Scientific Papers of Arthur Holly Compton

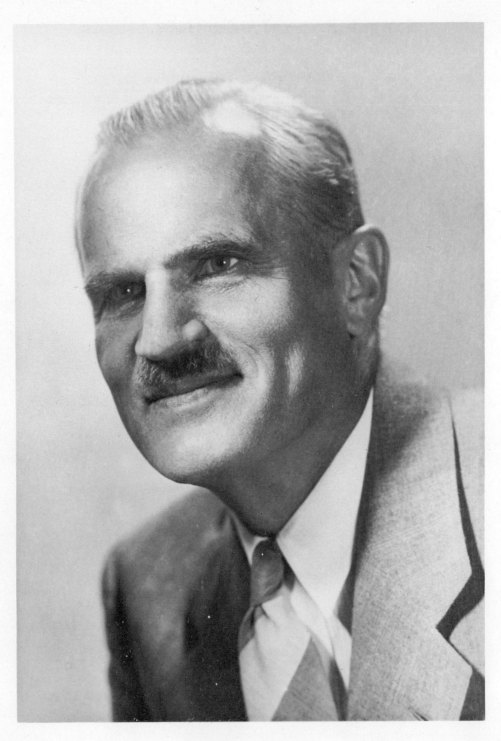

Arthur Holly Compton
Photograph by Jules Pierlow

Scientific Papers of Arthur Holly Compton

X-Ray and Other Studies

Edited and with an Introduction by

Robert S. Shankland

The University of Chicago Press

Chicago and London

ROBERT S. SHANKLAND is Ambrose Swasey Professor of Physics at Case Western Reserve University. He is a leading authority on acoustics and has published *Atomic and Nuclear Physics* and a number of papers on the history of modern physics.
[1973]

Publication of this volume was assisted by a grant from Washington University.

QC
71
C64
1973

The University of Chicago Press, Chicago 60637
The University of Chicago Press, Ltd., London
© 1973 by The University of Chicago
All rights reserved. Published 1973
Printed in the United States of America
International Standard Book Number: 0-226-11430-9
Library of Congress Catalog Card Number: 73-84189

Contents

(*Numbers in brackets indicate Bibliography entry numbers.*)

List of Illustrations / x
Acknowledgments / xi
Introduction / xiii

1913
A Laboratory Method of Demonstrating the Earth's Rotation [5] / 1

1915
A Determination of Latitude, Azimuth, and the Length of the Day
  Independent of Astronomical Observations [7] / 5
Watching the Earth Revolve [8] / 14
The Distribution of the Electrons in Atoms [11] / 16
The Variation of the Specific Heat of Solids with Temperature [12] / 18

1916
A Physical Study of the Thermal Conductivity of Solids [13] / 31
On the Location of the Thermal Energy of Solids [14] / 39
A Recording X-Ray Spectrometer, and the High Frequency Spectrum
  of Tungsten [16] / 45

1917
The Intensity of X-Ray Reflection, and the Distribution of the Electrons
  in Atoms [17] / 59
The Reflection Coefficient of Monochromatic X-Rays from Rock Salt
  and Calcite [18] / 88
The Nature of the Ultimate Magnetic Particle (with Oswald Rognley) [19] / 90

1918
The Size and Shape of the Electron [20] / 94
The Non-molecular Structure of Solids [23] / 105
Note on the Grating Space of Calcite and the X-Ray Spectrum of Gallium
  [24] / 135

1919
The Law of Absorption of High Frequency Radiation (Abstract) [26] / 138
The Size and Shape of the Electron: I. The Scattering of High Frequency
  Radiation [27] / 139

A Sensitive Modification of the Quadrant Electrometer: Its Theory and Use (with K. T. Compton) [28] / 163

The Size and Shape of the Electron: II. The Absorption of High Frequency Radiation [29] / 177

Radio-activity and Gravitation (E. Rutherford with A. H. Compton) (Letter) [30] / 190

1920

A Photoelectric Photometer [31] / 191
Cathode Fall in Neon (with C. C. Van Voorhis) [32] / 197
Radioactivity and the Gravitational Field [33] / 203
Is the Atom the Ultimate Magnetic Particle? (with Oswald Rognley) [34] / 207

1921

The Absorption of Gamma Rays by Magnetized Iron [35] / 220
Classical Electrodynamics and the Dissipation of X-Ray Energy [36] / 224
Possible Magnetic Polarity of Free Electrons [37] / 261
The Elementary Particle of Positive Electricity [38] / 264
The Degradation of Gamma-Ray Energy [39] / 265
The Wave-length of Hard Gamma Rays [40] / 286
The Magnetic Electron [41] / 294
Secondary High Frequency Radiation (Abstract) [42] / 305
The Polarization of Secondary X-Rays (with C. F. Hagenow) (Abstract) [43] / 306
A Possible Origin of the Defect of the Combination Principle in X-Rays (Abstract) [45] / 308
The Softening of Secondary X-Rays (Letter) [46] / 311

1922

The Width of X-Ray Spectrum Lines [47] / 313
The Spectrum of Secondary X-Rays (Abstract) [48] / 318
The Intensity of X-Ray Reflection from Powdered Crystals (with Newell L. Freeman) (Letter) [49] / 320
Secondary Radiations Produced by X-Rays [51] / 321
Radiation a Form of Matter (Letter) [52] / 378

1923

The Luminous Efficiency of Gases Excited by Electric Discharge (with C. C. Van Voorhis) (Abstract) [54] / 381
A Quantum Theory of the Scattering of X-Rays by Light Elements [55] / 382
The Total Reflexion of X-Rays [57] / 402
Recoil of Electrons from Scattered X-Rays (Letter) [58] / 413

Absorption Measurements of the Change of Wave-Length Accompanying the Scattering of X-Rays [59] / 414
The Spectrum of Scattered X-Rays [60] / 429
The Quantum Integral and Diffraction by a Crystal [61] / 434

1924
A Quantum Theory of Uniform Rectilinear Motion (Abstract) [62] / 438
Scattering of X-Ray Quanta and the J Phenomena (Letter) [63] / 439
A Measurement of the Polarization of Secondary X-Rays (with C. F. Hagenow) [64] / 441
The Recoil of Electrons from Scattered X-Rays (with J. C. Hubbard) [65] / 446
The Wave-length of Molybdenum $K\alpha$ Rays When Scattered by Light Elements (with Y. H. Woo) [68] / 457
The Scattering of X-Rays [69] / 460
A General Quantum Theory of the Wave-length of Scattered X-Rays [70] / 476

1925
The Effect of a Surrounding Box on the Spectrum of Scattered X-Rays (with J. A. Bearden) [74] / 485
Measurements of β-Rays Associated with Scattered X-Rays (with Alfred W. Simon) [75] / 488
The Density of Rock Salt and Calcite (O. K. DeFoe with A. H. Compton) [76] / 496
The Grating Space of Calcite and Rock Salt (with H. N. Beets and O. K. DeFoe) [77] / 499
On the Mechanism of X-Ray Scattering [78] / 504
Directed Quanta of Scattered X-Rays (with Alfred W. Simon) [79] / 508
X-Ray Spectra from a Ruled Reflection Grating (with R. L. Doan) [80] / 519
Light Waves or Light Bullets? [81] / 523

1926
Electron Distribution in Sodium Chloride (Abstract) [83] / 525

1927
X-Rays as a Branch of Optics [84] / 527
Coherence of the Reflected X-Rays from Crystals (G. E. M. Jauncey with A. H. Compton) [85] / 544

1928
On the Interaction between Radiation and Electrons [86] / 545
Some Experimental Difficulties with the Electromagnetic Theory of Radiation [88] / 552

The Spectrum and State of Polarization of Fluorescent X-Rays [89] / 576
The Corpuscular Properties of Light [90] / 581

1929

An Attempt to Detect a Unidirectional Effect of X-Rays (with K. N. Mathur and H. R. Sarna) [92] / 597
A New Wave-length Standard for X-Rays [93] / 601
The Efficiency of Production of Fluorescent X-Rays [94] / 613
What Things Are Made Of: I and II [95] / 630
Compton Effect [96] / 637

1930

The Determination of Electron Distributions from Measurements of Scattered X-Rays [98] / 640
Scattering of X-Rays and the Distribution of Electrons in Helium (Abstract) [99] / 654

1931

Electron Distribution in Argon, and the Existence of Zero Point Energy [102] / 656
The Optics of X-Rays [103] / 657
A Precision X-Ray Spectrometer and the Wave Length of Mo $K\alpha_1$ [105] / 672
The Uncertainty Principle and Free Will [106] / 684
Assault on Atoms [107] / 686

1934

The Appearance of Atoms as Determined by X-Ray Scattering (E. O. Wollan with A. H. Compton) [136] / 698

1935

Incoherent Scattering and the Concept of Discrete Electrons [142] / 703

1936

Scattering of X-Rays by a Spinning Electron [151] / 707

1938

An Alternative Interpretation of Jauncey's "Heavy Electron" Spectra [160] / 711

1940

What We Have Learned from Scattered X-Rays [167] / 712
Physical Differences between Types of Penetrating Radiation [168] / 721

1945

Modern Physics and the Discovery of X-Rays [178] / 727

1946

The Scattering of X-Ray Photons [179] / 732

1952

Man's Awareness and the Limits of Physical Science (Abstract) [180] / 737

1956

The World of Science in the Late Eighteenth Century and Today [181] / 738

1961

The Scattering of X-Rays as Particles [182] / 746

Appendix 1: A. H. Compton and O. W. Richardson / 751
Appendix 2: An Exchange of Letters between A. H. Compton and
   Gordon Ferrie Hull / 756
Appendix 3: The Compton Experiment *by Albert Einstein* / 759
Bibliography of Compton's Scientific Works / 763
Index / 785

Illustrations

frontispiece Arthur Holly Compton
following page xxix

1. Compton and other students with Sir Ernest Rutherford and Sir Joseph J. Thomson at the Cavendish Laboratory, 1919–20
2. The X-ray crystal spectrometer with large ionization chamber used by Compton for the X-ray scattering experiments that led to the discovery of the Compton effect
3. The 1927 Solvay Conference of Physics at Brussels
4. Compton and Albert A. Michelson in Compton's laboratory, 1927
5. Compton and O. W. Richardson at the Volta Electrical Congress at Como, Italy, 1927
6. Compton with the special ionization chamber that he designed and used for his world-wide cosmic-ray survey, 1931–33
7. Compton checking research apparatus at the University of Chicago, 1935

Acknowledgments

Acknowledgment is gratefully made to the following for permission to reprint the materials which comprise this volume: *American Journal of Physics; The American Journal of Roentgenology, Radium Therapy, and Nuclear Medicine; The Bulletin of the National Research Council; Indian Journal of Physics; Journal of the Franklin Institute; Journal of the Optical Society of America; Journal of the Washington Academy of Sciences; Nature; Philosophical Magazine; The Physical Review; Proceedings of the American Philosophical Society; Proceedings of the National Academy of Sciences; Radiology; The Review of Scientific Instruments; Science; Scientific American; Transactions of the Illuminating Engineering Society;* Washington University Press (Washington University Studies); J. A. Bearden; Mrs. O. K. Defoe; Lady Henrietta M. Richardson; H. O. W. Richardson; John D. Richardson; Mrs. Lillian M. Denisoff; Humanities Research Center, The University of Texas at Austin; Gorden F. Hull, Jr., and the Estate of Albert Einstein. Data relating to the original publication of the Compton papers are given in the Bibliography at the end of this book.

Many individuals have assisted in the preparation of this volume. Among these I wish especially to mention Mrs. Arthur Holly Compton and Luis W. Alvarez, J. A. Bearden, J. M. Benade, Ralph D. Bennett, Richard L. Doan, Carl Eckart, Newell S. Gingrich, Eric Rogers, Thomas H. Osgood, Alfred W. Simon, and E. O. Wollan, who, as Professor Compton's graduate students at Washington University or the University of Chicago, worked with him during the years of his greatest activity in X-ray research.

I wish also to thank the many other physicists throughout the world who were active during the years when Professor Compton made his greatest discoveries, and who answered our many inquiries about Compton's researches and their impact on the development of quantum physics. I am deeply appreciative for the advice given by S. K. Allison, J. W. Beams, Sir Lawrence Bragg, G. Breit, L. Brillouin, Louis de Broglie, E. U. Condon, W. D. Coolidge, K. K. Darrow, P. A. M. Dirac, P. P. Ewald, L. L. Foldy, S. Goudsmit, W. Heisenberg, H. Hönl, A. L. Hughes, E. C. Kemble, Paul Kirkpatrick, P. E. Klopsteg, W. B. Lewis, J. A. Prins, Arthur Ruark, B. W. Sargent, Sir George P. Thomson, G. E. Uhlenbeck, J. Valasek, J. H. Van Vleck, D. L. Webster, G. Wentzel, and J. A. Wheeler.

I am also indebted to Professor Roger H. Stuewer of the University of Minnesota and Professor Martin J. Klein of Yale University for helping to clarify historical facts related to Professor Compton's discoveries and their influence on the development of physics.

Finally, I wish to express my greatest thanks to Miss Marjorie Johnston for her detailed and expert contributions at all stages in the preparation of this volume. Having been Professor Compton's secretary during his years as chancellor of

Washington University in St. Louis, and having later had a major role in assembling and administering the Compton Archives at that university, she was able to contribute beyond measure to the preparation of this book.

R. S. SHANKLAND

Introduction by R. S. Shankland

It is generally agreed that the collected papers of a scientist are most useful when they are reprinted in chronological order. This has been done in the present volume, although serious consideration was given to dividing the X-ray papers into several groups corresponding to Arthur Compton's principal research interests. Such a division, however, would obscure important interrelationships among Compton's several X-ray experiments as, for example, when purely classical experiments, including those to determine the percentage polarization in scattering, were performed specifically to clarify the nature of the quantum scattering process now known as the Compton effect. Other relationships could be cited; but the reader will no doubt wish to follow for himself the various avenues of investigation carried through by Compton and revealed in these papers, from the X-ray work of his Ph.D. thesis at Princeton in 1916 to his final X-ray experiments in the early 1930s before his personal research interests changed to cosmic rays.

Arthur Compton's long study of X-rays began while he was still an undergraduate student at the College of Wooster in Ohio. His oldest brother, Karl, had stayed on at Wooster after graduation to earn a master's degree before going to Princeton University for his Ph.D. studies. He was Arthur's closest friend, and they shared many scientific interests. During Karl's year of graduate study and teaching at Wooster, Arthur worked with him almost constantly, using the X-ray apparatus that was available there, an experience that was of the greatest value for his own work when Arthur himself went to Princeton in 1913 for graduate studies.

During the Wooster and early Princeton days, Arthur had several research interests in addition to X-rays. His early delight in astronomy had led to his development of a novel device to measure the earth's rotation, and he had optimistic hopes of obtaining useful energy from this source. He was also greatly interested in gliders and airplanes and maintained an active concern in them for many years. At Princeton, however, X-rays soon became the central concern of his research efforts, first under the direction of Professor O. W. Richardson, who was to return to England at the end of Arthur's first term, but from whom he learned a great deal during this short period. Reading the Richardson-Compton papers now collected in the library at the University of Texas reveals clearly that an important impetus for Arthur Compton's pursuit of the basic science of X-rays came from Richardson. Professor Richardson had assembled a fine laboratory of modern X-ray equipment for his own use at Princeton; and when he returned to England at the end of 1913 and was unable to take it with him, he turned it over with his blessing to Arthur Compton—a most lucky happening for a new graduate student.[1] The excellent equipment received from Richardson included

1. See Appendix 1.

one of the early Snook-Roentgen Company X-ray tubes and high-voltage transformer circuits. After Richardson returned to England, Professor H. Lester Cooke supervised Compton's research work in X-rays. During Compton's graduate studies, additional guidance and inspiration were received from Deans William Magie and Andrew West, Professor Henry Norris Russell, the noted astronomer, and Professor Augustus Trowbridge.

Arthur began his work shortly after the Braggs had made their great discoveries in crystal structure with X-ray diffraction. Early references in Compton's personal correspondence reveal that he had planned to carry out similar experiments, but that the Braggs had anticipated him. Compton also closely followed Moseley's work on X-rays and, in fact, at one time thought he had found errors in Moseley's results, which, however, turned out not to be the case. Another important influence on Compton at this time was a visit to Princeton by Sir Ernest Rutherford, who showed considerable interest in Compton's X-ray experiments, as Moseley was then working in his laboratory at Manchester. Rutherford evidenced real enthusiasm for young Compton's experimental methods and for his grasp of research procedures.

There is little evidence, however, that physics research at Princeton in those years was greatly influenced by developments on the continent of Europe. Nevertheless, the Princeton curricula contained many important ingredients of classical physics that had been brought from England by Sir James Jeans, the most notable being his work in electricity and magnetism, mechanics, and the kinetic theory of gases. Some of Compton's early Princeton studies involved specific heats, and there is one reference in his correspondence to Debye's 1912 publication on this subject; but curiously, in his own 1915 paper on specific heats, his references to Einstein's earlier quantum theory are highly critical. His class notes on kinetic theory, however, indicate a thorough study of the classical aspects of specific heats, Planck's law of energy quantization, and Einstein's theory of the Brownian movement.

Compton's Ph.D. thesis, published in the *Physical Review* in January of 1917, was on X-ray diffraction by crystals. In his experiments he determined the coefficient for reflection of MoKα X-rays from the cleavage faces of a crystal of rock salt. Two crystals were employed, the first to select homogeneous X-rays from the tube, and the second to measure the coefficient of reflection. This was one of the first uses, with two crystals, of the Bragg X-ray spectrometer, and Compton was one of the pioneers in the development of this important instrument. He also repeated the experiment with crystals of calcite. These measurements were compared with two basic theoretical results that had been obtained by C. G. Darwin in 1914. In the first theory, based on reflection from a perfect crystal, the predicted intensity of X-ray reflection was much less than Compton and Moseley found experimentally. Darwin was in close touch with Moseley's work, as they were both in Rutherford's laboratory at Manchester. He thus soon learned that actual crystals are built up of numerous small blocks assembled by nature into what is called a mosaic. Darwin then developed a second theory of X-ray reflection from such an ideally imperfect mosaic crystal. This second theory was in close agreement with Compton's and Moseley's experiments and has been the basis of much quantitative X-ray work since that time.

The principal purpose of Compton's Ph.D. research was to use the intensities of X-ray reflections from crystals to determine the distribution of electrons within the crystal lattice. For this purpose he used a Fourier series representation, first introduced by W. H. Bragg in 1915, to specify the average electron density in various parts of the crystal lattice. Applications of the Fourier series analysis to electron densities in one dimension in a crystal were made in 1925 by W. Duane and by J. R. Havighurst. The method was then extended to two-dimensional analyses by W. L. Bragg (son of W. H. Bragg, and later Sir Lawrence Bragg) and his great school at Manchester, in whose hands it became a renowned tool for complex crystal structure analysis.

During his years of graduate study at Princeton, Compton seriously considered several alternative careers. He had had summer employment at Westinghouse Air Brake Company in Pittsburgh and was greatly interested in engineering. Earlier he had considered attending Case School of Applied Science as an undergraduate because of this engineering interest, and later, after only one year at Princeton, had planned to transfer to Cornell University to study engineering, with the view of working for Westinghouse. While at Princeton he also investigated the possibility of a research career at the General Electric Research Laboratory in Schenectady and had several interviews and correspondence with Dr. W. R. Whitney, General Electric's famed director of research. It seems very probable, however, that his brother Karl urged him to continue his studies in physics; for after finishing the planned year at Princeton, he stayed on there to complete the Ph.D. degree in physics instead of going to Cornell.

After he received his Ph.D. at Princeton, several opportunities presented themselves to Compton, including a fine Princeton fellowship for travel and study in Europe, which he declined because he planned to be married, and he could hold it only if unmarried. He was also offered teaching opportunities at both Lake Forest College and the University of Minnesota. Professor Anthony Zeleny, on leave from Minnesota, had spent the year 1915–16 at Princeton, where he had been impressed with Compton and had invited him to come to Minnesota. In the end, Compton did choose Minnesota, at a considerably smaller stipend than he would have received at Lake Forest, but with Zeleny's promise of a thousand-dollar's worth of X-ray equipment for his personal use. This decision was highly important for Compton's career, as he was thus able to continue his experimental work within an established research tradition.

At Minnesota in the college year 1916–17, Compton developed a balanced double crystal X-ray spectrometer for his experiments. The most important of these experiments was his demonstration, with O. Rognley, that the ferromagnetism in a magnetite crystal is not due to any effect of electron orbital motions around the nuclei in this ferromagnetic crystal. This was proved by demonstrating that no measurable change in X-ray diffracted intensities from a magnetite crystal was found as between the magnetized and unmagnetized state. He correctly concluded that ferromagnetism must be due to an inherent property of the electron's charge, a property now commonly called electron spin magnetic moment. In a paper presented to the AAAS in 1920 and published in the *Journal of the Franklin Institute* for August 1921, Compton stated: "I then conclude that the electron itself, spinning like a tiny gyroscope, is probably the ultimate magnetic particle and is responsible for ferromagnetism." In this article Compton not only asserted that the electron is the elementary magnetic particle but also that its spin is quantized and may be the cause of the unexplained gyromagnetic anomaly found by Stewart and Barnett.

It was not until 1925 that Goudsmit and Uhlenbeck introduced the generally accepted concept of electron spin based on detailed studies of atomic spectra. In their publication in *Nature* they state that Compton was the first to suggest quantized moments for the spinning electron. Their own great contribution was to show that the fourth quantum number, necessary to completely specify atomic states for spectral classifications, is the quantum number of the electron spin. They were able to explain in detail the experimental results of the anomalous Zeeman effect by providing a quantitative interpretation of the empirical rules that had been introduced by Landé.

During the two years 1917–19 Compton worked for the Westinghouse Electric and Manufacturing Company in Pittsburgh. His principal activity there during World War I was in the development of aircraft instruments for the Signal Corps. After the armistice his interest shifted to the physics and engineering that pioneered the sodium vapor lamp, to which he made basic contributions. Many years later this interest in lighting led to his participation in the development of the fluorescent lamp industry in the United States, as a consultant for the lamp development laboratory of the General Electric Company at Nela Park in Cleveland.

Compton's Westinghouse notebooks also reveal his continuing interest in X-rays. He did not have the opportunity in these years to do significant X-ray experiments; however, he spent a great deal of time on such matters as calculating the densities of calcite and rock salt and on related subjects that were an extension of his earlier work on X-rays at Princeton and Minnesota and a preparation for his later researches on the scattering of X-rays.

After two years at Westinghouse, he was awarded one of the first National Research Council fellowships. This he was eager to accept because it revived the opportunity for European study that he had earlier declined, and especially since he was now able to make arrangements for work at Cambridge with Ernest Rutherford, who had just moved to the Cavendish Laboratory from Manchester. Compton's move to Cambridge also indicates his final commitment to a career of research in basic physics rather than engineering, which for many years had been a very real alternative in all his plans. The Minnesota-Westinghouse years had afforded him ample opportunity to compare industrial and academic research, both of which had greatly appealed to him. The university career won out, however, and he therefore traveled, with his family, to England to work with Rutherford.

When Compton arrived in Cambridge for the fall term in 1919, he found the University, and especially the Cavendish Laboratory, filled with young men from all parts of the British Empire who had just returned from the war. Some young scientists, like Moseley, had not survived the great conflict, but many students, including P. M. S. Blackett, began their scientific careers at Cambridge with distinguished war records. It was a highly crowded and hectic time, and Compton found it impossible to do X-ray experiments of the type he had been planning at Westinghouse. Instead, Rutherford provided him with a laboratory and necessary equipment to study the scattering of gamma rays. During the year, Compton carried out the splendid gamma-ray scattering experiments that were such an important prelude to his later great work at Washington University in St. Louis. Several earlier workers, including D. C. H. Florance, A. S. Eve, and especially Professor J. A. Gray of McGill University, had already performed experiments on the scattering of gamma rays. Unlike the earlier X-ray scattering results of Professor C. G. Barkla, these gamma-ray scattering experiments had not yielded results consistent with J. J. Thomson's classical theory of X-ray scattering. Neither the absorption coefficient of scattered rays nor the angular dependence of the scattered intensity was in agreement with theory. More important still, the gamma-ray experiments indicated a progressive softening of the radiation as the scattering angle was increased—a result inconsistent with Thomson's theory. These results were especially evident in the series of carefully planned experiments carried out by Professor Gray, who deserves great credit for assigning scattering as the mechanism involved in the gamma-ray interaction that produces the change in wavelength.[2] However, Compton's Cambridge experiments were more definitive than any of the earlier work, and his detailed experience with X-ray absorption enabled him to make considerably improved determinations of the degree of softening of the gamma rays as a function of the scattering angle. These early experimental results obtained at Cambridge are entirely consistent with all later work on gamma-ray scattering.

It is by no means evident from a study of Compton's notebooks or correspondence exactly when the original idea came to him for pursuing this type of research. As with nearly every new idea, it is difficult to trace its exact origin; but certainly in the course of the years at Princeton, Minnesota, and Westinghouse, Compton became increasingly interested in the nature of the basic inter-

---

2. *Nature* 108 (1921): 435.

action between radiation and electrons and less concerned with the use of X rays as a tool to determine electron distributions in crystal structures. By the time he returned to the United States in 1920 to begin his work as head of the Physics Department at Washington University in St. Louis, he had formulated a detailed program of research directed principally to the X-ray scattering experiments that led him to his greatest discovery. On the cruise home in 1920 he outlined his proposed St. Louis researches, and there exists among his papers a most interesting sketch on the stationery of the ship Aquitania that shows the essential details of the proposed X-ray scattering experiment that he soon performed. It is interesting that this sketch incorporates the Snook-Roentgen type of X-ray equipment that he had already used so successfully at Princeton and Minnesota. Snook was a pioneer in the development of X-ray equipment in this country, and his excellent apparatus was an important factor in Compton's success.

When Compton assumed his duties as Wayman Crow Professor of Physics and head of the Department of Physics of Washington University in St. Louis, he immediately undertook the carefully planned program of research in X rays that he had been developing for several years. He concentrated first on studying the scattering of hard X rays from carbon as a function of scattering angle. He soon found a definite change in wavelength of the scattered radiation, closely analogous to his results obtained with gamma rays at Cambridge, and also to the results of J. A. Gray and earlier workers. However, Compton's X-ray results provided a major advance over all the earlier measurements. He employed a special Bragg X-ray crystal spectrometer, so that his observations gave the degree of softening of the scattered X rays with high precision, in contrast to all earlier work in which the wavelengths of the scattered radiation had been estimated by absorption techniques. During the next three years (1920–23) he constantly refined and improved these X-ray scattering experiments so that there could not be the slightest doubt that the scattering predicted by J. J. Thomson's famous electromagnetic theory could not explain them. Essential apparatus for these improvements also included special X-ray tubes, blown by Compton himself, in which the conventional glass bulb was eliminated to permit a closer proximity between the X-ray target and the carbon scatterer, and also the famous Compton tilted needle and adjustable quadrant electrometer which had been developed in collaboration with Karl Compton during a summer at Princeton. Both of these improvements greatly increased the sensitivity of detection of the scattered X-rays in Compton's experiments.

Compton tried every possible theoretical means he could devise to bring his results into conformity with a classical explanation. He made detailed scattering calculations with postulated electrons of finite size and special shapes, such as rings, to account for the observed dependence of X-ray intensities and frequencies on scattering angle. He also examined classical Doppler effects in which the electrons were assumed to recoil to produce the observed effects. His first attempt at a quantum explanation of the experiments was to investigate the consequences of a fluorescence type of interaction between X rays and electrons in the scatterer, in which the incident radiation excited an electron which then reradiated energy of a lower frequency.

None of these explanations proved to be satisfactory. At last, however, Compton arrived at the correct explanation by assuming that the interaction was between an X-ray quantum having both energy $h\nu$ and linear momentum $h\nu/c$, striking an essentially free electron, in a collision that obeyed the laws of conservation of energy and momentum for each individual event. The development of his ideas is rather completely recorded in papers in the *Washington University Studies*, January 1921,[3] and in the *Bulletin of the National Research Council*, October 1922.[4] For those interested in the detailed historical development, a careful study of these papers will reveal the great care with which Compton eliminated all classical explanations of his results before adopting a quantum interaction for his effect. Neither his notebooks nor personal letters reveal explicitly how he finally decided to treat the quantum as a particle having not only energy $h\nu$ but also linear momentum $h\nu/c$. There is no reference or hint that he was aware of Einstein's 1917 paper on this subject. He was, however, acquainted with Einstein's 1905 paper on the photoelectric effect, since his brother Karl had done his Ph.D. thesis at Princeton to check Einstein's predictions. The 1905 paper of Einstein discusses only the photon's energy, however, and not its momentum. It seems very probable that Compton arrived at his result by a different route, namely, from the classical electrodynamics that he had learned so thoroughly at Princeton during his graduate studies.[5] He was very proficient in this subject, and it was a natural step for him to posit the same relationship between the photon's energy and momentum as had long been known to exist between the energy transfer and momentum transfer for directed radiation in classical electrodynamics—namely, to divide the energy term by the speed of light to obtain the momentum term. In any case, when Compton assumed $h\nu$ and $h\nu/c$ for the X-ray quantum and took account of the change in mass of the recoiling electron with its speed, he found close agreement between his theory and his experiments for the change in wavelength as a function of scattering angle.

Strong corroboration for this view is to be found in the papers of O. W. Richardson, who was also a master of classical electrodynamics and had been Compton's first research advisor at Princeton.[6] Beginning in December 1921 Richardson had made a number of calculations to explain Compton's results, which apparently had been communicated to him privately. In the first of these calculations, Richardson assumed electrons of specific size and shape and then, by classical electrodynamics, attempted to explain the dependence of scattering on angle. These

---

3. See Bibliography entry 36. Text reproduced in this book, p. 224.
4. See Bibliography entry 51. Text reproduced in this book, p. 321.
5. See Appendix 2, Compton's letter to Hull.
6. See Appendix 1.

attempts were unsuccessful, as were all similar calculations made by Compton during the first years at Washington University.

Next Richardson set up the equations for the conservation of energy and the conservation of linear momentum for the interaction. Here he used $h\nu$ for the energy and $h\nu/c$ for the momentum, basing this relationship on classical electrodynamics. Many years later, when introducing Compton as the Guthrie lecturer (1935), he again referred to this choice as follows: "The interaction between radiation and a free electron is very simple and in fact the simplest interaction which radiation can undergo. Associated with the energy $h\nu$ according to electrodynamic theory there is momentum $h\nu/c$."

It appears that in the early 1920s the photon hypothesis was not well known to active research workers in England or in the United States, and it thus seems clear that Compton arrived at the correct theory for the phenomenon quite independently of earlier work along these lines.

On the European continent, however, the situation was entirely different. Einstein's special theory of relativity and his photon hypothesis were well known. P. Debye, in fact, published a theory of the Compton effect in an article appearing in *Physikalische Zeitschrift* in April of 1923,[7] in which he refers to Compton's earlier work reported in the *Bulletin of the National Research Council*.[8] However, Compton had presented his theory, including the important formula for the change in wavelength with scattering angle, at the December 1, 1922, meeting of the American Physical Society held at the University of Chicago. This and the earlier National Research Council publication clearly established his priority. At the Washington meeting of the American Physical Society on April 20, 1923, Compton gave the precise experimental value of 0.0242 Angstrom for what has since been known as the Compton wavelength of the electron. Then, in the May 1923 issue of the *Physical Review*,[9] his final summary paper was printed, giving both his theory and the experimental results that had been presented earlier at meetings of the American Physical Society. Compton's publications, starting in 1922, clearly establish his priority in both the theoretical explanation of the Compton scattering phenomenon and the precise experimental data upon which his theory was based. Since Debye's paper reports no experiments, and since his theoretical paper appeared after Compton had presented his at several meetings of the American Physical Society, there has been no question since that time that the discovery should properly be attributed to Arthur Compton.

An attempt was made by Bohr, Kramers, and Slater to explain Compton's experimental results without the necessity of treating the X-ray photon as a particle. They proposed that in individual events momentum and energy are not strictly conserved, but that these laws are obeyed only on a statistical average.

7. 24 (1923): 161.
8. See footnote 4 above.
9. See Bibliography entry 55. Text reproduced in this book, p. 382.

Their theory introduced the concept of virtual oscillators in the radiation field, and this aspect of the work continues to be important for quantum electrodynamics. However, other experiments, discussed below, soon proved that the Compton type of quantum interaction is strictly valid in each individual interaction between X rays and electrons.

In 1927 Erwin Schrödinger showed that wave mechanics could be used to calculate the change in wavelength of the Compton X rays scattered from free electrons when these are treated as De Broglie waves. However, the earlier corpuscular theory of Compton had already had a decisive influence on the development of quantum mechanics before Schrödinger's paper appeared, and it was only with the Dirac relativistic form of quantum mechanics that the true wave-corpuscle dualism in the Compton scattering process was clearly demonstrated.

In 1923, shortly after his discovery of the phenomenon that was soon to be called the Compton effect, Professor Compton moved to the University of Chicago, to the professorship formerly held by Professor Robert A. Millikan. During his first years in Chicago, working with a number of graduate students, he continued his experiments on the Compton effect, including a detailed study of X-ray scattering from many elements, both of the Compton modified line, which is due to scattering from essentially free electrons, and the unmodified Thomson scattering, which depends on the distribution of all the electrons in the atoms of the scatterer. Y. H. Woo made detailed studies of the relative intensity of the modified and unmodified lines from a wide range of scatterers, all confirming the predictions of Compton's theory. Because of the questions raised by the Bohr-Kramers-Slater theory as to whether a strict photon interaction obeying the conservation laws for energy and momentum was necessary to explain the Compton wavelength shift, several experiments were undertaken specifically to test this point. In Compton's laboratory R. D. Bennett used Geiger point-counters to observe coincidences between scattered photon and recoil Compton electrons. The experiment that was most decisive in settling this question, however, was one performed by Bothe and Geiger in Berlin, which revealed directly that the time coincidences between recoil electron and scattered photon pairs do occur as required by the conservation laws. Sixty-six coincidences within 0.001 sec. were recorded. Concurrently, Wilson cloud chamber experiments by Compton and A. W. Simon found the correct angular correlation between scattered photon and recoil electron as predicted by theory for 18 cases within 20° of the theoretical angles. Several years later, in 1934, when I was one of Professor Compton's graduate students, he suggested that I devise an experiment using Geiger-Mueller counters to show both the time coincidence and the angular correlation between scattered photon and recoil electron as predicted by theory. The early results of these experiments seemed to support the Bohr-Kramers-Slater hypothesis; but, after improvements of technique, it was shown that the time coincidence between scattered photon and recoil electron was within 10^{-4} sec., and at the same time the angular predictions of the Compton theory were verified within $\pm 10°$.[10]

Important evidence for the Compton process was furnished by excellent photographs taken by C. T. R. Wilson in 1923 in his famous cloud chamber at Cambridge. These revealed directly the Compton recoil electrons with the correct energies and angular dependence relative to the incident X-ray beam. Wilson, however, made no contributions to the theory. He shared the 1927 Nobel Prize for physics with Compton, not for the Compton effect, but rather for the invention and use of his cloud chamber.

This was a time (1922–27) of great interest in the Compton effect both in theory and experiment. Compton reported his findings at a number of scientific meetings during this period, and in the earlier years he was seriously criticized and attacked both for his experimental results and his theoretical interpretation. The distinguished Professor William Duane of Harvard University questioned the validity of the measured Compton change in wavelength on scattering, which he was unable to confirm in his own laboratory and which he attributed to tertiary radiation from carbon and oxygen in the enclosing box. This controversy reached its climax at the meeting of the British Association for the Advancement of Science at Toronto in August 1924, where Compton presented the latest results of his work. Interest was so great that the president of the British Association, Sir William H. Bragg, scheduled a special meeting at which Compton and Duane debated their views. This was a notable occasion, at which Compton answered all criticisms with clarity and good humor and convinced the large majority of those present that he was correct in both experiment and theory. There were important exceptions, however, including C. V. Raman of India and T. Lyman of Harvard, who were not willing to accept Compton's interpretation. It seems quite probable that the worldwide attention given to Compton's work at this Toronto meeting was a factor in his winning the Nobel Prize in 1927, especially after the "debate" was reported rather fully in *Nature*.[11]

Another eminent physicist who objected to Compton's experiments and their interpretation was Professor Charles G. Barkla of Edinburgh University. Barkla had done important pioneer work in X-rays, for which he was awarded the Nobel Prize in 1917. In a letter to *Nature* in 1923 Barkla wrote: "In a number of recent papers, Prof. A. H. Compton brings forward what purports to be a Quantum Theory of the scattering of X-rays. I venture to think that this theory—or more correctly system of rules—has little connection with the phenomena of X-ray scattering as I observed it nearly twenty years ago, and as I still know it." A reply was made to this by Compton in *Nature* in early 1924. After bringing forward arguments to support his point of view, he wrote: "I am thus wholly unable to

---

10. For details on these and later experiments of this type see R. S. Shankland, *Atomic and Nuclear Physics*, 2d ed. (New York: Macmillan, 1960), pp. 208-13.
11. 114 (1924): 627.

agree with Barkla's conclusion that 'Compton's formula holds neither for the apparent change of wave-length, nor for the energy of the recoil electrons.'"

Additional widespread interest in Compton's discovery was due to the enthusiasm of Albert Einstein in Berlin.[12] For a number of years Einstein had searched for crucial experimental evidence to confirm his 1917 and earlier papers on the photon hypothesis. In fact, at his suggestion, Bothe and Geiger had performed an experiment preceding Compton's that attempted to decide between classical and quantum properties of light. This experiment had not proved successful and is different from the well-known Bothe-Geiger experiment referred to above, which was a decisive factor in establishing the validity of the Compton effect. When Compton's theory and experiment became known to him, therefore Einstein discussed it with great enthusiasm and praised it highly; and it seems certain that his interest had a major influence in the ready acceptance of the result, especially on the European continent.

Einstein's discussion of the Compton effect in a Berlin newspaper and at the Berlin seminars not only emphasized its importance for the photon description of radiation, which Einstein had especially emphasized in his 1917 paper, but also strongly reminded physicists that the wave properties of X-rays and radiation are equally necessary and could not be ruled out after the great successes they had achieved in explaining a multitude of phenomena in interference, diffraction, polarization, and scattering of light. Compton, of course, had depended on the wave properties of X-rays in his experiments. First he had used the Bragg diffraction method to determine accurately the wavelengths of his scattered X-rays and hence the energies of the photons. He had also employed classical polarization techniques, as had Professor Gray, to show that the Compton effect is a direct scattering process and not a fluorescence phenomenon. Thus the discovery of the Compton effect made it essential for physicists to consider both the corpuscular and wave nature of X-rays and of radiation in general.

Compton's theory had been developed in terms of the old quantum theory and was adequate to predict the change in wavelength on scattering, the Compton wavelength of the electron (an early important example of a fundamental length), and the energy of the recoil electrons. However, the theory was soon shown to be inadequate for a quantitative explanation of the angular dependence of the scattered X-ray intensities, the total scattering cross-section as a function of energy, and the exact state of polarization of the Compton scattered X-rays. Thus the discovery of the Compton effect not only proved that both wave and corpuscular properties of radiation were required but also emphasized the need for a more basic quantum mechanics, soon developed by Heisenberg, de Broglie, Born, Schrödinger, and Dirac. Heisenberg, in fact, used the Compton interaction as an elegant means of illustrating his uncertainty principle, while the Dirac relativistic quantum mechanics gave a complete description of the Compton process, includ-

---

12. See Appendix 3.

ing the correct scattering cross-sections as calculated by Klein and Nishina and the dependence of the state of polarization at high energies.

The Compton effect was studied in several other laboratories besides those at the University of Chicago. At Stanford University, David L. Webster, Paul Kirkpatrick, and P. A. Ross beautifully confirmed Compton's findings when they were under serious question. Webster especially settled the matter as to whether the spurious "box effect" proposed by Professor W. Duane of Harvard was responsible for the observed changes in wavelength. The box effect was also disproved by J. A. Bearden, working at Chicago with Compton in an experiment done in the open air outside a third-floor laboratory window. The Stanford experiments not only supported Compton's findings but were extended to give greater detail and precision in the results both on the Compton modified line and also on the unshifted Thomson scattered X-rays. Soon Professor Duane, with the collaboration of Samuel K. Allison, discovered the cause of the disagreement between his experiments and Compton's and promptly acknowledged his confirmation at the next meeting of the American Physical Society.

At the California Institute of Technology, Professor J. W. M. DuMond and Harry Kirkpatrick conducted notable experiments on the Compton effect. DuMond developed his multicrystal X-ray spectrometer specifically for this work, and his detailed observations on the shape of the Compton scattered lines provided important information on the Fermi distribution of the nearly free electrons in the scatterer.

The Compton effect proved conclusively that radiation has a corpuscular nature and strongly suggested that it retains the properties of discrete energy $h\nu$ and linear momentum $h\nu/c$ while traveling through free space, and not simply in the processes of emission and absorption. However, the extreme corpuscular picture of radiation required modification when it was shown by G. I. Taylor and later by A. J. Dempster that normal diffraction effects are exhibited even by very weak light beams, thus proving that individual photons carry wave as well as corpuscular properties. Thus Compton's experiments and their interpretation clinched the argument on the reality of the corpuscular properties of radiant energy, as proposed by Planck and Einstein, and emphasized the impossibility of any easy way out of the paradox of waves and particles. Thereafter, any understanding of quantum physics by means of conventional concepts was impossible, and a telling argument was given for the need of a new theory of quantum mechanics that would incorporate within itself a unified description of both the wave and corpuscular nature of both radiation and matter.

Arthur Compton had a life-long interest in classical electrodynamics, as a result of the tradition at Princeton University established by Professors James Jeans and O. W. Richardson. He was always interested in the optics of X-rays and especially those classical phenomena that Roentgen had not discovered in his pioneering experiments. While at Washington University, he made the significant discovery that X-rays are totally reflected when they strike a surface within the critical angle for total reflection. He was, of course, familiar with the classical conditions for total reflection as given by dispersion theory, and was able to calculate from the available data the approximate expected angle for total reflection. On this basis he designed an experiment to test the phenomenon, and was able to demonstrate clearly the total reflection of X-rays from metallic surfaces and from glass. Not only did his results agree with the simple theory, but, by studying the angular dependence of reflected intensity near the critical angle, he took an important step in a more detailed understanding of the interaction between the X-rays and the bound scattering electrons in crystal lattices.

His interest in this subject continued at the University of Chicago, where he suggested to his graduate student, Richard L. Doan, that ruled gratings might be used to produce X-ray spectra by making use of total reflection. Doan undertook this project for his Ph.D. thesis and designed and produced on Michelson's ruling engine the special gratings of speculum metal having the necessary properties for use with X-rays near grazing incidence. Doan conducted this research most successfully and obtained the first X-ray grating spectra, while in the same laboratory, by the same method, T. H. Osgood became the first to bridge the gap in the electromagnetic spectrum between X-rays and ultraviolet optics.

This pioneer work of Doan, Osgood, and Compton led to much activity in this field of X-ray measurements. Compton's graduate student, J. A. Bearden, developed the work, both at Chicago and later at the Johns Hopkins University, into a method of high precision, and it was also carried forward at the University of Upsala in Sweden by Erik Bäcklin, a student of Professor Manne Siegbahn.

One of the notable results of the X-ray grating method developed at the University of Chicago has been its use to determine improved values for the electronic charge and Avogadro's number. Grating measurements give an absolute value for X-ray wavelengths, and these, when combined with Bragg reflections from crystals, yield lattice spacings. The two measurements lead to improved values of the Avogadro number, which, combined with the Faraday constant, give the electronic charge. The X-ray values for these fundamental atomic constants definitely differed from those obtained by Millikan using the oil drop method, which for many years had been accepted as standard. This disagreement led to considerable controversy, but in the end it was clearly demonstrated that the X-ray grating method gave much more reliable values. It is of interest to note that Professor J. A. Bearden, who contributed so much to the X-ray determinations of the electronic charge, also had a considerable share in locating the cause of error in the viscosity of air measurements that had introduced the principal uncertainty in Millikan's oil drop experiment. Bearden found that the torsion constant of the supporting wire used in the viscosity measurement for air must be determined under exactly the same load conditions as are used in the experiment and that an appreciable error had been introduced by assuming that Hooke's law was valid.

In the years 1928-31 Compton's graduate students were also extending the double crystal X-ray spectrometer methods into tools of high accuracy. This work was done principally by Newell S. Gingrich, W. W. Colvert, and Y. C. Tu. Gingrich was responsible for the earliest Chicago work with the two crystals used in the antiparallel position, a technique introduced by Professor Bergen Davis of Columbia University. Compton's own pioneer work with two crystals had used them in the parallel position, which gives no dispersion. The last Ph.D. thesis related to this work was performed by Luis W. Alvarez, who ruled a special optical grating suitable for use at grazing incidence. This experiment investigated the possibility that the X-ray grating method at grazing incidence does not yield reliable wavelength measurements. Alvarez found by an elegant and conclusive experiment that no anomalies occur when gratings are used at grazing incidence, thus supporting the validity of the X-ray determinations of the electronic charge.

Dr. Tu studied a variety of crystals, including some very perfect gem specimens loaned by the Peacock Jewelry Company in Chicago, to determine the degree to which mosaic structure might influence the X-ray results. He found that the X-ray determinations were not influenced to any appreciable extent by mosaic structure as had been claimed by critics of the X-ray methods.

A series of precision X-ray diffraction measurements by J. C. Stearns in 1930 gave the final proof of Compton and Rognley's Minnesota work of 1917 showing that the X-rays diffracted by ferromagnetic crystals are uninfluenced by the state of magnetization. Stearns' work confirmed precisely Compton's early conclusion that ferromagnetism is due to an inherent magnetic property of the spinning electron itself and not to orbital electronic motions in the crystal lattice, proving that alignment of electron spin magnetism is the basic cause of ferromagnetic behavior.

The question of the degree of polarization of scattered X-rays was an important topic throughout almost the entire period of Compton's X-ray work. At Washington University, with C. F. Hagenow, and later at the University of Chicago, he had shown that the modified X-rays scattered at 90° are almost completely polarized, thus proving that the Compton interaction is indeed a true scattering process. However, with the further development of quantum theory, and in particular the Dirac form of relativistic quantum mechanics, a more refined theory for the Compton scattering process predicted that at very high energies a small fraction of Compton scattered X-rays should be unpolarized. This prediction was accurately confirmed in the Ph.D. thesis of Eric Rodgers, who used the high energy X-rays from the General Electric machine at Mercy Hospital in Chicago.

The final X-ray experiments carried on at Chicago by Compton's students involved the determination of electron distributions in inert gas atoms. This interest originated in his Ph.D. research at Princeton, where he had pioneered methods to determine electron distributions in crystal lattices. The Chicago experiments were conducted principally by E. O. Wollan, who scattered X-rays from the noble gases, and from the measured angular distributions of the scattered radiation determined the electron distributions in the atoms with high precision. There was much interest in this work because of its relationship to the Thomas-Fermi atom model and also to the Hartree self-consistent field methods of calculation then being developed.

In the early 1930s Professor Compton's principal research interests changed from X-rays to cosmic rays. At first this was probably due to the fact that the interaction of high energy photons and electrons in the cosmic rays is one of the most important examples of the Compton scattering process. His work on cosmic rays is listed in the bibliography at the end of this volume.

It is clear that Compton's contributions to X-ray physics covered many aspects of the subject. It is also evident that his interest in classical X-ray physics, which began at Princeton, continued throughout all the years of his investigations. In fact, the title of his Nobel Prize lecture in 1927 was "X-Rays as a Branch of Optics." His work on polarization, total reflection, the double crystal X-ray spectrometer, X-ray diffraction gratings, and electron distributions in crystals and gaseous atoms all emphasize his continuing interest in the wave nature of X-rays.

His greatest achievement, however, was the discovery of the Compton effect, which made it necessary for him to adopt a corpuscular description of X-rays to explain his results. Compton did not use a quantum explanation for his results until he had exhausted all possible classical explanations. He resisted making any hasty claim based on the newer theories until he had followed every possible route of established physics. In studying this impressive series of X-ray papers in chronological order, the reader can learn much about actual methods of scientific research in addition to the physics that they contain.

In addition to his research papers on X-rays, Compton published two definitive books in this field. *X-Rays and Electrons*, which appeared in 1926, was one of the first textbooks dealing with modern physics to be published in this country. It included a detailed discussion of the growth of X-ray research, especially his own work that led to the Compton effect and other great advances. When a revision was urgently requested by the publisher, he collaborated with his younger colleague, Samuel K. Allison, to produce *X-Rays in Theory and Experiment*, which was published in 1935. Since that time this book has maintained its position as the standard textbook and reference work on X-rays.

A word is in order on the influence that the discovery of the Compton effect had on the development of quantum mechanics in the years 1922-28. It is true that Compton used the concept of the photon having localized energy and momentum and the special theory of relativity in the development of his theory. As emphasized above, however, he seems to have been singularly uninfluenced by the publications of Einstein on these subjects. His route was rather an extension of classical physics modified to include quantum effects. The importance of his work for the development of the new quantum mechanics seems to have been as follows. The Compton effect showed conclusively that no modification of classical wave electrodynamics, such as that proposed by Bohr, Kramers, and Slater, could be accepted, but that both the wave and corpuscular aspects of radiation must be recognized. On the other hand, impressive as were the predictions of the Compton theory based on the old quantum theory, the results revealed significant limitations; namely, that the total scattering cross-section as a function of energy and the differential scattering cross-section as a function of angle were not predicted accurately. There were other important phenomena that pointed toward revisions in theory, and soon the development of the new quantum mechanics was under way, and many new results rapidly appeared. The Compton effect certainly played an important role in stimulating this great advance, for it is such a basic interaction between radiation and matter that any acceptable theory must be precisely correct in explaining all its features. This explanation was finally achieved by several basic results obtained with the new quantum mechanics. Dirac's relativistic electron theory, including the effect of electron spin, led to the Klein-Nishina formula for the Compton scattering cross-section. Of even greater importance was Dirac's quantization of the radiation field, which provided the unified description of the wave and corpuscular aspects of light that the discovery of the Compton effect had demanded. Moreover, when this theory was applied by G. Wentzel to X-ray scattering by *bound* electrons, it explained both the "modified" and the "unmodified" Compton lines in all detail. The quantitative agreement of these theoretical results with experiment was in a sense the capstone proving the validity and importance of the Compton effect.

Because of its basic nature the Compton interaction has continued to be of central importance for quantum physics. Today high energy interactions between photons and electrons are seen to play a central role in astrophysical processes. Important applications, such as that used at the Stanford linear accelerator to produce monoenergetic gamma-ray beams by the interaction of intense laser light with the accelerator electron beam, remind us of the continuing importance of the Compton effect.

At the AAAS meeting in Cleveland in December 1930, R. A. Millikan, as president, delivered a major address on his cosmic-ray researches. In this talk he referred repeatedly to his use of the Klein-Nishina formula for the scattering of gamma rays, which he believed the cosmic rays to be. Compton was in the audience, and since the Klein-Nishina formula is a direct consequence of the Compton effect theory as extended by Dirac, it was only natural that Compton's interest was greatly aroused. As a result he shifted his own major research interest to cosmic rays. The bibliography of his scientific papers at the end of this book lists the many publications in this field by Compton and his students—work which was reluctantly terminated by his involvement in the nuclear energy program of World War II.

The most notable result of this research program was the proof that the primary cosmic rays are charged particles—mostly protons—and not gamma rays as Millikan had maintained. Compton organized a world-wide survey of cosmic-ray intensities which carried him to all parts of the world. Its results showed that the earth's magnetic field influenced the intensities in a major way, which would occur only if the primary cosmic rays were charged particles. Further work with his graduate student Luis W. Alvarez in Mexico City proved that an appreciable fraction of the primary cosmic rays were positively charged and thus probably protons.

The interesting account of Compton's cosmic ray work with L. W. Alvarez, R. D. Bennett, R. L. Doan, I. A. Getting, J. L. Hopfield, M. Schein, J. C. Stearns, R. J. Stephenson, R. N. Turner, and E. O. Wollan must be reserved for another occasion.

Fig. 1. Physics research students with Professors Ernest Rutherford and Joseph J. Thomson at the Cavendish Laboratory, Cambridge University, 1919–20. Compton is seated second from right in the middle row. It was while here that he began his famous experiments on the scattering of radiation.

Fig. 2. The X-ray chrystal spectrometer with large ionization chamber used by Compton at Washington University, St. Louis, in 1920–23 for his definitive X-ray scattering experiments which led to his discovery of the Compton effect. The X-ray tube is in the large lead-shielded box.

Fig. 3. The 1927 Solvay Conference of physics at Brussels, Belgium. Compton was the "American guest" at this conference and discussed his X-ray scattering experiments for which he won the Nobel Prize later that year. The participants shown in the photograph are, from left to right, as follows: *front*, Irving Langmuir, Max Planck, Madame Curie, H. A. Lorentz, Albert Einstein, Paul Langevin, C. E. Guye, C. T. R. Wilson, O. W. Richardson; *middle*, P. Debye, M. Knudsen, W. L. Bragg, H. A. Kramers, P. A. M. Dirac, Arthur Holly Compton, L. V. de Broglie, M. Born, Niels Bohr; *rear*, August Piccard, E. Henriot, P. Ehrenfest, E. Herzen, T. de Donder, Erwin Schrödinger, E. Verschaffelt, W. Pauli, Werner Heisenberg, R. H. Fowler, L. Brillouin.

Fig. 4. Compton and Albert A. Michelson examining an X-ray tube in Compton's laboratory in 1927 where Michelson had come to congratulate him on winning the Nobel Prize.

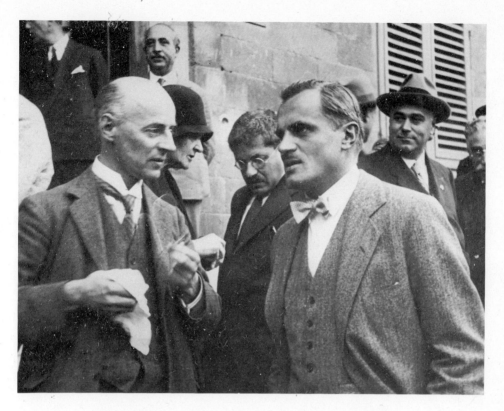

Fig. 5. Compton and O. W. Richardson at the Volta Electrical Congress at Como, Italy, in 1927. Richardson was Compton's first research advisor at Princeton University in 1913. Madame Curie and Paul Ehrenfest are in the background.

Fig. 6. Compton with the special ionization chamber which he designed and used for his world-wide cosmic-ray survey during 1931–33, which proved that cosmic rays are charged particles.

Fig. 7. Compton checking research apparatus in his laboratory at the University of Chicago in 1935.
Photograph by Stephen Deutch.

SPECIAL ARTICLES

A LABORATORY METHOD OF DEMONSTRATING THE EARTH'S ROTATION

THE two laboratory methods in general use for proving the rotation of the earth are Foucault's pendulum and gyroscope experiments. The first is inapplicable in many laboratories, because there is no convenient place to hang a sufficiently long and heavy pendulum, while the apparatus for the second is necessarily expensive. The following experiment is designed to provide a simple and convenient means by which the earth's rotation may be demonstrated in a small laboratory. The demonstration depends upon the fact that, if a circular tube filled with water is placed in a plane perpendicular to the earth's axis, the upper part of the tube with the water in it is moving toward the east with respect to the lower part. If the tube is quickly rotated through 180 degrees about its east and west diameter as an axis, the part of the tube which was on the upper side attains a relatively westward motion as it is turned downwards (since it is drawing nearer the earth's axis). But the water in this part of the tube retains a large part of its original eastward motion, and this can be detected by suitable means.

Since the east and west axis itself is rotating with the earth, only that component of the water's momentum which is parallel to this axis will have an effect in producing a relative motion when the tube is turned. If then a is the angular velocity of the earth's rotation, r the radius of the circle into which the tube is bent, and θ the angular distance of any small portion of the tube from the east and west axis, the relative velocity between the water and the tube when it is quickly turned from a position perpendicular to the earth's axis through 180 degrees is

$$\text{Velocity} = V = \frac{ar}{\pi} \int_0^{2\pi} \sin^2 \theta \, d\theta = ar.$$

In order to prevent convection currents, it is best to hold the ring normally in a horizontal position, in which case the relative motion is of course $ar \sin \phi$, where ϕ is the latitude of the experimenter.

To perform the experiment, glass tubing 1.3 cm. inside diameter was bent into a circular ring 99.3 cm. in radius, and a short glass tube closed with a rubber tube and screw

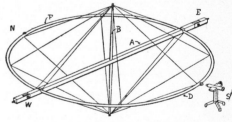

FIG. 1

clamp was sealed into it to allow for the expansion of the water and to provide a place for filling. The ring was fastened with tape into notches in the wooden rod A (Fig. 1), which served as the horizontal axis, and was supported by wires from the extremities of the cross rod B. The ends of the rod A were made adjustable perpendicularly to the plane of the ring, so that the ring might be made to swing on an axis parallel to its plane. The ends of the rod were swung in solid supports, adjustable to make the axis horizontal. In order that the motion of the water might be detected, a mixture of linseed oil and oil of cloves of the same density as water was prepared, and a few drops of the mixture were shaken up with the water with which the tube was to be filled. The globules of oil were observed at a point C, between the ends of the axis, through a micrometer microscope. Difficulties from the astigmatic refraction of the light by the water in the cylindrical glass tube were overcome by sealing a tubular paraffine cap, closed with a cover-glass and filled with water, on the part of the glass tube under the microscope, thus presenting a plane surface through which to make the observation. One side of the ring was weighted, so that on releasing a catch at the side of the observer the tube swung around through 180 degrees in a definite time, and was held again by the catch just under the microscope.

In taking a reading, the microscope was focused as nearly as possible on the center of the tube, and the ring was left in position until the oil globules had no appreciable motion. As soon as the catch which held the ring in position was released, the time was counted, with the aid of a metronome ticking half-seconds, until the tube had turned and an oil globule had been fixed upon to follow. The globule was followed through a measured length of time by turning the micrometer screw, and the distance through which it moved was recorded. Examples of these observations are given in the first three columns of Table I.

Variations in the readings arose from the fact that the part of the ring toward the east was near a cold wall, so that convection currents were produced as soon as the tube left the horizontal position in making a turn. This effect was made as small as possible by stirring the air with an electric fan. Other variations came from the fact that it was found impossible to adjust the horizontal axis so nearly parallel to the plane of the ring as to prevent a slight effect from turning the

TABLE I

Time from Releasing Catch to Following Water's Motion	Time of Following Water's Motion, Sec.	Distance Through which Water is Followed, Mm.	Time from Completion of Turn to Following Water's Motion, Sec.	Time on Curves of Completion of Turn	Initial Velocity, V, Mm. Sec.$^{-1}$
Case I. Weight on side D. Change from heavy to light side.					
7.5 secs.	22.5	+ .40	4.5	+21.2	+.041
7.0	23.0	+ .37	4.0	+22.1	+.039
Case II. Weight on side D. Change from light to heavy side.					
7.5	22.5	+1.57	4.5	− 1.0	+.160
8.0	22.0	+1.35	5.0	+ .5	+.155
Case III. Weight on side F. Change from heavy to light side.					
8.0	22.0	− .59	5.0	+13.4	−.067
7.5	22.5	− .70	4.5	+11.4	−.075
Case IV. Weight on side F. Change from light to heavy side.					
7.5	22.5	+ .37	4.5	+19.9	+.045
8.0	22.0	+ .67	5.0	+11.5	+.075

Average V: Case I. = .0434; Case II. = .1580; Case III. = − .0633; Case IV. = .0671.

tube. Errors from the first cause were corrected by reversing the direction of turning in alternate readings. Those from the latter cause were nullified by taking readings with one side of the ring weighted and then shifting the weight to the other side. In this manner ten readings of each of four different kinds were taken (Cases I., II., III. and IV.), and the fact that the predominant motion is positive, or toward the west as observed on the south side, shows that the earth is turning from the west to the east.

Calculation of the Initial Velocity

In order to make an accurate estimate of the velocity corresponding to any given reading, the rate of decrease of velocity of the water in the ring must be determined. If the retardation r is taken to be proportional to the velocity V for this low velocity,

$$r = \frac{dV}{dt} = CV,$$

$$\frac{dV}{V} = Cdt,$$

and

$$\log V = Ct + K$$

will express the value of the velocity at different times. In order to determine the constants C and K, the ring was held in a vertical position until the colder water near the east wall produced a considerable motion. It was then brought back to the horizontal and the time observed which was required to move successive quarter millimeters. A few

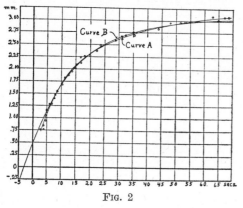

FIG. 2

such readings are given in Table II. From a large number of such observations an average curve was drawn, showing the relation of the distance covered to the time (Fig. 2, Curve A). The slope of this curve was taken at two

TABLE II

	.25	.50	.75	1.00	1.25	1.50	1.75	2.00	Distance in Mm.
1	2	4	7	11	16	22	30	42	Time in seconds.
2	2	2	5	7	10	15	21		
3	3	7	11	17	23	31	45	68	

of the most definite points, $t = 12.5$ and $t = 30$, and these values were substituted in equation (1) to determine the constants C and K. The curve in Fig. 3 was then drawn from the resulting formula, showing the velocity at any time. Curve B, Fig. 2, was then constructed by integrating this curve graphically with respect to t.

The water in the ring has its maximum velocity just before the turn is completed. The time required to make a complete turn was three seconds, and if this is subtracted from the time in column 1, Table I., it gives the length of time between the completion of the turn and the first observation of the motion (column 4, Table I.). Now if a portion of Curve B (Fig. 2) be taken, such that the distance represented on the curve in the time of any particular reading is the same as the distance in that reading, the beginning of that portion of the curve will correspond to the time at which the motion of the globules

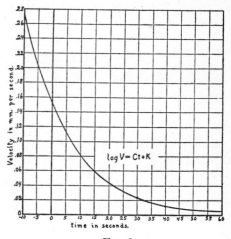

FIG. 3

was first observed (column 5, Table I.). So if the number of seconds in column four is subtracted from the time corresponding to the beginning of the reading, the time corresponding to the completion of the turn is obtained, and the velocity at that time can be read from the curve in Fig. 3. This value is given in column six, and is the velocity at the time of completing the turn. The velocities in each of the four cases are averaged separately, and the average of the four averages is taken as the true motion due to the earth's rotation.

The average of the velocities in these four cases is .0513 mm. per second. From the formula $V = ar \sin \phi$ derived above, we obtain $V = .0484$, a difference of 5 per cent. As a check upon the accuracy of the readings, it will be seen that the differences between the velocities in Cases I. and II. and between those in III. and IV., representing double the velocity due to the difference in density of the water in different parts of the tube, are about equal; also the differences between Cases I. and III., and II. and IV., representing the variation due to imperfect adjustment of the axis, are approximately the same. In order to show that there was no appreciable effect from convection currents while the ring was in a horizontal position, several readings were taken after the tube had remained at rest for some time, none of which showed a motion larger than .015 mm. per second.

In order to obtain the best possible results, the ring should be mounted as rigidly as possible in a room of equal temperature throughout, and the axis should be capable of accurate adjustment parallel to the ring. If the radius of the ring were made smaller, although the effect of the earth's rotation would be less, it would be easier to keep all parts of the tube at an equal temperature, and the ring could be turned more quickly. Moreover, since the motion would not be so great, the velocity of the water would diminish less rapidly, so that more accurate readings could be obtained. With a more mobile liquid the motion would of course continue longer. Even with the comparatively crude apparatus described above, however, it is not difficult to show that the earth revolves.

ARTHUR HOLLY COMPTON

PHYSICAL LABORATORY,
UNIVERSITY OF WOOSTER,
January 13, 1913

A DETERMINATION OF LATITUDE, AZIMUTH, AND THE LENGTH OF THE DAY INDEPENDENT OF ASTRONOMICAL OBSERVATIONS.

By Arthur H. Compton.

IN a previous paper[1] an experiment was described which afforded a means of measuring the component of the earth's rotation about a vertical axis. Assuming the latitude to be known, the rate of the earth's rotation could then be calculated. The present paper shows how the same method may be employed to measure also the components of the earth's rotation about two mutually perpendicular horizontal axes, so that the rate of the earth's rotation can be determined directly. From the ratio of the vertical component to the resultant rotation the latitude may be found, and from the ratio of the two horizontal components the azimuth may be determined.[2]

If a circular tube filled with liquid is placed in a plane perpendicular to the axis about which the rotation is to be measured, one side of the tube is, in general, moving with respect to the other side. If now the tube is quickly rotated through 180 degrees about an axis in its own plane, the part of the tube on one side of the axis will have its motion changed as it is shifted to the other side, while the liquid retains a large part of its original motion. For example, if a tube bent into a ring of radius r is placed in a plane perpendicular to the earth's axis and is then turned

---

[1] A. H. Compton. "A Laboratory Method of Demonstrating the Earth's Rotation," Science, N. S., Vol. 37, p. 803, 1913.

[2] By experiments with an Atwoods machine, such as those conducted by John G. Hagen (John G. Hagen, "How Atwoods Machine Shows the Rotation of the Earth even Quantitatively," International Congress of Mathematics, Aug., 1912) it is theoretically possible to determine the azimuth from the ratio of the deviation of the falling weight toward the south to that toward the east. The earth-rotation ring here described, however, is the only apparatus which has been shown capable of measuring the earth's angular velocity about both vertical and horizontal axes, which is necessary for a determination of the latitude and the length of the day independent of astronomical data.

half way around about a horizontal axis, the upper portion of the tube acquires a relative velocity toward the west when turned to a position nearer the earth's axis than originally equal to $2r\omega$, where ω is the angular velocity of the earth's rotation. The liquid in this part of the tube, however, will retain its original motion, and so will have a relative momentum toward the east. Since the pivots upon which the horizontal axis rests are constrained to follow the earth in its rotation, the component of the motion of the liquid parallel to the direction of the tube at these points is without influence on the relative motion, and only that component of the liquid's momentum which is parallel to the axis will have an effect in producing relative motion when the tube is turned. So if θ is the angular distance of any small portion of the ring from the axis about which it is turned, the mean momentum per unit length of the tube which tends to cause relative motion immediately after the ring is shifted from a position perpendicular to the earth's axis through 180 degrees is:

$$\rho V = 2r\omega\rho \frac{\int_0^{2\pi} \sin^2\theta d\theta}{\int_0^{2\pi} d\theta}$$

or

$$V = \omega r \qquad (1)$$

where ρ is the mass of the liquid per unit length of the tube, and V is the relative velocity between the liquid and the tube.

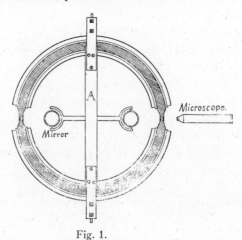

Fig. 1.
Showing the construction of the earth rotation ring.

The ring used to perform this experiment was made of one inch brass tubing bent into a circle eighteen inches in diameter. Where the windows were placed the tube was constricted to a diameter of about three eighths of an inch, as in Fig. 1, so as to increase the velocity of the liquid at the point of observation. In order to prevent convection currents as far as possible, the tube was covered with a quarter of an inch of asbestos, and enclosed in a concentric tin tube with an intervening air space. The protected tube was then mounted on a rigid rod A (Fig. 1), the ends of which were made adjustable perpendicular to the plane of the ring, so that the ring might be made to swing upon an axis accurately parallel to its plane. An iron framework was so constructed that the axis could be supported in either a horizontal or a vertical position in order to measure either the vertical or the horizontal components of the earth's rotation.

Carbon disulphide was first used to fill the tube, on account of its low viscosity. Its motion was made visible by shaking up with it an aqueous solution of calcium chloride of the same density, which formed small suspended globules whose motion was easily visible through the microscope. Because of its high coefficient of expansion, however, the convection currents due to slight differences in temperature in different parts of the tube rendered the use of this liquid impracticable. In fact, no liquid could be found whose coefficient of expansion was nearly as low as that of water, so this was finally used to fill the tube. A mixture of coal oil and carbon tetrachloride was prepared of the density of water at 4° Centigrade, at which temperature most of the measurements were made. The slight change in relative density due to a rise to room temperature did not noticeably affect the motion of the smaller globules of the oil when shaken up in the water.

When the ring was held in a horizontal plane, no particular pains were required to eliminate convection currents, since the only time that a difference in density in different parts of the tube could affect the motion of the water was while it was being turned over. When the ring was held in a vertical plane, however, in order to measure the horizontal components of the earth's rotation, the slightest variation in density in different parts of the tube was immediately noticed. Great precautions were taken to keep the ring at uniform temperature throughout. The whole apparatus was enclosed in an asbestos box, and the within was stirred by an electric fan, as in Fig. 2. A further asbestos shield prevented the observer's breath from striking the enclosing box, and the surrounding air was kept well stirred by an electric fan. By this means the difference in temperature of different parts of the tube was kept within 0.05 of a degree, but even this small difference at ordinary room temperature produced convection currents comparable in magnitude with the motion of the water due to the earth's rotation. In order to eliminate still further these currents, the apparatus was set in a constant temperature room and kept at 4° Centigrade by means of a thermostat. In this manner it was found possible to eliminate almost entirely the effect due to the convection currents.

In taking a reading, a microscope with an eye-piece scale was focused on the center of the tube under the glass window, and the ring was held in position until the oil globules had no appreciable motion. The ring was then quickly turned over, and the number of scale divisions passed by the globules between the fifth and fifteenth seconds after the ring was reversed was noted. A telegraph sounder actuated by the laboratory clock was used to measure the time during which the motion was followed. Immediately after the ring was turned over there was a large motion across the tube, but this soon died out, and the motion along the tube could be accurately measured.

TABLE I.

Reading Given in Scale Divisions.

	Case A.	Case B.	Case C.	Case D.
Setting I..........................	+13	+25	+10	+24
Axis vertical, Ring Approx.......	+12.5	+25	+11	+24.5
ÆNE×WSW...................	+14.8	+20.5	+13.8	+17.5
Ave. 9 r'd'gs:	+15.35	+21.19	+13.42	+19.71

X_1 = Average of 4 cases = + 17.41 divisions.

Setting II........................	− 7	−10	−14	− 8
Axis vertical, Ring approx.......	− 8.2	−13	−14	− 7.5
NNW×SSE...................	−10	− 9.5	− 9	− 6.5
Ave. 10 r'd'gs:................	− 9.70	− 9.59	−12.33	− 6.60

X_2 = Average of 4 cases = − 9.55 divisions.

Setting III.......................	+14.5	+ 8	+19.5	+17.5
Ring horizontal.................	+35	+20	+11	+30
	+10	+24	+32	+12
Ave. 10 r'd'gs:................	+15.35	+18.20	+23.15	+17.10

X_3 = Average of 4 cases = + 18.42 divisions.

Some typical readings thus obtained with the ring set in the three mutually perpendicular planes are shown in Table I. The four cases, A, B, C, D, represent the four different ways in which the ring may be turned. By taking the average of these four cases, differences in the readings due to slight convection currents and to inaccurate adjustment of the axis about which the ring is turned cancel out. The positive direction is taken as upward as seen in the microscope (a real downward motion) in the first two settings, and toward the right in the third. In the first setting observations were taken on the west side of the ring, and the fact that the water was moving relatively downward on this side after the ring was reversed indicates that the earth is revolving from west to east. Similarly the relatively upward motion observed on the

$$\varphi = \text{latitude} = \sin^{-1} \frac{X_3}{\sqrt{X_1^2 + X_2^2 + X_3^2}} = 42.8°.$$

$$\psi = \text{azimuth} = \tan^{-1} \frac{X_2}{X_1} = 28.7°.$$

east side in the second setting indicates the same sort of motion. The third setting showed a relative motion to the left on the side of the ring observed, which shows a rotation of the earth in a counter-clockwise direction about a vertical axis. Thus qualitatively the rotation of the earth about the three different axes is shown.

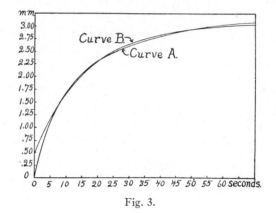

Fig. 3.

In order to make an accurate estimate of the angular velocity corresponding to any observed motion, it is necessary to find the law of motion of the water in the tube. If the motion were uniform the resistance would be proportional to the velocity, as assumed in the previous paper,[1] but this is not exactly true when the velocity changes with the time, if the motion of the water at the center of the tube be considered. For instance if the water in the tube is given an impulsive motion, immediately after the impulse the resistance at the center of the tube is zero, while the velocity is a maximum. However, if the greater part of the resistance occurs at definite points in the tube, as in the ring used in this experiment, the above law will hold more accurately than for a tube of

[1] Ibid., p. 805.

uniform diameter. An experimental test of the accuracy of this law for a uniform circular tube is afforded by a comparison of the curves A and B in Fig. 3, taken from my previous paper. Here curve A represents the motion of the water as determined by a large number of readings, while curve B represents the motion as it would be if the resistance were proportional to the velocity, showing a rather close agreement. We shall assume, therefore, that for our present purposes the resistance may be considered proportional to the velocity, that is:

$$\frac{d^2x}{dt^2} + c\frac{dx}{dt} = 0,$$

where x is the distance travelled along the tube, t is the time, and c is a constant depending upon the viscosity and density of the liquid and the dimensions of the tube. The solution of this equation may be put in the form:

$$X = \frac{V}{c}(e^{-ct_1} - e^{-ct_2}),$$

where X is the distance through which the motion of the water is followed, V is the initial velocity of the water after the ring is turned over, t_1 is the time of beginning, and t_2 that of ending the observation of the motion of the globules. Since in all the readings the times t_1 and t_2 were taken the same, the initial velocity of the water V is proportional to the distance X through which its motion is followed. But by equation (1)

$$V = \omega r,$$

so that X is proportional to ω, that is, the distance through which the water is followed is a measure of the component of the angular velocity about an axis perpendicular to the plane of the ring. Thus if we let α be the factor of proportionality,

$$\omega = \alpha X. \qquad (2)$$

Let ξ, η, and ζ be the components of the earth's angular velocity about axes perpendicular to the plane of the ring in settings I., II. and III. respectively, and X_1, X_2, X_3 be the average motions observed in the three settings. Then by equation (2)

$$\xi = \alpha X_1; \quad \eta = \alpha X_2; \quad \zeta = \alpha X_3.$$

If we call φ the latitude and ψ the angle between the ξ axis and the north, then from Table I.:

$$\left.\begin{aligned}\xi &= \omega \cos\varphi \cos\psi = +17.41\,\alpha, \\ \eta &= \omega \cos\varphi \sin\psi = -9{,}.55\,\alpha \\ \zeta &= \omega \sin\varphi = +18.42\,\alpha,\end{aligned}\right\} \qquad (3)$$

from which

$$\omega = 27.08\,\alpha,$$
$$\sin \varphi = 0.680,$$
$$\varphi = 42.8° = \text{latitude},$$
$$\tan \psi = 0.548,$$
$$\psi = 28.7° = \text{azimuth}.$$

A quantitative determination of the absolute magnitude of the earth's angular velocity from these data may be made if the constant α in equation (2) is evaluated. An attempt was made to determine this constant by placing the ring on a spectrometer table which was turned at the desired angular velocity by means of a driving clock. The ring was turned with known angular velocity until the motion of the ring became uniform, and was then stopped under the microscope, and the motion of the water observed. The constant as determined in this way is not strictly comparable, however, with that which enters when the ring is turned over just before the motion of the water is observed. It was found necessary, therefore, to use a different method for calibrating the tube.

The method employed was a direct determination of the angular velocity of the vertical component of the earth's rotation. This was done by placing the whole apparatus upon the table of the spectrometer, as shown in Fig. 4. The table was rotated by means of the driving clock at such speed that the readings taken on reversing the ring with the clock running were approximately equal but opposite in sign to those taken when the clock was stopped. So if X_4 be the observed motion of the water when the clock was stopped and X_4' that when running, the vertical component of the earth's angular velocity is

$$\zeta = \frac{-\chi}{\left(1 - \dfrac{X_4'}{X_4}\right)},$$

where χ is the angular velocity at which the spectrometer table was turning. As shown in Table II., the readings for determining X_4 and X_4' were taken in alternate sets of four, one under each of the four cases, every other set being made with the clock running. In this manner all systematic errors were eliminated, so that the accuracy with which X_4 and X_4' can be determined is a direct function of the number of readings taken. It may be noted that although X_4 and X_3 are both measures of the earth's rotation about a vertical axis, the two quantities are not directly comparable, since the temperature at which X_4 was measured

TABLE II.

Apparatus on Spectrometer Table. Readings in Scale Divisions.

Case A.		Case B.		Case C.		Case D.	
Clock: Off.	On.	Off.	On.	Off.	On.	Off.	On.
+24		+37		+38		−12	
	−22		−10		−24		−63
+38		+49		+21		+16	
	−49		−22		−17		−34

Average X_4 (28 readings) $= + 27.5$ divisions.
Average X_4' (28 readings) $= - 29.9$ divisions.
$\chi =$ angular velocity of spectrometer table relative to earth
$\quad = - 3°$ in 512.4 sidereal seconds,
$\quad = - 0.00585$ degrees/second.
$\zeta =$ vertical component of earth's angular velocity.

$$= \frac{-\chi}{\left(1 - \frac{X_4'}{X_4}\right)} = 0.673 \text{ revolutions/day}.$$

$\omega =$ earth's angular velocity.

$$= \frac{\zeta}{\sin \varphi} = 0.991 \text{ revolutions/day}.$$

was some twenty degrees higher than that for X_3, and the viscosity was correspondingly less. The quantity χ was measured directly, and was found to be $-0.00585°$ per sidereal second. Substituting the values of X_4 and X_4' as obtained in Table II., the vertical component of the earth's angular velocity becomes:

$$\zeta = \frac{0.00585}{\left(1 + \frac{29.9}{27.5}\right)}$$

$$= 0.00280 \text{ degrees/second,}$$

$$= 0.673 \text{ revolutions/day}.$$

But by equation (3)

$$\zeta = + 18.42 \, \alpha,$$

so that

$$\alpha = 0.03656$$

and

$$\omega = 0.991 \text{ revolutions/day}.$$

The length of the day is therefore 24 hours 12 minutes in sidereal time.

These values of ω, φ and ψ may be compared with their values as determined astronomically thus:

By Data from Earth-Rotation Ring.	By Astronomical Data.	Difference.
$\omega = 0.991$ revs/day	1.000 revs/day	0.9%
$\varphi = 42.8°$	40.4°	2.4°
$\psi = 28.7°$	30.1°	1.4°

The remarkable agreement of the two values of ω is only accidental, since if the true value of φ is used in determining ω by data from the earth-rotation ring,

$$\omega = \frac{\zeta}{\sin \varphi} = \frac{0.673}{0.650} = 1.034 \text{ revs./day},$$

which represents a difference of 3.4 per cent. Although the comparatively low degree of accuracy of these data renders them valueless for work which requires precision, it is interesting to find that these quantities can be determined without reference to astronomical observations.

In conclusion I wish to thank Professor Russell of the Department of Astronomy for his suggestions, and Professor Magie for his kind encouragement in carrying on this experiment.

PALMER PHYSICAL LABORATORY,
   PRINCETON, N. J.
      November 3, 1914.

Watching the Earth Revolve

An Apparatus That Enables the Movements of the Earth to be Directly Studied

By Arthur H. Compton

For most people the fact that the sun rises in the morning, travels slowly across the sky and sets in the evening is sufficient evidence that the earth goes around. Our ancestors, however, believed for the same reason that the sun and moon and stars all actually move across the sky while the earth itself stands still. Indeed, the attempt of Copernicus and Galileo to dispel this idea, which seemed so evident as to be almost axiomatic, was the cause of their bitter persecution. It is really impossible to prove definitely by means of observations on the heavenly bodies whether the earth really revolves while the stars remain fixed or whether it is the stars which revolve about the earth. Even though we may show that these bodies are millions or trillions of miles from us, we can still explain their apparent daily motion by keeping the earth at rest if good proof that the earth is actually revolving.[1] Even his experiment, however, did not show that all the apparent motion of the stars across the heavens is due to the turning of the earth. Since a pendulum swings in a vertical plane, it is only the part of the earth's rotation about a vertical axis which Foucault's apparatus was able to measure. Suppose that the pendulum is set up at the point O (Fig. 2) on the earth's surface. It is evident that there will be some rotation about the vertical axis OZ, but this will be less rapid than the rotation about an axis OP, parallel to the earth's axis. If the earth turns around the axis $O'P'$ once in 24 hours, there ought to be a rotation about a vertical axis at Paris, whose latitude is 49 degrees, at the rate of once in about 32 hours; and by means of his enormous pendulum Foucault showed that such a rotation the rotation about these three axes is measured, not only the length of the day, but also the position of the true north and the latitude can be calculated, and this wholly independent of astronomical observations.

The earth rotation ring shown in the photographs was made for the purpose of measuring these three components of the earth's rotation. The principle on which this apparatus works is comparatively simple. The instrument consists essentially of a circular tube filled with water and mounted on an axis in its own plane, as in Fig. 3. This apparatus is set in a plane perpendicular to the axis OC, about which the earth's rotation is to be measured. If the rotation is in the direction indicated by the solid arrows, it will be seen that the side A of the ring is moving toward the left relative to the other side, and after the ring has been stand-

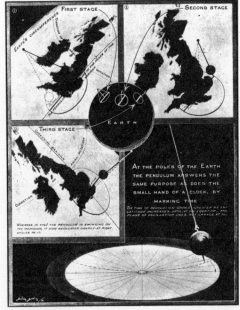

Fig. 1.—Foucault's pendulum, which was the first satisfactory means of showing that the earth actually revolves.

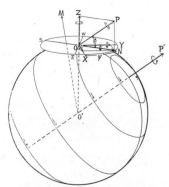

Fig. 2.—Foucault's pendulum was able to measure the earth's rotation only about a vertical axis OZ, while the earth rotation ring measures the rotation about the three axes OX, OY and OZ.

The actual length of the day can then be calculated, which was impossible from Foucault's experiment, and the latitude and the position of the true north can also be determined.

ing a few minutes the water within the tube has the same sort of motion. Now let the ring be quickly turned half way around about its axis, so that the part A comes to the nearer side, as shown by the dotted lines. It is evident that the water in that part of the tube will retain a large part of its original motion toward the left, so that there will be a relative motion between the water and the microscope, which turns with the earth. The speed of this relative motion will of course depend upon how fast the earth is revolving about the

we suppose that the stars are traveling through the heavens with a sufficiently great speed. In fact, this is the assumption on which Ptolemy based his theory of the universe.

It was not until the middle of the last century that Foucault performed his famous pendulum experiment in the Pantheon at Paris (Fig. 1), which was the first

[1] This experiment is described in the SCIENTIFIC AMERICAN, February 14th, 1914.

actually exists. But the fact that there is such a rotation about the vertical axis does not show what the real angular velocity of the earth is nor the direction of the axis about which the earth turns. For example, a comparatively small rotation about such an axis as OM would give the same effect on Foucault's pendulum as a much more rapid rotation about the axis $O'P'$. In order to show that all the apparent motion of the stars across the sky is due to the earth's rotation, it is necessary to determine, without observations on the stars, how fast the earth is revolving, and where its axis is located. This requires more data than are given by Foucault's experiment.

If we can measure the rotation about two horizontal axes, OX and OY, as well as about the vertical axis OZ, the earth's rotation will be completely determined. For by combining the rotation about the OX and the OY axes, the rotation about a north and south axis ON can be found, and combining this rotation with that about the vertical axis the true rate of the earth's rotation about OP can be calculated. It is evident that by comparing the relative magnitudes of the rotation about the OX and the OY axes the angle ψ, or the azimuth of the X axis can be obtained, and from the ratio of the rotations about ON and OP the angle ϕ, which is the latitude of the observer, can be determined. Thus, if

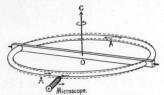

Fig. 3.—If the earth is revolving about the axis OC, when the ring is reversed there is a relative motion between the water and the microscope as shown by the dotted arrows.

Fig. 6.—Measuring the absolute magnitude of the earth's rotation about a vertical axis.

axis OC as well as upon the dimensions of the ring. With the apparatus here described the motion was usually about as fast as that of the minute hand of a watch, and could easily be seen through the microscope.

The ring used in these experiments was made of 1-inch tubing, bent into a circle a foot and a half in diameter. Where the windows were placed the tube was constricted somewhat so as to increase the velocity of the water which was being watched. The motion of the water which filled the tube was made visible by shaking up with it a mixture of coal oil and carbon

Fig. 4.—Watching the earth revolve. The apparatus is in a constant temperature room just above freezing point to avoid convection currents in the water.

tetrachloride of the same density as water, which formed small suspended globules whose motion was easily visible through the observing microscope. In order to avoid spurious motions due to differences in temperature in different portions of the tube, some parts of the experiment had to be performed with the apparatus boxed up in a cold room, as in Fig. 4, but in measuring the effect due to the vertical component of the earth's rotation, as in Figs. 5 and 6, no such particular precautions had to be taken.

When the ring was held in a vertical plane, as in Fig. 4, the oil globules are always seen to rise on the east side of the tube and go down on the west side, after the ring is reversed. This shows conclusively that the earth is turning over from West to East. Similarly, if the ring is in a horizontal plane, a motion to the left is always observed, which, as we saw above, indicates a motion of the earth in a counter clockwise direction about a vertical axis. It is an interesting experiment to project the motion of the oil globules through the microscopes onto a screen, with the apparatus set up as in Fig. 5. In this manner a room full of people can be shown a moving picture of the earth going around.

As an average of a number of readings, the ratio of the velocity observed about the OY axis to that about the OX axis indicated that the true north was 61.3 degrees from the OX axis, and when the motion about the vertical axis OX was determined, the latitude ϕ was found to be 42.8 degrees. In order to find out from these figures how fast the earth is going around, the apparatus was set up as in Fig. 6, keeping the ring in a horizontal position in order to measure the earth's rotation about a vertical axis. The spectrometer table upon which the apparatus was placed could be turned at any desired speed by means of the driving clock C. First a set of readings was taken with the clock stopped, and the motion of the globules to the left was measured. Then the clock was started, and was so adjusted that the globules moved just as fast toward the right as they had moved before toward the left. It is evident that the spectrometer table was then turning backward twice as fast, relative to the earth, as the earth itself was turning forward. The spectrometer table was turning at the rate of 1.346 times per day, which means that the earth is turning about a vertical axis at the rate of 0.673 revolutions per day. Since the rate of the rotation about this axis to that about OP was already known, it was easy to calculate that the rate of the earth's rotation about its axis is 0.991 revolutions per day. That is, the length of the day, according to these data is 24 hours and 12 minutes.

It is interesting to compare these values of the azimuth, the latitude and the length of the day with their values as determined astronomically, thus:

	By data from earth rotation ring.	By astronomical data.
Day	24.2 hours	24.0 hours
Latitude	42.8 degrees	40.4 degrees
Azimuth	61.3 degrees	59.9 degrees

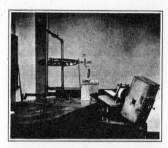

Fig. 5.—The apparatus set up with a projection lantern for showing real moving pictures of the earth's rotation upon a screen.

These figures show conclusively, within the limit of experimental error, that the earth turns about an axis which is identical with its astronomical axis, and that the rate of its rotation is that determined by astronomical observations. Thus it is evident that it is the earth alone which revolves, while the stars remain relatively fixed.

The Distribution of the Electrons in Atoms.

THE spectra which are obtained by the diffraction of X-rays by crystals are characteristic both of the substance which emits the X-rays and of the crystal which acts as the grating. If the lines of an accurately ruled plane grating are small in width compared with their distance apart, the intensities of the different orders of spectra are nearly the same. If, however, the lines have a width comparable with the grating constant, the intensities of the higher orders rapidly diminish. When a crystal diffracts a beam of X-rays, the different layers of atoms correspond to the lines of the ordinary transmission grating, so that the relative intensity of the higher orders of spectra will depend upon the ratio of the effective diameter of the atoms in scattering the X-rays to the distance between the successive layers of atoms.

There are good reasons for believing that it is the electrons in atoms which scatter the X-rays. On this assumption it may be shown that if the density of the space distribution of the electrons in each layer of atoms is some function $f(z)$ of z, where the z axis is taken normal to the reflecting planes, the ratio of the amplitude of the nth order spectrum to the amplitude it would have if all the electrons were in the same plane is:—

$$\frac{P}{P_0} = \frac{\int_a^b f(z) \cos(\beta + 2\pi nz/d) dz}{\cos\left(\beta + \pi n \frac{b+a}{d}\right) \int_a^b f(z) dz},$$

where $b - a = d$ is the grating space, and β is the phase angle of the reflected ray. If it is possible to find some function $f(z)$ which will lead to the values of P/P_0 as determined experimentally, an indication will be obtained of the distribution of the electrons in the atoms.

W. H. Bragg has published experimental results (*Phil. Mag.*, vol. xxvii., p. 895, 1914) showing the rate of variation of the intensity with the order when X-rays are reflected from rock-salt. It can be shown from his data that the intensities of the different orders cannot be accounted for by assuming that the atoms in the salt crystal are made up of single rings of electrons, or by assuming a uniform volume distribution of the electrons in spheres. A distribution which fits Bragg's data acceptably is an arrangement of the electrons in equally-spaced, concentric rings, each ring having the same number of electrons, and the diameter of the outer ring being about 0·7 of the distance between the successive planes of atoms.

If, as D. L. Webster assumes (*Phys. Rev.*, vol. v., p. 238, 1915) the trains of waves of the primary beam of X-rays are short compared with the distance which the rays penetrate the crystal, certain corrections have to be applied to the experimental data, and on this assumption it can be shown that the average distance of the electrons from the centre of the atom is small compared with the distance between the atoms.

Experiments are now in progress to test the validity of Webster's assumption and to determine more accurately the rate of variation of the intensity of the reflected beam with the order. It is hoped that it will be possible in this manner to obtain more definite information concerning the distribution of the electrons in the atoms.   ARTHUR H. COMPTON.

Palmer Physical Laboratory, Princeton, N.J.,
    April 29.

THE VARIATION OF THE SPECIFIC HEAT OF SOLIDS WITH TEMPERATURE.

By Arthur H. Compton.

THE very considerable success of the quantum hypothesis in explaining the variations of the specific heat of solids with temperature has been taken as a strong confirmation of that hypothesis. Before this evidence can be considered as conclusive, however, it is necessary to see if there may not be some other satisfactory solution of the problem of specific heat, which does not involve the conception of quanta. It has been pointed out by several writers[1] that the sharp decrease of the specific heat of solids at low temperatures can be qualitatively explained if it is assumed that at these temperatures the atoms become so intimately associated that degrees of freedom are lost. In fact Benedicks[1] has been able to obtain an empirical expression on this assumption which, with properly chosen constants, fits the experimental data acceptably. In the present paper an assumption is introduced which leads directly to an expression for the variation of the specific heat with temperature which will be shown to agree at least as well with experiment as the expressions derived from the quantum hypothesis.

The assumption on which the following work is based is:

If the relative energy between two neighboring atoms in a solid falls below a certain critical value, the two atoms become agglomerated[2] so that the degree of freedom between them vanishes; but as soon as the energy increases again above the critical value, the degree of freedom reappears.

The defence of this assumption from various lines of evidence will form the subject of a later paper.

Derivation of a Formula for the Specific Heat.

The energy content of the unagglomerated degrees of freedom may be written, according to equipartition,

$$U_u = nRT,$$

---

[1] F. Richarz, Zeitschr. f. anorg. Chem., 58, 356; 59, 146. J. Duclaux, Compt. Rend., 155, 1015. C. Benedicks, Ann. d. Phys., 42, 133.

[2] The term "agglomeration," suggested by Benedicks, is used to indicate any state of association of the atoms on account of which degrees of freedom for thermal motion disappear.

where n is the number of unagglomerated degrees of freedom, R the gas constant for a single molecule, and T the absolute temperature. In the case of the chemical combination of two atoms, when a degree of freedom vanishes its energy is usually transmitted to the other degrees of freedom in the form of heat of formation. In the case of endothermic substances, however, the vanishing degree of freedom absorbs a certain amount of energy. We may assume provisionally, therefore, that when two atoms in a solid agglomerate they retain a certain amount of potential energy. The amount of this energy will be,

$$U_a = \gamma(3N - n),$$

where γ is the potential energy of each agglomerated degree of freedom, and $3N$ is the greatest possible number of degrees of freedom. The total energy content of the solid is therefore,

$$U = U_u + U_a = nRT + \gamma(3N - n).$$

If we call P the probability that a certain possible degree of freedom shall actually exist, $P = n/3N$, and

(1) $$U = 3NRTP + 3N\gamma(1 - P).$$

In the case of a solid the kinetic energy of a degree of freedom is on the average equal to its potential energy. We may assume, therefore, that the probability for a certain value of the potential energy is equal to the probability for the same value of the kinetic energy. By Maxwell's distribution law, the probability that the relative velocity of two atoms along the line of their centers shall lie between $|u|$ and $|u + du|$ is

$$\frac{2}{\alpha\sqrt{\pi}} e^{-\frac{u^2}{a^2}} du,$$

where

$$\alpha = 2\sqrt{\frac{RT}{m}}.$$

This is therefore also the probability that the relative kinetic energy shall lie between $\tfrac{1}{2}mu^2$ and $\tfrac{1}{2}m(u + du)^2$. Since the probability for the potential energy is equal to that for the kinetic energy, the probability that the relative potential energy shall lie between $\tfrac{1}{2}mv^2$ and $\tfrac{1}{2}m(v + dv)^2$ is similarly,

$$\frac{2}{\alpha\sqrt{\pi}} e^{-\frac{v^2}{a^2}} dv.$$

Thus the probability that the total energy shall lie between $\tfrac{1}{2}m(u^2 + v^2)$ and $\tfrac{1}{2}m\{u^2 + v^2 + d(u^2 + v^2)\}$ is the product of these two expressions, or

$$\frac{4}{\alpha^2 \pi} e^{-\frac{u^2}{a^2}} e^{-\frac{v^2}{a^2}} du\, dv.$$

If we let ϵ be the critical value of the energy below which a degree of freedom remains agglomerated, the probability that the degree of freedom shall be agglomerated is:

$$P' = \int_0^{\frac{1}{2}m(u^2+v^2)=\epsilon} \frac{4}{\alpha^2 \pi} e^{-\frac{u^2}{a^2}} e^{-\frac{v^2}{a^2}} du dv,$$

$$= \frac{4}{\alpha^2 \pi} \int_0^{u^2=\frac{2\epsilon}{m}} e^{-\frac{u^2}{a^2}} du \int_0^{v^2=\frac{2\epsilon}{m}-u^2} e^{-\frac{v^2}{a^2}} dv,$$

or

$$= \frac{4}{\alpha^2 \pi} \int_0^{\beta} e^{-\frac{u^2}{a^2}} du \int_0^{\sqrt{\beta^2-u^2}} e^{-\frac{v^2}{a^2}} dv;$$

where

$$\beta^2 = \frac{2\epsilon}{m}.$$

This integral may be determined by a series method, the solution being,

$$P' = \frac{\beta^2}{\alpha^2} - \frac{1}{2!}\frac{\beta^4}{\alpha^4} + \frac{1}{3!}\frac{\beta^6}{\alpha^6} - \cdots.$$

This is the probability that a possible degree of freedom shall be agglomerated. The probability that it shall actually exist is therefore:

$$P = 1 - P'$$

$$= 1 - \frac{\beta^2}{\alpha^2} + \frac{1}{2!}\frac{\beta^4}{\alpha^4} - \frac{1}{3!}\frac{\beta^6}{\alpha^6} + \cdots.$$

But

$$e^{-\frac{\beta^2}{\alpha^2}} = 1 - \frac{\beta^2}{\alpha^2} + \frac{1}{2!}\frac{\beta^4}{\alpha^4} - \frac{1}{3!}\frac{\beta^6}{\alpha^6} + \cdots,$$

and by comparison of series,

$$P = e^{-\frac{\beta^2}{\alpha^2}}.$$

We may substitute for β^2/α^2 its equivalent,

$$\frac{\beta^2}{\alpha^2} = \frac{\frac{2\epsilon}{m}}{\frac{4RT}{m}} = \frac{\epsilon}{2RT} = \frac{\tau}{T}, \quad \text{where} \quad \tau = \frac{\epsilon}{2R},$$

then

(2) $$P = e^{-\frac{\tau}{T}}.$$

Substituting this value of P in equation (1) we have:

(3) $$U = 3NRTe^{-\frac{\tau}{T}} + 3N\gamma(1 - e^{-\frac{\tau}{T}}),$$

and the specific heat is:

(4) $$C_v = \frac{dU}{dT} = 3NRe^{-\frac{\tau}{T}} + \frac{3NR\tau}{T}e^{-\frac{\tau}{T}} - 3N\gamma\frac{\tau}{T^2}e^{-\frac{\tau}{T}}.$$

Since there is a T^2 in the denominator of the last term, it is evident that as long as γ has a finite value, T can be made so small that C_v will become negative. This is impossible, as it implies a condition of instability, so we must place $\gamma = 0$, and

(5) $$C_v = 3NRe^{-\frac{\tau}{T}}\left(\frac{\tau}{T}+1\right).$$

If $T = \infty$, $C_v = 3NR = C_\infty$,

(6) $$\therefore \frac{C_v}{C_\infty} = e^{-\frac{\tau}{T}}\left(\frac{\tau}{T}+1\right).$$

This is the expression for the variation of the specific heat with temperature to which our assumption leads.

Testing this Equation.

The curves of Fig. 1 show how this formula compares with that of Debye for the specific heat.[1] The solid line is plotted from equation (6)

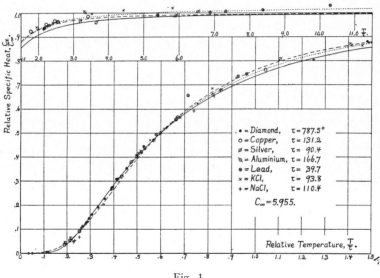

Fig. 1.

and the broken one from Debye's equation,

$$\frac{C_v}{C_\infty} = \frac{12}{x^3}\int_0^x \frac{\xi^3 d\xi}{e^\xi - 1} - \frac{3x}{e^x - 1},$$

where $x = \Theta/T$. The two curves have been made to coincide at $C_v/C_\infty = 0.5$. That they are in general good agreement is evident at a glance.

[1] Debye, Ann. d. Phys., 39, 789. Debye's formula is used because his has been shown to be the most accurate of any of the existing expressions (cf. E. H. Griffiths and E. Griffiths, Proc. Roy. Soc., A, 90, 558).

Their comparative accuracy has been tested with the data used by Debye and by Nernst and Lindemann[1] in testing their own formulæ. I have taken their calculations of C_v from the observed value of C_p. The value of γ for each of the substances is chosen so as to bring the experimental values as near as possible to the two curves where they cross.[2]

At lower temperatures, though the curves are not far apart, there is an evident preponderance of experimental evidence in favor of the solid curve. Above $T/\tau = 0.6$, however, the experimental values gradually rise from the solid curve, cross the broken one at about 2.25, and continue approximately parallel to the solid curve but at a higher level. The dotted curve is drawn as an approximate experimental mean. It is evident that the data fit neither formula accurately for these temperatures, but that the values of C_v/C_∞ approach a limit some two per cent. higher than unity. That these high values of C_v are not due to proximity to the melting point is evident when one notices that none of the experiments are made at a temperature closer than 191° to the melting point, while in the case of KCl the value of C_v/C_∞ rises above unity when the temperature is only .40 as high as the melting point. A possible explanation of the high values of the specific heat at these temperatures is that there may be more degrees of freedom in a solid than indicated by Rayleigh's formula, $C_\infty = 3R$. Whatever the cause of this discrepancy, however, it is necessary on any theory that C_v shall never become greater than C_∞, so we must assume a value of C_∞ greater than $3R = 5.955$, which is the value used in calculating the values of C_v/C_∞ in Fig. 1. The value 6.081 has therefore been chosen as an experimental limit which C_v seems to approach.

Fig. 2 shows the same theoretical curves and the experimental data plotted with this new value of C_∞. The curves have here been made to coincide at $C_v/C_\infty = 0.65$, in order to show more clearly the differences between the two formulæ. It will be seen that this correction makes the observed specific heats conform very well with the curve plotted according to my expression, while they vary consistently from that of Debye. Thus while the data can in no way be made to conform at all temperatures with Debye's formula, by making a correction which would seem necessary in any case, the data may be made to fit well the equation

---

[1] Zeitschr. für Electrochemie, 17, p. 817.

[2] In the following tables the values of C_v for Al, Pb, KCl and NaCl are taken from Nernst and Lindemann's paper (*loc. cit.*), those for Cu, Ag and diamond from that of Debye (*loc. cit.*). C_v is calculated from the observed values of C_p according to the formula, due to Nernst and Lindemann, $C_v = C_p - 0.0214 C_p^2 T/T_s$, where T_s is the melting point of the substance. C_∞ is taken to be 6.081 as explained in the text.

T	C_v Obs.	C_v Calc.	T	C_v Obs.	C_v Calc.
Diamond, $\tau = 794.0°$			Silver (continued)		
30°	0.00	0.00	45.5°	2.46	2.44
42	0.00	0.00	51.4	2.80	2.83
88	0.03	0.01	53.8	2.89	2.98
92	0.03	0.015	77.0	4.04	4.04
205	0.62	0.63	100	4.80	4.65
209	0.66	0.67	200	5.61	5.61
220	0.72	0.75	273	5.75	5.79
222	0.76	0.80	331	5.71	5.87
232	0.86	0.87	535	5.90	5.93
243	0.95	0.98	589	5.99	5.99
262	1.14	1.18	Aluminum, $\tau = 169.6°$		
284	1.35	1.41			
306	1.58	1.60	32.4°	0.25	0.19
331	1.83	1.87	35.1	0.33	0.28
358	2.11	2.13	83.0	2.40	2.40
413	2.64	2.62	86.0	2.51	2.51
1169	5.24	5.18	88.3	2.61	2.60
Copper, $\tau = 133.5°$			137	3.91	3.94
			235	5.17	5.09
23.5°	0.22	0.14	331	5.58	5.55
27.7	0.32	0.26	433	5.74	5.72
33.4	0.54	0.56	555	5.98	5.85
87	3.32	3.32	NaCl, $\tau = 113.5°$		
88	3.37	3.36			
137	4.53	4.54	25.0°	0.29	0.36
234	5.50	5.40	25.5	0.31	0.38
290	5.66	5.60	28.0	0.40	0.52
323	5.75	5.69	67.5	3.05	3.04
450	5.87	5.86	69.0	3.12	3.10
Lead, $\tau = 38.4°$			81.4	3.52	3.60
			83.4	3.72	3.71
23.0°	2.95	3.04	138	4.79	4.87
28.3	3.91	3.69	235	5.55	5.57
36.8	4.38	4.38	KCl, $\tau = 96.1°$		
38.1	4.43	4.45			
85.5	5.57	5.60	26.9°	0.76	0.78
90.2	5.63	5.67	33.7	1.25	1.35
200	5.91	5.98	39.0	1.83	1.80
273	5.99	6.02	52.8	2.79	2.78
290	5.99	6.02	63.2	3.34	3.34
332	6.03	6.03	76.6	4.08	3.91
409	6.15	6.05	86.0	4.33	4.22
Silver, $\tau = 92.0°$			137	5.18	5.15
			235	5.73	5.69
35.0°	1.58	1.58	331	5.93	5.87
39.1	1.90	1.93	416	6.02	5.95
42.9	2.26	2.25	550	6.09	5.99

just derived. In fact, all variations from the theoretical values seem to lie within the limits of experimental error.

Discussion of the Formula.

Some interesting aspects of formula (6) appear if the exponent of e is written in a different form. τ is defined as $\epsilon/2R$, where ϵ is the energy required to liberate an agglomerated degree of freedom. This quantity has the same dimensions as Planck's energy quantum $h\nu$, both being elementary units of energy. In fact Benedicks has shown,[1] from a consideration of the relation of hardness to the frequency of vibration of an atom, that ϵ is probably proportional to ν. If we call the factor of proportionality b, the probability becomes

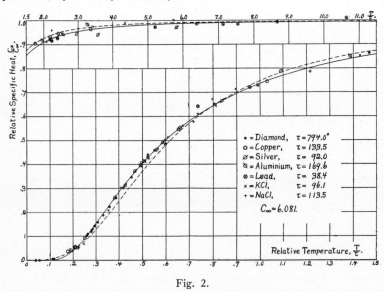

Fig. 2.

(7)
$$P = e^{-\frac{\tau}{T}} = e^{-\frac{\epsilon}{2RT}} = e^{-\frac{b\nu}{2RT}} = e^{-\frac{\beta\nu}{T}},$$

where

$$\beta = \frac{b}{2R}.$$

The value of β can easily be determined from the expression $\beta = \tau/\nu$, which follows from equation (7). The values of τ for the different substances considered are determined by their specific heats as shown in Fig. 2. The values of ν may be taken as those assigned by Nernst and Lindemann's formula for the specific heat, since their values of ν have been shown to be in very accurate agreement with the characteristic

[1] *Loc. cit.*

frequencies of those substances whose reststrahlen can be obtained. The different values of β as thus determined are given in the following table.

Substance.	τ	$\nu \times 10^{-12}$	$\beta \times 10^{11}$	Difference from Mean.
Lead..................	38.4	1.9	2.02	+0.3%
Silver.................	92.0	4.5	2.04	+1.4
KCl...................	96.1	4.73	2.03	+0.6
NaCl..................	113.1	5.8	1.96	−2.9
Copper................	133.5	6.6	2.02	+0.3
Aluminium	169.6	8.3	2.04	+1.4
Diamond..............	794.0	40.0	1.98	−1.4

β is thus determined just as accurately as ν is known, and since it remains constant over so great a range of frequencies, equation (7) may be considered valid, even though its derivation is not rigid.

It is interesting to see how the quantity b in this equation compares with the similar quantity h of Planck's expression. b is determined by the equation $b = 2R\beta$, in which the accepted value of R is about 1.35×10^{-16} erg deg.$^{-1}$, and $\beta = 2.01 \times 10^{-11}$ deg. sec. Thus

$$b = 5.44 \times 10^{-27} \text{ erg sec.},$$

while Planck's constant h has the value

$$h = 6.55 \times 10^{-27} \text{ erg sec.}$$

The value for b found here is in excellent agreement with certain values of h as determined by photoelectric methods: Richardson and Compton[1] 5.4×10^{-27}; Hughes[2] 5.6×10^{-27}; Cornelius[3] 5.7×10^{-27}; though Kadesch[4] and Millikan[5] very recently have obtained photoelectric values of h more nearly 6.55×10^{-27}.

Now that equation (7) has been established we are in a position to make a new application of Debye's theory which interprets the heat energy of solids in terms of their elastic vibrations. He considers a whole spectrum of frequencies, the number of vibrations between the frequencies ν and $\nu + d\nu$ being

(8) $$dn = \frac{9N}{\nu_m^3} \nu^2 d\nu,$$

---

[1] Phil. Mag., 24, 574.
[2] Phil. Trans. Roy. Soc., 212, 205.
[3] K. T. Compton, Phys. Rev., 1, 382.
[4] Phys. Rev., 3, 367.
[5] Phys. Rev., 4, 73.

where $3N$ is the total number of degrees of freedom in the solid, and ν_m is the maximum possible frequency, determined by the equation:

$$(9) \qquad \frac{1}{\nu_m^3} = \frac{4\pi V}{9N} \left\{ 2\left(\frac{2\rho(1+\sigma)\chi}{3(1-2\sigma)}\right)^{\frac{3}{2}} + \left(\frac{2\rho(1+\sigma)\chi}{3(1-\sigma)}\right)^{\frac{3}{2}} \right\},$$

where V is the volume, ρ the density, χ the compressibility, and σ Poisson's ratio, for the substance considered. The average energy of a degree of freedom of frequency ν is, by equations (3) and (7),

$$(10) \qquad \frac{3NRTe^{-\frac{\beta\nu}{T}}}{3N} = RTe^{-\frac{\beta\nu}{T}};$$

so the total energy in the solid is

$$U = \int_0^{\nu_m} \frac{9N}{\nu_m^3} \nu^2 \cdot RTe^{-\frac{\beta\nu}{T}},$$

$$= \frac{9NRT}{\nu_m^3} \left\{ \frac{2T^3}{\beta^3} - e^{-\frac{\beta\nu_m}{T}} \left(\frac{T}{\beta}\nu_m^2 + \frac{2T^2}{\beta^2}\nu_m + \frac{2T^3}{\beta^3}\right) \right\}.$$

Substituting $\tau = \beta\nu_m$,

$$U = 9NRT \left\{ \frac{2T^3}{\tau^3} - e^{-\frac{\tau}{T}} \left(\frac{T}{\tau} + \frac{2T^2}{\tau^2} + \frac{2T^3}{\tau^3}\right) \right\}.$$

According to Debye's assumptions ν_m, and hence also τ, are not functions of T, so the specific heat is:

$$C_v = \frac{dU}{dT} = 9RN \left\{ 8\frac{T^3}{\tau^3} - e^{-\frac{\tau}{T}}\left(8\frac{T^3}{\tau^3} + 8\frac{T^2}{\tau^2} + 4\frac{T}{\tau} + 1\right) \right\},$$

and

$$(11) \qquad \frac{C_v}{C_\infty} = 3\left\{ 8\frac{T^3}{\tau^3} - e^{-\frac{\tau}{T}}\left(8\frac{T^3}{\tau^3} + 8\frac{T^2}{\tau^2} + 4\frac{T}{\tau} + 1\right) \right\}.$$

Fig. 3 shows how this expression compares with equation (6). The solid line as before represents the first equation derived, while the broken one represents equation (11). Although the difference between these two curves is not large, it is evident, particularly at low temperatures, that this new formula does not represent the facts accurately. It is necessary to conclude, therefore, either that the energy of each degree of freedom is not accurately expressed by equation (10) or that some of the assumptions on which Debye's theory is based are not valid.

It is evident that if equation (11) is to be valid, ν_m must not vary with the temperature. That is, by equation (9), the quantities σ and χ must be independent of the temperature. While there is no evidence of any considerable variation in the value of σ, Grüneisen has shown[1]

[1] Ann. d. Phys., 33, 1239. Some of Grüneisen's results indicating such a variation of the compressibility with the temperature may be given:

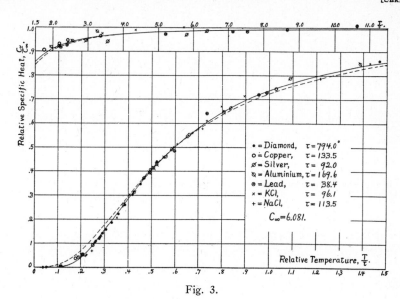

Fig. 3.

Compressibility × 10⁶.

T	Al	Fe	Cu	Ag	Pt	Sn	Pb
83°	1.32	0.606	0.718	0.709	0.374	2.1	2.5
290	1.46	0.633	0.773	0.763	0.392	3.1	3.2
373	1.70	0.652	0.801	0.820	0.398	—	—

that there is a general and decided increase in the compressibility χ with the temperature. It is evident, therefore, that account must be taken of the variation of ν_m with the temperature. If this is done, equation (11) becomes:

$$\frac{C_v}{C_\infty} = 3 \left\{ 8\frac{T^3}{\tau^3} - e^{-\frac{\tau}{T}}\left(8\frac{T^3}{\tau^3} + 8\frac{T^2}{\tau^2} + 4\frac{T}{\tau} + 1\right) \right.$$
$$\left. - \left[6\frac{T^4}{\tau^4} - e^{-\frac{\tau}{T}}\left(6\frac{T^4}{\tau^4} + 6\frac{T^3}{\tau^3} + 3\frac{T^2}{\tau^2} + \frac{T}{\tau}\right)\right]\frac{d\tau}{dT} \right\}.$$

The order of magnitude of $d\tau/dT$ is 5×10^{-4}, so the last term is negligible except at high temperatures; but instead of considering τ to be constant, its particular value which corresponds to the temperature for which C_v/C_∞ is evaluated must be employed in order to make equation (11) valid. In the case of aluminum, for example, the compressibility at 373° is 29 per cent. greater than at 83°. This corresponds to a difference in ν_m, and hence also in τ of 13.5 per cent., and in the specific heat at the lower temperature of about 45 per cent. Thus if the specific heat of aluminum at 83° is calculated from its elastic constants at 373°

the resulting value is some 45 per cent. greater than if calculated from its constants at the lower temperature. This shows conclusively that equation (11) is not valid if τ is considered independent of the temperature.

That the same argument applies to Debye's formula for the specific heat is shown by the form of his equation at low temperatures,

$$\frac{C_v}{C_\infty} = 77.93 \frac{R^3 T^3}{h^3 \nu_m^3}.$$

In this equation the value of the ratio C_v/C_∞ for aluminum varies 46 per cent. according as the value of ν_m is calculated from the elastic constants at 373° or at 83°. It is thus evident that the error introduced by Debye's assumption that ν_m is independent of the temperature is much too large to be neglected.

It can be shown, moreover, that the fundamental assumption on which Debye's theory is based, that the heat energy of a solid lies in the elastic vibrations of the body as a whole, does not represent the truth. It may be shown that the velocity of propagation of a thermal disturbance in a solid is directly proportional to the diameter of the elementary vibrator. If, as Debye assumes, the whole substance is capable of vibrating as a unit in its heat motion, the thermal conductivity should therefore be very great. This may be shown more clearly in the following manner. Consider two infinite parallel planes in an elastic medium, at one of which the medium is maintained with vibrational energy greater than at the other corresponding to a difference in temperature of one degree. The thermal conductivity, or rate at which energy is transmitted from the hotter to the colder plane per unit area will then be equal to the product of the difference in the energy content of the medium at the two planes by the velocity of transmission of the vibrational disturbance; i. e.:

$$\text{thermal conductivity} = V \cdot C(t_1 - t_2) = V \cdot C,$$

where C is the heat capacity per unit volume of the medium, and V is the mean velocity of propagation of a vibrational disturbance. For aluminum C is about 0.5 calories cm.$^{-3}$ deg.$^{-1}$, and V may be taken as about 5×10^6 cm. sec.$^{-1}$, so the thermal conductivity should be of the order of magnitude of 2.5×10^5 calories cm.$^{-2}$ sec.$^{-1}$ deg.$^{-1}$. This is wholly out of accord with the experimental value, which is 0.5 calories cm.$^{-1}$ sec.$^{-1}$ deg.$^{-1}$. In fact the dimensions themselves are different, the rate of heat transmission according to this assumption being independent of the distance apart of the two parallel planes. It is evident, therefore, that the heat energy of solids is not contained in the elastic vibrations of the body as a whole, but is contained in the motion of very much smaller elements. Since the lower limit of the frequencies which enter into Debye's theory is determined by the dimensions of the vibrator, the range of frequencies in which thermal motion occurs is thus limited. If the elementary vibrator is of atomic dimensions, the only frequencies to be considered are the natural frequencies of the atoms themselves.[1] Thus the assumption that thermal motion occurs in all the possible frequencies from o to ν_m cannot be accepted.

Debye's formula for the specific heat must therefore be considered to be largely empirical. In order to compare the results of this agglomeration hypothesis with those of the quantum hypothesis, equation (6) should be compared rather with Einstein's formula,[2]

$$\frac{C_v}{C_\infty} = \frac{h^2\nu^2}{R^2T^2} \frac{e^{h\nu/RT}}{(e^{h\nu/RT}-1)^2},$$

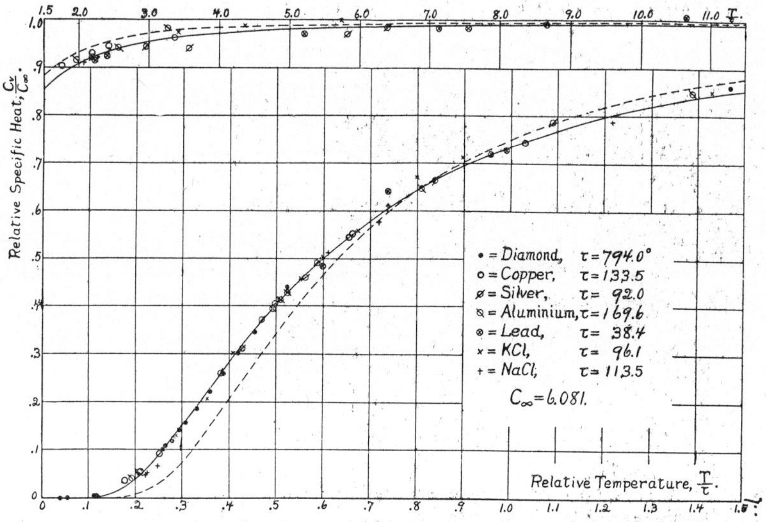

Fig. 4.

which is a valid deduction from the quantum hypothesis. Fig. 4 shows how these two expressions compare. The solid line represents my

---

[1] The great increase in thermal conductivity at low temperatures may be shown to indicate that the diameter of the elementary vibrator becomes comparatively large at low temperatures. Under these circumstances Debye's assumption of a whole spectrum of frequencies may be valid. This may account for the fact that at extremely low temperatures, $T/\tau = 0$ to 0.22, the observed specific heats are slightly larger than indicated by equation (6). If this be true, equation (11) should hold at these temperatures, and Debye's "third power law" would still be true.

[2] Einstein, Ann. d. Phys., 22, 180.

formula and the broken one that of Einstein. The experimental values are the same as in Fig. 2. It is evident that Einstein's expression is much the less accurate of the two. Thus from the standpoint of the variation of specific heat with temperature the assumption is strongly supported that at sufficiently low temperatures many of the atoms of a solid become so intimately associated that they lose degrees of freedom.

Summary.

The assumption has been made that a possible degree of freedom between two atoms in a solid actually exists only as long as the relative energy between the two atoms is greater than a certain critical value.

An expression for the variation of the specific heat of solids with temperature has been derived from this assumption, which seems to agree more satisfactorily with experiment than any of the existing formulæ.

Debye's expression for the variation of the specific heat with temperature has been found to be largely empirical; so to compare this agglomeration hypothesis with the quantum hypothesis my formula is rather to be compared with that of Einstein.

The strong support of experimental evidence in this case seems to indicate that this agglomeration hypothesis represents more accurately the condition of the atoms in a solid than does the quantum hypothesis.

I wish to express my thanks to Professor Adams and Professor Magie for their continued interest in this study.

PALMER PHYSICAL LABORATORY,
  PRINCETON, N. J.,
    May 5, 1915.

A PHYSICAL STUDY OF THE THERMAL CONDUCTIVITY OF SOLIDS.

By Arthur H. Compton.

SINCE the physical interpretation of the variation of the specific heat of solids with temperature has been found to be of considerable theoretical importance, it is to be expected that a study of the physical nature of the thermal conductivity of solids may also bring interesting results. The present paper may be considered as a start at such a study.

In order not to be disturbed by thermal conduction due to electrons, we shall consider only non-metallic substances. For the sake of further simplicity we shall consider in the first place the conduction of heat through a crystal, in which the atoms are closely packed. Consider two infinite, parallel planes in such a medium, at a distance x apart, and maintained at a constant difference of temperature, $T_1 - T_0$. Let δx be the average distance in the x direction through which the energy of an atomic collision is transmitted. The temperature drop along this distance δx will evidently be

$$\delta T = \frac{\delta x}{x}(T_1 - T_0).$$

This quantity represents the average difference in temperature between the atom which gives out energy at each collision, and the atom or atoms which absorb it. According to ordinary kinetic theory, if the collisions of the atoms are elastic, the average amount of energy transmitted at each collision may be shown to be

$$\delta \epsilon = \frac{4m_1 m_2}{(m_1 + m_2)^2} R \delta T,$$

where m_1 and m_2 are the masses of the atoms in collision, and R is the gas constant for a single molecule. This may be written

$$\delta \epsilon = b R \delta T,$$

where b is in most cases approximately equal to 1.

In the kind of a medium we are considering, the number of atomic collisions in unit time on unit area of the surface of a plane separating adjacent layers of atoms will be equal to the number of atoms per unit area multiplied by the frequency of oscillation of each atom. That is,

$$\eta = n^2 \nu,$$

where ν is probably equal to the natural frequency of the atoms as determined by optical methods, and n^2 is the number of atoms per unit area of the plane. n^2 will be equal to the 2/3 power of the number of atoms in unit volume in the case of cubic crystals, but not necessarily for other kinds. The rate of transfer of energy per unit area across a plane perpendicular to the x axis will therefore be

$$\frac{dE}{dT} = \eta \cdot \delta\epsilon = n^2\nu \cdot bR \cdot \frac{\delta x}{x}(T_1 - T_0);$$

and since there can be no piling up of energy at any point of the substance this is also the rate of transfer of energy from one boundary plane to the other. But by definition,

$$\frac{dE}{dT} = k\frac{T_1 - T_0}{x},$$

where k is the coefficient of thermal conductivity, so that

$$k = bn^2 R\nu\delta x,$$

or

$$\delta x = k/bn^2 R\nu.$$

According to the experiments of Eucken,[1] the thermal conductivity of crystals is approximately inversely proportional to the temperature. If we call the factor of proportionality, c,

$$k = c/T,$$

and the distance through which the energy of an atomic collision is transmitted is

(1) $$\delta x = c/T \cdot \frac{1}{bn^2 R\nu}.$$

That is, as the temperature falls, and the atoms become more and more closely packed, the distance through which the effect of an atomic collision is transmitted increases, as appears plausible.

Thus we have the distance δx for a crystal expressed in terms of known quantities. For rock-salt at 0° C. these quantities are:

$$c/T = 1.36 \times 10^6 \text{ erg cm.}^{-1} \text{ sec.}^{-1} \text{ deg.}^{-1},$$
$$b = 0.955,$$
$$n = 3.58 \times 10^7 \text{ cm.}^{-1},$$
$$\nu = 5.8 \times 10^{12} \text{ sec.}^{-1},$$
$$R = 1.35 \times 10^{-16} \text{ erg deg.}^{-1};$$

[1] A. Eucken, Ann. d. Physik, 34, 185 (1911).

and thus
$$\delta x = 1.42 \times 10^{-6} \text{ cm}.$$
But the distance between two neighboring atoms in rock-salt is $d = 2.80 \times 10^{-8}$ cm., or $\delta x = 51d$. The value of δx for other crystals is of the same order of magnitude. We find, therefore, that the energy of a collision between two atoms in a crystal is transmitted through a distance which is approximately inversely proportional to the temperature, and many times the distance between two neighboring atoms.

If it should happen that the crystals of which a substance is composed are of smaller dimensions than the average distance through which the effect of a thermal disturbance is transmitted, a decrease in temperature will no longer mean a decrease in the δx of equation (1), since this distance will evidently be limited by the size of the component crystals. On the other hand, it is probable that the frequency of the collisions between the atoms of neighboring crystals will diminish as the amplitude of the vibrations becomes less, since the atoms will not be closely packed at the juncture of the crystals. This would result in a decrease in the conductivity with temperature, such as is actually observed in the case of all amorphous substances.

That something of this kind actually occurs in amorphous substances becomes evident when we consider a definite example. In the case of fused quartz at $-190°$ C., for instance,

$$k = 6.61 \times 10^4 \text{ erg cm.}^{-1} \text{ sec.}^{-1} \text{ deg.}^{-1},$$

$$b = .95$$

$$\nu^6 = 1.82 \times 10^{13} \text{ sec.}^{-1} \text{ (mean value)},$$

$$n = 4.1 \times 10^7 \text{ cm.}^{-1},$$

and if we calculate δx according to equation (1) we obtain

$$\delta x = 1.7 \times 10^{-8} \text{ cm.} = 0.7d.$$

Thus if a collision occurred at each oscillation of an atom, we should have to conclude that the effect of the collision was transmitted through less than the distance between two atoms, which would be meaningless. It is evident, therefore, that some of the atoms in an amorphous substance must oscillate without affecting appreciably the atoms next to them, as suggested above.

The physical interpretation of this result, that the effect of each atomic collision in a crystal is felt at a considerable distance, is an interesting problem. According to the agglomeration theory of the variation of the specific heat of solids with temperature,[1] at low temperatures the

[1] C. Benedicks, Ann. d. Physik, 42, 133 (1913); A. H. Compton, Phys. Rev., 6, 377 (1915).

atoms of solids lose degrees of freedom in a manner very similar to that in which a degree of freedom is lost when two atoms combine chemically. It is evident that if large groups of atoms should thus be formed which would vibrate as units, the thermal conductivity would be greatly increased. Benedicks[1] has accordingly considered the sharp increase in the heat conductivity of crystals at low temperatures to be an indication of the formation of such groups, and hence as a support of the agglomeration hypothesis. If these atomic aggregations vibrate independently, the distance through which the effect of a collision will be transmitted will be equal to their average diameter. As we have seen above, for rock-salt at 0° C. this distance is some 50 times the distance between two neighboring atoms; but the probability that a degree of freedom shall be agglomerated is in this case only 1/3. Thus it is seen that the size of the groups formed by a possible agglomeration of the atoms is not of the proper order of magnitude to account for the thermal conductivity of crystals.

A more probable interpretation would seem to be that each atom transmits the greater part of the energy of every thermal disturbance which reaches it, retaining or scattering the remaining part, the atoms thus acting as if imperfectly elastic. If we assume that the energy of each collision is transferred from atom to atom, being diminished by a definite factor as it traverses each atom, it can be shown by reasoning similar to that above that the thermal conductivity of a crystal in which adjacent atoms are arranged in lines parallel to the axis of conduction is

(2) $$k = bn^2 dR\nu \cdot \frac{2-\alpha}{\alpha^2};$$

where d is the distance between successive atoms in the direction of conduction, and α is the fraction of the energy absorbed by each atom as the thermal disturbance traverses it. This absorbed energy is given out again at the next collision of the atom. For other kinds of crystals the function of α is somewhat different, but is in any case approximately proportional to $1/\alpha^2$ as long as α is small. For rock-salt at 0° C.

$$\alpha = 0.202.$$

According to this explanation the energy of an atomic collision is not transmitted through a definite distance, but dies out gradually as it passes through a substance. The *mean* range, however, must be the same as the δx which we have used above.

If we assume that the fraction of the energy of each thermal disturbance which an atom retains is proportional to its amplitude of vibration, and hence, according to ordinary kinetic theory, to the square root of the temperature, *i. e.*,

$$\alpha = c\sqrt{T},$$

we have,

(3) $$k = bn^2 dRv \cdot \frac{2 - cT^{\frac{1}{2}}}{c^2 T},$$

where c is determined by the conductivity at a given temperature. The following tables show how this expression compares with Eucken's experimental data. The first calculated values are found according to the expression

$$k = c/T,$$

and the second according to equation (3). The value of α at 273° K. is calculated according to equation (2), using the value of d determined by X-ray measurements,[1] and v for quartz in the mean value, 1.82×10^{13}, determined from specific heat data.[2] k is expressed in calories cm.$^{-1}$ sec.$^{-1}$ deg.$^{-1}$.

TABLE I.

T.	NaCl; $\alpha_{273} = 0.202$.			KCl; $\alpha_{273} = 0.175$.		
	k (Calc.1).	k (Obs.).	k (Calc.2).	k (Calc.1).	k (Obs.).	k (Calc.2)
373°	0.0122	0.0116	0.0119	0.0121	0.0118	0.0118
273	0.0167	0.0167	0.0167	0.0166	0.0166	0.0166
195	0.0233	0.0249	0.0238	0.0232	0.0248	0.0237
83	0.0548	0.0636	0.0577	0.0545	0.0502	0.0567
22	—	—	—	0.206	0.125	0.215
	SiO$_2$ ∥ ; $\alpha_{273} = 0.200$[3].			SiO$_2$ ⊥ ; $\alpha_{273} = 0.309$[3].		
373	0.0236	0.0215?	0.0230	0.0127	0.0133	0.0123
273	0.0325	0.0325	0.0325	0.0173	0.0173	0.0173
195	0.0455	0.0467	0.0465	0.0242	0.0241	0.0250
83	0.1067	0.1170	0.1124	0.0569	0.0586	0.0620
22	—	—	—	0.214	0.58	0.234

The data are not sufficiently consistent to fit any formula accurately, but it is apparent that equation (3) fits the data in general somewhat more closely than does the strictly linear law except for quartz ⊥, in which case equation (3) does not strictly apply.[3]

[1] Cf. W. H. Bragg and W. L. Bragg, "X-rays and Crystal Structure," pp. 88 and 160.

[2] The data used are those of Nernst (Ann. d. Physik, 36, 395) and F. Koref (Ann. d. Physik, 36, 64). v is calculated according to the formula $v = \tau/\beta$, where β is a universal constant and τ is determined by the specific heat at a given temperature. Cf. A. H. Compton, loc. cit.

[3] According to the structure of quartz as found by Professor Bragg, the arrangement of the atoms parallel to the axis is probably not greatly different from a linear distribution; but perpendicular to the axis the arrangement is more irregular. Thus while the function of β for quartz ∥ is probably nearly that given in equation (2), for quartz ⊥ this function may be different. The value of n^2 which should be used in this case is also doubtful, which could account for the large value of α for quartz ⊥.

We have so far been considering the average energy of a degree of freedom in a solid to be RT, instead of the somewhat different value which is found by a study of the specific heat. If we use instead the energy assigned by the quantum hypothesis,

$$h\nu \left\{ \frac{1}{e^{\frac{h\nu}{RT}} - 1} + \frac{1}{2} \right\},$$

two changes must be made. In the first place, the average amount of energy given out by an atom at each collision is no longer

$$\delta\epsilon = bR\delta T,$$

but is

$$\delta\epsilon = b\frac{\partial}{\partial T}\left\{ h\nu \left(\frac{1}{e^{\frac{h\nu}{RT}} - 1} + \frac{1}{2} \right) \right\} \delta T$$

$$= \frac{bh^2\nu^2}{RT^2} \cdot \frac{e^{\frac{h\nu}{RT}}\delta T}{\left(e^{\frac{h\nu}{RT}} - 1 \right)^2}.$$

In the second place, if α is proportional to the amplitude of vibration, or the square root of the energy, *i. e.*,

$$\alpha = c\left\{ \frac{1}{e^{\frac{h\nu}{RT}} - 1} + \frac{1}{2} \right\}^{\frac{1}{2}},$$

the expression for the thermal conductivity corresponding to equation (3) is

$$(4) \quad k = bn^2 d\nu \cdot \frac{h^2\nu^2}{RT^2} \cdot \frac{e^{\frac{h\nu}{RT}}}{\left(e^{\frac{h\nu}{RT}} - 1 \right)^2} \cdot \frac{2 - c\left\{ \frac{1}{e^{\frac{h\nu}{RT}} - 1} + \frac{1}{2} \right\}^{\frac{1}{2}}}{c^2 \left\{ \frac{1}{e^{\frac{h\nu}{RT}} + 1} + \frac{1}{2} \right\}}.$$

This expression makes the thermal conductivity approach zero at very low temperatures, giving values much smaller than are actually observed. This is shown clearly in the following table, for the cases of rock-salt and sylvine (calc.$_3$).

TABLE II.

T.	NaCl; $\alpha_{273} = 0.202$.			KCl α: $_{273} = 0.175$.		
	k (Calc.$_3$).	k (Obs.).	k (Calc.$_4$).	k (Calc.$_3$).	k (Obs.).	k (Calc.$_4$).
373°	0.0124	0.0116	0.0118	0.0127	0.0118	0.0117
273	0.0167	0.0167	0.0167	0.0166	0.0166	0.0166
195	0.0214	0.0249	0.0240	0.0223	0.0248	0.0240
83	0.0263	0.0636	0.0590	0.0323	0.0502	0.0577
22				0.0015	0.125	0.217

In order to obtain a satisfactory expression for the thermal conductivity, if the energy of a degree of freedom is that assigned by the quantum hypothesis, it is necessary to assume that α is at least approximately proportional to the expression

$$\left\{ \frac{h^2\nu^2}{RT} \cdot \frac{e^{\frac{h\nu}{RT}}}{\left(e^{\frac{h\nu}{RT}} - 1\right)^2} \right\}^{\frac{1}{2}},$$

which is the square root of the product of the heat capacity of a degree of freedom according to Einstein's formula by the absolute temperature. It is hard to conceive of any possible direct connection between this quantity and the amount of "inelasticity" α of an atom. It seems difficult, therefore, to obtain on the basis of the quantum hypothesis a justifiable formula for the heat conductivity.

The case is different, however, if use is made of the agglomeration hypothesis. According to this hypothesis, the number of existing degrees of freedom is less than the greatest possible number by the factor $e^{-\frac{\beta\nu}{T}}$, where β is a universal constant; but the degrees of freedom which actually exist have the amount of energy assigned by equipartition, that is RT. This being the case, the energy given out at each collision is, as originally,

$$\delta\epsilon = bR\delta T;$$

but the number of collisions is diminished, since the number of existing degrees of freedom is reduced, in such a manner that

$$\eta = n^2\nu e^{-\frac{\beta\nu}{T}}.$$

The average energy of each *atom* is, however, $3RTe^{-\frac{\beta\nu}{T}}$, and if as before we consider the fraction of the passing energy which each atom absorbs to be proportional to its amplitude of vibration, we have

$$\alpha = cT^{\frac{1}{2}}e^{-\frac{\beta\nu}{2T}}.$$

The expression for the thermal conductivity then becomes:

$$k = bn^2 dR\nu \cdot e^{-\frac{\beta\nu}{T}} \cdot \frac{2 - cT^{\frac{1}{2}}e^{-\frac{\beta\nu}{2T}}}{c^2 T e^{-\frac{\beta\nu}{T}}},$$

(4)
$$= bn^2 dR\nu \cdot \frac{2 - cT^{\frac{1}{2}}e^{-\frac{\beta\nu}{2T}}}{c^2 T}.$$

That this expression is practically equivalent to equation (2) will be seen on comparing the values of $k(\text{calc.}_4)$ in Table II, with the corresponding values of k(calc.$_2$) of Table I. Thus the agglomeration hypothesis leads directly to a satisfactory formula for the heat conductivity while the quantum hypothesis apparently does not.

The essential difference between these two hypotheses is that while according to the quantum hypothesis all of the possible degrees of freedom possess energy, but in an amount which differs from that assigned by equipartition, according to the agglomeration hypothesis only a part of the possible degrees of freedom possess energy, but those which have it carry the amount assigned by equipartition. The fact that the latter of these hypotheses leads naturally to a satisfactory formula for the thermal conductivity while the former seems to give such an expression only with the use of a highly improbable assumption may be taken as evidence against the quantum hypothesis, and as a verification of the law of equipartition of energy as applied to the degrees of freedom of solids.

Summary.

In the first place it has been shown that the average distance through which the energy of an atomic collision in a crystal is transmitted is approximately inversely proportional to the temperature, and is many times the distance between the atoms.

This result may be explained on the assumption that each atom in a crystal absorbs or scatters a small fraction of the energy of every thermal disturbance which traverses it and transmits the rest.

In order to obtain a satisfactory formula for the thermal conductivity on the basis of the quantum hypothesis it seems necessary to make certain improbable assumptions; the agglomeration hypothesis, on the other hand, leads to such a formula with much more plausible assumptions.

This study of the thermal conductivity therefore brings evidence against the quantum hypothesis, and supports the law of the equipartition of energy as applied to the degrees of freedom in a solid.

I am indebted to my brother Professor Karl T. Compton for his suggestions and to Professor E. P. Adams for his helpful criticism in connection with this paper.

PALMER PHYSICAL LABORATORY,
   PRINCETON UNIVERSITY,
      December 9, 1915.

ON THE LOCATION OF THE THERMAL ENERGY OF SOLIDS.

By Arthur H. Compton.

FROM a consideration of the variation of the specific heat of solids with temperature, Debye has suggested[1] that the thermal energy of a solid lies not in the independent motion of the individual atoms, as assumed by Einstein,[2] but in the elastic vibrations of the body as a whole. These include the fundamental tonic vibrations as determined by the dimensions of the body, and all the harmonics up to and including the natural vibrations of the atoms themselves. On this assumption he has been able to obtain an expression for the specific heat of solids, based on the quantum hypothesis, which fits the experimental data acceptably throughout a considerable range of temperature. A formula for the specific heat of solids based on an agglomeration hypothesis has also been recently obtained,[3] but this hypothesis is found to lead to an accurate result only on the assumption that the thermal energy of solids lies not in their elastic vibrations, but in the motions of the individual atoms. In the present paper an attempt will be made to find out whether the vibrations which have a greater period than that of the individual atoms affect in any important manner the specific heat of solids.

It has been pointed out in a previous paper[4] that Debye's assumption that the thermal energy of a solid is distributed throughout all its possible elastic vibrations leads to a value of the thermal conductivity which is of a much greater order of magnitude than that observed experimentally. This makes his assumption in its original form untenable. In order to account for the magnitude of the thermal conductivity, it has been found[5] that the energy of each atomic collision is transmitted through an average distance of some 50 times the atomic distances in the case of a crystal, and probably not many times the atomic distances in the case of an amorphous substance. It is evident, therefore, that there must be a maximum value of the wave-length of the elastic vibrations which can possess heat energy, which is determined by the distance a thermal

---

[1] P. Debye, Ann. d. Phys., 39, 789 (1912).
[2] A. Einstein, Ann. d. Phys., 22, 180 (1907).
[3] A. H. Compton, Phys. Rev., 6, 377 (1915).
[4] A. H. Compton, loc. cit., p. 387.
[5] A. H. Compton (Preceding paper on Thermal Conductivity).

disturbance can be transmitted through the substance. The existence of such a maximum wave-length changes somewhat the form of Debye's specific heat formula, and the amount of this change will be shown to be in general greater for amorphous than for crystalline substances. If, therefore, the vibrations of frequency less than the natural frequency of the atoms possess any considerable amount of thermal energy, we should expect to find a certain difference between the specific heats of amorphous and of crystalline forms of the same substance. If, on the other hand, only the independent motions of the individual atoms possess a considerable amount of energy, the distance through which the effect of an atomic collision is transmitted cannot affect the specific heat, and no difference is to be expected on this account between the specific heat of amorphous and that of crystalline substances.

We shall first consider the effect on Debye's formula for the specific heat of introducing a lower limit to the frequency of vibration which is comparable in magnitude with the upper limit. Debye has shown[1] that for a perfectly elastic sphere the total number of degrees of freedom for elastic vibrations of frequency less than ν, where ν is large compared with the fundamental frequency, may be taken as

$$Z\nu = FV\nu^3,$$

if V is the volume of the sphere, and F is a quantity determined by the elastic constants and the density of the substance. If ν_m is the upper and ν_l the lower limit of the frequencies to be considered, the total number of degrees of freedom is therefore

$$Z = Z\nu_m - Z\nu_l = FV(\nu_m^3 - \nu_l^3).$$

But by Rayleigh's formula, the total number of degrees of freedom in a solid is

$$Z = 3N,$$

and thus

$$FV = 3N/(\nu_m^3 - \nu_l^3).$$

The number of degrees of freedom between the frequencies ν and $\nu + d\nu$ is

$$dn = \frac{\partial Z\nu}{\partial \nu} d\nu = 3FV\nu^2 d\nu$$

$$= \frac{9N\nu^2 d\nu}{\nu_m^3 - \nu_l^3}.$$

If with Debye we assume that each degree of freedom has the amount of energy assigned by the quantum hypothesis,

$$h\nu/(e^{\frac{h\nu}{RT}} - 1),$$

[1] P. Debye, loc. cit.

the total energy content of the sphere is

$$U = \int_{\nu_l}^{\nu_m} \frac{h\nu}{e^{\frac{h\nu}{RT}} + 1} \cdot \frac{9N}{\nu_m^3 - \nu_l^3} \nu^2 d\nu,$$

and we find for the specific heat at constant volume,

$$C\nu = \frac{dU}{dT} = \frac{9NR}{x^3 - y^3} \left\{ 4 \int_y^x \frac{\xi^3 d\xi}{e^\xi - 1} - \frac{x^4}{e^x - 1} + \frac{y^4}{e^y - 1} \right\};$$

where

$$x = \frac{h\nu_m}{RT}; \quad y = \frac{h\nu_l}{RT}; \quad \xi = \frac{h\nu}{RT}.$$

The ratio of the specific heat to its value at very high temperatures is

$$\frac{C\nu}{C\infty} = \frac{C\nu}{3NR} = \frac{3}{x^3 - y^3} \left\{ 4 \int_y^x \frac{\xi^3 d\xi}{e^\xi - 1} - \frac{x^4}{e^x - 1} + \frac{y^4}{e^y - 1} \right\}.$$

On integration this becomes:

$$\frac{C\nu}{C\infty} = \frac{3}{x^3 - y^3} \left\{ \frac{y^4}{e^y - 1} - \frac{x^4}{e^x - 1} - 4 \sum_{n=1}^{\infty} \left(\frac{e^{-nx}}{n} \left[x^3 + \frac{3x^2}{n} + \frac{6x}{n^2} + \frac{6}{n^3} \right] \right.\right.$$
$$\left.\left. - \frac{e^{-ny}}{n} \left[y^3 + \frac{3y^2}{n} + \frac{6y}{n^2} + \frac{6}{n^3} \right] \right) \right\}.$$

which is identical with the expression obtained by Debye if y is put equal to zero. The difference between the value of $C\nu/C\infty$ according to this equation when y has a finite value and when y is zero may be shown to be

$$\left(\frac{C\nu}{C\infty} \right)_{y=y} - \left(\frac{C\nu}{C\infty} \right)_{y=0} = \frac{3}{x^3 - y^3} \left\{ \frac{y^4}{e^y - 1} \right.$$
$$\left. - 24 \sum_{n=1}^{\infty} \frac{1}{e^{ny}} \left(\frac{y^4}{4!} + \frac{ny^5}{5!} + \frac{n^2 y^6}{6!} + \cdots \right) \right\} + \frac{y^3}{x^3} \left(\frac{C\nu}{C\infty} \right)_{y=0}.$$

The value of this correction for small values of y at ordinary temperatures is approximately $- 3(y/x)^3$, and does not vary rapidly with the temperature.

If the vibrations which are longer than the natural period of the atoms possess heat energy, as Debye assumes, his formula for the specific heat must be modified by some expression such as this. If, however, only the vibrations of the individual atoms possess thermal energy, as assumed in the writer's agglomeration formula, no such correction is to be applied. If, therefore, there is found to be any effect on the specific heat due to the fact that the energy of an atomic collision is transmitted a relatively small distance through a solid, Debye's assumption in a somewhat modified form will be justified.

For rock-salt at 0° C. it is found[1] that the mean distance through which the energy of an atomic collision is transmitted is $\delta x = 1.42 \times 10^{-6}$ cm. The maximum wave-length to be considered will therefore be at least approximately $4\delta x$, or 5.6×10^{-6} cm. The velocity of transmission of a vibrational disturbance in rock-salt is about 3.5×10^5 cm. sec.$^{-1}$, so the lower limit of the frequency is of the order of magnitude of

$$\nu_l = 3.5 \times 10^5 / 5.6 \times 10^{-6} = 6 \times 10^{10} \text{ sec.}^{-1}$$

The maximum frequency, ν_m, is approximately the same as the natural frequency of the atoms, and this is

$$\nu_m = 5.8 \times 10^{12} \text{ sec.}^{-1}$$

The ratio y/x is thus,

$$y/x = \frac{h\nu_l}{RT} \bigg/ \frac{h\nu_m}{RT} = \nu_l/\nu_m = 10^{-2}$$

approximately. Thus for rock-salt at 0° C. the correction to be applied to Debye's formula is of the order of magnitude of 10^{-6}, and is entirely negligible.

It is found,[3] however, that the quantity δx in amorphous substances may be of the order of magnitude of the distance between the atoms, which would make the frequency ν_l approach the frequency of the vibration of the individual atoms. The value of δx is too uncertain for amorphous substances to calculate ν_l with sufficient accuracy to make a definite estimate of the magnitude of the correction to be applied in such a case, but this correction might be expected to be rather large. Such a correction would not necessarily mean that the specific heat of a substance should be less in the amorphous than in the crystalline state, since the frequency of the vibrations of the atoms in the two forms is generally different. It is possible, also, that similar atoms in an amorphous substance may be so arranged as to have widely different natural frequencies. Such a distribution of the frequencies of the atoms would not affect greatly the form of the specific heat curve at higher temperatures, but would tend to *increase* the relative value of the specific heat at low temperatures. The effect of a correction term such as we have been discussing would be, on the other hand, to change the specific heat curve in such a manner as to *decrease* the relative magnitude of the specific heat at low temperatures. Thus if experimental evidence is found to indicate a relatively smaller specific heat at low temperatures for amorphous than for crystalline substances, such a correction term may be inferred to exist; while if experiment indicates the opposite, no definite conclusion can be drawn as to the existence of such a correction term.

---

[1] A. H. Compton (Preceding paper on Thermal Conductivity).

[2] Ibid.

The following table will show how the specific heats of certain crystalline and amorphous substances compare for rather high relative temperatures. T_m is the arithmetic mean of the absolute temperatures between which the mean specific heat per gram molecule C_m is determined. T and C represent corresponding quantities determined at definite temperatures.

AgCl (Crystalline).[1]		AgCl (Amorphous).[1]		Silicon (Crystalline).[2]		Silicon (Amorphous).[2]	
T_m	C_m	T_m	C_m	T_m	C_m	T_m	C_m
301°	12.53	301°	12.29	297°	4.84	300°	5.08
235	12.31	234	12.89	234	4.10	234	4.26
137	10.83	137	10.52	138	2.44	138	2.58

Although for both of these substances the actual magnitude of the specific heat of the two forms is appreciably different, it can be shown that the specific heat curve of the amorphous and crystalline modifications are of practically the same form. The result is different, however, at low relative temperatures, as is shown in the case of quartz.

SiO$_2$ (Rock Crystal)		SiO$_2$ (Amorphous Quartz).		SiO$_2$ (Rock Crystal).		SiO$_2$ (Amorphous Quartz).	
T	C^3	T	C^3	T_m	C_m^4	T_m	C_m^4
25.8°	0.416	26.25°	0.637	137	5.30	137	5.32
28.75	0.520	29.4	0.644	234	8.81	235	8.87
31.2	0.527	35.2	0.844				
36.1	0.794	42.6	1.33				
84.3	3.03	84.0	3.14				
89.0	3.25						
92.6	3.39						

It will be seen at a glance that at low temperatures the specific heat of the amorphous form is decidedly greater than that of the crystalline form, while at higher temperatures the specific heats are nearly the same. This difference in the rate of the variation of the specific heat of the two modifications is, as we have seen above, of such a nature as to indicate that similar atoms in the amorphous form have different natural frequencies. If the fact that the energy of an atomic collision is transmitted through a shorter distance in an amorphous than in a crystalline substance has any effect on the relative specific heat of the two forms, this effect is entirely masked, at least in the case of quartz, by the effect of the differing frequencies of the atoms in the amorphous state. This test, therefore,

[1] R. Ewald, Ann. d. Phys., 44, 1213 (1914).
[2] A. S. Russel, Phys. Zeitschr., 13, 61 (1912).
[3] W. Nernst, Ann. d. Phys., 36, 395 (1911).
[4] F. Koref, Ann. d. Phys., 36, 64 (1911).

gives no evidence of the kind of a difference between the specific heat of amorphous and of crystalline substances that would be expected if the elastic vibrations of the solid possess thermal energy.

We cannot conclude from this that all the heat energy lies in the vibrations of the individual atoms, but the above result would indicate that the vibrations of longer period probably play no important part in the specific heat. Further measurements of the specific heat of amorphous substances at very low temperatures may very possibly lead to important information on this point.

Summary.

An explanation of the specific heat according to the quantum hypothesis implies that the heat energy of a solid lies in its elastic vibrations, while according to the writer's agglomeration hypothesis this energy lies in the vibrations of the individual atoms.

A previous study of the thermal conductivity has shown that the distance through which the energy of an atomic collision is transmitted is in a crystal many times the distance between neighboring atoms, but is in amorphous substances probably not many times the atomic distances. This implies that the maximum wave-length of the elastic vibrations which can possess thermal energy is greater for crystalline than for amorphous substances, which is shown to mean that if it is the elastic vibrations which possess the heat energy the specific heat of amorphous substances should be comparatively less at low relative temperatures than that of crystalline substances.

The fact that experiment shows no such difference in the specific heat of substances in the amorphous and in the crystalline form indicates that the vibrations of period greater than that of the individual atoms probably play no important part in the specific heat; though the evidence to this effect cannot be taken to be conclusive.

Palmer Physical Laboratory,
   Princeton University,
      December 10, 1915.

A RECORDING X-RAY SPECTROMETER, AND THE HIGH FREQUENCY SPECTRUM OF TUNGSTEN.

By Arthur H. Compton.

ALTHOUGH the photographic method of obtaining the spectrum of a beam of X-rays reflected from a crystal has led to most interesting results in the hands of Moseley,[1] de Broglie[2] and others, it has been found possible to examine such a spectrum more thoroughly by ionization methods.[3] The reason for this is that while by the photographic method the intensity of the different spectrum lines can at best be only qualitatively measured, the ionization method is capable of giving quantitatively the relative intensity of the reflected beam of X-rays at different angles. The ionization method as usually applied, however, is open to the objection that a very large number of separate observations are necessary to obtain accurately a complete spectrum. In the present paper an apparatus will be described which gives a continuous record of the intensity of the beam of X-rays at different angles, and a study of the X-ray spectrum of tungsten will be made to illustrate the manipulation of the instrument.

This apparatus differs from the well-known Bragg X-ray spectrometer in two essentials: (1) The ionization current due to the reflected beam of X-rays, instead of charging up an electroscope directly, goes to one pair of quadrants of an electrometer and is shunted to earth through a high resistance. The electrometer thus acts as a highly sensitive galvanometer, a steady ionization current producing a steady proportional deflection. (2) The angle of the crystal and of the ionization chamber are varied continuously in such a manner that the ionization chamber is always in position to receive the reflected beam of X-rays, and the electrometer deflections corresponding to each particular angle of the crystal are recorded on a moving roll of photographic paper. In this manner a record of the complete spectrum can be obtained with a minimum of trouble and of exposure to the X-rays.

The arrangement of the apparatus is shown diagrammatically in Fig. 1. The X-rays pass from the anticathode A, through the slits B and B',

---

[1] H. G. J. Moseley, Phil. Mag., 26, 1024 (1913).
[2] M. de Broglie, Compt. Rend., 157, 924 and 1413 (1913).
[3] Cf. W. H. Bragg and W. L. Bragg, X-rays and Crystal Structure, p. 66.

are reflected from the crystal C and pass through the slits D' and D into the ionization chamber I. The ionization chamber rests on an arm fastened to the spectrometer table S, and the table on which the crystal rests is geared to move with half the angular velocity of the ionization chamber, though its position can also be varied by means of the slow motion screw F. The ionization current is carried from the chamber I to one pair of quadrants of a highly sensitive electrometer E, and is shunted to the ground through a variable xylol-alcohol resistance R. The intensity of the primary beam is measured in a similar manner by

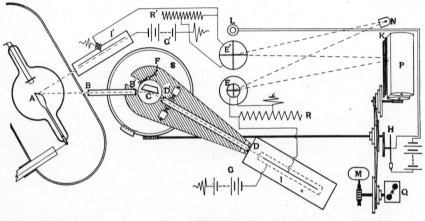

Fig. 1.

means of the ionization chamber I', the resistance R' and the electrometer E'. The mirrors of both electrometers reflect beams of light from the Nernst glower N, through a fine horizontal slit k onto a roll of bromide paper P. A motor-actuated driving clock MQ moves, by means of a system of pulleys and worm gears, both the spectrometer table and the roll of bromide paper at a constant, though adjustable, speed. The pointer H is so geared to the shaft which drives the spectrometer table that it makes one revolution for each degree through which the table turns, and by means of a mercury contact the lamp L is turned on for an instant at each degree or half degree. This illuminates the slit K and marks off the angle by a series of lines across the bromide paper, as is shown in the records, Figs. 4 to 11.

The construction of the system for measuring the reflected beam of X-rays is shown in more detail in Fig. 2. The ionization chamber I is closed at one end by a thin mica window m and at the other by a perforated brass cap covered with a plate of glass g. This arrangement allows the crystal to be adjusted optically to the angle of reflection.

The outside of the ionization chamber is grounded, but the wire framework f is raised to a sufficiently high potential to produce a saturation current to the wire w which is connected with the electrometer. The wire w passes through the ebonite plug e_1, is fastened to one terminal of the glass tube R containing a mixture of xylol and alcohol, and goes through the ebonite plugs e_2 and e_3 to the electrometer. The wire can be grounded directly by screwing up the mercury cup O until it touches a contact point, or it can be grounded through the desired high resistance by moving up one of the other mercury cups. If all the cups are down the system is insulated. The electric shielding consists of the rather heavy copper box c and the brass tube b. The electrometer E is placed

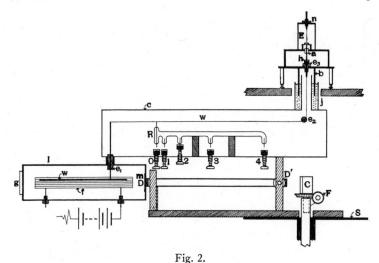

Fig. 2.

directly over the axis of the spectrometer, so that the joint in the shield at j permits free motion of the spectrometer arm, while an oil seal renders the shield airtight.

The electrometer E deserves particular mention, because it is so admirably adapted to this work. Its chief advantages lie in its flexibility, in its high sensitivity, in its perfect electrostatic shielding and in the use of a single piece of insulation for the sensitive quadrants. The quadrants of the electrometer are only 1.3 cm. in diameter. One pair of quadrants is supported by brass rods which fit into the electrometer case, and is thus permanently grounded; while the other pair is supported on a fine brass rod which passes through the amber insulation a and dips into the mercury cup h. The needle is suspended by a fine quartz fiber, and the whole system is sputtered with platinum. Most of the moment of inertia of the needle is due to the mirror, so this is made just large enough to reflect a sufficiently strong beam of light. The needle is "dead-beat," and comes approximately to rest in a few seconds. With ten volts on the needle the electrostatic control is small, and with the suspension used the sensitivity is about 1,000 divisions per volt. When the potential of the needle is above forty volts, the electrostatic control becomes the determining factor, and the sensitivity can be varied from 1,000 divisions to ∞ by adjusting the position of the quadrants and the height of the needle. The highest sensitivity that it has been found practicable to use is about 25,000 mm. per volt at a distance of a meter. The capacity of the electrometer is about 15 electrostatic units, so that it is capable of measuring a very small charge.

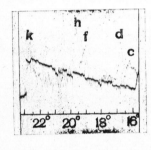
Fig. 3.

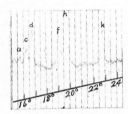
Fig. 4.

The Effect of Slight Radioactivity in the Ionization Chamber.—Fig. 3 shows an early attempt to obtain the X-ray spectrum of tungsten with this apparatus. The dark line represents the primary beam of X-rays and the light one the reflected beam. The electrometer in this case had a sensitivity of about 10,000 divisions per volt, and the shunting resistance was rather high. The electrostatic shielding at this time consisted of steel tubing and galvanized iron. Although the more prominent spectrum lines are easily distinguishable, there are so many spurious deflections of the electrometer which in no way correspond to variations of the primary beam of X-rays that it is impossible to make any accurate measurements. Supposing that these irregular motions were due to stray ionization currents within the electrostatic shield, a copper and brass shield was used instead, thus decreasing the contact potential difference between the shield and the wire. In this manner the relative size of the spurious deflections was much reduced, as is shown in Fig. 4, though these motions were still so large as to interfere seriously with the readings.

Investigation showed that the remaining irregular motions of the electrometer needle were due, in large measure at least, to real ionization currents in the ionization chamber, which were independent of the X-rays, as is shown in Fig. 5 in which the X-ray tube was not running. The magnitude of this current was found to be of the same order as the natural ionization current usually observed in brass vessels. It was also noticed that the deflections always occurred suddenly in the direction of the normal ionization current, and then died out gradually, which suggested ionization due to alpha particles.

Fig. 5.

As will be shown later (equation 4 below), if the shunt resistance and the sensitivity of the electrometer are known, the total ionization corresponding to each hump on a record can be calculated from the area under the hump. When such a calculation is made on the record shown in Fig. 5, it is found that on the average each hump corresponds to the production of about 8×10^4 ions. Not all of the alpha particles will escape from the walls of the ionization chamber with their full velocity, since most of them will start at some distance from the surface of the metal. If we assume that the alpha particles have a definite range in the metal, a simple calculation shows that the average range of the emitted particles will be half the maximum range. Since, however, the ionization is somewhat greater per centimeter path for the more slowly moving particles, the average ionization will be somewhat more than half, about 0.6 of that produced by a particle which starts from the surface (assuming the emitted particles to be completely absorbed by the gas in the ionization chamber). A particle starting from the surface therefore produces about $8 \times 10^4/0.6 = 1.3 \times 10^5$ ions. An alpha particle from radium would produce 1.5×10^5 ions, which agrees well enough with the observed value to show that these spurious deflections are doubtless due to alpha particles given off either by the walls of the ionization chamber or by the gas within it.[1]

---

[1] An effect exactly similar to that here described has been observed by Rutherford and Geiger (Proc. Roy. Soc., A, 81, 141 (1908)) while counting the alpha particles from radium by an ionization method. They attributed the effect to the natural radioactivity of the metal of which the ionization chamber is composed. Recently Shrader (PHYS. REV., 6, 292 (1915)) has found that if the electrode within the ionization chamber has a sharp point, and if the chamber is at high enough potential to produce ionization by collision, the "natural disturbances" depend only on the nature of the point, and can be practically eliminated if the point is properly treated. It is to be noticed, however, that the only disturbances which would affect his electroscope would be the ones which would ionize the air near the needle point of his electrode, where any ionization was greatly magnified by collision. It would seem that his apparatus was not sufficiently sensitive to be affected by the occasional production of alpha particles in other parts of the ionization chamber. His work cannot, therefore, be taken to indicate that the walls of his ionization chamber were not slightly radioactive, so that his results are not inconsistent with the writer's and those of Rutherford and Geiger.

A thorough cleaning of the ionization chamber with nitric acid seemed to reduce the amount of radioactivity but little. Its relative effect could be greatly diminished, however, by filling the ionization chamber with a dense gas, and thus increasing the ionization due to the X-rays. The spectrum shown in Fig. 6 was obtained in this manner, using a simple ionization chamber such as I', Fig. 1, filled with dry hydrogen iodide. The effect of the alpha particles was still fruther reduced by changing the ionization chamber to the type shown in Fig. 2. The ionization due to the X-rays is produced chiefly within the wire network f, which is kept at a sufficiently high potential to produce a saturation current to the wire w. The outer casing of the ionization chamber is grounded, and enough air-space is left between the casing and the network to absorb all the alpha particles which leave the walls. Thus the particles from the walls of the vessel do not affect the current flowing to the wire w; this is affected only by the radioactivity of the wire netting and of the gas within it. By this means the number of spurious deflections of the electrometer was reduced to about 1/4 of the number when the simpler ionization was used, while the ionization due to the X-rays remained about the same. All of the spectra after Fig. 6 were obtained with this new ionization chamber.

A Study of the High Frequency Spectrum of Tungsten.—As a source of X-rays for the examination of the spectrum of tungsten, a Coolidge X-ray tube with a tungsten anticathode was used. In order that the tube should remain steady over long runs it was found best to heat the thermionic filament by means of batteries of rather large capacity, 175 ampere hours. The source of high potential was a Snook-Roentgen machine. This consists of a 10-kilowatt step-up transformer with a commutating device in the secondary circuit, which makes it possible to obtain a direct though intermittent high potential current. This apparatus was easily capable of running the Coolidge tube at its maximum capacity, and was found to be very satisfactory. In most of the runs a current of about 30 milliamperes was sent through the tube, with an alternate spark gap between sharp points of about 10 cm.

Fig. 6 shows a complete spectrum of the X-radiation from tungsten as analyzed by a crystal of zinc-blende. As in the other figures the heavy line indicates the strength of the primary beam, and the light one of the reflected beam of X-rays, though of course the two curves are on greatly different scales. Each vertical line corresponds to one degree on the spectrometer table, or 1/2 degree of rotation of the crystal. The angles marked on all of the spectra indicate the glancing angle θ of the beam of X-rays which strikes the crystal. The broken base line represents the zero point for the electrometer measuring the reflected beam, and the solid line that of the one measuring the direct beam of X-rays. At the angle $\theta = 30.5°$ a zero reading was taken, by putting a lead screen between the slit B' and the crystal, to make sure that the zero point of the sensitive electrometer had not shifted. In obtaining this spectrum wide slits were used, the slit at D being about 1.0 cm. The shunting resistance was about 3.5×10^{11} ohms, and the sensitivity 1,050 mm. per volt. With this low resistance and wide slits it was possible to go over the complete spectrum in an hour and a half.

Owing to the comparatively great wave-length of the characteristic L-radiation from tungsten and to the closeness of the layers of atoms in

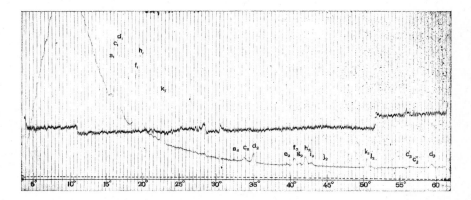

Fig. 6.

the cleavage planes of zinc-blende, it is impossible to obtain a complete third order spectrum in this case, but the first and second orders stand out very clearly. The line marked k is the single line observed in the spectrum of tungsten by Moseley,[1] and the prominent lines c, d, f, h and k correspond to the five lines observed by Moseley and Darwin[2] in the spectrum of platinum. Recently Barnes[3] has detected also the lines g and i in the tungsten spectrum by a photographic method. Lines at a and l can also be seen plainly in this photograph, l being visible only in the second order spectrum because of its proximity to the prominent line k. An indication is also given of possible lines at e and j. These are confirmed by other records, though the line j is rather doubtful. A series of four records, such as Fig. 7, which is a part of the first order spectrum from rock-salt using a slit at D only 0.4 mm. wide, indicates that the line c is really a very close double. A line also appears con-

[1] H. G. J. Moseley, Phil. Mag., 27, 703 (1914).
[2] H. G. J. Moseley and C. G. Darwin, Phil. Mag., 26, 211 (1913).
[3] J. Barnes, Phil. Mag., 30, 368 (1915).

sistently on a number of records between *a* and *c*. All of these lines are shown more or less distinctly in Fig. 8.

In order to make an accurate determination of the wave-length of the different lines, a crystal of calcite was used, as recommended by Professor Bragg, because of the extremely perfect faces obtainable. The X-ray tube was swung around until the rays which fell on the crystal left the target almost tangent to its face, the slits B and D were about 2 mm., and the slits B' and D' were made wide. The slit at D and the target of the X-ray tube were set as nearly as possible at equal distances, about 52 cm., from the crystal. With this arrangement it was found possible to measure the angles of reflection with considerable precision.

It was necessary first to determine the zero point accurately. This was done by taking a series of four records, two on either side of the zero point, one with the spectrometer moving toward and the other from zero. Fig. 8 is one of this series, taken with the spectrometer moving toward the zero point. The mean angle as measured from the three most prominent lines in the four records was taken as the true zero, and this could be determined with a probable error of less than ± 0.2'. The angle of the

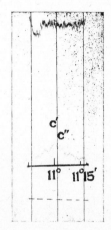

Fig. 7.

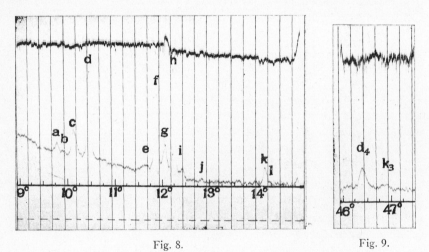

Fig. 8.   Fig. 9.

spectrum lines was then determined from a series of six records of the first order spectrum and two records showing the third order k and the fourth order c, d, f and h lines. Fig. 9 shows a part of one of these records, including the fourth order d and the third order k lines. The angles found for the higher order spectra did not correspond exactly with those calculated from the first order according to the formula

$$n\lambda = 2d \sin \theta.$$

The reason for the difference was traced to a combination of two causes: first, the front edge of the crystal projected about 0.5 mm. in front of the axis of the crystal table, and second, the slight penetration of the X-rays into the crystal made the observed angle slightly less than the true angle of diffraction. The latter effect is very small, being only about 0.2′ for tungsten rays diffracted by calcite. For X-radiation of short wave-length, however, and a crystal such as diamond, this error becomes of considerable magnitude, as has been noticed in the case of rhodium rays.[1] The correction due to the inaccurate adjustment of the crystal could be calculated accurately from a comparison of the position of the different orders of the same line. The sum of the two corrections was $3.15' \pm 0.10'$, being practically equal for the different lines. The corrected angles of the lines of the first order spectrum are given in column 2 of Table I. The error in the angle here given for the more intense lines

TABLE I.

Line.	θ	$\lambda \times 10^8$ Cm.	Barnes's Values.	Remarks.
a	9° 44.5′	1.0249		
b	9° 53.1′	1.0399		Line somewhat doubtful.
c'	10° 3.7′	1.0582·	1.082	Very close double.
c''	10° 7.7′	1.0652·		
d	10° 25.7′	1.0959·	1.113	
e	11° 36.3′	1.2185		
f	11° 49.9′	1.2420·	1.258	
g	12° 0.4′	1.2601·	1.277	
h	12° 11.4′	1.2787·	1.296	
i	12° 22.7′	1.2985·	1.312	
j	12° 44.7′	1.3363		Line uncertain.
k	14° 4.7′	1.4735·	1.477	Moseley gives 1.486.
l	14° 11.1′	1.4844·		

is probably less than $\pm 0.5'$, though for some of the fainter lines the probable error is as much as $\pm 1.0'$.

The distance between the successive atomic layers in the cleavage planes of calcite may be calculated according to the formula[2]

---

[1] W. H. Bragg, Phil. Mag. 27, 898 (1914).

[2] W. H. Bragg, Proc. Roy. Soc., A, 89, 468 (1914). "X-rays and Crystal Structure," p. 112. Professor Bragg uses the value $\phi(\beta) = 1.08$, which makes his value of d for calcite differ appreciably from that here obtained.

$$d = \left(\frac{1}{2}\frac{M}{\rho N \phi(\beta)}\right)^{\frac{1}{3}},$$

where M is the molecular weight of $CaCO_3$, ρ is the density of the crystal, N is the number of molecules per gram-molecule, and $\phi(\beta)$ is the volume of a rhombohedron the distance between whose opposite faces is unity and the angle between whose edges is β. This function may be shown to be

$$\varphi(\beta) = \frac{(1 + \cos\beta)^2}{\sin\beta(1 + 2\cos\beta)}.$$

For calcite, β is 101° 54′, which makes $\phi(\beta) = 1.0960$[1]. The density of the crystal used was carefully determined, and was found to be $\rho = 2.7116 \pm 0.0004$ g. cm^{-3}. (at 18°). Taking $M = 100.09$ and $N = 6.062 \times 10^{23}$ (Millikan), we find $d = 3.0279 \times 10^{-8}$ cm. From this value of d the wave-lengths can be immediately calculated according to the expression

$$\lambda = 2d \sin\theta.$$

Using the values of θ given in the second column of Table I., we obtain the wave-lengths given in the third column. For convenience in comparison the wave-lengths found by Barnes[1] are given in the fourth column.[2]

The Intensity of the Different Orders of Spectra.—It is difficult to obtain an accurate estimate of the relative intensity of the various lines in the same order of the spectrum, on account of the difference in the absorption of the corresponding wave-lengths. It is possible, however, to make a comparatively accurate determination of the relative intensity of the different order reflections of the same line. In order to do this, the slit at B' was made narrow, about 0.9 mm., slit D' was removed, and slit D was made 12 mm. wide, which was more than broad enough to take in all the rays reflected from the crystal. When the crystal and ionization chamber are moved at uniform speed past a spectrum line, the area under the curve representing the line on the resulting record is proportional to the integral of the ionization current produced by the line, and hence to the intensity of the line itself. Thus by comparing

---

[1] J. Barnes, loc. cit.

[2] Note added May 12, 1916: In a preliminary report on this work, made at the meeting of the American Physical Society on February 26, the wave-lengths of the spectrum lines were calculated using as the grating space for calcite 3.0695×10^{-8} cm. instead of the value here found. The results thus obtained agreed acceptably with Barnes's measurements. W. S. Gorton has since shown (PHYS. REV., Feb., 1916), by comparison with rock-salt, that d for calcite is rather about 3.028×10^{-8} cm., which called my attention to an error in my previous calculation of the grating space. The wave-lengths here obtained are in much better agreement with those found by Gorton than with those due to Barnes.

the areas under the humps on the record, the intensities of the different order lines may be compared.

Since the period of the needle and the time required for the electrometer to charge up to its maximum potential might also be expected to affect the area of the hump, it may be worth while to investigate the question in detail. The equation of motion of the spot of light reflected by the electrometer needle onto the photographic paper is

$$(1) \qquad I\frac{d^2y}{dt^2} + D\frac{dy}{dt} + My = \frac{M}{a}V,$$

where y is the deflection of the spot of light, I is the inertia term, D the damping term, M represents the restoring force, $1/a$ is the sensitivity of the electrometer, and V is the potential difference between the two pairs of quadrants. On integrating this expression between the times t_0 and t_1 we obtain,

$$(2) \qquad \frac{Ia}{M}\left[\frac{dy}{dt}\right]_{t_0}^{t_1} + \frac{Da}{M}\left[y\right]_{t_0}^{t_1} + a\int_{t_0}^{t_1} y\,dt = \int_{t_0}^{t_1} V\,dt.$$

The potential of the system at any time is given by the expression

$$V = Ri - RC\,dV/dt,$$

where R is the resistance of the shunt to ground, i is the ionization current, and C is the electrostatic capacity of the system. Integrating as before,

$$\int_{t_0}^{t_1} V\,dt = R\int_{t_0}^{t_1} i\,dt - RC\left[V\right]_{t_0}^{t_1}.$$

Combining with equation (1) and substituting in (2) we get

$$(3) \qquad \int_{t_0}^{t_1} i\,dt = \frac{a}{R}\int_{t_0}^{t_1} y\,dt + \left(\frac{aD}{RM} + aC\right)\left[y\right]_{t_0}^{t_1}$$
$$+ \left(\frac{aI}{RM} + \frac{aCD}{M}\right)\left[\frac{dy}{dt}\right]_{t_0}^{t_1} + \frac{aCI}{M}\left[\frac{d^2y}{dt^2}\right]_{t_0}^{t_1}.$$

Since the intensity of a line is proportional to the total ionization produced in the ionization chamber as it moves past the line, expression (3) is evidently a measure of the intensity if the times t_0 and t_1 are taken just before and just after the line is traversed. If the displacement, velocity and acceleration of the spot of light are the same at the times t_0 and t_1, equation (3) reduces to

$$(4) \qquad \int_{t_0}^{t_1} i\,dt = \frac{a}{R}\int_{t_0}^{t_1} y\,dt = \frac{a}{R}\frac{dt}{dx}\int_{x_0}^{x_1} y\,dx$$
$$= \frac{a}{R}\frac{dt}{dx}A,$$

where dx/dt is the velocity with which the roll of paper moves, and A is the area under the hump on the record. In practice it is easily possible to satisfy these conditions with sufficient accuracy, so that the intensity may be taken to be proportional to the area under the curve.

The records shown in Figs. 10 and 11 are examples of this method of comparing intensities. Fig. 10 shows the first and second order f and

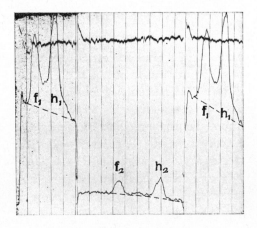

Fig. 10.

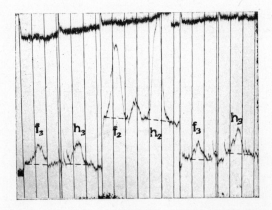

Fig. 11.

h lines as reflected from a cleavage plane of rock-salt, and in Fig. 11 the second and third orders of these lines are compared. In Fig. 10 the lighter line has been inked, in order to make a better reproduction. The areas under the humps were measured with a planimeter. The largest part of the error probably occurred in selecting the base line from which to measure the area. These records give the ratio of the intensities of the three orders as 100 : 18.9 : 3.9.

Bragg[1] has studied the relative intensities of the first three orders of reflection from rock-salt, using the rhodium line of wave-length 0.614×10^{-8} cm. He found that he could express his results within the probable error of his experiments by the formula

$$J_\theta = \frac{A(1 + \cos^2 2\theta) e^{-\beta \frac{\sin^2 \theta}{\lambda^2}}}{\sin^2 \theta},$$

where J_θ is the intensity of a given spectrum line when reflected at the angle θ, and A and B are constants. The factor $(1 + \cos^2 2\theta)$ was deduced theoretically by J. J. Thomson, and indicates a polarization of the reflected beam. Since in Bragg's measurements θ never became greater than about 20°, he was unable to decide experimentally whether this factor should really appear. In the writer's experiments the angle θ for the third order was 43°, so that the polarization should have been nearly complete. In the following table the first calculated values do not include the factor $(1 + \cos^2 2\theta)$, but this factor is included in the second calculation. The fact that the observed values agree much

TABLE II.

Order.	Mean Angle.	Intensities.		
		Calc.$_1$	Obs.	Calc.$_2$
1	13° 5′	100.0	100.0	100.0
2	26° 55′	21.6	18.9	16.2
3	42° 59′	7.5	3.9	4.2

more closely with the values calculated when this factor is considered shows that the crystal grating does polarize the reflected beam and that this polarization is approximately complete when the angle of reflection is 45°.

Summary.—A recording X-ray spectrometer has been described by means of which a continuous record can be obtained of the ionization produced by a beam of X-rays reflected from a crystal as the angle of the crystal is varied.

A spurious and irregular current in the ionization chamber has been observed, which has been found to be due to a slight radioactivity within the ionization chamber. Methods are described whereby the effect of this radioactivity may be considerably reduced.

[1] W. H. Bragg, Phil. Mag., 27, 893 (1914). also "X-rays and Crystal Structures," p. 198.

The X-ray spectrum of tungsten has been examined in some detail. In addition to the seven lines already known, an indication has been found of six others, two of which are, however, somewhat uncertain. The wave-lengths of the different lines have been carefully determined.

The relative intensities of the different order reflections of the same line have been shown to be proportional to the areas under the humps produced in the record by the spectrum line; and the relative intensities of the first three orders in the spectrum from rock-salt have been determined by measuring these areas.

The observed intensities of the different orders indicate that polarization occurs when a beam of X-rays is reflected by a crystal.

In conclusion I wish to thank Professor H. L. Cooke for many valuable suggestions in the design of the apparatus and for helpful advice in all parts of the work.

PALMER PHYSICAL LABORATORY,
    PRINCETON UNIVERSITY,
        February 16, 1916.

THE INTENSITY OF X-RAY REFLECTION, AND THE DISTRIBUTION OF THE ELECTRONS IN ATOMS.

By Arthur H. Compton.

IN the study of the spectra of X-rays as analyzed by crystal gratings, the remarkably low intensity of the higher orders of reflection has from the first attracted a considerable amount of attention. Preliminary measurements by Mr. W. L. Bragg[1] showed that, when corrected for temperature effects, the intensities of the different orders of reflection of a given X-ray spectrum line are approximately proportional to the inverse square of the order. A more detailed experimental investigation by Professor W. H. Bragg[2] showed that if X-rays of a definite wave-length are reflected at a glancing angle θ by a crystal in which the successive layers of atoms are similar and are similarly spaced, the energy in the reflected beam can be expressed with considerable accuracy by the formula

$$(1) \qquad E_r = \frac{C(1 + \cos^2 2\theta)}{\sin^2 \theta} e^{-B \sin^2 \theta}.$$

In this expression C is a constant depending upon the energy in the incident beam, the wave-length of the X-rays, and the nature of the crystal. The factor $e^{-B \sin^2 \theta}$ accounts for the effect of the thermal motion of the atoms. The constant B can be determined experimentally[2] or may be calculated from certain thermal properties of the crystal.[3, 4] The polarization factor $(1 + \cos^2 2\theta)$ was originally deduced by J. J. Thomson[5] for any case of the scattering of X-rays by electrons, and was first introduced into the formula for X-ray reflection by Darwin.[4] Professor Bragg was not able to verify this polarization factor, since in his experiments θ was always small. Using rays of longer wave-length, however, the writer has been able to measure[6] the reflection at sufficiently large glancing angles to obtain an appreciable effect due to this factor, and thus

---

[1] W. L. Bragg, Proc. Roy. Soc., A, 89, 468 (1914).

[2] W. H. Bragg, Phil. Mag., 27, 881 (1914), also W. H. Bragg and W. L. Bragg, X-rays and Crystal Structure, p. 195.

[3] P. Debye, Ann. d. Phys., 43, 49 (1914).

[4] C. G. Darwin, Phil. Mag., 27, 325 (1914).

[5] J. J. Thomson, Conduction of Electricity through Gases, p. 326.

[6] A. H. Compton, Phys. Rev., 7, 658 (1916).

to verify its existence. Since the sine of the glancing angle is proportional to the order of reflection, this formula includes the result found by W. L. Bragg, but is more general, as it expresses the intensity of the reflection from all possible planes in the crystal.

The theory of the intensity of X-ray reflection has been examined in considerable detail by Mr. C. G. Darwin,[4, 7] who finds that if all the electrons which are effective in scattering the X-rays are close to the centers of the atoms, the energy in the beam reflected at an angle θ should be proportional to

$$\frac{1 + \cos^2 2\theta \, e^{-B \sin^2 \theta}}{\sin \theta \cos \theta}.$$

This expression differs from Bragg's experimental formula by the factor $\tan \theta$, which must be explained, as Darwin pointed out, by the fact that the electrons are not all concentrated near the centers of the atoms, but that at least some of the electrons are at distances from the atomic centers which are of the same order of magnitude as the distance between the atoms. Since the relative intensity of the different orders of X-ray reflection is thus a function of the distribution of the electrons in the atoms of the crystal, it should be possible, knowing the relative intensity of the different orders, to obtain some definite idea of the manner in which these electrons are arranged.

The possibility of finding an arrangement of the electrons which will account in a satisfactory manner for the observed intensity of X-ray reflection at different angles was suggested first by Professor Bragg[8] and independently soon after by the writer.[9] Both of us were able to show the nature of the effect on the intensity of reflection due to certain different distributions of the electrons in the atoms of a crystal grating, but we both neglected to consider certain important factors that must seriously modify the conclusions at which we arrived. We based our arguments on the assumption that the reflected energy would be the same for all orders if all the scattering occurred at the centers of the atoms. This is indeed true for the intensity in the middle of the reflected line, if the crystal acts as a perfect grating, but since the effective width of the spectrum line can be shown to be proportional to $1/\sin \theta \cos \theta$, the reflected energy is reduced in the same ratio. Thus instead of a factor $1/\sin^2 \theta$ there is really, as pointed out above, a factor of only $1/\tan \theta$ to be accounted for by the assumed structure of the atom. In the present paper a more complete theory will be obtained of the dependence

[7] C. G. Darwin, Phil. Mag., 27, 675 (1914).
[8] W. H. Bragg, Bakerian Lecture, March 18, 1915; Phil. Trans., A, 215, 253 (July 13, 1915).
[9] A. H. Compton, Nature, May 27, 1915.

of the intensity of reflection upon the angle and upon the distribution of the electrons in the atoms of the reflecting crystal. The result obtained will be in exact accord with that found in a different manner by Darwin, but will be so expressed that it will be found possible to determine with some definiteness the distribution of the electrons in certain atoms by comparison with the observed intensities of X-ray reflection.

The Intensity of X-ray Reflection.

Let us therefore obtain a general expression for the energy of a beam of X-rays of wave-length λ which is reflected at a glancing angle θ from a crystal all of whose atoms are similar. For sake of simplicity we may consider the primary beam to be polarized in such a manner that the electric vector is perpendicular to the plane of reflection. This will eliminate the polarization factor $(1 + \cos^2 2\theta)/2$. The temperature factor, $e^{-B\sin^2\theta}$ may also be neglected if we consider the atoms to be in their positions of rest. These factors can be introduced later into the expression for the intensity of reflection without modifying the rest of the calculation.

Let X-rays thus polarized strike the crystal C (Fig. 1) in the elementary solid angle $d\sigma$ included between $\theta + \epsilon_1$, ϵ_3 and $\theta + \epsilon_1 + d\epsilon_1$, $\epsilon_3 + d\epsilon_3$, and consider the ray reflected at the angle $\theta + \epsilon_2, \epsilon_4$. Here θ is the angle

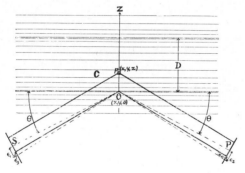

Fig. 1.

of maximum reflection, defined by the relation $n\lambda = 2D\sin\theta$, ϵ_1 and ϵ_2 are small angles measured in the plane of reflection and ϵ_3 and ϵ_4 are similar small angles taken perpendicular to this plane. We shall consider the effect of the reflected ray at the surface of a sphere of very large radius R. If $I_i \cdot d\sigma = I_i \cdot d\epsilon_1 d\epsilon_3$ is the intensity at the surface of the crystal due to this bundle of incident rays, the corresponding amplitude dA_i of the electric vector may be defined by the relation $I_i d\sigma = c(dA_i)^2$. The amplitude at the distance R of the ray scattered by a single electron may then be taken[4, 5] as $(e^2/mC^2R)dA_i$ where e is the charge and m the mass of the electron, and C is the velocity of light. If the electrons are held very rigidly in position, this expression might have to be slightly modified, but it seems probable that such effects are inappreciable for waves of the frequency of X-rays. The scattering due to the positive nucleus of the atom may be neglected on account of its comparatively great mass.

If we write $\phi = e^2/mC^2$, the amplitude of the reflected ray at the point P $(R, \epsilon_2, \epsilon_4)$ due to a single electron near the surface of the crystal is therefore $(\phi/R) \cdot dA_i$. If $f(z)$ is the volume density of distribution of the electrons at a distance z from the middle plane of a layer of atoms in the crystal, the electric displacement at P at the time t due to a volume element of an atomic layer near the surface of the crystal is therefore

$$f(z)dxdydz \cdot \frac{\phi}{R}dA_i \cdot \cos 2\pi \left(\frac{t}{T} - \frac{\delta + r}{\lambda}\right),$$

where T is the period of vibration, $2\pi\delta/\lambda$ is the phase angle at P due to a crystal element at $(x, y, 0)$ at the time $t = 0$, and r is the distance $2z\{\sin\theta + (\epsilon_1 + \epsilon_2)\cos\theta\}$. Neglecting the small term $(\epsilon_1 + \epsilon_2)\cos\theta$, and writing,

$$\beta = 2\pi\left(\frac{t}{T} - \frac{\delta}{\lambda}\right),$$

we find for the displacement at P due to an element of area of a layer of atoms near the surface of the crystal,

$$\frac{\phi}{R}dA_i \cdot dxdy \cdot \int_a^b f(z) \cos\left(\beta - \frac{4\pi z \sin\theta}{\lambda}\right) dz,$$

where $b - a$ is the thickness of the atomic layer, and represents the diameter of the atoms. If N is the number of electrons per unit volume of the crystal, and D is the distance between two successive atomic layers, the number of electrons in unit area of such a layer is ND. The function $f(z)$ may therefore be written $NDF(z)$, where $F(z)$ is the probability that a given electron shall be at a distance z from the mid-plane of the atomic layer to which it belongs. Since the function $F(z)$ can nowhere be greater than at the plane of symmetry, i. e., at $z = 0$, the displacement becomes a maximum when $\beta = 0$. The integral factor of the above expression then becomes,

$$\psi = \int_a^b F(z) \cos\left(\frac{4\pi z \sin\theta}{\lambda}\right) dz,$$

and the amplitude at P due to an element of area of the atomic layer considered may be written

$$\frac{\phi}{R}dA_i \cdot ND \cdot \psi \cdot dxdy.$$

If we consider an element of the crystal so small that the reflection from all of its atomic layers may be considered to be in the same phase, yet containing a large number dn of such layers, the amplitude of the reflected beam at P due to this element is evidently

$$\frac{\phi}{R} dA_i N D \psi dx dy \cdot dn.$$

Or since $dn = dz/D$, the amplitude due to a volume element of the crystal is

$$\frac{\phi}{R} dA_i N \psi dx dy dz = dA_0 \cdot dx dy dz,$$

where dA_0 is the amplitude of the reflected ray at P per unit volume of the crystal near its surface due to an incident beam of solid angle $d\sigma$.* When the rays are reflected from a depth z below the surface of the crystal, they travel a distance $2z/\sin\theta$ through the crystal before they emerge, so that the intensity is reduced by a factor $e^{2\mu z/\sin\theta}$, where μ is the absorption coefficient. The amplitude is thus reduced by a factor $e^{\mu z/\sin\theta}$, so the amplitude of the beam reflected from any part of the crystal is, per unit volume of the crystal,

(2) $$dA_r = dA_0 e^{-\frac{\mu z}{\sin\theta}} = dA_i N \frac{\phi}{R} \psi e^{-\frac{\mu z}{\sin\theta}}.$$

Perfect Crystal and Long Trains of Waves.—We shall first determine the energy in the reflected beam on the assumptions (1) that the crystal has no faults, but acts as a perfect grating, and (2) that the X-rays come in trains of waves which are long compared with the depth to which they penetrate the crystal. These are the assumptions on which Darwin's theory of X-ray reflection is based, although he considers also the general nature of the modification to be expected if the crystal is imperfect. On these assumptions reenforcement will occur from all parts of the crystal struck by the rays if all the ϵ's are zero, but when the ϵ's differ from zero the reflections from different parts of the crystal will in general be in different phases. The difference in the path of the ray reflected from an element of the crystal at p $(x, y,$

* It will be noted that this expression seems to make the amplitude of the reflected beam proportional to the number of electrons and hence approximately proportional to the atomic weight of the atoms of the crystal. Though this result is confirmed by some early experiments by Mr. W. L. Bragg,[1] later experiments give different results. In the case of calcite Professor Bragg finds[3] that the *intensity* of the beam reflected by an atom is more nearly proportional to the atomic weight than is the *amplitude*. He writes me that some of his results point one way and some the other. It seems very difficult to obtain a consistent reflection formula on the assumption that it is the intensity which is proportional to the number of electrons. The difficulty is probably to be explained by the nature of the function ψ, for if many of the electrons are at an appreciable distance from the center, the amplitude of the ray scattered by an atom will be considerably less than the sum of the amplitudes due to each electron.

z) and that reflected from an element at O (Fig. 2) is

$$x(\epsilon_2 - \epsilon_1) \sin \theta + y(\epsilon_4 + \epsilon_3) + z\{2 \sin \theta + (\epsilon_1 + \epsilon_2) \cos \theta\},$$

where second order terms in the ϵ's are neglected. The reflected rays

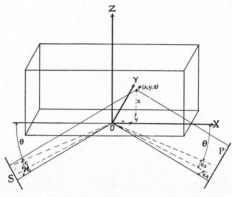

Fig. 2.

from all parts of the crystal are in the same phase when the difference in their paths is $2z \sin \theta$, so the difference in phase between the two reflected beams is

$$\delta = \frac{2\pi}{\lambda} \{x(\epsilon_2 - \epsilon_1) \sin \theta + y(\epsilon_3 + \epsilon_4) + z(\epsilon_1 + \epsilon_2) \cos \theta\}.$$

If the phase angle at P of the ray reflected from the point (0, 0, 0) is $2\pi\gamma/\lambda$, the displacement due to an element of volume of the crystal at p is, by equation (2),

$$dA_0 e^{-\frac{\mu z}{\sin \theta}} \cdot \cos \frac{2\pi}{\lambda} \{\gamma - x(\epsilon_2 - \epsilon_1) \sin \theta - y(\epsilon_3 + \epsilon_4) - z(\epsilon_1 + \epsilon_2) \cos \theta\} dx\,dy\,dz.$$

The displacement at P due to the reflection from the whole crystal is therefore

$$dA_0 \int_0^\infty dz \int_{-(k/2)}^{k/2} dy \int_{z \cot \theta - l/\sin \theta}^{z \cot \theta + l/\sin \theta} dx \cdot e^{-\frac{\mu z}{\sin \theta}} \cos \frac{2\pi}{\lambda} \{\gamma - x(\epsilon_2 - \epsilon_1) \sin \theta$$

$$- y(\epsilon_3 + \epsilon_4) - z(\epsilon_1 + \epsilon_2) \cos \theta\},$$

where l is the width and k the length of the slit through which the primary beam passes just before it strikes the crystal. This becomes when integrated,

$$\frac{4dA_0 BC}{gh} \cdot \frac{m \cos \frac{2\pi\gamma}{\lambda} - b \sin \frac{2\pi\gamma}{\lambda}}{m^2 + b^2},$$

where

$$B = \sin\left\{\frac{\pi l}{\lambda}(\epsilon_2 - \epsilon_1)\right\}, \quad C = \sin\left\{\frac{\pi k}{\lambda}(\epsilon_3 + \epsilon_4)\right\},$$

$$g = \frac{2\pi}{\lambda}(\epsilon_2 - \epsilon_1)\sin\theta, \quad h = \frac{2\pi}{\lambda}(\epsilon_3 + \epsilon_4),$$

$$m = \frac{\mu}{\sin\theta}, \quad b = \frac{4\pi}{\lambda}\epsilon_2\cos\theta.$$

If this expression is differentiated with respect to γ and equated to zero, it is found that the maximum displacement occurs when

$$\frac{2\pi\gamma}{\lambda} = \tan^{-1}\left(\frac{b}{m}\right),$$

i. e.,

$$\sin\frac{2\pi\gamma}{\lambda} = \frac{b}{\sqrt{m^2 + b^2}}, \quad \cos\frac{2\pi\gamma}{\lambda} = \frac{m}{\sqrt{m^2 + b^2}}.$$

Substituting these values in the above equation, we find for the amplitude of the reflected beam,

(3) $$dA_\epsilon = \frac{4dA_0 BC}{gh\sqrt{m^2 + b^2}}.$$

The corresponding intensity of reflection is therefore,

$$dJ_\epsilon = c(dA_\epsilon)^2 = \frac{16c(dA_0)^2 B^2 C^2}{g^2 h^2 (m^2 + b^2)},$$

where c has the value assigned above. This may be expanded into the form

$$J_\epsilon d\epsilon_1 d\epsilon_3 = \frac{cl^2 k^2}{\sin^2\theta} \cdot \frac{\sin^2\{n(\epsilon_2 - \epsilon_1)\}}{\{n(\epsilon_2 - \epsilon_1)\}^2} \cdot \frac{\sin^2\{p(\epsilon_3 + \epsilon_4)\}}{\{p(\epsilon_3 + \epsilon_4)\}^2} \cdot \frac{A_0^2 d\epsilon_1 d\epsilon_3}{m^2 + g\epsilon_2^2},$$

where $n = \pi l/\lambda$, $p = \pi k/\lambda$, and $q = (4\pi/\lambda)\cos\theta$, and dJ_ϵ and dA_0 are written as $J_\epsilon d\epsilon_1 d\epsilon_3$ and $A_0\sqrt{d\epsilon_1 d\epsilon_3}$ respectively. dJ_ϵ has its greatest value when all the ϵ's are zero, and is then

$$dJ_{(\text{max})} = \frac{c(dA_0)^2 l^2 k^2}{\mu^2}.$$

The maximum intensity of the reflected beam is thus independent of θ except for the function dA_0, a result which is in accord with that found by Webster[10] on similar assumptions.

It is not, however, this maximum intensity which is measured in the experiments. It is customary[2,6] rather to have the opening into the ionization chamber which measures the reflected X-ray beam large enough to receive the rays reflected at all angles ϵ_2 and ϵ_4. If the X-rays come from a point source, the angle ϵ_3 has all values between $-k/2r$ and

[10] D. L. Webster, Phys. Rev., 5, 241 (1915).

$+ k/2r$, where r is the distance from the anticathode to the slit, and ϵ_1 has a range of values $\delta\epsilon_1 = l/r$. The energy per unit time in the reflected beam which is due to the rays incident at angles between ϵ_1 and $\epsilon_1 + d\epsilon_1$ is therefore,

$$E_{\epsilon_1} d\epsilon_1 = \int_{-\infty}^{\infty} R d\epsilon_4 \int_{-k/2r}^{+k/2r} d\epsilon_3 \int_{-\infty}^{\infty} R d\epsilon_2 \cdot J_\epsilon d\epsilon_1.$$

If the crystal is rotated with a constant angular velocity $\omega = d\epsilon_1/dt$, it is exposed to rays incident at the angle ϵ_1 for a time $\delta\epsilon_1/\omega = \delta t$, and the total reflected energy due to the rays incident at the angle ϵ_1 is $E_{\epsilon_1} d\epsilon_1 \cdot \delta t$. If in the time Δt the crystal moves through all angles ϵ_1 at which any appreciable radiation of wave-length λ is reflected, the whole reflected energy is thus,

$$E_r = \delta t \int_{-\infty}^{\infty} E_{\epsilon_1} d\epsilon_1 = R^2 \delta t \int_{-\infty}^{\infty} d\epsilon_4 \int_{-k/2r}^{+k/2r} d\epsilon_3 \int_{-\infty}^{\infty} d\epsilon_2 \int_{-\infty}^{\infty} d\epsilon_1 \cdot J_\epsilon.$$

Substituting the value of J_ϵ given above, and integrating, this becomes,

$$E_r = \frac{cA_0^2 R^2 l^2 k^3}{r \sin^2 \theta} \cdot \frac{\pi^3}{mnpq} \delta t,$$

or

(4) $$E_r = \frac{cl^2 k^2 \lambda^3 \cdot N^2 A_i^2 \phi^2 \psi^2}{2r^2 \omega \mu \sin(2\theta)}.$$

The amplitude dA_i of the beam incident in the solid angle $d\sigma$ has been defined by the relation $c(dA_i)^2 = cA_i^2 d\sigma = I_i d\sigma$, where I_i is the intensity. The energy of the radiation in the solid angle $d\sigma = d\epsilon_1 d\epsilon_3$ which passes through the slit in unit time is therefore $klI_i d\sigma = cklA_i^2 d\epsilon_1 d\epsilon_3$, and as above, putting $\delta\epsilon_1 = l/r$, $\delta\epsilon_3 = k/r$, the total energy which passes through the slit in the time Δt is

$$E_i = klI_i \cdot \delta\epsilon_1 \delta\epsilon_3 \Delta t = \frac{k^2 l^2}{r^2} \cdot cA_i^2 \Delta t.$$

Writing $\Delta\theta = \omega \Delta t$ we find for the ratio of the reflected to the incident energy,

(5) $$R = \frac{E_r}{E_i} = \frac{N^2 \lambda^3 \phi^2 \psi^2}{2\mu \sin 2\theta} \cdot \frac{1}{\Delta\theta},$$

where $\Delta\theta$ is the angle through which the crystal is turned while making the observations. This result is in accord with that obtained by Darwin, but is worked out for somewhat different experimental conditions.

Imperfect Crystal.—It remains to determine the effect of changing our assumptions. Let us consider the case in which the crystal is not perfect, but is made up of a large number of small crystals, each of which acts as a separate unit. We shall have to determine the energy in the beam reflected by each little crystal, and sum up for all the component crystals.

Each component crystal will for convenience be considered to have the form of a rectangular parallelopiped of dimensions $\delta x, \delta y, \delta z$. If one of these little crystals is so small that the rays are not appreciably absorbed in passing through it, the phase of the beam reflected from the whole little crystal is the same as that of the ray reflected from its center. In a crystal whose center is at x_1, y_1, z_1 the amplitude contributed by a crystal element at that point is, by equation (2), $e^{-\mu z_1/\sin\theta} dA_0 dxdydz$. If δ is the phase difference between this ray and the one reflected from the point x, y, z, the amplitude contributed by a crystal element at x, y, z is therefore $e^{-\mu z_1/\sin\theta} dA_0 \cos\delta \cdot dxdydz$. Substituting the value of δ similar to that used above, we find for the amplitude at P of the beam reflected by the whole little crystal,

$$dA_\epsilon' = dA_0 e^{-\frac{\mu z_1}{\sin\theta}} \int_{z_1-(\delta z/2)}^{z_1+(\delta z/2)} dz \int_{y_1-(\delta y/2)}^{y_1+(\delta y/2)} dy \int_{x_1-(\delta x/2)}^{x_1+(\delta x/2)} dx \cdot \cos\frac{2\pi}{\lambda}$$

$$\times \{(x-x_1)(\epsilon_2-\epsilon_1)\sin\theta + (y-y_1)(\epsilon_3+\epsilon_4) + (z-z_1)(\epsilon_1+\epsilon_2)\cos\theta\}$$

$$= dA_0 e^{-\frac{\mu z_1}{\sin\theta}} \cdot \frac{\sin\left\{\frac{\pi\delta x}{\lambda}(\epsilon_2-\epsilon_1)\sin\theta\right\}}{\frac{\pi\delta x}{\lambda}(\epsilon_2-\epsilon_1)\sin\theta} \cdot \frac{\sin\left\{\frac{\pi\delta y}{\lambda}(\epsilon_3+\epsilon_4)\right\}}{\frac{\pi\delta y}{\lambda}(\epsilon_3+\epsilon_r)}$$

$$\times \frac{\sin\left\{\frac{\pi\delta z}{\lambda}(\epsilon_1+\epsilon_2)\cos\theta\right\}}{\frac{\pi\delta z}{\lambda}(\epsilon_1+\epsilon_2)\cos\theta} \cdot \delta x \delta y \delta z.$$

As before, the whole energy reflected by this crystal as it is turned through an angle $\Delta\theta$ at an angular velocity ω is given by

$$E_r' = R^2\delta t \int_{-\infty}^{\infty} d\epsilon_4 \int_{-(k/2r)}^{+(k/2r)} d\epsilon_3 \int_{-\infty}^{\infty} d\epsilon_2 \int_{-\infty}^{\infty} d\epsilon_1 \cdot c(A_\epsilon')^2.$$

Substituting the above value of $A_\epsilon' = dA_\epsilon'/d\epsilon_1 d\epsilon_3$ and integrating, we obtain

$$E_{r'} = \frac{cA_0^2 R^2 kl\lambda^3 e^{-\frac{2\mu z_1}{\sin\theta}}}{2r^2\omega \sin\theta \cos\theta} \delta x \delta y \delta z.$$

The energy reflected from the whole crystal is of course the sum of the energies reflected from all the component crystals, i. e., $E_r = \Sigma E_{r'}$, or replacing the summation sign by an integral,

$$(4a) \quad E_r = \int_0^\infty dz \int_{-(k/2)}^{+(k/2)} dy \int_{z\cot\theta-(l/2\sin\theta)}^{z\cot\theta+(l/2\sin\theta)} dx \cdot \frac{cA_0^2 R^2 kl\lambda^3 e^{-\frac{2\mu z}{\sin\theta}}}{2r^2\omega \sin\theta \cos\theta}$$

$$= \frac{cl^2 k^2 \lambda^3 N^2 A_i^2 \phi^2 \psi^2}{2r^2\omega\mu \sin 2\theta},$$

which is the same as was found in the case of a perfect crystal. Although this result has been obtained from the consideration of one particular type of imperfect crystal, it is evident that the same formula will hold in whatever manner the crystal is divided into its components.

Short Trains of Waves.—It has been pointed out by Webster[10] that if X-rays do not come in long trains like light waves, but come rather in trains which are short compared with the depth to which they penetrate the crystal, the intensity of the reflected beam at the angle of maximum reflection is proportional to $1/\sin^2 \theta$. He has since noted, however,[11] that the breadth of the reflected line increases with θ so that the total reflected energy does not obey the same law. The following analysis will show that the total energy reflected in this case is the same as with long trains of waves. If we consider the primary beam to be made up of trains M waves in length, the total number of reenforcing layers is M/n, where n is the order of the reflection. The thickness of the crystal which is effective in reenforcment at any instant is therefore

$$\delta z = \frac{M}{n} D,$$

where D is the distance between two adjacent layers of atoms. If the point x_1, y_1, z_1 is the center of the part of the crystal which has reflected the ray reaching P (R, ϵ_2, ϵ_4) at any particular instant, the amplitude of of the reflected wave at P at that instant is

$$dA_0'e^{-\frac{\mu z_1}{\sin \theta}} \int_{z_1 - (MD/2n)}^{z_1 + (MD/2n)} dz \int_{-(k/2)}^{k/2} dy \int_{z \cot \theta - (l/2 \sin \theta)}^{z \cot \theta + (l/2 \sin \theta)} dx$$

$$\times \cos \frac{2\pi}{\lambda} \left\{ \begin{array}{l} (x - x_1)(\epsilon_2 - \epsilon_1) \sin \theta \\ + (y - y_1)(\epsilon_4 + \epsilon_3) \\ + (z - z_1)(\epsilon_1 + \epsilon_2) \cos \theta \end{array} \right\}$$

$$= dA_0'e^{-\frac{\mu z_1}{\sin \theta}} \cdot \frac{klMD}{n \sin \theta} \cdot \frac{\sin u}{u} \cdot \frac{\sin v}{v} \cdot \frac{\sin w}{w},$$

where

$$u = \frac{\pi l}{\lambda}(\epsilon_2 - \epsilon_1), \quad v = \frac{\pi k}{\lambda}(\epsilon_3 + \epsilon_4), \quad w = \frac{2\pi MD}{n\lambda} \epsilon_2 \cos \theta,$$

and the corresponding intensity is

$$c(dA_0')^2 e^{-\frac{2\mu z_1}{\sin \theta}} \frac{k^2 l^2 M^2 D^2}{n^2 \sin^2 \theta} \cdot \frac{\sin^2 u}{u^2} \cdot \frac{\sin^2 v}{v^2} \cdot \frac{\sin^2 w}{w^2}.$$

Since by hypothesis the primary train of waves is short compared with the depth to which it penetrates the crystal, the length of the reflected train will depend, not on that of the incident one, but only upon the depth of penetration. In order to find the mean intensity of the reflected beam at this angle, we must therefore integrate this intensity over the whole length of the reflected train, and divide by the average distance between successive trains of waves. If this distance is ρ, the mean intensity of the reflected beam at this angle is thus,

$$\frac{c}{\rho}(dA_0')^2 \frac{k^2l^2M^2D^2}{n^2\sin^2\theta} \cdot \frac{\sin^2 u}{u^2} \cdot \frac{\sin^2 v}{v^2} \cdot \frac{\sin^2 w}{w^2} \int_0^\infty e^{-\frac{2\mu z}{\sin\theta}} \cdot 2\sin\theta\, dz$$

$$= \frac{c(dA_0')^2l^2k^2M^2D^2}{\rho n^2\mu} \cdot \frac{\sin^2 u}{u^2} \cdot \frac{\sin^2 v}{v^2} \cdot \frac{\sin^2 w}{w^2}.$$

Since $n\lambda = 2D\sin\theta$, this may be written:

$$J_\epsilon d\sigma = \frac{cM^2\lambda^2k^2l^2(dA_0')^2}{4\rho\mu\sin^2\theta} \cdot \frac{\sin^2 u \sin^2 v \sin^2 w}{u^2v^2w^2}.$$

When $u = v = w = 0$, this intensity is a maximum, and the resulting expression agrees with that obtained by Webster. To obtain the whole energy of the reflected beam, however, we must have as before,

$$E_r = R^2\delta t \int_{-\infty}^\infty d\epsilon_4 \int_{-(k/2r)}^{+(k/2r)} d\epsilon_3 \int_{-\infty}^\infty d\epsilon_2 \int_{-\infty}^\infty d\epsilon_1 \cdot J_\epsilon,$$

and substituting the above value of J_ϵ this becomes

$$E_r = \frac{cM\lambda^4k^2l^2(A_i')^2N^2\phi^2\psi^2}{2r^2\omega\rho\mu\sin 2\theta}.$$

The mean value of the square of the amplitude of the incident beam is evidently $A_i^2 = A_i'^2 \cdot M\lambda/\rho$, so the energy of the reflected beam is as formerly,

(4b) $$E_r = \frac{cl^2k^2\lambda^3N^2A_i^2\phi^2\psi^2}{2r^2\omega\mu\sin 2\theta},$$

where cA_i^2 now represents the mean intensity of the incident beam instead of its intensity at a given instant.

If there is any difference in the reflecting power of a crystal according to its degree of perfection or the nature of the incident rays, it must therefore be accounted for by a difference in the value of the absorption coefficient μ, since all the other factors in equation (4) have definitely defined values. It has been found by Darwin[7] that the value of this absorption coefficient does differ according to the degree of perfection of the crystal. This results from the fact that a sort of selective absorption occurs near the angle of maximum reflection, which is great only in the case of a crystal that is nearly perfect. Darwin concludes, however, that this effect may be accounted for by inserting a constant factor into the reflection formula.* We may therefore consider the expression here derived for the energy of the reflected beam of X-rays to hold for any crystal and any beam of X-rays if μ is taken as the *effective* absorption coefficient.

If a beam of X-rays of wave-length λ, polarized so that its electric vector is perpendicular to the plane of incidence, is reflected at an angle θ from a crystal in which all the atomic layers are equally spaced and whose atoms have no thermal motion, the energy in the reflected beam is therefore given by equation (5) as

$$E_r = \frac{E_i}{\Delta\theta} \frac{N^2 \lambda^3 \phi^2 \psi^2}{2\mu \sin 2\theta},$$

where E_i is the whole energy of the radiation of wave-length λ which strikes the crystal as it is turned with uniform angular velocity through an angle $\Delta\theta$, and $\Delta\theta$ is large enough to include all the angles at which rays of wave-length λ are reflected. N is the number of electrons per unit volume of the crystal, μ is the effective absorption coefficient of the X-rays in the crystal at the angle of reflection, and may differ by a constant factor from the coefficient of absorption at other angles. ϕ is the amplitude at unit distance of the ray scattered by a single electron, and has the value e^2/mC^2. ψ is a factor depending upon the arrangement of the electrons in the atomic layers, and has the value

$$\int_a^b F(z) \cos\left(\frac{4\pi \sin \theta}{\lambda} z\right) dz,$$

where $b - a$ is the diameter of an atom, and $F(z)$ is the probability that

---

* The statement that this correction for selective absorption at the angle of maximum reflection can be accounted for by the insertion of a constant factor into the reflection formula implies that this absorption is equally strong in all orders. Darwin has shown (loc. cit.) that this means a high degree of imperfection of the reflecting crystal. If the crystal is more perfect, the selective absorption must be large compared to the normal absorption and be proportional to $1/\sin 2\theta$. This would make the reflected energy E independent of the angle except for the factor ψ. Darwin considers the fact that the reflected energy at the larger angles is known to fall off rapidly to be sufficient proof that the reflecting crystals cannot be so perfect. It is to be noted, however, that since it is possible, as will be shown below, completely to explain the low intensity of the higher orders by assuming the proper distributions of the electrons in the atoms, the existence of this diminution cannot properly be used to prove the imperfection of the crystal. The fact that the absorption coefficient μ which is to be used in the reflection equation is found by Darwin to be of the same order of magnitude as the normal absorption coefficient indicates, however, that the selective absorption is not large, and hence that the crystal must be very imperfect. The writer has recently made a series of experiments comparing the rate of falling off of the higher orders when tungsten and rhodium rays are reflected from the same crystal. These experiments, which will be published in the near future, indicate rather definitely the existence of the term $\sin 2\theta$ in the denominator of the reflection expression, which would not occur if the crystal used were not sufficiently imperfect to make the absorption coefficient approximately constant.

a given electron will be at a distance z from the mid-plane of the atomic layer to which it belongs. This factor corresponds to the "excess scattering" factor introduced by Darwin.[4] As pointed out above, if the primary beam of X-rays is unpolarized, the factor $\frac{1}{2}(1 + \cos^2 2\theta)$ will have to be introduced, and to account for the thermal motion of the atoms the factor $e^{-B \sin^2 \theta}$ must also be included. Introducing these factors, writing $\sin 2\theta = 2 \sin \theta \cos \theta$, and expanding the terms ϕ and ψ, the complete expression for the energy of the reflected beam becomes:

$$(6) \quad E_r = \frac{E_i}{\Delta\theta} \cdot \frac{N^2 \lambda^3}{2\mu} \cdot \frac{1 + \cos^2 2\theta}{4 \sin \theta \cos \theta} \left(\frac{e^2}{mC^2}\right)^2 \times \left\{\int_a^b F(z) \cos \frac{4\pi z \sin \theta}{\lambda} dz\right\}^2 \cdot e^{-B \sin^2 \theta}.$$

It may be remarked that this same equation holds when the crystal remains stationary if the angular aperture $\Delta\theta$ of the slit as observed from the anticathode includes all the angles at which X-rays of wavelength λ are reflected.

The Distribution of the Electrons in Atoms.

By comparing this equation with Bragg's experimental formula (1) it should be possible to determine the form of the function $F(z)$, and hence to find the mean distances of the electrons from the centers of their respective atoms. We shall first attempt a direct solution of the problem, and shall find that there is no possible distribution of the electrons in atoms which will give Bragg's formula as it stands; but a slightly modified form of his law will be found to lead to a definite value of $F(z)$, thus determining the probable distance of an electron from the center of its atom. On comparison with experiment it will be shown, however, that the differences in the intensities of reflection by different crystals are such that no single distribution of the electrons can explain the reflection from all crystals. Arbitrarily chosen distributions which will explain satisfactorily the observed differences between the reflections from certain crystals will then be described, and these will be of considerable interest in connection with certain hypotheses concerning the structure of the atom.

Direct Method.—According to equation (6), if X-rays of wave-length λ are reflected by a certain crystal at two different angles, θ and θ_1, the ratio of the energy in the reflected beam in the two cases is

$$(7) \quad \frac{E}{E_1} = \frac{\dfrac{1 + \cos^2 2\theta}{\sin \theta \cos \theta} \left\{\int_a^b F(z) \cos\left(\dfrac{4\pi z \sin \theta}{\lambda}\right) dz\right\}^2 e^{-B \sin^2 \theta}}{\dfrac{1 + \cos^2 2\theta_1}{\sin \theta_1 \cos \theta_1} \left\{\int_a^b F(z) \cos\left(\dfrac{4\pi z \sin \theta_1}{\lambda}\right) dz\right\}^2 e^{-B \sin^2 \theta_1}}.$$

Bragg's experimental formula (1) gives this ratio as

$$\frac{E}{E_1} = \frac{\dfrac{1 + \cos^2 2\theta}{\sin^2 \theta} e^{-B \sin^2 \theta}}{\dfrac{1 + \cos^2 2\theta_1}{\sin^2 \theta_1} e^{-B \sin^2 \theta_1}} \qquad (8)$$

and on comparing these two expressions, we find

$$\frac{\left\{\int_a^b F(z) \cos\left(\dfrac{4\pi z \sin \theta}{\lambda}\right) dz\right\}^2}{\left\{\int_a^b F(z) \cos\left(\dfrac{4\pi z \sin \theta_1}{\lambda}\right) dz\right\}^2} = \frac{\tan \theta_1}{\tan \theta}. \qquad (9)$$

Extracting the square root, and substituting

$$x = \frac{4\pi \sin \theta_1}{\lambda} z, \qquad \phi(x) = F\left(\frac{\lambda x}{4\pi \sin \theta_1}\right) = F(z),$$

$$\alpha^2 = \frac{\tan \theta}{\tan \theta_1}, \qquad \beta = \frac{\sin \theta}{\sin \theta_1},$$

$$r = \frac{4\pi \sin \theta_1}{\lambda} a, \qquad s = \frac{4\pi \sin \theta_1}{\lambda} b,$$

this equation may be reduced to the form

$$\int_r^s \phi(x)\{\alpha \cos(\beta x) \pm \cos x\} dx = 0. \qquad (10)$$

From this equation we have to evaluate the function $\phi(x)$ and the limits of integration, r and s.

This seems to be a difficult form of integral equation to attack by direct mathematical methods. It is possible by graphical methods to show that there are an infinite number of solutions of the form $\phi(x) = c \sin px$, if $r = -s$, where the constants c and p can have any value. These solutions, however, imply that the probability function ϕ becomes negative at certain points, which is meaningless, so that they do not apply to this problem. There are also an infinite number of solutions of the type

$$\phi(x) = c_1 \sin^n p_1 x + c_2 \cos^m p_2 x + c_3,$$

if p_1 and p_2 are integers, r and s are integral multiples of π, β is an integer, c_1, c_2, n, m, may have any value, and c_3 is large enough to prevent the expression from becoming negative. Experiment shows, however, that β is not necessarily an integer, so that these solutions also do not apply. Other possible solutions are not obvious.

The necessary form of the function $F(z)$, and hence also of the function

$\phi(x)$ can be found, however, by certain physical considerations. If we consider the primary beam of X-rays to be plane polarized, and the reflecting crystal to be at a very low temperature, the experimental law expressed in equation (8) becomes $E/E_1 = \sin^2 \theta_1/\sin^2 \theta$. The ratio of the intensity of the reflected beam at two different angles thus depends only upon the angles at which reflection occurs, and is independent of the wave-length of the reflected beam. Thus with the layers of atoms at a distance a apart and with a beam of X-rays of wave-length λ, the ratio of the reflected energy at the two angles θ and θ_1 is the same as when the layers are at a distance ma apart using X-rays of wave-length $m\lambda$. Since, however, the ratio of these intensities is determined by the distribution of the electrons in the atomic layers, this means that this distribution shall be the same on the scale ma as it is on the scale a. That is, if the density of distribution is p times as great at $z = (1/q)a$ as at $z = (n/q)a$, it must be also p times as great at $z = (1/q)(ma)$ as at $z = (n/q)(ma)$, or

$$p = \frac{F\left(\frac{1}{q}a\right)}{F\left(\frac{n}{q}a\right)} = \frac{F\left(\frac{1}{q}ma\right)}{F\left(\frac{n}{q}ma\right)},$$

and in general

(11) $$\frac{F(z)}{F(nz)} = \frac{F(mz)}{F(nmz)},$$

where the constants n and m may have any value. The only type of function which will satisfy this relation may be shown to be

$$F(z) = b_1 z^{-p} \qquad [z > 0]$$
$$= b_1(-z)^{-p} \qquad [z < 0],$$

where b_1 and p are arbitrary constants. Since

$$z = \frac{\lambda x}{4\pi \sin \theta_1},$$

(12) $$\phi(x) = F(z) = b_1 \left\{\frac{\lambda x}{4\pi \sin \theta}\right\}^{-p} = bx^{-p} \quad [x > 0].$$

Equation (11) shows that the density of distribution of the electrons can vanish nowhere unless it is everywhere zero, so the limits of integration of equation (10) must be $-\infty$ and ∞. This of course corresponds to an atom with an infinite radius.

Substituting this value for $\phi(x)$, and remembering that the function under the integral sign is symmetrical on both sides of the plane $z = 0$, equation (10) may be written

(13) $$\int_0^\infty x^{-p}\{\alpha \cos \beta x \pm \cos x\} dx = 0$$

The integral is a known one,[12] the solution being

$(\alpha \beta^{p-1} \pm 1) \dfrac{\pi}{2 \Gamma(p)} \sec \left(\dfrac{p\pi}{2}\right)$ for $[0 < p < 1]$

∞ for $[p \gtreqless 1]$

indeterminate for $[p \lesseqgtr 0]$.

This integral can vanish only when

$$\alpha \beta^{p-1} \pm 1 = 0$$

or

$$\frac{\cos \theta}{\cos \theta_1} = \left\{\frac{\sin \theta}{\sin \theta_1}\right\}^{2p-1}.$$

Since θ and θ_1 can have any values, it is obvious that there is no constant value of p which will satisfy this equation. Thus there is no solution of the integral equation (10) which meets the conditions. We must conclude, therefore, that there is no possible distribution of the electrons in the atomic layers which will give Bragg's experimental law as it stands.*

If equation (1) is modified by inserting the factor $\cos \theta$ in the denominator, its value is not greatly changed, since θ is usually small. This modification changes the quantity on the right-hand side of equation (9) to $\sin \theta_1/\sin \theta$, and the coefficient α in equations (10) and (13) becomes $(\sin \theta/\sin \theta_1)^{1/2}$, the other quantities remaining unchanged. In this case the left hand side of equation (13) vanishes only when

$$\frac{\sin^{1/2} \theta}{\sin^{1/2} \theta_1} \cdot \left\{\frac{\sin \theta}{\sin \theta_1}\right\}^{p-1} = 1,$$

or

$$p = \tfrac{1}{2}$$

(14) whence

$F(z) = b_1 z^{-1/2}$ $[z > 0]$

$= b_1(-z)^{-1/2}$ $[z < 0]$.

Thus although Bragg's law in its original form is not given by any possible

[11] D. L. Webster, Phys. Rev., 7, 696 (1916).

[12] Bierens de Haan, Nouv. Tables d'integrales definies.

* On somewhat different assumptions from those used here, Professor Bragg has found a distribution of the scattering material which leads to the law $E/E_1 = n_1^2/n^2$, where n is the order of reflection.[8] The distribution he finds is, however, a function of the distance between the atomic layers, and so cannot truly represent the arrangement of the electrons in the atoms. In fact it is possible to show by reasoning similar to that used above that on his assumptions also there is no distribution of the electrons which will give his law for the intensity of reflection.

distribution of the electrons, it can be obtained in a slightly modified form if the density of distribution of the electrons is inversely proportional to the square root of the distance from the mid-plane of the atomic layer to which they belong.

The quantity $F(z)$ represents the probability that a given electron shall be at a distance z from the middle of its layer of atoms at any instant. It is evident, therefore, that the relation $\int_{-\infty}^{\infty} F(z)dz = 1$ or $\int_{0}^{\infty} F(z)dz = \frac{1}{2}$ must be satisfied. The constant b_1 in equation (12), representing the probability $F(z)$ at unit distance from the mid-plane is thus defined by the expression

$$b_1 \int_0^{\infty} z^{-p} dz = \tfrac{1}{2},$$

or
$$b_1 = 0,$$
since
$$\int_0^{\infty} z^{-p} dz = \infty$$

for all values of p. This means that there can be no appreciable density of distribution of the electrons in the atomic layer unless there are an infinite number of electrons in it. It is evident therefore that the function $F(z)$ cannot have the form $b \cdot z^{-p}$. According to the argument above, this means that the relative intensity of the beam of X-rays reflected at different angles must depend not only upon the angle of reflection, but also upon the wave-length of the incident beam of X-rays.

Empirical Method.—Since there is no possible distribution of the electrons in the atoms which will give a law of reflection of the form of Bragg's empirical expression, let us see if it is not possible to find some arbitrary distribution which will give a reflection formula fitting the experimenral data better than his law. In order conveniently to compare data obtained with different crystals and with X-rays of different wave-lengths, let us remember, according to equation (7), that the part of the ratio E/E_1 which is due to the diffuseness of the atomic planes is

$$U^2 = \frac{\left\{\int_a^b F(z) \cos\left(\frac{4\pi z \sin \theta}{\lambda}\right) dz\right\}^2}{\left\{\int_a^b F(z) \cos\left(\frac{4\pi z \sin \theta_1}{\lambda}\right) dz\right\}^2} = \frac{\psi^2}{\psi_1^2} = \frac{E}{E_1} \cdot \frac{\dfrac{1+\cos^2 2\theta_1}{\sin \theta_1 \cos \theta_1}}{\dfrac{1+\cos^2 2\theta}{\sin \theta \cos \theta}} \cdot \frac{e^{-B \sin^2 \theta_1}}{e^{-B \sin^2 \theta}}.$$

From any assumed distribution of electrons $F(z)$ the theoretical values of

(15) $$U = \frac{\psi}{\psi_1}$$

may be calculated for different angles of reflection, and by comparison with the corresponding experimental values as given by the equation

$$(16) \quad U = \left\{ \frac{E}{E_1} \cdot \frac{\frac{1 + \cos^2 2\theta_1}{\sin \theta_1 \cos \theta_1}}{\frac{1 + \cos^2 2\theta}{\sin \theta \cos \theta}} \cdot \frac{e^{-B \sin^2 \theta_1}}{e^{-B \sin^2 \theta}} \right\}^{1/2},$$

the accuracy of the assumed distribution may be tested.

For the cases of rock-salt and calcite the values of E/E_1 have been determined experimentally by Professor Bragg.[2,8] His results are given in Tables I. and II., columns 3. In the case of calcite the reflection at

TABLE I.
Rock-Salt.

Crystal Face.	$\frac{\sin \theta}{\sin \theta_1}$	$\frac{E}{E_1}$	U_{obs}	U_{calc_1}	U_{calc_2}
100	1.00	1.00	1.00	1.00	1.00
110	1.41	.41	.79	.83	.81
111	1.73	.244	.69	.75	.695
100	2.00	.187	.67	.70	.63
110	2.83	.0705	.55	.58	.56
100	3.00	.0625	.55	.56	.55
111	3.46	.042	.54	.51	.50

TABLE II.
Calcite.

Crystal Face.	$\frac{\sin \theta}{\sin \theta_1}$	$\frac{E}{E_1}$	U_{obs}	U_{calc_1}	U_{calc_2}
$1\bar{1}1$	.80	1.44	1.05	1.12	1.03
100	1.00	1.12	1.03	1.00	1.00
111	1.07	.95	.98	.97	.98
$1\bar{1}0$	1.24	.67	.88	.90	.95
110	1.59	.46	.86	.78	.87
$1\bar{1}1$	1.60	.48	.88	.78	.87
100	2.00	.257	.72	.70	.79
211	2.12	.286	.78	.68	.76
111	2.14	.297	.80	.68	.76
$1\bar{1}0$	2.46	.164	.65	.62	.69
100	3.00	.113	.61	.56	.60
110	3.22	.091	.57	.54	.58
$1\bar{1}1$	3.21	.108	.62	.54	.58
100	4.00	.034	.41	.48	.44
111	4.28	.036	.42	.46	.42
110	4.78	.021	.37	.43	.36
$1\bar{1}0$	4.91	.013	.30	.42	.35

different angles is produced by layers of different kinds of atoms, but the values here given have been corrected according to the method explained by Professor Bragg[3] for the effect of the differences in the atomic layers, so these figures may be taken as due to a crystal all of whose atoms are similar and whose atomic layers are uniformly spaced. In these tables θ_1 is taken as the angle of the first order reflection from the (1, 0, 0) face of the crystal. The values of U calculated from the observed values of E/E_1 are given in columns 4, and are plotted in Fig. 3 against the corresponding values of $\sin \theta / \sin \theta_1$.* The open circles in this figure represent the experimental values of U for rock-salt, and the solid ones for calcite. The agreement between the results of successive experiments indicates that these data may be taken as accurate within a probable error of about one or two per cent., although consistent errors of considerably greater magnitude may occur due to various causes.

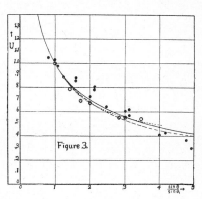

Fig. 3.

The values of U calculated from Bragg's law, equation (1), are given in columns 5 of the tables and are shown on Fig. 3 by the solid line. If the function $F(z)$ is assumed to have the form bz^{-p}, U has the values observed for the first and second order reflections from rock-salt when

* In calculating the values of U here given, the constant B in equation (16) has been evaluated by Debye's formula,[3]

$$B = \frac{1.142 \times 10^{-12}}{A\Theta\lambda^2} \frac{\phi(x)}{x},$$

where A is the atomic weight, Θ is Debye's "characteristic temperature," $x = \Theta/T$, where T is the absolute temperature, and $\phi(x)$ is a function which Debye evaluates. The value of B thus calculated has been found by Bragg[2] to account with considerable accuracy for the observed changes of the intensity of reflection with temperature. For rock-salt $\Theta = 260°$ K., $A = 29$, $x = .89$, $\phi = .80$, and using rhodium rays of wave-length $\lambda = .614 \times 10^{-8}$ we obtain $B = 3.6$. In the case of calcite $\Theta = 910°$ (determined from the specific heat at 298° K.), and taking $A = 20$, $x = 3.10$ and $\phi = .473$ we find $B = 0.25$. It is these values which have been used in calculating U. If a "Nullpunktsenergie" is assumed, Debye shows that the value of B is

$$B = \frac{1.142 \times 10^{-12}}{A\Theta\lambda^2}\left(\frac{\phi(x)}{x} + \frac{1}{4}\right).$$

In this case for rock-salt $B = 4.6$ and for calcite $B = .67$. This makes no very large difference in the values of U. In any case the effect of the temperature factor is not large at the angles at which these measurements are made, so that any slight error in the value of B will not greatly affect the results.

$p = 0.425$, which gives $U = (\sin \theta_1/\sin \theta)^{.575}$. This expression is plotted in the broken line of Fig. 3, and is about as accurate an expression for the reflected energy as can be obtained with this type of function. Professor Bragg has suggested[8] the use of a function of the form $F(z) = be^{-cz}$ to express the distribution of the electrons. The coefficient b in this case has a fiinte value, so although this function also implies an atom of infinite radius it is not *a priori* impossible. It can be shown, however, that the values of U obtained with such a distribution are considerably too large at moderately small angles, and decrease much too rapidly at large angles. Better agreement is obtained if a distribution of this type is combined with a certain concentration $F(O) = a$ of the electrons at the middle of the reflecting layers. In this case

$$U = \frac{a/b + c \Big/ \left\{ c^2 + \left(\frac{4\pi \sin \theta}{\lambda}\right)^2 \right\}}{a/b + c \Big/ \left\{ c^2 + \left(\frac{4\pi \sin \theta_1}{\lambda}\right)^2 \right\}}.$$

In order to make the values of U thus calculated agree with the observed values for the second and third orders of reflection from rock-salt we must put $a/b = 0.274 \lambda$ and $c = 1.51/\lambda$. The values of U thus obtained are plotted in the dotted curve of Fig. 3, and are seen to be in better agreement with the experiments on rock-salt than is the empirical law expressed by the solid curve. A comparison of the experimental data for rock-salt and calcite as shown in Fig. 3 shows at once that a curve which corresponds to the observed values of U for rock-salt cannot fit accurately the data for calcite. Professor Bragg has suggested[8] the possibility of the existence of such a difference, but it was a difficult thing to detect on account of the many disturbing factors. Since by equation (15) U is a function only of the distribution of the electrons in the atomic layers and of $\sin \theta/\lambda$, any observed differences in U for a given value of $\sin \theta/\lambda$ can be explained only by differences in the distributions of the electrons in the layers of atoms which do the reflecting. Thus we may conclude that the distribution is not the same in the atoms of calcite as it is in the atoms of rock-salt.* It appears, therefore, that

---

* Note added December 9, 1916: A more striking example of the difference in the value of U for different crystals is afforded by Vegard's recent determination of the intensity of X-ray reflection from silver (Phil. Mag., 32, 94, July, 1916). He finds the intensity of the first three orders of reflection from the (111) plane to be in the ratio of 1.00 : 0.45 : 0.11, respectively. The corresponding values of U are 1.00 : 1.03 : 0.73. In the case of rock-salt we have found the values of U for the first three orders of reflection from the (100) plane to be 1.00, 0.67 and 0.55. The difference is too great to be accounted for by experimental error. Since, as pointed out above, U depends only upon the distribution of the electrons in the atoms of the crystal, these differences must be taken as a strong confirmation of our

instead of considering the electrons to be distributed according to some general law such as we have been discussing, one should rather consider every atom to possess a finite number of electrons, each placed at a definite distance from its center.

Atoms with a Finite Number of Electrons.—If each atom of a certain kind has an electron at a distance a from its center, the average effect from a large number of such atoms will be the same as that due to a uniform distribution of the electrons over the surface of a sphere of radius a. The center of this equivalent spherical shell will be in the mid-plane of the atomic layer, and the probability that a given electron in the shell will be at a distance z from the middle of the layer may be shown to be

$$F(z) = c \quad [-a < z < a],$$

or, in virtue of the relation

$$\int_{-a}^{a} c\, dz = 1,$$

$$F(z) = \frac{1}{2a} [-a < z < a].$$

The value of ψ due to an electron in such a shell is therefore

(17) $$\psi' = \frac{1}{2a} \int_{-a}^{a} \cos\left(\frac{4\pi z \sin \theta}{\lambda}\right) dz = \frac{\sin\left(\frac{4\pi a \sin \theta}{\lambda}\right)}{\frac{4\pi a \sin \theta}{\lambda}},$$

and the value of ψ for a whole atom is

(18) $$\psi = \frac{1}{\nu}\sum_r \psi_r' = \frac{1}{\nu}\sum_r \sin\left(\frac{4\pi a_r \sin \theta}{\lambda}\right) \bigg/ \left(\frac{4\pi a_r \sin \theta}{\lambda}\right),$$

where ν is the number of electrons in the atom, and the summation extends over all the r's from 1 to ν. In any actual case this summation is most readily performed by plotting ψ' according to equation (17) and taking off its values corresponding to the desired values of $a \sin \theta/\lambda$. From the value of ψ thus obtained, the quantity U can be calculated as before by equation (15).

Attempts to obtain a suitable formula for U by thus adding the effects of a number of electrons placed at arbitrary distances a from the centers

fundamental assumption that the intensity of X-ray reflection depends upon the distribution of the electrons in the atoms of the reflecting crystal. If the values of U for silver are calculated on the basis of the assumptions used below in the case of calcite, giving to each electron an angular momentum of h/π, we obtain the values of U for the first three orders of reflection as 1.00 : 0.93 : 0.82. Though the agreement is far from perfect, it is much better than that obtained with the values 1.00 : 0.70 : 0.56, assigned by Bragg's law.

of the atoms soon convince one that the form of the resulting curve is very sensitive to changes in the assumed values of a. The results obtained above indicate that there must be a rather strong concentration of the electrons near the centers of the atoms, but it is difficult to select a distribution of the outer electrons which will give a reflection formula that agrees with the experimental data.

In the case of calcite it was found that a fairly acceptable expression could be obtained if the atoms of calcium, carbon and oxygen were assumed to have a number of electrons equal to their positive valence, *i. e.*, 2, 4 and 6, respectively, placed at distances from the centers of the atoms inversely proportional to the valence, together with an arbitrary number of electrons placed at the centers of the atoms. The use of this number of electrons in the outer part of each atom was suggested by well-known theories of valence, while some such spacing as this seemed necessary in order to obtain a suitable formula for U. This assumption concerning the spacing is, however, very nearly what is to be expected if these valency electrons revolve in rings about the centers of their respective atoms, all the electrons having the same angular momentum. For according to classical mechanics,[13] the centrifugal force $m\omega^2 a$, where m is the mass and ω the angular velocity of the electron and a is the radius of its orbit, must be balanced by the centripetal force $(M - \sigma_n)e^2/a^2$ where e is the electronic charge, M is the total charge in electronic units on the part of the atom inside the ring considered, and σ_n is a term which depends upon the mutual repulsion between the electrons in the same ring. With one electron in the ring $\sigma_1 = 0$, for two $\sigma_2 = .25$, and similarly $\sigma_4 = .96$, $\sigma_6 = 1.83$, $\sigma_7 = 2.31$ and $\sigma_8 = 2.81$. But if the angular momentum is constant, *i. e.*, $ma^2\omega = c$, we may write

$$ma \cdot \frac{c^2}{m^2 a^4} = \frac{e^2}{a^2}(M - \sigma_n),$$

or

$$(19) \qquad a = \frac{c^2}{me^2} \frac{1}{(M - \sigma_n)}.$$

Since for the outer ring of a neutral atom M is equal to the number of electrons in the ring, a is thus approximately inversely proportional to the number of valency electrons. This suggested, therefore, the assumption of the constancy of the angular momentum of the electrons in the atoms as a working basis.

It is of course possible to find any number of satisfactory arrangements unless the number of electrons in the atoms is defined. Barkla has shown[13] that the number of electrons in an atom which are effective in

[13] H. G. J. Moseley, Phil. Mag., 26, 1032 (1913).

scattering X-rays is approximately equal to half the atomic weight. His work in connection with that of Rutherford on the scattering of alpha particles[15] and of Moseley on the high frequency spectra of the elements[14] makes it very probable that if N is the atomic number, there are N electrons distributed about the atomic nucleus, and that all these electrons are effective in scattering X-rays. This, together with the assumption used above concerning the valency electrons, determines the number of electrons in both rings of the first row of elements in the periodic table, including carbon and oxygen. For the second and third rows it has been assumed, in accordance with Moseley's interpretation of his X-ray spectra,[13] that the inner ring contains 4 electrons. This leaves 6 electrons in the middle ring of the second row of atoms. There remain 8 electrons to be placed in the calcium atom, and it seemed reasonable to put these in a ring just inside of the valency electrons.

With the number of electrons in each ring thus determined, the relative distance of each electron from the nucleus can immediately be

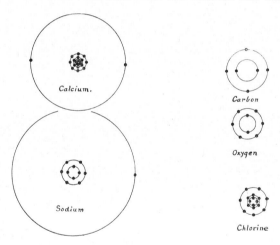

Fig. 4.

calculated by equation (19), and the corresponding values of U can be determined by equations (15) and (18). If the angular momentum of all the electrons is the same, their relative spacing in the atoms here used is as shown in Fig. 4. The value of U obtained for calcite on the basis of these assumptions is shown in the dotted curve of Fig. 5. Although the agreement is fair, it is evident that the theoretical values are too high in the low orders, and fall off too rapidly in the higher orders. It is to be noted that as there is but a single arbitrary constant, the

[14] C. G. Barkla, Phil. Mag., 14, 408 (1907).
[15] E. Rutherford, Phil. Mag., 27, 488 (1914).

angular momentum of an electron, in the equation for U, even this general agreement indicates that we are working along the right line. If the assumptions are modified to the extent of giving the electrons in the inner rings and in the second calcium ring half the angular momentum of those in the outer rings, the values of U given in column 6 of Table II. and shown in the solid curve of Fig. 5 are obtained. The equation for ψ is in this case,

$$\psi = \frac{1}{50} \left\{ \begin{bmatrix} 4\sin(.013k\sin\theta)/.013k\sin\theta + 6\sin(.018k\sin\theta)/.018k\sin\theta \\ +8\sin(.139k\sin\theta)/.139k\sin\theta + 2\sin(.571k\sin\theta)/.571k\sin\theta \end{bmatrix} \\ +[2\sin(.043k\sin\theta)/.043k\sin\theta + 4\sin(.329k\sin\theta)/.329k\sin\theta] \\ +3[2\sin(.032k\sin\theta)/.032k\sin\theta + 6\sin(.240k\sin\theta)/.240k\sin\theta] \right\}.$$

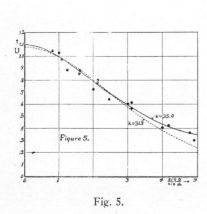

Fig. 5.

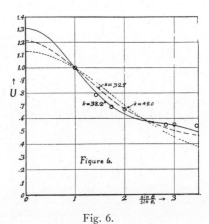

Fig. 6.

50 is the number of electrons in a molecule of calcium carbonate, and k is the constant $(4\pi/\lambda) \cdot (c^2/me^2)$ where c is the angular momentum of an electron in one of the outer rings. The agreement of the theoretical curve with the observed values is well within the probable experimental error, and is evidently better than that obtained with Bragg's empirical formula.

If the assumption that all the electrons have the same angular momentum is used in the case of rock-salt, a result is obtained which does not at all correspond with the observed intensities of reflection, as is shown in the dotted curve of Fig. 6. In order to account for the low value of U around the second order, it seems necessary that there shall be a ring of electrons about 1.00×10^{-8} cm. in radius which contains a considerable number of electrons, and that there shall be no other heavy ring of more than half as great diameter. This arrangement is obtained if we consider the electrons in the outer rings of chlorine to have 3/2 as great angular momentum as those of sodium, and if, as before, we assume the inner ring of each atom to have $\frac{1}{2}$ the normal angular momentum. According to the work of J. J. Thomson[16] it appears probable that the valency electron of sodium goes over to the chlorine atom and becomes a part of its outer ring. This would give 8 electrons in the outer ring of chlorine and none in that of sodium. The values of U given in Table I. and shown in the solid curve of Fig. 6 have been calculated on this basis. If the valency electron of sodium is not thus transferred, the broken curve in this figure is obtained on the same assumptions concerning angular momentum. It will be seen that the former hypothesis yields the better results. Although the accuracy of the experimental values of U is not great enough at present to warrant any definite conclusions concerning such details, the difference in the curves obtained by thus shifting the position of a single electron indicates the extreme sensitiveness of this method for determining the distribution of the electrons in atoms.

Conclusions and Discussion.

Thus we see that if certain definite distributions of the electrons in atoms are assumed, it is possible to explain in a satisfactory manner the rapid decrease in the intensity of the higher orders of reflection of X-rays by crystals. Although it is not possible thus to derive Bragg's empirical formula, expressions for the intensity of reflection can be derived which agree better with experiment than does his law. We have shown from a theoretical standpoint that any equation which is to express adequately the relative intensity of the beam of X-rays reflected at different angles must be a function not only of the angle of reflection, as is Bragg's law, but must involve also the wave-length of the incident beam of X-rays. Professor Bragg has suggested that the observed values of the intensities of the reflection from different crystals actually show differences which do not depend upon the angle alone. These differences are usually masked by the large general variations of the intensity with the angle of reflection which is to be expected with any crystal grating; in the cases of rock-salt and calcite, however, these general variations can be accounted for, and it has been found above that there remain certain decided differences between the spectra from these crystals. It has been shown theoretically that any such differences must be due to differences in the distribution of the electrons in the atoms of the respective crystals, and the arrangements of the electrons which have been assigned to the different atoms have been found capable of accounting for these variations. Thus our fundamental hypothesis that the intensity of the reflected beam is a function of this distribution is strongly confirmed.

[16] J. J. Thomson, Phil. Mag., 27, 757 (1914).

The scarcity and comparative uncertainty of the experimental data make it premature to arrive at any positive conclusions concerning the details of the distribution of the electrons in the atoms considered. In general, however, it may be said that a comparison of the different distributions of the electrons which have been found to give satisfactory results show that there is usually a rather strong concentration of the electrons near the centers of the atoms, but with a considerable number at appreciable distances from the atomic centers. The expression $F(z) = bz^{-\frac{1}{2}}$, which was found to give Bragg's law in a slightly modified form, implies that the volume of density of distribution of electrons is inversely proportional to the 5/2 power of the distance from the center of the atom, and the satisfactory distributions in the case of a limited number of electrons have followed this rule approximately.

From the value of the constant k which is determined by the experimental data, it is possible to calculate the radii of the different rings of electrons according to the formula

$$a = \frac{k\lambda}{4\pi(M + \sigma_n)}.$$

Using the values corresponding to the solid curve of Fig. 5 we thus find for the outer ring of calcium, $a = 0.97 \times 10^{-8}$ cm., for carbon, 0.56×10^{-8} cm., and for oxygen, 0.41×10^{-8} cm. The absolute magnitudes here given are of course subject to revision by further experiment, but if the number of electrons in the outer ring is taken equal to the positive valence of the atom, a number of unsatisfactory attempts to obtain expressions for U with other arrangements make it seem necessary that the ratio of the diameters of the rings shall be about that here given. This result is interesting in connection with the fact that while the carbon and oxygen atoms in calcite are only about 1.07×10^{-8} cm. apart, the calcium atom is about 2.95×10^{-8} cm. from its nearest neighbor. In the case of rock-salt, if the valency electron of sodium is in the outer chlorine ring, the radius of the next sodium ring is 0.36×10^{-8} cm., and of the outer chlorine ring is 1.00×10^{-8} cm. If the valency electron remains in the sodium atom, it is, according to our assumptions, 1.86×10^{-8} cm. from the center of the atom, and the chlorine ring is reduced to 0.89×10^{-8} cm. This is a possible arrangement, since the atoms are 2.81×10^{-8} cm. apart. As mentioned above, the diameter of the chlorine ring is probably determined with a fair degree of accuracy. The arrangement of the inner rings is by no means so definitely determined. A number of attempts were made, however, to fit the data for rock-salt assuming only 2 instead of 4 electrons in the inner rings, and this seemingly small change made the curve for U depart too far from the experimental values to be acceptable. Results such as these make it reasonable to suppose that in the case of the lighter elements it may be possible, with a sufficient amount of accurate experimental data, to determine positively the distance of each individual electron from the center of its atom.

Bearing on Theories of Atomic Structure.—While it is difficult by any such "cut and try" method to find the only possible arrangement of the electrons in atoms, it is evident that a study of the intensity of X-ray spectra thus affords an extremely sensitive test of any theory which assigns a definite distribution to the electrons in atoms. From the experimental data now available it may be said, for example, that unless some important factor has been neglected in our formula for X-ray reflection, it seems impossible to account for the rapid diminution of the intensity of the higher orders on any theory, such as Crehore's,[17] which would confine the electrons of an atom within a distance less than 10^{-10} cm. from its center. Such a distribution would make the quantity ψ of equation (6) approximately equal to unity, and this equation would then differ from the experimental equation (1) by the factor $\tan \theta$, which is much greater than the experimental error. Even if such a factor were introduced into the theoretical equation, it would still be unable to account for the variations in intensity which are characteristic of the individual crsytal used as a grating. It seems necessary, therefore, to reject Crehore's theory of the atom as an impossible hypothesis.

On the other hand, it seems possible to explain all the X-ray intensities on the basis of the type of atom suggested by Bohr.[18] The working assumption that we have used, the constancy of angular momentum, is the fundamental hypothesis of Bohr's theory. We may calculate the angular momentum of the electrons in the atoms here considered according to the relation

$$c = \sqrt{k \frac{\lambda m e^2}{4\pi}}.$$

Using the values $\lambda = 0.614 \times 10^{-8}$ cm., $m = 9.0 \times 10^{-28}$ gm. and $e = 4.77 \times 10^{-10}$ gm.$^{1/2}$cm.$^{3/2}$sec.$^{-1}$, for the outer rings of the atoms in calcite $k = 35.0$ and c becomes 1.87×10^{-27} gm. cm.^{2} sec^{-1}. For the inner rings our hypotheses make c half this value. According to Bohr the angular momentum of an electron in an atom is an integral multiple of $h/2\pi$, where h is Planck's constant. Using the value $h = 6.57 \times 10^{-27}$

---

[17] A. C. Crehore, Phil. Mag., 26, 25 (1913) and elsewhere.
[18] N. Bohr, Phil. Mag., 26, pp. 1, 476, 857 (1913), 27, p. 506 (1914).

gm. cm.^{2} sec.$^{-1}$, we find $c = 2(1 - .10) \cdot h/2\pi$. In the case of rock-salt it is the outer chlorine ring whose diameter is most definitely defined. In this case $k = 9/4 \cdot 38.2$, and $c = 2.93 \times 10^{-27} = 3(1 - .06) \cdot h/2\pi$. It may be only an accident that the angular momentum in these cases works out so nearly in accord with Bohr's hypothesis, but it is at least a most interesting accident. The difference between the coefficients of $h/2\pi$ and integers is probably too great to be due to experimental error in the determination of the radii of the electronic rings, but there are good reasons to believe that there may be forces acting on the electrons which have not been considered, and which would be sufficient to account for the difference.

SUMMARY.

In the first part of this paper an expression is derived for the energy of a beam of X-rays of definite wave-length which is reflected from a crystal. The result is in accord with that previously derived by Darwin, and shows that the intensity of the reflected beam depends not only upon the angle of reflection but also upon the arrangement of the electrons within the atoms of the reflecting crystal.

The form of the equation for the energy in the reflected beam is shown to be independent of the degree of perfection of the crystal and of the length of the wave-trains of which the X-rays consist.

In the latter part of the paper a study is made of the possible distributions of the electrons in atoms which will account for the observed energy in the reflected beam of X-rays. It is shown that there is no possible distribution of the electrons which will lead to Bragg's empirical law for the intensity of reflection; it is found rather that any formula for this intensity must depend not only upon the angle of reflection but also upon the wave-length of the incident rays.

Attention is called to the fact that Bragg's experimental data indicate differences in the reflection from certain crystals which can arise only from differences in the distribution of the electrons in the atoms of which the crystals are composed.

Assuming a number of electrons in each crystal equal to the atomic number, and making certain plausible assumptions concerning the arrangement of these electrons in rings, it is found possible to account in a satisfactory manner for the observed intensities of the X-ray spectra.

Although the particular distributions assigned to the electrons in the atoms of calcite and rock-salt may not be the only ones which will account for the observed intensities of the X-ray spectra, it seems probable that these distributions are not far from correct.

The results of this investigation seem to show conclusively that the electrons are not concentrated within a very small distance from the center of the atom, as is assumed in Crehore's theory of atomic structure. On the other hand, the conclusions arrived at are in good accord with the theory of the atom due to Bohr.

My thanks are due to Professor H. L. Cooke for his helpful interest in this research.

PALMER PHYSICAL LABORATORY,
  PRINCETON UNIVERSITY,
    June 19, 1916.

The Reflection Coefficient of Monochromatic X-Rays from Rock Salt and Calcite.[1]

By A. H. Compton.

ACCORDING to classical electrodynamics, the ratio of the energy of a beam of X-rays reflected from a crystal to that of a beam incident upon it is given by Darwin's formula (2):

$$(1) \quad R = \frac{E_r}{E_i} = \frac{1}{\Delta\theta} \cdot \frac{N^2\lambda^3}{2\mu} \cdot \frac{1 + \cos^2 2\theta}{4 \sin\theta \cos\theta} \left(\frac{e^2}{mC^2}\right) e^{-B\sin^2\theta}$$

In this expression E_r is the energy in the beam of X-rays of wave-length λ which is reflected at a glancing angle θ, while the crystal is rotated with uniform angular velocity through an angle $\Delta\theta$ which is large enough to include all angles at which any appreciable amount of rays of this wave-length are reflected. E_i is the total energy of wave-length λ which falls on the crystal during this time; N is the number of electrons per unit volume in the crystal. μ is the absorption coefficient of the X-rays in the crystal; e is the charge and m the mass of an electron and C is the velocity of light. The factor ψ depends upon the distribution of the electrons in the atoms of the reflecting crystal, its value being approximately 0.76 for the first order reflection from the cleavage planes of rock salt, and 0.75 for calcite.[3] The constant B depends upon the thermal motion of the atoms of the crystal. It may be taken to be 2.6 in the case of rock salt reflecting molybdenum α rays, and 0.18 in the case of calcite.

It is evident that by measuring the ratio of the reflected energy E_r of wave-length λ to the incident energy E_i of the same wave-length, a test of this formula may be made. In order to measure this ratio, monochromatic X-rays were obtained by the reflection of a beam of X-rays from a crystal mounted on a standard Bragg X-ray spectrometer. The source of X-rays was a Coolidge tube with a molybdenum target, kindly supplied by Dr. Coolidge, so that it was possible to obtain a comparatively intense beam of monochromatic X-rays of wave-length 0.721×10^{-8} cm. The monochromatic beam thus obtained was reflected in turn by a crystal mounted on a second spectrometer, and the intensity of the second reflection was determined by the ionization method. This was compared with the intensity of the beam

---

[1] Abstract of a paper presented at the Washington meeting of the Physical Society, April 20–21, 1917.

[2] C. G. Darwin, Phil. Mag., 27, 325 (1913) and 27, 675 (1914).

[3] A. H. Compton, Phys. Rev., 9, 29 (1917).

incident on the second crystal by removing the crystal and swinging around the ionization chamber, so as to receive directly the monochromatic beam.

The quantity E_r was measured by the total deflection of the electrometer when the second crystal was turned with constant angular velocity through an angle $\Delta\theta$ past the angle of maximum reflection. The corresponding value of E_i was the deflection produced by the monochromatic beam when it passed into the ionization chamber for a time equal to that required to move the crystal through the angle $\Delta\theta$. The average value of R obtained in this manner was 0.0050 ± 0.0003 deg.$^{-1}$ in the case of the reflection from a cleavage face of calcite, and 0.023 ± 0.001 deg.$^{-1}$ for a cleavage face of rock salt.

If in equation (1) N is calculated assuming each atom to possess a number of electrons equal to its atomic number, and μ is taken to be the usual absorption coefficient (calcite 23.5: rock salt 18) we obtain R for rock salt = 0.040 deg.$^{-1}$ and for calcite 0.058 deg.$^{-1}$. It will be seen that for rock salt the experimental value of this ratio is about one half the calculated value, and for calcite is less than one tenth as large. The reason for this discrepancy is doubtless due to the fact that at the angle of maximum reflection a selective absorption occurs, as has been predicted by Darwin[1] from theoretical considerations, and has been observed experimentally by W. H. Bragg[2] in the case of diamond. The plausibility of this explanation is increased by the fact that if the reflecting surface of a calcite crystal is roughened by grinding, the reflection coefficient is some three times as great as from a cleavage face. The grinding makes the surface of the crystal imperfect, and thus greatly reduces the selective absorption.[1] Experiments are in progress to make quantitative measurements of the effective absorption coefficient at the angle of maximum reflection.

UNIVERSITY OF MINNESOTA.

THE NATURE OF THE ULTIMATE MAGNETIC PARTICLE

It appears probable from various considerations that when a substance is magnetically saturated, the "molecular magnets" of which it is composed have their axes arranged parallel with the external magnetic field. On this assumption it is possible to investigate the validity of those theories, such as Bohr's which would explain the magnetic properties of an atom as due to electrons revolving about the atomic center in orbits all lying in the same plane.

It has been shown that the relative intensity of the different orders of an X-ray spectrum line depends upon the distance of the electrons from the middle planes of the atomic layers in the diffracting crystal.[1] Imagine X-rays to be reflected from the surface of a ferro-magnetic crystal composed of atoms of the type just described. When the crystal is unmagnetized the different atoms will have their electronic orbits distributed in all possible planes, so that on the average the electrons will be at an appreciable distance from the mid-planes of their atomic layers. If, however, the crystal is magnetically saturated perpendicular to the reflecting face of the crystal, the electronic orbits, being perpendicular to the magnetic axes of their atoms, will all lie parallel to the crystal face. The electrons will therefore now be *in* the mid-planes of the layers of atoms which are effective in producing the reflected beam. It can be shown that such a shift of the elec-

[1] A. H. Compton, *Phys. Rev.*, 9, 29 (1917).

2

trons must produce a very considerable increase in the intensity of the reflected beam of X-rays. On the other hand, if the crystal is magnetized parallel to the reflecting face, the turning of the orbits will carry the electrons farther, on the average, from the middle of their atomic layers, and a decrease in the intensity of reflection should result. Of course if the electrons are arranged isotropically in the atom, or if the atom is not rotated by a magnetic field, which would mean that it is the electron or the positive nucleus that is the ultimate magnetic particle, no such effect should be observed.

We have hunted in vain for such an effect on the intensity of the reflected beam of X-rays when the reflecting crystal is strongly magnetized. In our experiment a "null method" was employed. The ionization due to the beam of X-rays reflected from a crystal of magnetite was balanced against that due to a beam reflected from a crystal of rock-salt, so that a very small change in the relative intensity of either beam could be detected, while variations in the X-ray tube itself had little effect. By means of an electromagnet with a laminated core the magnetite crystal was magnetically saturated, and then demagnetized with an alternating current. The effect of magnetization perpendicular to the plane of the crystal face was investigated for the first four orders. On account of mechanical difficulties the test was made only in the third order when the crystal was magnetized parallel to the reflecting surface. In no case was any change observed in the intensity of the reflected beam when the crystal was magnetized or demagnetized, though the method was sufficiently sensitive to detect a variation in the intensity of less than 1 per cent.

A direct calculation shows that a displacement of the atoms of 0.004 of the distance between the successive atomic layers is sufficient to cause 1 per cent. change in the intensity of the fourth order spectrum. If there is any displacement of the atoms when a crystal is magnetized, it is therefore very small. This confirms the observation of K. T. Compton and E. A. Trousdale[2] that magnetization does not shift the atoms in a crystal sufficiently to change the general form of the space lattice in which they are arranged, and verifies their conclusion that the ultimate magnetic particle is not a group of atoms, such as the chemical molecule, but is the individual atom or something within the atom.

It can be shown further that if all the electrons in an atom are in the same plane, the effect on the intensity of the reflected X-ray beam of turning the atom will be greater than one per cent. unless the effective radius of the atom is less than 10^{-10} cm. Other considerations, however, prove that the radius of the atom must be much greater than this.

There is a relatively small number (26) of electrons in the iron atom, and it appears probable that 8 of these, as valence electrons, are at a considerably greater distance than the others from the center of the atom. It is therefore difficult, though perhaps not impossible, to imagine an arrangement of the electrons so isotropic that a rotation of the atom will not produce an appreciable change in the intensity of the reflected X-ray beam.

The most obvious explanation of our negative result is that it is not the atom which is the elementary magnet, but that it is either the positive nucleus, as suggested by Merritt, or the electron, as suggested by Parson.

If the ultimate magnetic particle is not rotated to any great extent by the magnetic field, no conclusions can be drawn from our experiments. It appears much more probable however, that the molecular magnet is capable

[2] K. T. Compton and E. A. Trousdale, *Phys. Rev.*, 5, 315 (1915).

of being turned through a large angle, and on this basis we may conclude that:

1. The ultimate magnetic particle is either the atom or something within the atom.

2. If the atom is the ultimate magnet, its electrons are not all distributed in the same plane, as assumed by Bohr, but are arranged very nearly isotropically.

3. Our experiments are in accordance with the hypotheses that the atomic nuclei or the electrons themselves are the ultimate magnetic particles.

In a subsequent paper we shall describe our experiment in greater detail, and shall discuss more fully the significance of our negative result.
ARTHUR H. COMPTON,
OSWALD ROGNLEY

UNIVERSITY OF MINNESOTA

PHYSICS.—*The size and shape of the electron.* ARTHUR H. COMPTON, Research Laboratory, Westinghouse Lamp Company. (Communicated by G. K. Burgess.)

The radius of the electron is usual y deduced from the energy of the electron in motion, assuming its magnetic energy to be identical with its kinetic energy. If the electron is a sphere, its radius must be, according to this assumption, about 1×10^{-13} cm. It is thus sufficiently small to act as a point charge of electricity even with the shortest γ-rays.

Calculating on the basis of such an electron, J. J. Thomson[1] has shown that the fraction of the energy of an electromagnetic wave incident upon an electron which is scattered by it is given by the expression

$$\frac{8\pi}{3} \frac{e^4}{m^2 C^4}.$$

This corresponds to a mass absorption coefficient due to a scattering of the primary beam equal to

$$\frac{\sigma}{\rho} = \frac{8\pi}{3} \frac{e^4 N}{m^2 C^4}, \tag{1}$$

where N is the number of electrons which contribute to the scattering in a gram of the absorbing medium, C is the velocity of light, and e and m have their usual significance. As Barkla has pointed out, there may be absorption due to other causes,

[1] THOMSON, J. J. Conduction of Electricity through Gases, 2d ed., p. 321.

such as the production of secondary photoelectrons or beta rays, and for other than waves of short length the rays scattered by the different electrons in an atom are nearly enough in the same phase to produce the phenomenon of "excess scattering," so that the absorption coefficient is in most cases considerably greater than the value given by this expression. If the electron acts as a point charge there is, however, no possible grouping of the electrons which can, according to classical theory, produce a smaller absorption than that calculated according to Thomson's formula.

Barkla and Dunlop[2] have shown that for a considerable range of wave-lengths of X-rays the mass scattering coefficients of the lighter elements are given accurately by equation (1) if the number of electrons in the atom is taken to be approximately half the atomic weight. For elements of high atomic weight the phenomenon of excess scattering occurs, except with the very shortest wave-lengths, and the absorption coefficient due to scattering becomes much greater than this value. For wave-lengths less than 2×10^{-9} cm., however, the absorption coefficient becomes very appreciably less than that theoretically calculated, falling as low as one-fifth as great for the shortest γ-rays. Soddy and Russell[3] and Ishino[4] have shown that for these shortest rays the amount of energy scattered by the different elements is accurately proportional to their atomic numbers, so that all the electrons outside the nucleus are effective in producing absorption. It is therefore impossible to account for this very considerable decrease in the absorption coefficient for very short electromagnetic waves if the electron is considered to be a point charge of electricity.

If, however, the diameter of the electron is comparable in magnitude with the wave-length of the incident wave, the radiation scattered by different parts of the electron will be so different in phase that the energy of the scattered rays will be materially reduced. If, for example, the charge on an electron

[2] Barkla and Dunlop. Phil. Mag., March, 1916.
[3] Soddy and Russell. Phil. Mag. **18**: 620. 1910; **19**: 725. 1910.
[4] Ishino. Phil. Mag. **33**: 129. 1917.

is supposed to be in the form of rigid spherical shell, incapable of rotation, a simple calculation shows that the mass absorption coefficient due to scattering is given by

$$\frac{\sigma}{\rho} = \frac{8\pi}{3} \frac{e^4 N}{m^2 C^4} \sin^4\left(\frac{2\pi a}{\lambda}\right) \bigg/ \left(\frac{2\pi a}{\lambda}\right)^4, \tag{2}$$

where a is the radius of the spherical shell and λ is the wavelength of the incident beam. For long waves this becomes identical with equation (1), but it decreases rapidly as the wavelength approaches the diameter of the electron, as is shown in curve I, figure 1. Such an assumption is therefore able to explain at least qualitatively the decrease in the absorption for electromagnetic waves of very high frequency.

It would appear more reasonable to imagine the spherical shell electron to be subject to rotational as well as translational displacements when traversed by a γ-ray. The scattering due to such an electron is difficult to calculate, but an approximate expression can be obtained if the electron is considered to be perfectly flexible, so that each part of it can be moved independently of the other parts. On this hypothesis it can be shown that the intensity of the beam scattered by an electron at an angle θ with an unpolarized beam of γ-rays is given by the expression

$$I_\theta = I \frac{e^4(1+\cos^2\theta)}{2r^2 m^2 C^4} \left\{ \left(\frac{\lambda}{4\pi a}\right)^2 \sin^2\left(\frac{4\pi a}{\lambda} \sin\frac{\theta}{2}\right) \bigg/ \sin^2\frac{\theta}{2} \right\}. \tag{3}$$

Here I is the intensity of the incident beam, r is the distance at which the intensity of the scattered beam is measured, and the other quantities have the same meaning as before. The mass absorption coefficient due to scattering by such an electron is therefore

$$\frac{\sigma}{\rho} = 2\pi N r^2 \int_0^\pi \frac{I_\theta}{I} \sin\theta \, d\theta. \tag{4}$$

This integral may be evaluated graphically or by expansion into a series. The values of σ/ρ in the case of aluminium, taking the numbers of electrons per atom to be 13, are plotted in curve II, figure 1, for different values of a/λ. The values for a rigid spherical electron which is subject to rotation should lie between curves I and II for the range of a/λ here plotted.

Unfortunately the experimental data are too meager to submit these formulae to accurate quantitative test. There are,

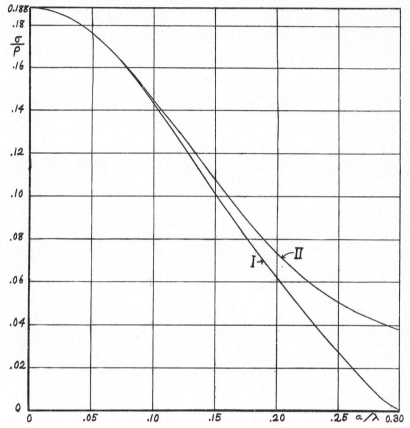

Fig. 1. Mass absorption coefficient for electrons (I) in the form of rigid spherical shells incapable of rotation, and (II) in the form of perfectly flexible spherical shells. The number of electrons per atom is taken as 13.

however, three points on the curve which are established with some accuracy. Barkla[5] has found that for relatively long X-rays the light elements scatter accurately according to equation (1), so that the part of the curves where a/λ is small is verified.

[5] BARKLA and DUNLOP. Phil. Mag.

Hull and Rice[6] have shown that for wave-lengths in the neighborhood of 0.17×10^{-8} cm. the value of σ/ρ for aluminium is about 0.12. From curve I this corresponds to an electronic radius of 2.2×10^{-10} cm., while curve II gives 2.3×10^{-10} cm. Ishino[7] finds that the value of σ/ρ, using the hard γ-rays from radium-C, is about 0.045. Taking the effective wave-length to be[8] 0.093×10^{-8} cm., curve I gives $a = 2.1 \times 10^{-10}$ and curve II gives $a = 2.5 \times 10^{-10}$ cm. Using either formula the agreement between the two values of the radius is within the limits of probable experimental error. The unusually low absorption coefficient for γ-rays can therefore be quantitatively explained on the hypothesis that the electron is a spherical shell of electricity of radius about 2.3×10^{-10} cm.

Another difficulty that is found in J. J. Thomson's simple theory is that it predicts that if a beam of X-rays is passed through a thin plate the intensity of the scattered rays on the two sides of the plate should be the same. It is well known, however, that the scattered radiation on the emergent side of the plate is much more intense than that on the incident side, both in the case of relatively soft X-rays and in the case of hard γ-rays. Barkla and Ayres[9] have shown that for rather hard X-rays and for those substances of low atomic weight whose absorption coefficient can be calculated accurately by equation (1) this prediction of Thomson's theory is also valid. In the case of the heavier atoms and the longer waves, however, the rays scattered at a small angle with the incident beam by the different electrons in the atom are so nearly in the same phase that the intensity is considerably increased, while at large angles the phase difference is much greater, and the intensity is much smaller. This explanation cannot, however, be applied to the excess scattering of γ-rays of short wave-length, since experiment shows[10] that for longer waves the light elements show no

[6] HULL and RICE. Phys. Rev. **8**: 326. 1916.
[7] ISHINO. Phil. Mag. **33**: 129. 1917.
[8] RUTHERFORD and ANDRADE. Phil. Mag. **28**: 263. 1914.
[9] BARKLA and AYERS. Phil. Mag. **21**: 271. 1911.
[10] BARKLA and AYERS. Phil. Mag. **21**: 271. 1911.

excess scattering on the emergent side, indicating that the electrons act independently, while for hard γ-rays the excess scattering is the same as for the heavier elements.

The same difficulty is present if instead of considering the electron as a point charge it is assumed to be a rigid spherical shell incapable of rotation, as this assumption also makes the scattered radiation symmetrical on the incident and the emergent sides. If, however, the electron is a spherical shell of electricity which can be rotated by a passing electromagnetic wave, it is

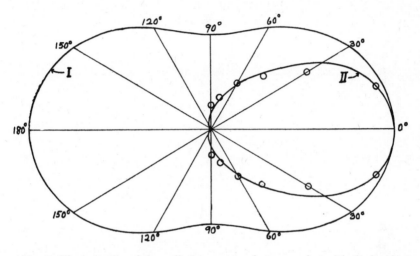

Fig. 2. The intensity of the radiation scattered at an angle θ with the incident radiation, the electron being a perfectly flexible spherical shell. I, radius of electron = 0; II, radius = $3\pi\lambda/4$.

capable of producing excess scattering on the emergent side for short γ-rays in much the same manner as groups of electrons in the atom produce excess scattering in the case of the longer X-rays. For purposes of calculation it is again simpler to consider the nearly equivalent case of the electron which is a flexible spherical shell. The intensity at any angle is then given by equation (3). When $a = 0$, this expression becomes identical with that calculated on Thomson's theory, and the corresponding values are plotted in curve I of figure 2. In curve II, I_θ/I is plotted for different values of θ, using the value $a = 3\pi\lambda/4$.

The circles are experimental values determined by D. C. H. Florance[11] using the γ-rays from radium bromide scattered by a plate of iron. Inasmuch as these rays are heterogeneous, and as the softer rays are scattered relatively more strongly at larger angles, the agreement of the experimental values with curve II is as good as can be expected.

A better quantitative test of this explanation is afforded by Ishino's observation[12] that the radiation scattered on the incident side of a plate struck by hard γ-rays from radium-C is about 15 per cent of that scattered on the emergent side. On the hypothesis of the electron as a flexible sphere this ratio is given by the relation

$$\frac{I_i}{I_e} = \int_{\pi/2}^{\pi} I_\theta \sin\theta\, d\theta \bigg/ \int_0^{\pi/2} I_\theta \sin\theta\, d\vartheta \tag{5}$$

The values of this ratio for different values of a/λ are plotted in figure 3. This curve explains beautifully the observation of Florance that the "incident" scattered rays are softer than the "emergent" and the primary rays, since it shows that the relative amount of the rays scattered backward is much greater for soft, or long wave-length, γ-rays than for hard rays. Rutherford and Andrade[13] have found the hard γ-rays from radium-C to consist of a strong line, $\lambda = 0.099 \times 10^{-8}$, and a weaker line, $\lambda = 0.071 \times 10^{-8}$ cm. Taking into account this selective effect, we may take the effective wave-length to be 0.095×10^{-8} cm. On this basis, and using $a = 2.3 \times 10^{-10}$ as determined above, the calculated value of the ratio of the incident to the emergent scattered radiation is 8 per cent. The agreement is hardly within the probable experimental error, but the calculated value is at least of the proper order of magnitude, which is a strong verification of a flexible or a rotatable electron.

According to electromagnetic theory it is obvious that the mass of an electron cannot be accounted for on the basis of a uniform distribution of electricity over the surface of a sphere

---

[11] FLORANCE. Phil. Mag. **20**: 921. 1910.
[12] ISHINO. Phil. Mag. **33**: 129. 1917.
[13] RUTHERFORD and ANDRADE. Phil. Mag. **28**: 263. 1914.

of the size here assumed. Much the same effect, so far as the scattering of γ-rays is concerned, results from the conception of the electron as a ring of electricity of diameter comparable with the wave-length of the incident beam. It has been shown by

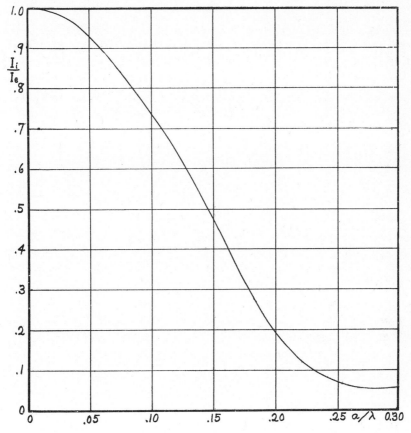

Fig. 3. Ratio of the intensity of the radiation scattered on the incident side to that of the radiation scattered on the emergent side of a plate. The electron is assumed to be a flexible sphere.

Davisson[14] and Webster[15] that this conception is compatible with the electromagnetic theory of the mass of the electron. I have not as yet been able to solve completely the problem of the scattering produced by such ring electrons. Approximate

[14] DAVISSON. Phys. Rev. **9**: 570. 1917.
[15] WEBSTER. Phys. Rev. **9**: 484. 1917.

methods show, however, that if the electron is a rigid ring whose plane is invariable, the scattered energy follows equation (2) rather closely, and is symmetrical on the incident and the emergent sides. If the electron is a flexible ring, or one capable of rotation about any axis, the scattering is more nearly that given by equation (4), but should be somewhat greater for large values of a/λ. The ratio of the incident to the emergent scattered radiation should also be appreciably larger than that given by expression (5). It seems probable, therefore, that the scattering of γ-rays and X-rays may be completely explained on the hypothesis that the electron is a ring of electricity of radius about 2×10^{-10} cm., if the ring is capable of rotation about any axis.

This hypothesis makes it possible to explain also the effect noticed by A. H. Forman[16] that the absorption coefficient of iron for a beam of X-rays is greater when the iron is magnetized parallel with the transmitted beam than when the iron is unmagnetized or magnetized perpendicular to the X-ray beam. Using an effective potential of 27,000 volts the effect was about 0.4 per cent, and with a potential of 81,000 volts it was 0.6 per cent. From X-ray spectra obtained under similar circumstances it can be shown that the effective wave-length used in the two cases was about 1.0×10^{-8} and 0.5×10^{-8} cm. respectively. If the ring electron acts as a tiny magnet, as suggested by Parson,[17] it may be turned by the magnetic field until its plane is perpendicular to the incident beam of X-rays This will make the rays scattered by the different parts of the electron more nearly in the same phase, so that the absorption due to scattered radiation will be increased. Moreover, since the incident rays can get a better hold on the electron in this position, its displacement will be greater than when unorientated, and absorption due to transformation of the X-rays into other types of energy will be greater. For the relatively long waves used by Forman the ratio of the absorption coefficient when

---

[16] FORMAN. Phys. Rev. **7**: 119. 1916.

[17] PARSON, A. L. Smithsonian Misc. Collections, Nov. 1915. Parson estimates his "magneton," or ring electron, to have a radius of 1.5×10^{-9} cm.

magnetized to that when unmagnetized should be approximately

$$k \left(\frac{2\pi a}{\lambda}\right)^2 \Big/ \sin^2\left(\frac{2\pi a}{\lambda}\right)$$

where a is the radius of the ring electron and k is the fraction of the electrons which are oriented by the magnetic field. Using the value $a = 2.3 \times 10^{-10}$ cm., this means that the change in the absorption due to magnetization for $\lambda = 1.0 \times 10^{-8}$ cm. is $0.7\ k$ per cent, and for $\lambda = 0.5 \times 10^{-8}$ cm. is $2.8\ k$ per cent. From the observed values of this difference we find that the fraction of the electrons oriented by the magnetic field is 0.6 and 0.26. The experimental basis of the latter value is much the more certain. Taking the number of electrons in the iron atom to be 26, this means that in order to explain Forman's effect in terms of ring electrons a number $0.26 \times 26 = 7$ of the electrons must be capable of being oriented by the magnetic field. This is what would be expected if it is the 8 valence electrons of iron which are responsible for its ferro-magnetic properties. Our hypothesis of a ring electron of radius 2.3×10^{-10} cm. is therefore capable of explaining satisfactorily Forman's effect.

It should be noted that Forman explains his effect as being due to an orientation of the *molecules* in the iron. The experiments of Rognley and the writer[18] on the effect of magnetizing a crystal on the intensity of the beam of X-rays reflected by it have shown that any orientation of the molecules, if it occurs at all, must be extremely small. It was found further that unless it is very nearly isotropic the atom also is not rotated by magnetization. Thus Forman's explanation of his effect is inadequate. The fact that his experiments can be explained in terms of an orientation of the electrons must be taken as a confirmation of the conclusion arrived at by Rognley and the writer that it is not the atom as a whole, but the electron itself that is the ultimate magnetic particle.

[18] COMPTON and ROGNLEY. Science (N. S.) 46: 415. 1917.

Summary. Ishino's experiments, showing that the scattering of hard γ-rays by different materials is strictly proportional to the number of electrons and is not proportional to the masses, proves that the electrons are responsible for practically all of the scattering, and that for these wave-lengths they act independently of each other. According to the classical electrodynamical theory, this means that if the electrons are sensibly point charges of electricity, the absorption coefficient due to scattering for these rays must be given by equation (1). Since this equation does not hold for these wave-lengths, we cannot consider the electron to be a point charge. In order to account for the small absorption coefficient of γ-rays *the electron must have an effective radius of about* 2.3×10^{-10} *cm.* In order to explain the fact that the emergent scattered radiation is more intense than the incident radiation, it is necessary to assume further that *the different parts of the charge of the electron can possess certain motions independently of each other*. It appears that these phenomena, together with the electromagnetic mass of the electron, can be quantitatively explained on the hypothesis that the electron consists of a ring of electricity subject to rotation about any axis and of radius about 2.3×10^{-10} cm. This hypothesis is confirmed by the fact that it explains satisfactorily Forman's effect of magnetization of iron upon its absorption coefficient, for which there is no other apparent explanation.

THE NON-MOLECULAR STRUCTURE OF SOLIDS.*

BY

ARTHUR H. COMPTON,

Research Laboratory, Westinghouse Lamp Company.

ONE of the most striking things in the structure of crystals as worked out by Prof. W. H. and Mr. W. L. Bragg † is the orderly manner in which the atoms of the different crystals are

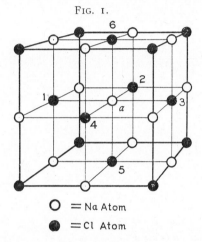

FIG. 1.

O = Na Atom
● = Cl Atom

arranged. The evidence of X-ray analysis points definitely to a structure in the simple crystals which have been examined in which there is no evidence whatever of any grouping of the atoms into molecules. It seems rather that the position of an atom depends equally upon all the neighboring atoms, so that the whole

---

* Communicated by the Author.

Contribution from the Research Laboratory of the Westinghouse Lamp Company, East Pittsburgh, Pa., January 5, 1918.

This paper was sent to three members of the Physics faculty at Princeton University on July 25, 1914. It was read before the Physics Colloquium of Princeton University on October 8 1914. Though not originally prepared for publication, the subject has become one of such general interest that it has been thought worth while to publish the article at this time without alteration.—THE WRITER.

† Proc. Roy. Soc., 1913–1914.

mass is a homogeneous bundle of atoms which have no tendency to form into molecules.

If we take, for example, a crystal of rock-salt, we find the atoms of sodium and chlorine arranged alternately in a simple cubic pattern (Fig. 1). Any one sodium atom a is equally strongly bound to any one of six chlorine atoms, 1, 2, 3, 4, 5, 6, or *vice versa*. It is evident that it is easy for the atoms to pair off into molecules on passing into the liquid state. In doing so, however, there are equal probabilities for an atom to combine with any one of six others, so that we cannot expect the same pair of atoms to form a molecule after the substance melts as were together before they assumed the solid state. Thus no particular

FIG. 2.

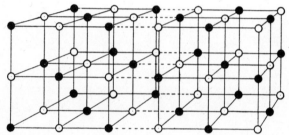

pair of atoms in the crystal can be said to belong to the same molecule.

In order to make more evident the absence of true molecular structure, let us consider a salt crystal to be made up of two cubes (Fig. 2), each cube having an odd number of atoms on its edge. Then the whole crystal has an equal number of sodium and chlorine atoms, but if it be split in the middle, one half will contain an excess of chlorine and the other of sodium atoms. Thus by the simple operation of splitting the crystal along a cleavage plane we must have performed the exceedingly delicate operation of mechanically severing a molecule into two parts—a chemical process—if there is a true molecular structure in the crystal. Exactly similar arguments show the non-molecular structure of the other kinds of crystals which the Braggs have determined.[1]

---

[1] So far as the writer is aware, the fact that the work of the Braggs thus indicates a non-molecular structure in the crystals which they examined was first pointed out by him in *The Scientific American Supplement*, July 4, 1914, page 5.

In these simple crystals, then, if the structure as determined by X-rays is correct, each atom is held in position equally by all the neighboring atoms. It is natural to suppose, therefore, that in substances which are chemically more complicated, as in these simple crystals, each atom is affected by forces equal, on the average on all sides, so that there will be no grouping into molecules. Thus by the definite indication by X-ray analysis of a non-molecular structure for certain simple crystals we are led to the idea that very possibly the structure of all compounds, however complicated, is non-molecular in the solid state.

We may therefore state the thesis that we hope to prove, thus: *Every atom in a solid has its position so determined by the other atoms that it cannot be said definitely to belong or not to belong to the same molecule as some other one atom.* That is, in the solid state the atoms of the substance are so intimately intermingled that particular molecules cannot be definitely defined. Since this proposition is apparently not in accord with the generally accepted molecular theory of solid matter, it is important to see how it is confirmed by a study of the general properties of matter as well as by the molecular theory itself. In this way we hope to form an argument which will show that *non-molecular structure is an essential attribute of all solid matter.*

ARGUMENT FROM THE PROPERTIES OF SOLID MATTER.

From an historical standpoint, the way in which the conception of molecules in the solid state was formed was by a deduction from the kinetic theory of gases. It was easy to explain the change from gas to liquid as due to a shrinking of molecular distances to something comparable with the diameter of the molecules, at which distance the molecules would be held together by intermolecular forces. Likewise the passage from the liquid to the solid state was explained as the settling of the molecules into positions of stable equilibrium. Since the molecules have a definite form in the gaseous state, it was natural to suppose that they would keep this form in the liquid and solid states. We have already seen, however, that there is no trouble with an explanation of the passage from the solid to the liquid state, even with a non-molecular structure in the solid. For example, NaCl, which is known to have a non-molecular structure, actually does melt, so that a molecular structure of the solid is not necessary. The explanation of this we saw was simply that the atoms in the salt crystal paired off when assuming the liquid state. Since, therefore, a non-molecular structure accords as well with the general kinetic theory of gases and liquids as does a molecular solid, there is no *a priori* reason why solids should be molecular in structure.

It will be convenient, however, to speak of the atoms in the solid state as belonging to molecules, meaning either the molecules to which they belonged before assuming the solid state or those to which they will belong if they assume again the liquid or the gaseous state. We shall then employ a sort of *reductio ad absurdum* method, and try to show that these groups of atoms cannot exist in the solid as definitely defined molecules.

In order to prove the thesis as stated above, it will be necessary to show that at least some of the atoms of every molecule are bound as strongly to atoms of other molecules as they are to their own. It is evident that this means a non-molecular structure, since there can be no definable boundary to the molecules in such a case. We shall attempt to prove this point by steps, showing:

1. That each atom in a solid oscillates about a definite position of stable equilibrium.

2. That the forces holding the atoms in their positions of stable equilibrium are of the same nature and are comparable in magnitude with the forces binding together the atoms of the same " molecule "; and

3. That at each of the frequent collisions the forces between the atoms of different " molecules " become so great that an atom has even chances to be torn from the atoms of its own " molecule," so that, on the average, it is bound equally to all the neighboring atoms.

The evidence for the fact that each atom has a definite position in a solid is well known. In the first place it is a generally accepted part of the molecular theory that in the solid state the molecules are in positions of stable equilibrium from which they hardly move. The fact that a true solid will not change its shape under small forces even for great periods of time shows that the molecules are not free to move from place to place. Also, the existence of the latent heat of fusion and of sublimation shows that in the solid state the molecules are in positions of minimum potential energy and hence of stable equilibrium.

But we can show, also, that the molecules in a solid are arranged in definite directions as well as in definite positions, which means, of course, that the atoms are in definite positions. We shall point out below that the structure of all solids is crystalline, so anything which we prove for crystals will hold for all solid matter. Now it is evident that to form any regular crystal structure the molecules must be arranged in a regular order. But it is generally accepted that the crystal form is some function of the chemical constitution. Unless, however, the atoms in the molecule are arranged in a definite order, it is hard to see how the crystal form can be influenced by the chemical composition. As a matter of fact, a definite arrangement of the atoms in crystals is assumed in the explanation of many of the properties of matter. For example, Lord Kelvin, in explaining pyro- and piezo-electricity (Poynting and Thomson, " Electricity and Magnetism "), assumes that the molecules in the crystals which show this effect have their parts arranged in a certain definite manner, as described below (p. 27). Likewise Bridgman, in explaining the phenomenon of polymorphism under high pressures, says (*Phys. Rev.*, iii, p. 199): " It seems natural that the crystalline form assumed under the free action of the orienting forces is one in which the potential energy of the attractive forces is a minimum; that is, that the local centres of attraction within the molecule have approached as close as possible to each other. If the molecules are forced to assume a different arrangement, even one occupying less volume, these centres of attraction must be pulled apart, and work done against the attractive forces while decreasing the volume. That is, the internal energy of the form with the smaller volume will be the greater." Also, it may easily be shown that the explanation of the expansion of certain substances on solidification depends upon the existence of local centres of attraction within the molecules. It is possible to imagine systems in which the arrangement which would bring the local centers of attraction nearest together would not be the arrangement of minimum volume, so that when the molecules would be in stable equilibrium the volume might be a maximum. Indeed, it is difficult to see how there can be any other explanation. These local centres of attraction, however, can be nothing other than atoms or groups of atoms. We shall see below (p. 11) that if the orienting forces are the attractions between the atoms, such a position of the molecules in the solid state is very easily accounted for. It seems evident, therefore, that in order to explain the crystal structure the molecules must be oriented in such a manner as to give to each atom a definite position.

Aside from this evidence that atoms in crystals have their positions defined, we have further evidence that the atoms in any solid oscillate about definite centres of stable equilibrium. According to the theory of the equipartition of energy, each degree of freedom of an atom in the solid state should possess the same average kinetic energy as a degree of freedom of an atom in the gaseous state at the same temperature. But if we give to each atom in the solid state three degrees of freedom, the specific heat of the atom calculated from the rate of increase of the kinetic energy is only half that which corresponds to Dulong and Petit's law for the atomic heat. Since an atom is supposed to have but three degrees of freedom, in order to account for this difference it is necessary to ascribe to each degree of freedom of the atom potential energy as well as kinetic energy, which increases as the temperature, to give the atom the proper amount of specific heat. But the existence of this potential energy means that the atom is vibrating about a position of stable equilibrium. As the kinetic energy of the atom accounted for half the specific heat, the potential energy must account for the other half, and must therefore be equal, on the average, to the kinetic energy. This indicates that the motion of an atom is, in general very nearly harmonic motion about a central position.

In the explanation of heat radiation, also, we have an indication that the atoms oscillate about centres of stable equilibrium. The explanation referred to is the common one that in solids heat waves and long light waves are produced chiefly by the oscillation of the atoms in harmonic motion about positions of stable equilibrium. Thus from two different lines of evidence we come to the conclusion that the atoms in any kind of a solid must oscillate about positions of stable equilibrium.

THE NATURE AND MAGNITUDE OF THE INTERMOLECULAR FORCES.

The next question is, What is the nature of the forces which hold the atoms in position? This will become evident as we discuss the phenomenon of the cohesion of solids. The accepted molecular theory explains cohesion as due to the settling of the molecules into positions of stable equilibrium, where they are held by the intermolecular forces. It is evident, on the kinetic theory, that the collisions of each molecule with the neighboring molecules will be very frequent in the solid state. But we can show that when two molecules are in "contact" the attraction between atoms of the different molecules is of the same nature and of the same order of magnitude as the attraction between the atoms of the same molecule. This is evident from a consideration of the forces which occur in chemical action. For instance, the fact that some compounds are dissociated when they are dissolved in water implies that the atoms in the water molecules exert a greater force on the atoms of the compound than these do on each other. Similarly when two molecules interact directly, as in an organic reaction in which dissociation takes no part, the atoms of one molecule must exert a greater force on certain atoms in the second molecule than do the other atoms of the second molecule. In fact, the universality of reversible reactions shows that in every case there is a delicate balance between the forces binding an atom to its own molecule and the forces tending to make it join the molecule with which it is in contact. There can be little doubt, therefore, that at least while two molecules are in contact the forces between their atoms are not only of the same nature but also of the same order of magnitude as the forces binding together the atoms of a chemical molecule. (It might be thought at first that this theory of attraction between the atoms of different molecules is opposed to the theory of chemical valency which would seem not to allow an atom to attract more other atoms than it has chemical bonds. According to Stark's theory (Campbell, "Modern Electrical Theory," p. 341), which is possibly as good as any explanation yet offered of the mechanism of valency, a chemical combination between two atoms is not a direct attraction of one atom for another, but a simultaneous attraction of both atoms for the same electron, which thus forms a bond between them. This common electron is one which originally belonged to the electropositive atom, but which was more strongly attracted by the electronegative atom, and on being brought nearer to this atom has the effect of giving it a negative charge and of giving the electropositive atom a positive charge. This method of binding may be illustrated by Fig. 3.

(According to Stark, the band spectra of substances represent the vibrations of the electrons concerned in chemical combination, which are, of course, these "valency" electrons. The fact that the spectrum bands widen as the pressure of the gas increases indicates that these electrons are influenced by neighboring atoms as well as by the atoms of their own molecule. This suggests that the valency electrons may not be fixed in position, but that their positions may be altered by the presence of other molecules.

FIG. 3.

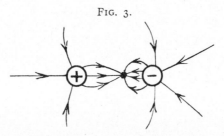

So, if another molecule should come near the first, we might expect them to be held together in some such manner as shown in Fig. 4.

(Since the electronegative atom acts as though negatively charged, and the electropositive atom as though positively charged, on account of the position of the valence electron, it is easy to see how a whole system of atoms may be held together, due to

FIG. 4.   FIG. 5.

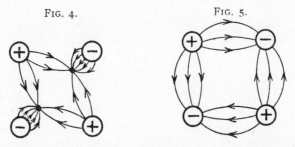

the electrostatic forces. For example, when two molecules composed of oppositely charged atoms come together, a position of stable equilibrium is as illustrated in Fig. 5. If two more come near, a position of stable equilibrium is as shown in Fig. 6, which is the elementary cube of an alkali halide. It seems probable that if the equivalent charge and the diameter of every atom were known it might be possible in this way to calculate the different crystalline forms of every compound.)

Since, therefore, we know that the forces between atoms of different molecules are of the same nature and of the same order of magnitude for at least a considerable portion of the time as those binding a molecule together it is natural to assume that these forces are always similar. Evidence that the forces between the atoms of different molecules are, in general, comparable with the forces between atoms of the same molecule results from a consideration of the atomic heat of substances in the solid state. We found above that in order to explain the specific heat of an atom in the solid state it was necessary to assign to it three degrees of freedom. This is true in whatever way the atom may be combined chemically, so long as the substance obeys Kopp's

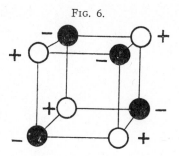

Fig. 6.

extension of the Dulong-Petit law. According to the Staigmüller-Boltzmann theory, however, in the gaseous state an atom has no degrees of freedom along its chemical bonds. In order, therefore, that in the solid state it shall have three degrees of freedom, the attractions of the atoms of the neighboring molecules must loosen the chemical bonds enough to allow other degrees of freedom. But in order to do this these attractions must be at least of the same order of magnitude as the forces binding the atom to its own molecule.

If, as we have assumed above from a consideration of the Dulong-Petit law, an atom in the solid state has degrees of freedom along its chemical bonds, an atom on the surface of a solid ought, according to the law of velocity distribution, to break away occasionally from the solid as in sublimation. A single atom subliming in this manner should carry with it either a positive or a negative change,[2] so that we should expect the vapor from a subliming solid to be partially ionized. The proportion of the ionized atoms would, of course, be relatively small, since it would be much easier for a group of atoms forming a gaseous molecule to leave the solid than for a single atom, which would be held back by an opposite charge in the solid. Moreover, even though an atom in the *interior* of the solid may have three degrees of freedom, it is easily possible that at the *surface* of the solid, since all the attractions are from one side, the atom might be held firmly to the other atoms so as not to be free to escape. If this be true, it should be possible to show that in the form of very thin foil a substance has smaller specific heat than when in lumps. But even in this case it would seem that an atom on the surface of a solid should not be held as firmly along its chemical bond as it would be in the gaseous state. So we should expect a vapor to become ionized at a lower temperature when in contact with its solid than when not in contact. Is it possible that this is the explanation of the production of thermions from hot salts?

There is evidence, also, from an altogether different source that the forces which hold the molecules of a solid in positions of stable equilibrium are the attractions of the neighboring atoms, and are comparable in magnitude with the forces between the atoms of the same molecule. A measure of the forces holding the atoms together in a molecule is the total heat of combination. Also, since a substance melts when its molecules attain kinetic energy enough to tear them away from the neighboring molecules, the melting-point may be taken as a rough measure of the firmness with which the molecules are held together in the solid. We should expect, therefore, if the firmness with which the molecules are held together depends upon the same properties of the atom as the firmness with which the atoms are held together in the molecule, that the melting-point should vary roughly as the heat of combination.

In order to test this prediction fairly it must be remembered that the melting-point of a solid must depend to some extent upon

---

[2] This statement, of course, assumes that the substance is a polar compound, in which case the atoms carry electric charges. If the combination is non-polar, no such evaporation of ions could occur. This is possibly the reason why the emission of thermions of atomic size is confined to the strongly polar compounds, such as the alkali and alkaline earth halides.

the arrangement of the atoms in the molecule as well as upon the magnitude of the attractions between the different atoms. Also, it seems very probable that at the melting-point only a definite proportion of the molecules escape from positions of stable equilibrium, the rest remaining in small groups forming "liquid crystals." (This is the explanation that is given for the expansion of water below four degrees. The high specific heat of liquids in general may be explained as due to the latent heat of fusion of these "liquid crystals" as they melt with rising temperature.) It is probable that the relative number of molecules which have sufficient kinetic energy to escape from the positions of stable equilibrium is different for different substances at their melting-points, which would also indicate that the melting point is only a rough measure of the stability of a molecule in a solid. Moreover, the heat of combination is not an absolute measure of the stability of the molecule, for, in the first place, this varies with the stability of the different atoms in the elemental state; e.g., CO_2 from amorphous carbon has a heat of formation of 97.3×10^3, while from diamond it is only 94.3×10^3. In the second place, the heats of formation of substances differ relatively at their melting-points and at room temperature. In view of these and other disturbing factors, we have no reason to expect more than a rough agreement between the atomic heats of formation and the melting-points of the different substances.

In testing this prediction I have grouped similar compounds together, since one would expect that the atomic arrangement of such compounds in the solid state would be similar (see Table I).

TABLE I.

Showing Relation Between the Melting-points (m.p.) and the Atomic Heats of Formation (A.H.F.) of Various Substances.

(A) ALKALI HALIDES, ETC.

Substance	A.H.F.	m.p.	Substance	m.p.	Substance	m.p.
KCl	53×10^3	770	KBr	750	KI	723
NaCl	49	801	NaBr	765	NaI	695
LiCl	47	600				
HgCl	15.5	500				
AgCl	14.5	460	AgBr	427	AgI	540
HCl	11	−111	HBr (4)	−86	HI (−3)	−51

(B) BICHLORIDES, BROMIDES AND IODIDES.

Substance	A.H.F.	m.p.	Substance	m.p.	Substance	m.p.
(BaCl$_2$		960)	BaBr$_2$	880	BaI$_2$	740
SrCl$_2$	62	825	SrBr$_2$	630		
CaCl$_2$	57	780	CaBr$_2$	760	CaI$_2$	740
MgCl$_2$	51	708				
BeCl$_2$		600	BeBr$_2$	601		
PbCl$_2$	28	447				
HgCl$_2$	18	290	HgBr$_2$	244	HgI$_2$	241
CuCl$_2$	17	498				

(C) TRI-HALIDES.

Substance	A.H.F.	m.p.	Substance	m.p.	Substance	m.p.
BiCl$_3$	23	227	BiBr$_3$	215		
SbCl$_3$	23	73	SbBr$_3$	93	SbI$_3$	167
AsCl$_3$		−18				
PCl$_3$		−112				
FeCl$_3$	24	301				
AlCl$_3$	24+10^3	301	AlBr$_3$	93	AlI$_3$	185
GaCl$_3$		73				
TlCl$_3$		25				

(D) ALKALI, ETC., SULPHATES, CARBONATES AND NITRATES.

Substance	A.H.F.	m.p.	Substance	A.H.F.	m.p.	Substance	A.H.F.	m.p.
K$_2$SO$_4$	49	1070	K$_2$CO$_3$	47	880	KNO$_3$	24	345
Na$_2$SO$_4$	47	884	Na$_2$CO$_3$	45	849	NaNO$_3$	22	313
Li$_2$SO$_4$	48	853	Li$_2$CO$_3$		710	LiNO$_3$	22	258
Tl$_2$SO$_4$	32	632						
H$_2$SO$_4$	27	10.4				HNO$_3$	8	−47
Ag$_2$SO$_4$	24	676				AgNO$_3$	6	218

(E) ALKALI-EARTH, ETC., SULPHATES, CARBONATES AND NITRATES.

Substance	A.H.F.	m.p.	Substance	A.H.F.	m.p.	Substance	A.H.F.	m.p.
BaSO$_4$		infus.	BaCO$_3$		795	Ba(NO)$_2$		575
CrSO$_4$		dec. w.ht.	SrCO$_3$		dec. 1160	Sr(NO$_3$)$_2$		645
CaSO$_4$	53	v. high	CaCO$_3$	54	dec. r. ht.	Ca(NO$_3$)$_2$	27	561
PbSO$_4$	38	937				Cu(NO$_3$)$_2$	9	114.5

(F) OXIDES.

Substance	A.H.F.	m.p.	Substance	A.H.F.	m.p.	Substance	A.H.F.	m.p.
MgO	71.5 ×10^3	over 2000	Cu$_2$O	14	rd. ht.	SO$_3$	26	14.8
SiO$_2$	60	1600	NO$_2$	−.6	−10	H$_2$O	23	0
P$_2$O$_5$	53	subl. r. ht.	N$_2$O	−6.3	−102	CO	14.5	−207
Al$_2$O$_3$	30	wht. ht.	CaO	70	infus.	Bi$_2$O	4	84
PbO	25	rd. ht.	B$_2$O$_3$	55	577	Cl$_2$O	−6	−19
SO$_2$	23	−76	CO$_2$	32	−65	NO	−11	−167

	(G) OXIDES AND SULPHIDES.					
Substance	m.p.	Substance	m.p.	Substance		m.p.
Sb_2O_3	rd. ht.	H_2O	(23) 0	Bi_2S_3		decomp.
Sb_2O_5	300	ZnO	V. high	CS_2	(−6)	−110
Bi_2O_3	860	Sb_2S_3	fusible	H_2S	(1)	−86
CO_2 (32)	−65	Sb_2S_5	fusible	ZnS		1050

The heat of formation used is that of a gramme-molecule, divided by the number of atoms, thus giving what may be called the "atomic heat of formation." It will readily be seen, by a glance at the tables, that in general the higher the heat of formation the higher the melting-point of a substance. There are a few exceptions to the rule, but the discrepancies are small, and there can be no doubt that the rule holds in general. For instance, if we take the alkali chlorides, the decrease in melting-point is regular, except for a slight variation in sodium, and so, also, as we pass from the chlorides to the bromides and iodides; similarly with the sulphates, carbonates, and nitrates. Among the oxides there are a great number of exceptions, but the predominance of high melting-points at the head of the list and of low melting-points at the foot shows that the general rule holds. Moreover, when the sulphides and the oxides are compared, the former invariably have the lower melting-point, as they should. In order to show that these uniform results are not confined to elements of the same periodic group, Table II and the graph with it have been made, showing all the compounds in the order of their

TABLE II.

The Melting-points of Different Substances in the Order of Their Atomic Heats of Formation.

—Substance	A.H.F.	Authority	m.p.	—Substance	A.H.F.	Authority	m.p.
			C°				C°
ThO_2	110 ×10³	3	infus	B_2O_3	55	1	577
UO_2	85	5	2176	TiO_2	54		1500
Al_2O_3	79	1	2040	$CaCO_3$	54	2	1289
MgO	72	1	>2000	$CaSO_4$	53	2	>1000
CaO	70	1, 2	1995	P_2O_5	53		*850
$BaCl_2$	66	2	960	KCl	53	1, 2	770
SrO	65	1, 2	3000	$MgCl_2$	51	2	708
V_2O_5	63	5	658	K_2SO_4	49	1, 2	1070
$SrCl_2$	62	1, 2	825	NaCl	49	1, 2	801
SiO_2	60	1	1600	$LiSO_4$	48	2	853
ZrO_2	59	7	2500	WO_3	48	7	*800
$CaCl_2$	57	2	780	Na_2SO_4	47	1, 2	884

*=approximate value.

TABLE II. (*Continued*)

Substance	A.H.F.	Authority	m.p. C°	Substance	A.H.F	Authority	m.p. C°
LiCl	47	2	600	CuO	19	2	1148
K_2CO_3	47		880	$HgCl_2$	18		290
MnO	45		*1500	$CuCl_2$	17		498
Na_2CO_3	45	1, 2	849	HgCl	15		500
MoO_3	45	5	759	CCl_4	15	1	-24
As_2O_5	42	2	960	Cu_2O	14		*800
Fe_3O_4	40	4	1538	Tl_2O	14		*700
Fe_2O_3	39	18	1458	AgCl	14	1, 2	460
$PbSO_4$	38	2	937	PbI_2	14	6	373
$CdSO_4$	37	2	>1000	CO	14		-207
LiOH	37		*850	C_2H_5COOH	11	2	-22
KOH	35	1, 2	360	HCl	11	2	-111
NaOH	34	1, 2	318	$Cu(NO_3)_2$	9		114
FeO	32	18	1419	AgI	8	13	500
K_2O	32		*800	HNO_3	8	1	-47
Na_2O	32		*800	CH_3OH	8	2	-95
Tl_2SO_4	32	2	632	Trifluorocresol	7	17	-5
$ZnCl_2$	32	1, 2	375	C_2H_5OH	7×10^3	2	-112
CO_2	32	1, 2	-65	CH_3CHO	7	2	-120
TeO_2	29	5	*600	$AgNO_3$	6	1, 2	218
Co_3O_4	28	5	750	Nitrofluorophenol	5	17	74
$PbCl_2$	28	2	447	$CHCl_3$	5	2	-70
$Ca(NO_3)_2$	27	1	561	$(C_2H_5)_2O$	5	2	-117
$SnCl_2$	27	2	249	Fluoronitracetanilide	4	17	96.5
H_2SO_4	27	2	10				
SO_3	26		15	Bi_2O_3	4		84
$SnCl_4$	26		-33	Parafluorphenol	4	17	28
PbO	25	2	*800	Orthofluorphenol	4	17	16
Ag_2SO_4	24		676	Metafluorphenol	4	17	14
TlCl	24	2	407	HBr	4	2	-86
KNO_3	24×10^3	1, 2	345	CH_4	4	1, 2	-184
$FeCl_3$	24	2	01	Metafluoracetanilide	3	17	85
$BiCl_3$	23	2	227				
$SbCl_3$	23	2	73	Difluorethylnitramine	3	17	45
SO_2	23		-76				
$NaNO_3$	22	1, 2	313	Nitrofluorphenetol	3	17	14
$LiNO_3$	22	2	258	Metatrinitrotoluol	3	17	-2.4
$CdCl_2$	21	2	568	Parafluorphenetol	3	17	-8.5
Tl_2O	21	2	300	C_2H_6	3	1, 2	-171
SeO_2	21	5	>260	C_3H_8	3	2	-195
$AlCl_3$	20		190	NH_3	3		-75
NOCl	-2	9	-60	Parafluornitrobenzol	2	17	27
HI	-3	1, 2	-51	Ethylenitramine	2	17	6
H_2Se	-5		-64	SiH_4	2		<0
N_2O_3	-5	1	-106	Fluornitraniline	1	17	96
Cl_2O	-6		-19	Dinitrofluorbenzol	1	17	26
CS_2	-6	1	-110	Metafluornitrobenzol	1	17	4
N_2O	-6		-112				
AsH_3	-9		-113				
HCN	-10		-14	H_2S	1	2	-86
NO	-11	2	-167	Metacresol	$+.3$	17	4
C_2H_2	-12×10^3	2	-82	C_7H_8	$-.1$	2	-97
C_2N_2	-19	1	-35	C_2H_4	$-.3$	2	-169
SbH_3	-22	1	-91	NO_2	$-.6$	1	-10
H_2O	19	1	0	C_6H_6	-1	2	5
$(NH_4)_2SO_4$	19	1, 2	140	Aniline	-1	2	-8

*= approximate value.

AUTHORITIES.

1. Berthelot, *Ann. d. Chim. and d. Phys.*, 1878.
2. Thomsen, "Thermochemistry."
3. E. Chauvenet, *Ann. Chim. Phys.*, **8**, 23, 425.
4. J. Ermiloff, *J. Soc. Metal. Rus.*, **4**, 1.
5. W. Mixter, *Am. J. Sc.*, **4**, 32, 202.
6. Koref, *Z. anorg. Chem.*, **66**, 88.
7. Weissl Kaiser, Nurmann, *Z. anorg. Chem.*, **65**, 345.
8. De Forcrand, *C. R.*, 150, 1399.
9. E. Briner and E. Pylkoff, *J. Chim. Phys.*, x, 671.
10. L. Rolla, *Ren d. Accad. Linc.*, v, 21, 281.
11. H. von Wartenberg, *Z. anorg. Chem.*, **79**, 71-87.
12. C. Matignon, C. R. 154, 1353.
13. Ulrich Fisher, *Z. anorg. Chem.*, **78**, 41-67.
14. F. Meyer, Thése, Paris, 1912, pp. 30, 31, 54.
15. O. Ruff and Gersten, *Ber. d. Chem. Ges.*, **45**, 63.
16. G. Darzens, C. R. 154, 1233.
17. Frederic Swarts. *Bl, Acad. Belg.*, 1912, 481-523.
18. Le Chatelier.

"atomic heats of formation." Here there is no chance for any other factor, such as atomic weight or place in the periodic table, to affect the melting-points consistently. I have not made any selection, but have used all the compounds for which the required data are available. In spite of occasional rather large discrepancies in this table, there can be no doubt that in general the melting-point rises with an increase in the atomic heat of formation.[3]

I have given so much space to this demonstration of the connection between the heat of combination and the melting-point because this seems to be a definite proof that the forces which bind the atoms of different molecules together in the solid state, measured by the melting-point, are similar to those which bind a chemical molecule together, measured by the atomic heat of combination. The similarity must be not only a likeness in nature, but also comparability in magnitude, since if the magnitude of the forces were not comparable there would be no reason to expect the relative melting-points to be comparable with the relative heats of formation.[4] We may consider it established,

---

[3] Table II and the graph representing it (Fig. 7) are more complete than those given in the original paper. They were shown in their present form before the American Physical Society at its meeting October 30, 1915.

[4] A calculation of the energy required to raise the temperature of a substance from absolute zero to its melting-point and melt it shows that, on the average, this energy is about one-fourth of its heat of formation. Thus the energy required to remove the atoms of a "molecule" in the solid form from the atoms of the neighboring "molecule" is of the same order of magnitude as that required to separate the atoms of the same molecule from each other.

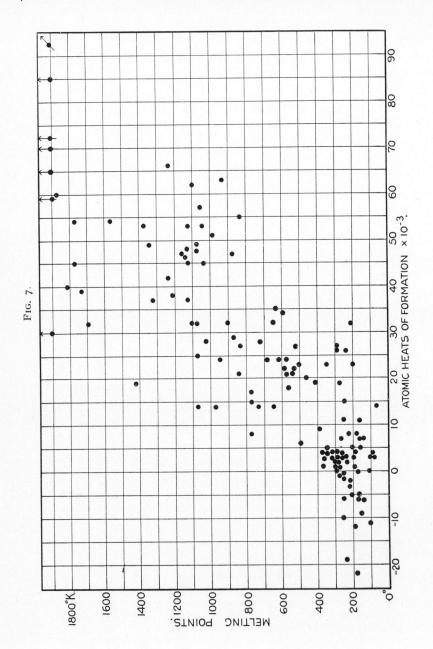

Fig. 7.

therefore, that in a solid the forces affecting an atom due to neighboring atoms of other molecules are both of the same nature as and comparable in magnitude with the forces due to the other atoms of its own molecule.

It thus appears that the forces of cohesion which hold the molecules in positions of stable equilibrium must be the mutual attractions of the atoms of neighboring molecules. In order, however, that each molecule should be in the most stable position with respect to the other molecules, it is evident that it must orient itself in such a manner that the atoms of different molecules which attract each other most strongly shall be next to each other. One result of this is that the molecules must evidently all be oriented and arranged in a definite, repeating order, which means a crystalline structure. Thus we see that the structure of all true solids must be crystalline. If this is true, the molecules in the so-called "amorphous solids" must not be in positions of most stable equilibrium under the atomic forces, and these substances should therefore have neither a definite melting-point nor perfect elasticity even for small forces. It seems probable that the "near solid state" in which these substances exist is caused by the agglomeration of large numbers of molecules into minute crystalline groups, similar to liquid crystals, which are so large compared with their distances apart that the viscosity is very great.

A second result of this arrangement of the molecules is that since an atom is nearest the other atoms of its own molecule which attract it most strongly, and since it comes next to the atoms of the neighboring molecules which attract it most strongly, it seems probable that, wherever possible, the arrangement of the atoms on all sides of it shall be similar. So if an atom should escape from the other atoms of its own molecule it would be held just as strongly by the atoms of the other neighboring molecules.

The forces on an atom due to the atoms of its own molecule and to those of other molecules, on the average, equal.

The next step in the argument to prove that there is no true molecular structure in the solid state will be to show that the forces between the atoms of the different supposed molecules are as great as those between the atoms of the same molecule. The first part of this proof is to show that the distance between the atoms in a solid is small compared with the "diameters" of the atoms themselves. That is, if the atoms are considered as spheres, the distance between the surface of two neighboring spheres is small compared to their diameters.

Possibly the most direct way to show this point is to compare the size of the molecules as deduced (1) from the viscosity of a gas, (2) from the thermal conductivity, and (3) from Van der Waals's equation, with the size of the molecules as deduced from the density of the densest known form of the substance, assuming the molecules to be spheres arranged in the order of closest packing. As we saw above, according to the Staigmüller-Boltzmann theory, the atoms of a molecule in the gaseous state are in such close contact that there is no freedom for motion along a chemical bond. So if it can be shown that the molecules are very nearly in contact, the atoms of different molecules must also be very nearly in contact. The following figures are taken from Kaye and Laby's tables, page 33:

Molecular diameter deduced from

	Viscosity	Thermal Conductivity	Van der Waal's Equation	Limiting Density
H_2	2.47 times 10^{-8}	2.40 times 10^{-8}	2.32 times 10^{-8}	2.92 10^{-8}
He	2.18		2.30	4.34
N_2	3.50	3.31	3.53	2.97
O_2	3.39	3.31		2.79
A	3.36		2.86	4.43
Kr			3.14	4.93
Xe			3.42	4.88
C_2H_4	4.55	4.68		5.26
CO_2	4.18	4.32	3.40	4.42
N_2O	4.27	4.20		4.58
H_2O	4.09			3.49

These data assume that the molecules are spherical, which undoubtedly introduces some error, but the uniformly close agreement of the values in the fourth column with those in the other three can leave but little doubt that the molecules in the solid state, and hence the atoms also, must be very nearly in "contact."

In support of this evidence for the proximity of the atoms in a solid we may mention the evidence which comes from the X-ray spectra from different crystals that the distance of the thermal motion is small compared with the diameter of the atoms themselves. This evidence comes from a consideration of the relative intensity of the different orders of spectra. At first the rapid falling off in intensity of the higher orders for such crystals as rock salt when compared with crystals like potassium ferrocyanide was explained as due to the greater heat motion of the lighter atoms in the former case. But the Laue photographs which de Broglie obtained at great extremes of temperature—from liquid air to red heat—showed a surprisingly small difference in the intensity of the spots, showing that the heat motion has very little effect. Apparently the only remaining explanation is that the X-rays are not reflected from single points in the atoms, but by electrons distributed throughout the atoms. Thus it is the size of the atom compared to the distance apart of the reflecting atoms which accounts for the decrease in intensity of the higher orders of spectra. (If a large proportion of the electrons are in the central ring about the nucleus, the intensity should not diminish as rapidly as if they are uniformly distributed throughout the sphere. It should be possible to test this point.[5]) But since the heat motion of the atom does not materially affect the intensity of the higher orders of spectra, while the diameter of the atoms does so affect it, the distance of the heat motion, and hence the distance between the atoms, must be small compared with the diameter of the atoms.

A more definite confirmation comes from the arrangement of the atoms in those simple crystals whose structure has been worked out. If we assume here the existence of molecules, the distance between the atoms within a molecule is the same as that between neighboring atoms of different molecules. The atoms of the same molecule are certainly very near together, however, so for these simple compounds, at least, any two neighboring atoms are very close together.

One further line of evidence toward this same point may be valuable. The expansion of solids on heating is due, of course, to the lengthening of the paths of the molecules as the temperature

---

[5] It was this fact that suggested the writer's study of the distribution of the electrons in atoms as indicated by the intensity of X-ray reflection (cf. *Nature*, May 27, 1915, and *Phys. Rev.*, **9**, 29, 1917). In the latter paper it was shown that the distance of the outer electrons from the atomic centres was comparable with the distance between the atoms. According to Debye's theory of the effect of heating on the intensity of X-ray spectra (*Ann. d. Phys.*, **43**, 49, 1914), which has been verified experimentally by Professor Bragg (*Phil. Mag.*, **27**, 881, 1914), the thermal motion is relatively small, and reduces to practically zero at very low temperatures, as is here suggested.

rises. It seems natural to suppose that at absolute zero the length of these paths would decrease until the molecules come in contact. Even if the coefficient of expansion were constant the volume of a solid would not decrease enough to indicate at ordinary temperatures a free path comparable with the diameter of the atoms. As a matter of fact, at low temperatures the coefficient of expansion approaches zero, which indicates an "agglomeration" into groups in which the atoms are all in contact. This is the same theory of agglomeration as the one used to explain the decrease of specific heat at low temperatures. If this theory is correct, at low temperatures all the atoms, whether of the same or of neighboring molecules, are held in "contact" with each other.[6] We may conclude, then, with considerable certainty that the atoms of neighboring molecules are very nearly in contact with each other, and that they are probably as near together as the atoms of the same molecules.

We showed above that the forces between atoms are of the same nature, whether due to atoms of the same molecule or to those of different molecules. It seems evident, therefore, that the attraction between atoms should be some function of the distance such that if the distances are equal the forces should also be equal. This is confirmed by the evidence of chemical action which we considered above. But, since we know that the distances between atoms of the same molecule and atoms of neighboring molecules are approximately equal, the forces should likewise be very nearly equal. Thus we see that the forces on all sides of an atom must, on the average, be at least very nearly the same.[8]

We can go still farther and show it to be probable that the forces on an atom due to the atoms of neighboring molecules are, on the average, exactly equal to the forces from the other atoms of its own molecule. From the discussion in the last paragraph we see that an atom is as strongly attracted to another atom with which it is in contact, though of a different molecule, as it is to a similar atom in its own molecule. When, therefore, due to the heat motion of a molecule in the solid, one of its atoms comes in contact with an unlike atom of a different molecule, there are even chances that it will remain fast to the other molecule. So

---

[6] The theory referred to is that suggested by C. Benedicks (*Ann. d. Phys.*, **42**, 133, 1913).

in a very large number of collisions it will belong just as much of the time to one molecule as it will to the other, and so cannot be said to be a part of either. As a matter of fact, since we know, from a study of the Dulong-Petit law, that an atom has three degrees of freedom in a solid, and since an atom cannot remain joined to another without losing a degree of freedom (Staigmüller-Boltzmann Theorie), it is evident that an atom cannot remain in contact for more than an infinitesimal interval with another atom, but must continually oscillate back and forth, exerting equal attractions on all the neighboring atoms.

The fact that Dulong and Petit's law is not accurately followed—i,e., that the atoms of solids do not always have three degrees of freedom, is a support rather than an objection to our hypothesis; for as the temperature decreases, the time interval during which an atom remains in contact with a neighboring atom will, on the average, increase, thereby decreasing the total number of degrees of freedom at any one instant. In the gaseous state, as we have just seen, each atom has three degrees of freedom minus the number of its chemical bonds. But if the gas is heated to a sufficiently high temperature it becomes ionized; that is, the chemical bonds become loosened so that parts of the molecule occasionally separate (Meyer, " Kinetic Theory of Gases," p. 132). According to our hypothesis, when a molecule assumes the solid state, the attractions of the neighboring atoms loosen the chemical bonds sufficiently to give each atom three degrees of freedom. Even in the solid state, however, each atom is strongly bound to the neighboring atoms, so that if the kinetic energy of the atom should become sufficiently low, one would expect it to be held in contact with some of the other atoms thus taking away its degrees of freedom, until it should be moved from its place by a sufficiently violent collision. This would be exactly analogous to the binding of the atoms together in a gas molecule until there is a sufficiently violent collision to ionize the gas. The only differences are that, since the atom in the solid state is not held so strongly by other atoms, the collisions would not have to be so violent, and that, since the atom in the solid state is surrounded by other atoms, it might have either zero, one or two degrees of freedom when held in contact with other atoms. It is easy to see that on this hypothesis, when the kinetic energy becomes sufficiently low, large numbers of the atoms will be held in contact with each other, thus losing their degrees of freedom, so that as absolute zero is approached the specific heat should also approach zero. This is, I believe, about the same as the "Agglomeration Theorie" on which Dean Magie reported to Physics Colloquium last winter.[7]

CONCLUSION.

THE STRUCTURE OF SOLIDS IS NOT MOLECULAR.

Since, therefore, each atom of a substance in the solid state exerts equal forces on all the neighboring atoms, it cannot be said definitely to belong or not to belong to the same molecule as some other one atom.[8] In other words, the atoms of a solid are so intimately intermingled that particular molecules cannot be definitely defined. Thus we can no longer speak of solid

---

[7] These were the considerations which led to the development of the theory of specific heat presented by the writer in the *Physical Review*, **6**, 377, 1915, where it was found that this agglomeration hypothesis is capable of explaining in a fairly satisfactory manner the variation of the specific heat of solids with temperature. It may be pointed out that the theories of the temperature variation of specific heat based on the quantum hypothesis (Einstein, Nernst-Lindemann, Debye) lead to the same conclusion as that here arrived at concerning the rigidity with which atoms are held in a solid. According to these theories, the energy corresponding to a degree of freedom depends upon the corresponding frequency of vibration. In the case of gases this frequency is very high, while in the solid state the frequency of the natural vibrations of the atom is relatively low. This means that the rigidity with which the atom is held in position is much less in the solid state than in the gaseous state, due doubtless to the fact that the forces due to the surrounding atoms are nearly equal on all sides.

[8] On account of the selective nature of the forces between atoms, we can state these forces as a function of the distance only when we consider but a single arrangement of the atoms. This argument therefore proves that the "bonds" on the different sides of an atom are of the same strength only for those atoms which have similar atomic groups arranged symmetrically about them. For this reason it is probable that there are some complicated compounds in which certain groups of atoms may retain their identity in the solid state. The writer has been unable to think of any substance, however, which could crystallize in such a manner that some one atom would not be equally strongly bound to two or more similar atomic groups. The molecule is therefore indeterminate, inasmuch as this particular atom cannot be said to belong to one atomic group rather than to another.

matter as having a molecular structure, but must consider it to be purely atomic, which is the thesis we started to prove.[9]

A CONSIDERATION OF OBJECTIONS.

As an immediate result of our hypothesis it must follow that matter in the solid state can have no properties which depend upon a molecular structure as distinguished from an atomic structure. It is important, therefore, to consider those explanations of properties of solid matter which involve a molecular structure. All

FIG. 8.

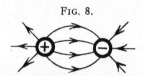

such explanations which I have been able to find are those which consider solid matter to be made up of electric or magnetic doublets. For instance, there is the common explanation of electric strain as due to molecules in the dielectric which are made up

FIG. 9.

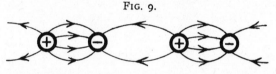

each of two portions, one positively, the other negatively electrified, and that the two portions are shifted by the forces accompanying the strain. Of course, in making this assumption chemical and electric attraction are identified, as we have assumed

---

[9] Langmuir's conclusion from similar considerations is rather that the crystal is a single molecule. The difference depends chiefly upon whether the molecule is defined as the smallest portion of a substance which has the characteristics of the substance or as a group of atoms held together by chemical forces. The above argument is based on the former definition, contending that these smallest portions cannot be definitely defined. Langmuir concludes, on the basis of the latter definition, that, since all the atoms in a crystal are similarly bound together, the whole crystal is a single molecule. Since, however, the forces of cohesion and adhesion are essentially identical with chemical forces, a similar argument would show that the whole earth, and perhaps even the whole universe, is a single molecule. The writer therefore prefers the former conception of the molecule as having a more definite meaning.

above. Following Poynting and Thomson ("Electricity and Magnetism," p. 59): "We may, for our present purpose, . . . conventionally represent a molecule as in the figure (Fig. 8) . . . If the dielectric is in a neutral condition, we must suppose that the molecular axes are distributed equally in all directions, . . . and

FIG. 10.

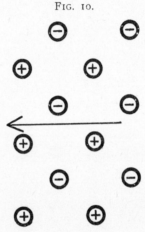

there is no resultant electric action outside." When an electric field is imposed, these molecules arrange themselves as in Fig. 9, which has the same effect as placing a charge on the surfaces of the dielectric.

FIG. 11.

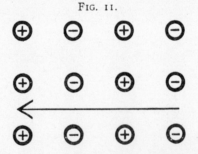

If this means that the molecules are capable of rotation through any angle, then in a crystal such as a KCl crystal, if the electric field were perpendicular to the (111) planes, as in Fig. 10, the supposed molecules would arrange themselves as in Fig. 11. Thus the (111) planes would become cleavage planes, which is evidently absurd, since there is no apparent change in the crystal structure. But if the theory means only that a molecule can be turned slightly from its stable position, there is nothing which requires a true molecular structure, since a slight shift in the position of the atoms will produce the same result. (If this shift in position is great enough, it should be possible to detect it in crystal dielectrics by obtaining X-ray spectra from them. For example, if the electric field is perpendicular to a cleavage plane of a KCl crystal, we might expect the atoms to be displaced as in Fig. 12. It may be shown that such a displacement would reduce the comparative intensity of the higher orders of

Fig. 12.

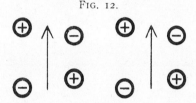

spectra, and this, by careful measurement, might be noticeable.)

Another explanation of electric phenomena which apparently depends upon a true molecular structure in a crystal is Lord Kelvin's explanation of pyroelectricity, which is also extended to piezoelectricity. As explained in Poynting and Thomson ("Electricity and Magnetism," p. 150), this explanation is: "The molecules are set in the crystal with their electric axis

Fig. 13.

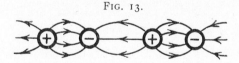

parallel to a certain direction, and there will therefore be a definite amount of strain passing from each positive element of a doublet to the negative element of the doublet next to it. The crystal, as a whole, will be the seat of an electric strain along the direction of the axis. The constitution corresponds to that of a saturated permanent magnet on the molecular theory of magnetism. The electric strain will manifest itself almost entirely at the ends of the crystal if the doublets are near enough together, and the crystal will produce a field outside equal to that which would be produced by charges respectively positive and negative at the two ends. . . . But . . . charges will in time gather on the surfaces at the ends of the axis, . . . entirely masking the existence of the strain, so that there is no external field. This is the ordinary condition of a neutral crystal.

"We have now to suppose that the electric strain passing forward from molecule to molecule depends on the temperature, and that it alters nearly in proportion to the change in tempera-

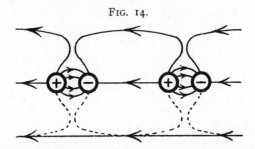

FIG. 14.

ture. Then when the temperature changes there will be an unbalanced strain proportional to the change, producing the same effect as a positive charge at one end and a negative charge at the other until conduction again does its work in bringing up masking charges."

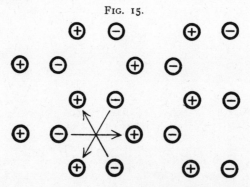

FIG. 15.

The mechanical model which Poynting and Thomson suggest as an illustration is a series of molecular doublets, arranged as illustrated in Fig. 13, which, when heated, assume the arrangement shown in Fig. 14 as the molecules become separated. Here certain of the lines of force combine, giving the effect of a free charge at either end. This model would explain pyroelectricity as due merely to molecular separations, and we ought to get a similar effect if the molecular separation is produced by tension, which is the piezoelectricity found by Curie.

An objection to this model is that it assumes that as the temperature rises the parts of the molecule come relatively closer together, while experiments in specific heat seem to indicate that with rising temperature the parts of a molecule become farther separated. A model to which this objection does not apply, which does not assume any molecular structure, and which accords fully with Kelvin's explanation, may easily be imagined. Suppose the atoms in a crystal to be arranged as in Fig. 15. It will be seen at once that such an arrangement will correspond to the " saturated permanent magnet on the molecular theory of magnetism." The left end will have the equivalent of a positive charge and the right end that of a negative charge. There will evidently be in this model three different axes along which there will be such an effect, which might correspond to the three electric axes of a quartz crystal. As the atoms move farther apart with a rise in temperature the strength of the doublets is increased, and, if the ends were neutralized before, a new charge will now appear.

Of course, the same result would occur here also if the atoms were separated by tension, which would explain piezoelectricity. If the structure of a quartz crystal can be determined as the Braggs are now trying to do, it will be an easy matter to see whether a true molecular structure is required for the explanation of pyro- and piezo-electricity.

There are, however, some difficulties with a complete explanation of the magnetic properties of solids, without assuming a molecular structure, that are not so easily met. One of the arguments for a molecular structure from magnetic properties comes from a consideration of the Heusler alloys. In these alloys no element by itself is very strongly dia- or para-magnetic, but when the different elements are alloyed the mixture is almost as paramagnetic as iron. From this it might be concluded that atoms from two different slightly magnetic substances can combine to form a molecule that is ferromagnetic.

We have, however, strong indications that magnetism is a purely atomic property. A good experimental indication of this comes from depositing iron electrolytically in a magnetic field. For instance, if a varnished silver wire with a thin scratch through the varnish is placed in a solution of ferric chloride, when the iron is deposited electrolytically an extremely small magnetic field will make of the deposited iron a saturated magnet. This indicates that the elementary magnetic doublets are easily turned around while the iron is being deposited. The iron is deposited, however, atom by atom, and, since the position taken by each atom is determined by the crystal form and not by the magnetic field, it must be the atoms and not the molecules which are oriented by the magnetic field.

Perhaps the strongest evidence that the magnetic effects are not molecular but atomic phenomena is the fact that the structure of the atom which the most generally accepted theories suggest offers an excellent explanation of magnetism as due to atomic properties. I refer to the theory that the atom consists of electrons revolving in orbits, held together by the appropriate positive charge. This, of course, makes each atom an elementary magnetic doublet. Langevin and Weiss have developed a complete theory on this assumption (Professor Richardson's lectures on "Electron Theory"), explaining both dia- and para-magnetism, which seems to be as satisfactory as any yet proposed. The fact that a theory based on this assumption explains magnetic phenomena so well without any reference to a molecular structure would indicate that these effects have no dependence on molecular structure.

However, we have good reason to believe that the environment of an atom greatly influences its magnetic properties. For instance, different samples of iron vary greatly in their magnetic properties, according to their temper, per cent. of carbon, amount of alloy, temperature, etc. The explanation of Heusler alloys would therefore seem to be that the environment of the atoms of the different elements is changed in such a manner by being alloyed that they are free to turn under a magnetic field.

There is another difficulty, however, which results from a purely atomic explanation of magnetism. This is an explanation of the heating effect when a piece of iron is placed in an alternating magnetic field. This alternating magnetic field, if intense enough, is supposed to turn the elementary doublets completely around. If these doublets are molecular, it is easy to see how their turning can impart kinetic energy to the system and so raise the temperature; but if they are atomic, since only the translational degrees of freedom of an atom are supposed to possess heat motion, any amount of rotation of the atom would not affect the temperature of the substance. Two possible explanations may be offered. In the first place, it is possible that, since an atom can no longer be considered to be a sphere, its rotary degrees of freedom actually do enter into the thermal motions. Or, in the second place, though the rotary degrees of freedom may not themselves enter into the heat motion, it is possible, since the atoms are not spheres, that their rotation may displace other atoms translationally, which would account for the heating effect. It must be confessed, however, that neither of these explanations appears wholly satisfactory, and that this heating effect presents a real difficulty to any non-molecular explanation of magnetic phenomena.

A test for this point would be to take Laue photographs through a crystal such as pyrrhotite, which shows magnetic effects, and see if there is any change in the internal structure when the crystal becomes saturated.[10]

SUMMARY.

To sum up our argument then: It is shown, in the first place, that the arrangement of the atoms in certain crystals, as determined by their X-ray spectra, indicates definitely that in these crystals there is no molecular structure.

In extending the argument to all solid matter it is pointed out, from the dependence of crystal form on chemical composition, from a consideration of the Dulong-Petit law and of the nature of cohesion, and from the evidence of X-rays as to certain crystals, that each atom in a solid oscillates about a definite position of stable equilibrium.

From a further examination of the nature of cohesion and of the forces concerned in chemical combination and especially from the general relation found between the atomic heat of formation of a substance and its melting-point it is found that the forces

---

[10] This test was suggested by my brother, K. T. Compton. It has been made by K. T. Compton and E. A. Trousdale (*Phys. Rev.*, **5**, 315, 1915) and repeated with greater care by the writer and Oswald Rognley (*Science*, N. S., **46**, 415, 1917). These experiments show that when magnetic crystals are saturated magnetically the atoms are not displaced by as much as 1/250 of the distance between the atoms. Magnetic effects are therefore not due to rotations of the molecules, but must be due to the atoms or something within the atoms.

holding the atoms in their positions of stable equilibrium are of the same nature and comparable in magnitude with the forces binding together a chemical molecule.

It is seen further that the atoms in a solid are very close together, so that they often come in contact. And, since an atom attracts equally all atoms of another kind which are in contact with it, an atom cannot remain combined for more than an infinitesimal interval with any other particular atom at ordinary temperatures.

Finally, it was shown that, since in the solid state each atom has three degrees of translational freedom and is strongly attracted by atoms other than those of its own "molecule," it must, on the average, exert equal attractions on all the neighboring atoms.

From this the conclusion is drawn that in the solid state the atoms are so intimately intermingled that particular molecules cannot be definitely defined.

When those properties of solid matter which have been explained by molecules are considered, nothing is found which indicates at all definitely a molecular structure.

We feel justified in concluding, therefore, *that the structure of solid matter is not molecular.*

July 25, 1914.

NOTE ON THE GRATING SPACE OF CALCITE AND THE X-RAY SPECTRUM OF GALLIUM.

By Arthur H. Compton.

IN a recent number of this journal[1] Uhler and Cooksey have described a method of measuring the angle of reflection of X-ray spectrum lines which seems to be remarkably free from systematic errors, and capable of high precision. They applied their method to the determination of the angle of reflection of the characteristic K lines of gallium from a crystal of calcite. In calculating the wave-length of these rays they obtained the grating space of calcite by comparing it experimentally with the grating space of rock-salt, which can be determined in terms of the known crystal structure. In making this comparison, however, they determined the angles of reflection from rock-salt by an "old" method which, as they point out, is liable to introduce appreciable errors. Their determination of the grating space of calcite and hence also of the wave-length of the characteristic X-rays from gallium, is therefore no more accurate than the measurements made by the "old" method which they criticize.

The reason assigned by Uhler and Cooksey for making this experimental determination of the grating space is "because a sufficiently satisfactory reduction factor [the ratio of the grating space of calcite to that of rock-salt] if present in the literature of the subject, has escaped our notice." It should be noted that the grating space in the case of calcite may be calculated from the known crystal structure as well as in the case of rock-salt. The formula to be used is given by W. H. Bragg[2] and the writer[3] as,

$$(1) \qquad d_1 = \left(\frac{1}{2} \frac{M_1}{\rho_1 N \phi(\beta_1)}\right)^{\frac{1}{3}},$$

where M_1 is the molecular weight of $CaCO_3$, ρ_1 is the density of the calcite crystal, N is the number of molecules per gram molecule, and $\phi(\beta_1)$ is the volume of a rhombohedron the distance between whose opposite faces is unity, and the angle between whose edges is β_1. This function is[4]

---

[1] Phys. Rev., 10, 645, 1917.

[2] W. H. Bragg, Proc. Roy. Soc. A., 89, 468 (1914). "X-rays and Crystal Structure," p. 112.

[3] A. H. Compton, Phys. Rev., 7, 655 (1916).

[4] A. H. Compton, *loc. cit.* Professor Bragg uses the value $\phi(\beta) = 1.08$, which makes his value of d for calcite differ appreciably from that here obtained.

$$\phi(\beta_1) = \frac{(1 + \cos \beta_1)^2}{\sin \beta_1 (1 + 2 \cos \beta_1)}.$$

For calcite $\beta_1 = 101° 55'$[1] which makes $\phi(\beta_1) = 1.0963$.

The corresponding expression for the grating space of rock-salt is

(2) $$d_2 = \left(\frac{1}{2} \frac{M_2}{\rho_2 N}\right)^{\frac{1}{3}},$$

the subscripts 2 indicating that the molecular weight and density are those corresponding to rock-salt. The reduction factor sought by Uhler and Cooksey is, therefore,

(3) $$R = \frac{d_1}{d_2} = \left(\frac{M_1 \rho_2}{M_2 \rho_1 \phi(\beta_1)}\right)^{\frac{1}{3}},$$

which gives the ratio of the grating space of calcite to that of rock-salt.

Bragg's expression[2] is not dependent upon the details of the arrangement of the atoms in the calcite crystal. It expresses only the fact that each elementary rhombohedron contains half a molecule of $CaCO_3$. The uncertainty of the applicability of this formula is thus no greater than in the corresponding case of rock-salt. In fact the calculated value of the grating space of calcite is probably the more accurate, since this crystal is more perfect and is less apt to contain inclusions than is rock-salt.

Substituting in formula (1) the values:

$M_1 = 100.075$,[3]

$\rho_1 = 2.7116$ g. cm.$^{-3}$,[4]

$N = 6.062 \times 10^{23}$ per gram molecule,[5]

$\phi(\beta_1) = 1.0963$,

we find for the grating space of calcite,

$$d = 3.0281 \times 10^{-8} \text{ cm}.$$

The greatest uncertainty in this value is due to N, whose probable error is ± 0.1 per cent. Since N occurs in the 1/3 power, the probable error in d is about .033 per cent. Thus the grating space of calcite is

$$d_1 = 3.0281 \pm .0010 \times 10^{-8} \text{ cm}.$$

The value determined by Uhler and Cooksey by comparison with rock-

---

[1] Calculated from Dana's value of 74°55′ for the dihedral angle.
[2] W. H. Bragg and W. L. Bragg, "X-rays and Crystal Structure," p. 110.
[3] International Atomic Weights 1917.
[4] A. H. Compton, loc. cit.
[5] R. A. Millikan, Phil. Mag., 34, 13 (1917).

salt, using $d_2 = 2.814 \times 10^{-8}$ cm. is $d_1 = 3.0307 \times 10^{-8}$ cm. Gorton[1] has determined the grating space of calcite by a similar comparison method, using the same value of d_2, and obtains $d_1 = 3.028 \times 10^{-8}$ cm., which agrees absolutely with the theoretical value. Millikan[2] gives for the grating space of calcite the value $(3.030 \pm .001) \times 10^{-8}$ cm., calculated by D. L. Webster[3] using Millikan's value of e. In this calculation Webster has made use of Bragg's value of $\phi(\beta_1) = 1.08$ instead of the true value 1.0963, which accounts for the difference between his value and that here given.

The wave-lengths of the characteristic X-rays from gallium given by Uhler and Cooksey require revision because of this error in their determination of the grating space of calcite. Their determinations of the angles of reflection from calcite were verified by their "new" method, and hence are not subject to the errors introduced when they determined the angles from rock-salt by their "old" method. Their values for the angles of reflection from calcite may thus be accepted without discount. Their values for the wave-length are given in the following table together with the corrected values using the value of $d_1 = 3.0281 \pm .0010 \times 10^{-8}$ cm.

Line.	Reflection Angle from Calcite.	$\lambda \times 10^8$ cm. Uhler & Cooksey.	$\lambda \times 10^8$ cm. Corrected.
α_2	12° 47′ 15″ ± 2″	1.34161 ± .00004	1.34046 ± .00045
α_1	12° 45′ 5″ ± 2″	1.33785 ± .00004	1.33673 ± .00044
β_1	11° 28′ 30″ ± 2″	1.25691 ± .00000	1.20482 ± .00041

The probable error in the wave-length is estimated by Uhler and Cooksey on the basis of their probable error in measuring the angle. It should be noted that a much larger error in the wave-length is introduced by the uncertainty of the grating space.

RESEARCH LABORATORY,
WESTINGHOUSE LAMP COMPANY,
January 22, 1918.

[1] W. S. Gorton, PHYS. REV., 7, 209 (1916).
[2] R. A. Millikan, loc. cit., p. 16
[3] D. L. Webster, PHYS. REV., 7, 607 (1916).

The Law of Absorption of High Frequency Radiation.[1]

By Arthur H. Compton.

OWENS' law for the absorption of X-rays, that

$$\mu = k_1 \lambda^3,$$

where μ is the mass-absorption coefficient, λ is the wave-length and k a constant, has been modified by Barkla to account for the scattering which becomes important at very high frequencies. In his expression,

$$\mu = k_2 \lambda^3 + \sigma_0$$

the mass-scattering coefficient σ_0 was originally considered as a constant, having the value 0.2, calculated by J. J. Thomson on the usual electron theory. Barkla and White have recently shown, however, that this quantity must diminish at short wave-lengths, but were unable to suggest any reason for the decrease.

If the electron is taken as a ring of electricity with a radius comparable with the wave-length of very hard X-rays and V-rays, the mass-scattering coefficient should vary according to the expression recently proposed by the writer,

$$\sigma = \sigma_0 \left(1 - 29.61 \frac{a^2}{\lambda^2} + 524.2 \frac{a^4}{\lambda^4} - 5396 \frac{a^6}{\lambda^6} + \cdots \right), \quad (1)$$

where σ_0 is the mass scattering coefficient as calculated by Thomson, and a is the radius of the electronic ring. At the same time the true or fluorescent absorption will depend upon the magnitude of the action of the incident radiation upon the electron. This will diminish for short wave-lengths, since in this case different parts of a wave may work against each other in trying to pry an electron loose from its fixed position. The energy absorbed in displacing an electron may be shown to be proportional to $\sqrt{\sigma}$. The mass absorption coefficient may therefore be represented by

$$\mu = k_3 \sqrt{\sigma} \lambda^3 + \sigma, \quad (2)$$

where σ is the function of a/λ represented by equation (1).

This expression is compared with the experimental values for the absorption coefficient of aluminium as given by Hull and Rice, Williams, and Bragg and Pierce, and is found to be satisfactory, especially for the short wave-lengths where the formulas of Owen and Barkla fail.

The value of *the radius of the electron* necessary to give best agreement with the experimental values *is 1.85×10^{-10} cm.* Unless there is some consistent error in the measurements of Hull and Rice, this value is determined within a probable error of about 5 per cent.

[1] Abstract of a paper presented at the New York meeting of the American Physical Society, March 1, 1919.

THE SIZE AND SHAPE OF THE ELECTRON.[1]

By Arthur H. Compton.

SYNOPSIS.—Attention is called to two outstanding differences between experiment and the theory of scattering of high frequency radiation based upon the hypothesis of a sensibly point charge electron. In the first place, according to this theory the mass scattering coefficient should never fall below about .2, whereas the observed scattering coefficient for very hard X-rays and γ-rays falls as low as one fourth of this value. In the second place, if the electron is small compared with the wave-length of the incident rays, when a beam of γ-rays is passed through a thin plate of matter the intensity of the scattered rays on the two sides of the plate should be the same, whereas it is well known that the scattered radiation on the emergent side of the plate is more intense than that on the incident side.

It is pointed out that the hypothesis that the electron has a diameter comparable with the wave-length of the hard γ-rays will account qualitatively for these differences, in virtue of the phase difference between rays scattered by different parts of the electron. The scattering coefficient for different wave-lengths is calculated on the basis of three types of electron: (1) A rigid spherical shell of electricity, incapable of rotation; (2) a flexible spherical shell of electricity; (3) a thin flexible ring of electricity. All three types are found to account satisfactorily for the meager available data on the magnitude of the scattering coefficient for various wave-lengths. The rigid spherical electron is incapable of accounting for the difference between the emergent and the incident scattered radiation, while the flexible ring electron accounts more accurately for this difference than does the flexible spherical shell electron.

It is concluded that the diameter of the electron is comparable in magnitude with the wave-length of the shortest γ-rays. Using the best available values for the wave-length and the scattering by matter of hard X-rays and γ-rays, the radius of the electron is estimated as about 2×10^{-10} cm. Evidence is also found that the radius of the electron is the same in the different elements. In order to explain the fact that the incident scattered radiation is less intense than the emergent radiation, the electron must be subject to rotations as well as translations.

I. THE SCATTERING OF HIGH FREQUENCY RADIATION.

THE radius of the electron is usually calculated from its kinetic energy when in motion, taking this to be identical with its magnetic energy. According to the customary assumption that the charge on an electron is uniformly distributed over the surface of a sphere, the radius of the sphere as thus calculated is about 1×10^{-13} cm. There are, however, a number of phenomena in connection with the scattering and absorption of high frequency radiation by matter, which appear to be

---

[1] A preliminary paper on this subject was read before the American Physical Society, December 28, 1917 (PHYS. REV., 11, 330, 1918). Cf. also J. Wash. Ac. Sci., 8, 1, 1918.

inexplicable according to classical electrodynamics if the dimensions of the electron are taken to be of this order of magnitude, whose explanation is obvious if the electron is assumed to be a flexible ring of electricity whose radius is comparable with the wave-length of short γ-rays. This paper is the first of a series of three, which will deal respectively with the scattering of high-frequency radiation, the absorption of high-frequency radiation, and the nature of the ultimate magnetic particle. The present discussion will deal with certain outstanding differences between experiment and the theory of scattering of high frequency radiation based upon the hypothesis of a sensibly point charge electron, and it will be shown that these differences may be explained on the basis of an electron of relatively large size. In order to preserve the directness of the argument, the details of the calculations will be reserved for the latter part of the paper.

PART I.

A. *The Scattering Coefficient of High Frequency Radiation.*

On the hypothesis that the electron is sensibly a point charge of electricity, Sir J. J. Thomson has shown[1] that the ratio of the energy of the electromagnetic radiation scattered by an isolated electron to the energy incident upon it is given by the expression

$$\frac{8\pi}{3} \frac{e^4}{m^2 C^4}.$$

Here e and m are respectively the charge and mass of the electron, and C is the velocity of light. If the electrons in any substance act independently of each other, the scattering coefficient per unit mass of the substance will therefore be

$$(1) \qquad \frac{\sigma}{\rho} = \frac{8\pi}{3} \frac{Ne^4}{m^2 C^4},$$

where σ is the ratio of the scattered to the incident energy per unit volume of the material, ρ is its density, and N is the number of electrons in unit mass of the substance.

Since this scattered energy is lost from the primary beam, the quantity σ/ρ represents also the part of the mass absorption coefficient which is due to scattering. As Barkla has pointed out,[2] there may be absorption due to other causes, such as the production of secondary beta or cathode rays, but this absorption due to scattering must always be present. Moreover, if the electrons in the absorbing material are grouped together in regions which are small compared with the wave-length of the incident

---

[1] J. J. Thomson, Conduction of Electricity through Gases, 2d ed., p. 325.
[2] C. G. Barkla and M. P. White, Phil. Mag., 34, 275, 1917.

beam, the electrons do not scatter independently. The rays scattered by the electrons in this case overlap in such a manner that a certain amount of "excess scattering" occurs. There is, however, no arrangement of the electrons which will result in less scattering than when they act independently. On the hypothesis of a point charge electron, it is possible for the scattering, and hence for the mass absorption coefficient, to be smaller than that predicted by Thomson's expression (1) only in case the electrons are held in position so firmly that their natural period of vibration is shorter than the period of the incident radiation.

In making the calculations from Thomson's theory, it may be assumed that the number of electrons in an atom which are effective in scattering the incident radiation is equal to the atomic number. This assumption is supported in the case of the lighter elements by the experiments of Barkla and Dunlop[1] when X-rays of ordinary hardness are used. It would seem possible that with the higher frequency γ-rays certain electrons might be effective in scattering which are too rigidly bound to scatter X-rays. Such an effect, however, would mean an increase rather than a decrease in the scattering for the shorter wave-lengths. That the atomic number is the number of effective electrons when γ-rays are used, is confirmed by the observations of Soddy and Russell[2] and of Ishino[3] to the effect that for the shortest rays the amount of energy scattered by atoms of the different elements is accurately proportional to their atomic numbers. This means that all the electrons outside of the nucleus are effective in producing absorption when hard γ-rays are used. If the electron is sensibly a point charge of electricity, the scattered energy should therefore be at least as great as the value assigned by equation (1).

Barkla and Dunlop[4] have shown that for a considerable range of wave-lengths of X-rays the mass scattering coefficients of the lighter elements are given accurately by equation (1) if the number of electrons in the atom is taken to be approximately half the atomic weight. For elements of high atomic weight, the scattering becomes greater than this value except for very short wave-lengths, indicating that the electrons are so closely packed that "excess scattering" occurs. For wave-lengths less than 2×10^{-9} cm., however, Barkla and White[5] have shown that the total mass absorption coefficient of the light elements is less than the value theoretically calculated for the mass scattering coefficient alone; and Soddy and Russell[6] have found that for the hard γ-rays from Radium

---

[1] C. G. Barkla and J. G. Dunlop, Phil. Mag., 31, 222, 1916.
[2] Soddy and Russell, Phil. Mag. 18, 620, 1910; 19, 725, 1910.
[3] M. Ishino, Phil. Mag., 33, 140, 1917.
[4] C. G. Barkla and J. G. Dunlop, *loc. cit.*
[5] C. G. Barkla and M. P. White, *loc. cit.*, p. 277.
[6] Soddy and Russell, *loc. cit.*

C the absorption of substances of lower atomic weight than mercury is only a small fraction of that required by Thomson's expression. Direct measurements of the scattering of hard γ-rays confirm the conclusions based on absorption measurements. Thus it has been found[1] in the case of radiation of very high frequency that the mass scattering coefficient falls as low as one fourth of the value predicted by Thomson's theory.

As has just been pointed out, it is impossible, according to classical electrodynamics, to account for this low scattering and absorption of radiation of very high frequency by matter if the electron is taken to be sensibly a point charge of electricity. If, on the other hand, the electron is considered to have a radius comparable with the wave-length of the incident radiation, a qualitative explanation of the phenomenon of low scattering for short wave-lengths is obvious. The effect of this hypothesis is to make an appreciable phase difference between the rays scattered by different parts of the electron. Thus the radiation scattered from A, Fig. 1, traverses a longer path between S and P than does the ray scattered from the part of the electron at B. If the wave-length is many times the diameter of the electron, the phase difference between these two rays will be negligible, and the reduction in the intensity of the scattered beam will be inappreciable; if, however, the difference in the two paths is comparable with the wave-length of the incident radiation, the phase difference will be such that the intensity of the ray scattered to P will be much reduced. The assumption of a relatively large electron is therefore capable of explaining qualitatively the observed decrease in the scattering of electromagnetic radiation when the wave-length becomes very short.

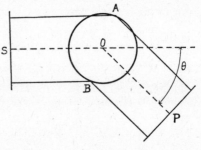

Fig. 1.

Calculation of the Scattering. I. *Rigid Spherical Electron.*—The exact manner in which the scattering will decrease with shorter wave-lengths will of course depend upon the form of electron considered. For example, taking the simplest case of the scattering due to a rigid, uniform, spherical shell of electricity, incapable of rotation, we find

$$(2) \qquad \frac{\sigma}{\rho} = \frac{8\pi}{3} \frac{e^4 N}{m^2 C^4} \sin^4\left(\frac{2\pi a}{\lambda}\right) \bigg/ \left(\frac{2\pi a}{\lambda}\right)^4,$$

[1] M. Ishino, *loc. cit.*

where a is the radius of the spherical shell and λ is the wave-length of the incident beam. The details of the derivation of this expression will be found in Part 2, Section 1. If $a = 10^{-13}$ cm., as usually assumed, this is practically identical with equation (1), even for radiation of the shortest known wave-length. The relative values of the scattering according to this expression are shown in curve I, Fig. 2, for different values of λ/a. In this diagram the value $\sigma/\sigma_0 = 1$ indicates the magnitude of the mass scattering if the electron were sensibly a point charge of electricity, and the calculated values are given in terms of this quantity.

2. *Flexible Spherical Electron.*—It would appear more reasonable to suppose that the spherical shell electron is subject to rotational as well as to translational displacements when traversed by a γ-ray. The scattering due to such an electron is difficult to calculate, but an approximate expression can be obtained if the electron is considered to be perfectly flexible, so that each part of it can be moved independently of the other parts. On this hypothesis it can be shown (cf. Part 2, Section 2) that the intensity of the beam scattered by an electron at an angle θ with an unpolarized beam of γ-rays is given by the expression,

$$(1) \quad I_\theta = I \frac{e^4(1 + \cos^2 \theta)}{2L^2 m^2 C^4} \left\{ \sin^2\left(\frac{4\pi a}{\lambda} \sin\frac{\theta}{2}\right) \Big/ \left(\frac{4\pi a}{\lambda} \sin\frac{\theta}{2}\right)^2 \right\}.$$

Here I is the intensity of the incident beam, L is the distance at which the intensity of the scattered beam is measured, and the other quantities have the same meaning as before. The mass absorption coefficient due to the scattering by such an electron is therefore,

$$(4) \quad \frac{\sigma}{\rho} = 2\pi N L^2 \int_0^\pi \frac{I_0}{I} \sin\theta d\theta.$$

This integral may be evaluated graphically or by expansion into a series (cf. *infra*, equation 17). The values of σ/σ_0 according to equation (4) are plotted in curve II, Fig. 2. The values for a rigid spherical electron which is subject to rotation should lie between curves I and II for the range of wave-lengths for which curve II is plotted.

3. *Ring Electron.*—According to electromagnetic theory it is obvious that the mass of an electron cannot be accounted for on the basis of a uniform distribution of electricity over the surface of a sphere of radius comparable with the wave-length of γ-rays. Much the same effect, so far as the scattering of high frequency radiation is concerned, results from the conception of the electron as a ring of electricity of relatively large diameter, similar in form to the "magneton" suggested by A. L. Parson.[1] It has been shown by Webster[2] and Davisson[3] that the assump-

[1] A. L. Parson, Smithsonian Misc. Collections, Nov., 1915.
[2] D. L. Webster, PHYS. REV., 9, 484, 1917.
[3] Davisson, PHYS. REV., 9, 570, 1917.

tion of such an electron is compatible with the electromagnetic theory of mass.

The exact calculation of the scattering produced by a thin ring of electricity is difficult. A chief factor in the complexity of the problem is the fact that the effective electromagnetic mass of a short arc of the ring differs according as it is accelerated parallel to the tangent to the arc, parallel to the axis of the ring, or parallel to a radius of the ring. The ratios of the effective masses along these three axes depends moreover upon the speed with which the electricity in the ring is rotating. In order to make the problem manageable, the assumptions have been made that the mass of an arc element is the same in all directions, and that the velocity of the electricity in the ring is small compared with the velocity of light. On the basis of these assumptions the mass scattering coefficient for a flexible electronic ring is found to be

$$(5) \quad \frac{\sigma}{\rho} = \frac{8\pi}{3} \frac{Ne^4}{m^2C^4} \left\{ 1 - a\left(\frac{a}{\lambda}\right)^2 + b\left(\frac{a}{\lambda}\right)^4 - c\left(\frac{a}{\lambda}\right)^6 + \cdots \right\},$$

where the coefficients $a, b, c, \cdots$ are constants which are evaluated below (cf. equation 21). The relative values of the scattering according to this expression are shown in curve *III*, Fig. 2. The scattering of γ-rays

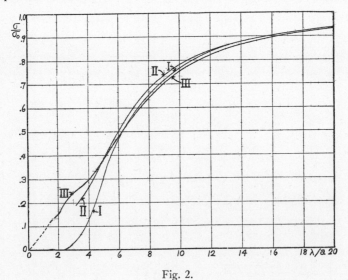

Fig. 2.

by a ring electron as thus calculated is an approximation which will doubtless correspond closely with the true value for relatively long waves, but which may differ appreciably for the shortest known radiation.

Unfortunately the experimental data are too meager to submit these formulæ to accurate quantitative test. There are, however, three points on the curve which have been established with some care. Barkla and Dunlop[1] have found that for relatively soft X-rays the light elements scatter according to equation (1), so that the part of the curves where λ/a is large is verified. Hull and Rice[2] have estimated from their absorption measurements that for X-rays and γ-rays whose wave-lengths are in the neighborhood of 0.15×10^{-8} cm. the value of σ/ρ for aluminium is 0.12. Taking the number of electrons in an aluminium atom to be 13, this gives for the relative scattering, 0.64. According to curve *I* this corresponds to an electronic radius of 1.9×10^{-10} cm. Curve *II* gives 2.0×10^{-10} cm., and curve *III*, 1.9×10^{-10} cm.[3] Ishino[4] finds that the value of σ/ρ for aluminium, using the hard γ-rays from radium C, is about 0.045, which means a value for the relative scattering of 0.24. The work of Rutherford and Andrade[5] shows that the principal part of the "homogeneous" radiations from radium C consists of a strong line $\lambda = 0.099$, and a weaker line $\lambda = .071$, $\times 10^{-8}$ cm. Both of these lines were prominent in Ishino's experiment, in which he filtered the γ-rays through a centimeter of lead. Rutherford has pointed out,[6] however, from a consideration of the velocities of the β-particles, that there must be a certain amount of radiation of much shorter wave-length. The existence of such extremely hard rays is confirmed by the fact that the absorption coefficient of the penetrating radiation of the atmosphere as determined at high altitudes is much smaller than that of the hard γ-rays from radium C, such as used by Ishino. The fact that it is impossible to detect these very short waves by crystal analysis, however, indicates that their effectiveness in the scattered beam is small compared with that of the two lines observed by Rutherford and Andrade. It seems reasonable, therefore, to take for the effective wave-length of the γ-rays used in Ishino's scattering experiments about $.08 \times 10^{-8}$ cm.[7] This gives for the value of the electronic radius, from curve *I*, 1.7, from curve *II*, 2.1 and from curve *III*, 2.7, $\times 10^{-10}$ cm.

[1] C. G. Barkla and Dunlop, *loc. cit.*

[2] A. W. Hull and M. Rice, PHYS. REV., 8, 326, 1916.

[3] In the second part of this paper, by using a more accurate formula for the mass absorption coefficient, the data of Hull and Rice will be shown to lead to a value of $(1.85 \pm .04) \times 10^{-12}$ cm., if the electron is taken to be a ring.

[4] M. Ishino, *loc. cit.*, p. 141.

[5] Sir E. Rutherford and Andrade, Phil. Mag., 28, 263, 1916.

[6] Sir E. Rutherford, Phil. Mag., 34, 153, 1917.

[7] In the paper last referred to, Rutherford estimated the effective wave-length of Ishino's γ-rays to be much shorter than the value here used. His estimate is based upon measurements of the absorption of high frequency X-rays filtered by means of a lead filter. He calculated the frequency of the X-rays according to the relation $h\nu = eV$, taking V to be the maximum voltage applied to the tube. As is apparent from the work of Rutherford, Barnes and Richardson (Phil. Mag., 30, 339, 1915), this relation does not express the *effective* fre-

The value of σ/ρ given by Hull and Rice is a mean over a relatively large range of wave-lengths, and Barkla is of the opinion[1] that Ishino's value of the scattering of the γ-rays from radium C is appreciably in in error because of a too high estimate of the true absorption. Thus, though the experimental values of the electronic radius agree best on the basis of the flexible spherical shell electron, as represented in curve II, the accuracy of the experiments is by no means sufficient to distinguish between the three hypotheses.

The important thing to notice is that if the electrons had dimensions comparable with 10^{-13} cm., as usually assumed, the scattering should be represented by the upper line of Fig. 2 where $\sigma/\sigma_0 = 1.0$. The fact that experiment gives consistently lower values when short wave-lengths are used is sufficient proof that the electron is not sensibly a point charge of electricity. On the other hand, it is possible to account for this reduced scattering within the probable experimental error if the electron has a radius of 2×10^{-10} cm.

B. *The Dissymmetry of the Scattering of Hard γ-rays on the Incident and Emergent Sides of a Plate.*

A second difficulty which is found with Sir J. J. Thomson's simple theory is that it predicts that if a beam of X-rays is passed through a thin plate of matter, the intensity of the scattered rays on the two sides of the plate should be the same. Barkla and Ayers[2] have shown that, for rather hard X-rays and for those substances of low atomic weight whose mass scattering coefficients can be calculated accurately by equation (1), this second prediction of Thomson's theory is also valid. On the other hand, it is well known that both in the case of relatively soft X-rays and in the case of hard γ-rays the scattered radiation on the emergent side of the plate is more intense than that on the incident side.

When heavy atoms and long waves are used, the dissymmetry between the emergent and the incident scattered radiation is accompanied by an increase in the total scattered energy. For this reason the phenomenon is described by the term "excess scattering." It is satisfactorily accounted for[3] by the fact that the electrons in the heavy atoms do not act independently in scattering the longer wave-length X-rays, since

quency of the filtered rays, especially when a lead filter is used. This doubtless accounts for the fact, which will be brought out in the following paper, that Rutherford's determinations of the absorption do not agree with those of Hull and Rice, who measured the absorption coefficient of homogeneous X-rays of known wave-length.

[1] C. G. Barkla and M. P. White, *loc. cit.*, p. 278.
[2] Barkla and Ayers, Phil. Mag., 21, 271, 1911.
[3] C. G. Darwin, Phil. Mag., 27, 329, 1914; D. L. Webster, Phil. Mag., 25, 234, 1913.

they are grouped so closely together that the rays scattered by the different electrons are in nearly the same phase. This has the effect of increasing the total scattering. But also, since the phase difference between the rays from the different electrons in an atom is less for the scattered rays which make small angles with the primary beam, there is greater reinforcement and hence greater intensity on the emergent than on the incident side of the scattering atom.

This explanation cannot be applied, however, to the case of the unsymmetrical scattering of very hard rays. This is clear for two reasons. In the first place, if an atom of medium weight is traversed by rays of increasing hardness, at first excess scattering occurs as described above; but as the wave-length becomes shorter the scattered radiation becomes nearly symmetrical until the scattered energy can be calculated according to Thomson's formula (1). The electrons now, therefore, are scattering independently, and must continue to do so for all shorter wave-lengths. Thus we see that the dissymmetry in the scattering which reappears as the wave-length becomes very short cannot be accounted for by the mutual action of the separate electrons. In the second place, the phenomenon of unsymmetrical scattering for very short waves is distinguished from the excess scattering which occurs with longer waves by the fact that in the former case the dissymmetry is accompanied not by an increase but by a decrease in the total scattering. If the phenomenon were due to the mutual action of the electrons, it would be accompanied by an increased total scattering, as before. It is thus evident that the unsymmetrical scattering of very short electromagnetic waves is due not to groups of electrons in the atoms, but to some property of the individual electrons.

The qualitative explanation of this phenomenon on the basis of our large electron hypothesis is at once apparent. Referring again to Fig. 1, it is obvious that if the diameter of the electron is comparable with the wave-length of the radiation, there will be an appreciable difference in phase between the rays scattered from different parts of the electron. Since this phase difference is greater for rays scattered at large than for those at small angles, the intensity of the incident radiation will be in the former case the more strongly reduced. In order to explain this phenomenon it is not sufficient, however, merely to assume that the electron is relatively large. For example, the hypothesis of the electron as a rigid spherical shell, incapable of rotation, though resulting in a reduced total scattering, would give rise to symmetrical scattering on the incident and emergent sides of a plate. To account for the observed dissymmetry, the further assumption must be made that the incident electromagnetic wave is capable of moving the different parts of the electron relatively to each other.

If the electron is sensibly a point charge of electricity, the intensity of the beam scattered by an electron at an angle θ with the incident beam is[1]

$$(6) \qquad I_\theta = I \frac{e^4(1 + \cos^2 \theta)}{2L^2 m^2 C^4}.$$

The corresponding expression for an electron in the form of a flexible spherical shell of electricity has already been given:

$$(3) \qquad I_\theta = I \frac{e^4(1 + \cos^2 \theta)}{2L^2 m^2 C^4} \left\{ \sin^2\left(\frac{4\pi a}{\lambda} \sin\frac{\theta}{2}\right)^2 \bigg/ \left(\frac{4\pi a}{\lambda} \sin\frac{\theta}{2}\right)^2 \right\}.$$

For the ring type electron we obtain

$$(7) \qquad I_\theta = I \frac{e^4(1 + \cos^2 \theta)}{2L^2 m^2 C^4} \left\{ 1 - \alpha \left(\frac{4\pi a}{\lambda} \sin\frac{\theta}{2}\right)^2 + \beta \left(\frac{4\pi a}{\lambda} \sin\frac{\theta}{2}\right)^4 - \cdots \right\},$$

where the values of the constants α, β, γ, etc., are those determined below (equation 20). When a/λ remains small, the scattering according to both expressions (3) and (7) approaches the value for a point charge electron (6).

D. C. H. Florance[2] has determined experimentally the values of the relative intensity of the radiation scattered at different angles when the hard γ-rays from radium bromide traverse a plate of iron. His values are indicated in Fig. 3 by circles. Taking the effective wave-length of these rays to be $.09 \times 10^{-8}$ cm., and using in equations (3) and (7) $a = 2 \times 10^{-10}$ cm. as above estimated, the relative scattering at different angles may be calculated. The intensity of the radiation scattered at different angles by a point charge electron is indicated in Fig. 3 by the outer solid curve, that due to the spherical shell electron by the inner solid curve, and that from the ring form electron by the broken curve. Inasmuch as the γ-rays used by Florance were heterogenous, and as the softer rays are scattered relatively more strongly at larger angles, the agreement of the experimental values with either of the inner curves is as good as can be expected. The important point to be noticed is, however, that the experimental values are entirely out of harmony with

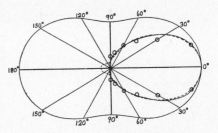

Fig. 3.

[1] J. J. Thomson, Conduction of Electricity through Gases, 2d ed., p. 326.
[2] D. C. H. Florance, Phil. Mag., 20, 921, 1910.

what is to be expected if the electron is sensibly a point charge of electricity.

A better quantitative test of this explanation of the dissymmetry of scattered γ-radiation is afforded by determinations of the ratio of the total radiation scattered on the incident side of a plate struck by hard γ-rays to that scattered on the emergent side. The theoretical value of this ratio is

$$(8) \quad \frac{I_i}{I_e} = \int_{\pi/2}^{\pi} I_\theta \sin\theta d\theta \bigg/ \int_0^{\pi/2} I_\theta \sin\theta d\theta.$$

The curves of Fig. 4 give the values of this ratio for different values of λ/a, the broken curve being calculated on the basis of the spherical

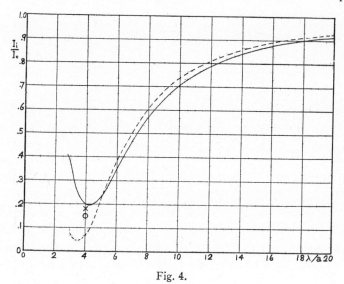

Fig. 4.

electron. These curves doubtless explain at least in part the observation of Florance[1] that the "incident" scattered rays are softer than the "emergent" and the primary rays, since they show that the relative amount of rays scattered backward is much greater for soft or long wavelength γ-rays than for the harder radiation.

The ratio of the incident to the emergent radiation has been determined for the hard γ-rays from radium C by Madsen,[2] who found the value 18 per cent., and Ishino,[3] who found 15 per cent. Assuming for the effective wave-length in this case $\lambda = 0.08 \times 10^{-8}$ cm., and for the radius of the electron 2×10^{-10} cm., as estimated above, Ishino's datum

[1] D. C. H. Florance, Phil. Mag., 27, 225, 1914.
[2] Madsen, Phil. Mag., 17, 423, 1909.
[3] M. Ishino, loc. cit., p. 138.

for this ratio is represented by the open circle in Fig. 4 and Madsen's datum by the cross. It is possible that neither of the theoretical values agree within the probable experimental error with these determinations, but in view of the approximate method of calculating the scattering by a ring electron the difference is not serious. The fact that the predicted values are of the proper order of magnitude is strong evidence that the dissymmetry in the scattering of γ-rays by matter is due to the interference of the rays scattered by different parts of the electron. Thus not only must the electron have a size comparable with the wave-length of γ-rays, but it must also be subject to rotations or be sufficiently flexible for γ-rays to move its different parts relatively to each other.

C. *Conclusions.*

As has been pointed out, therefore, according to classical electrodynamics the mass scattering coefficient for X-rays and γ-rays passing through matter should never fall below the value 0.18, as calculated on the basis of Thomson's theory, if an electron of the usual dimensions is postulated. Experiment shows, however, that for very high frequency radiation the scattering is much less than this. It is possible that certain assumptions regarding the conditions for scattering radiant energy, contrary to classical theory, might be made which would account for the observed low value of the scattering for very high frequencies. As long as the idea of the point charge electron is retained, however, no such assumptions can account for the observed dissymmetry between the incident and the emergent scattered radiations. Unless the theory that X-rays and γ-rays consist of waves or pulses is abandoned, the only possible explanation of this dissymmetry would seem to be that the scattering particles have dimensions comparable with the wave-length of the rays which they scatter. Since the scattering particles have been shown to be the electrons, the statement may therefore be made with confidence that *the diameter of the electron is comparable in magnitude with the wave-length of the shortest γ-rays.*

According to the best available values for the wave-length and the scattering by matter of hard X-rays and γ-rays, *the radius of the electron is about* 2×10^{-10} *cm.*

The fact that the scattering of hard γ-rays by atoms of the different elements is proportional to the atomic number shows that if the number of the electrons in an atom which are effective in scattering is equal to the atomic number, *the radius of the electron is the same in the different atoms.* This is clear, since if the electron had smaller dimensions in the atoms of one element, the scattering coefficient of this element would not decrease as rapidly for shorter wave-lengths, and the scattering by the different atoms would not be proportional to the number of electrons in the atoms.

In order to explain the fact that the emergent scattered radiation is more intense than the incident radiation, it is necessary to assume further that the different parts of the charge of the electron can possess certain motions independently of each other. That is, *the electron is subject to rotations as well as translations.*

PART 2.

1. *To calculate the energy scattered by a rigid spherical shell electron, incapable of rotation, whose diameter is comparable in magnitude with the wave-length of the incident radiation.*

Let us first derive an expression for the acceleration to which such an electron is subject when traversed by an electromagnetic wave, and then determine the energy scattered by integrating the intensity of the beam due to this acceleration over the surface of a sphere drawn with the electron at the center.

In Fig. 5, let us suppose that the γ-ray traverses the electron along the axis Z. We shall let A represent the amplitude of the incident wave, X its electric intensity at the plane z, and λ its wave-length. The radius of the electron we shall call a, and η will represent the charge on the surface of the electron between two planes z_1 and z_2 placed unit distance apart. As the electricity is by hypothesis distributed uniformly over the surface of the sphere, η is constant between $z = -a$ and $z = +a$, and the total charge on the electron is $e = 2a\eta$. The electric intensity at the plane z at any instant may be expressed by the relation,

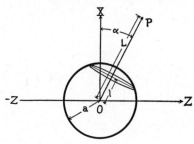

Fig. 5.

$$X = A \cos 2\pi \left(\frac{\delta - z}{\lambda} \right),$$

where $2\pi\delta/\lambda$ is the phase angle at $z = 0$ at that instant. The total force acting on the electron at this instant is therefore,

$$F = \int_{-a}^{a} A \cos 2\pi \left(\frac{\delta - z}{\lambda} \right) \eta dz$$

$$= \frac{A\eta\lambda}{\pi} \cos 2\pi \frac{\delta}{\lambda} \sin 2\pi \frac{a}{\lambda},$$

and the acceleration of the electron is

$$\ddot{x} = \frac{A\eta\lambda}{\pi m} \cos \frac{2\pi\delta}{\lambda} \sin \frac{2\pi a}{\lambda}. \tag{9}$$

Let us now calculate the intensity of the scattered ray at a distance L along the line OP which makes an angle α with the direction OX of the acceleration. According to classical theory the electric intensity at P due to a point charge electron at O subject to this acceleration would be

$$\frac{e\ddot{x} \sin \alpha}{LC^2} \tag{10}$$

where C is the velocity of light. Replacing e by ηdl and $\ddot{x}$ by its value as defined above, and integrating over the electron along the axis L from $l = -a$ to $l = +a$, the electric intensity of the scattered beam at L, α becomes:

$$R_{L,\,\alpha} = \frac{\sin \alpha}{LC^2} \int_{-a}^{a} \frac{A\eta\lambda}{\pi m} \sin 2\pi \frac{a}{\lambda} \cos \frac{2\pi}{\lambda} (\delta - L + l) \eta dl$$

$$= \frac{\eta^2 \lambda A}{\pi m L C^2} \sin \alpha \sin 2\pi \frac{a}{\lambda} \int_{-a}^{a} \cos \frac{2\pi}{\lambda} (\delta - L + l) dl$$

$$= \frac{\eta^2 \lambda^2 A}{\pi^2 L m C^2} \sin^2 2\pi \frac{a}{\lambda} \sin \alpha \cos 2\pi \frac{\delta - L}{\lambda}.$$

The amplitude of the electric vector of the scattered wave at the point L, α is therefore

$$A_{L,\,\alpha} = \frac{\eta^2 \lambda^2 A}{\pi^2 L m C^2} \sin^2 2\pi \frac{a}{\lambda} \sin \alpha,$$

or substituting for η its value $e/2a$,

$$A_{L,\,\alpha} = \frac{Ae^2}{LmC^2} \frac{\sin^2 (2\pi a/\lambda)}{(2\pi a/\lambda)^2} \cdot \sin \alpha.$$

The intensity of the radiation at this point is

$$I_{L,\,\alpha} = c A_{L,\,\alpha}^2 = \frac{cA^2 e^4}{L^2 m^2 C^4} \frac{\sin^4 (2\pi a/\lambda)}{(2\pi a/\lambda)^4} \sin^2 \alpha,$$

so that the total energy scattered by the electron is

$$E_s = \int_0^{\pi} I_{L,\,\alpha} \cdot 2\pi L \sin \alpha \cdot L d\alpha$$

$$= \frac{2\pi c A^2 e^4}{m^2 C^4} \cdot \frac{\sin^4 (2\pi a/\lambda)}{(2\pi a/\lambda)^4} \int_0^{\pi} \sin^3 \alpha \, d\alpha$$

$$= \frac{8\pi c A^2 e^4}{3 m^2 C^4} \cdot \frac{\sin^4 (2\pi a/\lambda)}{(2\pi a/\lambda)^4}.$$

The energy incident on unit area at the electron is, however, $I = cA^2$, so that the fraction of the incident energy scattered by the electron is

$$(11) \qquad E_s/I = \frac{8\pi e^4}{3m^2C^4} \frac{\sin^4(2\pi a/\lambda)}{(2\pi a/\lambda)^4}.$$

When a is small compared with λ this becomes

$$(12) \qquad \frac{e^4}{m^2C^4},$$

which is identical with the value given by Thomson[1] for a sensibly point charge electron.

If there are N electrons per unit mass of any substance, the mass scattering coefficient of the substance is therefore

$$(13) \qquad \frac{\sigma}{\rho} = \frac{8\pi}{3} \frac{e^4 N}{m^2 C^4} \frac{\sin^4(2\pi a/\lambda)}{(2\pi a/\lambda)^4},$$

where σ is the scattering coefficient per unit volume and ρ is the density of the scattering material.

2. *To calculate the energy scattered by an electron in the form of a flexible spherical shell of electricity.*

We shall treat this problem as if the mass of an element of the spherical shell were independent of the rest of the electron, being equal to $dm = m \cdot ds/s$, where s is the area of the surface of the electron, and m is its mass. As has been pointed out in part I. of this paper, the electro-magnetic mass of a spherical electron of the size here considered would be negligible. This form of electron is therefore only a convenient hypothesis to use in calculating the general effect on the scattering to be expected with any form of electron when the wave-length of the incident radiation approaches its largest dimensions.

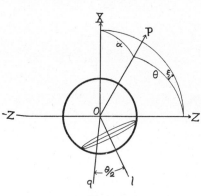

Fig. 6.

Let us suppose that the γ-ray strikes the electron when moving in the direction $-ZOZ$, and determine the intensity of the beam scattered in the direction OP. As shown in Fig. 6, OP makes the angle θ with the incident ray, and the angle α with the direction OX of the electric vector of the incident ray. The plane POZ is inclined at the dihedral angle ξ

[1] J. J. Thomson, *loc. cit.*

with the plane XOZ. We shall draw also the lines Ol and Oq in the plane POZ, line Oq being perpendicular to OZ, and Ol at an angle $\theta/2$ with Oq. The line Ol therefore bisects the external angle $-ZOP$. If now we consider the beam scattered in the direction OP, all the rays scattered from the element of the electron included between the planes l and $l + dl$ are in the same phase at P.

The amplitude of the electric vector at P due to the rays scattered by this element is, in accordance with expression (10),

$$\eta dl \cdot A \frac{\eta}{\mu} \cdot \sin \alpha / LC^2,$$

where η is the charge and μ the mass of the electron per unit distance along Ol, A is the amplitude of the electric vector of the incident beam, and as before L is the distance OP. If $2\pi\delta/\lambda$ is the phase of the ray at P scattered from the element of the electron at $l = 0$ at a given instant, the electric intensity at that instant of the ray scattered from any element to P is

$$\frac{A\eta^2}{\mu} \frac{dl}{LC^2} \sin \alpha \cdot \cos \frac{2\pi}{\lambda} \left(\delta - 2l \sin \frac{\theta}{2} \right),$$

and the electric intensity due to the whole electron is

$$R_{L,\,a,\,\theta,} = \frac{A\eta^2}{\mu} \frac{\sin \alpha}{LC^2} \int_{-a}^{a} \cos \frac{2\pi}{\lambda} \left(\delta - 2l \sin \frac{\theta}{2} \right) dl$$

$$= \frac{A\eta^2 \lambda}{2\pi\mu LC^2} \cdot \frac{\sin \alpha}{\sin \theta/2} \cdot \sin \left(\frac{4\pi a}{\lambda} \sin \frac{\theta}{2} \right) \cos \frac{2\pi\delta}{\lambda}.$$

Here, as before, a represents the radius of the electron. The amplitude is obviously the value of this quantity when $\cos (2\pi\delta/\lambda) = 1$, and the intensity of the ray scattered ot P by the electron is therefore

$$I_{L,\,a,\,\theta} = cA_{L,\,a,\,\theta}^2$$

$$= \frac{cA^2\eta^4\lambda^2}{4\pi^2\mu^2L^2C^4} \cdot \frac{\sin^2 \alpha}{\sin^2 \theta/2} \sin^2 \left(\frac{4\pi a}{\lambda} \sin \frac{\theta}{2} \right).$$

Since $2a\eta = e$, $2a\mu = m$ and $cA^2 = $ the intensity I of the incident beam, the first factor of this expression becomes

$$\frac{Ie^4}{m^2L^2C^4} \left(\frac{\lambda}{4\pi a} \right)^2.$$

When α is expressed in terms of θ and ξ we obtain

$$\cos \alpha = \sin \theta \cos \xi,$$

i. e.,

$$\sin \alpha = \sqrt{1 - \sin^2 \theta \cos^2 \xi}.$$

We may therefore write:

$$I_{L,\xi,\theta} = \frac{Ie^4}{m^2L^2C^4}\left(\frac{\lambda}{4\pi a}\right)^2 \sin^2\left(\frac{4\pi a}{\lambda}\sin\frac{\theta}{2}\right)\frac{1-\sin^2\theta\cos^2\xi}{\sin^2(\theta/2)}.$$

The intensity of the beam scattered by an electron at an angle θ with an unpolarized incident beam is the average of this quantity for all values of ξ, or

(14)
$$I_{L,\theta} = \frac{1}{2\pi}\int_0^{2\pi} I_{L,\xi,\theta}\,d\xi$$
$$= \frac{Ie^4}{2m^2L^2C^4}\left(\frac{\lambda}{4\pi a}\right)^2 \sin^2\left(\frac{4\pi a}{\lambda}\sin\frac{\theta}{2}\right)\frac{(\cos^2\theta + 1)}{\sin^2(\theta/2)}.$$

The total energy scattered in unit time by the electron is given by the quantity

(15)
$$E_s = \int_0^\pi I_{L,\theta}\cdot 2\pi L^2 \sin\theta\,d\theta$$
$$= \frac{\pi Ie^4}{m^2C^4}\left(\frac{\lambda}{4\pi a}\right)^2\cdot\int_0^\pi \sin^2\left(\frac{4\pi a}{\lambda}\sin\frac{\theta}{2}\right)\frac{(\cos^2\theta + 1)}{\sin^2(\theta/2)}\sin\theta\,d\theta.$$

This integral may be evaluated either graphically or by means of a series. To integrate by series, substitute $\theta/2 = x$ and $4\pi a/\lambda = b$. The integral factor then becomes:

$$\int_0^{\pi/2} \sin^2(b\sin x)\frac{4\sin^4 x - 4\sin^2 x + 2}{\sin^2 x}\cdot 2\sin x\cos x\cdot 2\,dx.$$

Writing $b\cdot\sin x = z$, this reduces to

$$8\int_0^b \sin^2 z\left(\frac{2z^3}{b^4} - \frac{2z}{b^2} + \frac{1}{z}\right)dz.$$

If $\sin^2 z$ is expanded into the series

$$\sin^2 z = z^2 - \left(\frac{1}{3!}+\frac{1}{3!}\right)z^4 + \left(\frac{1}{5!}+\frac{1}{3!3!}+\frac{1}{5!}\right)z^2 - \cdots,$$

each term may be integrated separately, the result being,

$$8b^2\{\alpha - \beta b^2 + \gamma b^4 - \cdots\},$$

where

$\alpha = \dfrac{1}{3}$ $= .33333,$

$\beta = \left(\dfrac{1}{4}-\dfrac{1}{3}+\dfrac{1}{4}\right)\left(\dfrac{1}{3!}+\dfrac{1}{3!}\right)$ $= .05556,$

$\gamma = \left(\dfrac{1}{6}-\dfrac{1}{4}+\dfrac{1}{5}\right)\left(\dfrac{1}{5!}+\dfrac{1}{3!3!}+\dfrac{1}{5!}\right)$ $= .00519,5,$

$\delta = \left(\dfrac{1}{8}-\dfrac{1}{5}+\dfrac{1}{6}\right)\left(\dfrac{1}{7!}+\dfrac{1}{5!3!}+\dfrac{1}{3!5!}+\dfrac{1}{7!}\right)$ $= .00029,10,$

$\epsilon = .00001, 076$ $\qquad \eta = .00000, 00053, 5,$

$\zeta = .00000, 0280$ $\qquad \theta = .00000, 00000, 77.$

The total energy scattered by the electron in unit time is therefore

$$(16) \qquad E_s = \frac{8\pi I e^4}{m^2 C^4} \left\{ \alpha - \beta \left(\frac{4\pi a}{\lambda} \right)^2 + \gamma \left(\frac{4\pi a}{\lambda} \right)^4 - \cdots \right\}.$$

When the wave-length is large compared with the radius of the electron, all terms after the first are negligible, in which case

$$(E_s/I)_{a/\lambda=0} = \frac{8\pi}{3} \frac{e^4}{m^2 C^4},$$

as it should. Writing as before N as the number of electrons per unit mass, σ as the scattering coefficient per unit volume and ρ as the density, the mass scattering coefficient of a substance composed of flexible spherical electrons is

$$(17) \qquad \frac{\sigma}{\rho} = \frac{8\pi e^4 N}{m^2 C^4} \left\{ \alpha - \beta \left(\frac{4\pi a}{\lambda} \right)^2 + \gamma \left(\frac{4\pi a}{\lambda} \right)^4 - \cdots \right\}.$$

3. *To calculate the energy scattered by an electron in the form of a thin, flexible, circular ring of electricity.*

In order to account for the electromagnetic mass of a ring electron, Webster[1] and Davisson[2] have shown that the ring must be very thin compared with its diameter. As a result of this fact, the inertia of any element of the ring is practically dependent only upon those parts of the ring immediately adjacent to it. Unless the wave-length is much smaller than the diameter of the electron, therefore, it is permissible to treat the mass of an element of the electronic ring as having a definite value.

Difficulties arise in this calculation, however, from the fact that the electromagnetic mass of an element differs according as the element is accelerated perpendicular or parallel to the tangent to the electronic ring at that point. The effective perpendicular mass depends also upon the speed with which the ring of electricity is rotating. For purposes of calculation I have assumed that the speed with which the electricity in the ring is moving is small compared with the velocity of light, and also that the mass of an element of the electron is independent of the direction of the acceleration. Admittedly the latter assumption makes the calculated value of the scattering only approximate, but it is probable that except possibly for the hardest γ-rays the approximation is close. I have further assumed, as in the case of the sphere, that the ring electron

---

[1] D. L. Webster, *loc. cit.*

[2] Davisson, *loc. cit.*

is flexible, *i. e.*, that the different parts of the charge are free to move relatively to each other. As was pointed out when the flexible sphere was considered, for comparatively long waves such an electron will scatter in practically the same manner as will a rigid electron which is free to rotate about any axis; for very short waves, however, the scattering by the two types of electron will not be exactly the same. The expression derived below for the scattering by a ring electron may therefore be relied upon for any except very short γ-rays.

In Fig. 7, imagine a beam of γ-rays going in the direction $-ZOZ$, and being scattered by an electron of radius a, represented by the heavy ring. Let us first determine the energy scattered in the direction OP, at

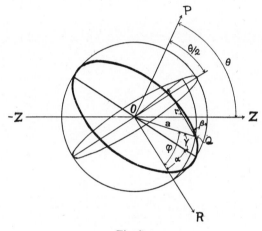

Fig. 7.

an angle θ with OZ. As is evident from the figure, the plane of the electronic ring makes a dihedral angle β with the plane ZOP, and the line of intersection of these two planes makes an angle α with the line OR bisecting the external angle $-ZOP$.

Consider first the energy scattered toward P due to the component of the electric vector which is perpendicular to the plane of the paper. If L is the distance, in the direction OP, at which the scattered radiation is evaluated, the electric displacement due to an element of the electron at $Q(a, \alpha, \beta, \gamma)$ is, in accordance with expression (10),

$$\frac{\eta a d\gamma \cdot A_1 \frac{\eta}{\mu}}{LC^2} \cdot \cos\frac{2\pi}{\lambda}\left(\delta - 2r\sin\frac{\theta}{2}\right).$$

In this statement, η is the charge and μ the mass per unit length of the circumference of the ring, A_1 is the amplitude of the electric vector perpendicular to the plane of the paper, r is the distance from Q to a plane drawn through O perpendicular to OR, and $2\pi\delta/\lambda$ is the phase angle of a ray scattered to P from this plane.

The displacement at P due to the whole electron is therefore

$$\frac{A_1\eta^2 a}{\mu L C^2} \int_0^{2\pi} \cos\frac{2\pi}{\lambda}\left(\delta - 2r\sin\frac{\theta}{2}\right) d\gamma.$$

This quantity is obviously a maximum when $\delta = 0$, so the amplitude at P due to the rays scattered by the whole electron is

$$\frac{A_1\eta^2 a}{\mu L C^2} \int_0^{2\pi} \cos\frac{4\pi}{\lambda}\left(r\sin\frac{\theta}{2}\right) d\gamma,$$

and the corresponding intensity is

(18) $$\frac{cA_1^2\eta^4 a^2}{\mu^2 L^2 C^4} \left\{ \int_0^{2\pi} \cos\frac{4\pi}{\lambda}\left(r\sin\frac{\theta}{2}\right) d\gamma \right\}^2.$$

This expression represents the intensity at P due to an electron with the particular orientation defined by the values of α and β. The probable intensity at P due to polarized rays scattered by an electron at O is the average of this quantity for all values of α and β; i. e.,

$$\frac{cA_1^2\eta^4 a^2}{\mu^2 L^2 C^4} \int_0^\pi \frac{d\alpha}{\pi} \int_0^\pi \frac{d\beta}{\pi} \left\{ \int_0^{2\pi} \cos\frac{4\pi}{\lambda}\left(r\sin\frac{\theta}{2}\right) d\gamma \right\}^2.$$

In a similar manner, the probable intensity at P due to the component of the incident electric vector which is parallel with the plane of the paper is

$$\frac{cA_2^2\eta^4 a^2 \cos^2\theta}{\mu^2 L^2 C^4} \int_0^\pi \frac{d\alpha}{\pi} \int_0^\pi \frac{d\beta}{\pi} \left\{ \int_0^{2\pi} \cos\frac{4\pi}{\lambda}\left(r\sin\frac{\theta}{2}\right) d\gamma \right\}^2,$$

where A_2 is the component of the incident amplitude parallel to the plane of the paper. Since on the average $A_1 = A_2$, and since the intensity of the unpolarized incident beam is $I = c(A_1^2 + A_2^2)$, we may write as the intensity of the beam scattered to P by an electron at O traversed by an unpolarized γ-ray,

$$I_{L,\theta} = \frac{I\eta^4 a^2}{\pi^2 \mu^2 L^2 C^4} \frac{(1+\cos^2\theta)}{2} \int_0^\pi d\alpha \int_0^\pi d\beta \left\{ \int_0^{2\pi} \cos\frac{4\pi}{\lambda}(r\sin\theta/2) d\lambda \right\}^2.$$

Remembering that $2\pi a\eta = e$ and $\eta/\mu = e/m$, the intensity of the beam scattered at the angle θ by an electron is

(19) $$I_{L,\theta} = \frac{Ie^4(1+\cos^2\theta)}{8\pi^4 m^2 L^2 C^4} \int_0^\pi d\alpha \int_0^\pi d\beta \left\{ \int_0^{2\pi} \cos\frac{4\pi}{\lambda}(r\sin\theta/2) d\gamma \right\}^2.$$

To evaluate this expression it is necessary to write r in terms of α, β and γ. In Fig. 7,

$$r = \alpha \cos\varphi$$

and
$$\cos \varphi = \cos \alpha \cos \gamma - \sin \alpha \cos \beta \sin \gamma.$$
Thus
$$r = a (\cos \alpha \cos \gamma - \sin \alpha \cos \beta \sin \gamma).$$
The first integral of equation (19) then becomes
$$F_1 = \int_0^{2\pi} \cos \frac{4\pi a}{\lambda} \left\{ \sin \frac{\theta}{2} (\cos \alpha \cos \gamma - \sin \alpha \cos \beta \sin \gamma) \right\} d\gamma.$$
Substitute
$$k = \frac{4\pi a}{\lambda} \sin \frac{\theta}{2} \cos \alpha,$$
and
$$l = \frac{4\pi a}{\lambda} \sin \theta/2 \sin \alpha \cos \beta.$$
Then
$$F_1 = \int_0^{2\pi} \cos(k \cos \gamma - l \sin \gamma) d\gamma.$$
We may write
$$k \cos \gamma - l \sin \gamma = m \sin (\gamma + \Delta),$$
where m is the maximum value of $k \cos \gamma - l \sin \gamma$, i. e.,
$$m = \sqrt{k^2 + l^2}$$
$$= \frac{4\pi a}{\lambda} \sin \frac{\theta}{2} \sqrt{1 - \sin^2 \alpha \sin^2 \beta},$$
and Δ is the appropriate phase angle. With this substitution,
$$F_1 = \int_0^{2\pi} \cos \{m \sin (\gamma + \Delta)\} d\gamma.$$
Since the integration extends from 0 to 2π, the value of Δ is immaterial, and may therefore be put to equal zero. The integral then becomes,
$$F_1 = \int_0^{2\pi} \cos (m \sin \gamma) d\gamma$$
$$= 2\pi \cdot \frac{1}{\pi} \int_0^{\pi} \cos (m \sin \gamma) d\gamma$$
$$= 2\pi \cdot J_0(m),$$
where
$$J_0(m) = 1 - \frac{m^2}{2^2} + \frac{m^4}{2^2 \cdot 4^2} - \frac{m^6}{2^2 \cdot 4^2 \cdot 6^2} + \cdots,$$
i. e., Bessel's J function of the zero order. Thus
$$F_1 = 2\pi J_0 \left(\frac{4\pi a}{\lambda} \sin \frac{\theta}{2} \sqrt{1 - \sin^2 \alpha \sin^2 \beta} \right).$$

The second integral of expression (19) may be written

$$F_2 = \int_0^\pi F_1^2 d\beta = \int_0^\pi \left\{ 2\pi J_0\left(\frac{4\pi a}{\lambda} \sin\frac{\theta}{2} \sqrt{1 - \sin^2\alpha \sin^2\beta}\right) \right\}^2 d\beta.$$

By substituting

$$\frac{4\pi a}{\lambda} \sin \theta/2 = k$$

and $\sin \alpha = l$ this is reduced to the form

$$F_2 = 4\pi^2 \int_0^\pi J_0^2(k\sqrt{1 - l^2 \sin^2\beta})d\beta.$$

The integral can be evaluated by expansion into a series of the form

$$J_0^2(x) = 1 - Ax^2 + Bx^4 - Cx^6 + \cdots,$$

and integrating term by term. In this series

$$A = \frac{1}{2^2} + \frac{1}{2^2},$$

$$B = \frac{1}{2^2 \cdot 4^2} + \frac{1}{2^2 \cdot 2^2} + \frac{1}{2^2 \cdot 4^2},$$

$$C = \frac{1}{2^2 \cdot 4^2 \cdot 6^2} + \frac{1}{2^2 \cdot 4^2 \cdot 2^2} + \frac{1}{2^2 \cdot 2^2 \cdot 4^2} + \frac{1}{2^2 \cdot 4^2 \cdot 6^2},$$

$$D = \frac{1}{2^2 \cdot 4^2 \cdot 6^2 \cdot 8^2} + \frac{1}{2^2 \cdot 4^2 \cdot 6^2 \cdot 2^2} + \frac{1}{2^2 \cdot 4^2 \cdot 2^2 \cdot 4^2} + \frac{1}{2^2 \cdot 2^2 \cdot 4^2 \cdot 6^2} + \frac{1}{2^2 \cdot 4^2 \cdot 6^2 \cdot 8^2},$$

etc.

Performing the integration we obtain

$$F_2 = 4\pi^3 \{M + Nl + Ol^2 + Pl^3 + \cdots\},$$

where

$$M = J_0^2(k),$$
$$N = \tfrac{1}{2}(Ak^2 - 2Bk^4 + 3Ck^6 - \cdots),$$
$$O = \tfrac{1}{2} \cdot \tfrac{3}{4}(Bk^4 - 3Ck^6 + 6Dk^8 - 10Ek^{10} + \cdots),$$
$$P = \tfrac{1}{2} \cdot \tfrac{3}{4} \cdot \tfrac{5}{6}(Ck^6 - 4Dk^8 + 10Ek^{10} - 20Fk^{12} + \cdots),$$

etc.

The third integral is

$$F^3 = \int_0^\pi F_2 d\alpha = 4\pi^3 \int_0^\pi (M + N \sin^2\alpha + O \sin^4\alpha + \cdots)d\alpha$$
$$= 4\pi^4 (M + \tfrac{1}{2} N + \tfrac{1}{2} \cdot \tfrac{3}{4} O + \tfrac{1}{2} \cdot \tfrac{3}{4} \cdot \tfrac{5}{6} P + \cdots).$$

Substituting the above values of M, N, O, etc., this may be written

$$F^3 = 4\pi^4(1 - \alpha k^2 + \beta k^4 - \gamma k^6 + \cdots),$$

where

$\alpha = (1 - \frac{1}{4})A$ = .37500,

$\beta = (1 - \frac{2}{4} + \frac{1}{4} \cdot \frac{9}{16})B$ = .06006, 0,

$\gamma = (1 - \frac{3}{4} + 3 \cdot \frac{1}{4} \cdot \frac{9}{16} - \frac{1}{4} \cdot \frac{9}{16} \cdot \frac{25}{36})C$ = .00498, 48,

$\delta = (1 - \frac{4}{4} + 6 \cdot \frac{1}{4} \cdot \frac{9}{16} - 4 \cdot \frac{1}{4} \cdot \frac{9}{16} \cdot \frac{25}{36} + \frac{1}{4} \cdot \frac{9}{16} \cdot \frac{25}{36} \cdot \frac{49}{64})D$ = .00025, 060,

$\epsilon = (1 - \frac{5}{4} + 10 \cdot \frac{1}{4} \cdot \frac{9}{16} - 10 \cdot \frac{1}{4} \cdot \frac{9}{16} \cdot \frac{25}{36} + 5 \cdot \frac{1}{4} \cdot \frac{9}{16} \cdot \frac{25}{36} \cdot \frac{49}{64}$
$\qquad - \frac{1}{4} \cdot \frac{9}{16} \cdot \frac{25}{36} \cdot \frac{49}{64} \cdot \frac{81}{100})E$ = .00000, 84241,

ζ = .00000, 02023, 6,

η = .00000, 00036, 51,

θ = .00000, 00000, 5056.

The scattering at the angle θ is therefore

$$(20) \quad I_{L,\theta} = \frac{Ie^4(1 + \cos^2 \theta)}{2m^2L^2C^4}\left\{1 - \alpha\left(\frac{4\pi a}{\lambda}\sin\frac{\theta}{2}\right)^2 + \beta\left(\frac{4\pi a}{\lambda}\sin\frac{\theta}{2}\right)^4 - \cdots\right\}.$$

The rate at which energy is scattered by an electron is obtained by integrating this expression over the surface of a sphere of radius L. That is

$$E_s = \int_0^\pi I_{L,\theta} \cdot 2\pi L^2 \sin\theta d\theta,$$

and since

$$\frac{\sigma}{\rho} = \frac{N}{I} E_s,$$

$$\frac{\sigma}{\rho} = \frac{Ne^4}{2m^2C^4} \int_0^\pi (1 + \cos^2\theta)\sin\theta\left\{1 - \alpha\left(\frac{4\pi a}{\lambda}\sin\frac{\theta}{2}\right)^2 + \beta\left(\frac{4\pi a}{\lambda}\sin\frac{\theta}{2}\right)^4 - \cdots\right\}d\theta.$$

If for $\sin^2(\theta/2)$ we substitute $1/2(1 - \cos\theta)$, this expression is immediately integrable, and we obtain for the mass scattering coefficient,

$$(21) \quad \frac{\sigma}{\rho} = \frac{8\pi}{3}\frac{Ne^4}{m^2C^4}\left(1 - a\left(\frac{a}{\lambda}\right)^2 + b\left(\frac{a}{\lambda}\right)^4 - c\left(\frac{a}{\lambda}\right)^6 + \cdots\right),$$

where

$a = \frac{3}{4}(8\pi^2)(1 + \frac{1}{3})\alpha$ = 29.60881,

$b = \frac{3}{4}(8\pi^2)^2(1 + \frac{1}{3} + \frac{1}{3} + \frac{1}{5})\beta$ = 524.1827,

$c = \frac{3}{4}(8\pi^2)^3(1 + \frac{1}{3} + \frac{3}{3} + \frac{3}{5})\gamma$ = 5,397.801,

$d = \frac{3}{4}(8\pi^2)^4(1 + \frac{1}{3} + \frac{6}{3} + \frac{6}{5} + \frac{1}{5} + \frac{1}{7})\delta$ = 35,619.04,

$e = \frac{3}{4}(8\pi^2)^5(1 + \frac{1}{3} + \frac{10}{3} + \frac{10}{5} + \frac{5}{5} + \frac{5}{7})\epsilon$ = 162,501.7,

$f = \frac{3}{4}(8\pi^2)^6(1 + \frac{1}{3} + \frac{1.5}{3} + \frac{1.5}{5} + \frac{1.5}{5} + \frac{1.5}{7} + \frac{1}{7} + \frac{1}{9})\zeta = 541,970.2,$

$g = 1,377,792,$ $\qquad n^1 = 3,717,000,$

$h = 2,757,220,$ $\qquad o^1 = 2,356,000,$

$i = 4,455,520,$ $\qquad p^1 = 1,334,000,$

$j = 5,935,500,$ $\qquad q^1 = 682,000,$

$k^1 = 6,632,700$ $\qquad r^1 = 318,000,$

$l^1 = 6,311,200,$ $\qquad s^1 = 136,000,$

$m^1 = 5,182,000,$ $\qquad t^1 = 54,000.$

The right hand member of this equation is convergent for all values of a/λ, but the convergence is very slow when λ approaches equality with a. The values of σ/ρ calculated according to this expression for different values of λ/a are shown in Fig. 2 by the solid part of curve *III*.

Elementary considerations suffice to determine the manner in which the scattering by a ring electron depends upon the wave-length when the frequency is very high. If we consider waves shorter than the diameter of the electron, it is apparent that the length of the arc of the ring which may be considered to vibrate as a unit due to the action of the incident beam will be proportional to the wave-length of the incident rays. The amplitude of the beam scattered by such a unit will therefore, by equation 10, be proportional to λe, and the intensity to $\lambda^2 e^2$. Since, however, the total number of such units in each electron will be inversely proportional to the wave-length, the intensity of the beam scattered by the whole electron will be proportional to $\lambda^2 e^2/\lambda$, *i. e.*, proportional to the wave-length. The solid part of curve *III* (Fig. 2) shows, as we should expect, that this relation holds approximately even for waves considerably longer than the diameter of the electron. I have therefore extrapolated curve *III* according to this law for values of λ/a too small for the practical application of formula 21, indicating these approximate values by the broken part of the curve.

RESEARCH LABORATORY,
WESTINGHOUSE LAMP COMPANY,
March 17, 1919.

[1] These values were determined by an approximation formula.

A SENSITIVE MODIFICATION OF THE QUADRANT ELECTROMETER: ITS THEORY AND USE.

By A. H. Compton and K. T. Compton.

Synopsis.

New Quadrant electrometer of high sensitivity, with tilted needle and movable quadrants. Principle. If the needle is given a slight tilt about its long axis and one pair of quadrants is raised or lowered a small distance with respect to the other pair, electrical control forces are set up which add to or subtract from the restoring torque from the suspension. *The theory* of these forces is discussed, the equation giving the sensitiveness in terms of the potential of the needle and the geometrical arrangement of the quadrants is derived and the method of utilizing electrostatic control to the best advantage is described. *Advantages:* High sensitivity, nearly independent of the deflection, and quick adjustment of sensitivity through a great range. *Results.* By using a small needle, 4.5 mm. in radius, with a slight tilt, sensitivities as high as 60,000 mm. per volt have been obtained. Adjustments are difficult at these extreme sensitivities, but the electrometer can usually be used up to 15,000 mm. per volt without undue trouble. Practical suggestions are given regarding needles, suspensions and mirrors.

Quadrant electrometer: needles of mica or aluminum. Technique.
Quadrant electrometer sensitivity; summary of elementary theory.

1. Introduction.

IF a quadrant electrometer is to be used for the quick and accurate measurement of small electric charges or small potential differences it should be very sensitive, it should be as nearly as possible equally sensitive over all parts of the scale, and the moving parts should come quickly to rest when a measurement is being made. If we consider certain possible means of securing high sensitiveness, we shall see that this may be obtained without unduly sacrificing either the uniformity of the scale deflections or the rapidity with which readings may be taken.

According to the simple theory of the quadrant electrometer, if the needle is at a potential V and the two pairs of quadrants are at potentials

v_1 and v_2 respectively, the angle through which the needle is turned is given by

$$k_2\theta = k_1(v_2 - v_1)\left(V - \frac{v_2 + v_1}{2}\right),$$

where $k_2\theta$ is the torque exerted on the needle by the suspending fibre when twisted through an angle θ and $k_1 = R^2/\pi h$, where R is the radius of the needle and h is the distance from the lower to the upper side of the quadrants. In practice, the term $(v_2 + v_1)/2$ may nearly always be neglected in comparison with V, so that the sensitiveness is directly proportional to the potential of the needle. As long as V remains comparatively small this expression is found to be accurate, but when the potential of the needle is large, another term, which is approximately proportional to $V^2\theta$, becomes of prominence. We have then[1]

$$k_2\theta = k_1V(v_2 - v_1) - k_3V^2\theta,$$

and the sensitiveness is

$$S = \frac{\theta}{v_2 - v_1} = \frac{k_1 V}{k_2 + k_3 V^2}. \tag{1}$$

With a given suspension the quantity k_2 is constant, but it is possible to vary k_3 over a large range of positive and negative values by changing the geometrical arrangement of the electrometer needle and quadrants. When k_3 is positive, the needle is held in more stable equilibrium, and there is said to be a "positive electrostatic control"; when k_3 is negative, the needle is held less firmly in position, and there is said to be a "negative electrostatic control." It is evident that, by adjusting the quantity k_3V^2 until it is opposite in sign and nearly equal to k_2, any desired degree of sensitiveness can be attained.

If, as is usually the case in practice, the two pairs of quadrants are not connected directly to a battery, but one pair is connected to an insulated system whose gain of charge is to be measured, there is another factor, called the "inductional electrostatic control," which has to be considered. This control is due to the difference of potential between the quadrants which is set up when the electrometer needle moves from one into the other. The torque produced by this effect is always in a direction to resist the motion of the needle, and may be written as $-k_4 \cdot V^2/C \cdot \theta$, where the constant k_4 depends on the construction of the electrometer and C is the capacity of the system connected with the

---

[1] *Cf.*, *e. g.*, R. Beatty, Electrician, 65, p. 729, 1910 and 69, p. 233, 1912; or Makower and Geiger, "Practical Measurements in Radioactivity," p. 6.

insulated quadrants.[1] The sensitiveness in this case is

$$S = \frac{\theta}{v_2 - v_1} = \frac{k_1 V}{k_2 + k_3 V^2 + k_4 \dfrac{V^2}{C}}, \quad (2)$$

where v_2 and v_1 refer now to the potentials of the quadrants before the needle moves from its zero position,—or their potentials if the needle were prevented from turning. Thus the sensitiveness when one pair of quadrants is insulated is always less than when both are at definite potentials. It is still possible to adjust k_3 in such a manner that the needle borders on instability and the sensitiveness becomes very great, but there is now the difficulty that when the insulated quadrants are grounded the term $k_4 \cdot V^2/C$ vanishes and, if $k_3 V^2$ is greater than k_2, the needle becomes unstable. This difficulty need not be serious, however, if it is possible to adjust k_3 quickly over a sufficiently large range. In any case, the sensitivity is increased by the introduction of the "negative electrostatic control" term $k_3 V^2$.

In order to use a quadrant electrometer at its maximum sensitiveness it is therefore essential to be able to vary the electrostatic control readily over wide limits. A number of practical means of obtaining an adjustable electrostatic control have been discussed by Beatty,[2] and one of these has been used by Parson[3] in his highly sensitive electrometer. When the needle of Parson's electrometer is in its zero position it rests over a slit, whose width can be varied at will. The slit has the effect of repelling the needle, so that by widening the slit the needle may be made to approach a condition of instability. The trouble with this arrangement is that the force on the needle, due to the slit, is not proportional to the angle through which the needle has turned, and the sensitiveness therefore varies over different parts of the scale. Thus, while the electrometer is well adapted to "detection" and use in measurements by a compensation or "null" method, it is not so suitable to use in the usual work where deflections are measured.

We have found that an electrostatic control may be obtained by slightly tilting the electrometer needle about its long axis, and at the same time moving one of the quadrants a little above or below the plane of the other quadrants. This type of control possesses the important advantage of giving nearly constant sensitiveness over a large range of deflections and of permitting quick adjustment to any desired amount of control, either positive or negative.

[1] J. J. Thomson, Phil. Mag., 46, p. 536, 1898; Makower and Geiger, *loc. cit.*
[2] *Loc. cit.*
[3] A. L. Parson, PHYS. REV., VI., p. 390, 1915.

2. Theory.

Let one pair of quadrants be displaced vertically a distance δ with respect of the other. The needle is suspended with its center at a distance p above the horizontal plane of symmetry and is tilted about its long axis so that its slope with respect to the horizontal plane is s. Evidently, as the needle deflects through an angle θ, it approaches the sides a and c (Fig. 1) and recedes from the sides b and d of the quadrants, the effect being to change the capacity of the system and therefore to give rise to forces tending to produce or resist further deflection. The magnitude of these forces, which combine to produce the electrostatic control, we shall proceed to calculate.

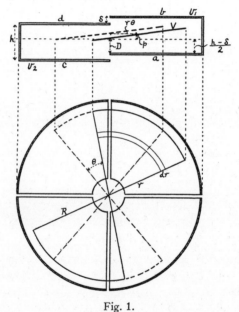

Fig. 1.

Consider first that portion of the needle lying between distances r and $r + dr$ from its center. The distance D between this element and the edge of the side a is

$$D = \frac{h - \delta}{2} + p - sr\theta,$$

where s is the slope $dD/rd\theta$ of the needle and θ is the angle of deflection of the needle from its zero position. The change in the charge induced on the side a by that part of the needle between r and $r + dr$ in turning through an additional infinitesimal angle $d\theta$ is $dq = \sigma r d\theta dr$. The surface density σ is, by Coulomb's law, $\sigma = (V - v_1)/4\pi D$, where V is the potential of the needle and v_1 that of the side a of the quadrant.

Thus
$$\frac{dq}{d\theta} = \frac{V - v_1}{4\pi \left(\frac{h}{2} - \frac{\delta}{2} + p - sr\theta\right)} r\,dr.$$

The change dW in the potential energy of this part of the system is $dW = \frac{1}{2}(V - v_1)dq$. Thus

$$dT_a = \frac{dW}{d\theta} = \frac{1}{2}(V - v_1)\frac{dq}{d\theta}$$

is the torque exerted on the portion of the needle considered by the side a. The total torque exerted on the needle by the side a is therefore

$$T_a = \int_0^R \frac{(V - v_1)^2 r\,dr}{4\pi(h - \delta + 2p - 2sr\theta)},$$

which, when integrated, takes the form of the rapidly converging series

$$T_a = \frac{R^2(V - v_1)^2}{16\pi}\left[\frac{4}{2}\frac{1}{(h - \delta + 2p)} + \frac{8}{3}\frac{Rs\theta}{(h - \delta + 2p)^2}\right.$$
$$\left. + \frac{16}{4}\frac{R^2 s^2 \theta^2}{(h - \delta + 2p)^3} + \cdots\right].$$

The torque due to the other faces may be similarly calculated, whence the total torque due to both ends of the needle is

$$T = 2(T_a + T_b + T_c + T_d),$$

which may be written in the most convenient form by grouping separately the coefficients of different powers of θ. Neglecting the term in $(v_2 + v_1)/2$ in comparison with those in V^2 and $V(v_2 - v_1)$, we have

$$T = \frac{R^2}{\pi h}\left\{V(v_2 - v_1)\left[\left(1 + \frac{\delta^2}{h^2} + 4\frac{p^2}{h^2} + \cdots\right) + \cdots\right]\right.$$
$$+ V^2\left[-\frac{4\delta p}{h^2}\left(1 + 2\frac{\delta^2}{h^2} + 8\frac{p^2}{h^2} + \cdots\right)\right.$$
$$+ \frac{8Rs\delta}{3h^2}\left(1 + 2\frac{\delta^2}{h^2} + 8\frac{p^2}{h^2} + \cdots\right)\theta$$
$$\left.\left.+ \frac{64R^3 s^3 \delta}{5h^4}\left(1 + 5\frac{\delta^2}{h^2} + 20\frac{p^2}{h^2} + \cdots\right)\theta^3 + \cdots\right]\right\},$$

which may be written

$$T = k_1 V(v_2 - v_1) - k' V^2 \delta p - k_3 V^2 \theta - k_5 V^2 \theta^3.$$

This equation for torque due to electrical forces includes all terms whose influence on the action of an electrometer of the type discussed may be detected. The values of the important constants k_3 and k_5 may be taken to be

$$k_3 = -\frac{8R^3s\delta}{3\pi h^3} \quad \text{and} \quad k_5 = -\frac{64R^5s^3\delta}{5\pi h^5}, \qquad (3)$$

and it is to be noted that both must have the same sign.

In equilibrium, this torque must be balanced by that due to the suspension, *i. e.*,

$$k_2\theta = k_1V(v_2 - v_1) - k'V^2\delta p - k_3V^2\theta - k_5V^2\theta^3. \qquad (4)$$

The term $-k'V^2\delta p$ is of interest because it shows the cause of the deflection from the zero position generally observed when the needle is charged while the quadrants remain grounded. Other possible causes are contact difference of potential between the quadrants due to solder or imperfect "finish" and a tilt of the needle about its short axis,—this being equivalent, as far as the equation is concerned, to a change of θ. With care in construction these latter causes are not effective. In order to use the electrometer, this zero deflection must be prevented, which can be done by reducing either p or δ to zero. It is advantageous to adjust p rather than δ to zero, for by so doing the values of the constants k_3 and k_5 are not appreciably affected. If the zero shift is thus eliminated for one potential V of the needle, it should not appear for any other needle potential unless one or both of the other causes of zero shift are effective. Therefore, when the height of the center of the needle is adjusted to its position ($p = 0$) of symmetry between the quadrants, as judged by the absence of zero shift, the term $k'V^2\delta p$ of equation (3) vanishes and we obtain for the sensitiveness

$$S = \frac{\theta}{v_2 - v_1} = \frac{k_1V}{k_2 + k_3V^2 + k_5V^2\theta^2}. \qquad (5)$$

If δ and s are of the same sign, as in Fig. 1, k_3 and k_5 are negative and the sensitiveness is increased, *i. e.*, there is a negative electrostatic control. Similarly if δ and s are of opposite sign, the electrostatic control is positive. Thus, with a given inclination of the needle, the control can be varied through wide limits by moving one pair of quadrants up and down. In practice we have used three of the quadrants in their position of symmetry, in which case the constants k_3 and k_5 have just half the values assigned in equations (3). With negative control, it is theoretically possible to obtain any sensitiveness between zero and infinity, while with positive control there is, for any given value of the constants, a maximum sensitiveness given by placing the derivative of S with respect to V equal to zero. If we neglect the term in k_5, which is very small and does not take a value until the needle has deflected, we obtain for the maximum sensitiveness under positive control

$$+S_{max} = \frac{k_1}{2k_2}\sqrt{\frac{k_2}{k_3}}$$

when (6)

$$V_m = \sqrt{\frac{k_2}{k_3}}.$$

Practical difficulties of a mechanical nature prevent the attainment of infinite sensitiveness under negative control. These difficulties will be discussed later.

When the electrometer is thus made very sensitive a relatively long time is required for the needle to come to its position of equilibrium. While the needle is in motion, the air resistance is proportional to the linear velocity of the needle, to its area, and approximately inversely to the distance between the needle and the quadrants. The torque due to air damping may therefore be written $k_6 \cdot R^4/h \cdot d\theta/dt$, where $d\theta/dt$ is the angular velocity of the needle. If I is the moment of inertia of the needle, the equation of its motion is therefore

$$k_1 V(v_2 - v_1) - (k_2 + k_3 V^2 + k_5 V^2 \theta^2)\theta - k_6 \frac{R^4}{h}\frac{d\theta}{dt} = I\frac{d^2\theta}{dt^2}.$$

By using needles made of thin mica sputtered with platinum (suggested to us by Professor Pegram of Columbia University) or made of thin aluminium leaf, the moment of inertia term may be made negligibly small in comparison with the other terms. In this case the time T required for the needle to return to zero after a deflection is evidently approximately proportional to

$$\frac{\theta}{\frac{d\theta}{dt}} = \frac{k_6 R^4}{h(k_2 + k_3 V^2 + k_5 V^2 \theta^2)}$$

$$= S\frac{k_6 R^4}{hk_1 V} = \frac{\pi k_6 R^2 S}{V},$$

by definition of the constant k_1. Thus

$$T \sim S\frac{R^2}{V}$$

and (7)

$$S \sim T\frac{V}{R^2}.$$

Equations (7) illustrate the well-known fact that the most advantageous combination of high sensitiveness with short period may be attained with small needles. The lower limit to which the needle may profitably be reduced is limited only by the fact that the needle must turn a mirror of sufficient size to give a well-defined spot of light on the scale. By taking into account the resolving power of the mirror, it can easily be shown that the minimum potential difference which can be detected is proportional to $1/R$ instead of to $1/R^2$, if dimensions of needle and mirror are maintained in a constant ratio. For detecting charge, there is further advantage in reducing the size of the needle and quadrants because the capacity of the electrometer is roughly proportional to the linear dimensions of the needle and quadrants.

3. Design.

Equations (5) and (7) give us information essential to the design of an electrometer to operate in the most satisfactory manner. We have used the electrometer in two sizes, one with a needle of 7 mm. radius and the other with a needle of 4.5 mm. radius. The former is rather easier to operate and adjust, while the latter gives greater sensitiveness for a given period. For very delicate work the size can be profitably reduced still further, and we understand that this has been done by Professor Pegram.

As to the use of the electrostatic control, there are a number of factors to be considered. The presence of the term $k_5 V^2 \theta^2$ in equation (5) shows that the sensitiveness is not the same for all values of θ, i. e., over all parts of the scale, though this term is not important except at very high sensitivities. It always tends to increase the control, whether positive or negative, as the electrometer deflects from the zero position. Thus, with positive control the sensitiveness decreases as the needle deflects from zero, and the needle cannot be put into unstable equilibrium. With negative control the sensitiveness increases as the needle deflects from zero and, if the deflection is too large, the needle may move past the point of instability and turn through an angle of 90° to a new position of equilibrium under positive control. Thus the size of the scale deflection which can be used decreases as the sensitiveness increases. This is a disadvantage, but it need not be serious if the design is guided by the following considerations.

Since these difficulties are due to the term in k_5, they are reduced to a minimum by reducing k_5 in relation to k_3. By equations (3) it is seen that the ratio

$$\frac{k_5}{k_3} = \frac{24R^2s^2}{h^2}.$$

It is therefore advantageous to make the tilt s of the needle small and the height h of the quadrants large. In other words, the desired amount of electrostatic control, given by the term k_3V^2, may be most advantageously obtained with a small tilt s of the needle and a relatively large displacement δ of the quadrant. In an actual case in which k_3 was about ten times larger than k_5, the action of the electrometer was apparently independent of the k_5 term for sensitivities below 15,000 mm. per volt, was not seriously disturbed up to 30,000 mm. per volt for deflections less than 100 mm., while at 60,000 mm. per volt the influence of this term was so serious that the electrometer could only be used as a "null" instrument.

Furthermore it is doubly advantageous to have both k_3 and k_5 small, so that the high sensitiveness arises largely from the term $k_1/k_2 \cdot V$. In the first place the high sensitiveness is then reached with little disturbance from the $k_5V^2\theta^2$ term and in the second place the greatest sensitiveness in relation to the period is obtained owing to the larger value of V attainable, as shown by equation (7). It is better, therefore, to use a suspending fiber whose constant k_2 is sufficiently small to permit the attainment of fairly high sensitiveness without electrostatic control and then to attain very high sensitiveness by applying a small negative control than to neutralize the torque of a strong fiber by a very strong control.

It is best, therefore, to use a fairly delicate suspension, to give the needle only a slight tilt and to displace the quadrants only enough to make the electrostatic control important for relatively high potentials of the needle. For example, we usually use a sputtered quartz suspension of such size that the little hook cemented on its end will oscillate with a period between 0.5 sec. and 1.0 sec. The needle is given just enough tilt to be detected by inspection. The movable quadrant is displaced about 0.15 mm. in the direction to produce negative control. Under these conditions extreme sensitivities are reached with needle potentials of 75 or 100 volts. If the electrometer is to be used at high sensitivity for work where the quadrants are to be insulated for considerable periods of time, as in getting a continuous photographic record of an X-ray spectrum, it is better to use a very fine suspension and slight positive control, for the zero position of the needle is then more stable.

The remaining features in the design of an electrometer of this type are obvious: a convenient means of adjusting the height of a quadrant from outside of the electrometer case and a means of adjusting the height of the needle while the potential source is connected with it. Both adjustments must be delicate and free from "lost motion."

4. Typical Results.

Figs. 2 and 3 show the action of the electrometer with various degrees of positive and negative control, respectively. Each represents results obtained with a given suspension and a given tilt of the needle—the various degrees of control being obtained by adjusting the moveable quadrant. If equation (1) is written in the form

$$S = \frac{aV}{1 - bV^2}, \tag{8}$$

the constant b determines the "electrostatic control."

Curves *1*, *2*, *3* and *4* are all coincident at the origin where, obviously, the "control" term vanishes. The straight dotted line represents a

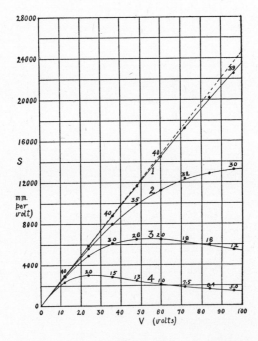

Fig. 2.

condition of zero electrostatic control in which the sensitiveness is strictly proportional to the potential of the needle. This condition was not quite reached in the measurements shown in curve *1*. Curve *2* was obtained by raising the moveable quadrant 0.08 mm., *i. e.*, a quarter turn of the adjusting screw, and readjusting the height of the needle. Curves *3* and *4* indicate further increased amounts of "control." The numbers above the curves represent the period of swing, being the time required to come to within 1 mm. of the resting point after a 50-mm. deflection. It is seen that the period is independent of the sensitiveness when there is no electrostatic control, but decreases as the control increases.

Curves 5, 6, 7, 8, 9 represent various degrees of "negative control," using a stronger suspension than was used in the experiments on "positive control." In this case curve 5 represents zero control, curve 6 the control obtained by lowering the moveable quadrant 0.08 mm., etc. In each case there is a critical potential V_c above which the needle cannot be kept in equilibrium and near which the sensitiveness is very high. This

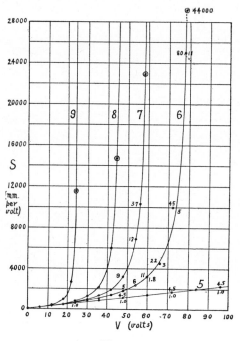

Fig. 3.

unstable condition is approached very rapidly when the control is strong as shown in curve 9, where an increase of one fourth volt in the needle potential 24 volts would double the sensitivity. In such a condition the needle is very hard to adjust and is sensitive to movements of the support. For this reason also it is best in practice to use relatively small electrostatic control and relatively high needle potentials, as in curve 6. The highest sensitiveness at which the zero point was sufficiently steady to permit measurements to be made is indicated by a circle in each case. The figures below the curves give the times required to make half the deflection, and show that most of the time of deflection is taken up with the final slow drift.

In both sets of curves the solid curves represent graphs of equation (8), with values of the constants indicated in Table I. The constants were

TABLE I.

Curve.	a.	b.	V_c, Volts.	Curve.	a.	b.	V_c, Volts.
1	245.0	0.000004		6	23.7	− 0.000152	81.2
2	245.0	0.000085		7	23.7	− 0.000279	59.7
3	245.0	0.000355		8	23.7	− 0.000475	45.8
4	245.0	0.001650		9	23.7	− 0.001650	24.6
5	23.7	0.000000	∞				

chosen to give best agreement with the experimental measurements, which are recorded as dots. The needle was of 4.5-mm. radius, in quadrants of 6.5 mm. radius. The capacity of the instrument was 10.4 cm. Exactly similar curves were obtained with a larger instrument with a needle of 7-mm. radius and a capacity of 12.5 cm., except that the periods of swing were somewhat longer.

5. Practical Hints.

Needles.—We have used mica and aluminium leaf needles, the former being more difficult to make but more satisfactory in operation because of greater rigidity and less air-damping. The mica needles are cut from the thinnest mica by scratching with a sharp needle around the outline of a brass template, cut to the desired shape and pressed firmly on the mica sheet. The fine wire stem is fastened to the needle with sealing wax and the whole system rendered conducting by sputtering beneath a platinum cathode *in vacuo*. While sputtering, the stem must be supported to prevent its falling over if the wax is softened by the discharge. An alternative procedure is to sputter the needle alone, then mount the stem with a touch of sealing wax, and make conducting contact across the wax by a tiny strip of gold leaf, cemented with india ink. The aluminium needles are easily cut from thin leaf held between a metal template and a flat piece of rubber, such as tire or shoe sole rubber, while a sharp-pointed knife blade is passed around the outline. The stem is mounted with soft wax, and is pressed against the needle so as to insure contact. This mounting and the adjustment of the "tilt" is facilitated by the use of a flat piece of metal which may be heated from underneath and against which the needle, held by the stem in a pair of tweezers, may be pressed. If the needle is irregularly warped, difficulties may be encountered in making adjustments and securing uniform scale deflections.

Suspensions.—Quartz fibers, 1 to 2 cm. long are mounted on platinum or silver hooks with sealing wax, and sputtered to obtain a conducting platinum coat. For use with "negative control," these fibers should be the finest ones that can be drawn in the oxy-gas flame by the bow and arrow method, or the coarsest ones that can be drawn out by the flame itself. For use with "positive control," the very finest fibers may be used, the limit being imposed only by the strength requisite for the support of the needle and mirror. The conductivity of the sputtered coat gradually diminishes and the suspensions usually require re-sputtering after about a year of service. If suspensions break, it is usually at the point of joining to the hook. They may then be easily repaired by dipping the end of the hook several times into india ink and then touching with the wet end the free end of the suspension. They are held together at once, and after drying for two or three minutes the suspension is apparently as strong as ever, and retains its conducting properties.

Mirrors.—When working at high sensitivity, the instrument is so much over damped that the moment of inertia may be made quite large without appreciably increasing the period. Thus the mirrors may be larger than might be supposed. We use mirrors with as much as 10 sq. mm. area on the 4.5 and 7.0 mm. instruments.

Ease of Adjustment.—The most serious sources of difficulty appear to be irregularities in the needle or quadrants and contact difference of potential between quadrants. Thus with some instruments it is easy to reach 10,000 or 15,000 mm. per volt sensitivity and in others difficult to reach 10,000 mm. per volt, due to accidental differences in construction. Among the seven or eight instruments of this type in use in our laboratories we have never had difficulty in obtaining 5,000 mm. per volt at the first trial, and usually count on being able to work between 5,000 and 10,000 mm. per volt without trouble. At higher sensitivities more trouble is experienced with zero drift. Difficulty due to contact difference of potential between quadrants may be identified by the failure to secure identical deflections from the zero, with quadrants earthed, when the sign of the potential of the needle is reversed, for this effect is proportional to the first power of the needle potential while all effects due to lack of symmetry depend on the square of the potential and are therefore independent of its sign. If such contact difference of potential is found to be troublesome, it may be removed by cleaning or compensated by a small potential permanently applied to the "earthed" quadrants.

By taking the precautions suggested in this paper we have been able to employ to advantage sensitiveness as high as 50,000 mm. per volt, at 1 meter scale distance.

Valuable suggestions in the design of the electrometer were made by Professor H. L. Cooke, to whom we are greatly indebted.

A. H. C., RESEARCH LABORATORY,
    WESTINGHOUSE LAMP COMPANY.
K. T. C., PALMER PHYSICAL LABORATORY,
    PRINCETON UNIVERSITY.

THE SIZE AND SHAPE OF THE ELECTRON.

By Arthur H. Compton.

Synopsis.

In this paper it is pointed out that Sir J. J. Thomson's explanation of the fluorescent absorption of X-rays on the basis of incident radiation consisting of short double pulses, taken in connection with Moseley's law relating atomic number and the wave-length of characteristic X-radiation, leads to Owen's experimental law for the fluorescent absorption of moderately soft X-rays, namely that $\tau/\nu = KN^4\lambda^3$, τ/ν being the atomic fluorescent absorption coefficient, N the atomic number, λ the wave-length of the incident radiation, and K a constant. Thus, though Thomson's explanation cannot be considered complete, it affords a provisional theoretical basis for Owen's law which has heretofore been lacking.

It is then shown that the law of absorption which holds for moderate frequencies must be modified in order to apply at very high frequencies in view of the fact, shown in a previous paper, that the wave-length becomes comparable with the radius of the electron. For absorbing elements in which excess scattering of the primary beam is inappreciable, the following law of absorption is derived: $\mu/\nu = K\varphi N^4\lambda^3 + \sigma/\nu$ where μ/ν is the atomic total absorption coefficient, σ/ν is the atomic scattering coefficient, and φ and σ are functions of the ratio of the wave-length to the radius of the electron which are evaluated in the paper.

It is found that this formula agrees satisfactorily with the experimental values for the absorption of high frequency radiation in aluminium, the only substance for which the requisite data are available to make an adequate test, if the electron is taken to be a ring of radius $(1.85 \pm .05) \times 10^{-10}$ cm. This result is in good agreement with the value 2×10^{-10} cm. previously estimated by the writer on the basis of measurements of the scattering of high frequency radiation.

II. The Absorption of High Frequency Radiation.

IN the first paper of this series[1] it was pointed out that the experimental observations on the scattering of high frequency radiation by matter could be explained only on the hypothesis that the radius of the electron is comparable with the wave-length of hard γ-rays. The phenomena of scattering were found to be quantitatively accounted for, within the probable errors of observation, if the electron was considered to be a flexible ring of electricity with a radius of 2×10^{-10} cm. The present discussion will deal chiefly with the modifications to be expected in the law of absorption of high frequency radiation if the electron is considered to have appreciable dimensions. Before studying these modifications, however, it will be instructive to consider the law of absorption for wavelengths which are long compared with the radius of the electron.

[1] A. H. Compton, Phys. Rev., 14, 20 (1919).

The absorption of high frequency radiation is due to two independent processes. The more important of these is usually the energy absorbed in exciting corpuscular and fluorescent radiation. There is always, however, a certain amount of energy removed from the primary beam because of scattering by the electrons in the absorbing screen. Thus the total absorption coefficient may be written as

$$\mu = \tau + \sigma,$$

where τ represents the fluorescent absorption, and σ is the absorption due to scattering. In order to compare directly the absorption by different elements, it is customary to consider the "atomic absorption coefficient," μ/ν, where ν is the number of atoms per unit volume of the absorbing screen and μ is the linear absorption coefficient. We may then write

$$\frac{\mu}{\nu} = \frac{\tau}{\nu} + \frac{\sigma}{\nu}.$$

Owen has shown[1] that the atomic fluorescent absorption coefficient follows the experimental law,

(1) $$\frac{\tau}{\nu} = KN^4\lambda^3,$$

where N is the atomic number of the absorber, λ is the wave-length of the incident rays, and K is a constant over certain ranges but changes abruptly when λ passes the critical wave-length required to excite a characteristic radiation in the absorber. The quantity σ/ν has been shown by Barkla and Dunlop[2] to be calculable for the lighter elements and for moderately hard X-rays according to Thomson's formula,

(2) $$\frac{\sigma_0}{\nu} = \frac{8\pi}{3}\frac{e^4 N}{m^2 C^4},$$

where e is the charge and m the mass of an electron and C the velocity of light. The total atomic absorption coefficient, for the elements of low atomic weight and for X-rays of moderately short wave-length, obeys therefore the experimental law,

(3) $$\frac{\mu}{\nu} = KN^4\lambda^3 + \sigma_0/\nu.$$

The absorption which is expressed by relation (1) is generally supposed to be due chiefly to a transformation of the incident energy into fluorescent radiation of longer wave-length. As in the similar case of ordinary light, no satisfactory explanation of this fluorescent absorption has been

[1] E. A. Owen, Proc. Roy. Soc., 94, 522, 1918.
[2] C. G. Barkla and J. G. Dunlop, Phil. Mag., 31, 222, 1916.

proposed. Sir J. J. Thomson has suggested a partial explanation, however, on the basis of an incident X-ray pulse of special wave-form, which leads to the experimental law for the variation of the fluorescent absorption with the wave-length of the incident rays and with the atomic number of the absorber. He shows[1] that if the pulse consists of an electric intensity X through a distance d followed by an intensity $-X$ for a distance d, after which the field due to the pulse vanishes, and if the distance $2d$ is short compared with the wave-length of the radiation excited by the absorbing electron, the atomic fluorescent absorption coefficient is

$$(4) \qquad \frac{\tau}{\nu} = 4\pi^3 \frac{e^2}{mC^2} d^3 \sum_k \frac{n_k}{\lambda_k^2}.$$

In this expression $2d$ may be taken as the wave-length of the incident radiation, and n_k is the number of electrons per atom of the type k, the wave-length of whose free radiation is λ_k.

Moseley has shown that the square root of the frequency of either the K or the L characteristic X-radiation is nearly proportional to the atomic number of the radiator. The same relation holds for Siegbahn's M radiation. We may therefore write $1/\lambda_k^2 = c_k N^4$, where N is the atomic number of the absorbing element, and the constant c_k has different values for the different types of characteristic radiation, but it is the same for all elements. The factor $\Sigma n_k/\lambda_k^2$ may therefore be written as $N^4 \Sigma c_k n_k$. Since the number of the electrons of any type k is probably the same for all the elements which have electrons of this type, and since those terms in the summation which are due to the most rigidly bound electrons are the most important, the factor $\Sigma c_k n_k$ may be considered to be practically constant for all elements except possibly the very light ones, which may not possess electrons of the same high frequency types as do the heavier elements. Writing also for the thickness $2d$ of the incident X-ray pulse its equivalent λ, Thomson's expression for the fluorescent absorption coefficient becomes:

$$(5) \qquad \frac{\tau}{\nu} = K_1 N^4 \lambda^3,$$

where

$$K_1 = \frac{2}{\pi^3} \frac{e^2}{mC^2} \Sigma c_k n_k.$$

As thus stated, Thomson's result is identical in form with the experimental law expressed by equation (1).

The incompleteness of this solution of the problem is evident, however,

[1] Sir J. J. Thomson, Conduction of Electricity through Gases, 2d ed., p. 325.

from the following considerations. In the first place, the solution by hypothesis does not apply if the natural period of any of the electrons in the absorbing material is comparable with or shorter than the period of the incident radiation. That is, the result is derived strictly for only the case where the wave-length of the incident radiation is considerably shorter than the wave-length of the K radiation characteristic of the absorbing material. In the second place, the value of the coefficient K_1 is too great by a factor of about 10.[1] In the third place, the basic hypothesis of incident X-rays consisting of very short pulses is inconsistent[2] with the fact that crystal analysis shows no X-radiation of frequency higher than that given by the relation $h\nu = eV$.

The exact similarity of form of the two expressions, and the fact that even the coefficient K_1 is not far from the proper order of magnitude, are nevertheless coincidences too remarkable not to have some physical significance. Thus, for example, we could hardly expect such an agreement if this absorption were not really due principally to fluorescent transformations. In any case this explanation affords a provisional theoretical basis for Owen's Law which has heretofore been lacking.

Expression (2) represents the scattering coefficient if the electrons in the atom are far enough apart to scatter independently, and if the wave-length is great enough for the electrons to act sensibly as point charges. The first of these conditions has been shown by the experiments of Barkla and Dunlop to be satisfied for X-rays of ordinary hardness in the case of absorbing elements of lower atomic weight than copper. For the heavier elements they find the scattering to be greater than thus calculated. This is probably due chiefly to a grouping of the electrons so close together in the heavy atoms that the waves scattered by the different electrons are nearly in the same phase. The increased scattering may also be partly due to the fact that there are in the heavier atoms certain electrons whose natural frequency is near that of ordinary X-rays. The second condition, that the electron must be small compared with the wave-length of the incident beam of X-rays, has been shown in the first paper of this series to hold for moderately hard radiation but not to hold if the wave-length of the incident rays is shorter than about 0.3 A.U.

In order to obtain an expression for the total absorption which will be valid at shorter wave-lengths, two modifications must be made in formula (3). In the first place, as just suggested, the value of the atomic scattering cannot be considered constant, but must decrease when the

---

[1] If, however, as suggested by Thomson's recent theory of atomic structure (Phil. Mag. 37, 419, 1919), not all the atoms possess electrons of a given type K, it is possible that K_1 may be equal to K.

[2] Cf. D. L. Webster, Phys. Rev., 7, 609, 1916.

wave-length becomes comparable with the diameter of the electron. If the electron is assumed to have the form of a flexible ring of electricity, it has been shown in the first paper of this series that the scattering will be expressed by the formula

$$(6) \quad \frac{\sigma}{\nu} = \frac{\sigma_0}{\nu}\left\{1 - 29.61\left(\frac{a}{\lambda}\right)^2 + 524.2\left(\frac{a}{\lambda}\right)^4 - 5{,}398\left(\frac{a}{\lambda}\right)^6 + \cdots\right\},$$

where σ_0/ν is the atomic scattering coefficient as expressed by equation (2), a is the radius of the electronic ring, and λ is the wave-length of the incident radiation. The values of the coefficients of the powers of a/λ have been given in detail in the previous paper.[1]

The second modification concerns the coefficient of λ^3 in the term representing the fluorescent absorption. In the derivation of Thomson's expression (4) for the fluorescent absorption, he shows that the energy absorbed from the incident ray by an electron is proportional to the square of the acceleration to which the electron is subject. It would seem that this relation must hold in whatever manner the fluorescent absorption is calculated. The acceleration of a comparatively large electron will be less than that of a small electron of the same mass, however, when both are traversed by X-rays of the same intensity, since in the former case the phase of the incident ray will not be the same at all parts of the electron. The fluorescent absorption due to a large electron will therefore be less than the value $KN^4\lambda^3$ by the factor,

$$\varphi = \left\{\frac{\text{acceleration of large electron}}{\text{acceleration of small electron}}\right\}^2.$$

If the large electron is assumed to have the form of a ring, its accelera-

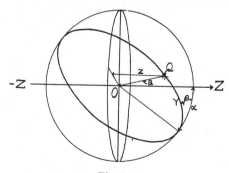

Fig. 1.

tion when traversed by an X-ray may be calculated in the following manner. Let us consider the motion of the electron represented by the heavy ring in Fig. 1, when traversed by an electromagnetic wave propa-

[1] A. H. Compton, *loc. cit.*, p. 42.

gated in the direction $-ZOZ$. The electric intensity at a given instant at the point $Q(a, \alpha, \beta, \gamma)$ on the ring is

$$X = A \cos 2\pi \left(\frac{\delta - Z}{\lambda}\right),$$

where A is the amplitude of the incident electric intensity, $2\pi\delta/\lambda$ is the phase angle at the point O and z is the Z coördinate of the point Q. But

$$z = a (\cos \alpha \cos \gamma + \sin \alpha \cos \beta \sin \gamma),$$

where a is the radius of the electron. The acceleration of the electron at this instant is therefore

$$\frac{Aa\eta}{m} \int_0^{2\pi} \cos \frac{2\pi}{\lambda} \{\delta - a (\cos \alpha \cos \gamma + \sin \alpha \cos \beta \sin \gamma)\} d\gamma,$$

where η is the charge per unit length along the circumference of the electron, and m is the electron's mass.

This integral is obviously a maximum when δ is zero. The amplitude of the acceleration is therefore,

$$f = \frac{Aa\eta}{m} \int_0^{2\pi} \cos \left\{ \frac{2\pi a}{\lambda} (\cos \alpha \cos \gamma + \sin \alpha \cos \beta \sin \gamma) \right\} d\gamma.$$

This quantity may be written

$$f = C_1 \int_0^{2\pi} \cos (k \cos \gamma + l \sin \gamma) d\gamma,$$

where

$$C_1 = Aa\eta/m,$$

$$k = 2\pi \frac{a}{\lambda} \cos \alpha,$$

$$l = 2\pi \frac{a}{\lambda} \sin \alpha \cos \beta,$$

or

$$f = C_1 \int_0^{2\pi} \cos \{M \sin (\gamma + \Delta)\} d\gamma,$$

where $M = \sqrt{k^2 + l^2}$ and Δ is the appropriate phase angle. Since the integration extends from 0 to 2π, the value of Δ is immaterial, and may therefore be put equal to zero. The integral then becomes:

$$f = C_1 \int_0^{2\pi} \cos (M \sin \gamma) d\gamma$$

$$= 2\pi C_1 J_0(M),$$

where

$$J_0 M = 1 - \frac{M^2}{2^2} + \frac{M^4}{2^2 4^2} - \frac{M^6}{2^2 4^2 6^2} + \cdots,$$

i.e., Bessel's J function of the zero order. Substituting

$$M = \frac{2\pi a}{\lambda}\sqrt{1 - \sin^2\alpha\cos^2\beta}$$

and

$$2\pi C_1 = \frac{A}{m} \cdot 2\pi a\eta = Ae/m,$$

where e is the total charge on the electron, we obtain

$$f = \frac{Ae}{m} J_0\left(\frac{2\pi a}{\lambda}\sqrt{1 - \sin^2\alpha\cos^2\beta}\right).$$

The ratio of the acceleration of the ring electron to that of a sensibly point charge electron is therefore

$$J_0\left(\frac{2\pi a}{\lambda}\sqrt{1 - \sin^2\alpha\cos^2\beta}\right).$$

The required value of φ is obviously the mean square of this quantity averaged over all angles α and β at which the ring electron may be oriented, *i. e.*,

$$\varphi = \frac{1}{\pi^2}\int_0^\pi\int_0^\pi J_0^2\left(\frac{2\pi a}{\lambda}\sqrt{1 - \sin^2\alpha\cos^2\beta}\right) d\alpha d\beta.$$

This is the same integral as that which enters into the calculation of the scattering by a ring electron, the solution being

$$\varphi = 1 - \alpha\left(\frac{2\pi a}{\lambda}\right)^2 + \beta\left(\frac{2\pi a}{\lambda}\right)^4 - \gamma\left(\frac{2\pi a}{\lambda}\right)^6 + \cdots,$$

where the constants $\alpha, \beta, \gamma \cdots$ have the values given on page (42) of the first paper of this series. For purposes of calculation this function may be more conveniently expressed as

$$\varphi = 1 - n\frac{a^2}{\lambda^2} + o\frac{a^4}{\lambda^4} - p\frac{a^6}{\lambda^6} + \cdots$$

where

$n = (2\pi)^2\alpha = 14.8044,$	$t = 545.4744,$	$z = .8111,$
$o = (2\pi)^4\beta = 93.6041,$	$u = 301.8218,$	$a = .1535,$
$p = \cdots = 306.6932,$	$v = 133.5450,$	$b = .0254,$
$q = 608.7241,$	$w = 48.3256,$	$c = .0037,$
$r = 807.8924,$	$x = 14.5661,$	$d = .0005,$
$s = 766.5255,$	$y = 3.7134,$	$e = ..0001.$

The values of φ, and of the ratio σ/σ_0 of the scattering by a ring electron to that of an electron of negligible size, are shown for different values of λ/a by curves *I* and *II* respectively of Fig. 2.[1]

[1] In a preliminary paper on the absorption of high frequency radiation (PHYS. REV., 13, 296, 1919), the quantity $\sqrt{\sigma}$ was used in place of the quantity φ. This was justified by the fact that for values of λ/a greater than 8, with which that paper was concerned, φ is almost exactly equal to $\sqrt{\sigma/\sigma_0}$.

The total absorption of high frequency radiation by the lighter elements may therefore be expressed by the formula,

$$(8) \qquad \frac{\mu}{\nu} = K\varphi N^4 \lambda^3 + \frac{\sigma}{\nu},$$

where σ/ν and φ have the values assigned by equations (6) and (7), and depend upon the radius of the electron. If this radius is taken as

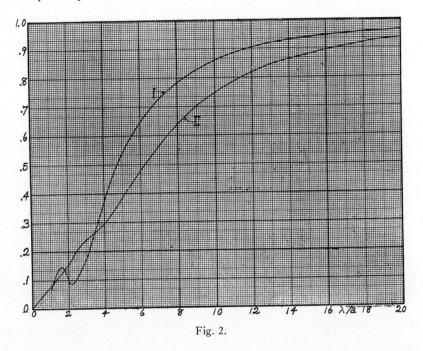

Fig. 2.

.02 A.U., as estimated from determinations of the scattering, K is the only arbitrary constant involved in this expression; though it must be admitted that the exponents of N and λ have no conclusive theoretical support.

The only reliable experimental data which are of value in testing expression (8) are those which refer to the mass absorption coefficients in aluminium.[1] In Fig. 3 are plotted the cube roots of the mass absorp-

[1] It is unfortunate that the measurements by Barkla and White (Phil. Mag., 34, 270, 1917) were not made with strictly homogeneous rays of known wave-length. They determine the effective wave-length of their rays by the absorption in copper, using the values given for this material given by Hull and Rice. Since, however, the variation of the absorption coefficient with wave-length of the light materials, aluminium, paper, paraffin, used by Barkla and White is much smaller than the variation in the case of copper, the effective wave-length of a heterogeneous beam will not be the same in the different materials. Moreover, the effect of inflection in the experimental absorption curves renders the true absorption for any definite wave-length uncertain. It would be of great interest to have the measure-

tion coefficients for this metal against the wave-length. The data include all the published measurements for strictly homogeneous rays of known wave-length which I have been able to find. The dotted line represents

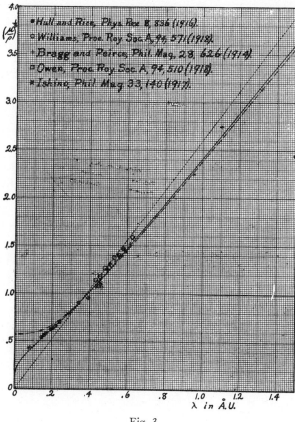

Fig. 3.

the simple cube law as expressed by equation (1), which neglects the absorption due to scattering. The broken line represents Owen's formula (3), which takes into account the scattering but is based on the hypothesis of a relatively small electron. The solid line represents equation (8), calculating the absorption on the basis of a ring electron of radius .02 A.U. as estimated from determinations of the scattering.[1] These curves are obtained from the values of μ/ν as calculated from expressions 1, 3 and 8 respectively, by multiplying by the factor ν/ρ, taking ν for alumin-

ments on these absorbers of low atomic weight repeated with strictly homogeneous radiation, since the predominance of the scattering over the fluorescent absorption by the light elements makes the measurements made on them of maximum value in determining the radius of the electron.

[1] Ibid.

ium to be 6.06×10^{22} and ρ to be 2.71. For the longer wave-lengths the three formulas differ but little, and the experimental variations are of little significance. It is at the shorter wave-lengths that the effects of the different hypotheses become evident. Here it is apparent how failure to take into account the scattering in equation (1) and the effect of the electron's size in equation (3) makes the dotted and the broken lines depart seriously from the experimental values, whereas the solid line shows a very fair agreement.

The mass absorption coefficients of aluminium for wave-lengths less than .35 A.U. are plotted in Fig. 4 in larger scale, in order to show the

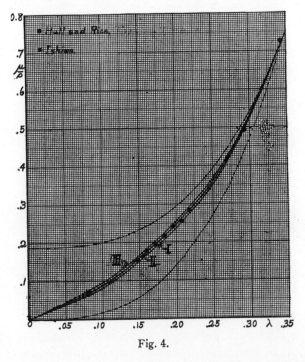

Fig. 4.

effect on the form of the theoretical curve of changing the value assumed for the radius of the electron. Here, as in Fig. 3, the dotted and the broken curves represent formulas (1) and (3) respectively, and both are obviously inaccurate for these very hard rays. It is of particular interest to note that the fact that the total absorption falls below the value of σ_0/ρ, required for the scattering coefficient alone on the basis of a small electron shows the hypothesis of a sensibly point charge electron to be untenable. Of the three solid lines, curve I is calculated on the basis of an electron of 2.0×10^{10} cm. radius, curve II for a radius of 1.85×10^{10} cm., and curve III for a radius of 1.7×10^{10} cm. Unless there is some consistent error in the experimental figures, we can, on the basis of the agreement of these curves, take the radius of the electron to be $(1.85 \pm .05) \times 10^{10}$ cm.[1]

[1] Attention has been called repeatedly (cf. Nature, 100, 510, 1918; H. S. Allen, Proc. Phys. Soc., 30, 143, 1918; 31, 53, 1919) to the fact that my original estimate of the radius of the electron (Jour. Wash. Ac. Sci., 8, 1, 1918) as 2.3×10^{-10} cm. was based on the value 9×10^{-10} cm. for the wave-length of the hard rays from radium C as determined by Rutherford and Andrade (Phil. Mag., 28, 263, 1914) instead of a value one third to one tenth as large more recently estimated by Rutherford (Phil. Mag., 34, 153, 1917) in light of measurements on the absorption of hard X-rays. In these latter measurements the X-radiation was obtained from a Coolidge tube operated by an induction coil, and the rays were filtered through considerable thicknesses of iron and lead. The wave-length of the end radiation was estimated according to the relation $\lambda = hC/eV$, where h is Planck's constant, C is the velocity of light, e the electronic charge, and V is the voltage across the X-ray tube. The value of V used in Rutherford's calculations was the *maximum* voltage across the tube, which obviously is considerably greater than the effective voltage. Furthermore, the filtering method of obtaining the radiation which has the shortest wave-length is uncertain in its results, as is apparent from the false indications obtained by this method by Rutherford, Barnes and Richardson (Phil. Mag., 30, 339, 1915). Both of these sources of error lead to a calculated value of the effective wave-length which is smaller than the true value.

That these errors are actually present in Rutherford's work is suggested by the following considerations. (1) For $\lambda = .086$ A.U. (maximum voltage = 144,000) he was unable to measure the intensity of the beam accurately after it had passed through 5 mm. of lead, whereas Rutherford and Andrade (Phil. Mag., 28, 266, 1914) find that homogeneous radiation shorter than 1.16 A.U. penetrates 6 mm. of lead without great loss.

(2) The absorption coefficient measured by Rutherford "is intermediate between μ and $\mu + \sigma$ (where μ is the true [fluorescent] absorption coefficient and σ the scattering coefficient), and probably closer to the former. The value of μ as given by Hull and Miss Rice corresponds to $\mu + \sigma$ in the above notation." That is, using the notation of the present paper, the data of Hull and Rice refer to the total absorption μ, while the figures given by Rutherford refer more nearly to the fluorescent absorption τ. Nevertheless, Rutherford points out that his measurements on both aluminium and lead agree with those of Hull and Rice for the wave-lengths which overlap. These wave-lengths are, however, in the neighborhood of .135 A.U., where, whether the electron is taken to be very small or of the size here estimated, by far the greater part of the total absorption in the case of aluminium is due to scattering. Rutherford's values for the absorption coefficient should therefore have been much smaller for aluminium than the values of Hull and Rice.

(3) That the minimum wave-length produced by the tube is not separated out by the filtering process is apparent from the following data on the absorption coefficient in lead for various thicknesses of the lead screen, as taken from Rutherford's paper (p. 154):

Max Voltage.	Range of Thickness in Lead, mm.	Absorption Coefficient.	Max Voltage.	Range of Thickness in Lead, mm.	Absorption Coefficient.
170,000	3.1–3.7	18	196,000	4.3–5.5	13
	3.7–4.3	17		5.5–6.4	12
	4.3–5.5	16		7.8–9.2	10
				8.8–10.0	8.5

If the absorption coefficients for 196,000 volts, for example, are plotted against the reciprocal of the mean thickness of the absorption screen, the curve shown in Fig. 5 is obtained. There is apparently no tendency for the absorption coefficient to approach a constant finite value for large thicknesses of the absorption screen. In fact these figures would rather lead to the con-

If a similar calculation is made in the cases of copper and lead, the absorption coefficients of which for very short wave-lengths have been determined by Hull and Rice, the agreement for long wave-lengths is satisfactory, but for shorter wave-lengths the calculated values are somewhat too low. The difference is greater in the case of lead than in the case of copper. This discrepancy is doubtless due at least in part to the fact, pointed out by Barkla and Dunlop,[1] that a considerable excess scattering occurs in these metals when traversed by X-rays, while

clusion that the absorption coefficient of the rays, after passing through a very thick lead screen, approaches zero. It is certain, at least, that the true absorption coefficient corresponding to the shortest wave-length produced at 196,000 volts is much less than the value 8.5 used by Rutherford.

It appears from these considerations that the wave-lengths assigned by Rutherford to correspond with the different observed absorption coefficients are considerably too short.

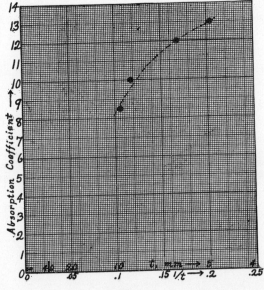

Fig. 5.

For this reason I have not considered his estimate of the wave-length of hard γ-rays, based on these absorption measurements, as reliable as the direct determinations of Rutherford and Andrade.

Note added September 3, 1919: After reading over these comments, Professor Rutherford still believes in the accuracy of his later measurements rather than in the earlier determinations by himself and Andrade. He writes me, "When I recall the faintness of the radium C lines and the difficulty of fixing them. . . . I am inclined to think that a mistake could easily arise. . . . I am inclined to give a good deal more weight to the Coolidge Tube experiments than you do." As Rutherford suggests, the question is one which must finally be answered by more refined measurements of the absorption of very hard homogenous rays of known wave-length.

[1] Barkla and Dunlop, *loc. cit.*

it does not occur to an appreciable extent in the case of aluminium for these wave-lengths. It is therefore impossible to test on the basis of measurements on lead and copper the validity of the assumptions underlying formula (8), which applies strictly to only those absorbing materials in which excess scattering does not occur.

Conclusions.—The excellent agreement of the absorption as calculated by equation (8) with the experimental values, in the only case (aluminium) in which the requisite data are available to make an adequate test, constitutes a strong support of the fundamental assumption of an electron of a size comparable with the wave-length of short X-rays. This agreement is the more significant since it is impossible to account for the low values of the absorption observed for very short X-rays if the electron is assumed to be sensibly a point charge of electricity.

These results must also be considered as a partial confirmation of the formula (8) proposed from theoretical considerations for the absorption coefficients of elements which do not show appreciable excess scattering.

Assuming the validity of this formula and the accuracy of the measurements of Hull and Rice on the absorption of hard X-rays in aluminium, and considering the electron to have the form of a flexible ring of electricity, the radius of the electron is calculated to be $(1.85 \pm .05) \times 10^{10}$ cm.

RESEARCH LABORATORY,
   WESTINGHOUSE LAMP COMPANY,
      May 24, 1919.

Radio-activity and Gravitation.

IN connection with the interesting letter of Prof. Donnan in NATURE of December 18, it may be of interest to mention some experimental results which have a bearing on this question. Some years ago Dr. Schuster suggested to one of us that it would be of interest to test whether the rate of transformation of radio-active substances was influenced by the intensity of gravitation. An accurate method of testing the rate of decay of radium emanation over a period of about a hundred days was developed, and it was intended to compare the rate of decay of samples which had been transported to suitable portions of the earth's surface. The outbreak of the war interfered with this plan.

Since, according to Einstein's theory, a gravitational acceleration is in no sense different from a centrifugal acceleration, experiments have been performed in the Cavendish Laboratory to test whether the rate of decay of radio-active substances is affected by subjecting them to the high centrifugal acceleration at the edge of a spinning disc. For the purpose of measurement the γ-ray activity was determined by a sensitive-balance method. Although the radio-active material was subjected to an acceleration of more than 20,000 times gravity, the change observed, if any, was certainly less than one part in a thousand.

This result is not in disaccord with the relation deduced by Prof. Donnan, for a simple calculation shows that his relation predicts an effect very much smaller than can be detected by measurements of this character.

E. RUTHERFORD.
A. H. COMPTON.

Cavendish Laboratory, December 19.

A PHOTOELECTRIC PHOTOMETER.*

BY ARTHUR H. COMPTON,
RESEARCH LABORATORY, WESTINGHOUSE LAMP COMPANY.

Abstract—This paper notes the difficulties which heretofore have prevented the use of the photoelectric cells in routine photometry. The use of a phermionic amplifier to increase current readings is suggested, also a filter to reduce the proportion of blue light and render the indications proportional to the photometric value. With such equipment, lamps are compared by varying their distances to the cell, until the deflection of the galvanometer is the same as that given by the standard lamp.

When light falls on the surface of an alkali metal, electrons are thrown off, the number liberated being proportional to the intensity of the light which strikes the metal. If an electric potential is supplied between the alkali metal and another electrode, with the alkali metal as cathode, the liberated electrons will therefore carry a current whose magnitude will be a measure of the intensity of the incident light. The photoelectric cell usually consists of a spherical glass bulb with an electrode placed as nearly as possible at its center. The bulb is highly evacuated, or filled to low pressure with an inert gas, and both the inner electrode and the inner surface of the bulb are coated, except for a small opening to admit the light, with a conducting layer of the distilled alkali metal. Either the outer surface or the inner electrode may then be used as cathode. A photoelectric cell made up in this manner responds to very faint illuminations, and has the advantage over the selenium cell of being practically instantaneous in its action. These characteristics have made it a valuable instrument for certain scientific purposes such as measuring the light from the stars and studying the energy in different parts of the spectrum.

It is at first sight surprising that no serious efforts have been made to apply the photoelectric cell to measurements of candlepower in industrial work. This may be accounted for by the fact that there are two essential difficulties which must be overcome before such an application of the instrument can be made. In

---
* Paper prepared for presentation before the Thirteenth Annual Convention of the Illuminating Engineering Society, Oct. 20 to 23, 1919, Chicago, Ill.

the first place the currents which pass through a photoelectric cell when illuminated to the extent customary in lamp photometry are of the order of only about 10^{-11} amperes, which is too small to be measured with an ordinary galvanometer. This makes necessary the use of some electrostatic method of detecting the current or the use of some method of magnifying the initial current. In the second place, the photoelectric cell has not the same relative sensitiveness in different parts of the spectrum as has the eye. Whereas the eye is most sensitive to yellow green light, the photoelectric cell is most sensitive to blue light, and is only slightly affected by red and yellow light. For this reason if a carbon and a tungsten lamp are of the same brightness as judged by the eye, the photoelectric cell will be affected considerably more strongly by the tungsten lamp because of its whiter color. It is necessary, therefore, to use in connection with the photoelectric cell an absorption screen which will correct for the selective action by radiation of short wave-length.

A customary method of increasing the current which passes through an illuminated photoelectric cell is to fill the cell with an inert gas at low pressure and apply across the tube a potential difference sufficiently great to produce ionization by collision. The current can thus be multiplied many fold. The formation of a hydride on the surface of the metal also increases greatly the photoelectric effect. The current as thus magnified can be readily measured by means of a portable electrometer of suitable design. Those who have used electrometers, however, will realize the difficulties which arise in warm, damp weather due to moist insulation. It seems almost impossible to avoid these troubles sufficiently to obtain in this manner reliable results in summer weather. For this reason I have found it advisable to magnify the photoelectric current still further by the introduction of a three electrode valve.

A diagrammatic sketch of the connections as thus used is shown in Fig. 1. The inner electrode C was used as the photosensitive cathode, while all the inner surface of the bulb, except for a small opening O acted as the anode A of the photoelectric cell. The cathode was connected directly to the grid G of the valve tube, and was shunted through a high resistance of about 1000 megohms to the negative terminal of the battery B_2. The grid potential V_1 and the plate potential V_2 were adjusted in such a manner as to obtain a plate current which was of a magnitude easily measurable on the galvanometer M when the photoelectric cell was illuminated, and as small a current as possible when the cell was dark. For the particular three electrode valve employed the most effective grid voltage was -20, while the plate voltage varied from 95 to 110 according to the intensity of the illumination used. The plate current can conveniently be made as large as several milliamperes, but more satisfactory results were obtained when the maximum current was adjusted to about half a microampere. Since all the connections between the cathode C, the grid and the high resistance can be completely enclosed and protected from moisture, all difficulty with regard to insulation leak and the measurement of the photoelectric current can thus be overcome. The only precaution necessary is to use batteries B_1 and B_2 of as constant voltage as possible.

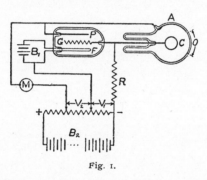

Fig. 1.

The color differences for which it is necessary to correct are the slight variations due to comparatively small changes in the temperature of the lamp filament. Unless caesium or rubidium is used for the photosensitive substance, it is impossible to duplicate the sensibility curve of the eye, since the other alkali metals are not sensitive to light in the extreme red. Sufficiently good results can be obtained, however, if only incandescent lamps near their normal operating temperature are to be measured, by the insertion of an absorption screen which will cut out the light of wave-length shorter than some certain value. If, for example, an unprotected photoelectric cell is used, since the maximum sensitiveness is in the blue, an increase in the temperature of the filament will produce a greater change in the photoelectric current than in the brightness of the lamp. On the other hand, if an absorption screen is inserted which permits only the red light to pass, the change in the photoelectric current will be less than the change in brightness. It is obvious that an absorption screen can be chosen to remove light shorter than some intermediate wave-length with the result that the change in the photoelectric current will be the same as the change in brightness. Experiment shows that if a potassium cell is used, a screen should be employed which transmits radiation longer than about 5300 A.° U. and absorbs the shorter wave-lengths. The selection of a suitable screen of this type is not at all difficult.

For many scientific purposes photoelectric cells using rubidium or caesium for the sensitive substance are desirable, because these cells give the greatest photoelectric current and are comparatively more sensitive to the longer wave-lengths than are cells using potassium or sodium. For industrial purposes, however, the low melting points of caesium and rubidium preclude their use, since the characteristics of the cells would change when the surface melts. For this reason potassium is the most satisfactory material to use. Its melting point is well above the maximum room temperature, and it is sufficiently sensitive to the action of red and yellow light to give measurable currents when the proper absorption screen is inserted.

If it were possible to use a high vacuum photoelectric cell the troublesome variations in the photoelectric current with changes in the voltage across the cell would be eliminated. Unfortunately in a good vacuum the vapor pressure of the potassium is sufficiently large to destroy rapidly the insulating properties of the glass. This makes it necessary to insert a certain amount of gas, and it is found that helium is satisfactory for the purpose. As mentioned above, this atmosphere serves the purpose also of magnifying the initial current by ionization, if a sufficiently large potential is applied to the tube.

A necessary precaution in the construction of a suitable photoelectric cell is that the bulb of the cell should be almost completely coated with a conducting layer of metal. If this is not done the insulated portions of the glass may collect charges which will modify the magnitude of the photoelectric current.

The characteristics of the three electrode valve are such that it is not practicable to calibrate a photoelectric photometer of this type permanently to read intensity of illumination. There are, however, a number of methods by which the instrument can be used to advantage. For example, in measuring horizontal candlepower a standard lamp at a fixed distance can be made to produce a certain deflection. The candlepower of other lamps can then be determined by measuring the distance from the photoelectric cell at which they will produce the same deflection.

For the intensity of illumination usually employed the plate current which is measured by the galvanometer M is approximately proportional to the square of the illumination. By help of this fact the photoelectric photometer may be used as a direct reading instrument in comparing lamps of about the same candlepower. If the galvanometer scale is divided according to this law, and if the reading is 100 per cent. when a standard lamp is used, the relative brightness of any other lamp can be read directly. This arrangement would be especially convenient where a number of lamps of the same rated candlepower are to be compared.

The same method of measurement can be used to advantage in determining the spherical power by means of an integrating sphere. The photoelectric cell with its orange screen may be placed at the opening in the sphere and be adjusted to give a deflection of 100 per cent. when a standard lamp is placed within the sphere. The relative total light emitted by any other lamp of about the same candlepower can then be read directly. Experiment has shown that it is not difficult by this means to compare the spherical candlepower of two similar lamps to within a probable error of 1/10 per cent.

Other electrical photometers, such as the selenium cell and the thermopile or bolometer with suitable absorption screens, have not been found adaptable to industrial photometry. The selenium cell, for the low illuminations which it is necessary to employ, is found to be slow in response and not very consistent in results. On the other hand, the thermopile and bolometer can hardly be made sensitive enough to respond to the illumination from lamps of low candlepower when a suitable absorption screen is employed. A satisfactory screen for use in connection with the thermopile is also a serious consideration, since it is necessary to cut out all the infra red radiation. This problem is much simplified when the photoelectric cell is used, since this instrument is not affected by radiation of wave-length longer than that of visible light.

As compared with the optical methods of photometry, the photoelectric photometer has all the advantages of a deflection instrument over one which requires the judgment of small differences of degree. It is not only easier but it also requires less time to read a deflection than to make an accurate judgment of relative brightness. It is obvious that an instrument of this kind cannot entirely replace the visual photometer, since the eye is the final authority in estimating illumination. But when the photoelectric photometer has been adjusted to make true measurements of intensity of illumination, it offers the advantages of greater speed, less fatigue and higher accuracy where a large number of measurements on similar lamps are to be made.

CATHODE FALL IN NEON.

By Arthur H. Compton and C. C. Van Voorhis.

Synopsis.

Systematic measurements have been made of the potential difference between the cathode and the beginning of the positive column when a discharge is passed through pure neon, using a number of different metals as cathode. The "normal cathode fall," or potential difference between the cathode and the cathode glow for normal current, was also determined for several metals. It was found that for normal current the potential difference between the cathode and the positive column is very nearly proportional to the potential difference between the cathode and the cathode glow when different metals are used as cathode. The values of the normal cathode fall in neon were found to be, with Pt cathode 152 volts, W 125 v., Tl 125 v., Al 120 v., Mo 115 (?) v., Mg 94 v., Ca 86 v., Na 75 v., and K 68 v. These values are in the order, as far as available data go, of the contact potential series, and are consistently slightly lower than the corresponding values in helium.

THE recent studies of H. A. Wilson[1] and C. A. Skinner[2] have emphasized the importance of a thorough experimental knowledge of the electrical characteristics of low pressure discharge tubes. The potential drop at the cathode, which is of especial theoretical significance, has been investigated by a large number of experimenters, including Hittorf,[3] Skinner,[4] Wilson,[5] Mey[6] and Defregger.[7] The results of the early experiments were often uncertain because of impurities in the gas and on the surface of the electrodes. Mey and Skinner have both made systematic studies of the cathode fall for a number of different metals in various gases. In most of this work, however, the gases hydrogen, nitrogen and oxygen were employed, which makes the significance of the results somewhat doubtful on account of danger of chemical action by the gases on the electrodes. Certain similar experiments by Mey, Defregger and Watson[8] have been performed in inert gases, but for the most part the tests were isolated ones, which makes an accurate comparison of the results for different metals of doubtful value. For

---

[1] H. A. Wilson, Phys. Rev., 8, 227, 1916.
[2] C. A. Skinner, Phys. Rev., 12, 143, 1918; 9, 97, 314, 1917.
[3] Hittorf, Wied. Ann., 20, 705, 1883; 21, 133, 1884.
[4] C. A. Skinner, loc. cit., and Wied. Ann., 68, 752, 1889; Phil. Mag., 8, 387, 1904.
[5] H. A. Wilson, Phil. Mag., 49, 505, 1900.
[6] K. Mey, Ann. d. Phys., 11, 127, 1903.
[7] R. Defregger, Ann. d. Phys., 12, 662 (1903).
[8] H. E. Watson, Proc. Camb. Phil. Soc., 17, 90, 1913.

this reason we undertook a systematic investigation of the cathode fall of a number of different metals in neon. This gas, besides being chemically inert, has the valuable characteristic of showing in its spectrum the presence of small traces of impurities, which makes it possible to be sure that uncontaminated gas is being used.

In any work of this character, the final value of the investigation depends upon the purity of the gas and the freedom of the surface of the cathode from any film of foreign material. The neon used was prepared by Claude and was very kindly loaned to us by the Bureau of Standards. It was enclosed in an all glass system, being circulated by the help of a mercury pump and a number of carefully constructed stopcocks. Though initially practically free from any impurity except water vapor and a trace of helium, the neon was subject to contamination by the gases given off from the electrodes of the discharge tubes. The

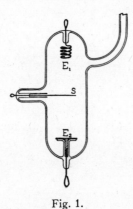

Fig. 1.

hydrogen thus introduced was burned by circulating the gas through a small bulb containing an oxidized copper wire heated electrically to a dull red. Final purification was secured by passing the neon through freshly-baked, activated cocoanut-charcoal, cooled in liquid air. All impurities other than helium were thus completely absorbed, and the pressure of the neon over the charcoal was so reduced that it was possible to pump back into the reservoir practically all the helium with the first portions of the neon, leaving behind gas which showed in a spectrum tube no impurity of any kind.

A typical form of discharge tube is shown in Fig. 1. It consists of a glass tube 3.5 cm. in diameter in which were sealed two electrodes E_1 and E_2 and a small sound wire S. It was found from previous experiments that in a discharge tube of this diameter the positive column begins about 4 cm. from the cathode when the pressure and current are such as to give the normal cathode fall. For this reason the tubes were made with electrodes 8 cm. apart and the sound wire placed midway between, so that either one could be used as cathode. These tubes were highly exhausted and baked for several hours at 500° C. before the neon was introduced. The special treatment required to secure clean metallic surfaces on the electrodes will be described below as each different metal is considered.

The electrical system employed may be explained from Fig. 2. The source of the current was a 500-volt direct current generator whose

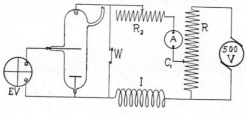

Fig. 2.

terminals were connected across the 2,500-ohm resistance R_1. By varying the voltage across the discharge tube with the sliding contact C_1, and by means of the widely variable resistance R_2, current varying in magnitude from 0.00001 to 0.2 ampere could be obtained. The resistance R_2 consisted of a variable length column of a solution of alcohol in xylol in series with a metallic high resistance. By momentarily shorting the circuit with the wire W and then breaking it again, the discharge could be started due to the action of the inductance I, which was the secondary of a small 2,200 volt transformer. The current strengths were measured by the milliammeter A, and the potential differences by means of an electrostatic voltmeter of special design EV.

Measurement of the Potential Drop from the Positive Column to the Cathode.

Preliminary experiments showed that there is a minimum potential difference between the cathode and the sound wire, placed as described above, for a certain value of the current. This minimum value is practically the same over a rather wide range of gas pressures, and the current strengths giving the minima vary approximately as the square of the pressures, just as for the cathode fall. Consequently for each different gas pressure in a given tube a series of readings of the potential drop was taken as the current strength was varied, and the minimum values were taken as the ones to be used in making comparison with other metals. The voltages as thus determined were practically independent of the pressure of the gas for pressures between 2.5 and 15 mm. of mercury. The values of the potential differences between the positive column and the cathode which are given below for each of the different metals are the average of a large number of measurements at different pressures in this range.

Aluminium, Magnesium, Calcium.—Thin flat discs of these metals were used as cathodes in bulbs similar to that shown in Fig. 1. In order to clean the surface and eliminate the absorbed hydrogen, these cathodes were maintained at a high temperature (about 600° C.), by means of a discharge in neon at low pressure, until the spectroscope revealed no impurities in the neon after continued operation. The mean value of the minimum potential drop from the beginning of the positive column to the cathode was found for aluminium to be 140 volts, for magnesium 115 volts and for calcium 102 volts.

Platinum, Tungsten, Molybdenum.—Sheets of pure platinum and tungsten, having been cleaned by raising them to a white heat in rarified neon, gave corresponding potential drops of 175 and 140 volts respectively. Cathodes of commercial tungsten wire (1 or 2 per cent. thorium oxide) and of pure molybdenum wire coiled in loose spirals were also employed. These gave for tungsten, 123.5 volts, and for molybdenum 133.5 volts. The difference between the two tungsten cathodes may perhaps be partly accounted for as due to the thorium oxide impurity in the wire, but is also probably due in part to the difference in shape of the two electrodes. If this is the case, the voltage observed in the case of molybdenum is not strictly comparable with that observed for the other metals.

Thallium.—A clean electrode of this metal was obtained by splashing the molten metal about in an atmosphere of pure neon. The minimum potential was found to be 145 volts.

Alkali Metals.—Clean cathodes of sodium, potassium and their alloy were obtained by vacuum distillation in a tube of the form shown in Fig. 3. The metals having been inserted in bulb B were distilled successively into bulbs D and F, and bulb D was sealed off at E. Neon was then admitted into bulb F, the alkali metal was melted and the neon was pumped out, bringing with it the clean molten metal. The minimum potential drop from the anode glow to the cathode in the case of sodium was 87 volts, for potassium 77 volts and for an alloy containing about equal weights of the two metals, 84 volts.

The Normal Cathode Fall.

In order that our results might be comparable with those of other observers, we measured directly the "normal cathode fall" from the cathode glow to the cathode for a number of different metals. For this purpose an extra sound wire was sealed into the discharge tube about 8 mm. from the cathode in such a manner that it was in the cathode glow when the normal current was passed through the tube. The current giving the minimum value for the cathode fall under a certain set of

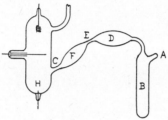

Fig. 3.

conditions was found to be practically the same as that which gave the minimum value for the potential drop between the cathode and the positive column. The values of the normal cathode fall C and of the minimum potential drop P from the positive column to the cathode are given in columns two and three respectively of Table I.

TABLE I.

Metal.	C.	P.	Ratio C/P.
Platinum	152 volts	175 volts	0.87
Magnesium	94	115	.82
Calcium	86	102	.84
Na-K Alloy	73.6	84	.88
Potassium	68	77	.88
			0.86 mean

It will be seen in the third column of this table that the ratio of the normal cathode fall to the fall in potential from the positive column is for these metals constant within the probable experimental error. It is therefore possible to calculate from our measurements of the minimum potential difference between the cathode and the positive column the value of the normal cathode fall for the different metals in neon by multiplying the former quantity by the factor .86. The values of the normal cathode fall as thus estimated are given in the second column of Table II.

TABLE II.

Metal.	Normal Cathode Fall.	
	In Neon.	In Helium.
Platinum....................	152	160 (D)
		165 (Dm)
Tungsten....................	125	
Thallium....................	125	
Aluminium..................	120	141 (D)
Molybdenum................	115(?)	
Magnesium..................	94	125 (D)
Calcium.....................	86	
Sodium......................	75	80 (M)
Na-K Alloy..................	73.6	78.5 (Dm)
Potassium...................	68	69 (M)

Discussion of Results.

We have no data available which determine the positions of tungsten, thallium, molybdenum and calcium in the contact potential series. For the remaining metals, however, our results are in accord with Skinner's hypothesis that the cathode fall varies inversely with the electropositiveness in the contact potential series.

The values quoted for the normal cathode fall in helium are those given by Defregger (D), Dember (Dm) and Mey (M). It will be seen that the cathode fall in neon is uniformly slightly lower than in helium. This is of interest in connection with the fact that the ionizing potential of neon is slightly lower than that of helium.

Research Laboratory,
    Westinghouse Lamp Company.

LXV. *Radioactivity and the Gravitational Field.* By ARTHUR H. COMPTON, *Ph.D.*, *National Research Fellow in Physics* *.

IT is well known that in order to account for the age and the present temperature of the earth, the average radioactivity of its component minerals must fall off rapidly a few miles below its surface. The high density of the radioactive minerals, however, makes it appear probable that they should occur more abundantly in the earth's interior than in the surface crust. Thus it appears that substances which at the earth's surface are radioactive may have practically no radioactivity in the earth's interior. These considerations suggest that the rate or energy of radioactive disintegration may be a function of the intensity or potential of the earth's gravitational field. This suggestion appears the more plausible since both radioactivity and gravitation are essential attributes of the atomic nucleus.

It has recently been pointed out by A. Donnan † that

---

* Communicated by Prof. Sir E. Rutherford, F.R.S.
† A. Donnan, 'Nature,' Dec. 17, 1919.

thermodynamic reasoning predicts a change in the energy evolved in radioactive disintegration when the potential of the gravitational field is varied. The following analysis shows, however, that this change is by no means large enough to account for the lack of radioactivity of the earth's interior. The cycle considered by Professor Donnan consists of the following four steps:—

1. A system at gravitational potential Z changes from state 1 to state 2, an amount of energy Q being liberated, which results in a change of mass from m_1 to m_2.

2. The system in state 2 is raised from potential Z to $Z+\delta Z$.

3. At potential $Z+\delta Z$ the system is changed back from state 2 to state 1.

4. The system in state 1 is lowered from potential $Z+\delta Z$ to potential Z. Being then in its original condition, the total energy evolved by the system is zero, and it possesses its original mass.

If the change δZ in the gravitational potential is small, the total work done by the system in performing this cycle is

$$Q - m_2 \delta Z - \left(Q + \frac{\partial Q}{\partial Z}\delta Z\right) + m_1 \delta Z = 0,$$

or

$$\frac{\partial Q}{\partial Z} = m_1 - m_2,$$

which is the expression obtained by Donnan. If now we consider Q as a function of Z only, we may write

$$\frac{dQ}{Q} = \frac{m_1 - m_2}{Q} dZ.$$

Putting R as the ratio between the energy evolved and the mass which disappears, we have

$$m_1 - m_2 = Q/R,$$

whence $\qquad dQ/Q = dZ/R.$

The difference in the gravitational potential between the surface and the centre of the earth is about 3×10^{11} cm.^{2} sec.$^{-2}$, and the ratio R is of the order of the square of the velocity of light, or 10^{21} cm.^{2} sec.$^{-2}$. Hence the decrease from this cause in the energy of radioactive disintegration, being less than one part in a billion, is wholly inadequate to account for the small amount of heat developed in the earth's interior.

This thermodynamic relation between the energy of radioactive disintegration and the gravitational potential does not, however, exclude the possibility of a connexion of an intimate character between, for example, the intensity of the gravitational field and the *rate* of radioactive disintegration. It is the latter type of relation which the present experiments have been designed to detect.

According to Einstein's generalized theory of relativity, a gravitational accelerational field is essentially the same as a field of centrifugal acceleration. We have therefore tested the effect due to a change in the gravitational field by subjecting the radioactive material to a strong centrifugal acceleration. It was obvious that the maximum centrifugal acceleration which could be attained at the edge of a rotating wheel would fall far short of the mean acceleration to which the atomic nucleus is subject due to the thermal agitation of the atoms. There was a chance, however, that a comparatively steady acceleration might have an effect different from that of the rapidly varying molecular accelerations.

The Experiments.—A small tube of radium emanation was placed in a hole near the circumference of a brass disk of 10 cm. radius. The gamma radiation from the emanation was measured when the disk was rotating slowly and when turning at approximately 250 revolutions per second. The acceleration was thus varied from about 1·5 to about 20,000 times the acceleration of gravity.

The gamma radiation was measured by a balance method. The ionization due to the gamma rays traversing a large ionization chamber was balanced against an adjustable current passing through a high resistance, a highly sensitive electrometer being used to detect any difference between the two currents. For the high resistance a Bronson resistance was at first employed. This was later discarded in favour of a resistance consisting of lampblack on sulphur. The latter resistance, though more subject to variations over long periods, has the advantage that it introduces no short period probability variations such as those due to the ionization by discrete alpha particles when the Bronson resistance is used. The probability variations in the ionization current due to the gamma rays, such as have been observed by Meyer[*], Laby[†] and others, were very noticeable in these measurements, and were the cause of practically the whole of the differences between successive measurements.

Altogether four extended series of measurements were

[*] Meyer, *Phys. Zeitschr.* xi. p. 1022 (1910).
[†] Laby, 'Nature,' lxxxvii. p. 144 (1911).

made, comparing the intensity of the radiation when the emanation was subject to small and to large accelerations. The results of these tests were as follows:—

1. An increase due to acceleration of $(\cdot 08 \pm \cdot 07)$ per cent.
2. A decrease due to acceleration of $(\cdot 17 \pm \cdot 09)$ per cent.
3. A decrease due to acceleration of $(\cdot 17 \pm \cdot 07)$ per cent.
4. An increase due to acceleration of $(\cdot 02 \pm \cdot 10)$ per cent.

The average of all these tests indicates a decrease in the intensity of the gamma radiation of $(\cdot 06 \pm \cdot 04)$ per cent. It is probable, therefore, that an acceleration of 20,000 times gravity produces no effect on the intensity of gamma radiation as large as one part in a thousand.

In order to explain the small degree of radioactivity of the earth's interior it would be necessary to assume a comparatively large change due to an increase in the gravitational acceleration. The negative result of this experiment therefore shows that we must look elsewhere for the cause of the confinement of the earth's radioactivity to its surface crust.

I wish to thank Professor Rutherford for proposing this problem to me, and for his helpful suggestions and encouragement as the work progressed.

Cavendish Laboratory,
    Cambridge University.
        Feb. 3, 1920.

[Reprinted from the Physical Review, N.S., Vol. XVI, No. 5, November, 1920.]

IS THE ATOM THE ULTIMATE MAGNETIC PARTICLE?

By Arthur H. Compton and Oswald Rognley.

Synopsis.

Effect of Magnetization of Crystal on Intensity of X-ray Reflection; Theory.—The intensity of reflection of X-rays from a crystal depends upon the arrangement of the atoms within the crystal and upon the arrangement of the electrons within the atom. If the ultimate magnetic particle is a group of atoms, such as the chemical molecule, magnetization of the crystal will change the orientation of the group and hence change the positions of the individual atoms. If the ultimate magnetic particle is the atom, magnetization will change the orientation of the atom and hence alter the arrangement of the electrons. In either case magnetization of a crystal should be accompanied by a change in the intensity of a beam of X-rays reflected from its surface.

Effect of Magnetization of Crystal on Intensity of X-ray Reflection; Experiment.—Such an effect was sought for by reflecting X-rays from a crystal of magnetite and measuring the intensity of the reflected beam by a sensitive balance method. The test was made on the first four orders of reflection from the natural (111) face when the crystal was magnetized perpendicular to its reflecting surface and on the third order when magnetized parallel with this surface. On magnetizing the crystal to 1/3 of saturation and on demagnetization, no change in intensity of the reflected beam was observed, though a variation of 1 per cent. would have been detected.

Molecule as Ultimate Magnetic Particle.—A displacement of the atoms due to magnetization by 1/300th of their distance apart would have produced a detectable effect. This experiment therefore affords a strong confirmation of the conclusion reached by K. T. Compton and E. A. Trousdale that the elementary magnet is not a group of atoms.

Motion of Elementary Magnet in Strong Magnetic Field.—An argument is presented which shows that when saturation occurs the elementary magnets very probably have their axes nearly parallel with the direction of magnetization.

Atom as Ultimate Magnetic Particle.—Subject to the validity of this conclusion, it is not found possible to explain the negative result of the experiment if the atom as whole acts as the elementary magnet. Certain other explanations are also discussed and found unsatisfactory.

Electron or Positive Nucleus as Ultimate Magnetic Particle.—Either of these conceptions is in accord with the result of the experiment, but auxiliary evidence favors the electron as the probable elementary magnet.

A FEW years ago K. T. Compton and E. A. Trousdale described in this journal[1] an experiment which led them to the conclusion that the ultimate magnetic particle is not any group of atoms, such as the chemical molecule, but is rather the atom or something within the atom. It has long been a favorite method of explaining the magnetic

[1] K. T. Compton and E. A. Trousdale, Phys. Rev., 5, 315 (1915).

properties of matter to interpret them in terms of the magnetic moment of electrons rotating in orbits. This hypothesis has seemed to receive support by the successes of the Rutherford-Bohr theory of atomic structure, which assumes the atom to consist of electrons revolving in orbits in such a manner that the atom must possess a large magnetic moment. Furthermore, the experiments of Barnett,[1] Einstein and de Haas[2] and J. Q. Stewart[3] have shown that the magnetization of iron is unquestionably accompanied by a change in angular momentum of the order of magnitude to be expected if the iron's magnetic properties are due to electrons revolving in orbits.

Experimental.—We have accordingly performed an experiment designed to test the hypothesis that the elementary magnet in a ferromagnetic substance is an atom consisting of rotating rings of electrons. In this experiment an attempt was made to detect a difference in the intensity of a beam of X-rays reflected from a crystal of magnetite when the crystal was magnetized and when unmagnetized.[4] It will be assumed for the present that all the elementary magnets of which a ferromagnetic substance is composed are arranged with their axes parallel with the magnetic field when the substance is magnetically saturated. If these ultimate magnetic particles are the individual atoms, the orientation of the atoms due to magnetization of the crystal will change the positions of the electrons of which the atoms are composed. In virtue of the fact that the intensity of a beam of X-rays reflected from a crystal face depends upon the arrangement of the electrons in the atoms which make up the crystal, such a shift of the electrons should make itself known by changing the intensity of the reflected beam.

Consider for example a crystal composed of atoms of the Rutherford type, each atom having all its electrons arranged in the same plane and perpendicular to its magnetic axis. When the crystal is unmagnetized, the axes of the electronic orbits will be oriented in all possible directions, so that most of the electrons will be at an appreciable distance from the mid-planes of their atomic layers. If, however, the crystal is magnetically saturated perpendicular to the reflecting face, the electronic orbits will all lie parallel to this face. The electrons will now, therefore, be *in* the midplanes of the layers of atoms which are effective in producing the reflected beam. Such a shift of the electrons should produce a very considerable increase in the intensity of the reflected beam of

---

[1] S. J. Barnett, PHYS. REV., 6, 240 (1915).
[2] Einstein and de Haas, Verh. d. deutsch. Phys. Ges., 17, 152 (1915).
[3] J. Q. Stewart, PHYS. REV., 11, 100 (1918).
[4] Cf. A. H. Compton and Oswald Rognley, Science, 46, 415 (1917); and PHYS. REV., 11, 132 (1918) for preliminary accounts of this work.

X-rays, since the rays scattered by all the electrons will now be in the same phase. If the crystal is magnetized parallel to the reflecting face, the turning of the orbits will carry the electrons farther on the average from the middle of their atomic layers, the phase difference between the rays scattered by the different electrons will be greater, and a decrease in the intensity of reflection should result. If, on the other hand, the electrons are arranged isotropically in the atom, or if the atom is not rotated by a magnetic field, which would be the case if it is the individual electron or the positive nucleus that is the ultimate magnetic particle, no such change in the intensity of the reflected beam should be observed.

In our search for this effect a null method was employed. The ionization due to the beam of X-rays reflected from a crystal of magnetite was balanced against that due to a beam of the same wave-length reflected from a crystal of rock-salt, so that a very small change in the relative intensity of either beam could be detected while variations in the X-ray tube itself had but little effect. The experimental arrangement is shown diagrammatically in Fig. 1. From the target A X-rays fell upon

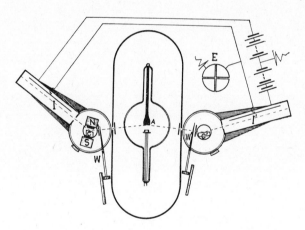

Fig. 1.

the crystals C and C', and were reflected into the ionization chambers I and I'. One of these chambers was at a positive and the other at a negative potential, and their electrodes were both connected to the same pair of electrometer quadrants. Thus if the ionization current was the same in both chambers there was no deflection of the electrometer. The beams could be balanced accurately against each other by use of the aluminium absorption wedges W and W', operated by means of micrometer screws. The crystal C was cemented to one pole of an electromagnet with a laminated core, so that by changing from direct to alternating current which could be gradually reduced to zero it was possible to magnetize or demagnetize the crystal at will.

For the source of X-rays we used a Coolidge tube with a molybdenum anticathode, kindly supplied by Dr. W. D. Coolidge. As the source of high potential, a Snook-Roentgen machine was employed. The first order of the molybdenum α line ($\lambda = .721$ A. U.) was always reflected into the ionization chamber I', while different orders of the same line were reflected by the magnetite crystal into the chamber I. It was possible with this arrangement to detect with certainty variations in the relative intensity of either reflected beam as small as 1 per cent. Unfortunately our magnet was not strong enough to saturate the magnetite crystal. The intensity of magnetization of the magnetite was about 150, which is just over 1/3 of the saturation intensity for magnetite as determined by du Bois.[1]

The effect of magnetization perpendicular to the plane of the crystal face was investigated in the first four orders. On account of mechanical difficulties, the test was made in only the third order spectrum when the crystal was magnetized parallel to the reflecting surface. In no case was a change as great as 1 per cent. observed in the intensity of the reflected beam when the crystal was magnetized or demagnetized. Our experimental results therefore may be summarized by the statement that *the intensity of the X-ray beam reflected from magnetite is not altered by as much as 1 per cent. in the first four orders when the crystal is magnetized to 1/3 of saturation perpendicular to the reflecting surface, nor in the third order by similarly magnetizing the crystal parallel to the reflecting face.*

The Orientation of the Elementary Magnets.—The conclusions which are to be drawn from this experiment depend upon the conception that we have of what occurs to the ultimate magnetic particles when a ferromagnetic substance is magnetized. It is usually supposed that these elementary magnets have their axes so oriented by the external magnetic field that at saturation they are very nearly parallel with this field. Indeed, the manner in which this conception accounts for the phenomenon of saturation was originally perhaps the strongest argument in favor of the so-called "molecular" theory of magnetism. It should be noted, however, that the phenomenon of saturation is not in itself sufficient evidence that the ultimate magnetic particles have their axes oriented along the magnetic lines of force. This phenomenon would also occur if the elementary magnets were turned through only a small angle, but from one position of stable equilibrium to another beyond which it might be incapable of turning.

[1] Du Bois, Phil. Mag., 29, 293 (1890).

Perhaps the most convincing argument in favor of a complete rotation of the elementary magnets by a strong magnetic field is that which was brought forward originally by J. Swinburne[1] in connection with Ewing's suggestion that the orientation of the "molecular magnets" is determined by magnetic rather than by frictional forces. Swinburne points out that on this hypothesis, if the magnetization of a piece of iron is reversed by a strong rotating field instead of by a field alternating through zero, the loss in energy should be little or nothing, for *if the molecules rotate with the field* no unstable movements are possible. Experiments by F. G. Baily[2] on iron, by R. Beattie[3] on nickel and cobalt, and especially by Weiss on pyrrhotite,[4] show that this is actually the case. For large fields, where saturation is approached, the hysteresis loss per cycle when a rotating magnetic field is applied is found in every case to be only a small fraction of the hysteresis loss per cycle due to an alternating magnetic field. There would seem to be no reason for this great reduction in the hysteresis if the axes of the elementary magnets are unable to follow closely the rotating field. If seems difficult, therefore, to avoid the conclusion that *the elementary magnet*, whatever it may be, *is free to be oriented in any direction*.[5] This conclusion will accordingly form the basis of our interpretation of the experiment just described.

[1] Cf. Encyclopædia Britannica, 11th ed., Vol. 17, p. 350.

The referee to whom the editors submitted this paper has offered the following comment: "It would seem that additional strong evidence is given by the experiments of Weiss and Kamerlingh Onnes in 'Researches sur L'Aimantation aux Tres Basses Temperatures,' *Journal de Physique*, vol. 9, pp. 555–584, 1910. In this research it is shown that the moment per c. c. resulting from complete orientation of the magnets differs by only 5 per cent in the case of nickel, 2 per cent in the case of iron, and 6 per cent in the case of magnetite from the experimentally measured saturation moment at ordinary temperatures."

It is indeed difficult to explain the very nearly complete saturation which this result indicates as due to anything other than an almost perfect alignment of the elementary magnets with the external field.

[2] Ibid.

[3] Ibid.

[4] Cf. E. H. Williams, The Electron Theory of Magnetism, p. 40.

[5] In any case the axis of the ultimate magnetic particle must describe a closed conical surface under the action of a rotating magnetic field in order to account for the reduced hysteresis. This conical surface is the locus of the possible stable orientations of the axis. If the elementary magnet is free to align itself with the external rotating field, the apex angle of this cone will be π; but if it is free to turn through only a small angle, the conical surface will have a sharp apex. The possibility suggests itself that the axis of this cone may represent some fixed axis in the elementary magnet, such as for example an electric axis, which may be slightly inclined to the magnetic axis and about which the magnetic axis may be free to turn. But any such rotation of one axis fixed in the particle about another means a rotation of the particle as a whole about the second axis. We cannot avoid in this manner, therefore, the conclusion that the elementary magnet is rotated by a rotating magnetic field.

In nature it is usual for an object to be capable of stable orientation in one direction,

The Molecule as the Ultimate Magnetic Particle.—It is obvious that if the elementary magnet consists of some group of atoms within the ferromagnetic material, the arrangement of the atoms in a crystal will be greatly altered when the groups of atoms are rotated by a magnetic field. One would therefore expect, as was pointed out by K. T. Compton and E. A. Trousdale, that magnetization would change the positions of the spots in Laue photographs taken through magnetic crystals, since these positions are determined by the arrangement of the atoms. As we have seen, the negative result obtained by these experimenters indicates that the arrangement of the atoms is not greatly changed by magnetization of ferromagnetic crystals.

The experiment which we have performed gives still more definite information with regard to the displacement of the atoms in a magnetic field. It has been shown by W. H. Bragg[1] that the minute changes in the positions of the atoms due to a relatively small rise in temperature is sufficient to affect appreciably the intensity of a reflected beam of X-rays. The negative result of our experiment shows clearly, therefore, that the atoms in a crystal are not moved as far by magnetization as they are by a change of temperature well within the melting point of the crystal. In fact direct calculation shows that a displacement of the atoms by 1/300th of the distance between the successive atomic layers would have caused a detectable change in the intensity of the 4th order spectrum. Our experiment therefore affords a very sensitive confirmation of the conclusion reached by K. T. Compton and E. A. Trousdale that *the ultimate magnetic particle is not a group of atoms, such as the chemical molecule, but is the individual atom or something within the atom.*

The Atom as the Ultimate Magnetic Particle.—If the atom as a whole is the ultimate magnetic particle, it is natural to assume that its magnetic moment is due to the presence of electrons revolving in orbits. The simplest example of this type is the Rutherford form of atom, in which all the electronic orbits lie in the same plane, perpendicular to the magnetic axis. Let us therefore calculate the order of magnitude of the effect due to magnetization on the intensity of a reflected beam of X-rays if the reflecting crystal is composed of atoms of this type.

Professor W. H. Bragg has shown[2] that the intensity of reflection of

or at most in a finite number of directions. The ability to possess stable orientation in any direction in the surface of a narrow cone would require a mechanism that is difficult to imagine. It is certainly a much more artificial hypothesis than the conception of freedom for orientation in any direction.

[1] W. H. Bragg and W. L. Bragg, X-rays and Crystal Structure, p. 195.
[2] W. H. Bragg, Phil. Mag., 27, 881 (1914).

X-rays from a crystal in which the successive atomic layers are similar and similarly spaced falls off approximately according to the law

$$(1) \quad E = \frac{C(1 + \cos^2 2\theta)}{\sin^2 \theta} e^{-B \sin^2 \theta}.$$

In this expression C and B are constants, and θ is the glancing angle at which the X-rays strike the crystal face. This law represents an average for a number of different kinds of crystals. Darwin has found,[1] however, from theoretical considerations, that if all the electrons are in the mid-planes of the atomic layers to which they belong, the intensity should fall off according to the law,[2]

$$(2) \quad E_m = \frac{C'(1 + \cos^2 2\theta)}{\sin \theta \cos \theta} e^{-B \sin^2 \theta}.$$

This expression corresponds to reflection from a crystal composed of atoms of the Rutherford type which is magnetically saturated perpendicularly to the crystal face. Expression (1) indicates the reflection from an unmagnetized crystal, in which the magnetic axes of the atoms are oriented at random. The ratio of the intensity of reflection from the magnetized to that from the unmagnetized crystal should therefore be

$$\frac{E_m}{E} = \frac{C'}{C} \tan \theta.$$

For our approximate calculations, since θ is never large we may substitute $\sin \theta$ for $\tan \theta$. Thus in virtue of the relation $n\lambda = 2D \sin \theta$ this ratio becomes,

$$\frac{E_m}{E} = \frac{C'}{C} \cdot \frac{n\lambda}{2D}$$

$$(3) \qquad\qquad = kn,$$

where $k = C'\lambda/2CD$ is a coefficient independent of the order of reflection n. By extrapolation for the intensity of the X-ray spectrum line of zero order, where theoretical considerations show that E_m/E must be unity, it has been shown by one of the writers[3] that for the first order reflection

---

[1] C. G. Darwin, Phil. Mag., 27, 675 (1914).

[2] This expression assumes that the effective absorption coefficient of the X-rays in the crystal is the same for all orders of reflection. It is possible that the effective value is less in the higher orders. This would mean even less rapid diminution of intensity for the higher orders, and the predicted change due to magnetization would be even greater.

[3] A. H. Compton, Phys. Rev., 9, 52 (1917). It should be noted that Bragg's expression (1) must of necessity break down when applied to estimate the intensity of the zero order spectrum line. The extrapolation here referred to is accordingly based upon certain particular arrangements of the electrons in the atoms. Any reasonable arrangement, however, will lead to values for this ratio which do not differ greatly from those here given, so these values cannot be much in error.

from a cleavage face of rock-salt the ratio E_m/E is about 1.5 and for calcite is about 1.2. The ratio is doubtless of the same order of magnitude for other crystals. Taking 1.3 as a mean value when $n = 1$, this gives in place of equation (3),

$$(4) \qquad E_m/E = 1.3n.$$

The intensity of reflection should therefore be increased by a factor of about 1.3 in the first order, 2.6 in the second order, 3.9 in the third order and 5.2 in the fourth order, if the crystal is composed wholly of atoms of the Rutherford type which are oriented with their axes parallel to the magnetic field.

It might be supposed that in the case of magnetite the iron atoms only would be oriented by the magnetic field while the oxygen atoms would remain unaffected. While this would reduce to some extent the change in intensity to be expected, it is obvious that the change would still be comparatively large. It is thus apparent that the hypothesis that atoms of the Rutherford type constitute the ultimate magnetic particles is incompatable with the negative result of our experiment.

Theoretical considerations seem to lead to the conclusion that, in order for an atom consisting of electrons revolving in orbits to be stable, the orbits must all lie in the same plane. It is therefore difficult to defend the hypothesis of a magnetic atom in which the electrons lie in different planes. It will be profitable, nevertheless, to consider the change to be expected in the intensity of the reflected X-ray beam if the atoms of the reflecting crystal have a more isotropic form.

Perhaps the most nearly isotropic form that has been suggested for the iron atom is that proposed by Hull[1] to account for the X-ray spectrum obtained from an iron crystal. This atom has two electrons near the center, the remaining 24 electrons being situated at the corners of three similarly oriented, concentric cubes whose diagonals are in the ratio $\frac{1}{4} : \frac{1}{2} : 1$. The diagonal of the outer cube, at whose corners the "valence electrons" are situated, he estimates as 1.25×10^{-8} cm. It is possible to calculate the order of magnitude of the effect due to the change in the orientation of such atoms by the help of the formula given by Darwin[2] and one of the writers,[3] in which the energy of the different orders of the reflected X-ray beam is expressed as a function of the distances of the various electrons from the mid-planes of the layers of atoms to which they belong and the distance between the successive layers of atoms. Since the value of D for the natural (111) faces of magnetite is known, 4.10

---

[1] A. W. Hull, Phys. Rev., 9, 85 (1917).
[2] C. G. Darwin, loc. cit.
[3] A. H. Compton, loc. cit.

$\times 10^{-8}$ cm., the intensity of the reflected beam can thus be determined for any given orientation of the atoms and for any order of reflection.

In Table I. we have shown the results of this calculation for different orders of reflection from magnetite. The numbers given in the second and third columns represent the ratio of the intensity when the crystal

TABLE I.

Order.	E_1/E_u.	E_2/E_u.
1	1.000	1.004
2	.96	1.03
3	.86	1.09
4	.51	1.09

is magnetized to its value when unmagnetized. The intensity E_u for the unmagnetized crystal is calculated for random orientation of the atoms. In the second column the intensity E_1 is calculated for the case in which the cube face of the atom is parallel with the crystal face, i.e., on the assumption that the magnetic axis of the atom is perpendicular to its cube face. The intensity E_2, used in the third column, is estimated for the atoms with their cube diagonals perpendicular to the crystal face. In performing these calculations it has been assumed that the oxygen atoms in magnetite have their electrons arranged in the same manner as in calcite[1] and that they are unaffected by the magnetic field. It is also supposed that parallel to the (111) planes of magnetite, which were those used in the experiment, all the atomic layers are similar and are similarly spaced. Whether or not these assumptions are strictly accurate, the change to be expected if the atoms are rotated by a magnetic field should be of the order of magnitude here estimated.

It is obvious from these calculations that if the iron atoms in a magnetite crystal are of the type suggested by Hull, a rotation of the atoms due to magnetization would have easily been detected in our experiments.

The surprisingly large calculated variation in the intensity of the reflected beam for a magnetic atom as nearly isotropic as that proposed by Hull makes it appear improbable that any reasonable fixed distribution of the electrons in the iron atom would be so isotropic as to make possible a rotation of the atoms without detection. There remains the possibility that the electrons may be arranged at random as a sort of atmosphere about the atomic nucleus. While this would result in an atom which would be on the average isotropic, it is obvious that such an atom would not as a whole have any polarity and could not therefore

[1] A. H. Compton, loc. cit.

as a whole act as an elementary magnet. Subject to some uncertainty with regard to the extent of the orientation of the ultimate magnetic particles, the conclusion may therefore be drawn from our experiment that *the elementary magnet is very probably not the atom as a whole.*

If one admits the possibility of only a very slight change in orientation of the elementary magnets, the effect of magnetization might have escaped notice even though the atoms are not isotropic. For example, if the iron atoms are built on Bohr's model, the magnetic moment per atom will be so great that the saturation intensity of magnetization of iron would require a maximum change of orientation of only 6 degrees. This would produce an effect which if an iron crystal had been used would have just been detectable in our experiment, but with the crystal of magnetite which we employed would have been considerably smaller than the errors of measurement. However, the argument given above for the nearly complete orientation of the elementary magnets seems to eliminate the necessity of considering this possibility.

Other Possible Explanations of Ferromagnetism.—There remain four possible explanations of the magnetic properties of iron: (1) Only a small fraction of the total number of atoms may be subject to orientation by an external magnetic field, these atoms possessing a very large magnetic moment. (2) The magnetic properties of the atom may be due to a few electrons revolving in very small orbits, a change in the orientation of these orbits occurring without any change in the orientation of the remainder of the atom. (3) The positive nucleus of the atom may be magnetic and subject to orientation by the magnetic field. And (4) the electron itself may be magnetic and subject to orientation by the magnetic field.

1. A minimum limit may be placed upon the number of atoms of the Rutherford type which may be turned around by the magnetic field without being detected in our experiment. It has been shown above that a complete orientation of all the atoms should increase the intensity of reflection in the fourth order by a factor of 5. Thus the orientation of 1/500th of the atoms would be sufficient to produce a change of 1 per cent. Since in our experiment we worked at only about .4 of the saturation intensity of the magnetite crystal, for complete saturation it is possible that 1 atom out of 200 might be turned by the magnetic field.

Considerations of symmetry make it appear improbable that in a homogeneous crystal of similarly arranged atoms so small a fraction of the atoms should be subject to orientation by a magnetic field. But perhaps a more serious difficulty with this hypothesis is the large magnitude of the magnetic moment which it is necessary to assign to the particular iron atoms which are subject to orientation. The magnetic moment per atom in magnetite is

$$\frac{\text{saturation intensity of magnetization}}{\text{number of atoms per cubic cm.}} = 5.1 \times 10^{-21} \text{ e.m.u.}$$

This value[1] is of the same order of magnitude as the moment 12×10^{-21} e.m.u. of an atom of oxygen,[2] for example, which can be calculated on the basis of Langevin's theory of a paramagnetic gas. If, however, only 1 in 200 of the atoms is subject to orientation by the external field, the magnetic moment of the mobile atoms must be $200 \times 5 \times 10^{-21}$ or 1×10^{-18} e.m.u. Suppose that all the electrons in this atom rotate in coplanar orbits, each with angular velocity $h/2\pi$, as assumed by Bohr. The magnetic moment per electron will then be 9.2×10^{-21} e.m.u.,[3] and for all 26 electrons will be $26 \times 9.2 \times 10^{-21} = .24 \times 10^{-18}$ e.m.u. This hypothesis would thus necessitate an atom with the apparently prohibitive magnetic moment of not less than 4 times that supposed by Bohr.

2. If the magnetic properties of magnetite are to be accounted for by a single electron in each atom rotating in an orbit whose plane can be altered without changing the orientation of the atom as a whole, it can be shown that the radius of this orbit must be less than 1×10^{-9} cm. in order to account for the negative result of our experiment. A study of the relative intensities of the different orders of an X-ray spectrum line shows, however,[4] that the outer electrons of an atom are at a distance of the order of 1×10^{-8} cm. from the center of the atom. If the mobile electrons are rotating about the nucleus, this hypothesis would therefore mean that the electrons responsible for the atom's magnetic properties are in one of the inner rings whose radius is less than 1/10 that of the atom. There are a number of effects, however, which are difficult to explain unless an atom's magnetic properties depend upon its surface electrons. Among these may be noted:[5]

1. The profound effect of chemical constitution on the magnetic properties of an atom,
2. The effect of temperature on magnetic properties,
3. The effect of mechanical jars in facilitating the orientation of the elementary magnets.

These phenomena make it appear improbable that the ultimate magnetic

[1] S. Dushman, Theories of Magnetism, p. 44.
[2] Ibid., p. 24.
[3] O. W. Richardson, Electron Theory of Matter, p. 395.
[4] A. H. Compton, Phys. Rev., 9, 52 (1917).
[5] Cf. K. T. Compton and E. A. Trousdale, loc. cit.

particle consists of a ring of electrons revolving as near the center of the atom as the result of our experiment requires.

It is possible, however, that these small orbits may be those of the outer valence electrons revolving about positions of equilibrium. If this is the case, there must be intense forces acting on these electrons of a kind concerning which we have as yet no knowledge. Thus the central force required to hold an electron in so small an orbit with angular momentum sufficient to account for the magnetic moment of the iron atom in magnetite must not be less than that due to a positive charge of 10 electronic units placed at the center of the orbit. In view of our ignorance concerning the character of the infra-atomic forces, it cannot perhaps be said that forces of this magnitude do not occur near the surface of the atom. We have, however, no other reason to suspect the existence of such forces, and the assumption that they exist would mean a very radical departure from our usual ideas of electrodynamics. One would therefore wish to consider this explanation of our experiment only as a last resort.

3. The considerations just brought forward as indicating that the magnetic properties of an atom depend in large measure upon its surface electrons are of equal weight as opposed to the hypothesis that the positive nucleus is the elementary magnet. Moreover the experiments of Barnett on magnetization[1] by rotation and by Stewart on rotation by magnetization[2] show that at least the major part of ferromagnetism is due to the motion of *negative* electricity. This is difficult to explain if the *positive* nucleus of the atom is the ultimate magnetic particle. Our experiment, however, brings forth no new evidence on this point.

4. It is clear that the result of our experiment is in accord with the hypothesis that the electron itself is the ultimate magnetic particle, unless the electron is assumed to have dimensions so great that a change in its orientation will produce an appreciable effect. Let us suppose with Parson[3] that the electron is a ring of electricity with a magnetic axis perpendicular to its plane, and in addition that on magnetization the axis of the electron is brought perpendicular to the reflecting surface of the crystal. If all the electrons are oriented by magnetization and if it is assumed that the different elements of the electron move under the action of the incident wave as if they were independent charged particles of definite mass, calculation then shows that the effect should not be noticeable in our experiment if the radius of the electron is less than 4×10^{-10} cm. Thus if the radius is 2×10^{-10} cm. as one of us has esti-

[1] S. J. Barnett, loc. cit.
[2] J. Q. Stewart, loc. cit.
[3] A. L. Parson, Smithsonian Misc. Collections, Nov., 1915.

mated on the basis of the scattering of X-rays and gamma rays,[1] the hypothesis that the electron is the ultimate magnetic particle is in accord with our experimental results.

Conclusions.—The fact that on magnetizing the crystal of magnetite the intensity of the reflected X-ray beam did not change by as much as 1 per cent. in any of the first four orders therefore supports the conclusion of K. T. Compton and E. A. Trousdale that the elementary magnet in a ferromagnetic substance is not a group of atoms.

Subject to the validity of our argument for the nearly complete alignment of the elementary magnets with the external magnetic field, it is also apparently impossible to explain this result on the hypothesis that the atom as a whole acts as the ultimate magnet, since an orientation of the atoms by the applied magnetic field would have made a noticeable change in the intensity of the reflected X-ray beam.

The hypotheses that only a small fraction of the atoms are turned by the magnetic field or that the magnetic effects are due to certain electrons whose orbits may change without affecting the remainder of the atom, have not appeared to be plausible explanations of our result.

Our experiment is in accord with the suggestion that *the magnetic properties of matter are due either to the nucleus of the atom or to the individual electrons.* Auxiliary evidence, however, indicates that *the electron is the more probable elementary magnet.*

This experiment was performed in the Physics Laboratory of the University of Minnesota during the winter of 1916–17. We take pleasure in thanking Professor K. T. Compton, of Princeton University, for his helpful suggestions in interpreting the results of this work.

                A. H. C.,
                        Cavendish Laboratory, Cambridge.
                O. R.,
                        Bureau of Standards, Washington.

May 22, 1920.

[1] A. H. Compton, PHYS. REV., 14, 31 (1919).

THE ABSORPTION OF GAMMA RAYS BY MAGNETIZED IRON.

By A. H. Compton.

Synopsis.

Ring Electron Theory of Magnetism; a Deduction Regarding the Absorption of Gamma Rays by Iron.—We should expect a ring electron to absorb more energy if its axis is parallel to a transversing beam of gamma rays than otherwise; therefore, if by magnetization the axes of the ring electrons are turned so as to be more nearly parallel to the gamma rays, the part of the absorption due to the transfer of energy to such electrons would be increased. The negative result given by the experiment carried out to test this deduction, however, does not disprove the theory but indicates merely that the absorption due to this cause is small.

Absorption of Gamma Rays by Iron.—The effect of magnetization was to change the absorption by less than 0.03 per cent. whether the magnetic field was parallel or perpendicular to the beam of hard gamma rays used.

THE recent experiments of Rognley and the writer[1] have indicated that the ultimate magnetic particle which is responsible for the property of ferromagnetism is probably the individual electron. If this is the case, it is natural to suppose that the electron may have the form of a ring, such as that proposed by Parson[2] in his "magneton" theory of atomic structure. It is clear, however, that the conceptions of a magnetic electron and a ring electron are not synonymous, since any form of electric charge in rapid rotation will exhibit magnetic polarity. The present investigation of the effect of magnetization on the absorption coefficient of gamma rays in iron has been carried out with the hope of detecting an effect due to the orientation by a magnetic field of anisotropic, magnetically polarized electrons.

Let us suppose that a ring electron has its axis parallel with an incident beam of gamma rays. The phase of the electromagnetic wave will in this case be the same over all parts of the ring. The acceleration of the electron produced by the electric vector of the incident wave will correspondingly be greater than if the plane of the ring is so inclined that the phase of the incident wave differs at different points on the ring. Because of its greater acceleration, the energy transferred to the motion of the electron as a whole will be greater when its axis is parallel with the incident beam than when oriented in any other manner. If by magne-

---

[1] A. H. Compton and Oswald Rognley, Phys. Rev., 16, 464 (1920).
[2] A. L. Parson, Smithsonian Inst. Pub. No. 2371 (1915).

tization the axes of some of the electrons in iron are turned until they are parallel with the transmitted gamma rays, any absorption which is due to motion imparted to the electron as a whole should be increased.[1] On the other hand, any energy absorbed in setting up internal or nutational oscillations of the electrons may or may not be affected by changing the electron's orientation, depending upon the manner in which the transfer of energy is effected.

When this experiment was undertaken, it was supposed on the basis of the work of Florance,[2] Ishino[3] and others that a large part of the total absorption of gamma rays is due to scattering, the magnitude of which depends upon the acceleration of the whole electron. Recent experiments by the writer,[4] however, have shown that if there is any true scattering of gamma rays, the energy thus dissipated is only a small fraction of that transformed into fluorescent radiation. These experiments have indicated further that the principal part of the fluorescent radiation produced is of a kind which is not characteristic of the particular absorbing element, thus differing from the fluorescent radiation excited by X-rays of ordinary hardness. While for X-rays most of the absorption due both to scattering and fluorescence is probably due to energy imparted to the motion of the electron as whole,[5] it is therefore not improbable that in the case of the much shorter gamma rays the principal part of the absorption is due to the transfer of energy to nutational and elastic oscillations of the electron itself. Because of this uncertainty as to the mechanism of absorption, while a positive effect on the absorption coefficient of gamma rays due to magnetization would presumably mean an anisotropic electron, the interpretation of no effect at all is ambiguous.

The Experiments.—In view of this ambiguity in the interpretation of the results it hardly seems worth while to record the details of the experimental procedure. It may suffice to say that extended series of observations were made by an accurate balance method, and all possible precautions, including check measurements with other metals substituted for iron, were taken to eliminate any errors due to such causes as the mechanical effects of the strong magnetic field. Though in the experiments on parallel magnetization the size of the beam of gamma rays was necessarily determined by the holes in the pole pieces of the magnet, a sufficiently large amount of radium emanation (about 75 millicuries) was employed to secure relatively intense ionization.

[1] The theory of this effect has been discussed in a preliminary manner by the writer in Journ. Wash. Acad. Sci., 8, 1 (1918).
[2] D. C. H. Florance, Phil. Mag., 27, 225 (1914).
[3] M. Ishino, Phil. Mag., 33, 140 (1917).
[4] A. H. Compton, Phil. Mag., in printer's hands.
[5] Cf. A. H. Compton, PHYS. REV., 14, 253 (1919).

The results of the experiment may be expressed as follows: When the iron was magnetized to saturation parallel with the transmitted beam of gamma rays, the observed increase in its absorption coefficient was (0.004 ± 0.019) per cent. For magnetization perpendicular to the gamma ray beam the absorption coefficient was increased by (0.023 ± 0.018) per cent. That is, for parallel magnetization the effect on the absorption coefficient is probably less than 1 part in 5,000, and for perpendicular magnetization the effect is probably less than 1 part in 3,000.

Interpretation of Results.—If the ratio of the wave-length of the gamma rays λ to the radius of the electron a is approximately 2.5, as was estimated by the writer on the basis of Ishino's experiments on the total scattering of gamma rays,[1] and if the scattering by a ring electron is calculated on the basis of the assumptions previously employed,[2] it can be shown that the part of the absorption coefficient due to scattering for (1) random orientation, (2) the axes of the ring electrons parallel and (3) the axes perpendicular to the transmitted gamma rays are in the ratio of about 1.0 to 1.5 to 0.8 respectively. Similarly the part of the fluorescent absorption which is due to energy transmitted to the motion of the electron as a whole may be calculated in the same general manner as that used by the writer in calculating the absorption of very hard X-rays.[2] The acceleration of a ring electron is so altered by its orientation that if $\lambda/a = 2.5$ this type of absorption in the three cases should be in the ratio 1.0 to 7.2 to 0.14 respectively. Since the writer's more recent experiments have indicated that these types of absorption do not play an important part in the total absorption of gamma rays, one does not feel justified in presenting the details of these calculations.

There is good reason to believe that when iron is saturated the elementary magnets of which it is composed are in nearly perfect alignment with the external magnetic field.[3] Thus if the elementary magnet is the ring electron, and if either of the two types of absorption just considered constitute any appreciable fraction of the total absorption of gamma rays, it is clear that magnetization of the iron should have produced a measurable effect upon the absorption coefficient. For if all the absorption were to be accounted for in this manner, the orientation of 1 electron in 20,000 by the applied magnetic field would hardly have escaped detection. If we accept the idea of nearly complete alignment of the elementary magnets with the external field when saturation occurs, it would therefore seem necessary to conclude either (1) that the ultimate magnetic particle is not a ring electron, or (2) that energy transmitted

---

[1] A. H. Compton, Phys. Rev., 14, 26 (1919).
[2] A. H. Compton, Phys. Rev., 14, 253 (1919).
[3] Cf., e.g., Compton and Rognley, loc. cit.

to the motion of the electron as a whole is not responsible for any considerable part of the total absorption of gamma rays.

If only one ring electron in each atom is oriented by the magnetic field, an effect of one part in 5,000 should have been produced on parallel magnetization if but 1 per cent. of the total absorption is due to scattering. Although experiment shows that by far the greater part of the absorption of gamma rays is due to fluorescence, the fact that for very hard X-rays scattering is still prominent makes it appear improbable that scattering should be as unimportant as this result would imply. Thus while it is not possible to draw any definite conclusion from this experiment until more information is available with regard to the mechanism of gamma ray absorption, the evidence seems rather opposed to the hypothesis that a ring electron is the ultimate magnetic particle.

It will be remembered, on the other hand, that A. H. Forman described in this journal several years ago[1] an experiment on the absorption of X-rays by magnetized iron which showed an increase in the absorption coefficient of about 10 times the probable error of measurement when iron was magnetized parallel with the transmitted rays. He found no effect for perpendicular magnetization. The writer has pointed out elsewhere[2] that the effect observed by Forman is of the order of magnitude to be expected if the electrons are rings of electricity which are oriented by the magnetic field. It is however rather surprising that while Forman found the change in the absorption coefficient due to magnetization to increase rapidly for shorter wave-lengths, the present experiment shows no effect for the still shorter wave-length gamma rays. In view of the apparently conflicting evidence, the writer does not feel justified in drawing any conclusions concerning the form of the electron from these experiments.

This work has been performed under the auspices of the National Research Council. I desire to thank Professor Rutherford for placing at my disposal the facilities of Cavendish Laboratory for carrying out the experiment.

CAVENDISH LABORATORY,
   CAMBRIDGE UNIVERSITY,
      AUGUST 20, 1920.

[1] A. H. Forman, PHYS. REV., 7, 119 (1916).
[2] A. H. Compton, Journ Wash. Acad. Sci., 8, 1 (1918).

Washington University Studies

Vol. VIII   JANUARY, 1921   No. 2

CLASSICAL ELECTRODYNAMICS AND THE DISSIPATION OF X-RAY ENERGY

ARTHUR H. COMPTON

Professor of Physics

The object of this paper is to find out whether it is possible to account for the known phenomena of X-ray absorption and scattering on the basis of the classical electrodynamics. In addition to its importance in connection with the theory of radiation, the answer to this question will determine whether we may use the phenomena of X-ray scattering as a basis for determining the arrangement of the mobile electricity within matter. We shall find that while our experimental knowledge of absorption does not find a ready explanation, the scattering of high frequency radiation is in good accord with the classical electrical theory.

I. THE SCATTERING OF HIGH FREQUENCY RADIATION

Let us consider first in a general way the assumptions that are necessary to explain the scattering of X-rays. The well known theory of J. J. Thomson[1] of the scattering of X-rays by matter is based upon four assumptions: 1. That the usual electro-magnetic theory is applicable to the problem. 2. That each electron in the scattering material acts independently of every other electron. 3. That there are no other forces

[1] J. J. Thomson, *Conduction of Electricity through Gases*, 2nd Ed., p. 321.

acting on the electrons which are comparable in magnitude with the forces due to the incident beam of X-rays. And 4, that the dimensions of the electron are negligible compared with the wave-length of the incident radiation. On the basis of these assumptions he showed that the scattering coefficient of any substance for electro-magnetic radiation should be

$$(1) \quad \sigma_o = \frac{8\pi}{3} \frac{e^4 n}{m^2 C^4},$$

where e is the charge and m the mass of the electron, n is the number of electrons per unit volume of the scattering material, and C is the velocity of light. He found further that the relative scattering at different angles θ with the incident beam should be proportional to $(1 + \cos^2 \theta)$.

For scattering materials of low atomic weight and X-rays of moderate hardness, this theory has been found to account fairly well for the experimental observations.[2] When, however, the conditions of the experiment are varied, the results depart widely from the predictions based upon these hypotheses. In the first place, when soft X-rays and matter of relatively high atomic weight are employed, the scattering becomes much greater than is to be expected according to formula (1).[3] Moreover, under these conditions a decided assymmetry occurs between the energy scattered at small angles and that scattered at large angles with the incident beam, the energy on the incident side being in some cases only a small fraction of that appearing on the emergent side of a scattering plate.[4] On the other hand, when radiation of very high frequency is employed, a difference of another kind appears. In this case apparently the same assymmetry between the incident and the emergent scattered radiation exists as was found for the comparatively long waves, but as the wave-

[2] Barkla and Ayers, *Phil. Mag.*, 21, 275 (1911).

[3] Crowther, *Proc. Camb. Phil. Soc.*, 16, 367 (1911); *Proc. Roy. Soc.*, 86, 478 (1912).
Barkla and Dunlop, *Phil. Mag.*, 31, 229 (1916).
Owen, *Proc. Camb. Phil. Soc.*, 16, 165 (1911).

[4] Crowther, *loc. cit.*

length is shortened the total scattered energy diminishes rapidly to a value very much smaller than that assigned by equation (1).[5]

The satisfactory agreement with Thomson's formula in the case of the scattering of X-rays of moderate hardness by elements of low atomic weight indicates that under these conditions the assumptions upon which his theory is based are justified. When radiation of the same wave-length is scattered by elements of high atomic weight, the assumption that the electron's size is negligible must continue to hold. If the increased and assymmetrical scattering by such elements is to be explained according to the classical electrodynamics, therefore, either the forces on the electrons in such atoms are no longer negligible, or the electrons must be grouped together in such a manner as to coöperate in their scattering. It is possible that both of these causes are effective in producing the excess scattering. The absence of any detectable refraction for X-rays, however, indicates that no considerable amount of resonance appears when X-rays traverse matter, such as should occur if the intra-atomic forces on the electron are comparable with the forces due to the incident X-ray beam. Moreover, while resonance due to such forces might explain the increase in the total scattering by heavy elements, it would not account for the fore and aft assymmetry of the scattered radiation.

It is possible, however, to give a satisfactory qualitative explanation of the phenomenon of excess scattering as due to coöperation between the electrons grouped close together near the centers of the atoms. In the atoms of the heavy elements it is probable that a considerable number of the electrons are within 10^{-9} cm. of the center of the atom. When X-rays 5×10^{-9} cm. wave-length are employed, which is a moderate hardness, the phase difference between the rays scattered by these inner electrons will not be large, so that the intensity

[5] Cf. e. g. Ishino, *Phil. Mag.*, 33, 129 (1917) and A. H. Compton, *Phil. Mag.* (in printer's hands).

of the scattered beam will be considerably greater than for the same electrons acting independently. Since the phase difference will be greater at large than at small angles with the incident beam, this concentration of the electrons accounts also for the relatively larger amount of energy scattered at the small angles.

Let us now consider the scattering by elements of low atomic weight of radiation of very short wave-length. We have seen that for longer waves these elements scatter in a manner which indicates that the electrons act independently of each other and that the intra-atomic forces are negligible. For very short waves, therefore, the electrons must continue to act independently, and the forces of constraint will presumably have even less effect than in the case of the lower frequency waves. Thus in the case of the scattering of gamma rays by light elements, of the four assumptions underlying Thomson's theory, the hypotheses of negligible positional forces on the electrons and of independent scattering by the different electrons are apparently justified. If the phenomena are to be explained on the usual electromagnetic theory, therefore, the assumption that the electron's size is negligible as compared with the wave-length of the incident radiation must cease to be valid for these very short waves. It has been pointed out by the writer[6] that the assumption of an electron of appreciable dimensions affords a satisfactory account of the scattering of radiation of very high frequency in virtue of the interference which occurs between the rays scattered by different parts of the electron.

It thus appears that if the scattering of high frequency radiation is to be explained on the basis of the classical electrodynamics, two general assumptions are necessary with regard to the distribution of the electricity within matter. Firstly, a considerable portion of the electrons in the atoms of the heavy elements must be at distances from the centers of the

[6] A. H. Compton, *Journ. Wash. Acad. Sci.*, Jan. 1, 1918. *Phys. Rev.* 14, 20 (1919), *Phil. Mag.* (in printer's hands).

atoms less than 10^{-9} cm. Secondly, the dimensions of the electrons themselves must be comparable with the wave-length of gamma rays. If we are to explain the scattering of high frequency radiation in a quantitative manner, therefore, it will be necessary to derive a scattering formula which will take into account such a twofold distribution of electricity inside the atom.

The Scattering of X-rays by Groups of Electrons. It is fortunate that the phase of a ray scattered by an electron of appreciable size is identical with that due to one of negligible dimensions placed at its center if the large electron consists of a symmetrical distribution of electricity. This enables us to calculate the scattering at any angle due to groups of electrons as if each electron were a point, and later to modify our result to take into account the distribution of the electricity within the electron itself.[7] C. G. Darwin has obtained an expression for the scattering of X-rays by groups of electrons in terms of the positions of the individual electrons.[8] This enabled him to account qualitatively for the phenomenon of excess scattering. But his formula is not applicable in a quantitative way, since it does not express the average intensity of the scattered beam but only the intensity at a particular instant. The general problem of determining the mean intensity of the beam scattered by any fixed distribution of electrons within the atom has been solved by Debye,[9] who has also calculated the scattering to be expected from an atom all of whose electrons are arranged in a single ring. The latter problem has been solved in a different manner by Schott,[10] with the same result as that previously obtained by Debye. General discussions of the excess scattering due to electrons

[7] This can be done strictly only in case the electrons all consist of the same isotropic distribution of electricity, since only under this condition will all the electrons scatter alike. Even if the electron is not isotropic, the error introduced in our calculation by the above assumption will, however, be exceedingly small.

[8] C. G. Darwin, *Phil. Mag.*, 27, 325 (1914).

[9] P. Debye, *Ann. d. Phys.*, 46, 809 (1915).

[10] G. A. Schott, *Proc. Roy. Soc.*, 96, 695 (1920).

arranged in groups have also been given by Barkla[11] and the writer.[12]

For our present purpose the general formula obtained by Debye is of the most interest. He finds that the average intensity of the radiation scattered at an angle θ with the incident beam by a single atom is given by the expression

$$(2) \quad I_\theta = I_1 \sum_1^N \sum_1^N \frac{\sin\left[\frac{4\pi}{2} s_{mn} \sin\frac{\theta}{2}\right]}{\frac{4\pi}{\lambda} s_{mn} \sin\frac{\theta}{2}},$$

where I_1 represents the intensity of the beam scattered by a single electron, N is the number of electrons per atom which contribute to the scattering, and s_{mn} is the distance from the mth to the nth electron. This expresssion can be applied strictly to any atom in which the electrons are held at fixed distances from each other, and thus includes static atoms and atoms all of whose electrons are in a single ring. For atoms which consist of several rings rotating at different angular velocities or in which the electrons move at random, however, the distances s_{mn} are not constant, which makes this formula inapplicable.

An attempt has been made by G. A. Schott[13] to obtain an expression which will describe the scattering by an atom consisting of a plurality of rings of electrons. In his discussion, however, he explicitly neglects the mutual coöperation in the scattering by the different rings. As a result, his formula gives for the maximum scattering per atom for long waves and small angles, the value

$$I_{\theta\,max} = I_1 \Sigma_s n_s^2,$$

where n_s is the number of electrons in the sth ring, and the

[11] Barkla and Dunlop, *Phil. Mag.*, 31, 229 (1916).

[12] A. H. Compton, *Phys. Rev.* 14, 20 (1919). Cf. also R. A. Houstoun, *Roy. Soc. Edinburgh Proc.*, 40, 43 (1920), of whose work I have as yet seen only an abstract. Sir J. J. Thomson has kindly loaned me a manuscript in which he has solved more completely than has Debye the problem of scattering from a fixed distribution of electrons.

[13] G. A. Schott, *loc. cit.*

summation is taken over all the rings. It will be seen, however from equation (2), as Barkla has also pointed out from general considerations,[14] that under these conditions the scattering should be

$$I_{\theta \, max} = I_1 N^2 = I_1 (\Sigma_s n_s)^2.$$

It can readily be shown that this discrepancy is due to Schott's neglect of the mutual action of the different electronic rings.

I have not attempted to solve explicitly the problem discussed by Schott. There are certain particular types of distribution of the electrons within the atoms, however, which lead to usable formulas for the intensity of the scattered radiation. One of these which is of particular interest is a distribution of pairs of electrons, the line joining the two electrons of each pair having a wholly random orientation, and the middle of the line coinciding with the center of the atom. This distribution includes as special cases the arrangement of the electrons in the hydrogen molecule, the helium atom and possibly the beryllium atom. Furthermore, if all the electrons are at the same distance from the center, the scattering will be very nearly the same as that due to a ring of electrons of the same radius. Thus a formula based upon this distribution enables us to calculate very approximately the scattering to be expected from atoms composed of rings of electrons. The details of the calculation in this case I shall reserve for a later paper,[15] but the result obtained for the scattering is:

$$(3) \quad I_\theta = I_1 \left\{ N + 2 \sum_1^{N/2} \left(\frac{\sin 2k_s}{2 k_s} - 2 \frac{\sin^2 k_s}{k_s^2} \right) + 4 \left(\sum_1^{N/2} \frac{\sin k_s}{k_s} \right)^2 \right\}$$

$$= I_1 \Psi, \text{ where } k_s = \frac{4\pi \rho_s}{\lambda} \sin \frac{\theta}{2},$$

and ρ_s is the distance of each of the sth pair of electrons from the center of the atom.

For very short wave-lengths, in which case the k's become very large, expression (3) approaches the value $I_\theta = I_1 N$. For long waves, when the k's become small, I_θ approaches the

[14] C. G. Barkla, *Phil. Mag.*, 31, 231 (1916).
[15] A. H. Compton (to be published soon in The Philosophical Magazine).

value $I_1 N^2$. This result therefore leads to the correct values for the scattering in the two limiting cases.

The Scattering of X-rays by Individual Electrons. The complete evaluation of the intensity of the beam of X-rays which will be scattered at any angle requires a knowledge of the quantity I_1, the scattering due to a single electron. If the electron has negligible dimensions, Thomson showed[16] that the intensity of the beam scattered by a single electron to a point at a distance L at an angle θ with the incident beam is,

$$(4) \quad I_\theta = I \frac{e^4 (1 + \cos^2 \theta)}{2 L^2 m^2 C^4},$$

where I is the intensity of the incident beam. If the dimensions of the electron are not negligible, it is necessary to take into account the phase difference between the rays scattered by different parts of the electron. The intensity of the scattered beam will then depend upon the particular form which is assigned to the electron. We shall consider the two cases of an electron in the form of a ring, and of a solid spherical electron, one of which will be found to account in a satisfactory manner for the experiments on the scattering of radiation of very high frequency.

In an earlier calculation of the scattering by a ring electron[17] a slight error was introduced by a faulty method of averaging the scattering over different orientations of the electron. This error has been pointed out by G. A. Schott,[18] who has derived a correct expression as a result of an intricate study of the limiting case of the scattering of X-rays by a ring containing a large number of electrons. The following much simpler analysis will avoid the error made in the writer's earlier calculation, and will lead to a result identical with Schott's.

In such a calculation it is necessary that we determine the acceleration to which each portion of an electron will be sub-

[16] J. J. Thomson, *Conduction of Electricity through Gases*, 2nd Ed., p. 326.
[17] A. H. Compton, *Phys. Rev.*, 14, 37 (1919).
[18] G. A. Schott, *loc. cit.*

ject when traversed by an electromagnetic wave. In a steady electric field the acceleration of the whole electron is of course Ee/m, where E is the electric intensity, and e and m have their usual significance. But it must be remembered that electromagnetic inertia of the electron is due to the back e.m.f. at each point on the electron due to the acceleration of the other parts. If the incident radiation is of so short wavelength that the electric intensity differs over different parts of the electron, the inertia of any portion of the electron will therefore differ from the corresponding inertia in a uniform electric field. The shortest wave-length radiation that we know is hard gamma rays, which almost certainly has effective wave-length greater than 10^{-10} cm. Supposing that the size of the electron is comparable in magnitude, those parts of the electron which are within say 1/20th of a wave-length or 5×10^{-12} cm. of the point under consideration will, however, have practically the same acceleration; while the parts of the electron at a distance greater than 5×10^{-12} cm. will have only a small effect on the effective mass of the element under consideration. For imagine a whole electronic unit of charge e of mass m placed 5×10^{-12} cm. from the element of charge whose motion we are studying; it may be shown in this case that the acceleration of the element of charge in a uniform electric field is not affected more than 1 part in 17 by the presence of the first charged particle. Thus the inertia of any element of an electron is practically dependent upon the parts of the electron within a radius of 5×10^{-12} cm., and is for the shortest wave-lengths with which we are acquainted sensibly independent of the wave-length. As long as we confine ourselves to the consideration of waves longer than 10^{-10} cm., therefore, it is legitimate to treat each element of the electron as possessing a definite ratio of charge to mass.

In an earlier calculation it was left uncertain whether the mass of an element of a ring electron is the same when accelerated perpendicular to or tangent to the ring. Following the considerations brought forward by Webster,[19] however, it will be seen that the principle of relativity demands that the mass shall be the same along every axis. For if the mass were greater for motions tangent to the ring, if the ring electron were moving freely in the direction of its axis, when turned edgewise to the direction of motion it would necessarily slow up in order that its kinetic energy might remain constant. It is clear that such variations in translational velocity according to the electron's orientation relative to its direction of motion would constitute a means of detecting a system's absolute velocity through the ether. Consequently the whole electron, and similarly each part of it, must have the same mass for motion in all directions.

For convenience in calculation the assumption will be made, as before, that the electron is flexible, i.e., that the positional forces on any element of it are negligible compared with the forces due to the incident radiation. This arbitrary assumption is justified in part by the fact that the scattering by such an electron is very similar, except for extremely short waves, to that of a rigid electron which is subject to rotation by the incident wave. Furthermore, if the different parts of the electron are incapable of any motion relative to each other, it is impossible to account for the fore and aft assymmetry of the scattering of very high frequency radiation.

In figure 1, imagine a beam of X-rays going in the direction -XOX, and being scattered by a ring electron of radius a, whose axis is at an angle γ with the bisector of the angle -XOP. If A_0 is the amplitude at P due to an electrified particle of charge e and mass m placed at O, and if e/m is the ratio of the charge to the mass for an element of the circumference of the ring electron, the amplitude at P due to a pair of elements at opposite sides of the ring will be,

$$2 A_0 \frac{ad\alpha}{2\pi a} \cdot \cos\left(\frac{2\pi}{\lambda} \cdot 2 z \sin \frac{\theta}{2}\right).$$

Here z is the distance of either element from a plane drawn

[19] D. L. Webster, *Phys. Rev.*, 9, 484 (1917).

through O perpendicular to the bisector OR, and a is the angle between a line drawn through the elements under consideration and the intersection of the plane just determined with the plane of the ring electron. It will be apparent from the figure that
$$z = a \sin a \sin \gamma,$$
whence the amplitude due to this pair of elements is
$$dA = A_0 \frac{da}{\pi} \cdot \cos\left(\frac{4\pi a}{\lambda} \sin a \sin \gamma \sin \frac{\theta}{2}\right).$$

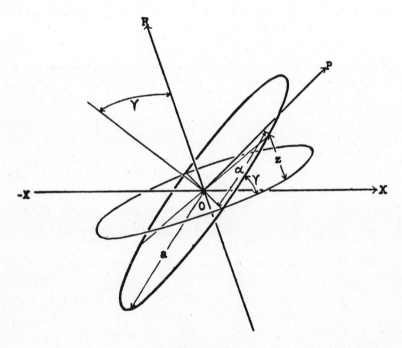

FIGURE 1

When this expression is integrated over all angles a between 0 and π, we find for the amplitude due to the ring electron for this particular orientation,
$$A_{\gamma, \theta} = A_0 \, J_0 \left[\frac{4\pi a}{\lambda} \sin \gamma \sin \frac{\theta}{2}\right],$$

where J_0 indicates Bessel's J function of the zero order. If we write $cA_0{}^2 = I_0$, we find for the corresponding intensity,

$$I_{\gamma,\theta} = c\,(A_{\gamma,\theta})^2 = I_0 J_0{}^2\left[\frac{4\pi a}{\lambda}\sin\gamma\sin\frac{\theta}{2}\right].$$

Assuming the axis of the electron to have entirely random orientation, the probability that the angle γ will lie between γ and $\gamma + d\gamma$ is $\tfrac{1}{2}\sin\gamma d\gamma$. Thus the average value of the intensity at P of the rays scattered by the ring electron is,

$$(5)\quad I_1 = I_0 \int_0^\pi J_0{}^2\left[\frac{4\pi a}{\lambda}\sin\gamma\sin\frac{\theta}{2}\right]\cdot\tfrac{1}{2}\sin\gamma d\gamma,$$

$$= I_0 \frac{1}{x}\sum_0^\infty J_{2n+1}(2x) = I_0\,\Phi_1(x),$$

where $x = \dfrac{4\pi a}{\lambda}\sin\dfrac{\theta}{2}$ and J_s represents Bessel's J function of the sth order.[20] This result may be put in the alternative form

$$I_1 = I_0 \sum_0^\infty \frac{(-1)^s}{(s!)^2(2s+1)}\,x^{2s},$$

which is that given by Schott. The form given in equation (5) is more readily calculable, however, for large values of x.

The scattering by a solid spherical electron may be calculated in a similar manner. The contribution to the amplitude due to the charge included in the shell whose inner and outer radii are r and $r + dr$ is the integral of the contributions of all the elementary rings of height between z and $z + dz$ above the plane through O perpendicular to OR (figure 2). It is clear that the rays scattered in the direction OP will be in the same phase from all parts of each elementary ring, and that the phase of the beam from the whole electron will be the same as that from a charged particle placed at its center. The volume of such a ring is $2\pi r dr dz$, so if ρ is the volume density of electricity at a distance r and e/m is the ratio of the charge

---

[20] This integral has been evaluated by Debye (*loc. cit.*) in obtaining an expression for the scattering by a ring containing an infinite number of electrons. Values of $J_s(2x)$ for different values of s and x may be found in Janke-Emde's *Function Tables*, pp. 147 et seq.

to the effective mass of any element, the amplitude at P due to the electricity in the ring is

$$A_o \frac{2\pi r\, dr\, dz \cdot \rho}{e} \cos\left(\frac{2\pi}{\lambda} 2z \sin\frac{\theta}{2}\right),$$

where as before A is the amplitude at P due to an electrified particle of charge e and mass m placed at O, and θ is the angle POX. The amplitude due to the whole spherical shell of radii r and $r + dr$ is of course the integral of this quantity between the limits $z = -r$ and $z = +r$, or

$$\frac{A_o\, \rho\, \lambda r}{\sin\theta/2} \sin\left(\frac{4\pi r}{\lambda} \sin\frac{\theta}{2}\right) dr.$$

Let us assume that the electrification is so distributed that

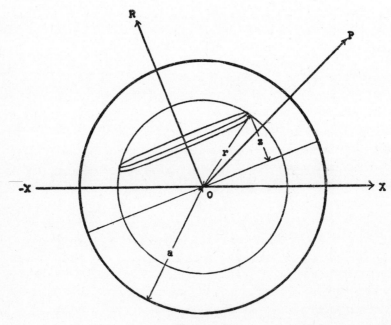

FIGURE 2

all concentric spherical shells of the same thickness, within the boundary of the electron, contain the same amount of electrification. The charge within a shell of thickness dr is then $e.dr/a$, where a is the radius of the electron, whence $\rho =$

$\frac{edr}{a}/4\pi r^2 dr = e/4\pi ar^2$. Substituting this value in the above expression, and integrating for all values of r between O and a, we find for the amplitude of the ray scattered by the whole electron,

$$A = A_0 \left\{ 1 - \frac{x^2}{3\cdot 3!} + \frac{x^4}{5\cdot 5!} - \frac{x^6}{7\cdot 7!} + \cdots \right\},$$

where $x = \frac{4\pi a}{\lambda} \sin \frac{\theta}{2}$. Writing as before $cA_0^2 = I_0$, we find for the corresponding intensity,

(6) $I = cA^2$

$$= I_0 \left\{ 1 - \frac{x^2}{3\cdot 3!} + \frac{x^4}{5\cdot 5!} - \cdots \right\}^2 = \Phi_2(x).$$

Scattering Formulas: The average intensity of the beam scattered by a single atom at an angle θ with the incident beam is therefore

(7) $\quad I_\theta = I_0 \cdot \Phi(x) \cdot \Psi,$

where

(8) $\begin{cases} I_0 = I \dfrac{e^4 (1 - \cos^2 \theta)}{2 L^2 m^2 C^4} \quad \begin{bmatrix} I = \text{intensity of incident beam,} \\ L = \text{distance at which scattered} \\ \quad \text{beam is evaluated.} \end{bmatrix} \\[2ex] \Phi(x) = \begin{bmatrix} \dfrac{1}{x} \sum\limits_0^\infty J_{2n+1}(2x) \text{ (for ring electron)} \\ \left\{ 1 - \dfrac{x^2}{3\cdot 3!} + \dfrac{x^4}{5\cdot 5!} - \cdots \right\}^2 \text{ (for solid spherical electron)} \end{bmatrix} \\[2ex] \qquad\qquad\qquad\qquad\qquad \begin{bmatrix} x = \dfrac{4\pi a}{\lambda} \sin \dfrac{\theta}{2} \\ a = \text{radius of electron} \end{bmatrix} \\[2ex] \Psi = \left\{ N + 2 \sum\limits_1^{N/2} \left(\dfrac{\sin 2k_s}{2k_s} - 2 \dfrac{\sin^2 k_s}{k_s^2} \right) + 4 \left(\sum\limits_1^{N/2} \dfrac{\sin k_s}{k_s} \right)^2 \right\}, \\[2ex] \begin{bmatrix} N = \text{atomic number of scattering element,} \\ k_s = \dfrac{4\pi \rho_s}{\lambda} \sin \dfrac{\theta}{2}, \text{ where } \rho/_s \text{ is the distance of the electrons} \\ \qquad\qquad\qquad \text{of the } s\text{th pair from the center of the atom.} \end{bmatrix} \end{cases}$

In these expressions I_0 represents the intensity of the scattered beam due to a particle of charge e and mass m but of neglible dimensions. The factor $\varphi(x)$ accounts for the distribution of the electric charge in the electron. The function

ψ sums up the effect due to the various electrons in the atom, taking into consideration the mutual coöperation in the scattering by the different electrons.

It is obvious that the total energy scattered by an atom is given by the expression

$$2\pi L^2 \int_0^\pi I_\theta \sin\theta d\theta.$$

The atomic scattering coefficient is therefore,

$$\frac{\sigma}{\nu} = \frac{\pi e^4}{m^2 C^4} \int_0^\pi (1 + \cos^2\theta) \sin\theta \cdot \phi(\mathrm{x}) \cdot \psi \cdot d\theta,$$

which by expression (1) is,

(9) $\quad \dfrac{\sigma}{\nu} = \dfrac{\sigma_0}{\nu} \cdot \dfrac{3}{8N} \int_0^\pi \phi(x) \cdot \psi \cdot (1 + \cos^2\theta) \sin\theta d\theta = \dfrac{\sigma_0}{\nu} \cdot \Theta.$

In any actual case this integral may be evaluated graphically if the distribution of the electricity within the atom is known.

It will be well at this point to review the various hypotheses upon which formulas (7) and (9) are based. Of the four assumptions underlying Thomson's formula, we have retained the idea of the applicability of the classical electrodynamics, because it is necessary as a working hypothesis. We have also retained the assumption that the positional forces on the various electrons are neglible as compared with the forces due to the incident radiation. This has been kept in view of the fact that our qualitative study of the phenomena of scattering did not disclose the presence of any such positional forces. In the factor ψ the assumption of the independence of the scattering by the different electrons has been eliminated. Our particular method of evaluating this factor has itself involved certain approximations, but it can be shown that the expression derived is sufficiently reliable for use in discussing the existing experimental determinations of the scattering. We have not retained Thomson's assumption that the dimensions of the electron are negligible, but have introduced the counter assumptions of certain special forms of electrons of comparatively large size. Expressions (7) and (9) do not, therefore, represent the most general formulas

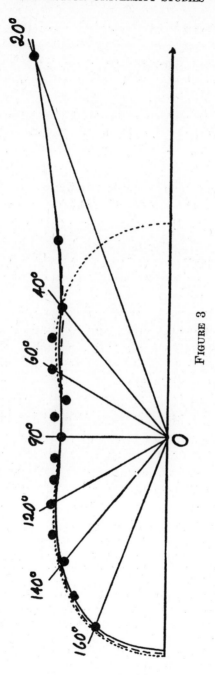

FIGURE 3

that can be given for the scattering according to the usual electromagnetic theory; but if we can thus explain the experiments in a satisfactory manner, the applicability of this theory to the scattering of high frequency radiation will be strongly confirmed.

Comparison with Experiment. First let us consider the application of this formula to the scattering of X-rays of comparatively long wave-length. The experiments of Barkla and Ayers[21] on the scattering of "moderately soft X-rays" by carbon are shown by the circles in figure 3. The dotted line represents the scattering to be expected according to Thomson's formula. In the broken curve this formula is modified by the insertion of the factor ψ, considering the carbon atom to consist of a pair of electrons at a distance 0.35 A. U. from the center, and two pairs at a distance of 0.6 A. U. from the center.[22] I have taken "moderately soft X-rays" to mean a uniform distribution of energy over wave-lengths from 0.45 to 0.9 A. U. The solid curve represents the scattering by the same kind of atom if the electrons are solid spheres of radius 0.05 A. U. This size of electron, which is determined by experiments on the scattering of hard X-rays and gamma rays as described below, will be used throughout the paper. The difference between the last two calculated values will be seen to be negligibly small for these relatively long waves, but this figure shows that for long wave-lengths the scattering even by light atoms cannot be calculated without considering the arrangement of the electrons within the atoms.

The same result is shown in figure 4, in which are plotted

[21] Barkla and Ayers, *Phil. Mag.*, 21, 275 (1911).

[22] It will be noted that these distances are very nearly 4 times those assigned by the assumption used by Bohr, namely that the angular momentum is $h/2\pi$, and that the radius of the orbit can then be calculated according to ordinary dynamics. In the case of the heavier elements the difference is even more marked. This result is in accord also with the distribution of the electrons in the atoms of calcite and rock-salt as calculated (A. H. Compton, *Phys. Rev.*, 9, 52, 1917) from the intensity of X-ray spectrum lines. The scattering predicted on the basis of the $h/2\pi$ relation is in all cases much greater than is actually observed.

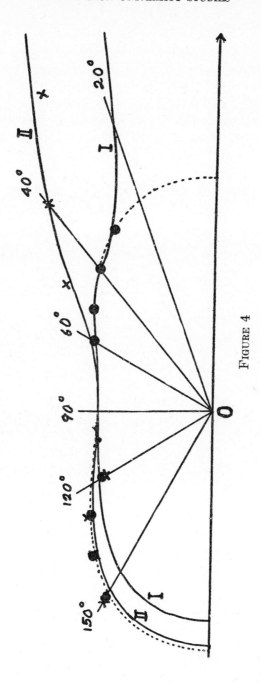

FIGURE 4

Owen's measurements [23] of the scattering of X-rays by filter paper ($C_6H_{10}O_5$). The crosses represent measurements with an equivalent spark gap of 2.5 cm. (which I have taken to mean a uniform distribution of energy between wave lengths 0.5 and 1.0 A. U.) and the circles correspond to a spark gap of 7 cm. (λ from 0.25 to 0.50 A. U.). As in the former figure, the dotted line is based upon Thomson's formula, while the solid lines are calculated for the two wave-lengths assuming the electrons in hydrogen to scatter independently, carbon to have the same structure as before, and oxygen to have a pair of electrons .26 A. U. from the center and three pairs at a distance of .42 A. U. It will be seen that the agreement of the experimental values with those calculated in both figures 4 and 5 is accurate within experimental error.

The scattering by elements of higher atomic weight has been examined by Crowther [24] and by Barkla and Dunlop. [25] Most of these tests have been made by examining the magnitude of the scattering in the neighborhood of 90 degrees. The results of the latter experimenters may be taken as typical. They determined the ratio of the scattering per unit mass by copper, silver, tin and lead respectively to that by aluminium for radiations of different penetrating power. I have shown their values of these ratios in figure 5, using the wave-length as estimated by the experimenters. The curves are calculated on the basis of the following arrangements of the electrons:

Aluminium		Copper		Silver and Tin		Lead	
No. of electrons	Distance A. U.	No.	Dist.	No.	Dist.	No.	Dist.
2	.12	2	.052	2	.036	2	.022
8	.26	10	.104	10	.073	10	.045
3	.7	8	.24	8	.17	16	.090
		8	.42	16	.34	16	.135
		1	1.05	8	.51	16	.202
				4½	.7	16	.31
						6	.6

[23] E. A. Owen, *Proc. Camb. Phil. Soc.*, 16, 165 (1911).
[24] Crowther, *Phil. Mag.*, 14, 670 (1907); *Proc. Roy. Soc.*, 86, 478 (1912).
[25] Barkla and Dunlop, *Phil. Mag.*, 31, 229 (1916).

It will be seen that these distributions lead to values of the scattering that agree with Barkla and Dunlop's measurements within the probable error of their experiments.

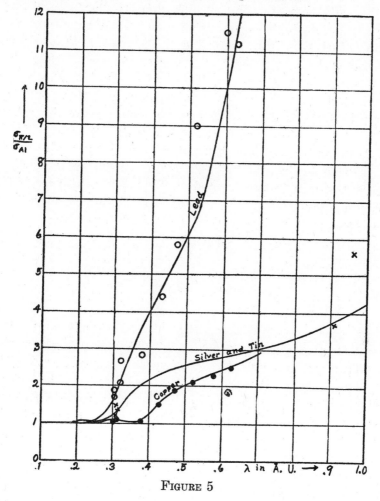

FIGURE 5

In an earlier paper on the scattering of high frequency radiation,[26] the writer discussed the experiments of Florance and Ishino on the "scattering" of gamma rays. More recent

[26] A. H. Compton, *Phys. Rev.*, 14, 20 (1919).

work[27] has shown, however, that in these experiments the secondary radiation measured was not truly scattered gamma rays. The only direct measurements that have been made of the scattering of radiation of higher frequency seem to be those of the writer[27] on the scattering of the hard gamma rays from radium C. These experiments showed that the scattered rays at large angles with the incident beam are less than 1/1000 as intense as one would expect on the hypothesis of a small electron. At smaller angles, however, there was apparently some scattering, 2 to 4 per cent of the theoretical value at 45°. For hard X-rays of wave-length about 0.15 A. U., Hull and Rice have estimated[28] from their absorption measurements that the total scattering coefficient is about 64 per cent of that to be expected on Thomson's theory. If for a rough calculation we take $\theta = 60°$ as the average angle at which the scattering is evaluated,* this datum gives:

$$\theta = 60°; \ x = \frac{4\pi a}{\lambda} \sin \frac{\theta}{2} = 0.42 \times 10^{10} a; \ I/I_o = 0.64.$$

Similarly, from the writer's experiments,

$$\theta = 120°; \ \lambda = c. \ 3 \times 10^{-10} \text{ cm.} \quad x = 3.63 \times 10^{10} a; \ I/I_o < 0.001.$$
$$\quad 45°; \quad \quad 1.3 \times 10^{-10} \quad \quad 1.61 \times 10^{10} a \quad = c.0.03.$$

For these very short waves the coöperation between the different electrons in the atom is negligible in comparison with the errors of experiment. The function ψ may therefore be taken as unity, and the ratio I/I_o is equal to ϕ. The experimental values of ϕ are accordingly plotted in figure 8 as functions of a/x. In this figure the broken line represents the value of ϕ as calculated for a ring electron according to equation (5), assuming the radius to be $a = 2.6 \times 10^{-10}$ cm., while the solid line is calculated for the spherical electron described above from equation (6) assuming $a = 5 \times 10^{-10}$ cm. It is clearly impossible to account for the very low value of the scattering of γ-rays on the basis of a flexible ring electron.

[27] A. H. Compton, *Phil. Mag.*, (in printers' hands).
[28] Hull and Rice, *Phys. Rev.*, 8, 326 (1916).
* Since there is more scattering forward than backward.

On the other hand, the spherical electron hypothesis gives a satisfactory explanation of the meager data that are available.

On account of the difficulty of making accurate measurements of the scattering of high frequency radiation, it is not as yet possible to give a more satisfactory test of the revised

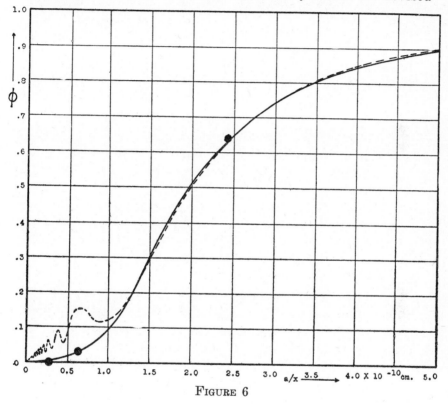

FIGURE 6

scattering formulas. The important point is, however, that the present hypotheses do afford an explanation of the scattering of gamma rays and X-rays which is qualitatively correct and quantitatively as reliable as our experimental knowledge. The comparison of the present theory with experiment must therefore be taken to confirm the applicability of the classical electrodynamics to the calculation of the scattering of high frequency radiation.

II. THE ABSORPTION OF HIGH FREQUENCY RADIATION

The experimental facts with regard to the absorption of X-radiation of not too high frequency are fairly well summed up by Owen's empirical formula,

$$(10) \quad \frac{\mu}{\nu} = K N^4 \lambda^3 + \sigma_0/\nu,$$

where μ/ν is the absorption coefficient for a single atom, N is the atomic number of the absorbing element, λ is the wave-length of the incident radiation, σ_0/ν is the atomic scattering coefficient as calculated according to Thomson's theory of scattering, and K is a constant for all elements whose fluorescent K radiation is excited by the incident X-rays, a different constant for those elements in which only the L and softer characteristic radiations are excited, etc. The first term in this expression represents the "true" or fluorescent absorption, and the second term the energy lost by the incident beam due to the scattering. This expression, though fairly satisfactory for X-rays which are not too hard, is not found to hold accurately for wave-lengths shorter than .3 A. U., the absorption for these very hard rays being usually less than indicated by Owen's formula.

The writer has shown elsewhere [29] that the term for the fluorescent absorption in Owen's law is identical with that calculated on the basis of J. J. Thomson's old theory of X-ray absorption [30] when this is combined with Moseley's law connecting frequency and atomic number. That is, Thomson's result is expressed in terms of the thickness of the incident X-ray pulse and the natural frequency of the absorbing electrons. The pulse thickness may obviously be interpreted as wave-length, whereas by Moseley's law the natural frequency of the absorbing electrons may be taken to be approximately proportional to the square of the atomic number. The appropriate substitutions were found to lead directly to Owen's term for the fluorescent absorption,

[29] A. H. Compton, *Phys. Rev.*, 14, 248 (1919).
[30] J. J. Thomson, *Cond. of Elec. through Gases*, 2nd Ed., pp. 326-8.

$$(11) \quad \frac{\tau}{\nu} = K N^4 \lambda^3.$$

The uncertainty of the theoretical basis for this fluorescent absorption term is obvious in view of the fact that it is an essential part of Thomson's theory that the incident X-ray shall have the form of a pulse, whereas it now appears highly probable that the X-rays consist of comparatively long trains of waves. For this reason the applicability of the classical electrodynamics to the calculation of X-rays must be left open until a more satisfactory theory of fluorescent absorption is developed. The provisional theoretical basis that has been found for the absorption term, however, serves to emphasize the physical existence of a type of fluorescent absorption which is at least approximately proportional to N^4 and λ^3.

A New Type of Absorption. If the absorption of high frequency radiation is to be explained according to the classical electrodynamics, there is another term which should appear in the expression for the total absorption. This arises from the fact that the incident radiation consists of wave-trains of finite length which presumably begin rather abruptly. When a single such train traverses an electron, it will set up a forced oscillation which will be compounded of a vibration of the frequency of the incident wave superposed upon a vibration whose frequency is that of the free oscillation of the electron. The first type of oscillation is that which results in the scattering, and persists only as long as the electron continues to be excited by the incident wave. The second type of oscillation represents a transformation of the energy of the primary beam into another form which will probably reappear as fluorescent radiation characteristic of the electron traversed.

The amount of energy thus transformed can be estimated in the following manner. Let us suppose that the electric vector of the primary wave-train is expressed by $A e^{i(p_1 t + \delta) - kt}$ where k represents the damping of the primary beam, δ is its phase at the time $t = 0$, and $p_1 = 2\pi C/\lambda$ where C is the velocity of light and λ is the wave-length. If l and $q_1 = 2\pi C/\lambda'$ are the corresponding damping and frequency terms for the natural oscillations of the absorbing electron, its equation of motion, considering it to act as a point charge, is

$$\frac{d^2x}{dt^2} + 2l\frac{dx}{dt} + q^2 x = A \cdot \frac{e}{m} \cdot e^{i(p_1 t + \delta) - kt},$$

where
$$q^2 = q_1{}^2 + l^2.$$

If the electron is at rest when $t = 0$, the solution of this differential equation is

$$x = A_1 e^{-kt} \cos(p_1 t + \Delta) + B e^{-lt} \cos(q_1 t - \beta),$$

where
$$\begin{cases} A_1 = \frac{Ae\sqrt{R}}{mp^2}; & \begin{cases} R = 1 / \left\{ \left[1 - \frac{q^2}{p^2}\right]^2 - 4\frac{k}{p^2}(k-l)\left[1 - \frac{q^2}{p^2}\right] \right. \\ \left. + \frac{4}{p^2}(k-l)^2 \right\}; \; p^2 = p_1{}^2 + k^2; \end{cases} \\ \Delta = \delta + \tan^{-1}\{2p_1(k-l)/(q^2-p^2) + 2k(k-l)\}; \\ B = A_1 \sqrt{1 + \left[\frac{p^2}{q^2} - 1\right]\sin^2 \Delta}; \; \beta = -\tan^{-1}\left[\frac{p}{q}\tan \Delta\right] \; [l \ll q]. \end{cases}$$

The energy which appears in the free oscillations of the electron is therefore

$$(11.5) \quad E_a = B^2 \frac{mq^2}{2} = \frac{A^2 e^2 R R'}{2 p^2 m},$$

where
$$R' = \frac{q^2}{p^2}\left\{1 + \left[\frac{p^2}{q^2} - 1\right]\sin^2 \Delta\right\}.$$

This energy depends upon the phase δ of the primary beam at the time $t = 0$. If the electron emitting the primary radiation is stimulated to vibration by the impact of a cathode particle, and if the duration of the impact is short compared with the period of the vibration, δ and hence also Δ will be approximately zero, so that $R' = 1$. If on the other hand the primary electron starts its oscillation by falling into a position of equilibrium from an outside point it may be that Δ should be taken more nearly as $\pi/2$, in which case $R' = q^2/p^2$. The mean value of R' for all values of δ is $\frac{1}{2}(1 + q^2/p^2)$. Since in any actual case for most of the electrons in the atom q^2/p^2

will be small compared with unity, R' will be very nearly independent of the wave-length of the primary rays, and will be comparable with but less than unity.

In order to calculate the corresponding term in the absorption coefficient, it is necessary to evaluate the total energy in the primary wave-train. This can be done if the damping of the electron giving rise to this wave-train is known. If we assume that this damping is r times that to which a single electron is subject due to its own radiation, the effective value of the quantity k is $\frac{1}{3} \frac{e^2 p^2}{mC^3}\ r$. The energy in the primary pulse per unit area at the electron is, however,

$$E_p = \int_0^\infty \frac{c}{8\pi} A^2 e^{-2kt} = \frac{C A^2}{16\pi k},$$

or

$$E_p = \frac{3}{16\pi r} \cdot \frac{m C^4 A^2}{e^2 p^2}.$$

Thus we should expect the absorption coefficient per electron due to this transformed radiation to be approximately

$$(12) \quad a_e = \frac{Ea}{E_p} = \frac{8\pi}{3} \frac{e^4}{m^2 C^4} \cdot RR'r.$$

If, as in calculating the scattering, we neglect resonance effects, the factor R becomes unity and the atomic "transformation" coefficient becomes

$$(12.5) \quad \frac{a}{\nu} = \frac{8\pi}{3} \frac{e^4 N}{m^2 C^4} \cdot R'r = R'r \cdot \frac{\sigma_0}{\nu}.$$

This is of the same order of magnitude as the scattering coefficient σ_0/ν as calculated according to Thomson's theory, and should be appreciable for very short wave-lengths where the ordinary fluorescent absorption becomes small. It is distinguished from the usual absorption as expressed by relation (11) in that it is very nearly independent of the wave-length and is proportional to the first power of the atomic number.

It is important to notice that the value of the "transformation" coefficient depends greatly upon the form of the front of the primary wave train. In the above calculation it has been assumed that the wave begins suddenly at the time t = o. In view of the catastrophic origin of X-rays and gamma rays, it does not appear improbable that this assumption may be nearly correct. If, however, instead of starting suddenly the amplitude of the incident wave gradually increases to a maximum and then dies down, the energy thus transformed into the natural vibration of the absorbing electron is very small. The wave-train reflected from a crystal must increase gradually in this manner as the waves scattered from the successive layers begin to reinforce each other. Furthermore, since energy is drawn from the first few waves of the train to set the absorbing electron in motion, it appears almost certain that after traversing an appreciable thickness of matter the front of the wave will be, as it were, smoothed off. As a result the absorption of this type will in any actual case probably be much less than the value given by equation (12.5).

To show the importance of the form of the wave front, consider a wave-train like that used above, which has been reflected from a perfect crystal composed of n layers of atoms equally spaced and similar in composition. We shall suppose that the number n is large, but not large enough to absorb any appreciable fraction of the incident radiation, nor large enough for the primary beam to decrease appreciably in amplitude due to damping over a distance of n wave-lengths. If $t = 0$ is the time when the ray reflected from the first layer of atoms reaches the transforming electron, the amplitude of the electric vector is zero until $t = 0$, is $\frac{Apt}{2\pi n} e^{i(pt + \delta)}$ between the times $t = 0$ and $t = 2\pi n/p$ and is $Ae^{i(pt + \delta) - kt}$ after $t = 2\pi n/p$. The equation of motion during the intermediate interval may be shown to be

$$x = -A_1 \frac{pt}{2\pi n} \cos(pt + \delta) + \frac{A_1 p}{2\pi nq} \sin qt \cdot \cos \delta,$$

where A_1 has the value used above, neglecting k and l during this short interval. Thus after $t = 2\pi n/p$, the equation of motion becomes

$$x = -A_1 e^{-k(t-t^0)} \cos(pt + \delta) + B_1 e^{-l(t-t^0)} \sin qt,$$

where $t_0 = 2\pi n/p$, and $B_1 = A_1 \dfrac{p}{q} \dfrac{\cos \delta}{2\pi n}$. The energy in the natural vibrations of the electron is now, therefore,

$$Ea' = B_1^2 \frac{mq^2}{2} = \frac{\cos^2 \delta}{8\pi^2} \cdot \frac{A^2 e^2 R}{mn^2 p2},$$

while that in the beam which traverses the electron is as before

$$Ep' = \frac{3}{16\pi} \frac{m C^4 A^2}{r e^2 p^2}.$$

The transformation coefficient for a single electron in this case is thus

$$a_e = Ea'/Ep' = \frac{2 \cos^2 \delta}{3\pi n^2} \frac{e^4 Rr}{m^2 C^4},$$

which, since n is large compared with 1, is a negligible fraction of the transformation coefficient for the beam before it was reflected from the crystal. It is apparent that measurements of the absorption of radiation of the same wave-length before and after reflection from a crystal will thus afford a means of testing for the existence of this absorption due to transformation.

The Revised Absorption Formula. We shall now consider the effect on the expression for the total absorption of introducing the assumptions which have been found necessary to account for the scattering. In the first place it is obvious that the atomic scattering coefficient is no longer given by Thomson's formula, but is given instead by expression (9). In the second place, any type of fluorescent absorption depends upon the energy which is transferred to an electron by the radiation which traverses it. If the energy thus absorbed results in motion of the electron as a whole, the amount of energy transferred will be proportional to the square of the maximum acceleration to which the electron is subject when traversed by radiation of unit intensity. This is true, for example, for the particular types of absorption considered in equations (11) and (12). But since the electric vector differs in phase over different parts of an electron of appreciable dimensions, the resultant acceleration is not as great as that to which an electron of negligible size would be subject. For this reason a factor must be introduced into expressions (11) and (12) which is the square of the ratio of the acceleration of the electron under consideration to that of a point charge electron when both are traversed by a beam of the same intensity.

The value of this factor for a ring electron has previously been calculated by the writer,[31] but the same error in averaging was introduced as that made in calculating the scattering. The correct expression may be shown to be

(13a) $\left\{\dfrac{\text{Acceleration of ring electron}}{\text{Acceleration of v. small electron}}\right\}^2 = \phi_1(z) =$

$$\dfrac{1}{z} \sum_0^\infty J_{2n+1}(2z),$$

where $z = 2\pi a/\lambda$, and ϕ is the same function as that used in equation (7). For a solid spherical electron of radius a, in which as before the volume density of electric charge is inversely proportional to the square of the distance from the center, the value of this ratio may be proved to be,

(13b) $\phi_2(z) = \left\{1 - \dfrac{z^2}{3\cdot 3!} + \dfrac{z^4}{5\cdot 5!} - \cdots\right\}^2,$

z being as before $2\pi a/\lambda$, and ϕ_2 having the same meaning as in equation (7).

The atomic absorption coefficient for primary radiation may therefore be written, according to the assumptions made in this paper,

(14) $\dfrac{\mu}{\nu} = K N^4 \lambda^3 \phi(z) + \dfrac{\sigma_0}{\nu} r R' \phi(z) + \dfrac{\sigma_0}{\nu} \Theta,$

where ϕ is given by equations (13), $R'r$ by equation (12.5) and Θ by equation (9). In the case of X-rays reflected from a

[31] A. H. Compton, *Phys. Rev.*, 14, 253 (1919).

crystal, the second term should be omitted. It will be understood that this formula does not rest upon as secure a theoretical foundation as does the expression given for the scattering. If absorption is to be accounted for on the basis of the usual electromagnetic theory, however, we should expect some absorption of each of the three types considered to appear.

Experimental Test. In order to test this formula, we may consider as typical the experimental values of the absorption of X-rays in aluminium, copper, and lead. In figures 7, 8, and 9 I have shown respectively all the experimental data that I

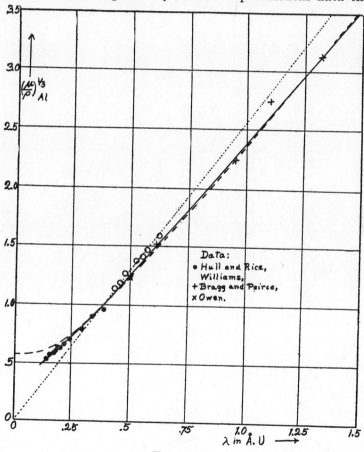

FIGURE 7

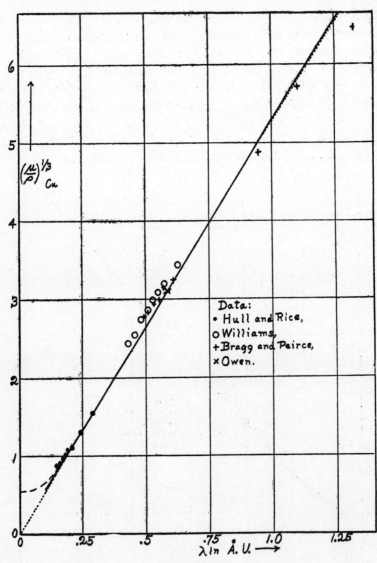

FIGURE 8

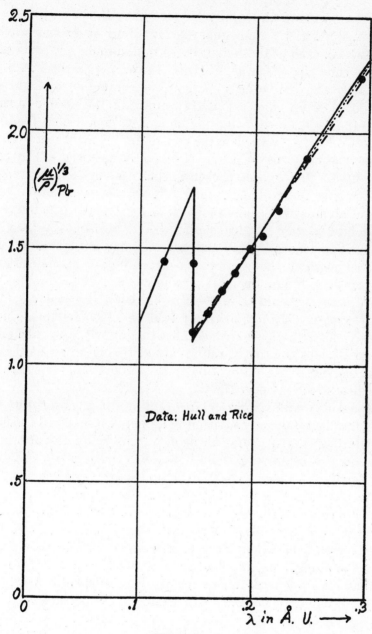

FIGURE 9

have been able to find which refer to the absorption by these metals for strictly homogeneous radiation of known wave-length. In each of these figures the dotted line represents the simple cube law which neglects the scattering, the broken line is calculated from Owen's original law, while the solid line is calculated from expression (14) omitting the second term, since the experiments were made on X-rays reflected from a crystal. For the wave-lengths considered, it makes little difference which type of large electron is assumed. The coefficients of λ^3 have been chosen in each case in such a manner as to secure the best agreement. It will be seen that in the case of aluminium the agreement with experiment is better for the solid line than for either of the other two lines, while for copper and lead all three curves are fairly satisfactory. It may be noted further that the coefficients of λ^3 in the case of copper and aluminium is 11.3, whereas according to the N^4 rule this ratio should be 10.8 — a rather good agreement. The same comparison in the case of lead is not so reliable, since there is but a single experimental datum on which to calculate this coefficient for the K radiation. This one gives us, however, a ratio to the absorption in aluminium of 174, whereas the theoretical value is 210.*

It is worth calling particular attention to the absorption of the highest frequency radiation in aluminium. It will be seen that in this case the *total* absorption falls below the absolute minimum of scattering alone if the electron is taken to have negligible dimensions. The only adequate explanation of these low values of the absorption coefficient on the basis of the classical electrodynamics seems to be that the size of the electron is comparable with the wave-length of the X-rays.

The total absorption of the hard gamma rays from radium C in aluminum, iron, and lead has been measured with care by Ishino, who finds the mass absorption coefficients to be respectively, .071, .068 and .076. Since his experiments were

* If the factor (Φ) is omitted, the ratio of the coefficients of λ^3 for lead and aluminium is 122, which is decidedly different from the theoretical ratio 210.

performed upon the primary radiation, all three terms of formula (14) must presumably be applied. In the case of the absorption by aluminium the fluorescent absorption t, being proportional to $\phi\lambda^3$, is theoretically negligible. The observed fluorescent absorption, .071, is therefore to be accounted for by the transformation term a/ρ and the scattering term σ/ρ. Since we have found that both of these types of absorption are, for short wave-lengths, proportional the number of electrons traversed, the corresponding mass transformation and scattering coefficients in iron and lead should be .068 and .057 respectively. Thus it appears that the "fluorescent" absorption coefficient τ/ρ is negligible in the case of iron and equal to .019 in the case of lead. From the theoretical expression that the atomic fluorescent absorption coefficient is

$$\frac{\tau}{\nu} = K\, N^4\, \lambda^3\, \phi_2\, (z),$$

and using the values of the constants determined from the X-ray measurements, it may be shown that the value $\tau/\rho = .019$ for lead corresponds to a wave-length $\lambda = .04$ A. U. This is in reasonable accord with the effective wave-length as estimated by other methods.

If the "true" absorption were estimated according to Owen's original expression, that in the case of lead would be 200 times as great as that for aluminium. The fact that the term which has been used to take into account the "transformed" radiation explains the likeness in magnitude of the absorption of gamma rays by the different elements may be considered as a partial justification of its introduction. Physically, however, it does not appear that this explanation of gamma ray absorption is adequate. Experiment indicates that a large part of the X-ray energy goes into the kinetic energy of high speed beta rays; and it is difficult to understand how any large part of the "transformed" energy, which by hypothesis is distributed nearly equally among all the electrons traversed, can show itself in the form of large amounts of energy on a comparatively small number of beta particles.

For this reason I am inclined to doubt whether the new type of absorption here introduced plays any considerable part in the absorption of gamma rays.

The Applicability of the Classical Electrodynamics. It has been shown that equations (7), (9) and (14) describe satisfactorily our present state of knowledge concerning the scattering and absorption of high frequency radiation by matter. The theoretical foundation of formula (14), which expresses the absorption, is not secure, since no satisfactory explanation has as yet been found for the principal fluorescent term. Until such a theory is forthcoming, it can merely be said that the absorption experiments afford no evidence against the classical electromagnetic theory. In deriving expressions (7) and (9), however, which describe the scattering, no assumptions have been introduced which in any way conflict with the classical electrodynamics. The experimental support of these formulas must therefore be taken to confirm the ordinary electrical theory as a basis for the calculation of the scattering of X-rays.

The Distribution of the Mobile Electricity within Matter. If this evidence for the adequacy of the classical electrodynamics is accepted, it is apparent that we have here a powerful means of investigating the inner structure of the atom. Expression (7) expresses the intensity of the scattered radiation in terms of no unknown quantities except the distribution of the mobile electricity within the atom. Experimental determinations of the scattering thus give direct evidence concerning this distribution. That the information is in usable form is indicated by the fact that we have found it possible to assign distributions to the electrons in the atoms on which experiments have been made with good success in accounting for their scattering. It is not claimed that there are no other arrangements of the electrons which would give as satisfactory results, but the true distributions are certainly not greatly unlike those here assigned. In particular such evidence may be of importance in the case of scattering by the lighter elements. For hydrogen and helium there is a single pair of electrons, so that a very few well chosen experiments should suffice to determine their distance apart. In fact such investigations should lead to fairly definite results for all of the lighter atoms. For the heavier atoms, on account of the much larger number of variables involved, it will probably be impossible to obtain by this means anything better than a general idea of the arrangement of the electrons.

Two other methods rather similar to this have been proposed for determining the positions of the electrons within the atom. The first of these is a study of the intensity of the different orders of reflection of an X-ray spectrum line from a crystal. The energy in such a reflected beam is a function of the distribution of the electrons in the atoms of the crystal.[32] Unfortunately, however, the energy depends also upon the absorption coefficient of the X-rays in the crystal and upon the thermal motion of the atoms. The effective absorption coefficient in this case is very difficult to determine, since a kind of selective absorption occurs at the angle of maximum reflection. Furthermore the heat motions of the atoms are not accurately known on account of the possibility of there being a motion corresponding to the absolute zero point of temperature. For these reasons this method is capable at best of giving only an approximate idea of the structure of the atom.

The second method is a study of the energy in the lines of the powdered crystal spectrograms of Debye and Scherrer and Hull. This method avoids the uncertainty due to the selective absorption of the X-rays, but the brightness of the spectrum lines in this case also depends upon the heat motions of the atoms. Though the effect of the zero point energy is not large, there is thus some uncertainty in the interpretation of these results as well as in the case of reflection from a large crystal. Since the intensity of the lines in the powdered crystal spectrum depends upon both the arrangement of the electrons and the heat motion of the atoms, while the intensity of

[32] C. G. Darwin, *Phil. Mag.*, 27, 325 (1914).

the general scattered radiation depends only upon the arrangement of the electrons, it is not impossible that a comparison of the results in the two cases might lead to a sufficiently accurate estimate of the heat motions of the atoms to test for the existence of the zero point energy.

SUMMARY

It has been shown that if the experiments on the scattering of high frequency radiation are to be explained on the basis of the classical electrodynamics, there must be a two-fold distribution of the mobile electricity within matter. In the first place, there must be a comparatively large number of electrons within a distance of .1 A. U. of the centers of the heavier atoms. In the second place the electrons themselves must consist of electricity distributed over a region comparable in radius with the wave-length of hard gamma rays.

On the basis of this two-fold distribution of electricity, a formula for the scattering of high frequency radiation by matter has been derived which has been found to be in satisfactory accord with experiment.

Owen's formula for the absorption of X-rays by matter has been modified to take into account this distribution of the electricity in matter, and an extra term has been inserted to take account of a certain type of fluorescent absorption which the usual electromagnetic theory demands. As thus modified, Owen's law is found to be in good agreement with experiment.

There appears, therefore, no evidence of the failure of the classical electrodynamics in connection with measurements of the absorption of high frequency radiation, while experiment confirms the view that the classical electrodynamics is applicable to calculation of the scattering of X-rays and gamma rays.

Assuming the validity of the classical electrical theory, it is pointed out that experiments on the scattering of high frequency radiation afford a powerful means of studying both the distribution of the electrons in atoms, and the structure of the electrons themselves.

Possible Magnetic Polarity of Free Electrons.
By Arthur H. Compton, *Ph.D.**

MY attention has recently been called by Mr. Shimizu to the fact observed by Mr. C. T. R. Wilson that the paths of beta and secondary cathode rays excited by X-rays in air usually terminate in converging helices. The comparatively uniform character of these curves shows itself clearly in certain of Mr. Wilson's beautiful stereoscopic photographs which have not been published but which he has very kindly allowed me to examine. For example, one photograph, which shows the complete tracks of 66 secondary cathode particles, reveals 52 tracks of helical form, only two or three tracks showing no general curvature of this type, the remaining 12 rays having paths too irregular to detect with certainty any helical curvature that may exist. These paths may have the form of either a right- or left-handed helix, and the axes of the different helices have a nearly random orientation. A detailed examination of a large number of tracks to prove that the observed curvature of the paths is not a random one, will necessarily involve much time and labour. In default, however, of a complete proof, it is of interest to see whether an explanation can be offered of the apparent tendency of the beta particles to move in a type of spiral orbit.

Though the spiral form of the paths would suggest a motion of the particles in a magnetic field, the chance orientation of the axes of the different helices shows that these axes are not determined by any external magnetic field, but are rather characteristic of the individual beta rays. More specifically, the axis of the helix must be parallel with some polarity of the beta particle which is relatively permanent in direction. It is apparent that a simple electric charge can possess no polarity whose orientation will remain constant in spite of numerous collisions with other charges unless it is in rapid rotation. Mr. Shimizu accordingly suggested that Mr. Wilson's photographs may be explicable on the assumption that the electron has a definite magnetic polarity which on account of gyroscopic action does not change rapidly in direction.

It is clear that a magnetic field whose direction is determined by the electron passing through it, is capable of producing the type of spiral track that is observed. But a beta particle which acts as a magnetic doublet as well as an electric charge is capable of producing such a magnetic field

* Communicated by Professor Sir E. Rutherford, F.R.S.

if the medium through which it passes is susceptible to magnetization. For the introduction of such a doublet will induce magnetization in the surrounding medium just as a bar magnet induces magnetization in a neighbouring mass of iron. If the atmosphere acts paramagnetically, the magnetic field at the doublet due to the induced magnetization has the same direction as the magnetization of the doublet. Conversely, the induced magnetic field due to a diamagnetic atmosphere is opposite in direction to the doublet's axis. This induced magnetic field will clearly have the same effect on the motion of the electron as would an externally applied field of the same intensity. That is, the beta ray will move in a helical path whose axis is parallel with the magnetic axis of the beta particle. The path may have the form of either a right- or left-handed helix, according as the north or south pole of the beta particle is foremost. This is in accord with observation, which shows paths of both kinds. Furthermore, if the induced magnetic field does not decrease while the velocity of the beta ray diminishes, the tracks of the particles will be *converging* helices such as appear in the photographs.

The intensity of the magnetic field induced at a beta particle due to the effect of its magnetic moment on the part of the medium at a distance greater than r from the doublet may be shown to be

$$H_r = \frac{8\pi\mu s}{3r^3},$$

where μ is the magnetic moment of the doublet and s is the effective susceptibility of the medium. If the cathode particle is moving at a speed corresponding to a drop through 10,000 volts, the minimum distance of its approach to an electron at rest is according to usual theory about 10^{-10} cm. Taking this as the value of r, using for s the value 3×10^{-8} of the magnetic susceptibility of air, and for μ the moment 10^{-20} of an electron with an angular momentum $h/2\pi$, this expression indicates that the intensity of the magnetic field induced at the electron should be of the order of 3000 gauss. This is approximately the field which would be required to produce the observed curvature.

The impulsive torque exerted by the beta ray doublet on a magnetic electron at rest when passing it at a distance of 10^{-10} cm. has, however, a period corresponding to X-ray frequency. It is clearly necessary, therefore, to take into account the inertia of the elementary magnetic doublets of which the medium is composed. Furthermore, the magnetic forces acting on an electron due to a beta particle at this close range will greatly exceed the restoring forces due to the other electrons in the atom which for weaker fields may make the atom diamagnetic. In fact, there is no reason for supposing that the effective susceptibility of the medium when subjected to such highly intense magnetic pulses of very short duration has any intimate relation to its susceptibility in steady and comparatively weak fields. It seems rather that the problem must be treated as a statistical one of encounters of a rapidly moving, electrically charged, magnetic doublet with a random distribution of similar doublets at rest. A preliminary investigation of this problem indicates, however, that the average motion of the beta ray should be rather similar to its motion when passing through a medium of uniform susceptibility. The above numerical discussion is therefore of value as showing that it is not unreasonable to expect a magnetic particle to induce in the surrounding medium a magnetization of the magnitude required to account for the observed helical paths.

If the obvious explanation of these spiral tracks is the correct one, their interpretation yields very valuable results. We have seen that the beta ray seems to act as a tiny gyroscope with a magnetic moment, which is capable of giving rise to torques on other electrons of a duration corresponding to the frequency of X-rays. The reaction, according to classical dynamics, must result in mutational oscillations of the spinning beta particle. This obviously supplies a mechanism for the production of high frequency radiation by a free electron, which has been suggested by Webster to account for Doppler effects at the target of an X-ray tube. The possibility that the electrons are magnetic doublets is also of great importance in connexion with our ideas of the structure of the atom and the nature of chemical combination.

Cavendish Laboratory,
    Cambridge University.
  August 16, 1920.

The Elementary Particle of Positive Electricity.

THE name "negative electron" was applied to the elementary particle of negative electricity after the experimental evidence for the variation of its mass with velocity had generally convinced physicists that its whole inertia was due to its electric charge. This meaning of the term "electron" was in accord with Dr. Johnstone Stoney's original use of the word to denote the elementary unit of electric charge. With the introduction of the principle of relativity it became clear that the variation of mass with velocity was no characteristic attribute of electrical inertia, and that therefore we have no proof that the negative electron's inertia is wholly electromagnetic in origin. In fact, the investigations of Abraham, Webster, and others have shown that there must be some mass present other than that due to the electron's electric field. If we abide by Dr. Stoney's original meaning of the word, it is therefore more than doubtful whether we are justified in calling this negatively electrified particle of matter an electron. Nevertheless, the term is now so well established in the literature that we use "electron" to denote this elementary particle regardless of our view concerning the origin of its mass.

The arguments for and against the electrical origin of the mass apply in exactly the same manner to the elementary particle of positive as to the corresponding particle of negative electricity. If the negative particle can legitimately be termed an "electron," it is thus equally legitimate to apply the term to the positive particle, since it likewise carries the fundamental unit of electric charge. Why not, therefore, denote both these elementary particles by the same generic term "electron," distinguishing the "positive" from the "negative" electrons when necessary, as several writers have long been accustomed to do?

It seems to me that the application of a distinctive name, such as "proton" or "hylon" or "hydrion," to the elementary particle of positive electricity can only suggest a distinction between the nature of the positive and negative electrons, which, so far as we are aware, does not exist. Thus, for example, when an atom of hydrogen is split into its two components the negative electron is just as really a hydrogen ion as is the positive electron. The fact that both components possess equally fundamental units of electric charge and are equally fundamental divisions of matter should suggest that the same generic name "electron" be applied to each.

ARTHUR H. COMPTON.

Washington University,
St. Louis, U.S.A.,
January 25.

From the PHILOSOPHICAL MAGAZINE, vol. xli. *May* 1921.

The Degradation of Gamma-Ray Energy. By ARTHUR H. COMPTON, *Ph.D., Wayman Crow Professor of Physics, Washington University* *.

IT has long been known that when matter is traversed by gamma rays, it becomes a source of secondary † gamma radiation. The relation between the primary and the secondary gamma rays, however, has not been definitely established. Although the secondary radiation is very appreciably less penetrating than the primary rays, it has usually been considered to be due principally to true scattering ‡. It is the purpose of the present paper to investigate the nature and the general characteristics of secondary gamma rays, and to study the mechanism whereby

* Communicated by Prof. Sir E. Rutherford, F.R.S.
† In this paper the term "secondary" gamma radiation is used to denote any radiation of the gamma type excited either directly or indirectly by the passage through matter of primary gamma rays. By "scattered" radiation is meant the radiation emitted by the electrons in matter (that due to the positive nuclei is theoretically negligible in comparison) due to the accelerations to which they are directly subjected by the primary rays. The term "fluorescent" radiation signifies as usual radiation of the energy absorbed from the primary beam and stored temporarily in the kinetic and potential energies of the electrons. Its frequency therefore depends jointly upon the frequency of the primary rays and the nature of the radiator.
‡ *Cf. e. g.* E. Rutherford, 'Radioactive Substances, etc.,' p. 282. J. A. Gray, Phil. Mag. xxvi. p. 611 (1913). D. C. H. Florance, Phil. Mag. xxvii. p. 225 (1914). K. W. F. Kohlrausch, *Phys. Zeitschr.* xxi. p. 193 (1920).

comparatively soft secondary radiation is excited by relatively hard primary radiation.

From theoretical considerations, both scattered and fluorescent radiation should undoubtedly be present in secondary gamma rays. According to J. J. Thomson's well-known theory [*], when the wave-length is so short that there is no appreciable co-operation in the scattering by the different electrons, the mass scattering coefficient should be about 0·2 for all elements and all wave-lengths (if the number of electrons per atom effective in the scattering is equal to the atomic number, as seems to be the case for hard X-rays). The magnitude of the scattering to be expected is considerably reduced if the wave-length approaches the size of the electron, and may, indeed, become very small if the ratio of the wave-length to the diameter of the electron approaches unity [†]. There is, however, on the basis of the classical electrodynamics, no means of eliminating completely the scattered radiation.

Fluorescent radiations of a comparatively soft type, presumably the characteristic K radiations, have been detected in the secondary gamma rays from elements of high atomic weights [‡]. But, in addition to this, there should be excited in all elements a harder fluorescent radiation by the impact of the high-speed beta particles liberated by the primary rays. The number of such electrons expelled by gamma rays is known to be much the same per unit mass for all elements, and it has been shown experimentally [§] that gamma rays in not greatly different amounts per unit mass are excited when beta rays fall upon different substances. Thus one would expect to find in the secondary gamma rays an appreciable amount of fluorescent radiation, which, like the scattered radiation, does not differ greatly according to the element used as radiator.

The usual method of distinguishing between scattered and fluorescent radiation is by comparing the absorption coefficients of the primary and secondary radiations. It is assumed that the scattered rays are of the same hardness as the primary rays, whereas all known high-frequency fluorescent radiations are of a less penetrating type. Gray [‡] and Florance [§], however, have rejected this criterion, for although they find that the secondary radiation excited by hard gamma rays is of a distinctly softer type than the

---

[*] J. J. Thomson, 'Conduction of Electricity through Gases,' 2nd ed. p. 325.
[†] A. H. Compton, Phys. Rev. xiv. p. 23 (1919).
[‡] J. A. Gray, *loc. cit.*
[§] D. C. H. Florance, *loc. cit.*

primary radiation, they conclude that the primary rays are truly scattered, but in the process of scattering are so modified as to become less penetrating. It is therefore important to determine under what circumstances, if any, the hardness of the scattered rays may differ from that of the primary rays.

If the scattering is due to electrons of negligible dimensions which are separated far enough to act independently of each other, there is no question but that the scattered ray will be exactly similar to the primary ray in every respect except intensity; for since the accelerations to which each electron is subject are strictly proportional to the electric intensity of the primary wave which traverses it, and since the electric intensity of the scattered ray (at a great distance) due to each electron is proportional to its acceleration, the electric vector of the scattered wave is strictly proportional to the electric vector of the primary wave. Thus the frequency, the wave-form, the damping, etc., will be the same in both beams. Radiations scattered by such electrons should, therefore, be identical in character with the primary waves.

Whatever type of scattering unit be assumed, it is also clear that, if the primary wave is perfectly homogeneous—*i. e.*, if it is an indefinitely long train of simple harmonic waves of constant frequency,—the scattered waves must also be homogeneous and of the same frequency. If, however, the scattering unit—whether a group of electrons or the individual electron—is of dimensions comparable with the wave-length of the incident radiation, theory demands that the scattering, especially at large angles, shall be less for short than for longer waves. This prediction is confirmed by measurements of the scattering of X-rays and gamma rays over a wide range of frequencies. If the primary beam consists of very short, highly-damped pulses, or of waves of some irregular form, it may, of course, be considered as the Fourier integral of a large number of long trains of waves of different wave-lengths. Thus, unless the primary beam consists of long trains of monochromatic waves, the scattered radiation will, in general, be softer than the primary rays, and the hardness will be greater at small than at large angles with the incident beam. This corresponds qualitatively with the properties of the secondary gamma rays*.

* This explanation of the difference in hardness, as well as an explanation of the distribution of the intensity of the scattered gamma radiation, has been discussed in detail, for the special case of scattering by a ring electron of comparatively large size, by the writer (Phys. Rev. xx. p. 30 (1919)).

While it is possible to account in this manner for the difference in penetrating power of the primary and secondary radiation if a sufficiently heterogeneous primary beam is postulated, it is clear that, as a result of scattering, there can be no transformation of radiation of one frequency into radiation of another frequency. That is, the scattered rays can be no softer than the softest components of the primary rays, and removal by filtering of the softer components of the primary radiation must harden also the secondary beam.

An experimental method of determining the relative amount of scattering and fluorescence has been applied to a study of the secondary radiation excited by the hard gamma rays from radium C. In figure 1 is shown diagrammatically the arrangement of the experiment. A source of hard

Fig. 1.

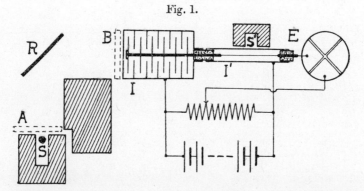

gamma rays S excites secondary radiation in a block R, and the intensity of this radiation is measured by an ionization chamber I, which is screened by heavy lead blocks from the direct beam of gamma rays, and which in the final experiments is surrounded on four sides by about 4 cm. of lead to keep out secondary radiation from the walls of the room. The ionization current when the radiator R is removed, is approximately balanced by the ionization current produced by an adjustable source of gamma rays S in a second chamber I'. The intensity of the secondary radiation is then measured by the difference in the readings of the electrometer E when the radiator R is in place and when removed.

The test for the presence of fluorescent radiation was made by comparing the intensity of the secondary radiation when an absorption screen was placed alternately in position A, in front of the source of gamma rays, and position B, in front of the ionization chamber. Supplementary tests showed that when the primary rays entered directly into the ionization chamber, the absorption was nearly the same whether the lead screen was at A or at B*. Let us suppose that the primary beam consists of any number of components of different wave-lengths λ_1, λ_2, ..., that I_1, I_2, ... are the intensities of these components in the primary beam, c_1, c_2, ... the fractions of each component scattered into the ionization chamber, k_1, k_2, ... the fractions of the respective energies transmitted through the absorption screen when placed at A, and k_1', k_2', ... the corresponding fractions when placed at B. Then it is clear that, when the absorption screen is placed at A, the intensity of the beam scattered into the ionization chamber is

$$I = c_1 k_1 I_1 + c_2 k_2 I_2 + \ldots = \Sigma c_s k_s I_s,$$

and, when placed at B, the intensity is

$$I' = c_1 k_1' I_1 + c_2 k_2' I_2 + \ldots = \Sigma c_s k_s' I_s.$$

But for all wave-lengths, k_s is very nearly equal to k_s'. Hence I is nearly equal to I': that is, for truly scattered radiation, the observed intensity of the secondary radiation should be approximately the same whether the absorbing plate is in the position of A or B.

If, on the other hand, the primary radiation excites in the radiator R a fluorescent radiation which is more readily absorbed than the primary rays, the observed intensity of the secondary radiation will be less when the absorption screen is in the position B; for if k_p is the fraction of the primary radiation transmitted through the absorption screen, while k_s is the corresponding transmission factor for the fluorescent radiation, the ratio I'/I of the intensity of the fluorescent radiation when the screen is at B to that when the screen is at A is obviously k_s/k_p. Thus the effect of any fluorescent radiation will be to make the fraction I'/I less than unity.

If all the secondary radiation is of the fluorescent type, the ratio k_s/k_p, and hence also of I'/I, should become indefinitely small as the thickness of the absorption screen is

* The supplementary experiment referred to showed that for the gamma rays from radium C filtered through 2 mm. of lead, and using an absorption screen of 1 cm. of lead, the value of k was 0·57 and of k' was 0·52. The difference is doubtless due to the difference in the amount of secondary radiation reaching the ionization chamber in the two cases. This difference will be relatively less important for softer radiation, but will be relatively somewhat more prominent for greater thicknesses of the screen.

increased. For wholly scattered radiation, as we have just seen, the value of this ratio should remain approximately unity for all thicknesses of the absorption screen. If the secondary radiation is a mixture of the two types, it will be seen that the ratio I′/I should approach, for large thicknesses of the absorption screen, the constant value

$$I'/I = c_s/(c_s + c_t), \quad \ldots \quad \ldots \quad (1)$$

where c_s is the fraction of the primary beam scattered into the ionization chamber with no absorption screen at B, and c_t is the corresponding fraction for the fluorescent radiation. Thus by measuring the ratio of the intensity of the secondary radiation when suitable absorption screens are placed alternately in front of the ionization chamber and the source, it is possible to determine the relative magnitude of the scattered and the fluorescent radiation.

The results of measurements of this ratio at three different angles with the primary beam are shown in Table I.

TABLE I.

Thickness of lead screen.	Ratios I′/I for iron radiator.		
	45°.	90°.	135°.
0	1	1	1
0·15 cm.	...	...	0·45
0·5	...	0·30	−0·02
1·0	0·52	0·13	−0·02
2·0	0·39	0·02	...
3·0	0·26	...	...
4·1	0·20	...	...

At each angle the measurements were continued until the intensity was too low for accurate determinations of I′/I. The probable error of the final measurements of the ratio at 135° was about 0·02, at 90° about 0·03, and at 45° about 0·04. On the basis of the above discussion, we may therefore conclude that for gamma rays which have traversed several centimetres of lead the secondary radiation at angles greater than 90° is, except for the small probable error, all of the fluorescent type.

At 45° it appears that the value of the ratio I′/I is approaching a constant value for large thicknesses of the absorption screen. The limiting value of this ratio would seem to be of the order of 5 or 10 per cent., which, according to expression (1), would represent approximately the fraction of the secondary radiation at this angle which is due to true scattering. It is clear, in view of the magnitude of the probable error, that such extrapolation for large thicknesses of the screen is precarious, and the evidence for any true scattering cannot be considered conclusive. Probably at least 90 per cent. of the secondary radiation at this angle is of the fluorescent type.

Though we thus find that there is very little of the penetrating primary gamma radiation present in the secondary rays, it is not impossible that some of the softer components of the unfiltered primary beam may be appreciably scattered, but be so strongly absorbed that they are not detected through the lead screens employed. An upper limit to the amount of such soft scattered radiation that may be present can be assigned in the following manner. If at any specified angle a certain screen suffices to make the ratio I'/I sensibly zero, it is clear that with this screen in position A there is no appreciable scattered radiation entering the ionization chamber, and practically all of the secondary radiation is fluorescent. The absorption coefficient for the primary rays which excite the fluorescent radiation may then be determined by measuring the decrease in ionization when additional screens are placed at A. Assuming that this absorption coefficient remains constant for all thicknesses of the absorption screen, which experiments on the primary beam show is very nearly the case, the intensity of the fluorescent radiation when no absorption screen is employed is

$$J_0 = J \cdot e^{\mu x},$$

where J is the observed intensity for a screen of thickness x, and μ is the linear absorption coefficient. Of the total secondary radiation I_0 observed when no screen is employed, the fraction J_0/I_0 at least consists of fluorescent radiation. Of the remainder, $(I_0 - J_0)/I_0$, a part may be fluorescent, since the true value of μ is presumably greater for small thicknesses of the absorption screen, and the rest will represent the truly scattered radiation. Thus the fraction $(I_0 - J_0)/I_0$ is an upper limit to the amount of scattered radiation which may be present in the secondary gamma rays when no absorption screens are employed.

Since theoretical considerations would lead one to expect the scattering to be greater at the smaller angles, an experimental determination of the value of this fraction was made for the secondary radiation at 45°. In this experiment the window of the ionization chamber consisted of 0·15 cm. of lead, and the same thickness of lead surrounded the source of gamma rays. This was necessary in order to cut out the beta rays. The results are shown in the following table :—

TABLE II.

x.	μ.	J/I_0.	J_0/I_0.	$(I_0-J_0)/I_0$.
3·1 cm.	0·57	0·17	0·98	0·02

The values in the fourth and fifth columns are calculated from the experimental data in the first three columns. After the primary gamma rays have passed through 3·1 cm. of lead, we have seen that probably not more than 10 per cent. of the secondary radiation is of the scattered type, but the intensity is so weak that the probable error of these measurements is necessarily rather large. It may be concluded, however, that at 45° probably not as much as 15 per cent. of the whole secondary radiation consists of scattered primary rays.

At the larger angles the results given in Table I. show that if there is any appreciable scattered radiation it must be of a very soft type, and since it is unable to penetrate a centimetre of lead, it cannot be identified with the hard gamma rays from radium C. It will be seen from this table also that the absorption coefficient of the fluorescent radiation is rapidly approaching that of the primary rays at the smaller angles. This fact, together with geometric difficulties which prevent securing intense secondary radiation at the smaller angles, makes very difficult any effort to separate the scattered and fluorescent radiation at angles much smaller than 45°. The question of the presence of scattered radiation at the smaller angles will be discussed in another paper on the basis of some experiments of a different type [*]. In this paper some positive evidence will be presented for the existence of true scattering at angles less than 15°.

In order to find out how far the actual scattering falls short of that to be expected from theoretical reasoning, a measurement was made of the relative intensity of the primary and secondary radiation at several angles. If the dimensions of the electron are negligible compared with the wave-length of the gamma rays, and if the electrons all scatter independently of each other, the usual theory [†] gives for the intensity of the scattered radiation

$$\frac{I_\theta}{I} = \frac{Ne^4(1+\cos^2\theta)}{2m^2l^2C^4}, \quad \ldots \quad (2)$$

[*] *Infrà*, p. 770.
[†] J. J. Thomson, *loc. cit.*

where I is the intensity of the primary beam at the ionization chamber when the radiator is replaced by the source of gamma rays, N is the number of electrons which are effective in scattering, e and m are the charge and mass respectively of the electron, θ is the angle with the primary beam at which the scattered beam is observed, l is the distance of the radiator from the source of gamma rays, and C is the velocity of light. Taking the number of electrons per atom as equal to the atomic number, and using the experimental values 10·3 cm. for l and 234 g. for the mass of the iron radiator, this expression gives for the ratio I_0/I at 90° the value 0·023. The experimental value of this ratio was 0·0017. But of this we have seen that less than 3 per cent. probably represents true scattering. The value of the ratio I_s/I, where I_s is the observed true scattering, is therefore less than 0·00005, only 2 per cent. of that required by theory. Similar results for the scattering at 45° and 135° are given in Table III. It will be seen from this table that at large angles, if there is any true scattering, it is probably less than a thousandth part of the amount predicted on the basis of the usual electron theory.

TABLE III.

Angle.	I_θ/I observed.	I_s/I observed.	I_s/I calculated.	$\dfrac{I_s \text{ obs.}}{I_s \text{ calc.}}$
45°	0·015	c. 0·001	0·035	c. 0·03
90°	0·0017	<0·00005	0·023	<0·002
135°	0·0008	<0·00002	0·035	<0·0005

It is not impossible to account for this very low value of the scattering on the basis of the classical electrodynamics, if suitable assumptions are made with regard to the wave-length of the primary gamma rays and the properties of the electron. Thus, for example, the writer has shown elsewhere* that if the electron is a rigid sphere which is not subject to rotational displacements by the primary beam, the scattering at all angles becomes negligible when the ratio of the wave-length to the radius of the electron is less than about 2·4. Certain other types of electron give a similar result for different values of this ratio. If this explanation is the correct one, the wave-length of these gamma rays must be considerably shorter than that of the hardest X-rays which have yet been studied, since for these rays the scattering, though somewhat smaller than that predicted by the usual theory, is apparently of the proper order of

* A. H. Compton, Phys. Rev. xx. p. 25 (1919).

magnitude *. It seems premature to attempt any detailed explanation of the failure of the usual electron theory until more definite information is available with regard to the wave-length of the hard gamma rays.

The Characteristics of the Fluorescent Radiation.

Let us now consider the properties of the fluorescent radiation excited by the hard gamma rays. The observed relative intensities of the secondary radiation from aluminium, iron, and lead at different angles with the primary beam, when gamma rays from radium C filtered through a centimetre of lead are employed, are shown in Table IV.

TABLE IV.

Secondary radiator.	30°.	45°.	60°.	75°.	Angle 90°.	120°.	135°.	150°.
Al......	...	6·2	4·0	1·7	(1·0)	0·7	0·4	0·3
Fe......	10	7·6	4·6	2·0	(1·0)	0·6	0·5	0·4
Pb......	11	6·8	4·3	2·2	(1·0)	0·8	...	0·7

In comparing the intensity of the scattered beam at any two angles, θ_1 and θ_2, the effect of absorption was eliminated as completely as possible by the well-known method of placing the radiating plate with its normal at an angle $(\theta_1 + \theta_2)/2$ with the primary beam. In this case the absorption is the same at the two angles at which the secondary radiation is compared. Since the window of the ionization chamber consisted of 0·15 cm. of lead, any soft fluorescent radiation was strongly absorbed.

Data similar to those given in this table have been published by Florance † and Kohlrausch ‡, except that in the present case the effect of absorption by the radiator has been largely eliminated, and the primary beam was rendered homogeneous by filtering through a suitable lead screen §.

* *C. f. e. g.* Hull & Rice, Phys. Rev. viii. p. 326 (1916). Barkla & White, Phil. Mag. xxxiv. p. 277 (1917).
† D. C. H. Florance, Phil. Mag. xx. p. 921 (1910).
‡ K. W. F. Kohlrausch, *loc. cit.*
§ It should be noted that on account of the differing hardness of the secondary radiation at different angles, the relative intensity observed at a given angle depends upon the fraction of the radiation absorbed by the ionization chamber. The ionization chamber used in the present experiments was so designed that it absorbed a large part of even the hard primary rays. This probably accounts for the fact that the writer's experiments show relatively more intense radiation at the smaller angles where the secondary rays are hard, than do the experiments of Florance and Kohlrausch, whose ionization chambers presumably absorbed only a small fraction of the incident radiation.

In common with these experimenters, it is found that the secondary radiation, which in the writer's work consisted almost wholly of fluorescent radiation, is very much more intense at small angles than at large angles with the incident gamma rays.

The relative amount of fluorescent radiation excited in different substances per unit mass is shown in Table V.

TABLE V.

Angle.	Relative fluorescence per unit mass:				
	Paraffin.	Al.	Fe.	Sn.	Pb.
135°	1·12	1·04	(1·00)	0·78	0·74
45°	1·7	0·9	(1·0)	0·8	0·9

The readings at 135° were taken for primary rays filtered through 0·5 cm. of lead, and those at 45° were with a 4·1 cm. lead filter. Thus it was made certain that practically all of the secondary radiation was of the fluorescent type. Sufficiently thin plates of the radiating materials were employed that the necessary corrections for the absorption of the primary and secondary radiation in the radiator were not large. The values here given therefore represent the amount of the fluorescent radiation excited in unit mass of the different radiators, which penetrates the 0·15 cm. lead window of the ionization chamber. It will be seen that the values do not differ greatly over a wide range of atomic weights.

The constancy is even more marked when the fluorescence per electron is calculated by multiplying each of the above values by the ratio (atomic weight)/(atomic number), as is done in the following table. At the angle 135° the constancy

TABLE VI.

Angle.	Relative fluorescence per electron:				
	Paraffin.	Al.	Fe.	Sn.	Pb.
135°	0·91	1·01	(1·00)	0·87	0·87
45°	1·4	0·9	(1·0)	0·9	1·1

of these values is somewhat accidental, since, as we shall see, the absorption coefficient of the fluorescent radiation at this angle is considerably greater for the radiation from the light than for that from the heavy elements. At 45°, however, the hardness of the fluorescent radiation is practically the same for the different radiators, so the constancy of the values at this angle is of real significance. The result expressed by this table is confirmed by the more quantitative experiments of Ishino*, who found that the magnitude of

* M. Ishino, *loc. cit.*

the total secondary radiation (which includes any scattered rays that may be present) per atom is more nearly proportional to the atomic number than to the atomic weight. The amount of the fluorescent radiation excited is therefore approximately proportional to the number of electrons traversed by the primary gamma rays.

It will be seen on examining Table I. that the fluorescent radiation at small angles with the primary beam is considerably harder than that at right angles. This matter was examined in greater detail for a number of different elements, with the results shown in Table VII. Care was taken in these experiments also to eliminate any possible soft scattered

TABLE VII.

Angle.	Mass absorption coefficients in Lead of the fluorescent radiation excited in different materials by hard gamma rays:				
	Paraffin.	Al.	Fe.	Sn.	Pb.
45°	0·10	0·10	0·11	0·09	0·05 (?)
90°	...	...	0·21	...	...
135°	0·78	0·50	0·50	0·32	0·15

radiation by interposing suitable absorption screens between the source of gamma rays and the radiating material. As a result, the intensity of the secondary radiation was so low that the values of the absorption coefficients obtained can be considered only approximate. The data suffice to show, however, that while the radiation from all substances is harder at small angles, the difference is less for elements of high atomic weight, so that whereas at large angles the radiation from the heavier elements is considerably more penetrating, at 45° the hardness differs but little from element to element[*].

Interesting information is obtained on examining the absorption coefficients of this penetrating fluorescent radiation in various materials. This was done for the secondary radiation from iron at 135° after the primary gamma rays had been filtered through 0·5 cm. of lead, with the results shown in the following table. With a similar geometric

TABLE VIII.

Mass absorption coefficents in different elements of the fluorescemt radiation at 135° excited in iron by hard gamma rays from radium C:			
Pb.	Sn.	Fe.	Al.
0·50	0·18	0·08	0·07

---

[*] The constancy observed for different elements at 45° is confirmed by the measurements of Florance, Phil. Mag. xx. p. 935 (1910).

arrangement, the mass absorption coefficient of the primary rays in lead was 0·062, which presumably means that this fluorescent radiation is of very appreciably longer wave-length than the primary gamma rays. On the other hand, the experiments of Hull and Rice* show that X-rays of wave-length 0.122×10^{-8} cm. have a mass absorption coefficient in lead of about 3·0, which indicates that even the softest part of this fluorescent radiation is of shorter wave-length than the critical wave-length 0.147×10^{-8} cm. required to excite the characteristic K radiation in lead. This conclusion is confirmed by the fact that the mass absorption of this fluorescent radiation is greater in lead than in tin, which is the reverse of the case for wave-lengths between the K radiation from lead and the radiation from tin. There can thus be no question but that the fluorescent rays under examination are of a distinctly harder type than the characteristic K radiation from even the heaviest elements.

In an experiment with a Coolidge tube operated by an induction coil at a maximum potential of 196,000 volts, Rutherford has obtained X-rays whose mass absorption coefficient in lead is as low as 0·75†. This is practically the same as the value observed for the fluorescent gamma radiation from paraffin at 135° (Table VII.). According to the quantum relation, $h\nu = eV$, the wave-length in Rutherford's experiment must have been greater than 0·063 Å.U. The wave-length of the softest part of this penetrating fluorescent radiation must therefore lie between 0·06 and 0·12 Å.U.

It is interesting to note that these secondary gamma rays bridge the gap which has existed between the hardest X-rays and the very penetrating gamma rays; for as we have just seen, the softest part of this secondary radiation falls within the wave-length of the hardest X-rays, while Table VII. shows that at small angles it is nearly as penetrating as the hard gamma rays from radium C.

The Origin of the Fluorescent Radiation.

Although the secondary gamma radiation under examination seems, without doubt, to be fluorescent in nature, it differs in several important respects from the characteristic fluorescent K and L radiations excited in matter when traversed by hard X-rays. In the first place, whereas these characteristic radiations differ greatly in hardness from

* Hull & Rice, *loc. cit.*
† E. Rutherford, Phil. Mag. xxxiv. p. 153 (1917).

element to element, the secondary gamma rays, especially at small angles with the incident beam, are of nearly the same hardness over a wide range of atomic numbers. And in the second place, while the characteristic radiations are found to be distributed uniformly with regard to intensity and quality at all angles with the primary beam, the fluorescent gamma rays show marked asymmetry in both quantity and quality in the forward and reverse directions. There is therefore good reason to suppose that the oscillators which give rise to this fluorescent radiation are radically different in character from those which are responsible for the K, L, and M characteristic radiations *.

An explanation of the origin of the fluorescent radiation which appears to be satisfactory is that the high-speed secondary beta particles liberated in the radiator by the primary gamma rays excite the secondary gamma rays as they traverse the matter of the radiator. On this view the fluorescent gamma rays should be identical in character with the so-called "white" radiation excited in the target of an X-ray tube by the impact of the cathode particles. Experiments have shown that when cathode rays or beta rays strike a target which is so thin that the particles are not greatly scattered and in which no considerable amount of characteristic radiation is excited, the X-rays emitted are more intense and harder in the general direction of the cathode ray beam than in the reverse direction †. This asymmetry is of the same kind as that observed for the secondary gamma rays, and though not so marked, is found to increase with the speed of the impinging electrons. For speeds comparable with those of fast beta rays the asymmetry may well become as great as that observed in the present experiments. But it is also known that the beta rays liberated by gamma rays are much more intense in the direction of the gamma ray beam than in the reverse

---

* The idea suggested itself that the secondary radiation which was being studied was a fluorescent radiation excited in the lead screens which surrounded the source of gamma rays, this fluorescent radiation being in turn scattered by the radiator into the ionization chamber. It is obvious that such a radiation would not be eliminated by placing additional lead screens over the source, while the ionization would be considerably reduced by placing screens over the ionization chamber. Considerations of the energy involved and of the characteristics of the secondary radiation rendered this suggestion improbable, but the possibility was definitely eliminated by removing all the lead screens and replacing them with iron. The phenomenon in this case was identical with that when lead screens were employed.

† G. W. C. Kaye, Proc. Camb. Phil. Soc. xv. p. 269 (1909). J. A. Gray, *loc. cit.*

direction. Indeed, "the results indicate that the beta particles initially escape in the direction of the gamma rays, and with the same speed for all kinds of matter"*. Thus the hypothesis that the fluorescent gamma radiation is due to the impact of the secondary beta particles accounts qualitatively for the observed asymmetry in the hardness and intensity of the secondary gamma rays.

With regard to the relative intensity of the fluorescence excited in different materials, attention may be called to the fact that the number of the beta particles excited by hard gamma rays is approximately the same per electron in different elements †. There is a somewhat larger number produced in the very heavy elements such as mercury and lead, which appears to be connected with the excitation of the characteristic K radiation in these elements. But since such radiation is too soft to have an appreciable effect in the present investigation, it seems probable that the number of *effective* beta particles excited in these elements does not differ greatly from that for the lighter elements. If, therefore, the simple assumption is made that the amount of secondary gamma rays excited depends only upon the number of electrons traversed by the secondary beta particles, our hypothesis gives a satisfactory account of the fact that the amount of secondary gamma radiation per electron is practically the same for all elements.

The greater scattering of beta particles by elements of high atomic weight means that in these elements a relatively larger number of the beta rays move in a direction opposed to the primary beam. For this reason we should expect, as Table VII. shows is actually the case, that at large angles with the incident beam the fluorescent radiation from the heavier elements will be more penetrating than that from the light ones.

An estimate of the relative energy in the secondary rays can be obtained by integrating over the surface of a sphere the observed relative intensity of the scattered beam at various angles. A rough summation of this kind, using the data of Tables III. and IV. and extrapolating by the help of Kohlrausch's data for the very small angles, shows that the ratio of the energy (as measured by the ionization) of the secondary radiation from iron to the total energy absorbed from the primary beam is about 0·69. This result is in good agreement with Ishino's estimate that the "scattering" by iron accounts for 62 per cent. of the total absorption. A

* E. Rutherford, 'Radioactive Substances, etc.,' p. 276.
† Eve, Phil. Mag. xviii. p. 275 (1909).

correction must be applied to this value to make allowance for the fact that while the greater part of the secondary radiation which enters the ionization is absorbed, in the present experiment only about half of the primary beam was thus absorbed. Taking this correction factor to be about 0·7, we find that approximately 50 per cent. of the absorbed primary rays is transformed into radiation of sufficiently high frequency to penetrate 0·15 cm. of lead*. The efficiency of transformation of the energy is therefore of a much higher order than that observed in an X-ray tube operating at usual potentials, in which case not as much as 1 per cent. of the energy of the cathode rays appears as X-rays. It is possible that this difference is to be accounted for by an excitation of gamma rays when the secondary beta particles are liberated in addition to that produced when they collide with other electrons.

A question of great theoretical importance is—What kind of oscillator can give rise to radiation which not only is more intense in one direction than in another, but also differs in wave-length in different directions? Since the secondary radiation differs in frequency from the primary rays, it would seem impossible to invoke any interference between the radiation from the different oscillators to account for this phenomenon. Such an explanation is rendered the more difficult by the fact that to explain the different hardness of the rays in different directions, oscillators of different frequencies would have to be present, between which there could be no fixed phase relations. An obvious means of accounting for the observed phenomenon is to suppose that the radiator which gives rise to the secondary rays is moving at high speed in the direction of the primary beam. In this case, both the intensity and the frequency of the fluorescent radiation will be greater in the forward than in the reverse direction, as is demanded by the experiments.

A rigid calculation of the relative intensity of the fluorescent radiation at different angles, according to this hypothesis, is not at present possible, because the scattering of the beta particles results in an irregular distribution of their velocities. It will nevertheless be instructive to consider the relative energy radiated in different directions by an oscillator moving at a speed comparable with that of light. It can be

---

* The estimate here made of the efficiency of transformation is, of course, based upon the assumption that unit energy of one frequency produces the same total number of ions as unit energy of another frequency. Though this assumption has not been tested over the range of frequencies here considered, it does not appear probable that any error thus introduced can change the order of magnitude of the result.

shown that for an electron whose acceleration is unpolarized relative to the observer*, and which is travelling at a velocity βC, the mean square of the electric vector at a great distance r is †

$$E^2 = \frac{e^2 \overline{\Gamma^2}}{16\pi^2 C^4 r^2}$$
$$\times \frac{(1-\beta\cos\theta)^2 - \tfrac{1}{3}(1-\beta^2)\sin^2\theta + \tfrac{1}{3}\cos\theta(2\beta - \cos\theta - \beta^2\cos\theta)}{(1-\beta\cos\theta)^6}, \quad (3)$$

where Γ^2 is the mean square of the acceleration relative to the observer at the moment the pulse under observation left the electron, e is the charge of the oscillator, C is the velocity of light, and θ is the angle between the direction of motion of the particle and the observed beam. Thus the ratio of the intensity of the radiation at an angle θ_1 to that at an angle θ_2 is given by the expression

$$R = \frac{I_1}{I_2} = \left\{\frac{1-\beta\cos\theta_2}{1-\beta\cos\theta_1}\right\}^6 \frac{(1-\beta\cos\theta_1)^2 - \tfrac{1}{3}(1-\beta^2)\sin^2\theta_1}{(1-\beta\cos\theta_2)^2 - \tfrac{1}{3}(1-\beta^2)\sin^2\theta_2}$$
$$\frac{+\tfrac{1}{3}\cos\theta_1(2\beta - \cos\theta_1 - \beta^2\cos\theta_1)}{+\tfrac{1}{3}\cos\theta_2(2\beta - \cos\theta_2 - \beta^2\cos\theta_2)}. \quad (4)$$

Assuming that all the radiating particles are moving in the same direction, the ratio of the intensity of the fluorescent radiation at the angle 45° to that at 135° has been calculated from this expression, with the results shown in figure 2. It will be seen that it is possible on this view to account for any reasonable degree of asymmetry of the secondary radiation. In the case of paraffin, in which the least scattering of the beta particles occurs, the observed ratio of the intensities at 45° and 135° was about 20. In addition to the effect of the scattering of the secondary beta rays, experimental errors arise because much of the soft radiation at 135° is absorbed before it enters the ionization chamber, while a considerable part of the hard radiation at 45° traverses the ionization chamber without being absorbed. The rapid increase of R with β, however, makes it reasonably certain, on the present view, that the average speed of the oscillators which emit the secondary gamma radiation does not differ greatly from half the speed of light.

* Of course such an oscillator will not be unpolarized relative to an observer moving with it. A slight polarization will not, however, make any great difference in the value of the ratio (4).
† The values of the three components of the electric vector from which this expression is derived may be found, e.g. in O. W. Richardson's 'Electron Theory,' p. 256.

Since the speed of even the swiftest alpha particles is only about one-tenth that of light, it is clear that the radiating particles cannot have mass comparable with atoms, but must

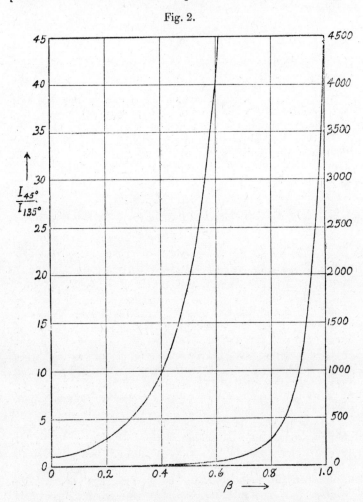

Fig. 2.

be individual electrons. We are thus led to the idea that it is the vibrations of the secondary beta particles themselves which give rise to the fluorescent gamma rays*.

* The corresponding hypothesis that the "white radiation" from an X-ray tube is due to vibrations of the cathode particles has been suggested on the basis of similar considerations by D. L. Webster, Phys. Rev. xiii. pp. 303–305 (1919).

Substituting the value $\beta = 0.5$ in equation (4), we can calculate the relative scattering to be expected at various angles. The result is shown in the solid curve of figure 3.

Fig. 3.

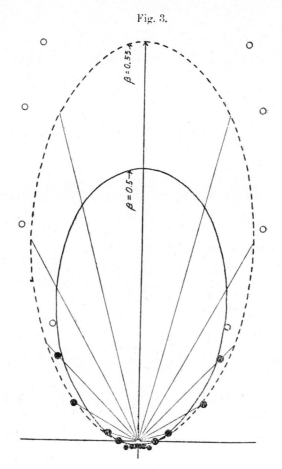

Curves showing intensity at different angles with motion of oscillator whose velocity is βC.
Solid circles writer's, open circles Kohlrausch's values of relative intensity of gamma rays.

The solid circles in this figure represent the writer's observations on aluminium, as given in Table IV. The open circles at the small angles show the results of Kohlrausch referred to the writer's value at 45°. It is not impossible that the large amount of secondary radiation at these small angles is due in part to the presence of some true scattering. However this may be, the generally satisfactory form of the theoretical curve suggests that we are working along the right line.

If this view of the origin of the fluorescent radiation is the correct one, we are supplied with a means of estimating roughly the wave-length of the primary gamma rays. It has been shown above that the wave-length of the softest part of the fluorescent radiation lies between 0·06 and 0·12 Å.U., and probably nearer the former. But according to the Doppler principle, if the oscillators producing the fluorescent radiation are moving in the direction of the primary rays, the ratio of the wave-length at an angle θ_1 to that at an angle θ_2 is

$$\frac{\lambda_1}{\lambda_2} = \frac{1-\beta\cos\theta_1}{1-\beta\cos\theta_2} \quad \cdots \cdots \quad (5)$$

Thus, if we take λ_2 to be about 0·08 Å.U. at $\theta_2 = 135°$, and the value of β to be 0·5, the wave-length of the penetrating fluorescent radiation at 45° is about 0·04 Å.U. This result does not vary greatly with different values of β; but the extreme hardness of the fluorescent radiation at 45° indicates that it is more nearly similar to the primary rays than to the soft secondary radiation which appears at the larger angles. Thus we shall probably not be far wrong in assigning a value 0·02 to 0·03 Å.U. as the wave-length of the most effective part of the hard gamma rays from radium C. This result is not in disaccord with the calculations of Rutherford* based upon the quantum hypothesis.

Summary.

The principal experimental results of this investigation may be summarized as follows:—

By far the greater part of the secondary gamma radiation from matter traversed by the hard gamma rays from radium C is fluorescent in nature. If any truly scattered radiation is present, at 45° it probably amounts to less than 15 per cent., and for angles greater than 90° to less than 3 per cent. of the secondary rays.

* E. Rutherford, Phil. Mag. xxxiv. p. 153 (1917). According to the quantum relation, an electron must have a velocity $\beta = 0·8$ in order to excite radiation of wave-length 0·04 Å.U. On the present view, therefore, the radiating beta particle must already have lost a large part of its energy of translation.

At large angles with the primary beam the scattered energy is probably less than 0·001 of that required by the usual electron theory.

The secondary fluorescent radiation is found, in accord with observations by others on the whole secondary radiation, to be harder and more intense at small angles with the incident beam than at large angles, and tables are given showing the manner of this variation.

While at large angles the radiation from heavy elements is somewhat more penetrating than that from the light elements, at small angles both the hardness and the intensity of the fluorescent radiation are approximately the same from elements covering a wide range of atomic numbers.

A study of the absorption coefficients of this radiation in various elements shows that the softest parts of it, though of shorter wave-length than the K radiation from lead, are not harder than the most penetrating X-rays. The hardest parts approach in penetrating power the primary gamma rays from radium C.

It is pointed out that the very small scattering observed is not incompatible with the classical electrodynamics, if the wave-length of the gamma rays and the diameter of the electron are of the same order of magnitude.

A satisfactory qualitative explanation of the observed fluorescent radiation is found in the gamma rays produced by the impact of the secondary beta particles liberated in the radiator by the primary gamma rays.

The observed asymmetry in the intensity and hardness may be accounted for if the oscillators which give rise to the fluorescent radiation are electrons moving in the direction of the primary beam with about half the speed of light.

The wave-length of the softest part of the observed fluorescent radiation is shown to lie between 0·06 and 0·12 Å.U., probably nearer the former value, while the wave-length of the hardest part is probably about half as great. By a comparison of absorption coefficients, the effective wave-length of the hard gamma rays from radium is estimated as about 2 or 3×10^{-10} cm.

The writer performed these experiments at the Cavendish Laboratory as National Research Fellow in Physics. He desires to express his appreciation of the interest which Professor Rutherford has shown in the work.

Washington University,
St. Louis,
September 24th, 1920.

The Wave-Length of Hard Gamma Rays. By ARTHUR H. COMPTON, *Ph.D., Physics Laboratory, Washington University* *.

THE only recorded attempt to measure directly the wavelength of hard gamma rays is apparently that of Rutherford and Andrade †, using the method of reflexion from a crystal of rock-salt. In these experiments spectrum lines were observed at angles as small as about 44 minutes, corresponding to a wave-length of about 0·07 Å.U. It was thought that this line, as well as one of wave-length 0·10 Å.U., could be detected through a 6-millimetre screen of lead, which would make it appear that these lines represent the hard gamma rays from radium C. Professor Rutherford informs me, however, that the appearance of these lines through the lead screen was doubtful. His more recent measurements of the absorption of X-rays of very high frequency ‡ have indicated rather that radiation, whose wavelength is about 0·08 Å.U., has an absorption coefficient in lead that is very much greater than that of the hard gamma rays from radium. Thus, while the crystal reflexion measurements show that radium gives off gamma rays of wavelengths 0·07 Å.U. and longer, the very penetrating radiation which it emits probably has a much shorter wave-length.

Various lines of theoretical reasoning suggest that there are in hard gamma rays components ranging in wave-length from 0·01 to 0·04 Å.U. Rutherford has pointed out § that radium C gives off beta rays with an energy corresponding to a fall through from 5 to 20×10^5 volts. According to the quantum relation, $h\nu = eV$, the limiting wave-length produced by the slower of these electrons would be about 0·03 Å.U., while that due to the fastest ones would be as short as 0·007 Å.U. In the second place, using an absorption formula which is satisfactory for hard X-rays of known wave-length, it is found ‖ by extrapolation that the absorption coefficient of hard gamma rays corresponds to a wave-length of about 0·04 Å.U. And finally, knowing approximately the

---

* Communicated by Prof. Sir E. Rutherford.
† Rutherford and Andrade, Phil. Mag. xxviii. p. 263 (1914).
‡ E. Rutherford, Phil. Mag. xxxiv. p. 153 (1917).
§ *Ibid.*
‖ A. H. Compton, Washington University Studies, Scientific Series, Jan. 1921.

wave-length of the "incident" secondary gamma radiation, and calculating from this the wave-length of the "emergent" secondary radiation on the hypothesis that the difference in wave-length is a Doppler effect due to motion of the particles emitting the secondary radiation, the wave-length of the primary gamma rays can be estimated, since the absorption coefficient of the primary and the "emergent" secondary radiation is nearly the same. This method leads to a value of between 0·02 and 0·03 Å.U. for the effective wave-length of the hard gamma rays from radium *.

In the present paper a new method of measuring the wave-length of high frequency radiation will be proposed, and the method will be applied to the determination of the wave-length of gamma rays. Instead of studying the spectrum lines reflected by a grating composed of regularly arranged atoms in a crystal, this method consists in observing the diffraction pattern due to the individual atoms. To consider an optical analogy, if the reflexion of X-rays from a crystal is compared with the spectrum from a ruled grating, the method of atomic diffraction corresponds to a study of the diffraction pattern due to a large number of parallel lines ruled at random distances. The distance between the different order lines in the spectrum is determined by the grating space between the lines ruled on the grating, while the distance between the bands of the diffraction pattern is determined by the breadth of the individual lines. The advantage of the method as applied to gamma-ray measurements lies in the fact that the effective diameter of the atom is much smaller than the distance between two atoms in a crystal, so that the effective width of the diffraction band is much greater than the distance between two spectrum lines. Thus, whereas the spectrum of hard gamma rays from a crystal grating would have to be studied at angles less than 1/2 degree, atomic diffraction measurements may be made at angles in the neighbourhood of 10 degrees. In order to use the method quantitatively, it is of course necessary to know the effective diameter of the atom. This may be determined, in a manner that will be described below, by measurements with X-rays of known wave-length.

Debye has shown that if an atom is composed of N electrons, and if at any instant the distance between the mth and the nth electron is s_{mn}, the probable intensity of the X-rays

---

* A. H. Compton, Phil. Mag. *suprà*, p. 749.

scattered at an angle θ with the primary beam whose intensity is I is*,

$$I_1 \sum_1^N \sum_1^N \frac{\sin\left[\frac{4\pi}{\lambda} s_{mn} \sin \frac{\theta}{2}\right]}{\left[\frac{4\pi}{\lambda} s_{mn} \sin \frac{\theta}{2}\right]},$$

where λ is the wave-length and I_1 is the intensity of the rays scattered by a single electron. This expression supposes that the forces holding the electrons in position are negligible in comparison with the forces due to the traversing radiation—an hypothesis supported by experiments on the scattering of X-rays. If $p_{mn}.ds$ is the probability that the distance s_{mn} will lie between s and $s+ds$, the average value of the intensity for all possible arrangements of the electrons in the atoms is,

$$I_\theta = I_1 \sum_1^N \sum_1^N \int_0^\infty \frac{\sin\left[\frac{\sin\theta/2}{\lambda}.4\pi s_{mn}\right]}{\frac{\sin\theta/2}{\lambda}.4\pi s_{mn}} p_{mn}\, ds_{mn},$$

or

$$\frac{I_\theta}{I_1} = F(N, p, \sin\theta/2/\lambda).$$

Since for any particular atom the quantities N and p remain constant, for an atom of atomic number N this ratio may be written,

$$\frac{I_\theta}{NI_1} = \psi_N(\sin\frac{\theta}{2}/\lambda).$$

If the quantity $\sin(\theta/2)/\lambda$ is sufficiently large, it will be seen that co-operation in the scattering by different electrons will be almost wholly a matter of chance, and the "excess scattering" function ψ will become practically unity. On the other hand, for very small values of this quantity co-operation between the electrons will be almost complete and the value of the function ψ will approach N. For intermediate values of $\sin(\theta/2)/\lambda$ the function will have a different value for every atom, since for no two atoms will the probabilities p_{mn} be identical. The experimental values of $\psi = I_\theta/I_1 . N$ for different materials, as measured by Barkla and Dunlop †, together with the values calculated for certain

* P. Debye, *Ann. d. Phys.* xlvi. p. 809 (1915).
† Barkla and Dunlop, *Phil. Mag.* xxxvii. p. 222 (1916).

arbitrary arrangements of the electrons in the atoms of the different elements *, are shown in fig. 1. The measurements for the different wave-lengths were all made at an angle $\theta = 90°$.

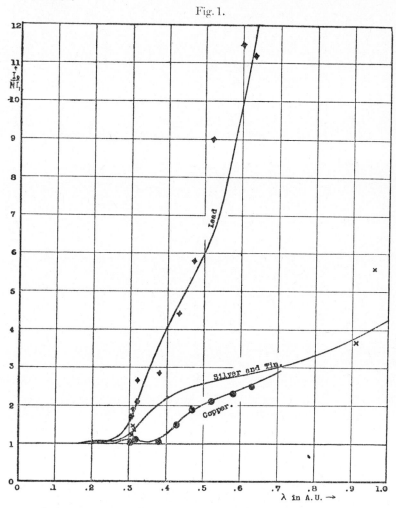

Fig. 1.

Let us suppose that according to these experiments the ratio of the value of ψ for lead to its value for copper is R when $\sin(\theta/2)/\lambda = c$. Then, if for some unknown wave-length λ' the value of this ratio becomes R at an angle θ', it

* Cf. A. H. Compton, Phil. (soon to be published).

is clear that $\sin(\theta'/2)/\lambda' = c = \sin(\theta/2)/\lambda$, whence

$$\lambda' = \frac{\sin \theta'/2}{\sin \theta/2} \lambda. \quad \ldots \quad \ldots \quad (2)$$

Thus, if it is possible to find an angle at which gamma rays scattered from lead and copper are in the same ratio as the X-rays scattered at 90° in Barkla and Dunlop's experiments, we have the data necessary to calculate the effective wave-length of the gamma rays.

Scattering of Hard Gamma Rays at Small Angles.—An element of uncertainty is introduced into the application of this method of determining the wave-length of gamma rays by the fact that recent measurements have shown[*] that only a very small part, if any, of the secondary gamma rays observed at large angles with the primary beam is truly scattered radiation. At an angle of 45° these measurements indicated that perhaps 5 or 10 per cent. of the secondary radiation consisted of scattered primary rays, though the absorption coefficient of the fluorescent secondary radiation was so nearly the same as that of the primary rays that it was not possible to establish with certainty the existence of any scattered rays. With the hope of placing the present wave-length experiments on a more certain footing, careful examination of the character of the secondary radiation at 22°·5 was made, using, with some refinements, the same general method as that employed in the earlier experiments. It was found that, though for small thicknesses of the absorption screen the absorption coefficient of the secondary radiation differed by only about 8 per cent. from that of the primary beam, even after traversing 5·6 cm. of lead the two absorption coefficients were still measurably different. Thus at least 50 per cent. of the radiation at this angle is certainly fluorescent in nature, but the fluorescent radiation is absorbed at so nearly the same rate as the primary rays that it was impossible to decide what part of the remaining 50 per cent. was fluorescent and what part might be scattered radiation. The experiments are, however, consistent with the view that for angles smaller than 30° with the primary beam a considerable portion of the secondary gamma radiation consists of truly scattered radiation.

Several investigators have found that at angles greater than 30° the ratio of the secondary radiation from one element

[*] A. H. Compton, *suprà,* p. 749.

to that from another is practically independent of the angle *. An examination was therefore made of the relative intensity of the secondary radiation from copper and lead at angles less than 30°. The experimental arrangement is shown in fig. 2. A strong source of hard gamma rays S (usually about

Fig. 2.

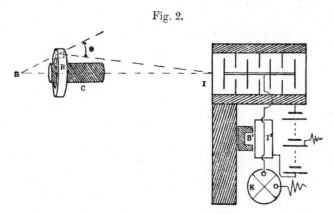

100 millicuries) is placed at a point on the axis of the ionization chamber I, and a lead cylinder C is placed between to cut off the primary gamma rays. The sample of copper or lead under examination is in the form of a ring R supported coaxially on the lead cylinder by a piece of cardboard. It is clear that all parts of the ring will scatter gamma rays into the ionization chamber at approximately the same angle. In order to secure the greatest possible intensity, the dimensions of the apparatus were so adjusted that the maximum differences in the angle θ made with the primary beam by the secondary rays entering the ionization chamber were between $\theta/2$ and $3\theta/2$. By this arrangement ionization due to the secondary radiation from the ring R can be obtained which is very considerably greater than that due to stray rays from the walls of the room, etc. It was found convenient, however, to balance the stray radiation against ionization produced by a small source S′ of gamma rays in a second chamber I′. The secondary radiation was then measured by the difference in the readings of the electrometer when the radiator R was in place and when removed.

The results of the experiments are shown in the following table. The values of the observed ratios of the intensity of

* *Cf.*, *e. g.*, M. Ishino, Phil. Mag. xxxiii. p. 140 (1914); K. W. F. Kohlrausch, *Phys. Zeitschr.* xxi. p. 193 (1920); A. H. Compton, *suprà*, p. 749.

the secondary rays from lead to that from copper, shown in the second column, are the averages of large numbers of readings:—

TABLE I.

Angle.	Observed Ratio. Intensity for Pb. Intensity for Cu.	Relative Intensity per electron.
30°	·605 ± ·012	·86 ± ·02
20°	·601 ± ·009	·85 ± ·01
15°	·638 ± ·008	·89 ± ·01
10°	·677 ± ·023	·95 ± ·03

In the third column the observed intensities are corrected for the difference in absorption of the rays in traversing the lead and copper rings, and the ratio of the relative intensity per electron is calculated, taking the number of electrons per atom as equal to the atomic number. It will be seen that while at all angles at which measurements were made the value of this ratio is slightly less than the theoretical value unity, the ratio shows a tendency to increase at the smaller angles. It is unfortunate that at still smaller angles the energy of the secondary radiation was so low that no satisfactory measurements could be made.

It may be mentioned that Kohlrausch * has recently made measurements similar to these at angles as low as 10°, and that his measurements do not show this tendency for the intensity of the secondary rays from lead to increase at the small angles more rapidly than that from the lighter elements. While Kohlrausch used several times as strong a source of gamma rays as that employed in these experiments, his apparatus was not specially adapted to taking measurements at small angles, and it would appear that his probable error at these angles was greater than that of the present measurements. In view of the consistency of the experiments here recorded, it appears probable that the observed increase in the ratio at 10° is not the result of chance.

The Wave-Length of the Gamma Rays.—A comparison of the values given in Table I. with the data in fig. 1 shows that the present experiments have not been carried to angles sufficiently small to give values in I_{Pb}/I_{Cu} overlapping Barkla

* K. W. F. Kohlrausch, *loc. cit.*

and Dunlop's experimental values for the scattering of X-rays at 90°. Since their measurements included wave-lengths as short as 0·3 Å.U., we may conclude from equation (2) that the effective wave-length of the gamma rays here employed is less than

$$\frac{\sin 5°}{\sin 45°} \times 0\cdot 3 = 0\cdot 037 \text{ Å.U.}$$

If Barkla and Dunlop's results are extrapolated according to the theoretical curve for lead shown in fig. 1, it will be seen that the ratio I_{Pb}/I_{Cu} at 90° begins to increase appreciably for wave-lengths in the neighbourhood of ·2 to ·25 Å.U. If from Table I. we take 10° as the angle at which the ratio begins to increase for gamma rays, the effective wave-length of these waves is by equation (2) between 0·025 and 0·030 Å.U. According to the present experiment this may therefore be taken as the approximate wave-length of hard gamma rays which have traversed about 8 mm. of lead.

The three principal elements of uncertainty which enter into this calculation are : (1) the wave-length 0·2 Å.U. at 90° is an extrapolated value, (2) a possible error in the experiments, and (3) the lack of positive evidence that the radiation measured in these experiments contains an appreciable fraction of truly scattered rays. None of these difficulties seem sufficiently serious to render improbable the correctness of the result as to order of magnitude. Indeed this method of estimating the wave-length of hard gamma rays is perhaps the most direct one that has been employed, and in as far as its results are in agreement with the predicted values considered at the beginning of the paper, they may be taken as a support of the theoretical bases of these predictions.

The writer performed this experiment at Cavendish Laboratory as National Research Fellow. He desires to thank Professor Sir E. Rutherford for the free use of the laboratory facilities and for valuable suggestions with regard to the experimental procedure.

Washington University,
  Saint Louis, U.S.A.
    December 1, 1920.

THE MAGNETIC ELECTRON.*

BY

ARTHUR H. COMPTON, Ph. D.,

Washington University, St. Louis.

THE evidence brought forward by the speakers who have preceded me has shown that many magnetic phenomena find a satisfactory explanation on the hypothesis that matter contains a large number of minute elementary magnets. The theories of para- and ferro-magnetism as developed by Langevin, Weiss and others, though based upon the hypothesis of such ultimate magnetic particles, make no assumptions concerning their nature. The explanation of diamagnetism, on the other hand, is based upon the view that this effect owes its origin to the circulation of electricity in resistanceless paths. The success of these theories in explaining the principal characteristics of magnetism gives us confidence in the real existence of these magnetic particles. Let us see, therefore, if it is possible to identify these elementary magnets with any of the fundamental divisions of matter.

The original investigations of ferromagnetism which led to the hypothesis of an elementary magnetic particle credited molecules with the properties of small permanent magnets. This view finds some support in the profound effect of heating, mechanical jarring, etc., on the ease of magnetization of iron. The dependence of magnetic permeability upon the chemical condition of a substance suggests the same view. But perhaps the strongest

---

* Based on a paper read before Section B of the American Association for the Advancement of Science, December 27, 1920.

[NOTE.—The Franklin Institute is not responsible for the statements and opinions advanced by contributors to the JOURNAL.]

COPYRIGHT, 1921, by THE FRANKLIN INSTITUTE.

argument that has been brought forward in support of the idea of molecular magnets has been the discovery of the Heusler alloys, in which by melting together elements which are only slightly magnetic an alloy with ferromagnetic properties is produced. It is, however, difficult to imagine what mechanism could reasonably give to a group of atoms, such as the chemical molecule, the properties of a single magnetic particle. Moreover, if on magnetization such a group of atoms should actually turn around within a crystal, as the elementary magnets are supposed to do, the resulting change in the positions of the atoms composing the molecule should produce a change in the crystal form; since, as we know, the form of the crystal is dependent upon the arrangement of its component atoms. It is, however, a matter of common observation that a magnetic field effects no such change in the form of a magnetic crystal.

Perhaps the most natural, and certainly the most generally accepted view of the nature of the elementary magnet, is that the revolution of electrons in orbits within the atom give to the atom as a whole the properties of a tiny permanent magnet. Support of this view is found in the quantitative explanation which it affords of the Zeeman effect. It seems but a step from the explanation of this effect to Langevin's explanation of diamagnetism as another result of the induced electronic currents within the atom. On Langevin's view the electronic orbits act as resistanceless circuits in which an external magnetic field induces changes of current. By Lenz's law these induced currents will always be in the direction to give the electronic orbit a magnetic polarity opposite to the applied field, thus accounting for the atom's diamagnetic properties. This theory offers a satisfactory qualitative explanation of diamagnetism, and accounts for the fact that diamagnetism is independent of temperature. But quantitatively it is inadequate. For, in order to explain the magnitude of the observed diamagnetic susceptibility on this view, one must suppose either that the atom possesses a number of electrons equal to several times its atomic number, or the distance between the electrons in the atom must be several times as great as is estimated by more direct methods. Moreover, the experiments of Barnett [1] and J. Q. Stewart [2] show that the ratio of charge to mass of the

---

[1] S. J. Barnett, *Phys. Rev.,* **6**, 240 (1915).
[2] J. Q. Stewart, *Phys. Rev.,* **11**, 100 (1918).

elementary magnet, though of the same order of magnitude, is appreciably greater than one would expect if the magnetic moment is due solely to electrons revolving in orbits. But perhaps a more serious difficulty with the usual electron theory of diamagnetism is that the induced change in magnetic moment of the electronic orbit involves also a change in its angular momentum. It is obvious, according to the classical electrical theory, that any electron revolving in an orbit will soon radiate its energy. Any angular momentum induced by an applied magnetic field will, on this theory, therefore, rapidly disappear so that diamagnetism should be merely a transient effect. Let us then assume with Bohr that if each electron has some definite angular momentum such as $h/2\pi$, no radiation occurs. On this view the electrons in the normal atom will all possess the requisite angular momentum, and when an external magnetic field is applied the induced change in angular momentum will put the electrons in an unstable condition. On this view also, therefore, the additional rotational energy induced by an applied magnetic field will not be permanent, but will soon be dissipated. In fact, the theory of atomic structure has yet to be proposed according to which diamagnetism, accounted for by the induced magnetic moment of electrons revolving in orbits, can be more than a transient phenomenon.

Besides the molecule and the atom we have the other two fundamental divisions of matter, the atomic nucleus and the electron. The sign of the Richardson-Barnett effect indicates that it is negative electricity which is chiefly responsible for magnetic effects, which makes the view that the positive nucleus is the elementary magnet difficult to defend. On the other hand, many of the magnetic properties of matter receive a satisfactory explanation on Parson's hypothesis,[3] that the electron is a continuous ring of negative electricity spinning rapidly about an axis perpendicular to its plane, and therefore possessing a magnetic moment as well as an electric charge. Thus, for example, the fact that such a ring can rotate without radiating enables this hypothesis to account for diamagnetism as a permanent instead of a transient effect. While retaining Parson's view of a magnetic electron of comparatively large size, we may suppose with Nicholson that instead of being a ring of electricity, the electron has a more nearly isotropic form with a strong concentration of electric charge near the centre

---

[3] A. L. Parson, Smithsonian Misc. Collections, 1915.

and a diminution of electric density as the radius increases. It is natural to suppose that the mass of such an electron is concentrated principally near its centre and that the ratio of the charge to the mass of its external portions will be greater than that for the electron as whole. While the explanation of the inertia of such a charge of electricity is perhaps not obvious, it is at least consistent with our usual conceptions and it has the advantage of offering an explanation for the large value of e/m observed in Barnett and Stewart's experiments. It also makes possible an explanation of the relatively large induced currents required to account for diamagnetism without introducing the assumption of a prohibitively large radius for the electric charge.

A series of experiments has recently been performed, designed to determine which of these fundamental divisions of matter is identical with the elementary magnet in ferromagnetic substances. The first of these, due to K. T. Compton and E. A. Trousdale,[4] had for its object the detection of any displacement of the atoms of a substance on magnetization. If the elementary magnet consists of a group of atoms such as the chemical molecule, the rotation of this elementary magnet into alignment with an applied external field will cause a displacement of the individual atoms. It is known, however, that the position of the spots on a Laue photograph depends upon the arrangement of the atoms within the crystal employed. If then, such a photograph is taken with a magnetic crystal, the character of the diffraction pattern should change when the direction of magnetization of the crystal is altered. In these experiments, however, no effect of this character was found. The obvious conclusion is that the ultimate magnetic particle does not consist of any group of atoms such as the chemical molecule.

The second of these experiments, performed by Mr. Rognley and myself,[5] was based upon the fact that the intensity of reflection of X-rays from the surface of a crystal depends not only upon the arrangement of the atoms within the crystal, but also upon the distribution of the electrons within the atoms. Let us suppose that the atom acts as a tiny magnet due to the orbital motion of its component electrons. Magnetization of the crystal will orient these atomic magnets and in so doing will change the planes of

---

[4] K. T. Compton and E. A. Trousdale, *Phys. Rev.*, **5**, 315 (1915).
[5] A. H. Compton and O. Rognley, *Phys. Rev.*, **16**, 464 (1920).

revolution of the electrons. This change in the electronic distribution should, therefore, affect the intensity of reflection of a beam of X-rays from the crystal's surface. An attempt was made to detect such a change in the intensity of X-ray reflection from a crystal of magnetite when strongly magnetized. Apparatus sufficiently sensitive to detect a change in intensity of less than one per cent. was employed, but magnetization of the crystal failed to produce any measurable effect. The following table shows in the first column the order of the X-ray spectrum line which was being studied; in the second column

TABLE I.

Order	E_1/E_u	E_2/E_u	E_3/E_u
1	1.05	1.000	1.004
2	1.27	0.96	1.03
3	1.48	0.86	1.09
4	1.70	0.51	1.09

the calculated ratio of intensity from the magnetized to that from the unmagnetized crystal, supposing the atom to have the Rutherford form; and the third and fourth columns represent the similar ratios as estimated from a cubic form of atom. In the third column it is supposed that the magnetic axis is perpendicular to a cube face, and in the fourth column that the magnetic axis is along the cube diagonal. According to experiment the value of these ratios was always unity, at least within one per·cent. It is clear that none of the types of atoms considered could be oriented by a magnetic field without producing a noticeable effect. In fact, it is difficult to imagine any form of magnetic atom which would be so nearly isotropic that it would have given no effect in our experiment. It is, therefore, difficult to avoid the conclusion that the elementary magnet is not the atom as a whole.

Since neither the molecule nor the atom gives a satisfactory explanation of these experiments, the view suggests itself that it is something within the atom, presumably the electron, which is the ultimate magnetic particle. Let us see then if we can find any positive evidence for the existence of an electron with a magnetic moment.

On the basis of the classical dynamics we should expect the electron, whatever its form, to possess thermal energy of rotational motion, equal on the average to that of a molecule or atom at the same temperature. On Planck's more recent quantum hypothesis, however, which is perhaps the more reasonable view, at the absolute zero of temperature each particle of matter—including the electron—should retain an average amount of energy $\tfrac{1}{2}h\nu$ for each degree of freedom for motion. For a rotating system this corresponds to an angular momentum of $h/2\pi$. Thus whatever view we adopt, the thermal motions of the electron will give to it an appreciable magnetic moment. For a particle of the small moment of inertia of the electron, the frequency of rotation corresponding to an angular momentum $h/2\pi$ will be exceedingly high, and the corresponding energy $\tfrac{1}{2}h\nu$ will be large compared with the additional energy which it may acquire due to an increase in temperature. Thus the angular momentum, and hence also the magnetic moment of the electron, will be nearly the same at different temperatures—a property characteristic of the elementary magnets. It is interesting to notice, also, that the magnitude of the magnetic moment of an electron spinning with an angular momentum $h/2\pi$ is of the proper order to account for ferromagnetic properties, being about one-third the magnetic moment of the iron atom.

If an electron with such an angular momentum is to have a peripheral velocity which does not approach that of light, it is necessary that the radius of gyration of the electron shall be greater than 10^{-11} cm. While such an electron is much larger than the spherical electron of Lorentz, recent experiments on the scattering of X-rays and gamma rays indicate the electron's diameter may be even greater than the minimum value thus required to explain magnetic properties. Experiment shows that the scattering of very high frequency radiation is considerably less than theory demands if the electron is supposed to have negligible dimensions. In the case of hard gamma rays, indeed, I have found the scattering at certain angles to fall below $1/1000$, the intensity predicted on the usual theory.[6] The only adequate explanation of these experiments seems to be that interference occurs between the rays scattered from the different parts of the same electron. Such an explanation clearly implies that the diameter of the electron is comparable with the wave-length of the radiation employed, which means that the effective radius of the electron is of the order of 10^{-10} cm. Considerations of the size of the electron, therefore,

---

[6] A. H. Compton, *Phil. Mag.* (in printer's hands).

support rather than oppose the view that the electron may have an appreciable magnetic moment.

Further evidence that the electron possesses properties other than those of an electric charge of negligible dimensions is afforded by a study of the white X-radiation emitted at the target of an X-ray tube. It was noticed by Kaye that the X-rays emitted in the direction of the cathode ray beam are harder and more intense than those traveling in the opposite direction. The difference in both hardness and intensity of the radiation at different angles is in good accord with the view proposed by D. L. Webster that the particles emitting the radiation are moving in the direction of the cathode-ray beam, giving rise to a Doppler effect. Indeed, it is very difficult to give any other explanation of the difference in wave-length of the radiation in different directions. But, on this view, in order to account for the difference in hardness observed in the case of gamma rays, the radiating particles must have a velocity of about one-half the speed of light. Since the highest known speeds at which atoms travel is only about one-tenth the velocity of light, as observed in the case of alpha particles, the swiftly moving radiators giving rise to this high-frequency X-radiation must therefore be free electrons. If this view is correct, it follows, as Webster has pointed out, that the electron must be a system capable of emitting radiation, and is therefore, not a mere charge of electricity of negligible dimensions. On the present view we may well suppose that the electron is spinning like a gyroscope and on traversing matter is set into mutational oscillations, resulting in the observed radiation.

Strong evidence that the electron possesses a magnetic moment is afforded by H. S. Allen's recent explanation of the rotation of the plane of polarization by optically active substances.[7] You will remember in Drude's classical work it is found that optical rotation may be explained if the electrons, when made to oscillate by a passing electric wave, do not move exactly in the plane of the electric vector. He supposes rather that there is a component of motion at right angles to the electric vector and finds that such a motion will account for the observed rotation. Allen shows that the motion perpendicular to the electric vector which Drude assumes is a natural consequence of the view that the electron is magnetic and has an appreciable diameter. It would take us too

---

[7] H. S. Allen, *Phil. Mag.*, **40**, 426 (1920).

far afield to discuss the details of this work, but the significance of the result is obvious, since it has heretofore been difficult to give a reasonable account of the type of motion postulated by Drude.

Finally, I wish to discuss a phenomenon, first noticed by C. T. R. Wilson and brought to my attention by Mr. Shimizu, which, if its obvious explanation is correct, gives direct evidence that free electrons possess magnetic polarity. Suppose that a magnetic electron is placed in a homogeneous paramagnetic medium. Every part of the medium will be slightly magnetized in the direction of the lines of force, and the magnetic field at the electron due to the magnetic moment of each portion of the medium will have a positive component in the direction of the electron's magnetic axis. Thus the magnetization induced in the surrounding medium will give rise to a magnetic force at the electron in the direction of its own magnetic axis. The case is exactly analogous to placing a bar magnet in a field of iron filings. The iron filings will be magnetized by induction in the direction of the lines of force and if the bar magnet is removed, there still exists a magnetic field where the magnet was because of the magnetization of the surrounding iron filings. If now the electron is in motion, this induced magnetic field will produce the same effect as would an externally applied field of the same intensity. That is, the force due to the magnetic field from the surrounding medium acting on the moving electric charge will make it follow a curved instead of a straight path.

If, because of its gyroscopic action, the axis of the electron does not change its direction, the induced magnetic field will always be in the same direction, and the electron will describe a helical orbit. In any actual medium, composed of discreet particles and therefore not homogeneous on an electronic scale, this spiral motion will be superposed upon an irregular motion due to collisions, and the axis of the electron will not remain fixed in direction. Thus any spiral motion that may appear should be rather broken. A rough calculation, assuming an electron to be projected into air with a speed corresponding to a drop through 10,000 volts, which is about that of the secondary cathode rays produced by ordinary X-rays, and having a magnetic moment corresponding to the angular momentum $h/2\pi$, indicates that the induced magnetic field at the electron should be of the order of 3000 gauss, if the permeability of the medium is that of ordinary air. This field is strong enough to produce a very decided curvature in the electron's path, so in spite of the irregularities in the electron's motion we might hope to observe experimentally the predicted helical tracks.

Below are a few of C. T. R. Wilson's photographs of the tracks of secondary cathode rays and beta particles. In the first figure are seen the tracks of the cathode rays ejected by a comparatively intense beam of X-rays. Let me call your attention particularly to the two tracks marked by arrows. You see here paths in the form of almost perfect helices. Most of the tracks are too

FIG. 1.

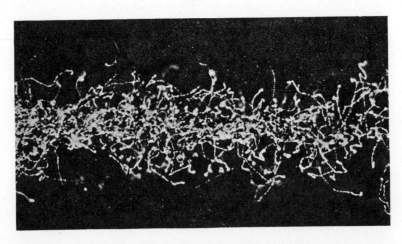

irregular and too confused with each other to trace so perfect a spiral form; but you will notice that in almost every case, the track terminates in a close spiral. The tracks can be examined more satisfactorily if we use a photograph showing a smaller number. In the next figure I have called attention particularly to three tracks. It is unfortunate that one cannot show these paths on the screen in three dimensions. Mr. Wilson showed me some remarkable stereoscopic photographs, as yet unpublished, which he obtained of X-rays passing through air. In one of these, showing altogether about 66 complete tracks, all but about 14 seemed to be of a spiral form. Of these fourteen 12 were too irregular to detect with certainty any spiral tendency that might exist, and the remaining two were for the most part straight. But to me there seemed no doubt, nor did there to others who examined them carefully, but that there was a real tendency to spiral motion in the tracks of these secondary cathode particles.

The beta rays from radium show the same consistent curvature. Notice particularly the path shown in Fig. 3 with its almost uniform curvature. If one would calculate the probability of such a curvature on the basis of chance collisions, each as likely to deflect the particle in one direction as in another, this type of path would be declared impossible.

Fig. 2.

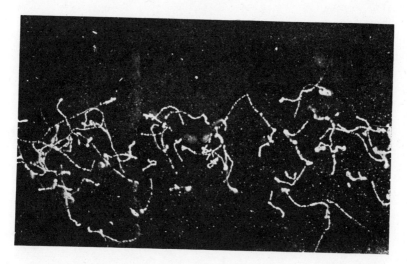

Examining again the tracks of the secondary cathode rays, let us see how their form compares with that to be expected for gyroscopic magnetic electrons. In the first place we find that the tracks exhibit a helical curvature of the kind we should anticipate. In the second place the axis of the helix is different for each beta particle, which we should anticipate since each beta particle induces its own magnetic field and the direction of the field is coincident with its own axis. And, finally, we notice changes in the direction of curvature such as might well result from sudden precessions of the electron's gyroscopic axis. If the obvious explanation of these spiral tracks is the correct one, we have here positive evidence for our hypothesis that the electron acts as a tiny magnet as well as an electric charge.

Let us then review the different lines of evidence that have given us information concerning the nature of the elementary magnet. In the first place, the Richardson-Barnett effect shows that magnetism is due chiefly to the circulation of negative electricity whose ratio of charge to mass is not greatly different from that of the electron. In the second place, experiments on the diffraction

FIG. 3.

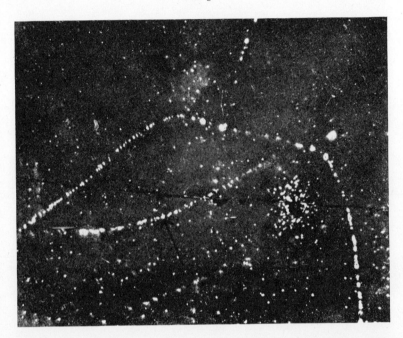

of X-rays by magnetic crystals indicate that the elementary magnet is not any group of atoms, such as the chemical molecule, nor even the atom itself; but lead rather to the view that it is the electron rotating about its own axis which is responsible for the ferromagnetism. And finally, positive evidence in favor of the hypothesis of some form of magnetic electron is supplied by a consideration of the curvature of the tracks of beta rays through air. May I then conclude that the electron itself, spinning like a tiny gyroscope, is probably the ultimate magnetic particle.

Secondary High Frequency Radiation.

By Arthur H. Compton.

The experiments performed by Florance and Ishino on the scattering of gamma rays showed a considerable difference in hardness between the primary and the secondary radiation, an effect difficult to explain if the secondary radiation was truly scattered. An experimental method of distinguishing between truly scattered and fluorescent radiation was therefore devised. This consisted of placing an absorbing screen alternately in the path of the primary and of the secondary beam. Any truly scattered radiation should remain of the same intensity for both positions of the absorbing screen, while the fluorescent rays, being in general softer than the primary rays exciting them, would be more strongly absorbed when the screen is in the path of the secondary radiation. The ratio of the truly scattered to the total secondary rays is the limit of the ratio of the intensity observed with the screen in the path of the secondary to that when the screen is in the path of the primary beam as the thickness of the screen is increased.

The experiments showed that with gamma rays the amount of truly scattered radiation at angles greater than 90° as determined by this criterion is a negligible part of the total secondary rays. At 45° about 3 per cent. of the radiation seems to be of the scattered type.

Similar experiments were performed with x-rays of wave-lengths between 0.12 and 0.50 Ä.U. using a modified Bragg spectrometer with a recording device to register the intensity of the secondary radiation for different angles of the ionization chamber. These rays showed an effect identical in character with that observed with gamma rays, but not so prominent. With $\lambda = 0.12$ Ä.U. the scattered rays form about 15 per cent. at 90° with the primary beam and 70 per cent. of the total radiation at 30°. For $\lambda = 0.50$ the scattering is relatively much more important, but the fluorescent radiation is still a large part of the total radiation. These results, both in the case of the gamma rays and of the x-rays, seem to be independent of the material used as radiator except in those cases where a perceptible amount of characteristic fluorescent radiation enters the ionization chamber.

The experiments therefore indicate the existence of a type of fluorescent high frequency radiation whose wave-length is independent of the particular substance used as radiator, depending only on the frequency of the exciting primary rays.

The measurements made by Barkla, Crowther, Owen and others represented the total secondary radiation, only a part of which, according to the present measurements, represents truly scattered rays. A direct measurement of the absolute value of the scattering shows that at small angles the scattering of x-rays approaches that predicted according to Thomson's theory if the number of effective electrons per atom is equal to the atomic number. At large angles however, the scattering is much less than predicted by this theory, especially for the shorter wave-lengths, being less than 1/1,000th of the theoretical value in the case of gamma rays. The results are in much better accord with the hypothesis that the electron is of sufficient size (about 3×10^{-10} cm.) to give rise to interference effects between the rays scattered from its different parts.

WASHINGTON UNIVERSITY,
SAINT LOUIS, April 21, 1921.

THE POLARIZATION OF SECONDARY X-RAYS.

BY A. H. COMPTON AND C. F. HAGENOW.

IN the previous paper by one of us, evidence is presented for the existence of a type of fluorescent radiation whose wave-length is independent of the material of the radiator, being a function only of the wave-length of the primary radiation. The explanation suggests itself that this general fluorescent radiation may be the x-rays excited by the impact of the secondary cathode rays liberated in the radiator by the action of the primary radiation. While such an explanation accounts for many of the characteristics of this radiation, quantitative measurements showed that at least in some cases as much as 50 per cent. of the energy of the primary rays appeared as general fluorescent radiation. Since the efficiency of production of x-rays by cathode rays is less than 1 per cent., it does not seem that this can be the real origin of the secondary rays. It occurred to us that some information on this point might be obtained from a careful measurement of the polarization of the fluorescent radiation.

Barkla, in his classic measurement of the polarization of secondary x-rays, found that at 90° the rays were approximately 80 per cent. polarized. The remaining 20 per cent. might be accounted for in part by experimental error, and the remainder Barkla ascribed as due possibly to a real lack of polarization due to forces acting on the scattering electrons. We employed much more sensitive methods of measuring the tertiary radiation than those used by Barkla and his followers. The first radiator was placed about 12.5 cm. directly above the second radiator, which was in turn placed on the crystal table of a Bragg spectrometer. The degree of polarization could then be measured by noting the scattering from the second radiator parallel and at right angles with the primary beam. The geometrical conditions were such that at the 90° position about 4 or 5 per cent. of the radiation should appear to be unpolarized (the apertures being so large that the scattering angles were not exact right angles).

For thick blocks of scattering material our results confirmed those of Barkla, It was found, however, that when thin sheets of the radiators were employed. the polarization became much more complete. As one might have anticipated, multiple scattering in the case of the thick radiators was responsbile for a considerable lack of polarization. Extrapolating our results to zero thickness we find the ratio of the intensity of the tertiary rays at 90° to that at 0° to be for a paper radiator, 0.05, for aluminium, 0.04 and for sulphur, 0.05. When these results are corrected for the imperfection of the geometrical conditions, it means that the polarization is complete within experimental error.

Since under the conditions of our experiment approximately 70 per cent. of the total secondary radiation at 90° is of the general fluorescent type, this result means that the fluorescent as well as the scattered rays are completely polarized. Barkla's experiments show that the general radiation from an x-ray tube is only partly polarized, which makes it difficult to support the view that the general fluorescent radiation is excited in exactly the same manner as the primary x-rays. The hypothesis that this radiation is emitted at the instant of liberation of the secondary cathode rays from the atoms, however, seems to offer a possible explanation of both the degree of polarization and the efficiency of production of these fluorescent rays.

WASHINGTON UNIVERSITY,
SAINT LOUIS, April 21, 1921.

A Possible Origin of the Defect of the Combination Principle in X-Rays.

By Arthur H. Compton.

The original Kossel relation, which represents an application of Ritz's combination principle to x-ray spectra, and which follows directly from Bohr's theory, states that

$$K_{a_1} + L_{a_1} = \frac{1}{h}(W_{M_1} - W_K) = K_\beta,$$

where K_a, L_a, K_β represent the frequencies of these lines respectively, W_{M_1} is the energy of an electron in the M_1 orbit and W_K its energy in the K orbit. It has been found experimentally, on the other hand, that

$$K_{a_1} + L_{a_1} = \frac{1}{h}(W_{M_1} - W_K) \neq K_\beta,$$

and that the difference

$$\Delta = K_{a_1} + L_{a_1} - K_\beta$$

is of the order of $+\frac{1}{2}$ per cent. of K_β. This difference has constituted a formidable difficulty for the Bohr-Sommerfeld theory of the origin of spectrum lines.

Perhaps the most obvious explanation of this discrepancy, proposed independently by Kossel[1] and Duane,[2] is that the K_β line may not be due to an electron falling from the same M orbit as the L_{α_1} line. According to Rubinowicz's principle of selection, indeed, no electrons should fall from the M_1 orbit to the K ring, but only from the $M_{2, 3 \text{ or } 4}$ orbits. On this view, therefore, K_β should equal

$$\frac{1}{h}(W_{M_{2, 3 \text{ or } 4}} - W_K).$$

A critical examination of the wave-lengths by Sommerfeld[3] and particularly by Birge[4] indicates, however, that this explanation probably does not account completely for the observed difference between K_β and $K_{\alpha_1} + L_{\alpha_1}$. In fact the difference Δ should on Kossel's view be approximately proportional to the 4th power of the atomic number, whereas Sommerfeld finds the experimental values of Δ to be nearly a linear function of N.

An alternative, or perhaps supplementary, explanation of this difference Δ is that as the electron passes from the M to the K orbit, traversing the L orbit, it sets the electrons of the L orbit into oscillation, thus being able itself to radiate less energy than $(W_K - W_M)$ when it reaches the K orbit. We know from a study of the tracks of β-rays and cathode rays through air that energy is lost in thus traversing groups of electrons. If the electron which gives rise to the radiation moves gradually from the M to the K orbit, the electrons of the L ring will assume their new positions gradually, and the work done on these electrons will be the same as that done by successive transfers from M to L, producing the L_α line, and from L to K, producing the K_α line. In this case the combination principle should therefore hold. If, however, the radiating electron were transferred instantaneously from the M to the K

---

[1] Kossel, Zeitschr. f. Phys., 1, 126, 1920.
[2] Duane and Stenström, Nat. Ac. Sci. Proc., 6, 484, 1920.
[3] Sommerfeld, Zeitschr. f. Phys., 1, 142, 1920.
[4] Birge, Phys. Rev., 16, 371, 1920.

orbit, the electrons of the L ring would be set into oscillation with an energy equal to their difference in energy for p and for $p-1$ electrons in the inner ring. Such energy of oscillation must be taken from the falling electron, and can be shown to represent a loss in frequency of this electron's radiation by approximately

$$(1) \qquad \Delta = \frac{1}{h}\Delta W = \frac{2qR}{2^2}(N - p - \varphi_q),$$

where p and q are the numbers of electrons in the K and L rings respectively, R is the Rydberg constant, N is the atomic number, and φ_q is Moseley's correction for the repulsion between the electrons in the L ring. The energy spent by the radiating electron in setting up oscillations in the ring of electrons through which it passes will therefore lie between zero and the value represented by expression (1). The fact that this view makes Δ a linear function of N is in accord with Sommerfeld's interpretation of the experiments, and the numerical values also are of the right order.

WASHINGTON UNIVERSITY,
SAINT LOUIS, MO.

The Softening of Secondary X-rays.

A NUMBER of experimenters have noticed that when a beam of X-rays or γ-rays traverses any substance, the secondary rays excited are less penetrating than the primary rays. Prof. J. A. Gray (Franklin Institute Journal, November, 1920) and the present writer (*Phil. Mag.*, May, 1921, and *Phys. Rev.*, August, 1921) have shown that the greater part of this softening is not due, as was at first supposed, to a greater scattering of the softer components of the primary beam, but rather to a real change in the character of the radiation. My conclusion was that this transformation consisted in the excitation of some fluorescent rays of wave-length slightly greater than that of the primary rays. Prof. Gray, on the other hand, showed that if the primary rays came in thin pulses, as suggested by Stokes's theory of X-rays, and if these rays are scattered by atoms or electrons of dimensions comparable with the thickness of the pulse, the thickness of the scattered pulse will be greater than that of the incident pulse. He accordingly suggests that the observed softening of the secondary rays may be due to the process of scattering.

It is clear that if the X-rays are made to come in long trains, as by reflection from a crystal, the scattering process can effect no change in wave-length. On Gray's view, therefore, if X-rays reflected from a crystal are allowed to traverse a radiator, the incident and the excited rays should both have the same wave-length and the same absorption coefficient. If, on the other hand, the softening is due to the excitation of fluorescent rays, as I had suggested, reflected X-rays should presumably be softened by scattering in the same manner as unreflected rays. An examination of the absorption coefficient of reflected X-rays before and after they have been scattered should therefore afford a crucial test of the two hypotheses.

The double reduction in intensity which occurs when the X-ray beam is first reflected by a crystal and then scattered by the radiator made Gray's preliminary attempts to perform this experiment unsuccessful. In the September (1921) issue of the *Philosophical Magazine*, however, Mr. S. J. Plimpton describes a successful attempt to measure the absorption of the K lines from rhodium and molybdenum after being scattered by paraffin and water. He observed no change in the absorption coefficient of the rays after being scattered by the paraffin. Apparently his measurements were made on the secondary rays at comparatively small angles, and this, together with the relatively long wave-lengths employed, form the conditions under which the least change in hardness occurs when unreflected X-rays are used. I accordingly repeated Mr. Plimpton's experiments, using the K lines from tungsten (effective $\lambda = 0.196$ Å.) reflected from rock-salt, and measured the absorption coefficient of the secondary radiation excited by these rays in paraffin. The absorption coefficient of these rays was found to be considerably greater, by about 52 per cent. at 90° and 22 per cent. at 30°, than that of the beam incident on the paraffin.

In order to compare my results with those of Mr. Plimpton, a molybdenum Coolidge tube was then substituted for the tungsten one, and the K_α line ($\lambda = 0.708$ Å.) was employed. An increase in the absorption coefficient of the secondary rays excited in paraffin was again observed, though it amounted to only 29 per cent. at 90° and only 6 ± 1.2 per cent. at 20° with the primary beam.

The softening thus observed when reflected X-rays are scattered is substantially the same as that found when unreflected rays of the same hardness are employed. Mr. Plimpton's negative result is apparently due to the fact that his experiment was performed under unfavourable conditions of wave-length and scattering angle. The conclusion seems necessary, therefore, that the softening of secondary X-rays is due, not to the process of scattering, but to the excitation of a fluorescent radiation in the radiator.

ARTHUR H. COMPTON.

Washington University, St. Louis, U.S.A.

THE WIDTH OF X-RAY SPECTRUM LINES.

By Arthur H. Compton.

Synopsis.

Width of X-ray Spectrum Lines.—(1) *Four causes* are discussed: (*a*) the width of the slit employed, (*b*) angular imperfections of the crystal, (*c*) finite resolving power of the crystal grating, and (*d*) lack of homogeneity of the x-rays. The effect of the first two causes is independent of the angle of reflection and therefore the same for all orders and all lines; the effect of the third is shown to vary as $1/\sin\theta\cos\theta$ and therefore to decrease as θ increases up to 45°; while the effect of the last is equal to $(d\lambda/\lambda)\tan\theta$ and therefore increases with θ. (2) *Measurements of two tungsten lines,* λ 1.242 and 1.279 Å, in the first four orders from the cleavage faces of calcite and rock-salt, are given. The fact that in both cases the width is least for the second order and greatest for the fourth order shows that both the third and fourth causes contribute measurably to the observed width, as well as the first two.

Lack of homogeneity of tungsten x-ray lines computed from the above measurements comes out greater than $0.0007\lambda \pm .00013\lambda$, which is about 0.5′ for the first order. This cannot be explained as a Doppler effect or as due to the damping of the electronic motion; however it is in accord with the complexity of the lines as predicted by Sommerfeld's theory of elliptic electronic orbits.

SEVERAL years ago the writer performed a series of experiments designed primarily to measure the relative intensities of the different orders of x-ray spectrum lines reflected from crystals of rock-salt and calcite. A recording x-ray spectrometer, which has previously been described,[1] was employed, the intensities of the lines being taken proportional to the area under the curves on the record representing the different spectrum lines. These intensity measurements, in which cleavage faces of the crystals were employed, were later found to be valueless as a test of the theory of reflection, because of the selective absorption which occurs at the angle of maximum reflection from a cleavage face.[2] The results show, however, an interesting broadening of the lines in the higher orders.

The width of the spectrum lines was measured at a point midway between the peak of the curve and a base-line representing the intensity of the general radiation. The average values thus found for the tungsten lines $\lambda = 1.242$ and $\lambda = 1.279$ Å.U. are as follows:

---

[1] A. H. Compton, Phys. Rev., 7, 658 (1916).

[2] A. H. Compton, Phys. Rev., 10, 95 (1917); W. L. Bragg, James and Bonsanquet, Phil. Mag., 41, 309 (1921).

TABLE I.

Width in Minutes of Arc at Mid-point of Line.

Crystal.	First Order.	Second Order.	Third Order.	Fourth Order.
Calcite................	16.3 ± .6	16.0 ± .36	17.3 ± .7	18.3 ± .12
(Calc.)................	15.4	(16.0)	16.7	(18.3)
Rock-salt.............	20.1 ± .5	18.9 ± .3	20.3 ± .6	21.2 ± 1.2
(Calc.)................	18.3	(18.9)	19.8	21.5

The size of the slits remained unchanged throughout the experiments. The measurements on the second and fourth orders of reflection from calcite are the most valuable, because these two lines are of nearly equal intensity, which tends to eliminate any consistent errors.

The observed width of a spectrum line may result from four different causes, (a) the angular aperture of the slits, (b) angular faults in the crystal grating, (c) the finite resolving power of the grating, and (d) the lack of homogeneity of the incident x-rays.[1] The width due to the size of the slits will obviously be the same at all angles. That due to angular faults in the crystal may vary somewhat according to the part of the crystal which is exposed to the x-rays, but should on the average be the same for all angles of reflection. It is clear that no crystal will act as a perfect grating, and that the resolving power will depend upon the size of the portions of the crystal which are effectively perfect. It can be shown[2] that if these perfect portions are assumed to be parallelopipeds of width along the crystal face δx, height δy and depth into the crystal δz, the effective width of the line (area/height) in radians is:

$$(1) \qquad \delta\theta_c = \frac{\lambda}{2\delta x \sin\theta} \quad \text{for} \quad \frac{\delta x}{\delta z}\tan\theta < 1,$$

$$\delta\theta_c = \frac{\lambda}{2\delta z \cos\theta} \quad \text{for} \quad \frac{\delta x}{\delta z}\tan\theta > 1,$$

where λ is the wave-length and θ the glancing angle. In crystals of rock-salt and calcite δx and δz are doubtless on the average about equal, so that $\delta\theta_c$ will be a minimum for $\theta = \pi/4$, increasing symmetrically for smaller and larger angles. Because of chance variations in δx and δz

---

[1] In the experiments here described, the beam incident upon the crystal was collimated by a pair of narrow slits, and the opening at the ionization chamber was wide enough to admit all the reflected rays. With Bragg's method of using narrow slits near the x-ray tube and at the ionization chamber and a wide slit at the crystal, the width also depends upon a fifth factor, the penetration of the x-rays into the crystal.

[2] This follows (after considerable reduction) from the results given by the writer in THE PHYSICAL REVIEW, 9, 37 (1917).

for the different component crystals, the change from $\delta\theta_c \propto 1/\sin\theta$ to $\delta\theta_c \propto 1/\cos\theta$ will be gradual. Thus it appears that the width of the line due to this finite resolving power should vary approximately according to the expression

(1a) $$\delta\theta_c \propto 1/\sin\theta\cos\theta.$$

The angular width due to a given range of wave-lengths $\delta\lambda$ in the incident x-rays may be found by differentiating $\lambda = 2D\sin\theta/n$ with respect to θ. Thus we find that

(2) $$\delta\theta_d = \frac{n\delta\lambda}{2D\cos\theta} = \frac{\delta\lambda}{\lambda}\tan\theta.$$

The total width of the line is a rather complicated function of $\delta\theta_a$ due to the width of the slits, $\delta\theta_b$ due to angular imperfections of the crystals, and $\delta\theta_c$ and $\delta\theta_d$ given by equations (1) and (2) respectively. Thus we may write

$$\delta\theta = F(\delta\theta_a, \delta\theta_b, \delta\theta_c, \delta\theta_d).$$

The function F will increase with every increase in any variable $\delta\theta_r$, in such a manner that $\delta\theta$ will be greater than any one of its components but will be less than the sum of all four. The form of this function cannot be determined unless we know the exact manner in which the intensity varies with the angular distance from the center of the line on account of each factor contributing to the width. Lacking such definite knowledge, a simple graphical analysis shows that an increase in $\delta\theta_r$ will result in an increase in $\delta\theta$ of the same order of magnitude but smaller than the change in $\delta\theta_r$, i.e.,

$$1 > \frac{\partial\delta\theta}{\partial\delta\theta_r} > 0.$$

Furthermore by Taylor's theorem, to a first approximation,

(3) $$\delta\theta = \delta\theta_0 + \frac{\partial\delta\theta}{\partial\delta\theta_a}\Delta\delta\theta_a + \frac{\partial\delta\theta}{\partial\delta\theta_b}\Delta\delta\theta_b + \frac{\partial\delta\theta}{\partial\delta\theta_c}\Delta\delta\theta_c + \frac{\partial\delta\theta}{\partial\delta\theta_d}\Delta\delta\theta_d,$$

where $\Delta\delta\theta_r$ represents a change in $\delta\theta_r$.

In the experiments using a calcite crystal, $\delta\theta_a$ and $\delta\theta_b$ remain constant, $\delta\theta_c$ decreases with θ (up to 45°), and $\delta\theta_d$ increases with θ. The observed increase in the width of the lines in the higher orders therefore means that $(\partial\delta\theta/\partial\delta\theta_d)\Delta\delta\theta_d$ is more prominent than $(\partial\delta\theta/\partial\delta\theta_c)\Delta\delta\theta_c$. If we neglect the term in $\delta\theta_c$ and compare the value of $\delta\theta$ at two angles θ_1 and θ_2, it follows from expression (3) that

$$\delta\theta_2 - \delta\theta_1 = \frac{\partial\delta\theta}{\partial\delta\theta_d}(\delta\theta_{d_2} - \delta\theta_{d_1})$$

whence according to equation (2)

$$(4) \qquad \frac{\delta\lambda}{\lambda} = \frac{\partial \delta\theta_d}{\partial \delta\theta} \frac{\delta\theta_2 - \delta\theta_1}{\tan\theta_2 - \tan\theta_1}.$$

Substituting the values of $\delta\theta$ observed in the second and fourth orders from calcite, noting that the angles θ_2 and θ_4 are 24.2° and 55.2° respectively and that $\partial\delta\theta_d/\partial\delta\theta$ is of the order of but greater than unity, we find that $\delta\lambda/\lambda$ is of the order of but greater than $0.0007 \pm .00014$. The values calculated in the second row of Table I. result from the further approximation that $\partial\delta\theta_d/\partial\delta\theta$ is constant for all values of $\delta\theta_d$. If we take this constant as unity, we find 0.0007 for the minimum value of $\delta\lambda/\lambda$, and 0.5′ for the minimum value of the width of the first order spectrum line due solely to lack of homogeneity of the x-rays. The observations on rock salt lead to values of the same order of magnitude.

The widths calculated for the lines reflected from calcite agree well with the experiments except in the first order. This line appears to be too broad, as we have noticed should result from the small resolving power if the crystal is sufficiently imperfect. The imperfection of the rock-salt crystal as compared with calcite shows itself both in the greater value of the constant $\delta\theta_b$ for rock salt, indicated by a greater width in all orders, and by the greater difference between the breadth of the first and second orders. According to equation (1a) the broadening due to this imperfection is nearly proportional to $1/\sin\theta\cos\theta$, i.e., to 2.5, 1.3, 1.0 and 1.1 in orders 1, 2, 3 and 4 respectively. A difference in breadth of 1.8′ between the first and second orders of rock salt[1] will therefore correspond to a difference of only 0.3′ between the second and fourth orders. The unmistakable broadening of the higher orders can apparently be accounted for only by a true lack of homogeneity of the x-ray spectrum line. Thus it appears that all the terms of expression (3) have values that are appreciable in experiment.

It is of interest to consider the possible origin of the observed non-homogeneity of the x-rays. Let us consider first the width to be expected on the basis of the Doppler effect. The average velocity of the tungsten molecules in the target at 2000° K. is only about 10^{-6} times the velocity of light, and could therefore account for only a negligible part of the non-homogeneity of the x-rays. The velocity of the thermal motion of an electron, according to the principle of equipartition of energy, would give rise to a Doppler effect of the required order of magnitude; but the fact that the spectrum lines are characteristic of the tungsten

---

[1] This difference corresponds to a value of $\delta\theta_c > 3.2′$ in the first order. Thus by equation (1), $\delta x < c. \; 2700\lambda = 4.5 \times 10^{-5}$ cm.

atom indicates that it is not a free electron which emits the radiation, but one which is a part of the atom.

The damping of the electron's linear oscillations due to its own radiation, on the other hand, results in a width which is not negligible. The damping factor for a single oscillating electron is

$$\frac{4\pi^2}{3} \frac{e^2}{m\lambda^2 c},$$

where e, m and C have their usual significance. For $\lambda = 1.25$ Å.U., this means that the wave will be damped to $1/e$ of its amplitude in 3,400 vibrations. By an application of Fourier's integral, Professor Jauncey finds[1] that the effective width of the line due to this damping is about 0.06 minute of arc in the first order, which is 12 per cent. of the minimum width found experimentally.

It appears probable that the chief cause of the non-homogeneity observed in these experiments is that the lines themselves are complex. Thus the line $\lambda = 1.279$ is known to have a faint companion 0.008 Å.U. distant which was not separated in the first and second orders by the slits employed; and according to Sommerfeld's theory of the fine structure of spectrum lines, the line 1.242, on which most of the fourth order measurements were made, should have faint satellites at distances of about 0.0007, 0.002 and 0.006 Å.U.

The experiments described in this paper were performed at Palmer Physical Laboratory, Princeton University.

WASHINGTON UNIVERSITY,
SAINT LOUIS,
July 18, 1921.

[1] Cf. G. E. M. Jauncey elsewhere in this number of the PHYSICAL REVIEW.

The Spectrum of Secondary X-rays.

By Arthur H. Compton.

An examination of the secondary rays excited in different materials when x-rays rendered nearly homogeneous by filtering were employed, showed that the secondary radiation was of a softer type than the primary rays which struck the radiator.[1] More recent experiments have shown that this phenomenon is not confined to heterogeneous x-rays, but occurs also when the rays incident upon the radiator have been reflected from a crystal.[2] The most obvious interpretation of these results was that in addition to scattered radiation there appeared in the secondary rays a type of fluorescent radiation, whose wave-length was nearly independent of the substance used as radiator, depending only upon the wave-length of the incident rays and the angle at which the secondary rays were examined.

In order to obtain more definite information with regard to the characteristics of the secondary x-radiation, a study has been made of the spectrum of the secondary rays excited in various substances by the x-rays from a Coolidge tube having a molybdenum target. A small piece of radiating material, such as celluloid or aluminium, placed in front of the first slit of the spectrometer, was illuminated by incident x-rays at approximately 90° with the secondary beam under investigation. The spectrum was studied by means of a calcite crystal grating, using both the ionization and photographic methods.

The spectra obtained show lines identical in wave-length with the primary K lines from molybdenum, thus proving that a part of the secondary radiation is truly scattered and unchanged in wave-length. In addition to these lines, a general radiation is observed which is more prominent in the secondary than in the primary beam. When the x-rays incident upon the radiator were unfiltered, the general secondary radiation had a broad intensity maximum at a wave-length slightly under 1 Å. U. On introducing a zirconium filter between the x-ray tube and the radiator, thus giving a primary beam consisting principally of the $K\alpha$ line from molybdenum together with some fluorescent K rays from zirconium, a much sharper maximum in the secondary fluorescent

---

[1] Phys. Rev., 18, 96 (1921).
[2] Nature, Nov. 17, 1921.

radiation was observed. This result has been verified by means of photographic spectra, which show a maximum of the general radiation at about 0.95 Å.U., which is about 35 per cent. greater than the wave-length of the exciting ray.

The energy in this general radiation is roughly 30 per cent. as great as the energy of the scattered K rays. Previous experiments have shown that when shorter wave-lengths are employed, the energy in the fluorescent rays may be even more prominent than the truly scattered rays. If we suppose that the incident x-ray beam ejects electrons moving forward with a kinetic energy hC/λ, where λ is the wave-length of the exciting ray, and if the ejected electron is oscillating at such a frequency that as observed in the direction of motion the wave-length is λ, on account of the Doppler effect the wave-length of the radiation at right angles with the primary beam will be very close to that of the fluorescent rays observed in these experiments.

WASHINGTON UNIVERSITY,
    SAINT LOUIS.

The Intensity of X-ray Reflection from Powdered Crystals.

In the May number of the *Philosophical Magazine* which has just reached us, Mr. C. G. Darwin has presented a most valuable discussion of the reflection of X-rays from imperfect crystals. He shows that, on account of the difficulty in determining the effective extinction coefficient of the X-rays in such crystals, it is very difficult to calculate with accuracy the intensity of the reflected beam. Hence he is unable to make a satisfactory comparison between the theoretical formulæ and the existing experiments on the intensity of X-ray reflection. This result is in general agreement with the conclusion reached by one of us (*Physical Review*, July 1917) on the basis of somewhat similar considerations.

Mr. Darwin concludes that a more satisfactory test might be made on powdered crystals, since in this case the only factor contributing appreciably to the extinction is the ordinary absorption of the X-rays in the powder, which can be measured directly. We had arrived at the same conclusion, and have made quantitative measurements of the intensity of the X-rays scattered by powdered crystals.

In our most recent experiments, the $K\alpha$ line from molybdenum ($\lambda = 0.708$ Å.), after reflection from a crystal of rock-salt, was allowed to fall upon a plate of powdered sodium chloride. The first order reflection from the [100] faces of the powdered crystals then entered the ionisation chamber. The method was thus similar to that employed by W. H. Bragg (Proc. Phys. Soc., Lond., 33, 222, 1921) except that the primary rays were homogeneous. The ratio of the energy reflected into the ionisation chamber due to this first order line to that incident upon the plate was 2.94×10^{-4}, with a probable error of about 10 per cent. The theoretical intensity of the line was calculated from a formula identical in significance with Darwin's formula (10.4) (*loc. cit.*), except that correction was made for the absorption of the X-rays in the crystal mass. We thus obtained the value 2.7×10^{-4}, which is in satisfactory agreement with the experimental measurement. Thus, at least to a close degree of approximation, the theory of X-ray reflection based upon the classical electrical theory gives accurate results.

This comparison of theory with experiment may be viewed in another manner. Any formula for the intensity of X-ray reflection must depend upon the value of a function ψ, the magnitude of which is determined by the distribution of the electrons in the atoms. The theoretical value 2.7×10^{-4} mentioned above is based upon the value $\psi^2 = 0.59$, i.e. upon the assumption that the intensity is 59 per cent. as great as it would be if all the electrons in sodium and chlorine were grouped together at the centres of their respective atoms. This value was estimated by one of us (*loc. cit.*) on the basis of some of W. H. Bragg's measurements of the relative intensity of the different orders of X-ray reflection from rock-salt. The corresponding value of ψ^2 as determined by the measurements of W. L. Bragg, James and Bosanquet is 0.43 (*Phil. Mag.*, July 1921). To obtain our experimental value 2.9×10^{-4} for the intensity of reflection from powdered crystals, the value of ψ must, however, be 0.64. The difference between the latter two values of ψ^2 supports Darwin's suggestion that the method employed by Bragg, James and Bosanquet for studying the intensity of X-ray reflection is not wholly trustworthy.

We hope in the near future to be able to report experimental results of a considerably higher degree of accuracy than those described above.

Arthur H. Compton.
Newell L. Freeman.

Washington University, Saint Louis, May 30.

BULLETIN

OF THE

NATIONAL RESEARCH COUNCIL

Vol. 4, Part 2   OCTOBER, 1922   Number 20

SECONDARY RADIATIONS PRODUCED BY X-RAYS,

AND SOME OF THEIR APPLICATIONS TO PHYSICAL PROBLEMS[1]

By Arthur H. Compton

Professor of Physics, Washington University

Contents

	Page
I. SECONDARY X-RAYS	1
a. The Scattering of X-rays	3
b. The Fluorescence of X-rays	14
II. PHOTOELECTRONS EXCITED BY X-RAYS	21
III. THE ABSORPTION OF X-RAYS	31
IV. THE REFRACTION OF X-RAYS	48
V. SOME APPLICATIONS OF SECONDARY X-RAYS TO PHYSICAL PROBLEMS	52

I. Secondary X-rays

By the term "secondary" X-rays is meant any radiation of the X type excited by the passage through matter of primary X-rays. "Scattered" radiation signifies the radiation emitted by the electrons in matter (that due to the positive nuclei is theoretically negligible in comparison) due to accelerations to which they are directly subjected by the primary rays. "Fluorescent" radiation is radiation of the energy absorbed from the primary beam and stored temporarily in the kinetic and potential energies of the electrons. In certain cases, as will be seen, it is difficult to distinguish experimentally between these two types of secondary radiation.

---

[1] This monograph is the third and last of a series which forms the report of the Committee on X-Ray Spectra, of the Division of Physical Sciences, of the National Research Council. This committee consists of the following members: William Duane, Harvard University, Chairman; Bergen Davis, Columbia University; A. W. Hull, General Electric Research Laboratory; D. L. Webster, Leland Stanford Junior University; Arthur H. Compton, Washington University.

Methods of Studying Secondary X-rays.—The usual method of investigating secondary X-rays may be explained by reference to Figure 1. Radiation from the target S of an X-ray tube, or from some other source of X- or γ-rays, is allowed to traverse a radiator R. This radiator is then found to emit X-rays in all directions. These rays may be investigated by means of an ionization chamber I which is carefully screened from the primary beam.

If the radiator consists of a plate of matter so thin that the X-rays are not appreciably diminished in intensity on traversing it, the intensity I_s of the secondary beam as it enters the ionization chamber may be written as

$$I_s = r_\theta \cdot I\, V/l_2^2,$$

where I is the intensity of the primary beam at R, V is the volume of the radiator, l_2 is the distance from the radiator to the ionization cham-

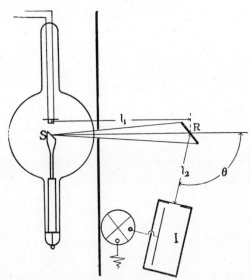

FIGURE 1. X-rays from the target S of the X-ray tube excite in the radiator R secondary rays which are measured by the ionization chamber I.

ber, and r_θ is a constant of proportionality which may be called the "radiating coefficient for the angle θ." Experiment shows that this coefficient is a function of the wave-length of the incident rays, the composition and physical state of the radiator, and the angle θ.

In order to learn the physical significance of the secondary radiation, it is important to know what part of it is scattered and what part is fluorescent. Two methods of making this distinction have been employed. The first depends upon Stokes' law, according to which the wave-length of the fluorescent radiation should be greater than that of the primary radiation which excites it. While this law has not been found to hold universally in optics, no indication of its failure has appeared in the X-ray region. The truly scattered rays, however, are presumably of the same wave-length as the primary rays (cf. infra, p. 18). Thus, since the longer wave-lengths of X-rays are the more strongly absorbed, it is possible by a comparative study of the absorption of the primary and the secondary rays to determine what part of the secondary beam is identical in character with the primary beam. This part constitutes the truly scattered rays, while the remaining less penetrating part of the secondary rays is according to this criterion fluorescent.

A more direct and certain method of separation consists in comparing the spectrum of the secondary with that of the primary X-rays. If the primary ray is homogeneous, and if the electrons traversed are at rest, so that no Doppler effect occurs, the scattered beam will be homogeneous and of the same wave-length; whereas the fluorescent rays will differ in wave-length from the primary. This method, while affording a very definite means of testing for the existence of the two types of secondary rays, is not readily adaptable to quantitative measurement of their relative intensities.

a. The Scattering of X-rays

By electrons acting as point charges.—It is a direct consequence of the electromagnetic theory of X-rays that scattering should occur. For every electron in matter traversed by primary X-rays will be subject to accelerations by the electric vector of the rays, and in virtue of its acceleration will itself radiate energy. On the basis of the assumptions:

1. That the classical electrodynamics is applicable,
2. That the forces of constraint on each electron in matter are negligible,
3. That the electrons scatter independently of each other, and
4. That the size of the electron is negligible, J. J. Thomson showed[1] that the intensity of the X-rays scattered per unit volume of matter traversed by the X-rays, to a point at distance L and at an angle θ with the primary beam, is

$$I_s = N_n I_0 = I \frac{N_n \ e^4(1+\cos^2\theta)}{2m^2L^2c^4}, \quad (1)$$

where N is the number of negative electrons in the atom, n the number of atoms per unit volume, I_0 is the intensity of the scattered beam due to each electron, I the intensity of the primary beam, e and m the charge and mass of the electron, and c the velocity of light. Using moderately soft X-rays and carbon for the radiating material, Barkla and Ayers[2]

---

[1] J. J. Thomson, "Conduction of Electricity through Gases," 2nd Ed., pp. 321 et seq.
[2] Barkla and Ayers, Phil. Mag., 21, 275 (1911).

found that the factor $(1+\cos^2 \theta)$ accounted satisfactorily for the relative intensity of the secondary rays at angles θ greater than 40°; and later work by Barkla[1] showed that this formula expressed quantitatively the intensity of the secondary rays at 90° from the lighter elements if N is taken as about half the atomic weight.[2] Since the view that the number N is identical with the atomic number has recently been strongly confirmed, this means that under certain conditions equation (1) gives an adequate quantitative expression of the scattering. Under these conditions, of moderately soft X-rays and scattering material of low atomic weight, the assumptions employed by Thomson therefore seem to be justified.[3]

Scattering by Groups of Electrons.—Using soft X-rays and scattering material of higher atomic weight, Owen,[4] Crowther,[5] Barkla and Dunlop[6] and others have found that the rays scattered in the forward direction are more intense than those scattered backward. This phenomenon of "excess scattering" is illustrated in Figure 2, in which the squares represent the scattering of soft X-rays by filter paper as

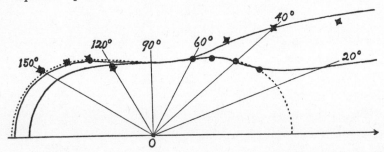

FIGURE 2 Scattering of Soft X-rays (★) and moderately hard X-rays (●) by filter paper, according to Owen. The dotted curve represents theoretical values for a random grouping, the solid curves for an empirical arrangement of the electrons in the atoms.

observed by Owen, and the dotted line shows the relative scattering at different angles calculated from Thomson's relation. The excess of the scattering over the theoretical value increases with the atomic number of the scattering material and with the wave-length of the X-rays employed. This is clearly shown by certain experiments due to Barkla and Dunlop,[6] shown in figure 3. Here the ratio of the intensity of the rays scattered at 90° per unit mass by copper, silver, tin and lead re-

---

[1] C. G. Barkla, Phil. Mag. 21, 648 (1911).
[2] Recent work by A. H. Compton, indicates that part of the secondary rays studied by Barkla and his collaborators may have been fluorescent in nature; but Compton's experiments lead to the same value for N as those of Barkla.
[3] A. H. Compton, Phys. Rev. 18, 96 (1921).
[4] Owen, Proc. Camb. Phil. Soc., 16, 165 (1911).
[5] Crowther, Proc. Camb. Phil. Soc. 16, 367 (1911); Proc. Roy. Soc., 86, 478 (1912).
[6] Barkla and Dunlop, Phil. Mag. 31, 229 (1916).

spectively to that of the rays scattered by aluminium is plotted against the effective wave-length. Expression (1) indicates that the relative mass scattering by the different elements should be proportional to the number of electrons per gram, that is, that I/I_{Al} should be nearly unity in all cases. It appears from these experiments that this relation may be exact for sufficiently short wave-lengths, but does not hold for the heavier elements at moderate wave-lengths.

The quantitative support of Thomson's theory in the special cases first considered, gives confidence in the application of the classical electrodynamics to the problem of scattering. Any effect due to the forces

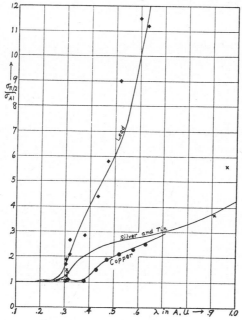

FIGURE 3. Relative scattering of X-rays at 90° per gram of various metals compared with that by aluminium. Data, Barkla and Dunlop; curves, theory based on empirical distribution of the electrons in the atoms.

of constraint on the scattering electrons would be to modify the amplitude of their motion, thus changing the intensity equally in all directions, and therefore could not account for the asymmetrical scattering in the forward and backward directions. If the size of the electron were appreciable, the difference in phase of the rays scattered by its different parts would result in a reduced instead of an increased intensity such as is observed. Thus in order to account for the excess scattering of soft X-rays by heavy elements, Thomson's original assumption that the electrons in matter act independently of each other must be modified.

The suggestion that the electrons in the heavy elements co-operate in their scattering seems to have been made first by Webster,[1] and was first stated in a satisfactory qualitative form by Darwin.[2] Debye[3] and Thomson[4] have solved independently the problem of the scattering of X-rays by atoms (or groups of atoms) consisting of electrons arranged at fixed distances from each other, taking into account the phases of the rays scattered by the different electrons. Their result may be put in the form

$$I_\theta = I_1 \sum_{1}^{N}{}_m \sum_{1}^{N}{}_n \frac{\sin\left(\frac{4\pi S_{mn}}{\lambda} \sin\frac{\theta}{2}\right)}{\frac{4\pi S_{mn}}{\lambda} \sin\frac{\theta}{2}}, \qquad (2a)$$

where I_θ is the intensity of the ray scattered at an angle θ by the group of electrons, I_1 is the intensity due to a single electron (according to Debye and Thomson identical with the I_θ of equation 1), N is the number of electrons in the group, and S_{mn} is the distance from the mth to the nth electron.

The more general problem of the scattering by atoms composed of electrons in relative motion was investigated by Schott[5] with unsatisfactory results.[6] Glocker and Kaupp,[7] however, have recently calculated the scattering by atoms composed of two or three coplanar rings of electrons revolving at different speeds. Glocker[8] has also calculated the scattering to be expected from Lande's pulsating tetrahedronal carbon atom, and finds a result practically the same as that for Bohr's plane carbon atom. This confirms the conclusion which had been reached by the writer,[9] that the scattering by groups of electrons in the atom depends principally upon the distances of the electrons from the enter, and only slightly upon their spatial distribution. The approximation is thus justified of calculating the scattering on the assumption that the electrons are arranged at random on the surface of shells of radii ρs. On this basis the intensity of the beam scattered by an atom is,[9]

$$I_\theta = I_1 \left\{ N + 2 \sum_{1}^{N/2} \left(\frac{\sin 2k_s}{2k_s} - 2 \frac{\sin^2 k_s}{k_s^2} \right) + 4 \left(\sum_{1}^{N/2} \frac{\sin k_s}{k_s} \right)^2 \right\}, \qquad (2b)$$

---

[1] D. L. Webster, Phil. Mag. 25, 234 (1913).
[2] C. G. Darwin, Phil. Mag. 27, 325 (1914).
[3] P. Debye, Ann. d. Phys. 46, 809 (1915).
[4] J. J. Thomson, manuscript read before the Royal Institution in 1916, and loaned to the writer.
[5] G. A. Schott, Proc. Roy. Soc. 96, 695 (1920).
[6] Cf. A. H. Compton, Washington University Studies, 8, 98 (1921).
[7] Glocker and Kaupp, Ann. d. Phys., 64, 541 (1921).
[8] R. Glocker, Zeitschr. f. Phys., 5, 54, (May 10, 1921).
[9] A. H. Compton, Washington U. Stud., 8, 99 (January, 1921).

where N is the atomic number and
$$k_s = \frac{4\pi \rho_s}{\lambda} \sin \frac{\theta}{2}.$$

This formula is much simpler in its application than those of Debye and Glocker, and it leads to equally reliable information concerning the distances of the electrons from the centers of the atoms. If sufficiently refined measurements of the scattering can be made, however, it may be possible to distinguish between the spatial arrangements considered in the different formulae.

The solid lines of Figures 2 and 3 are calculated* from formula (2b), assuming particular arrangements of the electrons in the atoms concerned.[1] These figures show that by taking into account the phase relations between rays from the various electrons in the atoms, a satisfactory explanation may be given of the excess scattering of soft X-rays by heavy atoms.

Scattering by electrons of appreciable dimensions.—When X-rays of very short wave-length are scattered by the lighter elements, the intensity of the scattered beam is less than that demanded by expression (1). For example, Ishino has found[2] that the total secondary gamma radiation averaged over different angles is less than one fourth of the energy calculated by expression (1) for the scattered radiation. Barkla and White pointed out[3] that similar reduced scattering occurs when sufficiently hard X-rays are used, when they observed a total absorption in paraffin less than the energy which should be lost according to Thomson's theory due to scattering alone. These results are supported and extended by A. H. Compton's measurements[4,5] of the part of the

---

[1] In figure 2, the squares represent measurements with an equivalent spark gap of 2.5 cm., which has been taken to mean a uniform distribution of energy over wave-lengths between 0.5 and 1.0 Å. U. The circles correspond to a spark gap of 7 cm. (λ from 0.25 to 0.50 Å. U.). In figure 3, the wave-lengths estimated by the experimenters were used in the calculations. The radius of the electron was taken to be 2.6×10^{-10} cm., as described below, but this size is almost negligible for the wave-lengths considered. The numbers of electrons at different distances from their atomic centers as employed in the calculations are as follows:

Hydrogen	Oxygen		Carbon		Aluminium		Copper		Silver & Tin		Lead	
No.	No.	Dist. (A.U.)	No.	Dist.	No.	Dist.	No.	Dist.	No.	Dist.	No.	Dist.
1	2	.26	2	.35	2	.12	2	.052	2	.036	2	.022
	6	.42	4	.6	8	.26	10	.104	10	.073	10	.045
					3	.7	8	.24	8	.17	16	.090
							8	.42	16	.34	16	.132
							1	1.05	8	.51	16	.202
									4½	.7	16	.31
											6	.6

While these exact distributions are of little value, because of the unreliability of the scattering measurements on which they are based, yet as to order of magnitude the results can hardly be wrong, and they represent the most direct experimental determinations of these distances that have so far been made.

[2] M. Ishino, Phil. Mag., 33, 129 (1917).
[3] Barkla & White, Phil. Mag., 34, 275 (1917).
[4] A. H. Compton, Phil. Mag., 41, 749 (1921).
[5] A. H. Compton, Phys. Rev., 18, 96 (1921).
* See Note 6, page 6.

8 SECONDARY RADIATIONS: COMPTON

secondary radiation which has the same wave-length as the primary X-rays (cf. infra p. 16). On the view that these rays only are truly scattered, his data, shown by the solid lines in Figure 4 for the case of paraffin, indicate that at small angles with the primary beam the scattering of hard X-rays by light elements is approximately that given by Thomson's expression (1). At larger angles, however, the intensity of the scattered beam is decidedly less than the theoretical value, the difference increasing for shorter wave-lengths. Indeed, for hard gamma rays he finds the "truly scattered" rays at large angles less than .001 of Thomson's theoretical value.[1]

There appears to be no possibility of accounting for this reduced scattering on the basis either of any grouping of electrons within the

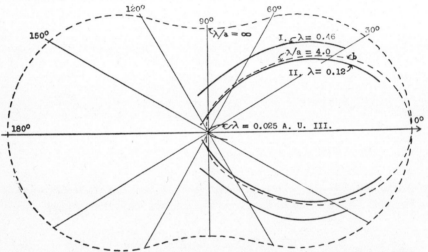

FIGURE 4. "True" scattering of hard X-rays and gamma rays by paraffine. Solid lines, experiment (Compton); broken lines theory for electron of radius a.

atoms or of any forces of constraint upon the electrons.[2] Unless the classical electrodynamics ceases to be applicable to these very high frequency rays, it therefore seems necessary to modify Thomson's assumption that the electron acts as a simple point charge of electricity.[3] It is obvious that if the electron is comparable in diameter with the wave-length of the X-rays, partial interference will occur between the rays scattered from its different parts, reducing the intensity of the

---

[1] For the data on γ-rays, cf. note 4, p. 7 and A. H. Compton, Phil. Mag. 41, 770 (1921). In figure 4 the *absolute* values of the scattering are compared with Thomson's formula, assuming that the number of effective electrons is equal to the atomic number. Figure 2 shows merely the *relative* scattering at different angles.

[2] A. H. Compton, Phys. Rev. XIV. p. 31 (1919); Washington U. Studies, 8, 96 (1921).

[3] For a possible modification of electrodynamics which may account for this reduced scattering as well as certain other phenomena, cf. infra, pp. 18 and 19.

scattered beam, and that this interference will be more nearly complete for rays scattered at the larger angles.

The theory of the scattering of high frequency radiation by matter containing electrons of appreciable size, taking into account this interference between the rays scattered by the different parts of the same electron, has been discussed in a series of papers by A. H. Compton[1] and also by G. A. Schott.[2] The magnitude of the scattering in this case is a function of the form of the electron, its rigidity, and the ratio of the charge to the effective mass of its different parts, as well as the wave-length of the X-rays. In the case of a spherical shell electron of negligible rigidity, the intensity of the ray scattered at an angle θ by a single electron is found to be*

$$I_1 = \frac{I_0 \sin^2 x}{x^2} \qquad \left[x = \frac{4\pi a}{\lambda} \sin \frac{\theta}{2}\right], \qquad (3a)$$

where I_0 is given by expression (1), a is the radius of the electron, and λ is the wave-length of the X-rays. Similarly if the electron is in the form of a ring of negligible rigidity, the scattering is*†

$$I_1 = I_0 \cdot \frac{1}{x} \sum_0^\infty J_{2n+1}(2x), \qquad (3b)$$

where J_s is Bessel's J function of the sth order, and the other quantities have the same significance as before. A solid sphere, without rigidity, with volume density of electric charge inversely proportional to the square of the distance from the center, and of equal ratio e/m of charge to mass everywhere, should scatter according to the expression,*

$$I_1 = I_0 \left\{1 - \frac{x^2}{3\cdot 3!} + \frac{x^4}{5\cdot 5!} - \cdots\right\}^2. \qquad (3c)$$

The scattering by rigid electrons and by electrons whose parts have different ratios of charge to mass has not been examined mathematically. It can be shown, however, that if the electron possesses rigidity, the scattering is not a function of x alone, but is a more complicated function of λ and θ.

In Figure 5, curves a, b and c show respectively the values of I_1 as calculated according to expressions 3a, 3b, and 3c, plotted as functions of a/x. That these formulae describe satisfactorily the relative scattering at different angles is shown by the dotted curve b of Figure 4, which is calculated for $\lambda/a = 4.0$ from expression (3a). But when the experimental data represented in Figure 4 by the solid lines, I, II, and III

---

[1] A. H. Compton, Journ. Wash. Acad. Sci., Jan. 1, 1918; Phys. Rev. 14, 20 (1919); Washington U. Stud. 8, 93 (1921).
[2] G. A. Schott, Proc. Roy. Soc. 96, 695 (1920).
  * See Note 1, page 9.   † See Note 2, page 9.

are transferred to Figure 5 (represented in this figure by broken lines), we see that the agreement is not perfect. Though the general variation of I_1/I_0 with λ and θ is in accord with the calculated values, it seems that the relative scattering cannot be described accurately as a function of x. This is shown by the fact that if, as in the figure, curve II for $\lambda = .12$Å coincides with the theoretical curves, curve I for $\lambda = .45$Å departs rather widely from the theory. According to the results of the

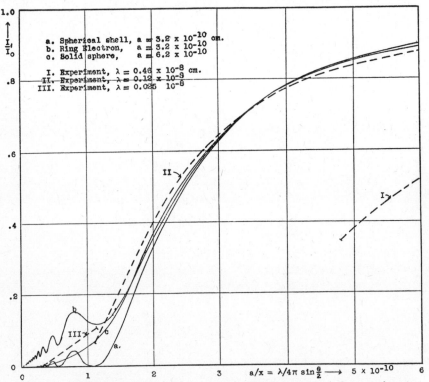

FIGURE 5. Scattering of hard X-rays, showing the departure of the experiments (broken lines) from $I^1/I_0 = 1$, and the approximate agreement with the theory (solid lines) based on the view that the electron's radius is comparable with the wave-length.

last paragraph, the fact that I_1/I_0 is not a function of x only would indicate that the electron has appreciable rigidity. It is doubtful, however, if the assumption of rigidity in the electron would modify the theoretical scattering in the manner demanded by the experiments. This is an important matter for further theoretical and experimental investigation, for if this discrepancy is real it would not seem possible to reconcile the results with the classical electrodynamics.

The important thing, however, is that if the electrons had dimensions comparable with 10^{-13} cm., as usually assumed, the classical theory requires scattering represented by the upper line of Figure 5, where $I_1/I_0 = 1.0$. On this basis, the fact that experiment gives consistently lower values when short wave-lengths are used, indicates that the electron is not sensibly a point charge of electricity. We find, indeed, that this reduced scattering for small values of a/x can be accounted for by interference between the rays from different parts of the electron, if its radius is of the order of 4×10^{-10} cm.

Scattering by Small Crystals.—An important special case of the scattering by groups of electrons is that of the scattering by a mass of minute crystals with random orientation. This includes, for example, the scattering by the ordinary inorganic solids. Debye and Scherrer[1] first pointed out that in such a random assemblage of crystals some will be so oriented as to reflect X-rays from every plane which can be drawn parallel to layers of atoms in the crystal. Thus a beam of X-rays passing through such a mass of crystals is partially scattered in a series of coaxial cones, the semi-apex angle θ of each cone being determined by the relation

$$n\lambda = 2D \sin \frac{\theta}{2},$$

where D is the distance between the layers of atoms giving rise to the reflection, and n is the order of reflection. Since every possible layer of atoms thus gives rise to a reflected beam which may be recorded on a photographic plate, these "powdered crystal spectrograms" are of great value as a means of studying crystal structure. Extensive applications of this method have been made by Debye and Scherrer[2] and especially by Hull[3] in determining the arrangement of the atoms in substances which cannot readily be obtained in the form of large crystals.

If motions of the electrons are neglected, the intensity of the beam scattered by such a mass of minute crystals is given by Debye's formula (2a), when the double summation is performed over all the electrons in each component crystal. This process leads, however, to an expression containing so many terms that it is quite unmanageable. A more satisfactory procedure is to start with the equation for the reflection of X-rays by crystals as given by Darwin[4] and A. H. Compton,[5] which applies if (as is the case experimentally) the crystals are large enough to give a spectrum line whose angular width is small compared with θ. By the application of this method it can be shown[6] that the total energy

---

[1] Debye and Scherrer, Phys. Zeits. 17, 277 (July 1, 1916).
[2] Debye and Scherrer, Loc. cit. and ibid., v. 18, 291 (1917), v. 19, 23 (1918) et al.
[3] A. W. Hull, paper Amer. Phys. Soc., Oct. 25, 1916; Phys. Rev. 9, 85 (1917), 10, 661 (1917) et al.
[4] C. G. Darwin, Phil. Mag., 27, 325 (1914).
[5] A. H. Compton, Phys. Rev. 9, 29 (1917).
[6] It is hoped to publish this calculation soon in the Phys. Rev.

scattered by a layer of crystals of thickness δx in the cone whose semi-apex angle θ is defined by $n\lambda = 2D \sin \frac{\theta}{2}$ is

$$E_\theta \, \delta x = E_i \cdot \frac{nN^2\lambda^3}{2 \sin \theta/2} \frac{\rho'}{\rho} \varphi^2 P \psi^2 D \cdot \delta x, \qquad (4)$$

where:

$E_i =$ incident energy of wave-length λ,

$n =$ number of planes in crystal giving rise to reflection at angle θ (e. g. for [100] planes $n=3$, for [110] $n=6$, for [111] $n=4$, etc.)

$N =$ number of electrons per unit volume in each component crystal,

$\frac{\rho'}{\rho} =$ ratio of density of mass of crystals to density of individual crystals,

$\varphi = e^2/mc^2$, $\qquad P = 1 + \cos^2\theta$,

$\psi =$ function of the arrangement of the electrons in the atoms,

$$= \int_a^b F(z) \cos\left(\frac{4I \sin \theta/2}{\lambda}\right) dz,$$

where

$b - a =$ diameter of the atom,

$F(z) =$ probability that a given electron will be at a distance z from the mid-plane of the layer of atoms to which it belongs,

$D =$ factor accounting for the thermal agitation of the atoms,

$$= e^{-\beta \frac{\sin^2\theta/2}{\lambda^2}},$$

where β is a constant characteristic of the crystal, which has been evaluated theoretically by Debye[1] (cf. also this report, vol. 1, p. 420).

If, as is usually the case in practice, the thickness δx of the mass of crystals is great enough to produce appreciable absorption, the reduction in intensity of the scattered beam due to this cause must also be taken into account.

For investigations of the arrangement of the atoms in crystals, in which the different angles θ at which scattered rays appear are of first importance, the scattering of X-rays by powdered crystals has been studied only by the photographic method. It has been customary to have the scattering material in the form either of a thin plate or of a narrow cylinder placed perpendicular to the incident beam, between the target and the photographic plate. W. H. Bragg has coated a flat plate with a layer of the powdered crystals, and has used this plate in place of the crystal of an X-ray spectrometer.[2] When the two arms of the spectrometer are equal, X-rays are scattered into the slit of the ionization chamber from all parts of the plate at approximately the angle θ.[3] With

---

[1] P. Debye, Ann. d. Phys., 43, 49 (1914); A. H. Compton, Phys. Rev. 9, 47 (1917).
[2] W. H. Bragg, Proc. Phys. Soc., 33, 222 (1921).
[3] Cf. e.g. Bragg's "X-rays and Crystal Structure," p. 26.

this arrangement he has succeeded in obtaining sufficient intensity to measure the scattered beam by the ionization method. The theoretical scattered energy in this case is*

$$E_s = E_\theta \cdot \frac{\left(l\left(1 - e^{\frac{-2\mu z}{\sin\theta/2}}\right)\right)}{4\pi L \mu \sin\theta}, \tag{5}$$

where E_θ is given by equation (4), E_i being the total energy falling on the layer of powdered crystals; L is the distance of the ionization chamber from the plate, l is the vertical height of the slit in the ionization chamber, μ is the absorption coefficient of the X-rays in the crystal mass, and z is the thickness of the layer of crystals. This formula supposes that the slit in the ionization chamber is broad enough to receive all the rays scattered at the angle θ, and that its height is small enough so that the curvature of the image of the scattered beam will not be great.

Expression (5) gives the intensity of the scattered rays in terms of known quantities, except for the function $F(z)$ of the arrangement of the electrons in the atoms.[1] Experimental determinations of the intensity E_s for different angles θ should therefore be of great value in determining the distribution of the electrons, and Bragg's preliminary measurements indicate that the method is feasible. Work of this kind with powdered crystals has a distinct advantage over similar work on the reflection from large crystals in that the absorption coefficient involved in equation (5) is directly measureable, whereas the absorption coefficient of X-rays in a large crystal at the angle of maximum reflection is very difficult to determine. It would seem that studies of the intensity of the scattering by such groups of crystals may afford most valuable information concerning the arrangement of electrons in atoms.

An interesting theoretical problem in this connection which awaits solution is to determine the relation between the scattering by atomic groups of electrons as expressed by equation (2) and the scattering by crystalline groups of electrons, as expressed by equations (4) and (5). It is found that a large part of the scattering of homogeneous X-rays by solid substances occurs in definitely defined cones, as is to be expected if these substances consist of finely divided crystals.[2] But the work of Barkla and others with heterogeneous rays indicates, as we have seen, scattering at all angles in accord with equations (2) based on independent

---

[1] The temperature factor D is uncertain to some extent, depending upon the question of the existence of a zero-point energy of thermal oscillation. This question, however, may possibly be answered by measurements of the scattering of X-rays by crystals at different temperatures.

[2] Recent experiments by G. E. M. Jauncey (Phys. Rev., 19, 435, 1922), however, show that even for the best crystals much more energy is spent in diffuse scattering than in crystalline reflection.

* See Note 6, page 11.

scattering by the different atoms. Thus it would appear that the energy in the spectrum lines from the fine crystals, when averaged over relatively large angles, must equal the energy scattered by the substance in an amorphous form.

An important difference between the scattering by small crystals and that by independent molecules lies in the fact that there can be no crystalline reflection at an angle smaller than that defined by $\lambda = 2D_{max} \sin\frac{\theta}{2}$, or

$$\theta_{min} = 2 \sin^{-1}(\lambda/2D_{max}),$$

where D_{max} is the largest grating space for any set of planes in the crystal, and λ is the wave-length of the incident radiation. Thus one might have predicted Hewlett's observation[1] that the scattering of X-rays of wave-length about .7 A.U. by powdered diamond and graphite becomes very small[2] at angles less than 5°. For molecules arranged at random, however, expressions (2) indicate very strong scattering at small angles, approaching N times the normal value (equation 1) as θ approaches zero. Hewlett finds,* however, even in the case of a liquid (mesitylene, $C_3H_6(CH_3)_3$) that the intensity of the scattered beam approaches zero at small angles. It follows that there must be even within the liquid sufficiently large groups of regularly arranged atoms to produce nearly complete interference at small angles. In order that formulas (2) may be valid at all angles, it would therefore seem possible to apply them only to the case of gases.

b. The Fluorescence of X-Rays

There exist several types of "characteristic," fluorescent radiation, consisting of a number of nearly homogeneous rays characteristic of the substance employed as radiator. There is some evidence also for the existence of a "general" fluorescent radiation, whose character depends principally upon that of the primary rays. These two types of fluorescent rays correspond closely to the characteristic and general radiations respectively from the target of the X-ray tube.

Characteristic Fluorescent X-rays

The characteristic fluorescent X-rays, discovered by Barkla and Sadler,[3] consist of the K, L, etc. series radiations from the substance traversed by the primary rays. These rays, therefore, have properties identical with those of the same radiations when excited at the target of an X-ray tube, which have been described in the other sections of this

---

[1] C. W. Hewlett, paper before Am. Phys. Soc. Nov. 26, 1921.
[2] Some recent experiments by A. R. Duane and W. Duane show that the energy scattered at an angle less than θ_{min} is about 1 per cent for aluminium and less than 5 per cent for water at an angle a degree or so larger than θ_{min}. They have been using this principle to measure the distances between crystal planes.
[3] Barkla & Sadler, Phil. Mag. 16, 550 (1908).
* See Note 1, page 14.

committee's report.[1] The characteristic fluorescent rays, are, however, usually mixed with much less general radiation than is the case with primary rays, and have therefore been much used as a source of comparatively homogeneous rays.

Experiments by Sadler[2] have shown that the K-rays from one element will excite the K-rays from all lighter elements, but not from the same or heavier elements. His results strongly suggest that the K, L_1, L_2 L_3, etc. series radiations characteristic of any particular element are excited only by rays of wave-length shorter than the critical K, L_1, etc. absorption wave-lengths (cf. this report Vol. I, p. 386) characteristic of the element. It appears, however, that no direct quantitative experiments have been performed to test this point.

The experiments of Barkla and his collaborators have shown that the characteristic fluorescent radiation emitted by an atom is unpolarized and is distributed uniformly in all directions. Apparently no quantitative measurements have been made of the intensity of this characteristic radiation. When it does occur, however, it is usually much more intense than the scattered radiation. The energy which goes into the radiation comes from that absorbed by the atom. Since the absorbed energy is closely proportional to λ^3 (cf. infra. p. 36), it is probable, if the incident waves are shorter than the critical wave-length, that the intensity of the characteristic fluorescent radiation is proportional to λ^4.

General Fluorescent X-rays

The softening of Secondary X-rays.—Under this head we shall discuss the fact that, even when no appreciable characteristic fluorescent radiation is excited, the secondary radiation is always somewhat less penetrating than the primary radiation which excites it. Secondary γ-rays are much softer than the primary γ-rays,[3] the backward secondary rays being both less penetrating and less intense than the rays in the forward direction. As a result also of careful experiments on the secondary radiation excited in carbon by the characteristic X-rays from different substances, Sadler and Mesham[4] found that this radiation was less penetrating than the primary beam, and that this difference in quality was greater the harder the primary rays employed.

It has been generally supposed[5] that this softening of the secondary radiation is due to the fact that the softer components of the beam were

---

[1] W. Duane, This Bulletin, vol. I, p. 383.
[2] C. A. Sadler, Phil. Mag. 18, 107 (1909).
[3] Eve., Phil. Mag. 8, 669 (1904); R. D. Kleeman, Phil. Mag. 15, 638 (1908); Madsen, Phil. Mag. 17, 423 (1909). D. C. H. Florance, Phil. Mag. 20, 921 (1910), et al.
[4] Sadler and Mesham, Phil. Mag. 24, 138 (1912).
[5] e.g. Florance, loc. cit.; Oba, Phil. Mag. 26, 601 (1914); A. H. Compton, Phys. Rev., 14, 20 (1919); K. W. F. Kohlrausch, Phys. Zeit. 21, 193 (1920), et al.

16 SECONDARY RADIATIONS: COMPTON

more strongly scattered than the harder ones. This may account for a part of the effect when γ-rays are employed, but Sadler and Mesham showed that in their experiments it was the harder components which were the more strongly scattered. In the case of γ-rays also, Gray[1] established the fact that only a small part of the softening was due to this cause, the greater part of the effect being due to a real change in

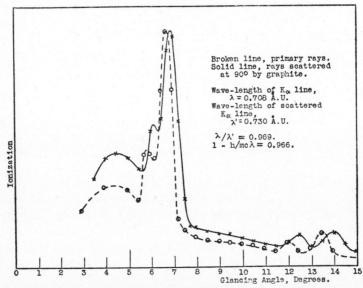

FIGURE 6. Spectrum of scattered X-rays, showing an increase in the wave-length of a spectrum line when it is scattered.

the character of the radiation as the secondary rays were formed. This conclusion has been experimentally confirmed in the X-ray region by Laub,[2] Gray,[3] Compton,[4] and Crowther,[5] and in the region of γ-rays by Florance[6] and Compton.[7]

Recent spectroscopic measurements by the writer show that the secondary rays have suffered a distinct change in wave-length. Thus in Figure 6 the spectrum of the molybdenum rays after being scattered at 90° by graphite show the K lines at angles distinctly greater than those at which they occur for the primary beam.

Absorption measurements indicate that the secondary rays are composed of two parts, one having the same wave-length as the primary

[1] J. A. Gray, Phil. Mag. 26, 611 (1913).
[2] J. Laub, Ann. d. Phys. 46, 785 (1915).
[3] J. A. Gray, Frank. Inst. Jour., Nov., 1920, p. 643.
[4] A. H. Compton, Phys. Rev. 18, 96 (1921); Nature, 108, 366 (1921).
[5] J. A. Crowther, Phil. Mag. 42, 719 (1921).
[6] D. C. H. Florance, Phil. Mag. 27, 225 (1914).
[7] A. H. Compton, Phil. Mag. 41, 749 (1921).

beam, and the other a slightly greater wave-length.‡ In view of the uncertainty of absorption experiments in deciding such a question, we shall have to wait for more precise spectroscopic measurements before we can say definitely whether any of the secondary rays have the same wave-length as the primary rays. Preliminary spectroscopic work, however, seems to support the absorption measurements in showing the existence of both the unchanged and the longer wave-lengths in the secondary beam. In any case the spectrum shown in Figure 6 leaves no doubt but that a large part of the secondary X-rays have suffered a real change in wave-length. According to the writer's absorption measurements, over the range of primary rays from .7 to .025, Å. U., the wave-length of the secondary X-rays at 90° with the incident beam is roughly 0.03 Å. U. greater than that of the primary ray which excites it.

Interpretation of the softening.—Three possible explanations of this effect have been suggested. Laub* and Crowther° account for the greater effective wave-length of the secondary beam on the assumption that there exist homogeneous fluorescent radiations characteristic of the radiator, but of higher frequency than the characteristic K radiations ("characteristic J-radiation"). Gray§† and Florance¶ conclude that the X-rays are truly scattered, but in the process of scattering are so modified as to become less penetrating. Compton has suggested‡ ‖ that the primary rays excite in the radiator a type of "general fluorescent radiation," similar in character to the general or "white" radiation emitted by an X-ray tube.

Analogy with the characteristic K and L radiations seems to be the chief reason for Laub and Crowther's view that the observed softening of the secondary ray is due to a fluorescent radiation characteristic of the radiator. The more exhaustive work of Sadler and Mesham+ had indicated, however, that these softer components of the secondary beam increased continuously in hardness with increasing hardness of the primary beam over a wide range. This result has been verified by Gray† and Compton‡ ‖, who have found that the secondary radiation not only from carbon but also from such elements as aluminium, copper, tin, and lead, may have any wave-length between 0.7 and 0.04 Å. U., according to the wave-length of the exciting rays and the angle at which the secondary rays are studied.[1] Evidence that the secondary radiation is in any way characteristic of the particular element employed as radiator is thus completely lacking.

---

* See Note 2, page 16.   ¶ See Note 6, page 16.
° See Note 5, page 16.   ‡ See Note 4, page 16.
§ See Note 1, page 16.   ‖ See Note 7, page 16.
† See Note 3, page 16.   + See Note 4, page 15.

[1] It is hoped to publish in the near future a full account of the experiments in the X-ray region of which reference 4, page 16 is a brief abstract.

Gray* has shown that if the primary rays consist of thin pulses, as suggested by Stokes' theory of X-rays, and if these rays are scattered by atoms or electrons of dimensions comparable with the thickness of the pulse, the thickness of the scattered pulse will be greater than that of the incident pulse. Thus the effective wave-length of the secondary rays will be greater than that of the primary beam. Since it has been shown that the sharp upper limit to the frequency of the primary X-ray beam from a tube operated at constant voltage is inconsistent with the pulse theory of X-rays,[1] this scattered pulse hypothesis is difficult to defend. But such an explanation has been definitely eliminated by Compton's observation[2] that X-rays which are reflected from a crystal, and which are therefore known to come in long trains of waves, excite in light elements a secondary radiation which is softened to about the same degree as that excited by ordinary X-rays.

If the incident X-rays are homogeneous, as in the writer's experiment, the scattered rays must be homogeneous and of the same wave-length † unless a Doppler effect is present. But in order to account for the observed softening of the secondary rays as due to a Doppler effect, the scattering particles would have to be moving in the direction of the primary beam at a speed comparable with that of light. This is not possible on the classical theory, which supposes that all the electrons in the radiator are effective in scattering. Thus the classical electrical theory appears irreconcilable with the view that the part of the secondary rays that are of greater wave-length than the primary beam are truly scattered. The assumption of a general fluorescent radiation is the obvious and apparently the only alternative. On this view, only that part of the secondary radiation whose wave-length is identical with that of the primary beam is truly scattered, and that of greater wave-length is fluorescent. If this result be correct, it is of far-reaching importance, since as a consequence of their neglect of this type of fluorescence little weight could be attached to the quantitative measurements of scattering made by the earlier investigators (see, however, note 2, page 4).

The Doppler Effect in Secondary X-rays.—On the basis of the quantum theory a different hypothesis may be formed. Let us suppose that each electron when it scatters X-rays receives a whole quantum of energy and reradiates the whole quantum in a definite direction. The momentum which the scattering electron receives from the radiation will then be $h\nu/c$, where h is Planck's constant, ν is the frequency and c is the velocity of light. This will result in a velocity in the forward direction which

---

[1] Cf. D. L. Webster, Phys. Rev. 7, 609 (1916).
[2] A. H. Compton, Nature, 108, 366 (1921).
* See Note 3, page 16. † See Note 5, page 16.

will produce a Doppler effect as the scattered rays are observed at different directions. In addition, as the electron radiates a quantum of energy toward the observer, the conservation of momentum principle demands that the electron shall recoil with a momentum $h\nu'/c$, where ν' is the average frequency of the scattered radiation. For cases in which the resulting velocity of the electron is small compared with the speed of light, it can be shown on this basis that the ratio of the average frequency of the rays scattered at 90° to that of the incident rays should be $1-h/mc\lambda$. In the case of the molybdenum Kα line ($\lambda = .708$ Å) this calculated ratio is 0.966, while the value of the ratio taken from the experiments shown in figure 6 is 0.969. This close numerical agreement would suggest that we should consider scattering as a quantum phenomenon instead of obeying the classical laws of electricity as assumed in the first section of this report.

The view that much of the secondary radiation comes from electrons moving at high speed is supported by the apparent Doppler effect observed in the case of secondary γ-rays. The writer found[1] that when hard γ-rays from RaC were used, the secondary rays at 135° with the primary beam had a wave-length of about 0.08 Å. U., while those at less than 20° were only very slightly softer than the incident rays. Using the value 0.025 Å. U. for the effective primary wave-length,[2] the wave-length of the secondary rays at 20° may be taken as about 0.03 Å. U. According to the Doppler formula,

$$\frac{\lambda_1}{\lambda_2} = \frac{1-\beta \cos \theta_1}{1-\beta \cos \theta_2},$$

the observed wave-lengths at the two angles 20° and 135° indicate a velocity of the source of $\beta = 52$ per cent the speed of light, if we suppose with Rutherford[3] that the secondary β-rays are initially ejected in the forward direction. This rapid motion of the radiator will result in a greater radiant energy in the forward than in the backward direction. A comparison of the observed assymmetry of the energy of the secondary γ-rays with that calculated for different velocities of the radiator leads to a value[4] of $\beta =$ about 55 per cent the speed of light. Thus the asymmetry of both the wave-length and the energy of the secondary γ-rays indicates radiation from a particle moving with slightly over half the speed of light. This is in good accord with Eve's observation[5] that the aver-

---

[1] A. H. Compton, Phil. Mag. 41, 749 (1921).
[2] A. H. Compton, Phil. Mag. 41, 770 (1921). This value of the wave-length is confirmed by Ellis' recent determination of the wave-lengths of the γ-rays from RaC from the magnetic spectra of secondary β-rays (Proc. Roy. Soc. A., 101, 6, 1922).
[3] E. Rutherford, "Radioactive Substances, etc." p. 276.
[4] A. H. Compton, Phil. Mag. 41, 767 (1921).
[5] Eve, Phil. Mag. 8, 669 (1904).

age speed of the secondary β-rays excited by the hard γ-rays from radium is somewhat greater than half the speed of light.

Polarization.—The view that these softened secondary X-rays are really scattered is apparently confirmed by a study of their polarization. Barkla, in his classic measurement of the polarization of secondary X-rays,[1] found that at 90° the secondary rays from carbon were approximately 80 per cent polarized. The remaining 20 per cent might be accounted for in part by experimental error, and the remainder Barkla ascribed as due possibly to a real lack of polarization because of forces acting on the electrons. Recent experiments by Compton and Hagenow[2] however, have shown that, when the multiple scattering within the radiator is eliminated, the polarization of the secondary X-rays is complete within an experimental error of about 1 per cent. Thus if any fluorescent radiation exists, it must be nearly completely polarized.

Conclusion.—While these properties of the secondary radiation are in accord with the view that it is truly scattered, one must not lose sight of the fact that the classical electron theory demands that scattered rays shall be of the same wave-length as the primary. Moreover, the very satisfactory explanations of excess scattering and of X-ray crystal reflection that have been given require that a considerable number of electrons shall co-operate in their scattering in their proper phases. Such co-operation is contrary to the quantum hypothesis of scattering which has been introduced to account for the change in wave-length of the secondary rays. But if we adhere to the classical theory, we must invoke a general fluorescent radiation to account for the observed softening. Both this suggestion of general fluorescent radiation and the idea of quantum scattering are difficult to reconcile with the very small intensity of the secondary radiation observed at small angles with the incident beam (supra, p. 14), since neither hypothesis supplies the mechanism necessary for destructive interference.

Thus we see that the classical electrodynamics succeeds in explaining quantitatively many of the phenomena of secondary X-radiation, supposing that a considerable part of this radiation is truly scattered. The change in wave-length of the rays as they are transformed from primary to secondary X-rays seems to be in accord rather with quantum principles. But it has not been possible to account on either basis for all the observed phenomena. The theory of secondary X-rays is thus at present in an unsatisfactory form. The close overlapping of the classical and the quantum principles as applied to this problem, however, suggests that here may be a most profitable field for studying the connection between these two points of view.

---

[1] C. G. Barkla, Proc. Roy. Soc. 77, 247 (1906).
[2] A. H. Compton and C. F. Hagenow, Phys. Rev. 18, 97 (1921).

II. Photoelectrons Excited by X-Rays

It was observed by Perrin[1] and by Curie and Sagnac,[2] early in the history of X-rays, that when these rays fall on solid screens a type of secondary radiation is emitted which is nearly completely absorbed in 1 mm. of air. Dorn[3] showed that this radiation consisted of negatively charged corpuscles which could be deflected by a magnetic field; and assuming the same ratio of e/m as that of the cathode rays, he found that the velocities of these secondary particles were of the order of 1/10th the velocity of light. We shall call these high speed electrons "photoelectrons," whether liberated by the action of light, X-rays or γ-rays.

Methods of Experimental Investigation

The presence of these photoelectrons can be detected by allowing X-rays to fall on a plate insulated in a good vacuum. The plate is then found to acquire a positive charge, due to the emission of the secondary electrons. The effect is thus strictly analogous to the photoelectric effect.

A second method of investigation is to make use of the ionization produced by the photoelectrons. Thus, it is found that if X-rays strike a solid substance placed in a gas, as in Figure 7, the ionization in the neighborhood of the solid is much more intense than that elsewhere in the gas. The region of intense ionization, being determined by the range of the photoelectrons, may be varied by changing the pressure of the gas. Thus, since the ionization due to the absorption of the X-rays in the gas is proportional to the pressure P, the total ionization I, if the secondary electrons are completely absorbed, is given by

$$I = CP + I_e,$$

where the constant of proportionality C can be determined by experiment, and I_e represents the ionization due to the photoelectrons from the solid. Thus

$$I_e = I - CP.$$

Theoretically this method is open to the objection that it does not distinguish between photoelectrons and secondary X-radiation of very soft type. Under ordinary conditions, however, the ionization due to the electrons is so much greater than that due to the very soft secondary X-rays that no confusion is apt to arise. This method is a convenient one, and has been much used.

In many respects the most satisfactory method of studying these secondary electronic rays is the beautiful one employed by C. T. R. Wilson,[4] in which the tracks of the individual particles are rendered

---

[1] Perrin, Ann. de Chim. et Phys. (7), vol. 2, p. 496 (1897).
[2] Curie & Sagnac, Jour. de Phys. (4), vol. 1, p. 13 (1902).
[3] Dorn, "Lorentz Jubilee Volume," p. 595 (1900).
[4] C. T. R. Wilson, Proc. Roy. Soc. 87, 277 (1912).

visible by condensing water droplets on the ions formed along their paths. By this means it is possible to count accurately the number of secondary electrons emitted, study their distribution, and make measurements of their range in air. If two simultaneous photographs are taken at right angles with each other, by the method described by Shimizu,[1] the exact shape and total length of the paths may also be determined.

For investigating the velocities of the photoelectrons excited by X-rays, the method of photographing their magnetic spectrum has given the best results. For this purpose, the arrangement employed first by Robinson and Rawlinson[2] is very satisfactory. This arrangement, suggested by Rutherford and Robinson's[3] method of photographing the magnetic spectra of primary β-rays, is illustrated in Figure 8.

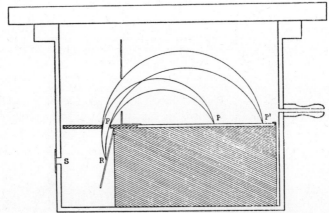

FIGURE 8. Magnetic photoelectron spectrometer. Photoelectrons leaving the radiator R with different speeds are bent by the magnetic field to different points PP^1 on the photographic plate.

A flat, air-tight, brass box having a window S for the admission of the primary X-rays, is evacuated and placed between the poles of a large electromagnet. Secondary electrons from the radiator R go out in all directions, and those passing through the slit F have their paths bent around by the magnetic field to some point P on a photographic plate. The geometrical arrangement is such that all electrons emitted with the same speed from a certain point on R, and passing through the slit F will fall on the same line at P. From the position of this line the radius of curvature can be determined, and the velocity of the electrons responsible for the line may be calculated from the formula,

$$v = RH\frac{e}{m},$$

---

[1] T. Shimizu, Proc. Roy. Soc. 99, 425 (1921).
[2] Robinson & Rawlinson, Phil. Mag. 28, 277 (1914).
[3] Rutherford & Robinson, Phil. Mag. 26, 717 (1915).

where R is the radius of curvature, H is the effective strength of the magnetic field, and e and m have their usual significance. A very good description of the details of this method is given by de Broglie in his paper on "The Corpuscular Spectra of the Elements."[1]

Distribution of the Photoelectrons

Longitudinal Asymmetry.—Mackenzie[2] first showed that the photoelectrons excited by γ-rays traversing a thin plate are much stronger in the direction of the γ-ray beam than in the opposite direction. The same effect is observed, though to a less degree, in the case of the photoelectrons produced by X-rays[3] and those produced by ultra-violet light.[4] The difference in the amount of the emergence and the incidence photoelectric emission from various metals as excited by γ-rays has been examined by Bragg.[5] He found that for light substances the emergent radiation is very much greater than for the incident (by a factor as great as 20 for the photoelectrons excited in carbon by hard γ-rays). For heavy metals the difference is far less marked, probably because of the much greater scattering of electronic rays by the heavy atoms. Rutherford[6] considers the experiments in accord with the view that the particles initially escape in the direction of the incident γ-rays. It is found that when X-rays are used, the degree of asymmetry does not differ for thin screens of different substances, and that the physical state has no appreciable effect.[8]

As has just been shown, the degree of asymmetry is a function of the wave-length of the incident radiation. However, Bragg finds * the asymmetry to be nearly the same with soft γ-rays as with hard γ-rays. Also Cooksie,[7] Stuhlmann† and Owen[8] find that the degree of asymmetry of the secondary radiation is practically the same, about 7 per cent greater in the forward direction, for wave-lengths varying from 3000 to 0.5 Å. U. There must therefore exist a region of wave-lengths between 0.5 and 0.05 Å. U. where the asymmetry of the photoelectric emission rapidly increases with decreasing wave-length.

The experiments so far considered were performed by ionization methods similar to that described above. It is rather surprising that photographs taken by Wilson's method show no appreciable asymmetry in the direction of the X-ray beam of the photoelectrons

---

[1] M. de Broglie, Jour. de Phys. et Rad. 2, p. 265 (1921).
[2] Mackenzie, Phil. Mag. 14, p. 176 (1907).
[3] C. D. Cooksie, Nature, 77, 509 (1908).
[4] O. Stuhlmann, Nature, May 12, 1910; Phil. Mag. 22, 854 (1911). R. D. Kleeman, Nature, May 19, 1910; Proc. Roy. Soc. (1910).
[5] W. H. Bragg, Phil. Mag. 15, 663 (1908).
[6] E. Rutherford, "Radioactive Substances and Their Radiations," p. 276 (1913).
[7] C. D. Cooksie, Phil. Mag. 29, 37 (1912).
[8] E. A. Owen, Phys. Soc. Proc. 30, 133 (1918).

* See Note 5.   † See Note 4.

liberated in air. The effect does appear when the X-rays traverse a plate of copper (Fig. 7), though here, as Mr. Wilson has pointed out in conversation with the writer, the predominance in the forward direction appears to be due to a bending of the paths of the electrons rather than to an initial asymmetry in their emission.

In order to account for the asymmetrical distribution of the photoelectrons, W. H. Bragg at one time suggested[1] that the primary X-rays exciting these electrons may consist of uncharged material particles, traveling with a speed approaching that of light. He supposed that these "neutrons," on colliding with electrons, transferred to them their energy and momentum. This would obviously account in a qualitative manner for the observed asymmetry of the electronic rays. The discovery† of the same type of asymmetry in the photoelectrons emitted under the action of ultra violet light, however, made this view difficult to defend,[2] and the later discovery of the interference and diffraction of X-rays has made the neutron hypothesis untenable.

An alternative explanation of the phenomenon has been put forward by Richardson.[2] He supposes that as the electron absorbs a quantum $h\nu$ of energy, the momentum of the absorbed radiation is also transferred to the electron, causing a resultant motion in the forward direction. If the kinetic energy after emission is $\frac{1}{2}mv^2 = h\nu$, the average ratio of the forward component of the velocity to the total velocity is shown to be

$$\frac{u}{v} = \frac{1}{2}\frac{v}{c}.$$

When v approaches the velocity of light c, the ratio u/v approaches $\frac{1}{2}$,* which indicates a decided preponderance in the forward direction, whereas for electrons liberated by ultra-violet light the ratio u/v is only about 1/500. This view thus accounts qualitatively for the observed increase in asymmetry of the photoelectric emission with increase in frequency of the exciting radiation. It seems to indicate, however, a greater and more uniform variation of asymmetry with wave-length than is actually observed.

It is interesting to note that the region of wave-lengths 0.5 to 0.05 Å. U. in which the asymmetry of distribution of the photoelectrons rapidly increases is just the region within which the secondary X-radiation becomes strongly asymmetrical. This suggests that both phenomena

---

[1] W. H. Bragg, Nature, Jan. 23, 1908; Phil. Mag. 16, 918 (1908).

[2] O. W. Richardson, Phil. Mag. 25, 144 (1913); The Electron Theory of Matter, pp. 478-481 (1914).

*This result is obtained without correcting for the variation of mass with velocity. When this correction is made, the limiting value of u/v for very high frequencies becomes unity. That is, the photoelectrons should be emitted in the direction of the incident rays.

† See Note 4, page 23.

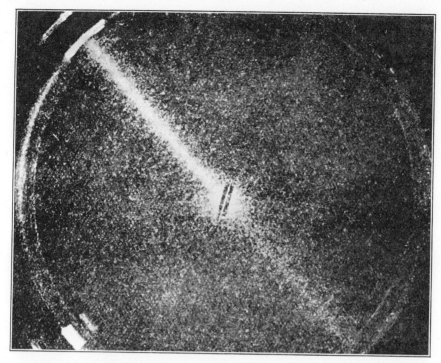

FIGURE 7. C. T. R. Wilson's photograph of X-rays traversing a thin plate of copper, showing the photoelectrons ejected from the copper and the absorption of the X-rays on passing through.

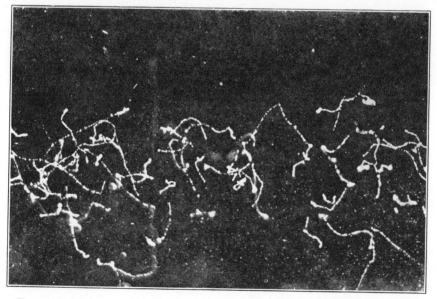

FIGURE 9. Path of X-rays through air, showing tracks of the photoelectrons produced.

may have a common origin, and tends to support the view that the secondary X-rays are emitted by electrons which are moving forward at high speed.

Lateral Asymmetry.—An examination of the electronic ray tracks as photographed by Wilson seems to reveal a marked tendency for these particles to be ejected nearly perpendicular to the direction of propagation of the X-ray beam. For example Figure 8 shows 31 tracks[1] whose point of origin can be distinguished with some certainty. Measurements on this photograph show 3 tracks starting within 45° of the horizontal, 7 whose initial direction is close to 45°, and 21 which start within 45° of the vertical. If the direction of ejection were a matter of chance, the horizontal and vertical directions would from considerations of symmetry be equally probable. Thus it appears that most of the photoelectrons are ejected nearly in the plane of the electric and magnetic vectors of the incident X-rays.

It would be of great interest to examine by this method the initial direction of the electrons ejected by polarized X-rays. It seems probable that the direction of ejection should be close to that of the electric vector.

Velocity of the Photoelectrons.—The maximum velocity of the photoelectrons liberated from a metal is given by the well known photoelectric equation

$$\frac{1}{2}mv^2 = h\nu - w_0, \tag{15}$$

where h is Planck's constant, ν is the frequency of the incident rays, and w_0 is the work done by the electron in escaping from the metal. This equation was first proposed by Einstein[2] as a deduction from the view that radiant energy occurs in discrete quanta, but was shown by Richardson[3] to be a direct consequence of Planck's radiation formula. As applied to light, its accuracy has been established by the researches of Richardson and K. T. Compton,[4] Hughes,[5] and particularly by the accurate experiments of Millikan[6] and his collaborators. In the X-ray region the studies of Innes,[7] Sadler,[8] Beatty,[9] Whiddington[10] and Moseley[11]

---

[1] This figure is C. T. R. Wilson's photograph, Proc. Roy. Soc. 87, plate 8, No. 4 (1912). I believe the lateral assymmetry of the photoelectrons here referred to has been noticed by Wilson, though I am unable to find any published statement to this effect.

[2] A. Einstein, Ann. d. Phys. 17, 145 (1905).

[3] O. W. Richardson, Phys. Rev. 34, 146 (1912); Phil. Mag. 24, 570 (1912).

[4] O. W. Richardson & K. T. Compton, Phil. Mag. 24, 575 (1912).

[5] A. L. Hughes, Phil. Trans; A, 212, 205 (1912).

[6] R. A. Millikan, Phys. Rev. 7, 18 & 355 (1916).

[7] P. D. Innes, Proc. Roy. Soc. 79, 442 (1907).

[8] Sadler, Phil. Mag. 19, 337 (1910).

[9] Beatty, Phil. Mag. 20, 320 (1910).

[10] Whiddington, Proc. Roy. Soc. 86, 360, 370 (1912).

[11] H. G. J. Moseley, Phil. Mag. 27, 703 (1914).

taken together showed[1] that the *maximum* energy of electrons liberated by X-rays of frequency ν is given very closely by

$$\frac{1}{2}m v^2 = h\nu. \qquad (16)$$

This is evidently in accord with the photoelectric equation, since the fastest electrons will come from the surface of the atom, where w_0 is negligible as compared with $h\nu$ for X-rays.

It is thus apparent that the fastest secondary electrons are emitted with kinetic energy equal to one quantum of the incident radiation. We are, however, equally interested in the energy relations of the slower electrons. Barkla and Shearer[2] were led to the conclusion that all the secondary electrons, whatever their origin, have on leaving the atom the speed corresponding to a whole quantum of the incident radiation. This conclusion was, however, shown by Richardson[3] to be questionable from a theoretical standpoint. Recently Simons,[4] from a study of the absorption of the secondary electrons in thin screens, suggested that different groups of velocities were present, corresponding to different energy losses by the electrons ejected from different parts of the atom. Finally a series of beautiful experiments by de Broglie[5] in the X-ray region and by Ellis[6] in the region of γ-rays, both using the magnetic spectrum method, has shown that at least a large part of the electrons emitted from different energy levels within the atom absorb one quantum of the incident energy, and emerge with their kinetic energy diminished only by the work required to leave the atom. For photoelectrons with these high velocities de Broglie finds it necessary to express the kinetic energy according to the relativity formula

$$T = 1/2 m_0 v^2 \left(1 + \frac{3}{4}\beta^2 + \frac{5}{8}\beta^4 + \cdots\right), \quad \left[\beta = \frac{v}{c}\right] \qquad (17)$$

where m_0 is the electron's mass at low speeds. The photoelectric equation as applied to high frequency radiation thus assumes the form

$$T = h\nu - w_P \qquad (18)$$

where ν is the frequency of the radiation which gives rise to the photo-electron and w_P is the energy required to remove the electron from its initial position in the P (K, L, M or N) energy level.

The experimental evidence indicates that the kinetic energy T, calculated by equation (18), is the maximum which may be possessed

---

[1] O. W. Richardson, "Electron Theory of Matter," Chapter XIX.
[2] Barkla and Shearer, Phil. Mag. 30, 745 (1915).
[3] O. W. Richardson, Proc. Roy. Soc. 94, 269 (1918).
[4] L. Simons, Phil. Mag. 4, 120, (1921).
[5] M. de Broglie, C. R. 172, pp. 274, 527, 746 & 806 (1921); Journ. de Phys. & Rad. 2, 265 (1921).
[6] C. D. Ellis, Proc. Roy. Soc. 99, 261 (1921).

by an electron ejected from the P energy level. The magnetic spectrum lines of the electronic rays as obtained by de Broglie, though relatively sharp on the high velocity side, shade off gradually on the side of the low velocities. In spite of the fact that thin foils were used, this shading is doubtless due in part to the loss in energy by some of the electrons before they reach the surface of the foil. Whether any electrons possess a smaller amount of kinetic energy as they leave the atom is therefore not answered by these experiments.

The fact should be mentioned that in the experiments of both de Broglie and Ellis a consistent tendency was noticed for the energy losses w_P to appear somewhat greater than the energy corresponding to the P energy level in the atom. This effect was most pronounced when w_P was small. It is possible, however, that this difference is due to consistent experimental errors, and the fact that de Broglie failed to notice such a difference in the experiments which he considered most reliable makes its existence appear questionable. In any case it may be stated that equation (18) holds within 2 or 3 per cent for the swiftest electrons leaving each energy level in the atom.

It is remarkable that in Ellis's paper no magnetic spectrum lines are recorded which are due to the expulsion of electrons from the outer rings. This result is confirmed by de Broglie's photographs, though since in his experiments X-rays are employed, the number of photo-electrons ejected from the outer rings is much smaller than that from the inner rings. In Ellis's experiments, however, γ-rays were employed, so that the greater part of the absorption was due to these outer electrons, as is shown by the fact that both the absorption per atom[1] and the number of electrons ejected per atom[2] is for the γ-rays nearly proportional to the number of electrons in the atom. We should therefore have expected these outer electrons, for which w_P is negligible compared with $h\nu$, to give rise to a strong line for which $T = h\nu$. Ellis finds, however, lines due only to the electrons in the K-rings, and these only for the heavy elements.

If, as has been suggested above, an electron scatters a whole quantum of energy at a time, and receives the momentum of the incident quantum, the average momentum in the forward direction of an electron which has scattered rays of wave-length λ is

$$M = \frac{h\nu}{c} = \frac{m_0 v}{\sqrt{1-\beta^2}},$$

where $\beta = v/c$. Thus the velocity of the electron is given by

$$\beta^2 = 1/(1+m^2c^2\lambda^2/h^2).$$

---

[1] Cf. M. Ishino, Phil. Mag. 33, 140, 1917; also infra, p. 45.
[2] Cf. infra, pp. 28 and 29.

For gamma rays with an effective wave-length of 0.025 Å.U., this means a forward velocity of about .7 the speed of light. Obviously such electrons should appear as photoelectrons, though with a velocity much less than that of an electron possessing a whole quantum of kinetic energy. It does not seem unreasonable to suppose that it may be such scattering electrons which constitute the photoelectrons excited by γ-rays in the lighter elements. The fact that the number of such electrons should be proportional to the atomic number and that the calculated velocity is of the observed order of magnitude confirms this view. Thus Ellis's failure to observe photoelectrons with the maximum kinetic energy $h\nu$ seems to support the hypothesis that most of the secondary β-rays excited by γ-rays are not a result of fluorescent absorption, but are rather a by-product of the scattering process.

Range of the Photoelectrons.—This question is identical with that of the range of the primary cathode rays, which has been discussed in an earlier portion of this report.[1] In one respect the problem is simplified however: it is possible to photograph the track of the electron by Shimizu's method, and to learn definitely both the distance to which the particle can penetrate, and the total length of its irregular path. While the former quantity is found to differ greatly for different rays, there is no evidence that the total length of path differs appreciably for electrons starting with the same initial velocity.

Number of Photoelectrons.—A series of experiments by W. H. Bragg,[2] Barkla[3] and their collaborators suggested strongly that the true absorption (as opposed to scattering) of X-rays is due solely to the excitation of the secondary electronic rays. This conclusion was supported by C. T. R. Wilson's photographs[4] of the path of an X-ray beam through air, which showed no ionization along the path of the X-rays except that due to the action of the high speed electrons which were liberated. On this view, X-ray energy can be dissipated in only two ways, either by scattering or by the excitation of photoelectrons.

It is possible that there may be two types of photoelectrons, those whose liberation excites the characteristic fluorescent radiation, and those which recoil after scattering a quantum of energy. According to the results of de Broglie and Ellis, each electron of the first type represents the absorption of one quantum of energy $h\nu$ from the primary beam. If the second type of photoelectron exists, it also should represent a whole quantum of energy, the greater part of which appears as scattered radiation and the remainder as kinetic energy of the recoiling electron. The energy of the second type will ordinarily be small compared with that of

---

[1] Bulletin of the National Research Council, vol. 1, p. 424 (1920).
[2] W. H. Bragg, Phil. Mag. 20, 385 (1910) et al.
[3] C. G. Barkla, Phil. Mag. 20, 370 (1910) et al.
[4] C. T. R. Wilson, Proc. Roy. Soc. 87, 288 (1912).

the first type. The evidence is thus consistent with the view that each photoelectron represents the removal of one quantum of energy from the primary beam, and that no other energy is lost except perhaps through true scattering.

It follows as a result of this conclusion that the energy absorbed per centimeter path of the X-ray beam should be

$$Nh\nu = E_i\tau,$$

where N is the number of β-particles liberated, E_i is the energy of the X-rays incident upon the substance and τ represents the fluorescent absorption coefficient (cf. infra, p. 37). Thus the number of photoelectrons liberated per centimeter path of the X-ray beam should be

$$N = E_i\tau/h\nu. \qquad (20)$$

Partial support of this relation (20) is given by the fact that the number of electrons ejected from an atom is independent of its state of chemical combination,[1] as is also the energy absorbed by the atom. Moore has shown[1] also that the number of photoelectrons emitted by different light atoms traversed by X-rays is proportional to the fourth power of the atomic weight. This corresponds exactly with Owen's law (cf. infra, p. 38) that the fluorescent absorption per atom under similar circumstances is proportional to the fourth power of the atomic number. It follows therefore that the number of photoelectrons is proportional to the X-ray energy truly absorbed, as stated by equation (20). Although no direct experimental determination of the factor of proportionality has been made, there seems no reason to doubt that this factor is the energy quantum $h\nu$.

In the region of γ-rays the number of secondary photoelectrons emitted per atom is more nearly proportional to the first power than to the fourth power of the atomic number, except for the very heavy elements in which appreciable characteristic X-rays are excited.[2] The number is, however, closely in accord with the total absorption of the γ-rays. Thus Hackett and Eve[2] find that γ-rays traversing plates of lead, iron and aluminium liberate β-particles in the ratio of 1.00 to 0.70 to 0.75 respectively. But Ishino[3] finds the total absorption per electron in these elements to occur in the ratio of 1.00 to 0.76 to 0.77 respectively. This close correspondence can leave little doubt that equation (20) applies to the electrons liberated by gamma-rays as well as those liberated by X-rays. Since these photoelectrons represent energy absorbed from the γ-ray beam, however, it follows that even in

---

[1] H. Moore, Proc. Roy. Soc. 91, 337 (1915).
[2] Cf. E. Rutherford, "Radioactive Substances, etc." pp. 275–276.
[3] M. Ishino, Phil. Mag. 33, 140 (1917).

the light elements by no means all of the absorbed energy can reappear in the secondary γ-rays, as has occasionally been argued.[1] The result is rather in accord either with the writer's conclusion[2] that the greater part of the energy removed from a γ-ray beam is fluorescently absorbed, or with the quantum conception of scattering suggested previously (p. 18).

Form of the Photoelectron Tracks.—From a study of the electron tracks photographed with his expansion apparatus, C. T. R. Wilson found, "The rays show two distinct kinds of deflection as a result of their encounters with the atoms of the gas—Rutherford's 'single' and 'compound' scattering. The gradual or cumulative deviation due to successive deflections of a very small amount is evidently, however, in this case much the more important factor in causing scattering, all the rays showing a large amount of curvature, while quite a small proportion show abrupt bends. When abrupt deflections occur they are frequently through large angles, 90° or more."[3]

Several people have noticed that the electron tracks in Wilson's photographs show a uniform curvature over considerable distances such as is not to be expected if the deflections are fortuitous. Many of the tracks have the form of converging helices,[4] such as might be due to motion in a strong magnetic field. But the axes of these helices have different orientations for the different electrons (nearly random orientations). It follows that the axis of each helix must be determined by some polarity of the photoelectron whose orientation remains nearly fixed as the particle moves through several centimeters of air and traverses several thousand atoms. This suggested to Shimizu[2] a gyroscopic action of the electron, and the writer has shown[5] that a spinning electron will induce magnetization in the surrounding atmosphere which will result in a strong induced magnetic field at the electron. Using for the angular momentum of the electron the value $h/2\pi$, it is found that one can thus account reasonably satisfactorily for the observed forms of the electron tracks.

An alternative suggestion of the origin of these helical tracks has recently been made by Glasson.[6] He supposes that a chance orientatation of the magnetic molecules in air gives rise to the magnetic fields which result in the curved tracks of the photoelectrons. It follows, however, from Weiss's theory of ferromagnetism that spontaneous magnetisation is not to be expected in a gas; and even should perfect alignment of the molecular magnets occur, the resulting field would be only

---

[1] Cf. e.g. Barkla and White, Phil. Mag. 34, 278 (1917).
[2] A. H. Compton, Phil. Mag. 41, 757 (1921).
[3] C. T. R. Wilson, Proc. Roy. Soc. 87, 289 (1912).
[4] Cf. A. H. Compton, Phil. Mag. 41, 279 (1921).
[5] A. H. Compton, ibid., and Frank. Inst. Journ. p. 154 Aug. (1921).
[6] J. L. Glasson, Nature, 108, 421 (Nov. 24, 1921).

a small fraction of that (greater than 1000 Gauss) required to account for the observed curvatures. This suggestion therefore does not help us.

If the writer's explanation of the spiral tracks is the correct one, it is of great interest. For it means that the electron acts as a tiny magnet as well as an electric charge, and that it is a dynamical system, which, by nutational or elastic oscillations, may radiate energy even though separate from an atom.

III. THE ABSORPTION OF X-RAYS

If radiation in traversing a thin layer of substance is reduced in intensity by a constant fraction μ per centimeter of the substance traversed, the intensity of the radiation after penetrating to a depth x is

$$I = I_0 \, e^{-\mu x},$$

where I_0 is the intensity at the surface. The quantity μ is called the "absorption coefficient." Similarly μ/ρ, the "mass absorption coefficient," is the fraction of a beam 1 cm.^{2} cross section absorbed per gram of substance traversed; and μ/ν, the "atomic absorption coefficient," where ν is the number of atoms per cm.,^{3} is the fraction of such a beam absorbed by each atom of the substance.

In order to obtain consistent results in the absorption measurements, the beam of X-rays passing through the absorbing material must be narrow, and the opening into the ionization chamber small, so that no appreciable amount of secondary rays will pass with the primary rays into the ionization chamber. This condition has not always been met in the earlier measurements on the absorption of X-rays and γ-rays, which has made much of this work of doubtful value. Measurements made on the X-rays reflected from crystals, however, have nearly always met this condition.

The following tables give the absorption coefficients of various wavelengths in different representative elements:

The values here given for lithium, carbon, oxygen, aluminium (H), iron and water (H) are due to Hewlett,[1] and those for aluminium (R), copper, molybdenum, silver, lead and water (R) are due to Richtmyer,[2] except for two measurements on lead due to Hull and Rice,[3] one each on aluminium and copper due to Duane[4] and those for wave-length .025*

---

[1] C. W. Hewlett, Phys. Rev. 17, 284 (1921).
[2] F. K. Richtmyer, Phys. Rev. 18, 13 (1921).
[3] A. W. Hull and M. Rice, Phys. Rev. 8, 836 (1916).
[4] W. Duane, Proc. Nat. Acad., March, 1922.
* $\lambda = .025$ represents the γ-rays from RaC, using the wave-length as measured by A. H. Compton, Phil. Mag. 41, 770 (1921). This value of the effective wavelength is confirmed by Ellis's recent results (Proc. Roy. Soc. A 101, p. 6, 1922). He finds homogeneous gamma-rays from RaC of wave-lengths .045, .025, .021 and .020 A. Line .020 is the strongest.

SECONDARY RADIATIONS: COMPTON

TABLE I. *Mass Absorption Coefficients, μ/ρ.*

Å.U.	Li 3	C 6	N 7	O 8	Al 13	Fe 26	Cu 29	Mo 42	Ag 47	Pb 82	H₂O (H)	H₂O (R)
.025	....	....	....	....	.071	....	....	....	....	.076	....	....
.100	....	.144	....	....	.162 (R)	.068 ← Duane	.32	....	....	....	....	.168
.125	....	.154	....	.146	(H) .174 / .181	.399	.46	1.35	Hull ⟶	3.0	.161	.176
.150	....	.163	.163	.163	.198	.585	.79	1.96	....	1.57	.180	.185
.175	....	.166	.171	.174	.221	.820	1.13	2.83	3.69	2.55	.195	.195
.20	....	.173	.177	.183	.259	1.06	1.56	4.02	6.00	4.60	.199	.204
.25	.172	.187	.193	.207	.358	1.88	2.77	7.42	11.4	8.48	.219	.229
.30	.188	.202	.224	.243	.370	3.09	4.50	12.7	18.2	14.2	.246	.261
.35	.208	.219	.251	.289	.517 .532	4.77	6.95	19.1	27.2	22.6	.283	.301
.40	.245	.240	....	.336	.719 .764	7.02	10.1	26.7	38.6	33.6	.334	.354
.50	.306	.304	....	.488	.982 1.06	13.86	18.8	48.6	11.5	60.6	.491	....
.60	.403	.394	....	.730	1.86 1.92	22.6	31.6	80.7	19.6	....	.711	....
.70	....	.532	....	1.08	3.05 3.23	35.3	48.8	18.8	....	....	1.023	....
.80	....	.706	....	1.53	4.84 5.05	50.7	....	27.2	87	⟵ Bragg	1.475	....
1.00	....	1.27	....	....	7.26 13.80	90.2	....	53.	....	....	2.70	....

TABLE II. Atomic Absorption Coefficients $\times 10^{23}$. $\frac{\mu}{\nu} = \frac{\mu}{\rho} \times \frac{W}{N}$.

Å. U.	H 1	Li 3	C 6	N 7	O 8	Al 13	Fe 26	Cu 29	Mo 42	Ag 47	Pb 82	H₂O (H)
.025	....	....	....	....	....	.317	....	....	....	....	....	....
.100	.04⁶	....	.285	....	....	.724	.625	....	....	....	2.60	.478
.125	.05	....	.305	.376	.385	.792	3.67	3.3	21.3	....	103.	.534
.150	.06	....	.323	.395	.430	.889	5.38	4.8	31.0	....	53.6	.578
.175	.05	....	.329	.409	.459	1.04	7.55	8.3	44.7	66.5	86.1	.591
.20	.05	....	.343	.446	.482	1.19	9.75	11.8	63.5	107	157	.650
.25	.04⁵	.197	.370	.518	.546	1.62	17.3	16.4	117	203	290	.730
.30	.04	.215	.400	.580	.641	2.34	28.4	29.0	201	323	485	.840
.35	.05	.238	.433	....	.763	3.31	43.9	47.2	302	483	772	.992
.40	.05	.280	.475	....	.886	4.56	64.5	72.9	422	686	1150	1.458
.50	.08	.280	.602	....	1.29	8.44	127	106	769	204	2070	2.11
.60	.09	.350	.780	....	1.92	14.0	208	197	1277	348	....	3.04
.70	.10	.462	1.052	....	2.85	22.1	325	332	297	....	....	4.38
.80	.17	....	1.40	....	4.03	32.4H	466	512	430	....	....	8.01
1.00	....	....	2.51	....	....	61.6H	830	....	838	2000	....	....

which are due to Ishino.[1] The writer has interpolated between the values given in the original papers to obtain the values for the wave-lengths desired. For additional data regarding the absorption coefficients of homogeneous X-rays reflected from crystals, the following authorities may be consulted: Bragg and Peirce (Phil. Mag. 28, 626, 1914) give data for the elements Al, Fe, Ni, Cu, Zu, Pd, Ag, Sn, Pt, Au, for wave-lengths between .491 and 1.32 Å.U. Hull and Rice* have studied the absorption by Al, Cu and Pb of short wave-lengths. Glocker (Phys. Zeitshr. 19, 66, 1918) gives a valuable discussion of absorption coefficients on the basis of data at that time available. Owen (Proc.

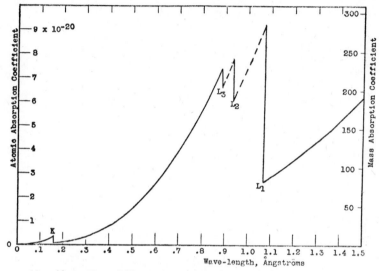

FIGURE 10. Absorption of X-rays by platinum, showing the rapid increase in absorption with increasing wave-length, and the critical K and L absorption limits.

Roy. Soc. 94, 510, 1918) gives absorption coefficients for 24 different elements of atomic numbers less than 35 for $\lambda = .586$ Å.U. Most of the work by Williams (Proc. Roy. Soc. A, 94, 571, 1918) has been repeated with greater care in the measurements from which Table I has been taken.

The most prominent characteristics of the absorption coefficients as functions of the wave-length and the atomic number are shown in Figures 10 and 11. In Figure 10 is shown the manner in which a given element, in this case platinum, absorbs radiation of different wave-lengths. In general the absorption coefficient increases rapidly with an increase of wave-length. There exist, however, certain critical regions in which for a slightly increased wave-length there is a sudden decrease in absorption. The wave-lengths at which such sudden changes

---

[1] M. Ishino, Phil. Mag. 33, 140 (1917).
* See Note 3, page 31.

occur are known as the critical absorption wave-lengths. It is found that if the wave-length of the radiation is shorter than the shortest of these critical wave-lengths, the complete X-ray spectrum of the absorbing element is excited, including the characteristic K-radiation. A slightly longer wave will excite only the characteristic fluorescent L, M, etc. radiations, but not that of the K-type. Similarly there are at least three critical absorption wave-lengths associated with the L series, at each of which a separate portion of the emission spectrum of the L series disappears, until at wave-lengths longer than 1.07 Å.U. no fluorescent L-radiation is excited. Experiment shows that the critical absorp-

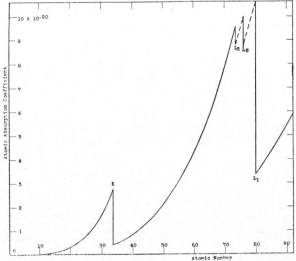

FIGURE 11. Absorption of X-rays of 1 A.U. wave-length per atom of different elements.

tion wave-length associated with any X-ray spectral series is very slightly shorter than the shortest emission wave-length of the series. Thus any element is especially transparent to its own characteristic radiation.

Figure 11 shows the absorption per atom of X-rays of wave-length 1.00 Å.U. in the different elements. The rapid increase of the absorption with the atomic number is prominent. But here again there exist the critical points at which sudden decreases in the absorption occur. Thus arsenic, of atomic number 33, absorbs this wave-length much more strongly than selenium, of number 34, corresponding with the fact that rays of 1 Å.U. wave-length will excite the characteristic K-radiation of arsenic but not of selenium. Similarly there exists a critical atomic number for the L series in the region of platinum ($N=78$).

Critical absorption wave-lengths have been observed corresponding not only to the K and L series of the absorber but to its M series as well in the case of the very heavy elements.

Some discussion has arisen with regard to the existence of a critical absorption of a shorter wave-length than the K-radiation, which could be ascribed to a possible "J" radiation. Several experimenters, including Barkla and White,[1] Williams,[2] Owen[3] and Dauvillier,[4] have obtained evidence which they have taken to indicate the existence of such critical wave-lengths; and Laub[5] and Crowther[6] have observed penetrating fluorescent radiation which they have attributed to this source. We have seen above, however (supra, p. 17), that this fluorescent radiation has no definite wave-length characteristic of the radiating element, and is probably ascribable either to a general fluorescent radiation similar to the "white" radiation from an X-ray tube or to scattering by moving electrons. Furthermore, there is no agreement between the values given by different observers for their critical J wave-lengths. The careful measurements of Richtmyer and Grant[7] and those of Hewlett[8] on the absorption of X-rays by light elements have shown no indication whatever of these supposed critical wave-lengths. And finally an examination of the radiation from an X-ray tube with an aluminium target led Duane and Shimizu to conclude[9] that "aluminium has no characteristic lines in its emission spectrum between the wave-lengths $\lambda = .1820$ Å. and 1.259 Å. that amount to as much as 2 per cent of the general radiation in the neighborhood." The evidence is thus strongly against the existence of a characteristic J radiation.

A very satisfactory discussion of these critical absorption wave-lengths, with tables of their values for the different elements, is given by Duane in the first part of this report.[10]

An empirical formula which has been found to express fairly satisfactorily the absorption by all elements of atomic number greater than 5 for wave-lengths between 0.1 and 1.4 Å.U.* is

$$\frac{\mu}{\nu} = KN^4\lambda^3 + .8N\sigma_0 \qquad (21)$$

---

[1] Barkla and White, Phil. Mag. 34, 270 (1917).
[2] Williams, Proc. Roy. Soc. 94, 567 (1918).
[3] E. A. Owen, Proc. Roy. Soc., 94, 339 (1918).
[4] Dauvillier, Ann. de Phys., 14, 49 (1920).
[5] J. Laub, Ann. d. Phys. 46, 785 (1915).
[6] J. A. Crowther, Phil. Mag. 42, 719 (1921).
[7] Richtmyer and Grant, Phys. Rev. 15, 547 (1920).
[8] Hewlett, Phys. Rev. 17, 284 (1921).
[9] Duane and Shimizu, Phys. Rev. 13, 288 (1919); 14, 389 (1919).
[10] W. Duane, Bull. Nat. Rsch. Coun. 1, 386 (1920).

* This is equivalent to a similar formula employed by Richtmyer,* except that through a typographical error he gives the value of K as 2.29×10^{-27}. The first use of the factors N^4 and λ^3 that I find are by Bragg and Peirce (Phil. Mag. 28, 626, 1914) and by Duane and Hunt (Phys. Rev. 6, 166, 1915) respectively. The term $N\sigma_0$, representing the scattering, was employed by Barkla and Collier (Phil. Mag. 29, 995, 1912); but Hull and Rice,† Hewlett§ and Richtmyer* have found a term equivalent to $.8 N\sigma_0$ to be more satisfactory, especially at very short wave-lengths with the lighter elements. Glocker (Phys. Zeitsch. 19, 66, 1918) suggests that somewhat better agreement may be obtained if slightly different values of the exponents of N and λ are used, though this is doubtful (cf. Richtmyer*).

---

* See Note 7, page 31.   † See Note 3, page 31.   § See Note 1, page 31.

Here λ is the wave-length of the X-rays employed, N is the atomic number of the absorber, K is a universal constant having the value 2.29×10^{-2} for wave-lengths shorter than the critical K absorption wave-length, if λ is expressed in cm., and a value $.33 \times 10^{-2}$ when λ is between the critical K and L absorption wave-lengths. The quantity σ_0 is given by the expression

$$\sigma_0 = \frac{8\pi}{3} \frac{e^4}{m^2 c^4}, \qquad (22)$$

and has the value 6.63×10^{-25}. It represents the total energy scattered by a single electron, calculated by integrating Thomson's formula (1) over the surface of a sphere.

The extent of the agreement of this expression (21) with the experimental values for the representative elements carbon, aluminium, iron silver and lead is exhibited for wave-lengths between 0.1 and 1.0 Å. U. in Figure 12. The logarithms of the atomic absorption coefficients are plotted against the logarithms of the wave-lengths. It is remarkable that a formula with but 4 arbitrary constants is able to express so accurately the absorption by some 80 elements of radiation over so wide a range of wave-lengths. It would suggest that the relation is of some physical significance. Nevertheless, the formula is unsatisfactory for extrapolation to shorter wave-lengths, since the minimum absorption that it can give, $0.8\ N\sigma_0$, corresponds to a mass absorption coefficient of about 0.16. This is not in agreement with the mass absorption coefficient about 0.07 observed for all elements when hard gamma rays are employed. A theoretical formula which describes the absorption more satisfactorily is given below (equation 39).

Theory of X-ray Absorption.—The absorption of high frequency radiation is due to at least two independent processes. The more important of these is usually the energy spent in exciting photoelectrons, and resulting in fluorescent radiation. There is always, however, a certain amount of energy removed from the primary beam by scattering. Thus the total absorption coefficient may be written as

$$\mu = \tau + \sigma,$$

where τ represents the "true" or "fluorescent" absorption, and σ the energy lost by scattering.[1]

The Fluorescent Absorption.—This part of the absorption represents the energy which, as we have seen (supra p. 29), is transformed into kinetic energy of the photoelectrons in accord with the quantum rela-

---

[1] On the classical theory, the energy of the scattered X-rays should equal the energy σ removed from the primary beam to produce the scattered rays. On the quantum theory (supra, p. 18) a part of the energy σ goes into the kinetic energy of recoil of the scattering electrons. In either case the energy σ removed in the process of scattering presumably follows a different law of variation with λ than does the energy τ spent in exciting the fluorescent radiation.

tion, $\frac{1}{2}mv^2 = hc/\lambda$. Experiment shows also,[1] in accord with our empirical formula (21), that it is the sum of a series of terms which may be written very approximately thus:

$$\frac{\tau}{\nu} = [K_K N^4 \lambda^3]^{\lambda < \lambda_K} + [K_L N^4 \lambda^3]^{\lambda < \lambda_L} + \cdots. \tag{23}$$

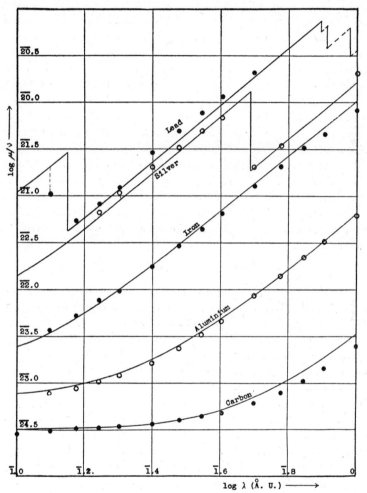

FIGURE 12. Comparison of the experimental values of X-ray absorption for different wave-lengths and atomic numbers with the values predicted by the empirical formula (21).

The first term in this expression presumably represents the absorption by the electrons in the K ring, and is to be used when the incident wavelength λ is less than the critical K absorption wave-length λ_K. Similarly

[1] This result was first reached empirically by E. A. Owen, Proc. Roy. Soc. 94, 522 (1918), and represents a more accurate statement of the absorption law first proposed by him in 1912. (Proc. Roy. Soc. A. 86, 434).

the other terms represent the absorption by the electrons in the L, M, etc. rings respectively. These facts, especially the reappearance of the absorbed energy in quanta of kinetic energy, would suggest that X-ray absorption is essentially a quantum phenomenon, to which presumably the classical electrodynamics may not be successfully applied. It is therefore surprising to find that quantum principles have offered no suggestion as to the significance of Owen's simple law (23)* whereas the classical electrical theory may be applied to the problem with striking results.

Equation (23) has been derived theoretically by the writer[1] in a simple though uncritical manner, making use of J. J. Thomson's old hypothesis of X-ray pulses. Such a solution of the problem is unsatisfactory, since the basic hypothesis of incident X-rays consisting of very short pulses is inconsistent with the fact that X-ray spectrum lines are very sharp.[2] But the fact that Owen's empirical formula can thus be derived theoretically, strongly suggests that the law may be of real physical significance.

Formal derivation of Owen's law.—Considering the general case in which both the emitting and the absorbing electrons are electrically charged particles capable of executing damped harmonic oscillations, the equation of motion of the absorbing electron when traversed by the incident wave is[3]

$$x = A_1 e^{-r_1 t} \cos(q_1 t + \delta_1) + A_2 e^{-r_2 t} \cos(q_2 t + \delta_2). \tag{24}$$

This motion consists of a forced oscillation whose frequency $q_1/2\pi$ and damping r_1 are those of the incident wave, combined with a free oscillation of the absorbing electron's natural frequency $q_2/2\pi$ and damping r_2. The energy removed from the primary beam in producing this motion is the initial energy of the electron's resulting free oscillation plus the energy spent in executing the forced oscillation against the resistance due to damping.

The energy corresponding to the second term of this expression is derived from the first wave of the incident train. It has therefore the effect of quickly smoothing off the front of the train of waves, so that after passing through a thin sheet of matter the incident wave train no longer starts suddenly, but gradually reaches a maximum and then dies down. A similar smoothing of the wave-front also results from reflection by a crystal. For this reason, in absorption measurements as usually made, the amplitude A_2 is so small as to make the corresponding part of the absorption negligible.[4] The absorption actually mea-

---

[1] A. H. Compton, Phys. Rev. 14, 249 (1919).
[2] Cf. e.g. D. L. Webster, Phys. Rev. 7, 609 (1918), and G. E. M. Jauncey, Phys. Rev. October (1921).
[3] A. H. Compton, Washington University Studies, 8, 117 (1921).
[4] A more complete discussion of this part of the absorption is given by A. H. Compton, Washington University Studies, 8, pp. 116-120 (1921).
* Cf., however, p. 43.

sured is accordingly due to the motion corresponding to the first term of expression (24) against the resistance due to damping.

The value of the absorption per atom due to the K electrons calculated on this basis may be shown to be,[1]

$$\left(\frac{\tau}{\nu}\right)_K = 8\pi N_K \frac{e^2}{mc} \cdot \frac{r^2}{p_1^2} \bigg/ \left\{\left(1-\frac{p_2^2}{p_1^2}\right)^2 + \frac{4r_1}{p_1^2}(r_2-r_1)\left(1-\frac{p_2^2}{p_1^2}\right) \right.$$
$$\left. + \frac{4}{p_1^2}(r_2-r_1)^2\right\}, \quad (25)$$

where $(\tau/\nu)_K$ is the part of the atomic absorption coefficient due to the N_K electrons in the K ring, e, m and c have their usual values, and $p_1^2 = q_1^2 + r_1^2$; $p_2^2 = q_2^2 + r_2^2$, where the q's and r's are as in equation (24).

Experiment shows* that the damping coefficient r_1 of the primary rays is very small. If it is considered negligible, there are two special cases of interest. First we may assume that the natural period of the absorbing electrons is very great compared with the period of the incident X-rays. Substituting $p_1 = 2\pi c/\lambda$, equation (25) then reduces to

$$\left(\frac{\tau}{\nu}\right)_K = \frac{2N_K}{\pi} \frac{e^2}{mc^2} \frac{r_2\lambda^2}{1+\frac{4r_2^2}{p_1^2}}. \quad (26)$$

If r_2 is small compared with p_1, this means that the absorption due to the K electrons is equal to a constant times $r_2\lambda^2$. In order for Owen's law to hold, i. e. $\tau/\nu = KN^4\lambda^3$, r_2 must therefore be proportional to $N^4\lambda$. It is accordingly possible to account for the observed absorption of X-rays on this view if the appropriate value of the damping of the absorbing electrons' motion is postulated.

As our second case we may consider the K absorption band to be due to the presence of electrons whose natural frequency may be anywhere between p_K and ∞, where p_K is the angular frequency corresponding to the critical absorption limit. If the damping of these electrons' forced oscillations is small compared with p, and is the same for different wavelengths, it can be shown that the number of electrons responsible for the K absorption band is related to the absorption coefficient thus:[2]

$$N_K = \frac{mc^2}{\pi e^2} \int_0^{\lambda_K} \left(\frac{\tau}{\nu}\right)_K \frac{d\lambda}{\lambda^2}. \quad (27)$$

Let us suppose that the absorption is proportional to some power of the wave-length, say to λ^{x+1}. Then we may write

---

[1] A somewhat similar calculation has been performed by R. A. Houstoun, Proc. Roy. Soc. Edinburgh, 40, 34 (1920).

[2] This value for N_K is obtained by a calculation similar to that performed by R. A. Houstoun, Proc. Roy. Soc. Edinburgh, 40, 34 (1920). It is hoped to treat this subject more fully in an early paper.

* See Note 2, page 39.

$$\left(\frac{\tau}{\nu}\right)_K = C_K \lambda^{x+1}$$

for wave-lengths between 0 and λ_K. Using this value in equation (27) we obtain,

$$N_K = \frac{mc^2}{\pi e^2 x} C_K \lambda_K^x. \qquad (28)$$

The unknown quantities x and N_K can be determined as follows. First, choose two elements such that the critical K absorption wave-length λ_{K1} of one is equal to the critical L absorption wave-length λ_{L2} of the other. It follows then from equation (28) that

$$\frac{N_L}{N_K} = \frac{C_{L2}}{C_{K1}}. \qquad (29)$$

That is, the ratio of the number of electrons responsible for the L and the K absorption bands is equal to the ratio of the atomic absorption coefficients for the same wave-length by the two elements chosen as specified above.[1] The weighted mean of the three critical L wave-lengths for lead is about 0.90 Å.U. The element with a critical K wave-length of this value should have an atomic number of 35.35. Experiment shows that the L absorption by lead is 4.2±.3 times as great as the K absorption by such an element[2], which means that there are about 4.2 times as many electrons in the L as in the K ring. This result is in exact accord with modern theories of atomic structure, which would assign 2 electrons to the K ring and 8 to the L orbits.

The value of the exponent x may be calculated if two elements are so chosen that the L absorption per atom due to one is equal to the K absorption by the other, i.e. $C_{L2} = C_{K1}$. Then by equation (28),

$$\frac{N_L}{N_K} = \frac{\lambda_{L2}^x}{\lambda_{K1}^x},$$

or

$$x = \log\left(\frac{N_L}{N_K}\right) \Big/ \log\left(\frac{\lambda_{L2}}{\lambda_{K1}}\right). \qquad (30)$$

The elements gold and silver satisfy within experimental error the condition imposed. λ_K for silver is .485, and the weighted mean λ_L for

---

[1] This formula (29) may be derived from much more general assumptions than those used above. It is necessary only to assume that the average absorption by a K electron is equal to that by an L electron if both have the same critical absorption limit. The formula may obviously be extended to the M etc. rings, and gives perhaps the most reliable experimental means now available for estimating the relative number of electrons in these rings.

[2] For the L absorption by lead, use is made of Richtmyer's experimental values, noting according to Bragg and Peirce (X-rays and Crystal Structure, p. 181), that 3/4 of this absorption is due to the L electrons. The K absorption by element 35.35 is calculated from the empirical law (21), remembering that a fraction $(2.29-.33)/2.29$ of this absorption is due to the K electrons.

gold is about 0.99. Using the experimental value $N_L/N_K = 4.2$, relation (30) becomes,

$$x = \log 4.2/\log (.99/.485) = 2.0.$$

Thus within experimental error the absorption should be proportional to the cube of the wave-length.

Equation (28) may now be written in the form

$$C_K = \frac{2\pi N_K e^2}{mc^2 \lambda_K^2} = DN_K \Big/ \lambda_K^2, \qquad (31)$$

where

$$D = 2\pi e^2/mc^2. \qquad (32)$$

Thus, introducing the similar terms representing the L, M, etc. absorption, we obtain the theoretical absorption law

$$\frac{\tau}{\nu} = D\lambda^3 \left\{ \Big[N_K/\lambda_K^2 \Big]^{\lambda < \lambda_K} + \Big[N_L/\lambda_L^2 \Big]^{\lambda < \lambda_L} + \cdots \right\}. \qquad (33)$$

In view of Moseley's approximate relation, $N^2 \alpha 1/\lambda_K$, this is approximately equivalent to Owen's formula. But now the formula does not involve a single arbitrary constant.[1]

The agreement of the experiments with this formula while good is not perfect. The accuracy of the λ^3 and the N^4 relations are exhibited in Figure 11. When $1/\lambda_K^2$ is substituted for N^4, the agreement is of about the same order of accuracy, though small consistent variations appear in the opposite sense as the atomic number of the absorber is varied. According to equation (32), the constant D should have the value 1.77×10^{-12}. Its average experimental value in the K absorption region, taking N_K to be 2, is 1.4×10^{-12}, and in the L region is 1.65×10^{-12}. The agreement as to order of magnitude is gratifying, and suggests that we are working along the right line. Nevertheless the differences between the theoretical and the experimental values are too great to be accounted for as experimental error, showing that our assumptions must in some way be modified.

The only considerable assumption that has been made in deriving equation (33) is that slightly damped electronic oscillators exist of natural frequencies distributed between the critical absorption frequency and infinity. It must be confessed that the assumption does not appear very plausible. Nevertheless, the fact that without introducing the quantum concept we can thus obtain a formula which is in many ways so satisfactory, strongly suggests the X-ray absorption is not a quantum phenomenon. The results support rather the view that X-ray absorption is a continuous process, obeying the usual laws of electrodynamics.

---

[1] N_K, N_L etc. are given by atomic theory as 2, 8, etc., the exponent of λ is determined by wave-length measurements, and the other quantities have fixed, known values.

Quantum theory of X-ray absorption.—Since the above discussion was written, there has appeared a theory of X-ray absorption by L. de Broglie which, although based upon widely different assumptions, leads to nearly the same result as that expressed by equation (33). On the assumptions, 1. that Wien's energy distribution law holds for black body radiation for X-ray frequencies, 2. that Kirchhoff's law relating the emission and absorption coefficients of a body is valid, and 3. that Bohr's hypothesis connecting the energy of the electron's stationary states with the corresponding critical absorption frequencies is correct, de Broglie shows that the atomic fluorescent absorption should be[1]

$$\frac{\tau}{\nu} = \frac{1}{8\pi kcT} \cdot \Sigma \eta_p A_{ip}^{\;n} E_p \cdot \lambda^3. \tag{34}$$

Here k is the Boltzmann constant, c the velocity of light, T the absolute temperature, $A_{ip}^{\;n}$ the probability that an atom ionized by the loss of an electron from the pth shell will return to its normal condition in unit time, E_p is the energy required to remove an electron from the pth orbit, and η_p is an undetermined constant whose value is probably not far from unity.

In view of the soundness of his assumptions, the theoretical basis for the λ^3 law of absorption seems very strong. While some of the experiments have seemed to throw doubt on the exact validity of the third power relation, those in the neighborhood of the critical wave-lengths are performed under adverse conditions, and those at short wave-lengths are difficult to interpret because of the unknown magnitude of the scattering. In general the experiments afford a satisfactory confirmation of the cube law over a wide range of wave-lengths.

Since the experiments show that the absorption is independent of the temperature and nearly proportional to N^4 or to E_p^2, de Broglie assumes that

$$A_{ip}^{\;n} = \alpha E_p T = \alpha(\epsilon_n - \epsilon_p)T = \alpha h\nu_p T,$$

where $\epsilon_n - \epsilon_p$ is the energy radiated as the electron returns from the state of p ionization to its normal condition. The constant of proportionality α is evaluated with the help of the principle of correspondence, considering what its value would be for low frequency oscillators in a body at high temperature.[2] He thus finds,

$$\alpha = \frac{8\pi^2}{c^3} \frac{e^2}{m} \frac{k}{h^2}. \tag{35}$$

The atomic absorption within the K absorption band is accordingly

$$\frac{\tau}{\nu} = \frac{\pi}{c^4} \frac{e^2}{m} \lambda^3 \left[\eta_K N_K \nu_K^2 + \eta_L N_L \nu_L^2 + \cdots \right],$$

---

[1] L. de Broglie, Jour. de Phys. et Rad. 3, 33 (1922), cf. also C. R. 171, 1137 (1920).
[2] L. de Broglie, C. R. 173, 1456 (1921).

or substituting the value of D given in equation (32),

$$\frac{\tau}{\nu} = \tfrac{1}{2}D\lambda^3 \left\{ \left[\eta_K N_K/\lambda_K^2 \right]^{\lambda<\lambda_K} + \left[\eta_L N_L/\lambda_L^2 \right]^{\lambda<\lambda_L} + \cdots \right\}. \quad (36)$$

In his papers, de Broglie has assumed that $\eta=1$, which would make this result differ from equation (33) by a factor of $\tfrac{1}{2}$. But, in a letter to the writer, he has pointed out that his theory does not definitely determine the value of this constant, so that there is no positive contradiction between the two results. It is certainly remarkable that results so nearly identical should be obtained on the basis of wholly different assumptions.

Absorption of very short wave-lengths.—Especial interest attaches to the study of the absorption of the highest frequency radiation, because the modifications necessary to fit this case involve important assumptions concerning the nature of X-rays and the properties of the electron. According to any expression based upon the usual electron theory, the atomic absorption coefficient can never fall below Thomson's theoretical value $N\sigma_0$ for the total scattering. Experimentally, however, it falls to a considerably lower value. For example, in the case of carbon $N\sigma_0 = 4.0 \times 10^{-24}$; whereas for 0.1 A.U., $\mu/\nu = 2.8 \times 10^{-24}$ and for hard gamma rays ($\lambda = 0.025$ Å) $\mu/\nu = 1.5 \times 10^{-24}$, less than half of the theoretical value of the scattering alone. The most obvious explanation of this low value of the absorption is that the electron is large enough for destructive interference to occur between the rays scattered by its different parts. Thus we are led to the same hypothesis of an electron of appreciable size that was found necessary to explain the direct experiments on the scattering of X-rays.[1]

In the above discussion (pp 4-11) it was found necessary to take into account the grouping of the electrons within the atom and the size of the electron itself in order to express adequately the scattering of X-rays. Consequently the quantity $N\sigma_0$ which assumes the electrons to act as point charges independent of each other, must be replaced by a term $\Omega N\sigma_0$ where the coefficient Ω is defined by the expression for the total energy scattered by an atom, thus:[2]

$$\frac{\sigma}{\nu} = \int_0^\pi \frac{I_\theta}{I} \cdot 2\pi L^2 \sin\theta d\theta \equiv \Omega N \int_0^\pi \frac{I_0}{I} 2\pi L^2 \sin\theta d\theta \equiv \Omega N \sigma_0. \quad (37)$$

Here I is the intensity of the incident beam which strikes the atom, I_θ is the intensity of the beam scattered by the atom to a distance L at an angle θ with the incident beam (cf. equations 2 and 3), and I_0 is the corresponding intensity scattered by a point charge electron

---

[1] It appears possible that the hypothesis of scattering one quantum by each electron may offer an alternative explanation of this reduced scattering. But no quantitative calculations along this line have as yet been made.

[2] A. H. Compton, Phys. Rev. 14, 250 (1919) and Washington University Studies, 8, 107 (1921).

(equation 1). Thus σ/ν can be determined experimentally by integrating the observed scattered intensity over the surface of a sphere. As thus measured, for wave-lengths less than 0.1 Å.U. the atomic scattering rapidly diminishes, until for hard gamma-rays it is only a small fraction of $N\sigma_0$.

A third type of absorption.—There is, however, another type of absorption which becomes prominent at very short wave-lengths. Ishino has shown[1] that the total absorption coefficient of hard gamma-rays, except for the very heavy elements in which appreciable characteristic radiation is excited, is proportional to the atomic number of the absorber. It might be thought that this absorption is due to scattering of the gamma rays; but Ishino and others have shown that only about half of the absorbed energy reappears as secondary rays. The writer has shown that even of these secondary rays only a very small part is of the same wave-length as the primary γ-rays.[2] That a large part of the energy spent by γ-rays is truly absorbed is confirmed by the fact that the number of high-speed secondary β-rays excited in the absorber is also closely proportional to the atomic number, and that they possess energy which is at least a considerable part of the absorbed energy.[3] Thus in the case of gamma-rays there exists a true absorption which is per atom proportional to the atomic number.

The remarkable simplicity of Owen's $N^4\lambda^3$ relation for fluorescent absorption of X-rays suggests strongly that the energy thus taken into account is dissipated by a single process, directly connected with the excitation of the characteristic fluorescent radiations. The true absorption of gamma-rays, which is proportional to the first power of N and in which the characteristic radiations are not concerned, is therefore presumably due to a different process.

Two alternative hypotheses of the origin of this true absorption of γ-rays may be presented. 1. We may think of this energy as spent in exciting a form of general fluorescent radiation through the medium of the secondary β-rays. Or 2. we may suppose, on the quantum idea of scattering, that when an electron receives a quantum of γ-ray energy, the momentum of the quantum gives to the scattering electron a considerable momentum, so that the scattered energy is equal to the energy absorbed less the kinetic energy of recoil of the scattering electron. On either view, the secondary rays will possess only a fraction of the energy removed from the primary beam, the remaining energy appearing as kinetic energy of the secondary photoelectrons or β-rays.

In any case, the difference between the energy removed from the γ-ray beam and the total energy in the secondary rays is a type of

---

[1] M. Ishino, Phil. Mag. 33, 140 (1917).
[2] Cf. Supra, p. 7.
[3] Cf. Supra, p. 29.

truly absorbed energy. We may call this the "momentum" absorption, since it seems to be due to the momentum carried by a quantum of radiation. The corresponding "momentum absorption coefficient" may be designated by the letter ω. This quantity is similar to the fluorescent absorption coefficient τ in that they both represent true absorption, and is similar to the scattering coefficient σ in that both are nearly proportional to the atomic number, and are probably both associated with the scattering process.

While the existence of this momentum absorption has been experimentally established only in the case of γ-rays, there is no reason to doubt its existence also in the case of X-rays. Indeed, the observation that the secondary X-rays are of longer wave-length than the primary (supra p. 17) would seem to require its existence. Its magnitude is presumably a function of the wave-length which increases rapidly with the frequency as the frequency of γ-rays is approached. It appears probable that it should again decrease for still higher frequencies. Thus we may write for the atomic momentum absorption

$$\frac{\omega}{\nu} = R(\lambda) \cdot N, \qquad (38)$$

where $R(\lambda)$ is probably always less than σ_0.

The general formula for the absorption of X-rays is accordingly,[1]

---

[1] From theoretical considerations, the writer (Phys. Rev. 14, 247, (1919), has suggested a modification of the principal absorption term τ/ν. For if (as assumed above) the energy spent in fluorescent absorption is transferred to the translational vibrations of the absorbing electrons, the absorption will be proportional to the square of the acceleration to which the electron is subject by a given electric field. But when the wave-length is comparable with the electron's diameter, the phase of the wave will differ over the electron's different parts, and the maximum acceleration will be less than for a smaller electron of the same mass. A factor should then be introduced into the fluorescent absorption term which is

$$\varphi = \left\{ \frac{\text{acceleration of real electron}}{\text{acceleration of point charge electron}} \right\}^2.$$

The value of this ratio for a spherical shell electron, for example, is

$$\varphi = \sin^2\left(\frac{2\pi a}{\lambda}\right) \Big/ \left(\frac{2\pi a}{\lambda}\right)^2,$$

where a is the radius of this electron. Using the value 2×10^{-10} cm. for a, as determined by scattering measurements, for $\lambda = 0.1$Å, $\varphi = 0.57$, and for $\lambda = 0.025$Å, $\varphi = .04$. According to the theoretical formula

$$\tau/\nu = K \varphi N^4 \lambda^3,$$

we should therefore expect the values of τ/ν to depart markedly from the λ^3 law at .1 Å.U., and to become almost negligible at 0.025 Å.U. The fact that Richtmyer's and Hewlett's measurements at .1 Å.U. show no such variation might perhaps be due to compensating changes in ω or σ. But if the writer's value of 0.025 is correct for the wave-length of the gamma rays from RaC, the characteristic fluorescent absorption of these rays by lead agrees accurately with the value extrapolated from longer wave-lengths *if φ is unity*.

Since our knowledge of the mechanism of absorption is not as complete as that of scattering, however, this contradiction does not mean that our assumed size of the electron is wrong, but that we don't know as much about the absorption process as we thought we did. It would appear from de Broglie's theory of absorption that the cube law should hold whatever the size of the electron.

$$\frac{\mu}{\nu} = \frac{\tau}{\nu} + \frac{\omega}{\nu} + \frac{\sigma}{\nu},$$
$$= KN^4\lambda^3 + RN + \Omega N \sigma_0. \tag{39}$$

Here K, N, λ and σ_0 are defined as in the empirical formula (21). Ω, determined by equation (37), is a function of the wave-length and the size of the electrons, as also presumably is the function R. The first term on the right hand side represents the energy which is used in exciting the characteristic radiations of the absorber, the second accounts for that used in producing the general fluorescent X-rays or in giving momentum to the scattering electrons, and the third indicates the X-ray energy truly scattered by the atom[1].

Conclusions.—While the experiments seem to have established the physical existence of these distinct types of absorption, the theory of X-ray absorption presents some fascinating unsolved problems. Thus, the "momentum absorption" ω has almost escaped detection, and from our present meager knowledge it is not possible to come to a definite conclusion concerning its origin. The absorption due to scattering is apparently in accord with the direct measurements of the scattered rays, and is explicable on the basis of the classical electrodynamics if the radius of the electron is of the order of 4×10^{-10} cm. From an experimental stand-point the principle absorption term $KN^4\lambda^3$ is best established. Its form and order of magnitude are in accord with the classical electrodynamics. The sudden changes in the value of K at certain definite wave-lengths are, on the other hand, apparently determined by quantum conditions. But there is some unknown process of damping the absorbing electrons' oscillations, which transforms radiant energy into the quanta of kinetic energy carried by the photoelectrons. The fact that the usual electron theory goes so far in explaining absorption phenomena, leads one to believe that here may be a vantage point for attacking the problem of the connection between the electron and the energy quantum.

---

[1] It is worth calling particular attention to the fact that the magnitude of the true absorption term ω *increases* with the frequency. The total absorption probably always diminishes with shorter wave-lengths, since increase in ω is accompanied by decrease in the scattered energy σ. But for light elements and wave-lengths so short that τ is small, the true absorption is due almost wholly to the momentum term ω, and hence (at least until extremely short wave-lengths are reached) increases as the wave-length is shortened.

In much practical work, as for example in the therapeutic use of X-rays, broad beams are employed, so that much of the scattered radiation is present with the transmitted rays. Under such conditions the effective absorption coefficient lies somewhere between the total absorption μ and the true absorption $\tau + \omega$. It appears highly probable that in such cases the increase in ω might for light elements more than balance the decrease in τ and the part of σ lost to the beam, as very short wavelengths are approached. There should accordingly be an optimum wave-length which, under given conditions, should be more penetrating than longer or shorter waves. It is possible that hard gamma rays, for which ω is about equal to σ, are on the short side of this optimum wave-length. If this is the case, X-rays with a somewhat longer wave would be preferable in therapeutic work because they would have greater penetration.

IV. The Refraction of X-rays

Theory.—According to Lorenz's theory of dispersion[1], if the frequency ν of the radiation transmitted by a substance is high compared with the natural frequency of the electrons in the substance, its index of refraction μ is approximately

$$\mu = 1 - \frac{ne^2}{2\pi m\nu^2}, \qquad (40)$$

where n is the number of electrons per unit volume, and e and m have their usual significance. If X-rays of ordinary hardness ($\lambda = 0.5$ Å.U.) traverse a substance of density 3, this expression indicates an index of refraction which is less than unity by about 1×10^{-6}. With the softest X-rays that have been examined spectroscopically ($\lambda =$ about 10 Å.U.), the refractive index should, however, differ from unity by as much as 4×10^{-4}. A difference of this magnitude should be measurable.

Webster and Clark[2] have calculated the refraction index in the neighborhood of an X-ray absorption band on the basis of the view, discussed above, that the absorption of X-rays is due to a frictional resistance to the forced oscillations of the electrons. Assuming that the energy thus dissipated is wholly accounted for by the characteristic fluorescent absorption τ, they find that, except near a critical absorption frequency, the index of refraction is expressed by

$$\mu = 1 - \Sigma_K \frac{\tau_K \lambda_K}{4\pi^2} \left\{ \frac{\nu_K^2}{\nu^2} + \frac{\nu_K^4}{\nu^4} \log \left|1 - \frac{\nu^2}{\nu_K^2}\right| \right\}. \qquad (41)$$

Here λ_K and ν_K are the critical wave-length and frequency respectively of the K absorption band, τ_K is the increase in the linear absorption coefficient as the critical K absorption frequency is passed, and the summation is taken over the absorption limits associated with the K, L, M etc. radiations. Except in the neighborhood of a critical absorption frequency, the effect on the refractive index of the electrons responsible for the characteristic fluorescent absorption as thus calculated is not as great as that due to the remaining electrons, as calculated from equation (40).

At the critical absorption frequency Webster and Clark[143] show that there occurs a maximum refractive index given approximately by

$$\mu = 1 + \frac{\tau_K \lambda_K}{4\pi^2} \log\left(\frac{\nu_K}{\Delta \nu_K}\right), \qquad (42)$$

where $\Delta \nu_K$ is the range of frequency through which occurs the sudden rise in absorption near the critical frequency ν_K. In the case of a rhodium prism, $\tau_K \lambda_K / 4\pi^2$ is about 10^{-7}. Thus unless $\Delta \nu_K$ is extraor-

---

[1] H. A. Lorentz, "The Theory of Electrons," 2nd Ed., p. 149.
[2] D. L. Webster and H. Clark, Phys. Rev. 8, 528 (1916).

dinarily small, even here the refractive index will differ only slightly from unity.

Experiments.—In his original examination of the properties of X-rays, Roentgen tried unsuccessfully to obtain refraction by means of prisms of a variety of materials such as ebonite, aluminium and water. Previous to the use of homogeneous rays reflected from crystals, perhaps the experiment conducted under conditions most favorable for measurable refraction was one by Barkla[1]. In this work X-rays of a wavelength which excited strongly the characteristic K-radiation from bromine were passed through a crystal of potassium bromide. The accuracy of his experiment was such that he was able to conclude that the refractive index for a wave-length of 0.5 Å.U. probably differed from unity by less than 5×10^{-6}.

A very satisfactory test of the refraction of homogeneous X-rays has been made by Webster and Clark.* They found that the refractive index for the different K lines of rhodium, transmitted by a rhodium prism, differed from unity by less than about 3×10^{-4}. These negative results are in accord with the electron theory of dispersion outlined above.

Variations from Bragg's law.—A direct test of the refraction of very soft X-rays is difficult, because of their strong absorption. It has been observed, however, by Stenström[2] that for wave-lengths greater than about 3 Å.U. reflected from crystals of sugar and gypsum, Bragg's relation

$$n\lambda = 2D \sin \theta$$

does not give accurately the angles of reflection. He interprets the difference as due to an appreciable refraction of the X-rays as they enter the crystal. Stenström's interpretation of his experiments has been criticized by Knipping[3], who tried to explain the discrepancy as due to a particular spatial arrangement of the atoms in crystals; but a more careful analysis by Ewald,[4] has shown that such an hypothesis is inadequate to explain the result. In fact, Ewald's calculations show a quantitative agreement between the discrepancies observed by Stenström and those calculated on the hypothesis of refraction.

The fact that Bragg's law cannot be strictly true seems to have been pointed out first by Darwin,[5] who gives for the difference between the observed glancing angle θ and the angle θ_0 anticipated from Bragg's formula,

$$\theta - \theta_0 = (1-\mu)/\sin \theta \cos \theta.$$

---

[1] C. G. Barkla, Phil. Mag. 31, 257 (1916).
[2] Stenström, dissertation, Lund (1919).
[3] P. Knipping, Zeits. f. Phys. 1, 40 (1920).
[4] P. P. Ewald, Phys. Zeits. 21, 617 (1920).
[5] C. G. Darwin, Phil. Mag. 27, 318 (1914).
* See Note 2, page 48.

a more useful expression for determining the index of refraction from these measurements is

$$1-\mu = \frac{\lambda_1 - \lambda_2}{\lambda_1} \cdot \frac{n_2^2}{n_2^2 - n_1^2} \sin^2\theta_1, \qquad (43)$$

where λ_1 and λ_2 are the apparent wave-lengths as measured in the n_1 and n_2 orders respectively. If the index of refraction of the crystal is known, the true wave-length can be calculated from the formula

$$\lambda = 2D \frac{\sin\theta}{n}\left(1 - \frac{1-\mu}{\sin^2\theta}\right). \qquad (44)$$

Duane and Patterson[1] and Siegbahn[2] have noticed that even with ordinary X-rays the wave-lengths observed in different orders do not agree. Thus for the tungsten $L\alpha$ line, $\lambda = 1.473$ Å., Duane and Patterson find $\lambda_1 - \lambda_2 = .00015$, whence $1-\mu$ for the calcite crystal used is $+8 \times 10^{-6}$. Similar measurements on $\lambda = 1.279$ and 1.096 Å. give $1-\mu = 10$ and 3×10^{-6} respectively. From equation (40) the corresponding theoretical values of $1-\mu$ are about 9, 7 and 5×10^{-6} respectively. Thus the index of refraction is less than unity and is close to the theoretical value.

Total Reflection of X-rays.—Since the refractive index is less than unity, a beam of X-rays striking a plane surface at a sufficiently large angle of incidence should be totally reflected. The critical glancing angle is given by

$$\cos\theta = \mu$$

or

$$\sin\theta = \sqrt{2}\sqrt{1-\mu}. \qquad (45)$$

For $\lambda = 1$ Å., the value of $1-\mu$ for a substance of density 3 is given by equation (40) as about 4×10^{-6}, in which case $\theta = 9.6$ minutes of arc—a readily measurable deflection.

The writer has tried the experiment of reflecting the tungsten line $\lambda = 1.279$ Å from surfaces of glass and of silver coated with lacquer.[3] The results are shown in Figure 13, which shows the intensity of the reflected beam at different angles θ. The theoretical values of the critical angle are calculated by equation (45), using the index of refraction given by Lorentz's formula in its more exact form. In view of the difficulties in measuring these small angles and the uncertainties with regard to the densities of the surfaces, these experiments are in surprising accord with the theory. The experimental values for $\lambda = 1.279$ are for crown glass, density 2.52, $1-\mu = 5.0 \times 10^{-6}$, and for a thin silver film on glass, $1-\mu = 20.9 \times 10^{-6}$.

Further experiments showed that within the critical angle the re-

---

[1] Duane and Patterson, Phys. Rev. 16, 532 (1920).
[2] M. Siegbahn, C. R. 173, 1350 (1921); 174, 745 (1920).
[3] A. H. Compton, paper before American Phys. Soc. Apr. 22, 1922.

flection was indeed specular and nearly total, and that the quantity $1-\mu$ for wave-lengths greater than .5 Å. U. is at least roughly proportional to the square of the wave-length. Thus the experiments on the refraction of X-rays are all in accord with the classical electron theory.

If the number of effective electrons per atom is assumed equal to the atomic number, equation (40) gives a moderately accurate means of

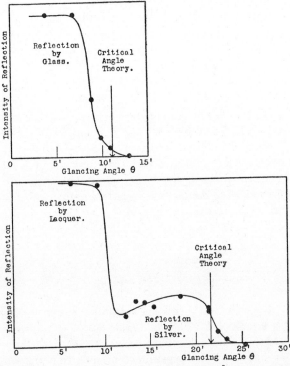

FIGURE 13. The total reflection of wave-length 1.279 Å. from various surfaces, showing the theoretical and experimental values of the critical angle.

measuring the wave-length, which is independent both of any crystalline grating space and of the quantum hypothesis. Or assuming the wave-length to be known, the measurements of the critical angle show that the number of electrons per atom affected by X-rays is probably within 5 per cent of the atomic number. This estimate has a very decided advantage over similar estimates made from X-ray scattering, since the intensity of scattering is affected by the grouping and size as well as the number of the electrons, while the refractive index depends only upon the number of electrons per unit volume (if no appreciable resonance occurs). These X-ray refraction measurements therefore afford a valuable confirmation of our estimates of the wave-length of X-rays and of the number of electrons in the atoms.

V. Some Applications of These Secondary Radiations to Physical Problems

1. That the secondary radiations produced by X-rays constitute a most effective tool for examining the structure of matter is apparent when one recalls that the very definite knowledge we have concerning the number of the electrons in matter and the arrangement of atoms in crystals came first from this source. The basis of the power of this tool lies in the fact that the wave-length of X-rays is comparable with the dimensions of the portions of matter we are studying. It follows that a study of scattered X-rays is capable of giving much the same information about atoms and electrons as may be given about dust particles by a study of the light which they scatter. The information is almost as definite as if we could look into the atom with a super-microscope which, since it employs X-rays or γ-rays instead of light, has a resolving power 10,000 to 100,000 times that of the best optical microscope. In studying the nature of radiation, the secondary X-rays present an equally valuable method, since for these high frequencies the magnitude of an energy quantum is relatively large. It is consequently possible in the X-ray region to distinguish more clearly between those phenomena which involve quantum principles and those which are to be explained on the basis of the classical electrodynamics.

It would be impossible in the compass of such a report as this to discuss adequately the information afforded concerning these major physical problems by the study of secondary X-rays. Nevertheless, certain problems which may be attacked with some success by this method may profitably be mentioned in order to give an idea of what can be accomplished. Since the applications of secondary X-rays to the problem of the structure of matter assume a knowledge of the nature of the X-rays, we shall first discuss the nature of radiation from the standpoint of X-rays.

2. *Nature of Radiation.*—Let us recall that it has been found possible to account satisfactorily for many of the experiments on the scattering of X-rays on the basis of the classical electrodynamics (cf. supra, pp. 4 to 11). The inadequacy of the quantum conception as applied to the scattering of radiation may be shown in the following manner. Experiment shows that a cathode electron at impact may give rise to one quantum of radiant energy of frequency $\nu = Ve/h$, where V is the potential applied to the tube, or to several quanta of lower frequency. Let us consider the case where $\nu > Ve/2h$, so that not more than one quantum of energy of this frequency can be radiated at the impact of each electron. We find that the resulting X-ray when scattered by matter shows the phenomenon of excess scattering. But this phenomenon is to be

accounted for by the fact that the incident X-ray *excites secondary radiation of the same frequency from several electrons in the same atom*, with the result that these radiations cooperate in the forward direction and partially interfere with each other at larger angles with the incident beam.

If, however, it is assumed that each electron both absorbs and emits radiation in quanta, the primary ray can excite radiation from only one electron, and these interference effects become inexplicable. If the radiation is gradually absorbed by all the electrons traversed and if each electron radiates when its energy exceeds the quantum, it is possible that a single quantum in the primary ray might occasionally liberate several quanta of secondary energy (though of course on the average not more than one quantum could be liberated). The chance that these several quanta would all be radiated by electrons in the same atom would, however, be so small as to be wholly inadequate to account for the observed interference effects. The same difficulty arises in explaining the reflection of X-rays by crystals, where in order to account for the appearance of the reflected beam as a sharp line it is necessary to suppose that a large number of electrons in the crystal emit radiation in their proper phases when excited by a single quantum of incident radiation. Thus the interference phenomena occurring in the scattering and reflection of X-rays are inconsistent with the view that an electron always emits scattered radiation in quanta.

There remains the possibility that the radiation itself always occurs in quanta, but that when scattering occurs the quantum of energy is radiated not by a single electron but by some group of electrons, affected by the incident wave. On this view it is a matter of probability whether or not an incident ray shall pass through a scattering body. If it passes through, it remains undiminished as a whole quantum; if it is scattered, some group of electrons in the body are effective, and the whole quantum is scattered, none passing through. This view also requires radiation of energy in discrete quanta, though now the energy quantum may be radiated by any number instead of by a single electron

3. It is important therefore to point out that a quantum of radiant energy cannot always retain its integrity, and that its parts may be separately scattered and absorbed. Hence scattered energy is not necessarily radiated in quanta, nor is radiation necessarily absorbed in integral quanta.

Perhaps the most convincing example of the division of radiation into parts of less than a quantum is that of the Michelson interferometer. We may suppose that one quantum of energy of wave-length λ strikes the semi-reflecting mirror, half being reflected to the movable mirror and half transmitted to the fixed one. The returning waves recombine at the semi-silvered surface, and in order that the energy quantum may remain complete, we may suppose that both components proceed together, either toward the eye or in the direction of the source. If the mirror is exactly half-silvered, the probability will be equal for the two directions. It is essential, however, that the initial division of the quantum occur at the half mirror, for if all the radiation during any given period went to either the movable or the fixed mirror, there would be nothing with which it could interfere on returning to the half mirror. It seems that the only possible interpretation of this interference is that while part of the energy quantum goes to the movable mirror, another part is simultaneously going to the fixed one. Thus neither part of the divided beam can carry the whole quantum of energy.

It remains to show that such a part of the original quantum can by itself be absorbed or scattered. Let us suppose the silvering of the mirrors is accurate, so that the interference of the rays at the eye is complete. Then imagine a thin absorption screen placed in the path of one of the divided beams. This absorption might for example be due to a slight tarnish on one of the mirrors. That absorption actually occurs is made evident by the incomplete interference of the recombined beam. But we have seen that neither part of the divided light beam possesses a whole quantum of energy. There accordingly appears to be no escape from the conclusion that radiation may be absorbed in amounts less than a quantum.

In a similar manner the light may be reflected or scattered from the surfaces of an interferometer mirror. It seems that in this scattering process we have an example of radiation of energy in smaller units than a quantum. The only escape from this conclusion would be to imagine a gradual absorption of energy by the electrons on the mirror surface, until a whole quantum is accumulated, and then a simultaneous emission of the radiation by the electrons each in its proper phase to produce interference effects. Such a view presents very grave difficulties.

A consideration of these and similar interference phenomena, whose importance must not be minimized, seems to lead with certainty to the conclusion that under certain conditions radiation does not occur in a definite direction, nor in definite quanta; that radiation may be absorbed in fractions of a quantum; and that, in the process of scattering at least, radiation may be emitted in fractions of a quantum. Most of the experiments on X-ray scattering and reflection, on the other hand, receive satisfactory explanation if the X-rays spread over a wide enough solid angle to excite oscillations of a large number of electrons in their proper phases. This study therefore supports the view that radiation occurs in waves spreading throughout space in accord with the usual electrical theory.

The experiments described above (p. 16), showing that the wave-length of the scattered X-ray is greater than that of the incident X-ray, present, however, a serious difficulty to this conclusion. This change in wave-length was found to receive quantitative explanation on the view that the radiation was received and emitted by each scattering electron in discrete quanta. No alternative explanation has as yet suggested itself. Nevertheless, the cogency of the argument based on interference phenomena is so great that it seems to me questionable whether the quantum interpretation of this experiment is the correct one

4. If then radiation may under certain conditions be emitted in infinitesimal fractions of a quantum, and if absorption is a continuous process, the question arises, has the quantum any real physical significance? To this our study of X-rays gives a definitely affirmative answer. Thus we have seen that de Broglie's and Ellis' experiments (supra p. 26) are in accord with the view that each photoelectron leaves its normal position in the atom with a kinetic energy $h\nu$, where ν is the frequency of the incident rays. Conversely, Duane and Hunt's experiments indicate that if the whole kinetic energy of a cathode ray is transformed to radiation at a single impact, then $E_{kinetic} = h\nu$. A similar relation also expresses quantitatively the frequency of the rays emitted as an electron falls from one energy level to another within the atom. It thus appears that the quantum law may describe a reversible mechanism whereby energy may be interchanged between radiation and the kinetic energy of an electron.

This mechanism is presumably that which is responsible for the fluorescent absorption of X-rays, since it is the energy thus absorbed which appears again as the kinetic energy of the photoelectrons. But the energy dissipated in scattering is not thus transformed, and need not therefore have any dependence upon the quantum mechanism. On this view there is no reason to question the application of the classical electrodynamics to the problem of scattering; so calculations on this basis may be used in studying the structure of matter. Such calculations may be employed with the greater confidence since we have found them capable of explaining the principal phenomena of X-ray scattering.

5. *The Structure of Matter.*—The advances in our knowledge of the structure of matter which have resulted from a study of secondary X-rays can merely be mentioned. Barkla in 1911 estimated the number of electrons in the atom necessary to account for the intensity of X-ray scattering on the basis of Thomson's theory. We have seen (supra, p. 4) that while his method calculation may be questioned, a more critical examination of the problem leads to about the same result: the number of electrons per atom is equal to about half the atomic weight.

Following Laue's discovery of X-ray diffraction, the work of the Braggs' and others has given us definite information concerning the arrangement of the individual atoms in crystals. Besides its value in crystallography, this study has thrown new light on the nature of cohesion, chemical valence, and other problems connected with the solid state. A most valuable method of pursuing this investigation has been developed by Debye and Scherrer,[1] Hull[2] and their co-workers, in which powdered crystals instead of large ones are employed. This method has made possible the study of many substances whose structure could not have been determined by the Bragg method.

The application of these secondary radiations to the study of atomic structure has hardly commenced. Experiments such as those of de Broglie and Ellis on the speed of the photoelectrons ejected by X-rays (supra, p. 26), tell us the energy with which the different electrons are held in the atom. Measurements of absorption coefficients and critical absorption wave-lengths make possible a determination of the number of electrons at each energy level in the atom (cf. supra, p. 41). The writer has shown[3] that a study of the intensity of X-ray spectra is capable of supplying rather definite information concerning the arrangement of the electrons in atoms; and recently Bragg, James and Bosanquet have made a rather extensive investigation from this standpoint of the distribution of the electrons in rock-salt.[4] The intensity of X-ray scattering by amorphous materials should lead to somewhat more definite results concerning this distribution; but such an investigation has not yet been seriously attempted. However, unless some unforeseen difficulty presents itself, it seems safe to predict that such studies will give us within a decade information concerning the arrangement of the electrons in the lighter atoms as definite as our present knowledge of the positions of the atoms in crystals.

It is difficult to overestimate the power of this new tool which is supplied us for the study of the structure of matter. The use of the tool has as yet hardly begun. But the successes already achieved give us hope that our knowledge of the nature of matter may thus be rapidly increased.

---

[1] Debye and Scherrer, Phys. Zeits. 17, 277 (1916) et. al.
[2] A. W. Hull, Phys. Rev. 10, 661 (1917) et. al.
[3] A. H. Compton, Phys. Rev. 9, 29 (1917).
[4] Bragg, James & Bosanquet, Phil. Mag. 41, 309 (1921); 42, 1 (1921).

RADIATION A FORM OF MATTER

To the Editor of Science: One sees the statement frequently made that, if one accepts Einstein's conclusion that the mass of a body is proportional to the total energy which it possesses, the principle of the conservation of matter must be abandoned. For if during any change energy is gained or lost by the body through radiation, there should be a corresponding gain or loss of mass. It has been calculated that in the case of radioactive disintegration the energy thus lost (or gained) through radiation represents an appreciable fraction of the total mass of the radioactive material. If, however, one takes the point of view that radiation is a form of matter, and that the amount of this matter is measured by the mass or inertia of the radiation, the total mass of the body plus that of the radiation emitted is unaltered by such changes. On this view the principle of the conservation of mass is strictly valid, being, as has been remarked, a corollary of the energy principle.

It is perhaps surprising to notice that according to the definitions of matter usually given electromagnetic radiation must be classed as matter. It is admittedly difficult to find a satisfactory definition. "Matter is that which occupies space," "matter is that which possesses mass or inertia," "matter is that which affects the senses," are, however, common statements. But radiation certainly occupies space; that it possesses mass is shown by the momentum which it imparts to a body which it strikes, producing radiation pressure; and who would deny that sunlight affects the senses? Unless, therefore, we change our idea of what is meant by the word "matter," this word includes not only solids, liquids and gases, but also the less tangible electromagnetic radiation.

The inclusion of radiation as a form of matter has important bearings in addition to the fact that it renews the validity of the principle of the conservation of matter. Thus, for example, we can no longer say that matter is composed wholly of positive and negative electrons, for the form of matter known as radiation includes no such electric charges. The statement that matter is composed of positive and negative electrons and electromagnetic radiation is, on the other hand, more complex than is required. We see rather that the fundamental thing in matter is not the electric charge but the electromagnetic field, for the electromagnetic field includes both the electrons and the radiation.

If the further simplification is made of considering the magnetic field as due to the electric field in motion, we may describe all forms of matter in terms of the intensity of the electric field at different points. The mass or inertia of the matter is proportional to the integral through the volume considered of the square of the electric intensity and of the magnetic intensity resulting from the motion of the electric field, whether this electric field is due to the presence of electrons or to the existence of electromagnetic radiation. The electric charge in an element of volume is proportional to the divergence of the electric intensity at the point. Thus all the fundamental properties of matter are determined if the intensity of the electric field throughout space and time is known. While the electrons can not be considered the fundamental elements which make up all matter, we have thus the intensity of the electric field as that which can be thought of as composing both the electrons and the radiation. Electric intensity, then, may be considered as that of which all matter is composed.

According to this point of view, matter is perfectly continuous. It is true that there are certain perhaps limited regions, the electrons, from which electric intensity diverges; but whether or not these regions of divergence are limited, the mass of the matter is associated with the electric intensity and is hence distributed through all space. Similarly, radiation propagated through space, as for example light coming from the sun to the earth, is on this view a continuous series of waves of matter. The old argument for the existence of an ether because some medium is necessary to transfer the radiant energy from the sun to the earth has accordingly no weight. For we now see that the radiation may be its own medium, somewhat as the stream of water from a hose acts as the medium for a wave if the nozzle is shaken.

Perhaps the only new thing in this letter is that, according to the common significance of the word, radiation must be considered a form of matter. But it has seemed to me that a consideration of this fact shows more clearly than we have seen before that matter is essentially continuous, and that the fundamental thing in matter is not the positive and negative electrons but is rather electric intensity.

ARTHUR H. COMPTON

WASHINGTON UNIVERSITY,
  ST. LOUIS, MISSOURI

41. The luminous efficiency of gases excited by electric discharge. ARTHUR H. COMPTON, Washington University, and C. C. VAN VOORHIS, Westinghouse Lamp Company.—A systematic investigation has been made of the light emitted by *twenty-four elementary gases* per unit energy consumed, when excited by an electric discharge at low pressure. The light was measured from a limited portion of the vacuum tube, and the corresponding electrical energy was determined by measuring the potential drop and the current in this part of the tube. The efficiencies vary from 0 in the case of arsenic vapor to 17 candles per watt in the case of sodium vapor. The maximum observed efficiencies of neon, mercury, and sodium are 1.8, 10, and 17 mean spherical candles per watt, respectively. These three are the only elements which give as high luminous efficiency as do present-day commercial illuminants. It is shown that it is very improbable that any gas composed of polyatomic molecules will have high efficiencies. A consideration of the spectral characteristics shows why the luminous efficiency of mercury vapor increases with the current density, while that of sodium vapor decreases when the current rises above a certain optimum value.

A QUANTUM THEORY OF THE SCATTERING OF X-RAYS BY LIGHT ELEMENTS

By Arthur H. Compton

Abstract

A quantum theory of the scattering of X-rays and γ-rays by light elements.—The hypothesis is suggested that when an X-ray quantum is scattered it spends all of its energy and momentum upon some particular electron. This electron in turn scatters the ray in some definite direction. The change in momentum of the X-ray quantum due to the change in its direction of propagation results in a recoil of the scattering electron. The energy in the scattered quantum is thus less than the energy in the primary quantum by the kinetic energy of recoil of the scattering electron. The corresponding *increase in the wave-length of the scattered beam* is $\lambda_\theta - \lambda_0 = (2h/mc) \sin^2 \frac{1}{2}\theta = 0.0484 \sin^2 \frac{1}{2}\theta$, where h is the Planck constant, m is the mass of the scattering electron, c is the velocity of light, and θ is the angle between the incident and the scattered ray. Hence the increase is independent of the wave-length. *The distribution of the scattered radiation* is found, by an indirect and not quite rigid method, to be concentrated in the forward direction according to a definite law (Eq. 27). The total energy removed from the primary beam comes out less than that given by the classical Thomson theory in the ratio $1/(1 + 2\alpha)$, where $\alpha = h/mc\lambda_0 = 0.0242/\lambda_0$. Of this energy a fraction $(1 + \alpha)/(1 + 2\alpha)$ reappears as scattered radiation, while the remainder is truly absorbed and transformed into kinetic energy of recoil of the scattering electrons. Hence, if σ_0 is the *scattering absorption coefficient* according to the classical theory, the coefficient according to this theory is $\sigma = \sigma_0/(1 + 2\alpha) = \sigma_s + \sigma_a$, where σ_s is the true scattering coefficient $[(1 + \alpha)\sigma/(1 + 2\alpha)^2]$, and σ_a is the coefficient of absorption due to scattering $[\alpha\sigma/(1 + 2\alpha)^2]$. Unpublished experimental results are given which show that for graphite and the Mo–K radiation the scattered radiation is longer than the primary, the observed difference ($\lambda_{\pi/2} - \lambda_0 = .022$) being close to the computed value .024. In the case of scattered γ-rays, the wave-length has been found to vary with θ in agreement with the theory, increasing from .022 A (primary) to .068 A ($\theta = 135°$). Also the velocity of secondary β-rays excited in light elements by γ-rays agrees with the suggestion that they are recoil electrons. As for the predicted variation of absorption with λ, Hewlett's results for carbon for wave-lengths below 0.5 A are in excellent agreement with this theory; also the predicted concentration in the forward direction is shown to be in agreement with the experimental results, both for X-rays and γ-rays. This remarkable *agreement between experiment and theory* indicates clearly that scattering is a quantum phenomenon and can be explained without introducing any new hypothesis as to the size of the electron or any new constants; also that a radiation quantum carries with it momentum as well as energy. The restriction to light elements is due to the assumption that the constraining forces acting on the scattering electrons are negligible, which is probably legitimate only for the lighter elements.

Spectrum of K-rays from Mo scattered by graphite, as compared with the spectrum of the primary rays, is given in Fig. 4, showing the change of wave-length.

Radiation from a moving isotropic radiator.—It is found that in a direction θ with the velocity, $I_\theta/I' = (1 - \beta)^2/(1 - \beta \cos \theta)^4 = (\nu_\theta/\nu')^4$. For the total radiation from a black body in motion to an observer at rest, $I/I' = (T/T')^4 = (\nu_m/\nu_m')^4$, where the primed quantities refer to the body at rest.

J. J. Thomson's classical theory of the scattering of X-rays, though supported by the early experiments of Barkla and others, has been found incapable of explaining many of the more recent experiments. This theory, based upon the usual electrodynamics, leads to the result that the energy scattered by an electron traversed by an X-ray beam of unit intensity is the same whatever may be the wave-length of the incident rays. Moreover, when the X-rays traverse a thin layer of matter, the intensity of the scattered radiation on the two sides of the layer should be the same. Experiments on the scattering of X-rays by light elements have shown that these predictions are correct when X-rays of moderate hardness are employed; but when very hard X-rays or γ-rays are employed, the scattered energy is found to be decidedly less than Thomson's theoretical value, and to be strongly concentrated on the emergent side of the scattering plate.

Several years ago the writer suggested that this reduced scattering of the very short wave-length X-rays might be the result of interference between the rays scattered by different parts of the electron, if the electron's diameter is comparable with the wave-length of the radiation. By assuming the proper radius for the electron, this hypothesis supplied a quantitative explanation of the scattering for any particular wave-length. But recent experiments have shown that the size of the electron which must thus be assumed increases with the wave-length of the X-rays employed,[1] and the conception of an electron whose size varies with the wave-length of the incident rays is difficult to defend.

Recently an even more serious difficulty with the classical theory of X-ray scattering has appeared. It has long been known that secondary γ-rays are softer than the primary rays which excite them, and recent experiments have shown that this is also true of X-rays. By a spectroscopic examination of the secondary X-rays from graphite, I have, indeed,

---

[1] A. H. Compton, Bull. Nat. Research Council, No. 20, p. 10 (Oct., 1922).

been able to show that only a small part, if any, of the secondary X-radiation is of the same wave-length as the primary.[1] While the energy of the secondary X-radiation is so nearly equal to that calculated from Thomson's classical theory that it is difficult to attribute it to anything other than true scattering,[2] these results show that if there is any scattering comparable in magnitude with that predicted by Thomson, it is of a greater wave-length than the primary X-rays.

Such a change in wave-length is directly counter to Thomson's theory of scattering, for this demands that the scattering electrons, radiating as they do because of their forced vibrations when traversed by a primary X-ray, shall give rise to radiation of exactly the same frequency as that of the radiation falling upon them. Nor does any modification of the theory such as the hypothesis of a large electron suggest a way out of the difficulty. This failure makes it appear improbable that a satisfactory explanation of the scattering of X-rays can be reached on the basis of the classical electrodynamics.

The Quantum Hypothesis of Scattering

According to the classical theory, each X-ray affects every electron in the matter traversed, and the scattering observed is that due to the combined effects of all the electrons. From the point of view of the quantum theory, we may suppose that any particular quantum of X-rays is not scattered by all the electrons in the radiator, but spends all of its energy upon some particular electron. This electron will in turn scatter the ray in some definite direction, at an angle with the incident beam. This bending of the path of the quantum of radiation results in a change in its momentum. As a consequence, the scattering electron will recoil with a momentum equal to the change in momentum of the X-ray. The energy in the scattered ray will be equal to that in the incident ray minus the kinetic energy of the recoil of the scattering electron; and since the scattered ray must be a complete quantum, the frequency will be reduced in the same ratio as is the energy. Thus on the quantum theory we should expect the wave-length of the scattered X-rays to be greater than that of the incident rays.

The effect of the momentum of the X-ray quantum is to set the

[1] In previous papers (Phil. Mag. 41, 749, 1921; Phys. Rev. 18, 96, 1921) I have defended the view that the softening of the secondary X-radiation was due to a considerable admixture of a form of fluorescent radiation. Gray (Phil. Mag. 26, 611, 1913; Frank. Inst. Journ., Nov., 1920, p. 643) and Florance (Phil. Mag. 27, 225, 1914) have considered that the evidence favored true scattering, and that the softening is in some way an accompaniment of the scattering process. The considerations brought forward in the present paper indicate that the latter view is the correct one.

[2] A. H. Compton, loc. cit., p. 16.

scattering electron in motion at an angle of less than 90° with the primary beam. But it is well known that the energy radiated by a moving body is greater in the direction of its motion. We should therefore expect, as is experimentally observed, that the intensity of the scattered radiation should be greater in the general direction of the primary X-rays than in the reverse direction.

The change in wave-length due to scattering.—Imagine, as in Fig. 1A,

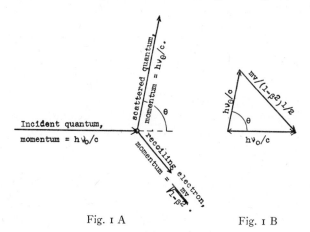

Fig. 1 A   Fig. 1 B

that an X-ray quantum of frequency ν_0 is scattered by an electron of mass m. The momentum of the incident ray will be $h\nu_0/c$, where c is the velocity of light and h is Planck's constant, and that of the scattered ray is $h\nu_\theta/c$ at an angle θ with the initial momentum. The principle of the conservation of momentum accordingly demands that the momentum of recoil of the scattering electron shall equal the vector difference between the momenta of these two rays, as in Fig. 1B. The momentum of the electron, $m\beta c/\sqrt{1-\beta^2}$, is thus given by the relation

$$\left(\frac{m\beta c}{\sqrt{1-\beta^2}}\right)^2 = \left(\frac{h\nu_0}{c}\right)^2 + \left(\frac{h\nu_\theta}{c}\right)^2 + 2\frac{h\nu_0}{c}\cdot\frac{h\nu_\theta}{c}\cos\theta, \quad (1)$$

where β is the ratio of the velocity of recoil of the electron to the velocity of light. But the energy $h\nu_\theta$ in the scattered quantum is equal to that of the incident quantum $h\nu_0$ less the kinetic energy of recoil of the scattering electron, i.e.,

$$h\nu_\theta = h\nu_0 - mc^2\left(\frac{1}{\sqrt{1-\beta^2}} - 1\right). \quad (2)$$

We thus have two independent equations containing the two unknown quantities β and ν_θ. On solving the equations we find

$$\nu_\theta = \nu_0/(1 + 2\alpha\sin^2\tfrac{1}{2}\theta), \quad (3)$$

where
$$\alpha = h\nu_0/mc^2 = h/mc\lambda_0. \tag{4}$$

Or in terms of wave-length instead of frequency,
$$\lambda_\theta = \lambda_0 + (2h/mc)\sin^2 \tfrac{1}{2}\theta. \tag{5}$$

It follows from Eq. (2) that $1/(1-\beta^2) = \{1 + \alpha[1-(\nu_\theta/\nu_0)]\}^2$, or solving explicitly for β
$$\beta = 2\alpha \sin \tfrac{1}{2}\theta \frac{\sqrt{1+(2\alpha+\alpha^2)\sin^2 \tfrac{1}{2}\theta}}{1+2(\alpha+\alpha^2)\sin^2 \tfrac{1}{2}\theta}. \tag{6}$$

Eq. (5) indicates an increase in wave-length due to the scattering process which varies from a few per cent in the case of ordinary X-rays to more than 200 per cent in the case of γ-rays scattered backward. At the same time the velocity of the recoil of the scattering electron, as calculated from Eq. (6), varies from zero when the ray is scattered directly forward to about 80 per cent of the speed of light when a γ-ray is scattered at a large angle.

It is of interest to notice that according to the classical theory, if an X-ray were scattered by an electron moving in the direction of propagation at a velocity $\beta'c$, the frequency of the ray scattered at an angle θ is given by the Doppler principle as
$$\nu_\theta = \nu_0 \bigg/ \left(1 + \frac{2\beta'}{1-\beta'}\sin^2 \tfrac{1}{2}\theta\right). \tag{7}$$

It will be seen that this is of exactly the same form as Eq. (3), derived on the hypothesis of the recoil of the scattering electron. Indeed, if $\alpha = \beta'/(1-\beta')$ or $\beta' = \alpha/(1+\alpha)$, the two expressions become identical. It is clear, therefore, that so far as the effect on the wave-length is concerned, we may replace the recoiling electron by a scattering electron moving in the direction of the incident beam at a velocity such that
$$\bar{\beta} = \alpha/(1+\alpha). \tag{8}$$

We shall call $\bar{\beta}c$ the "effective velocity" of the scattering electrons.

Energy distribution from a moving, isotropic radiator.—In preparation for the investigation of the spatial distribution of the energy scattered by a recoiling electron, let us study the energy radiated from a moving, isotropic body. If an observer moving with the radiating body draws a sphere about it, the condition of isotropy means that the probability is equal for all directions of emission of each energy quantum. That is, the probability that a quantum will traverse the sphere between the angles θ' and $\theta' + d\theta'$ with the direction of motion is $\tfrac{1}{2}\sin\theta' d\theta'$. But the surface which the moving observer considers a sphere (Fig. 2A) is

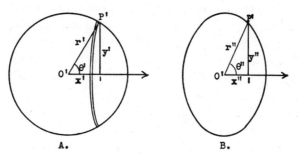

Fig. 2

considered by the stationary observer to be an oblate spheroid whose polar axis is reduced by the factor $\sqrt{1-\beta^2}$. Consequently a quantum of radiation which traverses the sphere at the angle θ', whose tangent is y'/x' (Fig. 2A), appears to the stationary observer to traverse the spheroid at an angle θ'' whose tangent is y''/x'' (Fig. 2B). Since $x' = x''/\sqrt{1-\beta^2}$ and $y' = y''$, we have

$$\tan \theta' = y'/x' = \sqrt{1-\beta^2}\, y''/x'' = \sqrt{1-\beta^2}\, \tan \theta'', \tag{9}$$

and

$$\sin \theta' = \frac{\sqrt{1-\beta^2}\, \tan \theta''}{\sqrt{1+(1-\beta^2)\tan^2 \theta''}}. \tag{10}$$

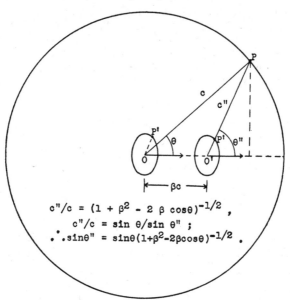

$c''/c = (1 + \beta^2 - 2\beta \cos\theta)^{-1/2}$,
$c''/c = \sin\theta/\sin\theta''$;
$\therefore \sin\theta'' = \sin\theta(1+\beta^2-2\beta\cos\theta)^{-1/2}$.

Fig. 3. The ray traversing the moving spheroid at P' at an angle θ'' reaches the stationary spherical surface drawn about O, at the point P, at an angle θ.

Imagine, as in Fig. 3, that a quantum is emitted at the instant $t = 0$, when the radiating body is at O. If it traverses the moving observer's sphere at an angle θ', it traverses the corresponding oblate spheroid, imagined by the stationary observer to be moving with the body, at an angle θ''. After 1 second, the quantum will have reached some point P on a sphere of radius c drawn about O, while the radiator will have moved a distance βc. The stationary observer at P therefore finds that the radiation is coming to him from the point O, at an angle θ with the direction of motion. That is, if the moving observer considers the quantum to be emitted at an angle θ' with the direction of motion, to the stationary observer the angle appears to be θ, where

$$\sin \theta / \sqrt{1 + \beta^2 - 2\beta \cos \theta} = \sin \theta'', \tag{11}$$

and θ'' is given in terms of θ' by Eq. (10). It follows that

$$\sin \theta' = \sin \theta \frac{\sqrt{1 - \beta^2}}{1 - \beta \cos \theta}. \tag{12}$$

On differentiating Eq. (12) we obtain

$$d\theta' = \frac{\sqrt{1 - \beta^2}}{1 - \beta \cos \theta} d\theta. \tag{13}$$

The probability that a given quantum will appear to the stationary observer to be emitted between the angles θ and $\theta + d\theta$ is therefore

$$P_\theta d\theta = P_{\theta'} d\theta' = \tfrac{1}{2} \sin \theta' d\theta',$$

where the values of $\sin \theta'$ and $d\theta'$ are given by Eqs. (12) and (13). Substituting these values we find

$$P_\theta d\theta = \frac{1 - \beta^2}{(1 - \beta \cos \theta)^2} \cdot \tfrac{1}{2} \sin \theta d\theta. \tag{14}$$

Suppose the moving observer notices that n' quanta are emitted per second. The stationary observer will estimate the rate of emission as

$$n'' = n' \sqrt{1 - \beta^2},$$

quanta per second, because of the difference in rate of the moving and stationary clocks. Of these n'' quanta, the number which are emitted between angles θ and $\theta + d\theta$ is $dn'' = n'' \cdot P_\theta d\theta$. But if dn'' per second are emitted at the angle θ, the number per second received by a stationary observer at this angle is $dn = dn''/(1 - \beta \cos \theta)$, since the radiator is approaching the observer at a velocity $\beta \cos \theta$. The energy of each quantum is, however, $h\nu_\theta$, where ν_θ is the frequency of the radiation as received by the stationary observer.[1] Thus the intensity, or the energy per unit area per unit time, of the radiation received at an angle θ and a distance R is

$$I_\theta = \frac{h\nu_\theta \cdot dn}{2\pi R^2 \sin\theta d\theta} = \frac{h\nu_\theta}{2\pi R^2 \sin\theta d\theta} \frac{n'(1-\beta^2)^{3/2}}{(1-\beta\cos\theta)^3} \frac{1}{2}\sin\theta d\theta \quad (15)$$
$$= \frac{n'h\nu_\theta}{4\pi R^2} \frac{(1-\beta^2)^{3/2}}{(1-\beta\cos\theta)^3}.$$

If the frequency of the oscillator emitting the radiation is measured by an observer moving with the radiator as ν', the stationary observer judges its frequency to be $\nu'' = \nu'\sqrt{1-\beta^2}$, and, in virtue of the Doppler effect, the frequency of the radiation received at an angle θ is

$$\nu_\theta = \nu''/(1-\beta\cos\theta) = \nu'[\sqrt{1-\beta^2}/(1-\beta\cos\theta)]. \quad (16)$$

Substituting this value of ν_θ in Eq. (15) we find

$$I_\theta = \frac{n'h\nu'}{4\pi R^2} \frac{(1-\beta^2)^2}{(1-\beta\cos\theta)^4}. \quad (17)$$

But the intensity of the radiation observed by the moving observer at a distance R from the source is $I' = n'h\nu'/4\pi R^2$. Thus,

$$I_\theta = I'[(1-\beta^2)^2/(1-\beta\cos\theta)^4] \quad (18)$$

is the intensity of the radiation received at an angle θ with the direction of motion of an isotropic radiator, which moves with a velocity βc, and which would radiate with intensity I' if it were at rest.[2]

It is interesting to note, on comparing Eqs. (16) and (18), that

$$I_\theta/I' = (\nu_\theta/\nu')^4. \quad (19)$$

[1] At first sight the assumption that the quantum which to the moving observer had energy $h\nu'$ will be $h\nu$ for the stationary observer seems inconsistent with the energy principle. When one considers, however, the work done by the moving body against the back-pressure of the radiation, it is found that the energy principle is satisfied. The conclusion reached by the present method of calculation is in exact accord with that which would be obtained according to Lorenz's equations, by considering the radiation to consist of electromagnetic waves.

[2] G. H. Livens gives for I_θ/I' the value $(1-\beta\cos\theta)^{-2}$ ("The Theory of Electricity," p. 600, 1918). At small velocities this value differs from the one here obtained by the factor $(1-\beta\cos\theta)^{-2}$. The difference is due to Livens' neglect of the concentration of the radiation in the small angles, as expressed by our Eq. (14). Cunningham ("The Principle of Relativity," p. 60, 1914) shows that if a plane wave is emitted by a radiator moving in the direction of propagation with a velocity βc, the intensity I received by a stationary observer is greater than the intensity I' estimated by the moving observer, in the ratio $(1-\beta^2)/(1-\beta)^2$, which is in accord with the value calculated according to the methods here employed.

The change in frequency given in Eq. (16) is that of the usual relativity theory. I have not noticed the publication of any result which is the equivalent of my formula (18) for the intensity of the radiation from a moving body.

This result may be obtained very simply for the total radiation from a black body, which is a special case of an isotropic radiator. For, suppose such a radiator is moving so that the frequency of maximum intensity which to a moving observer is ν_m' appears to the stationary observer to be ν_m. Then according to Wien's law, the apparent temperature T, as estimated by the stationary observer, is greater than the temperature T' for the moving observer by the ratio $T/T' = \nu_m/\nu_m'$. According to Stefan's law, however, the intensity of the total radiation from a black body is proportional to T^4; hence, if I and I' are the intensities of the radiation as measured by the stationary and the moving observers respectively,

$$I/I' = (T/T')^4 = (\nu_m/\nu_m')^4. \qquad (20)$$

The agreement of this result with Eq. (19) may be taken as confirming the correctness of the latter expression.

The intensity of scattering from recoiling electrons.—We have seen that the change in frequency of the radiation scattered by the recoiling electrons is the same as if the radiation were scattered by electrons moving in the direction of propagation with an effective velocity $\bar{\beta} = \alpha/(1 + \alpha)$, where $\alpha = h/mc\lambda_0$. It seems obvious that since these two methods of calculation result in the same change in wave-length, they must also result in the same change in intensity of the scattered beam. This assumption is supported by the fact that we find, as in Eq. 19, that the change in intensity is in certain special cases a function only of the change in frequency. I have not, however, succeeded in showing rigidly that if two methods of scattering result in the same relative wave-lengths at different angles, they will also result in the same relative intensity at different angles. Nevertheless, we shall assume that this proposition is true, and shall proceed to calculate the relative intensity of the scattered beam at different angles on the hypothesis that the scattering electrons are moving in the direction of the primary beam with a velocity $\bar{\beta} = \alpha/(1 + \alpha)$. If our assumption is correct, the results of the calculation will apply also to the scattering by recoiling electrons.

To an observer moving with the scattering electron, the intensity of the scattering at an angle θ', according to the usual electrodynamics, should be proportional to $(1 + \cos^2 \theta')$, if the primary beam is unpolarized. On the quantum theory, this means that the probability that a quantum will be emitted between the angles θ' and $\theta' + d\theta'$ is proportional to $(1 + \cos^2 \theta') \cdot \sin \theta' d\theta'$, since $2\pi \sin \theta' d\theta'$ is the solid angle included between θ' and $\theta' + d\theta'$. This may be written $P_{\theta'} d\theta' = k(1 + \cos^2 \theta') \sin \theta' d\theta'$.

The factor of proportionality k may be determined by performing the integration

$$\int_0^\pi P_{\theta'} d\theta' = k \int_0^\pi (1 + \cos^2 \theta') \sin \theta' d\theta' = 1,$$

with the result that $k = 3/8$. Thus

$$P_{\theta'} d\theta' = (3/8)(1 + \cos^2 \theta') \sin \theta' d\theta' \qquad (21)$$

is the probability that a quantum will be emitted at the angle θ' as measured by an observer moving with the scattering electron.

To the stationary observer, however, the quantum ejected at an angle θ' appears to move at an angle θ with the direction of the primary beam, where $\sin \theta'$ and $d\theta'$ are as given in Eqs. (12) and (13). Substituting these values in Eq. (21), we find for the probability that a given quantum will be scattered between the angles θ and $\theta + d\theta$,

$$P_\theta d\theta = \tfrac{3}{8} \sin \theta d\theta \frac{(1 - \beta^2)\{(1 + \beta^2)(1 + \cos^2 \theta) - 4\beta \cos \theta\}}{(1 - \beta \cos \theta)^4}. \qquad (22)$$

Suppose the stationary observer notices that n quanta are scattered per second. In the case of the radiator emitting n'' quanta per second while approaching the observer, the n''th quantum was emitted when the radiator was nearer the observer, so that the interval between the receipt of the 1st and the n''th quantum was less than a second. That is, more quanta were received per second than were emitted in the same time. In the case of scattering, however, though we suppose that each scattering electron is moving forward, the nth quantum is scattered by an electron starting from the same position as the 1st quantum. Thus the number of quanta received per second is also n.

We have seen (Eq. 3) that the frequency of the quantum received at an angle θ is $\nu_\theta = \nu_0/(1 + 2\alpha \sin^2 \tfrac{1}{2}\theta) = \nu_0/\{1 + \alpha(1 - \cos \theta)\}$, where ν_0, the frequency of the incident beam, is also the frequency of the ray scattered in the direction of the incident beam. The energy scattered per second at the angle θ is thus $nh\nu_\theta P_\theta d\theta$, and the intensity, or energy per second per unit area, of the ray scattered to a distance R is

$$I_\theta = \frac{nh\nu_\theta P_\theta d\theta}{2\pi R^2 \sin \theta d\theta}$$

$$= \frac{nh}{2\pi R^2} \cdot \frac{\nu_0}{1 + \alpha(1 - \cos \theta)} \cdot \frac{3}{8} \cdot \frac{(1 - \beta^2)\{(1 + \beta^2)(1 + \cos^2 \theta) - 4\beta \cos \theta\}}{(1 - \beta \cos \theta)^4}.$$

Substituting for β its value $\alpha/(1 + \alpha)$, and reducing, this becomes

$$I = \frac{3nh\nu_0}{16\pi R} \frac{(1 + 2\alpha)\{1 + \cos^2 \theta + 2\alpha(1 + \alpha)(1 - \cos \theta)^2\}}{(1 + \alpha - \alpha \cos \theta)^5}. \qquad (23)$$

In the forward direction, where $\theta = 0$, the intensity of the scattered beam is thus

$$I_0 = \frac{3}{8\pi} \frac{nh\nu_0}{R^2} (1 + 2\alpha). \tag{24}$$

Hence

$$\frac{I_\theta}{I_0} = \frac{1}{2} \frac{1 + \cos^2\theta + 2\alpha(1 + \alpha)(1 - \cos\theta)^2}{\{1 + \alpha(1 - \cos\theta)\}^5}. \tag{25}$$

On the hypothesis of recoiling electrons, however, for a ray scattered directly forward, the velocity of recoil is zero (Eq. 6). Since in this case the scattering electron is at rest, the intensity of the scattered beam should be that calculated on the basis of the classical theory, namely,

$$I_0 = I(Ne^4/R^2m^2c^4), \tag{26}$$

where I is the intensity of the primary beam traversing the N electrons which are effective in scattering. On combining this result with Eq. (25), we find for the intensity of the X-rays scattered at an angle θ with the incident beam,

$$I_\theta = I \frac{Ne^4}{2R^2m^2c^4} \frac{1 + \cos^2\theta + 2\alpha(1 + \alpha)(1 - \cos\theta)^2}{\{1 + \alpha(1 - \cos\theta)\}^5}. \tag{27}$$

The calculation of the energy removed from the primary beam may now be made without difficulty. We have supposed that n quanta are scattered per second. But on comparing Eqs. (24) and (26), we find that

$$n = \frac{8\pi}{3} \frac{INe^4}{h\nu_0 m^2 c^4 (1 + 2\alpha)}.$$

The energy removed from the primary beam per second is $nh\nu_0$. If we define *the scattering absorption coefficient* as the fraction of the energy of the primary beam removed by the scattering process per unit length of path through the medium, it has the value

$$\sigma = \frac{nh\nu_0}{I} = \frac{8\pi}{3} \frac{Ne^4}{m^2 c^4} \cdot \frac{1}{1 + 2\alpha} = \frac{\sigma_0}{1 + 2\alpha}, \tag{28}$$

where N is the number of scattering electrons per unit volume, and σ_0 is the scattering coefficient calculated on the basis of the classical theory.[1]

In order to determine the total energy truly scattered, we must integrate the scattered intensity over the surface of a sphere surrounding the scattering material, i.e., $\epsilon_s = \int_0^\pi I_\theta \cdot 2\pi R^2 \sin\theta d\theta$. On substituting the value of I_θ from Eq. (27), and integrating, this becomes

$$\epsilon_s = \frac{8\pi}{3} \frac{INe^4}{m^2 c^4} \frac{1 + \alpha}{(1 + 2\alpha)^2}.$$

[1] Cf. J. J. Thomson, "Conduction of Electricity through Gases," 2d ed., p. 325.

The *true scattering coefficient* is thus

$$\sigma_s = \frac{8\pi}{3}\frac{Ne^4}{m^2c^4}\frac{1+\alpha}{(1+2\alpha)^2} = \sigma_0\frac{1+\alpha}{(1+2\alpha)^2}. \qquad (29)$$

It is clear that the difference between the total energy removed from the primary beam and that which reappears as scattered radiation is the energy of recoil of the scattering electrons. This difference represents, therefore, a type of true absorption resulting from the scattering process. The corresponding *coefficient of true absorption due to scattering* is

$$\sigma_a = \sigma - \sigma_s = \frac{8\pi}{3}\frac{Ne^4}{m^2c^4}\frac{\alpha}{(1+2\alpha)^2} = \sigma_0\frac{\alpha}{(1+2\alpha)^2}. \qquad (30)$$

Experimental Test.

Let us now investigate the agreement of these various formulas with experments on the change of wave-length due to scattering, and on the magnitude of the scattering of X-rays and γ-rays by light elements.

Wave-length of the scattered rays.—If in Eq. (5) we substitute the accepted values of h, m, and c, we obtain

$$\lambda_\theta = \lambda_0 + 0.0484 \sin^2 \tfrac{1}{2}\theta, \qquad (31)$$

if λ is expressed in Angström units. It is perhaps surprising that the increase should be the same for all wave-lengths. Yet, as a result of an extensive experimental study of the change in wave-length on scattering, the writer has concluded that "over the range of primary rays from 0.7 to 0.025 A, the wave-length of the secondary X-rays at 90° with the incident beam is roughly 0.03 A greater than that of the primary beam which excites it."[1] Thus the experiments support the theory in showing a wave-length increase which seems independent of the incident wave-length, and which also is of the proper order of magnitude.

A quantitative test of the accuracy of Eq. (31) is possible in the case of the characteristic K-rays from molybdenum when scattered by graphite. In Fig. 4 is shown a spectrum of the X-rays scattered by graphite at right angles with the primary beam, when the graphite is traversed by X-rays from a molybdenum target.[2] The solid line represents the spectrum of these scattered rays, and is to be compared with the broken line, which represents the spectrum of the primary rays, using the same slits and crystal, and the same potential on the tube. The primary spectrum is, of course, plotted on a much smaller scale than

[1] A. H. Compton, Bull. N. R. C., No. 20, p. 17 (1922).
[2] It is hoped to publish soon a description of the experiments on which this figure is based.

the secondary. The zero point for the spectrum of both the primary and secondary X-rays was determined by finding the position of the first order lines on both sides of the zero point.

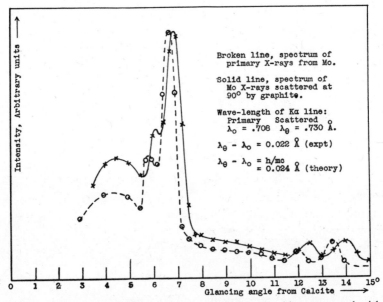

Fig. 4. Spectrum of molybdenum X-rays scattered by graphite, compared with the spectrum of the primary X-rays, showing an increase in wave-length on scattering.

It will be seen that the wave-length of the scattered rays is unquestionably greater than that of the primary rays which excite them. Thus the Kα line from molybdenum has a wave-length 0.708 A. The wave-length of this line in the scattered beam is found in these experiments, however, to be 0.730 A. That is,

$$\lambda_\theta - \lambda_0 = 0.022 \text{ A} \quad \text{(experiment)}.$$

But according to the present theory (Eq. 5),

$$\lambda_\theta - \lambda_0 = 0.0484 \sin^2 45° = 0.024 \text{ A} \quad \text{(theory)},$$

which is a very satisfactory agreement.

The variation in wave-length of the scattered beam with the angle is illustrated in the case of γ-rays. The writer has measured [1] the mass absorption coefficient in lead of the rays scattered at different angles when various substances are traversed by the hard γ-rays from RaC. The mean results for iron, aluminium and paraffin are given in column 2 of Table I. This variation in absorption coefficient corresponds to a

[1] A. H. Compton, Phil. Mag. 41, 760 (1921).

difference in wave-length at the different angles. Using the value given by Hull and Rice for the mass absorption coefficient in lead for wave-length 0.122, 3.0, remembering [1] that the characteristic fluorescent absorption τ/ρ is proportional to λ^3, and estimating the part of the absorption due to scattering by the method described below, I find for the wave-lengths corresponding to these absorption coefficients the values given in the fourth column of Table I. That this extrapolation is very

TABLE I

Wave-length of Primary and Scattered γ-rays

	Angle	μ/ρ	τ/ρ	λ obs.	λ calc.
Primary..........	0°	.076	.017	0.022 A	(0.022 A)
Scattered.........	45°	.10	.042	.030	0.029
"	90°	.21	.123	.043	0.047
"	135°	.59	.502	.068	0.063

nearly correct is indicated by the fact that it gives for the primary beam a wave-length 0.022 A. This is in good accord with the writer's value 0.025 A, calculated from the scattering of γ-rays by lead at small angles,[2] and with Ellis' measurements from his β-ray spectra, showing lines of wave-length .045, .025, .021 and .020 A, with line .020 the strongest.[3] Taking $\lambda_0 = 0.022$ A, the wave-lengths at the other angles may be calculated from Eq. (31). The results, given in the last column of Table I., and shown graphically in Fig. 5, are in satisfactory accord with the measured values. There is thus good reason for believing that Eq. (5) represents accurately the wave-length of the X-rays and γ-rays scattered by light elements.

Velocity of recoil of the scattering electrons.—The electrons which recoil in the process of the scattering of ordinary X-rays have not been observed. This is probably because their number and velocity is usually small compared with the number and velocity of the photoelectrons ejected as a result of the characteristic fluorescent absorption. I have pointed out elsewhere,[4] however, that there is good reason for believing that most of the secondary β-rays excited in light elements by the action of γ-rays are such recoil electrons. According to Eq. (6), the velocity of these electrons should vary from 0, when the γ-ray is scattered forward, to $v_{max} = \beta_{max} c = 2c\alpha[(1 + \alpha)/(1 + 2\alpha + 2\alpha^2)]$, when the γ-ray quantum

[1] Cf. L. de Broglie, Jour. de Phys. et Rad. 3, 33 (1922); A. H. Compton, Bull. N. R. C., No. 20, p. 43 (1922).
[2] A. H. Compton, Phil. Mag. 41, 777 (1921).
[3] C. D. Ellis, Proc. Roy. Soc. A, 101, 6 (1922).
[4] A. H. Compton, Bull. N. R. C., No. 20, p. 27 (1922).

is scattered backward. If for the hard γ-rays from radium C, $\alpha = 1.09$, corresponding to $\lambda = 0.022$ A, we thus obtain $\beta_{max} = 0.82$. The effective velocity of the scattering electrons is, therefore (Eq. 8), $\bar{\beta} = 0.52$. These results are in accord with the fact that the average velocity of the

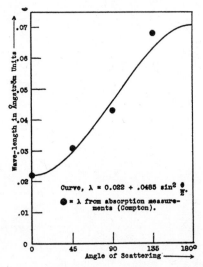

Fig. 5. The wave-length of scattered γ-rays at different angles with the primary beam, showing an increase at large angles similar to a Doppler effect.

β-rays excited by the γ-rays from radium is somewhat greater than half that of light.[1]

Absorption of X-rays due to scattering.—Valuable information concerning the magnitude of the scattering is given by the measurements of the absorption of X-rays due to scattering. Over a wide range of wave-lengths, the formula for the total mass absorption, $\mu/\rho = \kappa\lambda^3 + \sigma/\rho$, is found to hold, where μ is the linear absorption coefficient, ρ is the density, κ is a constant, and σ is the energy loss due to the scattering process. Usually the term $\kappa\lambda^3$, which represents the fluorescent absorption, is the more important; but when light elements and short wave-lengths are employed, the scattering process accounts for nearly all the energy loss. In this case, the constant κ can be determined by measurements on the longer wave-lengths, and the value of σ/ρ can then be estimated with considerable accuracy for the shorter wave-lengths from the observed values of μ/ρ.

Hewlett has measured the total absorption coefficient for carbon over a wide range of wave-lengths.[2] From his data for the longer wave-

[1] E. Rutherford, Radioactive Substances and their Radiations, p. 273.
[2] C. W. Hewlett, Phys. Rev. **17**, 284 (1921).

lengths I estimate the value of κ to be 0.912, if λ is expressed in A. On subtracting the corresponding values of $\kappa\lambda^3$ from his observed values of μ/ρ, the values of σ/ρ represented by the crosses of Fig. 6 are obtained.

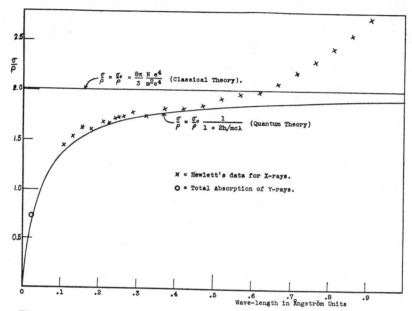

Fig. 6. The absorption in carbon due to scattering, for homogeneous X-rays.

The value of σ_0/ρ as calculated for carbon from Thomson's formula is shown by the horizontal line at $\sigma/\rho = 0.201$. The values of σ/ρ calculated from Eq. (28) are represented by the solid curve. The circle shows the experimental value of the total absorption of γ-rays by carbon, which on the present view is due wholly to the scattering process.

For wave-lengths less than 0.5 A, where the test is most significant, the agreement is perhaps within the experimental error. Experiments by Owen,[1] Crowther,[2] and Barkla and Ayers[3] show that at about 0.5 A the "excess scattering" begins to be appreciable, increasing rapidly in importance at the longer wave-lengths.[4] It is probably this effect which results in the increase of the scattering absorption above the theoretical value for the longer wave-lengths. Thus the experimental values of the absorption due to scattering seem to be in satisfactory accord with the present theory.

True absorption due to scattering has not been noticed in the case of

[1] E. A. Owen, Proc. Camb. Phil. Soc. **16**, 165 (1911).
[2] J. A. Crowther, Proc. Roy. Soc. **86**, 478 (1912).
[3] Barkla and Ayers, Phil. Mag. **21**, 275 (1911).
[4] Cf. A. H. Compton, Washington University Studies, **8**, 109 ff. (1921).

X-rays. In the case of hard γ-rays, however, Ishino has shown [1] that there is true absorption as well as scattering, and that for the lighter elements the true absorption is proportional to the atomic number. That is, this absorption is proportional to the number of electrons present, just as is the scattering. He gives for the true mass absorption coefficient of the hard γ-rays from RaC in both aluminium and iron the value 0.021. According to Eq. (30), the true mass absorption by aluminium should be 0.021 and by iron, 0.020, taking the effective wave-length of the rays to be 0.022 A. The difference between the theory and the experiments is less than the probable experimental error.

Ishino has also estimated the true mass scattering coefficients of the hard γ-rays from RaC by aluminium and iron to be 0.045 and 0.042 respectively.[2] These values are very far from the values 0.193 and 0.187 predicted by the classical theory. But taking $\lambda = 0.022$ A, as before, the corresponding values calculated from Eq. (29) are 0.040 and 0.038, which do not differ seriously from the experimental values.

It is well known that for soft X-rays scattered by light elements the total scattering is in accord with Thomson's formula. This is in agreement with the present theory, according to which the true scattering coefficient σ_s approaches Thomson's value σ_0 when $\alpha \equiv h/mc\lambda$ becomes small (Eq. 29).

The relative intensity of the X-rays scattered in different directions with the primary beam.—Our Eq. (27) predicts a concentration of the energy in the forward direction. A large number of experiments on the scattering of X-rays have shown that, except for the excess scattering at small angles, the ionization due to the scattered beam is symmetrical on the emergence and incidence sides of a scattering plate. The difference in intensity on the two sides according to Eq. (27) should, however, be noticeable. Thus if the wave-length is 0.7 A, which is probably about that used by Barkla and Ayers in their experiments on the scattering by carbon,[3] the ratio of the intensity of the rays scattered at 40° to that at 140° should be about 1.10. But their experimental ratio was 1.04, which differs from our theory by more than their probable experimental error.

It will be remembered, however, that our theory, and experiment also, indicates a difference in the wave-length of the X-rays scattered in different directions. The softer X-rays which are scattered backward are the more easily absorbed and, though of smaller intensity, may produce an

[1] M. Ishino, Phil. Mag. **33**, 140 (1917).
[2] M. Ishino, loc. cit.
[3] Barkla and Ayers, loc. cit.

ionization equal to that of the beam scattered forward. Indeed, if α is small compared with unity, as is the case for ordinary X-rays, Eq. (27) may be written approximately $I_\theta/I_\theta' = (\lambda_0/\lambda_\theta)^3$, where I_θ' is the intensity of the beam scattered at the angle θ according to the classical theory. The part of the absorption which results in ionization is however proportional to λ^3. Hence if, as is usually the case, only a small part of the X-rays entering the ionization chamber is absorbed by the gas in the chamber, the ionization is also proportional to λ^3. Thus if i_θ represents the ionization due to the beam scattered at the angle θ, and if i_θ' is the corresponding ionization on the classical theory, we have $i_\theta/i_\theta' = (I_\theta/I_\theta')(\lambda_\theta/\lambda_0)^3 = 1$, or $i_\theta = i_\theta'$. That is, to a first approximation, the ionization should be the same as that on the classical theory, though the energy in the scattered beam is less. This conclusion is in good accord with the experiments which have been performed on the scattering of ordinary X-rays, if correction is made for the excess scattering which appears at small angles.

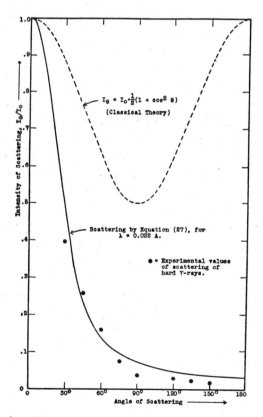

Fig. 7. Comparison of experimental and theoretical intensities of scattered γ-rays.

In the case of very short wave-lengths, however, the case is different. The writer has measured the γ-rays scattered at different angles by iron, using an ionization chamber so designed as to absorb the greater part of even the primary γ-ray beam.[1] It is not clear just how the ionization due to the γ-rays will vary with the wave-length under the conditions of the experiment, but it appears probable that the variation will not be great. If we suppose accordingly that the ionization measures the intensity of the scattered γ-ray beam, these data for the intensity are represented by the circles in Fig. 7. The experiments showed that the intensity at 90° was 0.074 times that predicted by the classical theory, or 0.037 I_0, where I_0 is the intensity of the scattering at the angle $\theta = 0$ as calculated on either the classical or the quantum theory. The absolute intensities of the scattered beam are accordingly plotted using I_0 as the unit. The solid curve shows the intensity in the same units, calculated according to Eq. (27). As before, the wave-length of the γ-rays is taken as 0.022 A. The beautiful agreement between the theoretical and the experimental values of the scattering is the more striking when one notices that there is not a single adjustable constant connecting the two sets of values.

Discussion

This remarkable agreement between our formulas and the experiments can leave but little doubt that the scattering of X-rays is a quantum phenomenon. The hypothesis of a large electron to explain these effects is accordingly superfluous, for all the experiments on X-ray scattering to which this hypothesis has been applied are now seen to be explicable from the point of view of the quantum theory without introducing any new hypotheses or constants. In addition, the present theory accounts satisfactorily for the change in wave-length due to scattering, which was left unaccounted for on the hypothesis of the large electron. From the standpoint of the scattering of X-rays and γ-rays, therefore, there is no longer any support for the hypothesis of an electron whose diameter is comparable with the wave-length of hard X-rays.

The present theory depends essentially upon the assumption that each electron which is effective in the scattering scatters a complete quantum. It involves also the hypothesis that the quanta of radiation are received from definite directions and are scattered in definite directions. The experimental support of the theory indicates very convincingly that a radiation quantum carries with it directed momentum as well as energy.

Emphasis has been laid upon the fact that in its present form the

---

[1] A. H. Compton, Phil. Mag. **41**, 758 (1921).

quantum theory of scattering applies only to light elements. The reason for this restriction is that we have tacitly assumed that there are no forces of constraint acting upon the scattering electrons. This assumption is probably legitimate in the case of the very light elements, but cannot be true for the heavy elements. For if the kinetic energy of recoil of an electron is less than the energy required to remove the electron from the atom, there is no chance for the electron to recoil in the manner we have supposed. The conditions of scattering in such a case remain to be investigated.

The manner in which interference occurs, as for example in the cases of excess scattering and X-ray reflection, is not yet clear. Perhaps if an electron is bound in the atom too firmly to recoil, the incident quantum of radiation may spread itself over a large number of electrons, distributing its energy and momentum among them, thus making interference possible. In any case, the problem of scattering is so closely allied with those of reflection and interference that a study of the problem may very possibly shed some light upon the difficult question of the relation between interference and the quantum theory.

Many of the ideas involved in this paper have been developed in discussion with Professor G. E. M. Jauncey of this department.

WASHINGTON UNIVERSITY,
SAINT LOUIS,
December 13, 1922

The Total Reflexion of X-Rays. By ARTHUR H. COMPTON, Ph.D., *Wayman Crow Professor of Physics, Washington University, Saint Louis.*

1. *Deviations from Bragg's Law.*

THOUGH direct attempts to measure the index of refraction of different substances of X-rays have hitherto failed, deviations from Bragg's relation,

$$n\lambda = 2D \sin \theta,$$

have been observed which have been ascribed to an appreciable refraction of the X-rays as they enter the crystal. These deviations were first observed by Stenström[†], using wave-lengths greater than 2·5 Å.U. Duane and Patterson[‡] and Siegbahn[§] have also noticed that even with ordinary X-rays the wave-lengths calculated from measurements in the different orders of reflexion do not exactly agree with each other. Stenström's conclusion that these experiments indicate appreciable refraction was criticized by Knipping[∥]; but a more careful analysis by Ewald[¶] shows a quantitative agreement between the discrepancies observed by Stenström and those calculated on the refraction hypothesis.

The fact that Bragg's law cannot be strictly true seems to have been pointed out first by Darwin[**], who gives for the difference between the observed glancing angle θ and the angle θ_0 anticipated from Bragg's formula,

$$\theta - \theta_0 = \delta/\sin\theta \cos\theta.$$

Here
$$\delta = 1 - \mu,$$

where μ is the index of refraction of the X-rays in the crystal. An expression which gives the index of refraction from measurements of the glancing angle for different orders of reflexion is

$$\delta = \frac{\lambda_1 - \lambda_2}{\lambda_2} \cdot \frac{n_2^2}{n_2^2 - n_1^2} \sin^2 \theta_1, \quad \ldots \quad (1)$$

where λ_1 and λ_2 are the apparent wave-lengths as measured

[†] W. Stenström, Dissertation, Lund (1919).
[‡] Duane and Patterson, *Phys. Rev.* xvi. p. 532 (1920).
[§] M. Siegbahn, *Comptes Rendus*, clxxiii. p. 1350 (1921); pp. 174, 745 (1922).
[∥] P. Knipping, *Zeits. f. Phys.* i. p. 40 (1920).
[¶] P. P. Ewald, *Phys. Zeits.* xxi. p. 617 (1920).
[**] C. G. Darwin, *Phil. Mag.* xxvii. p. 318 (1914).

in the n_1 and n_2 orders respectively, and θ_1 is the glancing-angle for the n_1 order*. If the index of refraction of the crystal is known, the true wave-length can be calculated from the formula,

$$n\lambda = 2\mathrm{D} \sin \theta \left(1 - \frac{\delta}{\sin^2 \theta}\right). \quad \ldots \quad (2)$$

Duane and Patterson † find an apparent difference in the wave-length of the tungsten Lα line, $\lambda = 1\cdot473$ Å., as observed in the first and second order from calcite, of $\lambda_1 - \lambda_2 = 0\cdot00015$ Å. According to equation (1), this means an index of refraction *less* than unity by $\delta = 8 \times 10^{-6}$. Their similar measurements on $1\cdot279$ and $1\cdot096$ Å. give $\delta = 10$ and 3×10^{-6} respectively.

2. *Drude-Lorentz Theory of Dispersion.*

According to the Drude-Lorentz theory of dispersion ‡, the index of refraction of a medium for radiation of very high frequency is given approximately by

$$\delta = 1 - \mu = \Sigma \frac{n_r e^2}{2\pi m (\nu^2 - \nu_r^2)} \quad \ldots \quad (3)$$

Here n_r represents the number of electrons per unit volume which vibrate with the natural frequency ν_r, e and m have their usual significance, ν is the frequency of the radiation, and the summation is taken over the different types r of the electrons. If, as in the case of ordinary X-rays traversing matter composed of light atoms, the frequency of the

---

* Darwin's investigation shows that, in spite of the fact that the wave-length is less than the distance between the individual atoms, refraction problems may be treated in the usual manner. If then θ and λ are the glancing angle and wave-length respectively outside the crystal, and θ' and λ' the corresponding angle and wave-length within the crystal,

$$\frac{\lambda}{\lambda'} = \mu; \quad \text{and} \quad \frac{\cos \theta}{\cos \theta'} = \mu, \quad \text{or} \quad \frac{\sin \theta'}{\sin \theta} = \frac{1}{\mu}\left(1 - \frac{1-\mu}{\sin^2 \theta}\right).$$

But Bragg has shown that

$$n\lambda' = 2\mathrm{D} \sin \theta';$$

whence

$$n\lambda = 2\mathrm{D} \sin \theta \left(\frac{1-\mu}{\sin^2 \theta}\right),$$

which is equation (2). This gives the true wave-length in terms of the known grating space and the observed glancing angle. To determine the index of refraction, the glancing angles θ_1 and θ_2 for two different orders n_1 and n_2 must be measured. This gives two equations (2), from which the unknown wave-length λ may be eliminated, giving, after some reduction, equation (1).

† Duane and Patterson, *loc. cit.*

‡ *Cf., e.g.*, H. A. Lorentz, 'The Theory of Electrons,' 2nd ed. p. 149.

radiation is considerably higher than the natural frequency of the electrons, this becomes very nearly,

$$\delta = \frac{ne^2}{2\pi m\nu^2}, \quad \ldots \ldots (4)$$

where n is the total number of electrons per unit volume of the medium.

If X-rays of wave-length 1·473 Å. traverse calcite, this expression (4) indicates an index of refraction which is less than unity by about 8×10^{-6}. For $\lambda = 1·279$, $\delta = 6 \times 10^{-6}$, and for $\lambda = 1·096$, $\delta = 4·5 \times 10^{-6}$. The close correspondence between these theoretical values and the values calculated from Duane and Patterson's experiments can leave little doubt but that the consistent changes in apparent wave-length with order which they observe are due to refraction.

3. Observation of Total Reflexion of X-rays.

Since the refractive index is less than unity, a beam of X-rays striking a plane surface at a sufficiently large angle of incidence should be totally reflected. The critical glancing angle is given by

$$\cos\theta = \mu,$$

or

$$\sin\theta = \sqrt{2\delta}.$$

For 1·279 Å., the value of δ for crown glass of density 2·52 is given by equation (4) as $5·2 \times 10^{-6}$, in which case $2\theta = 22$ minutes of arc, which is a readily measurable deflexion*.

Fig. 1.

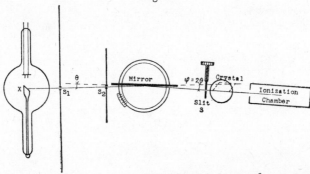

In searching for the totally reflected beam, the apparatus was set up as shown diagrammatically in fig. 1. The beam of X-rays passing through the slits S_1 and S_2 had an effective

* In this calculation, the composition of crown glass was assumed to be $CaO \cdot Na_2O \cdot 2SiO_2$. However, for a given density, variations in the composition affect only very slightly the index of refraction as calculated from equation (4).

width of about 2 minutes of arc. The mirror was a piece of ordinary plate glass, of density 2·52, mounted upon the prism table of a spectrometer. The vernier of the spectrometer was used only to check the accuracy of a lamp and scale system, which was set at such a distance from the mirror that a rotation of the mirror through an angle of $\theta = 1$ minute gave a deflexion of 1 mm. The slit S could be moved by a micrometer screw through any desired angle ϕ, while the opening into the ionization chamber was wide enough to admit rays coming from the mirror through a range of about 1·5 degrees.

The zero position of the slit S was determined by removing the mirror and adjusting the slit S for maximum deflexion. The zero position of the mirror could not be determined by the maximum deflexion, since, as we shall see, the beam reflected from the surface is nearly as intense as the primary, so that the maximum deflexion is not given when the front surface of the mirror just bisects the primary ray, but when part of the primary ray is reflected from its front surface. A more satisfactory method of determining the mirror's zero point was to turn it until the visual image of the slit S_1 coincided with the real slit.

In searching for the predicted total reflexion, the slit S was moved several minutes of arc from the zero position, and the ionization current noted for various positions of the mirror. The results thus obtained, for different positions of the movable slit, are shown in fig. 2. For example, when the slit was at an angle $\phi = 12'$, on turning the mirror from zero to $12'$ the intensity of the X-rays entering the chamber increased from zero at small angles to a maximum at $6'$ and then fell to zero at larger angles. If the X-rays entering the chamber were diffusely scattered, they should have been of maximum intensity when the mirror was at the smallest glancing angle at which all the primary beam would impinge upon its front surface, *i.e.* at about $4'$. Thus the fact that the X-rays are in every case of maximum intensity when the angle $\phi = 2\theta$ shows that there is an approximately specular reflexion of the X-rays at these small angles. The increasing breadth of the reflected beam at the larger angles shows that the reflexion is not perfectly regular. It appears rather that the glass surface reflects the X-rays in much the same manner that the air above a heated street reflects light to form a mirage. The fact that no appreciable reflexion is observed at angles as great as $\phi = 18'$ means that for the greater part of the X-ray beam the critical glancing angle θ is less than $9'$.

Fig. 2.

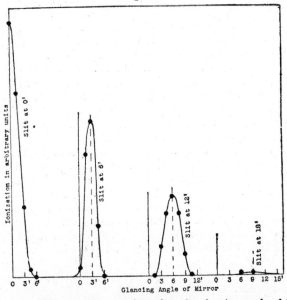

The maximum intensity occurring when the glancing angle of the mirror is half that of the slit indicates specular reflexion.

Fig. 3.

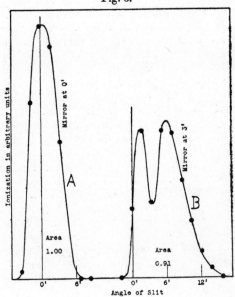

The reflected energy is nearly equal to the incident energy, indicating total reflexion.

If the observed phenomenon is really one of total reflexion, the energy in the reflected beam should be nearly as great as that in the primary beam. A measure of the total energy in each beam was obtained by integrating the observed ionization as the slit S was moved across the beam. Curve A of fig. 3 was thus obtained when the mirror was set at zero, and curve B when at 3′. This small glancing angle was necessary in order not to exceed the critical angle for the shorter wave-lengths. In order to satisfy this condition with greater certainty, the tube was operated at the comparatively low potential of about 12,000 peak volts. But the angle was so small that part of the primary beam was not intercepted by the mirror, giving rise to a sharp peak at 1′·5 in curve B. The ratio of the areas of the two curves, 0·91, shows that the greater part of the incident energy is indeed reflected.

4. *The Index of Refraction for X-rays.*

In order to test quantitatively the theoretical dispersion formula for these very high frequencies, it was necessary to determine the critical glancing angle for a known wavelength. For this purpose the calcite crystal C, on the crystal table of a Bragg spectrometer, was placed in the path of the reflected beam, so that its spectrum could be studied. The procedure then consisted in measuring the intensity of a given spectrum line in the beam reflected from the mirror, and in finding the angle $\phi = 2\theta$ beyond which the line fails to appear. The results of these experiments are shown in fig. 4. Curve A shows the intensity of the reflexion of the tungsten L line $\lambda = 1\cdot279$ Å. by the plate-glass mirror at various angles $\phi = 2\theta$. It will be seen that the line has almost disappeared from the reflected beam at a critical glancing angle of about 20′. Curve B shows the intensity of the same line as reflected from the surface of a silvered glass mirror coated with a thin film of lacquer. It is apparent that the critical angle for the lacquer is passed at about 22′, while the line finally disappears not far from the angle 43′ predicted by Lorentz's theory as the critical glancing angle for silver. From these curves we may take the critical glancing angle for glass as $\theta = \phi/2 =$ about 10′, for lacquer as about 11′, and for silver as about 22′·5. These values correspond to indices of refraction less than unity by $4\cdot2 \times 10^{-6}$ in the case of glass, $5\cdot1 \times 10^{-6}$ in the case of lacquer, and $21\cdot5 \times 10^{-6}$ in the case of silver.

The spectrum of the rays reflected from glass at the angle

$\phi = 8'$ is shown in fig. 5, curve A, as compared with the spectrum of the primary beam, curve B. The ordinates of the spectrum of the reflected rays are magnified by a factor of 3. It will be noticed that whereas the spectrum of the primary rays begins at $\theta = 2°\cdot1$, corresponding to a wavelength of 0·22 Å., the spectrum of the reflected rays begins rather gradually at about 4°·5, and apparently, if the slits

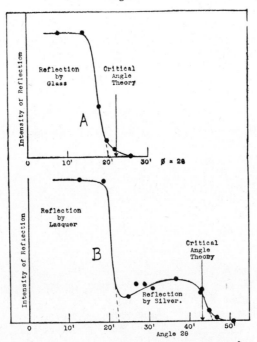

Fig. 4.

Total reflexion and critical angles for $\lambda = 1·279$ Å.

were sufficiently narrow, would begin at about 5°·0. Thus $\phi/2 = 4'$ is the critical angle for a wave-length of about ·52 Å., corresponding to a value of $1 - \mu = 0·9 \times 10^{-6}$. This value is not, however, as reliable as the values obtained from the spectrum line.

5. Comparison of the Experiments with Theory.

The calculation of the index of refraction of crown glass is not difficult, since the frequency of the radiation employed is considerably higher than that of the K absorption band of the heaviest element, calcium, in the glass. Consequently, the simple formula (4) can be applied with very little error. In the case of silver, however, the wave-length employed, 1·279 Å., is intermediate between the critical K and L absorption frequencies. It is thus necessary to take into account the natural frequencies of the electrons in both

Fig. 5.

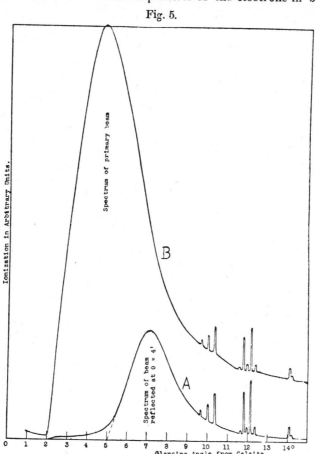

of the inner rings. It is surprising, however, how small a difference the corresponding degree of resonance makes. Thus if the middle of the K absorption band is taken as ·39 Å., that of the L series as 2·9 Å., the number of electrons in the K ring as 2, and in the L ring as 8, the value of δ, according to equation (3), is $19 \cdot 8 \times 10^{-6}$; but the approximate equation (2) gives the corresponding value as $20 \cdot 3 \times 10^{-6}$, a difference of only 2 per cent.

In Table I. are given the results of the experimental measurements of the index of refraction as compared with the theoretical values for glass, silver, and lacquer.

TABLE I.

Substance.	Density.	Wave-length Å.U.	Critical angle θ, expt.	$\delta = 1-\mu$, experiment.	$\delta = 1-\mu$, theory.
Glass......	2·52	1·279	10'	$4·2 \times 10^{-6}$	$5·2 \times 10^{-6}$
Glass......	2·52	·52	4'	$0·9 \times 10^{-6}$	$0·7 \times 10^{-6}$
Silver ...	10·5	1·279	22'·5	$21·5 \times 10^{-6}$	$19·8 \times 10^{-6}$
Lacquer..	—	1·279	11'·0	$5·1 \times 10^{-6}$	—

It is at first sight surprising that the value of δ should be greater for lacquer than for glass. Some recent experiments by Miss Laird have indicated*, however, that the density of thin films of celluloid increases considerably as the films become very thin, rising from the value of 1·41 for thick films to greater than 2·5 for the thinnest films measured. In view of the hydrogen content of these organic substances, it thus appears that the electron density in these dense thin films is greater than in glass. Thus in every case the experimental values of the index of refraction are in satisfactory accord with the usual electron theory of dispersion †.

6. *Application to Physical Problems.*

This establishment of the theoretical dispersion formula makes it possible to calculate with accuracy the index of refraction for these very high frequencies by substances containing only the lighter atoms. There remains some uncertainty with regard to the natural frequencies of the K and L electrons in the heavier elements; but as we have seen in the case of silver, errors from this cause can hardly be serious. Fortunately, the crystals most used in X-ray spectroscopy are composed of relatively light elements. In the case of calcite, formula (4) gives the value of $1-\mu$ as $3·67\lambda^2 \times 10^{-6}$, where λ is expressed in Ångström units. As compared with formula (3), this expression is accurate to 1 per cent. for wave-lengths less than 2 Å., and is accordingly reliable for corrections of the wave-length throughout the usual X-ray spectrum. The corresponding expression for

---

* E. R. Laird, Phys. Rev. xix. p. 384 (1922).

† There is an apparent tendency for the experimental value of $1-\mu$ for glass to fall below its theoretical value, though the difference is hardly more than the experimental error. If the difference is real, it may indicate the existence of a thin layer of matter of low density, such as water or oxygen, on the surface of the glass.

rock-salt is $3·04 \lambda^2 \times 10^{-6}$, which should also be accurate to 1 per cent. over the same range. Using these values for the index of refraction, the true wave-length can be calculated from the glancing angle according to equation (2).

A matter of considerable interest is the prominence of the L series spectrum lines in the spectrum of the totally reflected rays. In the primary beam, operating the tube at about 70,000 peak volts and using a wide opening into the ionization chamber, the continuous background is more prominent than even the strongest of the L series lines. But when wave-lengths shorter than $0·5$ Å. are absent, as they are in the reflected beam, the spectrum lines stand out much more prominently. In fact, the continuous background has almost disappeared. It is obvious from this result that the principal part of the continuous background in this region consists of the higher orders of reflexion of the shorter wave-lengths, and that comparatively little energy of wave-lengths greater than 1 Å. occurs in the general radiation.

This experiment illustrates the use of the phenomenon of total reflexion as a means of filtering the *short* wave-lengths from a beam of X-rays, a result which cannot be secured by absorption filters. Unfortunately the method can be applied only in the case of very narrow beams of X-rays, having a breadth of the order of a few minutes of arc. But in the case of such beams, the reflexion is so nearly total that the reduction in intensity is not great, while the critical glancing angle is so well defined that the shorter wave-lengths are rather sharply cut off. For use as such a filter, a mirror with a platinum surface is to be recommended, because its index of refraction differs so greatly from unity that the critical angles are comparatively large.

The existence of an index of refraction of the magnitude predicted by the classical electron theory would seem irreconcilable with the view that one quantum of X-rays can affect but a single electron. For there would seem to be no possibility of refraction unless the ray can spend a part of its energy in setting in vibration some of the electrons over which it passes so as to give rise to a secondary ray which will combine with the primary train. But this involves the idea that parts of the quantum are capable of affecting the traversed electrons. It is not easy to reconcile this result with such experiments as that of the softening of scattered X-rays[*], which seem to demand that each electron affected by the primary rays receives both the momentum and the energy of a whole quantum.

[*] *Cf.* A. H. Compton, Bulletin Nat. Research Council, No. 20, p. 16 (1922).

It is of value to note that equation (4) supplies a means of determining with fair accuracy the wave-length of an X-ray beam from an experimental measurement of its index of refraction, if the number of electrons per unit volume is known. If the number of effective electrons per atom is taken equal to the atomic number, as X-ray scattering experiments show is at least approximately the case, the measured value $\lambda = 4 \cdot 2 \times 10^{-6}$ for glass means a wave-length of $1 \cdot 18$ Å. This value is independent of any knowledge of crystalline structure or of the energy quantum, and so gives a very valuable confirmation of the wave-length $1 \cdot 28$ Å. calculated on the basis of other more accurate though less fundamental methods.

Perhaps of greater importance is the fact that, knowing the wave-length, it is possible to calculate from the observed index of refraction the number of electrons per atom which are affected by the incident X-rays. In calculating the theoretical value of the index of refraction from formula (3), the number of electrons per atom has been taken equal to the atomic number. From the accuracy with which the index as thus calculated agrees with the experiments, it follows that this assumption is probably correct to within about 5 per cent.

The two most direct methods hitherto employed for determining the number of electrons in the outer part of the atom are the study of the scattering of alpha rays by atomic nuclei and the measurement of the intensity of scattered X-rays. The first method determines the charge on the atomic nucleus, and it is not certain *à priori* that the charge thus measured is identical with the charge on the electrons which are responsible for high frequency radiation. The intensity of the scattered X-rays, on the other hand, though determined by the electrons exterior to the nucleus, is a function not only of their number but also of their arrangement and size. The index of refraction is theoretically independent of the arrangement and size of the electrons, and depends upon their resonance only to the same extent as does the scattering. Thus an estimate of the number of mobile electrons in an atom based upon a measurement of the index of refraction is more reliable than a similar estimate based upon these other methods. It is accordingly gratifying that such a measurement confirms the result of the earlier work, indicating that the number of electrons per atom which are affected by high frequency radiations is equal to the atomic number.

Washington University,
St. Louis,
Dec. 6, 1922.

Letters to the Editor.

Recoil of Electrons from Scattered X-Rays.

IN a recent paper before the Royal Society (as reported in NATURE, July 7, p. 26), C. T. R. Wilson announced that in his cloud expansion pictures of secondary β-rays produced by X-rays shorter than 0.5 Å, tracks of very short range appear. These electrons, he says, "are ejected nearly along the direction of the primary X-rays."

A quantum theory of the scattering of X-rays, devised primarily to account for the change in wave-length which occurs when X-rays are scattered, led me to predict (Bulletin National Research Council, No. 20, pp. 19 and 27, October 1922) that electrons should be ejected from atoms whenever X-rays are scattered. The idea is that a quantum of radiation is scattered in a definite direction by an individual electron. The change in momentum of the radiation, due to its change in direction, results in a recoil of the electron which deflects the ray. The direction of recoil is not far from that of the primary beam, in accord with Wilson's observation on his short tracks.

Corresponding to this momentum acquired by the electron, it has kinetic energy which varies from 0 when the scattered X-ray proceeds forward, to a maximum value $h\nu \cdot 2a/(1+2a)$, when the ray is scattered backward (P. Debye, Phys. Zeitschr. 24, 161, Apr. 15, 1923; A. H. Compton, Phys. Rev. 21, 486, May 1923). Here $a = \gamma/\lambda$, where $\gamma = h/mc = 0.0242$ Å, and λ is the incident wave-length. The ratio of the maximum energy of a photoelectron excited by an X-ray to the maximum energy of such a recoil electron would thus be $(1+2a)/2a$. But Wilson finds the length of the trails proportional to the square of the energy. The track due to the photoelectron should therefore be $(1+2a)^2/4a^2$ times that of the longest recoil electron tracks.

Taking Wilson's datum that a track of 1 cm. corresponds to 21,000 volts, the equation $Ve = hc/\lambda$ indicates that a ray of wave-length 0.5 Å will eject a photoelectron with a path of 1.4 cm. The recoil electron, taking $a = 0.0242/0.5$, should accordingly have a range of 0.11 mm., which should just be visible. For his harder X-rays, with a wave-length for example of 0.242 Å ($a = 0.1$), the recoil tracks on Wilson's photographs should be as long as 1.7 mm. The quantum idea of X-ray scattering thus leads to recoil electrons moving in the right direction and possessing energy which is of the same order of magnitude as that possessed by the electrons responsible for C. T. R. Wilson's very short tracks.

ARTHUR H. COMPTON.

University of Chicago,
August 4.

From the PHILOSOPHICAL MAGAZINE, vol. xlvi. *November* 1923.

Absorption Measurements of the Change of Wave-Length accompanying the Scattering of X-Rays. By ARTHUR H. COMPTON, *Wayman Crow Professor of Physics, Washington University, Saint Louis.*

IN some recent papers[*] the writer has described spectroscopic experiments which have shown that when the characteristic X-rays from molybdenum are scattered by graphite, the wave-length of the X-rays is increased. While these spectroscopic investigations have been made for only two wave-lengths, ·708 and ·630 Å., a quantum theory of the phenomenon has been developed[†] which predicts that a similar change in wave-length should occur whatever the wave-length of the primary beam. Absorption measurements on scattered γ-rays have indicated a change in wave-length of about the theoretical amount [‡], but interferometer measurements on light scattered by paraffin have failed to show any effect of this character [§]. Apparently, therefore, the change in wave-length due to scattering depends in some way upon the wave-length of the primary radiation used. The present experiments, in which the change in wave-length was measured by an absorption method, have as their primary object to test the theory over a wider range of wave-lengths, and for a greater variety of scattering materials than could be done conveniently by the spectroscopic method.

The quantum theory of this change in wave-length is based upon the hypothesis that each quantum of primary X-rays is scattered by an individual electron. If the frequency of the incident quantum is ν_0, its energy is $h\nu_0$, and its momentum is $h\nu_0/c$, where c is the velocity of light. Due to the change in direction of the quantum on scattering, its momentum is altered, resulting in a recoil of the scattering electron. Equating the momentum of recoil of the electron to the change in momentum of the quantum, we have,

$$\left\{\frac{m\beta c}{\sqrt{(1-\beta^2)}}\right\}^2 = \left(\frac{h\nu_0}{c}\right)^2 + \left(\frac{h\nu_\theta}{c}\right)^2 + 2\frac{h\nu_0}{c}\frac{h\nu_\theta}{c}\cos\theta. \quad . \quad (1)$$

---

[*] A. H. Compton, Bulletin National Research Council, xx. p. 16 (Oct. 1922): Paper before American Physical Society, April 28, 1923; Phys. Rev. June 1923; and Phys. Rev. xxii. (1923).
[†] A. H. Compton, Bull. N. R. C. xx. p. 18 (Oct. 1922); Paper before Am. Phys. Soc., Dec. 1, 1923; Phys. Rev. xxi. p. 207 (Dec. 1923) & xxi. p. 483 (May 1923). P. Debye, *Phys. Zeitschr.* xxviii. p. 161 (April 15, 1921).
[‡] A. H. Compton, Phil. Mag. (May 1921); Phys. Rev. xxii. (1923).
[§] P. A. Ross, Science, lvii. p. 614 (1923).

Here βc is the velocity with which the electron recoils, and ν_θ is the frequency of the rays scattered at the angle θ. But the energy of the scattered quantum is less than that of the incident quantum because of the energy spent in setting the scattering electron in motion. Thus,

$$h\nu_0 - h\nu_\theta = mc^2 \left\{ \frac{1}{\sqrt{(1-\beta^2)}} - 1 \right\} \quad \ldots \quad (2)$$

Combining these two equations we find,

$$\nu_\theta = \nu_0/1 + \alpha(1-\cos\theta), \quad \ldots \ldots \quad (3)$$

where
$$\alpha = h\nu_0/mc^2 = h/mc\lambda_0, \quad \ldots \ldots \quad (4)$$

or
$$\delta\lambda = \lambda_\theta - \lambda_0 = \gamma(1-\cos\theta), \quad \ldots \ldots \quad (5)$$

where
$$\gamma = h/mc = 0{\cdot}0242 \times 10^{-8} \text{ cm.}* \quad \ldots \quad (6)$$

Typical results of the spectrum measurements of this change in wave-length are shown in fig. 1, which represents, for slits of two different widths, the spectra of the Kα ray from molybdenum : (A) the primary ray, (B) as scattered by graphite at 45°, (C) as scattered at 90°, and (D) as scattered at 135° †. The line P is drawn in each case at the position of the primary line, and the line T at the theoretical position of the scattered line as given by equation (5). It will be noticed that, within a comparatively small probable error, the wave-length of one component of the scattered beam is exactly that predicted by this quantum theory. There remains, however, a part of the scattered beam which is unchanged in wave-length.

Experimental Method.

For the measurement of the difference in absorption coefficient between the primary and the scattered ray, a balance method was employed, as is shown diagrammatically in fig. 2. Two beams of X-rays from the target X of a Coolidge tube came through separate windows in the lead box. One of them was scattered by a radiator R into an ionization chamber I, and the other went directly through a slit of variable width S into a second ionization chamber I'.

---

* Since the mass of a quantum is $h\nu/c^2 = h/\lambda c$, the mass of a quantum of radiation of wave-length γ is $h/(h/mc)c = m$; *i.e.* a quantum of radiation of wave-length γ has a mass equal to that of the electron. This fact was pointed out to me by Dr. Eldridge through Prof. A. Sommerfeld.

† *Cf.* A. H. Compton, Phys. Rev. xxii. (1923).

One ionization chamber was kept at a positive and the other at a negative potential, so that with equal ionization currents

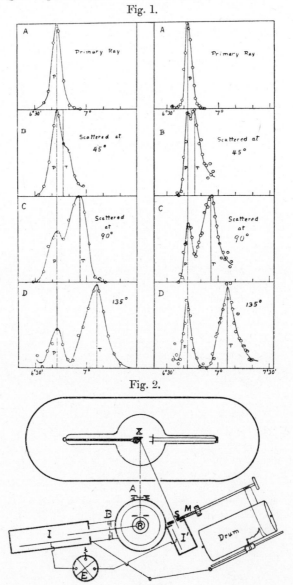

Fig. 1.

Fig. 2.

in each there was no deflexion on the electrometer E. The variable slit S was opened by a micrometer screw M.

The shaft driving this screw had wrapped about it a metal cord which was wrapped also about a drum in such a manner that the drum was rotated when the slit was opened by the micrometer screw. A recording pen moving along the drum was actuated by a metal cord fastened to the movable ionization chamber. Thus each position of the pen corresponded to a particular angle of the ionization chamber.

The chamber I was rotated through a range of angles of about 75° by a motor driving a worm gear. As the chamber was moving, the micrometer screw was turned so as to keep the electrometer at its zero position. In this manner a record was obtained on a sheet of paper placed around the drum showing the width of the slit for every angle of the

Fig. 3.

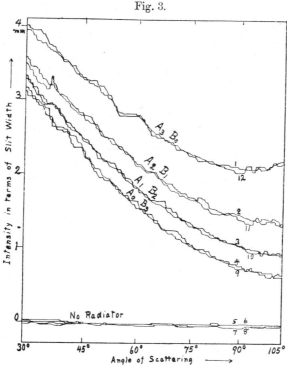

The scattering of hard X-rays by paraffin. In curve A_2B_1, 2 mm. of copper are in the path of the primary beam and 1 mm. in the path of the scattered beam, and similarly for the other curves. The fact that the intensity is greatest with the absorbing screen in the path of the primary beam shows that the wave-length of the X-rays is increased by scattering. The curves were made in the order of their numbers.

chamber I. A record of this type, copied in indian ink for reproduction, is shown in fig. 3, which represents the scattering of hard X-rays by a block of paraffin. In this figure the ordinates represent the width of the slit-opening, and the abscissæ represent the angle of the ionization chamber.

To interpret the graphs thus obtained, it was necessary to determine how the ionization in the chamber I' varied with the width of the slit. This was done by means of a sector disk made of lead which cut off a known and adjustable fraction of the primary X-ray beam. A calibration curve taken in this manner is shown in fig. 4, where the intensity

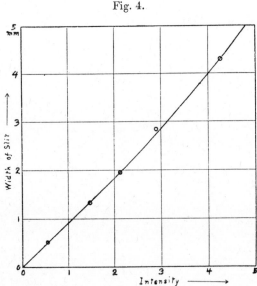

Fig. 4.

of the ionization is plotted against the width of the slit in millimetres. It will be seen that for slit-widths less than 2 mm. the calibration curve is sensibly a straight line, and that the departure from such a line is not large even for the greater widths.

The X-rays scattered into the chamber I were filtered through an absorption screen placed at A or B. It was necessary to place a similar absorption screen in the path of the rays entering I' in order that slight variations in the voltage applied to the X-ray tube should not destroy the balance between the two ionization currents. Of course, the beam entering the chamber I was much less intense than that entering the chamber I'. This difference in intensity was balanced by making the chamber I much larger and filling it with methyl-iodide vapour.

The change in absorption coefficient, corresponding to the change in wave-length of the scattered beam, was measured by observing the relative intensity of the scattered beam when an absorbing screen was transferred from position A to position B. If I_p is the intensity of the primary beam whose wave-length is λ and whose absorption coefficient in the absorbing screen is μ, and if S is the fraction of the energy of this beam which is scattered into the ionization chamber when no absorption screens are present, then the intensity of the scattered beam when a screen of thickness x is placed at A is $I_A = I_p S e^{-\mu x}$. If the absorption coefficient of the scattered ray is μ', the intensity of the scattered ray when the screen is placed at B is $I_B = I_p S e^{-\mu' x}$. Thus

$$\frac{I_A}{I_B} = e^{(\mu' - \mu)x}, \quad \ldots \ldots \ldots (7)$$

whence

$$\delta\mu = \mu' - \mu = \frac{1}{x}\log(I_A/I_B). \quad \ldots (8)$$

It is clear from equation (7) that if μ' were equal to μ, I_A would be equal to I_B. If the absorption coefficient of each component of a complex beam of X-rays were unchanged by scattering, its intensity should therefore be unaltered by moving the absorbing screen from A to B. Hence the intensity of the whole beam should also remain unchanged. Thus any change observed in the intensity of the scattered ray when the absorption screen is shifted from position A to position B represents a change in the absorption coefficient of the component rays when the X-rays are scattered *. The fact, which is shown clearly by fig. 3, that the intensity of the rays is greater when the screen is at A therefore means that μ' is greater than μ †, that is, that each component of the X-ray beam is softened during the scattering process.

* This conclusion is strictly true only in case the fraction of the rays which is scattered from the absorber placed at A onto the radiator is the same as the fraction scattered from the absorber when placed at B into the ionization chamber. The apparatus was so designed that this condition was at least very nearly fulfilled. However, even if uncorrected, the error from this source would have been very small.

† It is by no means always true, when heterogeneous X-rays are used, that the scattered ray is softer than the primary ray. For if the radiator is of considerable thickness, the more penetrating parts of the primary beam are scattered by the whole radiator while the softer components

Knowing the absorption coefficient of the primary beam, and the change in absorption coefficient due to scattering, one can determine the wave-length of both the primary and the scattered beam. For this purpose I have employed the absorption date for different wave-lengths given by Hewlett, Richtmyer, and Duane*.

Experimental Results.

The results of these absorption measurements are collected in Table I., under the head of observer 2, and the corresponding wave-lengths are exhibited in Table II. In each case the scattering material was in the form of a cylinder, 2 cm. in diameter, with walls of such thickness that more than half of the X-rays were transmitted. The wave-length changes are very consistent with each other in every case except that of the secondary radiation from lead for the effective primary wave-length $0\cdot14$ Å. In this case the change in wave-length, especially at the small angles, is considerably greater than the theory predicts. The difficulty obviously lies in the fact that the fluorescent K radiation from lead is being excited in large amounts, and that this secondary radiation is always softer than the ray which excites it. The measurements of the scattering by the other elements for wave-lengths $0\cdot12$ and $0\cdot13$ Å. are also slightly affected by the fluorescent K radiation from the lead slits, the tendency being to make the observed change in wave-length greater at small angles and less at large angles with the primary beam, just as in the case of the scattering by lead. These fluorescent rays are not excited appreciably when the effective wave-length of the primary beam is

---

are absorbed before they have penetrated very deeply. Because of this filtering process, it usually happens that the scattered ray is more penetrating than the primary. One cannot help but feel that this process may account in part for the small magnitude of the change in absorption observed by Barkla and Miss Sale (Phil. Mag. xlv. p. 758, 1923), even though they took the precaution of using thin sheets of paper as radiators. Changes in absorption due to scattering similar to those described here have been observed for γ-rays by Eve, Phil. Mag. viii. p. 669 (1904); R. D. Kleeman, Phil. Mag. xv. p. 638 (1908); Madsen, Phil. Mag. xvii. p. 423 (1909); D. C. H. Florance, Phil. Mag. xx. p. 921.(1910), xxvii. p. 225 (1914); J. A. Gray, Phil. Mag. xxvi. p. 611 (1913); A. H. Compton, Phil. Mag. xli. p. 749 (1921); *et al.* For X-rays the change has been observed by Sadler and Mesham, Phil. Mag. xxiv. p. 138 (1912); J. Laub, *Ann. de Phys.* xlvi. p. 785 (1915); J. A. Gray, Frank. Inst. Jour. p. 643, Nov. (1920); A. H. Compton, Phys. Rev. xviii. p. 96 (1921); Nature, cviii. p. 366 (1921); and J. A. Crowther, Phil. Mag. xlii. p. 719 (1921).

* These data are collected in Bulletin N. R. C. no. xx. p. 32 (1922).

904   Prof. Compton : *Absorption Measurements of Change*

TABLE I.

Mass Absorption Coefficients of Primary and Scattered X-Rays.

Radiator	Absorber	μ/ρ Primary	δ(μ/ρ), Scattered Ray.								Observer
			30°	45°	60°	75°	90°	105°	120°	135°	
Paraffin	Pb	·073	...	·027	...	...	...	...	...	·71	1
Aluminium	Pb	·073	...	·027	...	...	...	...	...	·43	1
Iron	Pb	·073	...	·037	...	...	...	...	...	·43	1
Tin	Pb	·073	...	·017	...	...	...	...	...	·25	1
Lead	Pb	·073	...	(?)	...	...	...	...	...	·08	1
Paraffin	Cu	·45	·043	·08	·12	·20	·245	·32	...	...	2
Graphite	Cu	·46	·08	·10	·12	·19	·15	·33	...	...	2
Aluminium	Cu	·45	·05	·08	·13	·19	·24	·30	...	...	2
Copper	Cu	·59	·04	·07	·10	·15	·19	·26	...	...	2
Lead	Cu	·64	·21	·27	·31	·32	·33	·34	...	...	2
Paraffin	Cu	·76	·065	·11	·17	·26	·34	·42	·50	·59	2
Graphite	Cu	·76	...	·105	·16	·23	·36	·46	·52	·59	2
Aluminium	Cu	·76	...	·10	·14	·21	·295	·35	·42	·53	2
Copper	Cu	·76	...	·06	·15	·20	·23	·32	·39	...	2
Tin	Cu	1·03	...	·07	·12	·17	·23	·36	...	...	2
Lead	Cu	1·05	...	·10	·14	·14	·19	·20	...	...	2
Paraffin	Cu	1·74	·09	·16	·29	·405	·52	·63	...	...	2
Paraffin	Cu	2·84	·15	·25	·33	·51	·60	·73	...	...	2
Paraffin	Cu	5·25	·17	·29	·50	·77	1·00	1·16	...	...	2
Paper	Al	·70–·90	...	...	...	...	·22	...	...	...	3
Paper	Al	1·15–1·83	...	...	...	...	·22	...	...	...	3
Paper	Al	1·90–4·42	...	...	...	...	·18	...	...	...	3
Paper	Al	4·8–10·3	...	...	...	...	·07	...	...	...	3

greater than 0·15 Å. The fluorescent K-rays from the copper absorbing screen would also have been a source of difficulty had not the differential absorption coefficients δ(μ/ρ) been measured after the scattered rays had been

TABLE II.

Change in Wave-Length Accompanying the Scattering of X-rays.

$\delta\lambda$ for Scattered Ray, A.U.

Radiator.	λ, A.U., Primary.	30°	45°	60°	75°	90°	105°	120°	135°	Theory.	Observer.
Free electron	All	·003	·007	·012	·018	·024	·030	·036	·041	·041	
Paraffin	·024		·006						·051		1
Aluminium	·024		·006						·039		1
Iron	·024		·006			·021			·029		1
Tin	·024		·007						·029		1
Lead	·024		·004						·013		1
Paraffin	·12	·005	·010	·013	·020	·024	·031				2
Graphite	·12	·010	·013	·015	·021	·027	·033				2
Aluminium	·12	·006	·010	·015	·020	·025	·030				2
Copper	·13	·004	·007	·010	·014	·017	·022				2
Lead	·14	·016	·021	·024	·025	·025	·026				2
Paraffin	·15	·004	·008	·012	·018	·023	·027	·033	·039		2
Graphite	·15		·007	·012	·017	·025	·031	·034	·038		2
Aluminium	·15		·007	·010	·016	·021	·025	·028	·035		2
Copper	·15		·005	·011	·015	·017	·023	·027			2
Tin	·17		·004	·007	·010	·013	·020				2
Lead	·17		·005	·008	·008	·011	·012				2
Paraffin	·21	·004	·008	·014	·018	·023	·027				2
Paraffin	·25	·005	·008	·012	·017	·020	·024				2
Paraffin	·32	·003	·006	·010	·015	·020	·023				2
Paper	·30–·33					·047					3
Paper	·41–·50					·023					3
Paper	·50–·69					·015					3
Graphite	·63					·020					4
Graphite	·71		·003			·018			·031		4
Paper	·70–·90					·003					3

filtered through at least ·5 mm. of copper. This precaution was sufficient also to eliminate the effect of the K-rays from tin when it was used as radiator.

For sake of completeness, I have included in Tables I.

and II. the results of some earlier experiments on γ-rays * (Observer 1), Barkla and Sale's recent experiments on the change in absorption coefficient of soft X-rays scattered by paper † (Observer 3), and my spectrum measurements on the change of wave-length of molybdenum K-rays ‡ (Observer 4). In the case of the γ-rays, the wave-lengths are calculated from the absorption coefficients according to the equation $\mu/\rho = \tau/\rho + \sigma/\rho$, where § $\tau/\rho = 1\cdot 64 \times 10^3 \lambda^3$ and ∥ $\sigma/\rho = \cdot 151/(1+\cdot 0485/\lambda)$. In Barkla and Sale's work, it did not seem possible to reproduce the results in successive series of experiments. I have accordingly averaged their results obtained for certain arbitrarily chosen ranges of wave-lengths, and have estimated the wave-lengths from Hewlett's data for the absorption of different wave-lengths in aluminium. The wave-length changes estimated from the spectrum measurements are the weighted mean values of the modified and the unmodified rays.

From Table II. it is apparent that in order that the scattered ray shall be changed in wave-length by the amount predicted by equation (5), X-rays of very short wave-length and radiators of low atomic number must be employed. These facts are exhibited in figs. 5 and 6. In fig. 5 is plotted the change in wave-length observed when X-rays of widely differing wave-lengths are scattered by paraffin ¶. For both wave-lengths 0·024 and 0·15 Å., the observed change is very nearly that predicted by the theory (as represented by the solid curve); but the change for $\lambda = 0\cdot 32$ Å. is slightly less, and that for $\lambda = 0\cdot 71$ is still less than the theoretical value. Similarly in fig. 6, whereas the change in wave-length of the rays scattered by carbon is within experimental error that demanded by theory, the wave-length change for the heavier elements becomes less and less as the atomic number becomes greater. The difference here shown between the rays

* The value of μ/ρ for the primary γ-rays in lead is that of M. Ishino, Phil. Mag. xxxiii. p. 140 (1917), and for the scattered rays is from A. H. Compton, Phil. Mag. xli. p. 760 (1921).

† C. G. Barkla & Rhoda Sale, Phil. Mag. xlv. p. 748 (1923).

‡ For $\lambda = 0\cdot 63$ (MoKβ-line, cf. A. H. Compton, Phys. Rev. xxi. p. 495 (1923). For $\lambda = 0\cdot 71$ (MoKα-line), cf. fig. 1 of this paper.

§ A. W. Hull & Marion Rice, Phys. Rev. viii. p. 836 (1916).

∥ The mass absorption due to scattering, according to the writer's quantum theory (Phys. Rev. xxi. p. 493, 1923), is $\sigma/\rho = \sigma_0/\rho(1+2\alpha)$, where σ_0/ρ is the mass scattering calculated on the classical theory, and has the value 0·151.

¶ In the measurements on γ-rays ($\lambda = \cdot 024$ Å.), as plotted in fig. 5, the mean wave-length change for paraffin, aluminium, and iron is used, in order to reduce the probable experimental error. All of these elements may be considered as of low atomic number when γ-rays are employed.

Fig. 5.

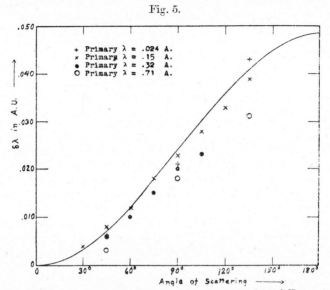

The change in wave-length accompanying the scattering of X-rays by paraffin, when different primary wave-lengths are employed.

Fig. 6.

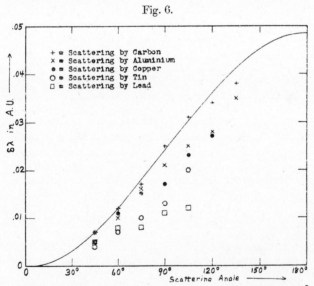

The change in wave-length when X-rays of wave-length ·15 ·17 Å. are scattered by different radiators, showing a smaller change for the heavier elements.

scattered by different elements when hard X-rays are used appears also when hard γ-rays are employed, as is shown in Table II. In this case, however, the difference between the different elements does not become apparent for elements lighter than iron.

I do not feel that these absorption experiments are sufficiently accurate to make from them any reliable estimate of the homogeneity of the scattered X-rays. The spectrum measurements, however, such as those shown in fig. 1, indicate clearly the existence of both a *modified ray*, whose wave-length is changed by the theoretical amount, and an *unmodified* ray of unchanged wave-length. It would seem possible to explain all of the present results on the assumption that only these two rays exist in the scattered beam, but that the energy distribution between the two rays varies with the wave-length, the angle of scattering, and the atomic number of the radiator. According to this view, for short wave-lengths and low atomic numbers nearly all of the energy lies in the modified ray, while for long waves and high atomic numbers the unmodified ray has the greater energy.

These experiments therefore suggest that for such comparatively great wave-lengths as those used in optics, when the usual materials are used as radiators, the unmodified ray should predominate, and the effective change in wave-length due to scattering should be very much less than that which occurs when X-rays are scattered. This is in accord with the negative result of Ross's experiment, in which he attempted to detect a change in wave-length when light was scattered by paraffin.

The Limb-Effect.

If the electrons were really free, which would correspond to atoms of zero atomic number, the present experiments suggest that the change in wave-length predicted by the theory should occur for even very long waves. J. Q. Stewart has recently presented an argument which suggests strongly that there exists about the sun a comparatively dense atmosphere of free electrons [*]. If this is the case, we should expect, in addition to the spectrum lines transmitted directly through this atmosphere [†], to find some scattered light from

[*] J. Q. Stewart, Nature, cxi. p. 186 (1923); Phys. Rev. xxii. (1923).
[†] Of course, the solar lines are absorption, rather than emission lines. The change in wave-length should occur, however, in exactly the same manner. For the continuous background on either side of a dark line should be shifted toward the red, which would shift the centre of gravity of the dark line itself.

the atmosphere which would be of greater wave-length than the direct ray. Since the thickness of the atmosphere traversed is greater near the limb, the amount of scattering, and hence the effective increase in wave-length, should be greater at the limb. We might thus expect the mean wave-length of a spectrum line from the sun's limb to be slightly greater than that of the same line from the middle of the photosphere. The difference should probably be less than 0·024 Å., since even at the limb the direct ray would probably be responsible for a large part of the spectrum line.

An effect of exactly this character is found in the solar spectrum, and is known as the "limb effect." Dr. C. E. St. John writes me that the wave-lengths from the limb are greater than those from the centre by from 0·004 to about 0·010 Å. in passing from the violet to the red. This neglects the very strong lines, which show no change in wave-length, and which presumably originate above the denser part of the electron atmosphere. The observed limb effect is thus of the right sign and of the right order of magnitude. At first sight it seems difficult to account for the fact that the red lines are shifted more than the violet. It is very possible, however, that the violet light is the more rapidly absorbed by the sun's atmosphere, so that the violet light reaching us traverses a thinner stratum of electrons. Our ignorance of the relative amount of the primary and the scattered light in the solar lines makes it impossible at present to give this explanation of the limb effect a quantitative test; but qualitatively it seems to be satisfactory.

Possible Origin of the Unmodified Ray.

Two different hypotheses suggest themselves to account for the presence of the unmodified ray. The first is that if the momentum of the light quantum is insufficient to impart to the scattering electron enough kinetic energy to eject it from the atom, the electron is held so firmly that it cannot recoil. Since in this case no energy is lost by recoil, the frequency of the scattered ray is the same as that of the incident ray. According to this view, many of the electrons in the heavier elements would be so tightly bound that they could recoil only from quanta possessing great energy, whereas for low energy or long wave-length quanta, even in the lighter elements some of the electrons would not recoil, thus giving rise to unmodified scattering. This is in accord with experiment. Quantitatively, however, the hypothesis is not so satisfactory. Thus the kinetic energy of an electron recoiling with the impulse imparted by a molybdenum Kα-ray when deflected through 135° is greater than the critical ionizing energy (280 volts) of the K electrons in carbon. We should therefore expect that no unmodified ray should appear when these rays are scattered at 135° by carbon. The spectra exhibited in fig. 1, however, show that the unmodified ray is present under these conditions.

The second hypothesis is based upon the view that when interference occurs, two or more electrons must scatter the same quantum. The theory upon which equation (5) is based, however, supposed that each quantum is scattered by a single electron. The change in wave-length is proportional to $1/m$, where m is the mass of the body which scatters the ray. If, then, the ray is scattered simultaneously by two electrons, the change in wave-length should be $1/2$ the maximum value, and if interference occurs between the rays scattered by a large number of electrons, as in the case of crystal reflexion, the change in wave-length should be negligible. According to the wave-theory, partial interference should always occur when more than one electron is traversed by an electromagnetic wave. Experimentally, however, we have no evidence that the rays scattered by small groups of electrons, such as those in an atom of low atomic number, interfere with each other except when the phases of the rays scattered by the individual electrons are nearly identical. This is the condition, for example, under which excess scattering of X-rays occurs. There is accordingly some justification for the assumption that an electron scatters independently only when removed from other electrons by a distance greater than some fraction of a wave-length of the incident ray. If the electrons are closer than this, they will cooperate in their scattering, and in view of their large total mass, no appreciable change in wave-length will result. This hypothesis therefore leads also to an unmodified ray which possesses greater relative energy as the wave-length of the primary ray and the atomic number of the radiator are increased.

On the latter hypothesis, there should be no change of wave-length when X-rays are regularly reflected from a crystal, or when light is reflected by the free electrons of a metallic mirror, whereas according to the former hypothesis such a change might have been anticipated. The fact that experiment seems not to show any wave-length change in these cases * is a point in favour of the view that the

* The test on light reflected by a mirror has been made by P. A. Ross (*loc. cit.*); that on the wave-length of reflected X-rays is being made in this laboratory.

unmodified ray results from scattering by groups instead of by single electrons.

Summary.

The present absorption measurements on hard X-rays, when combined with the writer's earlier measurements on γ-rays and his spectrum measurements on soft X-rays, show that over the range of primary wave-lengths from 0·7 to 0·024 Å., there occurs a change in wave-length during the scattering process.

For light elements and short wave-lengths the effective wave-length change is very near the theoretical value $\delta\lambda = 0\cdot024\,(1-\cos\theta)$, but is less for long wave-length X-rays and for radiators composed of heavy elements.

It appears probable that in each case the scattered ray consists of two portions, an unmodified ray for which $\delta\lambda = 0$, and a modified ray for which $\delta\lambda = 0\cdot024\,(1-\cos\theta)$. The effective wave-length change then depends upon the distribution of energy between these two rays.

Two different hypotheses are suggested to account for the existence of the unmodified ray.

The limb effect, or difference in wave-length of solar lines between the centre and the limb of the photosphere, receives a satisfactory qualitative explanation on the view that it is due to a change in wave-length as the light is scattered by an electron atmosphere around the sun.

Washington University,
    Saint Louis, U.S.A.
June 23, 1923.

THE SPECTRUM OF SCATTERED X-RAYS[1]

By Arthur H. Compton

Abstract

The spectrum of molybdenum Kα rays scattered by graphite at 45°, 90° and 135° has been compared with the spectrum of the primary beam. A primary spectrum line when scattered is broken up into two lines, an "unmodified" line whose wave-length remains unchanged, and a "modified" line whose wave-length is greater than that of the primary spectrum line. Within a probable error of about 0.001 A, the difference in the wave-lengths $(\lambda - \lambda_0)$ increases with the angle θ between the primary and the scattered rays according to the quantum relation $(\lambda - \lambda_0) = \lambda(1 - \cos \theta)$, where $\lambda = h/mc = 0.0242$ A. This wave-length change is confirmed also by absorption measurements. The modified ray does not seem to be as homogeneous as the unmodified ray; it is less intense at small angles and more intense at large angles than is the unmodified ray.

An x-ray tube of small diameter and with a water-cooled target is described, which is suitable for giving intense x-rays.

THE WRITER has recently proposed a theory of the scattering of x-rays, based upon the postulate that each quantum of x-rays is scattered by an individual electron.[2,3] The recoil of this scattering electron, due to the change in momentum of the x-ray quantum when its direction is altered, reduces the energy and hence also the frequency of the quantum of radiation. The corresponding increase in the wave-length of the x-rays due to scattering was shown to be

$$\lambda - \lambda_0 = \gamma(1 - \cos \theta) \quad (1)$$

where λ is the wave-length of the ray scattered at an angle θ with the primary ray whose wave-length is λ_0, and

$$\gamma = h/mc = 0.0242 \text{ A}$$

where h is Planck's constant, m is the mass of the electron and c the

---

[1] A report on this work was presented before the American Physical Society, Apr. 21, 1923 (Phys. Rev. **21**, 715, 1923).

[2] A. H. Compton, Bull. Nat. Res. Coun., No. 20, p. 18 (October 1922); Phys. Rev. **21**, 207 (abstract) (Feb. 1923); Phys. Rev. **21**, 483 (May, 1923).

[3] Cf. also P. Debye, Phys. Zeitschr. **24**, 161 (April 15, 1923).

velocity of light. It is the purpose of this paper to present more precise experimental data than has previously been given regarding this change in wave-length when x-rays are scattered.

Apparatus and method. For the quantitative measurement of the change in wave-length it was clearly desirable to employ a spectroscopic method. In view of the comparatively low intensity of scattered x-rays, the apparatus had to be designed in such a manner as to secure the maximum intensity in the beam whose wave-length was measured. The arrangement of the apparatus is shown diagrammatically in Fig. 1. Rays proceeded from the molybdenum target T of an x-ray tube to the graphite

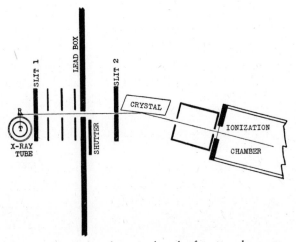

Fig. 1. Measuring the wave-length of scattered x-rays.

scattering block R, which was placed in line with the slits 1 and 2. Lead diaphragms, suitably disposed, prevented stray radiation from leaving the lead box that surrounded the x-ray tube. Since the slit 1 and the diaphragms were mounted upon an insulating support, it was possible to place the x-ray tube close to the slit without danger of puncture. The x-rays, after passing through the slits, were measured by a Bragg spectrometer in the usual manner.

The x-ray tube was of special design. A water-cooled target was mounted in a narrow glass tube, as shown in Fig. 2, so as to shorten as much as possible the distance between the target T and the radiator R. This distance in the experiments was about 2 cm. When 1.5 kw was dissipated in the x-ray tube, the intensity of the rays reaching the radiator was thus 125 times as great as it would have been if a standard Coolidge tube with a molybdenum target had been employed. The electrodes for this tube were very kindly supplied by the General Electric Company.

THE SPECTRUM OF SCATTERED X-RAYS

In the final experiments the distance between the slits was about 18 cm, their length about 2 cm, and their width about 0.01 cm. Using a crystal of calcite, this made possible a rather high resolving power even in the first order spectrum.

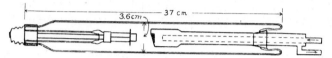

Fig. 2. X-ray tube.

Spectra of scattered molybdenum rays. Results of the measurements, using slits of two different widths, are shown in Figs. 3 and 4. Curves A represent the spectrum of the Kα line, and curves B, C and D are the spectra of this line after being scattered at angles of 45°, 90° and 135° respectively with the primary beam. While in Fig. 4 the experimental

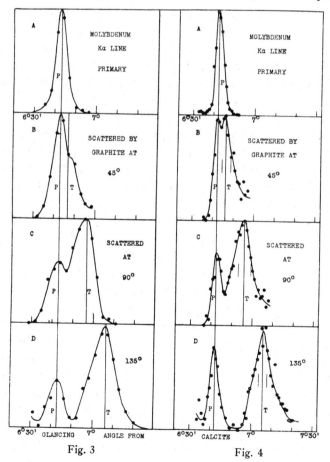

Fig. 3 Fig. 4

points are a little erratic, it may be noted that in this case the intensity of the x-rays is only about 1/25,000 as great as if the spectrum of the primary beam were under examination, so that small variations produce a relatively large effect.

It is clear from these curves that when a homogeneous x-ray is scattered by graphite it is separated into two distinct parts, one of the same wave-length as the primary beam, and the other of increased wave-length. Let us call these the *modified* and the *unmodified* rays respectively. In each curve the line P is drawn through the peak of the curve representing the primary line, and the line T is drawn at the angle at which the scattered line should appear according to Eq. (1). In Fig. 4, in which the settings were made with the greater care, within an experimental error of less than 1 minute of arc, or about 0.001 A, the peak of the unmodified ray falls upon the line P and the peak of the modified ray falls upon the line T. The wave-length of the modified ray thus increases with the scattering angle as predicted by the quantum theory, while the wave-length of the unmodified ray is in accord with the classical theory.

There is a distinct difference between the widths of the unmodified and the modified lines. A part of the width of the modified line is due to the fact that the graphite radiator R subtends a rather large angle as viewed from the target T, so that the angles at which the rays are scattered to the spectrometer crystal vary over an appreciable range. As nearly as I can estimate, the width at the middle of the modified line due to this cause is that indicated in Fig. 4 by the two short lines above the letter T. It does not appear, however, that this geometrical consideration is a sufficient explanation for the whole increased width of the modified line, at least for the rays scattered at 135°. It seems more probable that the modified line is heterogeneous, even in a ray scattered at a definite angle.

The unmodified ray is usually more prominent in a beam scattered at a small angle with the primary beam, and the modified ray more prominent when scattered at a large angle. A part of the unmodified ray is doubtless due to regular reflection from the minute crystals of which the graphite is composed. If this were the only source of the unmodified ray, however, we should expect its intensity to diminish more rapidly at large angles than is actually observed. The conditions which determine the distribution of energy between these two rays are those which determine whether an x-ray shall be scattered according to the simple quantum law or in some other manner. I have studied this distribution experimentally by another method, and shall discuss it in another paper;[4] but the reasons underlying this distribution are puzzling.

Experiments with shorter wave-lengths. These experiments have been performed using a single wave-length, $\lambda = 0.711$ A. In this case we find for the modified ray a change in wave-length which increases with the angle of scattering exactly in the manner described by Eq. (1). While these experiments seem conclusive, the evidence would of course be more complete if similar experiments had been performed for other wave-lengths. Preliminary experiments similar to those here described have been performed using the K radiation from tungsten, of wave-length about 0.2 A. This work has shown a change in wave-length of the same order of magnitude as that observed using the molybdenum $K\alpha$ line. Furthermore, as described in earlier papers,[5] absorption measurements have confirmed these results as to order of magnitude over a very wide range of wave-lengths. This satisfactory agreement between the experiments and the theory gives confidence in the quantum formula (1) for the change in wave-length due to scattering. There is, indeed, no indication of any discrepancy whatever, for the range of wave-length investigated, when this formula is applied to the wave-length of the modified ray.

WASHINGTON UNIVERSITY,
SAINT LOUIS,
May 9, 1923.

[4] A. H. Compton, Phil. Mag. (in printer's hands)
[5] Cf. e.g., A. H. Compton, Phys. Rev. **21**, pp. 494–6 (1923)

THE QUANTUM INTEGRAL AND DIFFRACTION BY A CRYSTAL

By Arthur H. Compton

Ryerson Physical Laboratory, University of Chicago

Communicated, October 5, 1923

Duane has recently pointed out[1] that if the momentum of a crystal grating perpendicular to the crystal face is nh/a, where n is an integer, h is Planck's constant and a is the distance between successive atomic layers, and if the momentum of the incident radiation quantum is $h\nu/c$, then Bragg's diffraction formula $n\lambda = 2a \sin\theta$ is a necessary consequence. It is worth while to point out that the general statement of the quantum postulate, $\int p\, dq = nh + \eta$, leads directly to the result that the momentum of the crystal changes by integral multiples of h/a as Duane assumes.

Let us express our quantum postulate in the form

$$\bar{p} \equiv \frac{\int p\, dq}{\int dq} = \frac{nh}{q_1} + \gamma, \tag{1}$$

where $\bar{p}$ is the displacement average of the momentum, $q_1 = \int dq$ is the displacement necessary to bring the system back to its original condition, and the constant γ, corresponding for example to the zero point energy of an oscillator according to Planck's second radiation formula, represents the minimum value of the average momentum. Applying this expression to the case of a beam of infinite plane waves of wave-length λ, it is clear that after the beam has propagated itself through a complete wave-length it is again in its original condition. Thus $q_1 = \lambda$. The momentum of the beam in the direction of propagation is therefore

$$p = nh/\lambda + \gamma.$$

This corresponds, according to the relativity theory, to an energy

$$\epsilon = pc = nhc/\lambda + \gamma c = nh\nu + \gamma c.$$

If $n = 1$ and $\gamma = 0$, we thus have $\epsilon = h\nu$, which is in accord with the results of photoelectric experiments. Thus the momentum of the light ray is

$$p = h/\lambda. \tag{2}$$

Let us consider in a similar manner the motion of an infinite three dimensional grating, such as a crystal. If a_x is the distance between the layers of atoms in the X-direction, the condition of the grating after it has moved a distance a_x is indistinguishable from its original condition, since layers of atoms again occupy the original positions. Thus in equation (1), $q_1 = a_x$, and hence

$$\bar{p}_x = n_x h/a_x + \gamma_x = p_x, \qquad (3)$$

since $p_x = \bar{p}_x$ for uniform motion. Just as in the case of the light ray, where p was the momentum of the whole ray, so here p represents the momentum of the whole crystal. The constant γ_x in this equation assumes different values according to the motion of the axes relative to which the momentum of the crystal is measured. This equation states that the momentum of the crystal along the X-axis changes by integral multiples of h/a_x, i.e., that

$$\delta p_x = \frac{h}{a_x} \delta n_x, \qquad (4a)$$

where δp_x is the change in the X component of the momentum, and δn_x is an integer. Similarly, if a_y and a_z are the distances between the layers of atoms along the Y and Z axes, respectively,

$$\delta p_y = \frac{h}{a_y} \delta n_y \quad \text{and} \quad \delta p_z = \frac{h}{a_z} \delta n_z. \qquad (4b, c)$$

These expressions (4) ascribe a momentum to the crystal which is quantized in precisely the manner asumed by Duane. He had shown that the dimensions of the equations demanded a length in the denominator of the right-hand members, and the lengths a were the only ones which appeared suitable; but the constant of proportionality, unity, remained arbitrary. We now see that this quantized momentum of the crystal is a direct consequence of the fundamental quantum postulate.

We shall now proceed with the discussion of the passage of radiation through the crystal according to the method suggested by Duane, though in somewhat greater detail. Let l_x, l_y, l_z be the direction cosines of the incident ray, and l_x', l_y', l_z' the direction cosines of the diffracted ray. If λ is the wave-length of the incident ray and λ' that of the diffracted ray, the increase in the X component of the momentum of the ray by the diffraction is $hl_x'/\lambda' - hl_x/\lambda$. By the principle of the conservation of momentum, this change in the momentum of the radiation must be balanced by the change in the momentum of the crystal, i.e., $hl_x'/\lambda' - hl_x/\lambda + h\delta n_x/a_x = 0$. Thus

$$\frac{l_x'}{\lambda'} - \frac{l_x}{\lambda} + \frac{\delta n_x}{a_x} = 0, \quad \frac{l_y'}{\lambda'} - \frac{l_y}{\lambda} + \frac{\delta n_y}{a_y} = 0, \quad \frac{l_z'}{\lambda'} - \frac{l_z}{\lambda} + \frac{\delta n_z}{a_z} = 0. \qquad (5)$$

The total change in momentum of the crystal is $(\delta p_x^2 + \delta p_y^2 + \delta p_z^2)^{1/2}$, or

$$\delta p = h \left\{ \left(\frac{\delta n_x}{a_x}\right)^2 + \left(\frac{\delta n_y}{a_y}\right)^2 + \left(\frac{\delta n_z}{a_z}\right)^2 \right\}^{1/2}. \quad (6)$$

From the principle of the conservation of energy it follows that the change in energy of the radiation is balanced by the change in kinetic energy of the crystal. In order to avoid changes of wave-length due to the Doppler effect, we must suppose that the initial velocity of the crystal is zero, and hence also that the initial value of $p = 0$. The energy equation is then

$$\frac{hc}{\lambda'} - \frac{hc}{\lambda} + \frac{(\delta p)^2}{2M} = 0. \quad (7)$$

Substituting the value of δp given in equation (6), and using in this equation the values of $\delta n/a$ given by expression (5), equation (7) becomes

$$\frac{hc}{\lambda} - \frac{hc}{\lambda'} = \frac{h^2}{2M} \left\{ \left(\frac{l_x}{\lambda} - \frac{l_x'}{\lambda'}\right)^2 + \left(\frac{l_y}{\lambda} - \frac{l_y'}{\lambda'}\right)^2 + \left(\frac{l_z}{\lambda} - \frac{l_z'}{\lambda'}\right)^2 \right\}. \quad (8)$$

On multiplying both sides by λ/hc we obtain

$$1 - \frac{\lambda}{\lambda'} = \frac{h/\lambda c}{2M} \left\{ \left(l_x - l_x'\frac{\lambda}{\lambda'}\right)^2 + \left(l_y - l_y'\frac{\lambda}{\lambda'}\right)^2 + \left(l_z - l_z'\frac{\lambda}{\lambda'}\right)^2 \right\}. \quad (9)$$

In this expression $h/\lambda c = h\nu/c^2$ is the mass of the incident quantum of radiation, which is very small indeed compared with the mass M of the crystal. Hence we have almost exactly

$$1 - \lambda/\lambda' = 0, \text{ or } \lambda' = \lambda. \quad (10)$$

When the value of λ' given by equation (10) is substituted in equations (5), we obtain

$$l_x - l_x' = \delta n_x \frac{\lambda}{a_x}, \quad l_y - l_y' = \delta n_y \frac{\lambda}{a_y}, \quad l_z - l_z' = \delta n_z \frac{\lambda}{a_z}. \quad (11a, b, c)$$

These expressions are exactly those obtained on the theory of interference for the angles at which the ray of wave-length λ may be diffracted by a crystal. Here, however, they are based upon the energy and momentum principles and the quantum postulate.

In order to put this result in the more familiar form known as Bragg's law, let us suppose that the incident ray lies in the XY plane, and that momentum is imparted only along the X axis, i.e., we assume $l_z = 0$, $\delta n_y = 0$ and $\delta n_z = 0$. It follows from equation (11c) that $l_z' = 0$, meaning that the diffracted ray also lies in the XY plane. If we call θ the glancing angle of incidence of the ray as it strikes the YZ plane of the crystal and θ' the glancing angle of emergence, we have $l_x = \sin\theta$, $l_x' = -\sin\theta'$, $l_y = \cos\theta$ and $l_y' = \cos\theta'$. Equation (11b) thus becomes

$$\cos\theta - \cos\theta' = 0,$$

whence $\theta' = \pm\theta$. If $\theta' = -\theta$, we find from (11a) that $\delta n_x = 0$, whence if the ray is undeflected no momentum is imparted to the crystal. If, however, a number $\delta n_x = n$ quanta of momentum are imparted to the crystal, we must have $\theta_1 = \theta$, and equation (11a) becomes

$$n\lambda = 2a_x \sin\theta. \tag{12}$$

This is identical with Bragg's expression, derived from the usual interference considerations, in which n represents the order to the diffracted beam.

It will be noted that this derivation of equations (11) and (12) has assumed infinitely long trains of waves, and a diffracting crystal which is infinite in extent. The same assumptions are also used (at least implicitly) when these expressions are derived on the interference theory. In both cases the modifications for finite wave-trains and finite crystals may be made by considering these finite quantities as the Fourier integrals of infinite wave-trains or gratings. The equations thus resulting from the quantum postulate have been given by G. Breit;[2] though in accord with the viewpoint of the present paper, we should consider the momentum of the crystal itself to be quantized rather than Breit's suggestion of some disturbance traversing the crystal.

The argument leading to equation (9) is precisely similar to that used by the writer in calculating the change in wave-length when X-rays are scattered by individual electrons,[3] except that in the latter case the mass $h/\lambda c$ of the radiation is comparable with the mass m of the scattering electron, so that the change in wave-length becomes appreciable. The fact that equation (9) indicates no measurable change in wave-length for the diffracted ray suggests that it is scattering by large groups of electrons, such as atoms or minute crystals, which gives rise to the scattered X-ray of unmodified wave-length.[4]

Attention may well be called to the fact that the present quantum conception of diffraction is far from being in conflict with the wave theory. In fact we were able to quantize the incident radiation only in view of the fact that it repeats itself at regular space intervals. Thus even from the quantum viewpoint electromagnetic radiation is seen to consist of waves.

[1] W. Duane, *Proc. Nat. Acad. Sci.*, May, 1923.
[2] G. Breit, *Ibid.*, July, 1923.
[3] A. H. Compton, *Physic. Rev.*, May, 1923.
[4] A. H. Compton, *Ibid.*, July, 1923; P. A. Ross, *Proc. Nat. Acad. Sci.*, July, 1923.

48. A quantum theory of uniform rectilinear motion. ARTHUR H. COMPTON, University of Chicago.—For uniform rectilinear motion, the quantum postulate $\int p\, dq = nh$ states that the momentum of a system is $p = nh/q_1$, where $q_1 = \int dq$ is the displacement required to bring the system back to its initial condition. The fact that a thing in uniform rectilinear motion repeats its initial condition at regular space intervals makes it in the general sense a train of waves, for which $\lambda = q_1$. Using Bohr's correspondence principle, each value of n is identified with the order of a harmonic component of the wave. For the nth harmonic, $\lambda_n = q_1/n$. Thus in general, for sine wave, $p = h/\lambda$. But the momentum of a wave-train of energy E and velocity v is $p = E/v$. Thus $E = hv/\lambda = h\nu$. The application of these equations to electromagnetic radiation is confirmed by the change of wave-length of x-rays when scattered and by the photo-electric effect. Thus also on the quantum theory radiation consists of trains of waves. Considering a moving diffraction grating of grating space D as a train of waves, its momentum is similarly nD/h, which is the basic hypotheses of Duane's quantum theory of diffraction.

Scattering of X-ray Quanta and the J Phenomena.

In NATURE of November 17, p. 723, Prof. Barkla discusses the transformations of X-rays during scattering and transmission through matter, with particular reference to my recently proposed quantum theory of X-ray scattering. Permit me to present my apology to Prof. Barkla for not referring to the work done on this subject by himself in 1904 (*Phil. Mag.* 7, 550), and by Beatty in 1907 (*Phil. Mag.* 14, 604) working in his laboratory. This early work was not mentioned because of later statements of his, such as, "the scattered radiation differs inappreciably in penetrating power from the primary radiation, that is to say, there is no appreciable degradation accompanying the process of scattering" (Barkla and Ayers, *Phil. Mag.* 21, 271, 1911). These statements led me to think that Barkla judged the differences between the primary and the scattered X-rays observed in these earlier experiments to be smaller than the experimental error.

Prof. Barkla and I agree upon the fact that the secondary X-rays from light substances are usually less penetrating than are the primary rays. Our difference lies chiefly in the interpretation of this softening. Barkla adheres to Thomson's classical theory of scattering, according to which the wavelength of the scattered beam is identical with that of the primary beam. To account for the observed difference between the primary and the scattered X-rays, he supposes that some undefined transformation occurs, subsequent to the scattering, during the transmission of the scattered radiation through the radiating substance and through the absorbers. On this view the change in penetrating power of the secondary rays is not immediately produced by the scattering, but becomes evident only after the scattered rays have traversed an appreciable thickness of matter.

My interpretation of the softening differs fundamentally from Barkla's in that the change of wavelength is considered to occur when the ray is scattered. If a quantum of primary X-rays is scattered in some other direction by a single electron, the change in momentum of the X-ray results in a recoil of the electron, which takes up a part of the energy of the ray quantum. Thus the energy, and hence also the frequency of the deflected radiation quantum, is less than that of the primary ray, and the wavelength is correspondingly greater. If, however, the primary quantum is scattered by a group of electrons the combined mass of which is relatively great, no appreciable energy is taken up in their recoil, and no appreciable change in wave-length results.

My recent spectroscopic measurements of the X-rays scattered by graphite (paper Am. Phys. Soc., Apr. 28, 1923; *Phys. Rev.* Nov. 1923) separate the scattered molybdenum K_α ray into two components, one of accurately the same wave-length as the primary, and the other of increased wave-length, due presumably to scattering by groups of electrons and by single electrons respectively. The wave-length change of the modified ray varies from 0·007 Å at 45° to 0·041 Å at 135°, in accurate accord ($\pm 0·001$ Å) with the quantum theory. These results have been beautifully confirmed and extended by the photographic spectra obtained by P. A. Ross (Proc. Nat. Acad. July 1923; *Phys. Rev.* Nov. 1923). These spectra show both the modified and the unmodified lines for the rays scattered at various angles by elements over a wide range of atomic numbers, and in each case a wavelength change for the modified line in accurate agreement with the theoretical formula.

As an explanation of the change in penetrating power, Barkla's "J" transformation is obviously less complete than is this quantum theory, since it says nothing regarding the mechanism of the transformation and makes no prediction regarding the magnitude of the change.

Barkla attempts to disprove the existence of the recoil electrons which result from my theory of scattering by a consideration of the relative ionisation in hydrogen and air. He quotes Shearer's estimate (*Phil. Mag.* 1915) of 0·0016 as an upper limit of this ratio when the K-rays from tin are used, and calculates from my formula a value "of the order of 0·01," which does not agree well. Using Hewlett's absorption data (*Phys. Rev.* 1921) for calculating the ionisation of air, and taking $\lambda_{K_\alpha} = 0·487$ Å as the wave-length of the tin rays, I calculate from my formula the ratio of the energy spent in producing ionisation in hydrogen to that spent in air, per unit volume, has the value 0·0040. If low-speed electrons in hydrogen produce the same ionisation per unit energy as do high-speed electrons in air, this figure represents also the relative ionisation to be expected in the two gases. A similar calculation for copper ($\lambda = 1·484$ Å) gives the ratio 0·00005.

As an average of all his measurements Shearer found the values of this ratio to be for tin K rays, 0·0035, which is close to the theoretical value; and for copper, 0·0018. Suspecting that a considerable part of the observed ionisation in hydrogen was due to impurities, Shearer estimated the ratio 0·0016 for the tin rays on the basis of the three lowest observed values. Since Shearer's experimental values for the tin rays vary between 0·0009 and 0·0063, and for the copper rays between 0·0003 and 0·0031, it appears that the differences between the values calculated from the quantum theory and those observed in Shearer's experiments are within the probable experimental error.

On the other hand, very direct evidence for the existence of the recoil electrons is afforded by Wilson's and Bothe's recent cloud-expansion photographs. Wilson concludes (Proc. Roy. Soc. 104, 24, 1923), referring to his "fish" tracks, that "Their direction and range, and the value of the minimum frequency of the radiation which is required to produce them, are in agreement with the suggestion made by A. H. Compton, that a single electron may be effective in scattering a quantum of radiation, and that in so doing it receives the whole momentum of the quantum." Moreover, the relative number of "fish" tracks and of long-range tracks which appear in these photographs is in accord with the view that each fish track represents a quantum of scattered rays (cf. paper by Prof. Hubbard and myself to appear soon in the *Physical Review*). In support of these conclusions, I have received a letter from W. Bothe in which he states (my translation): "I have made precise measurements on the recoil rays [by the Wilson photograph method], and I find that their velocity is in satisfactory accord with your (and Debye's) theory." He writes that the paper describing this work is in the press.

In view of the fact that there was no evidence for the existence of these recoil electrons at the time this theory was presented, their existence and the quantitative agreement with the predictions as to their number and velocity constitute a strong support of the fundamental hypotheses of the quantum theory of scattering.

I am thus wholly unable to agree with Barkla's conclusion that "Compton's formula holds neither for the apparent change of wave-length, nor for the energy of the recoil electrons."

Prof. Barkla, however, emphasises the fact that the

scattered rays obtained when one uses soft primary rays, thin radiators, and thin absorbers are nearly identical in penetrating power with the primary rays producing them. My own spectroscopic and absorption measurements, so far as they go, are in accord with this observation, showing (*Phil. Mag.*, Nov. 1923) that " the effective wave-length change . . . is less for long wave-length X-rays." Here again our differences are chiefly in the interpretation of the experiments.

The experiments of Ross (*Phys. Rev.*, Nov. 1923) and myself (*Phil. Mag.*, Nov. 1923) show that the smaller effective wave-length change observed with the longer wave-lengths does not mean a smaller change for the modified portion of the scattered ray, but rather that a larger fraction of the scattered ray is of the unmodified type. This fact is in accord with the approximate rule which I have suggested, that if the wave-length of the incident ray is greater than the distance between adjacent electrons in the scattering material, a group of electrons will co-operate in their scattering, giving rise to an unmodified line ; whereas if the wave-length is less than the distance between adjacent electrons, each quantum will be scattered by a single electron and give rise to a modified ray. On this view, if Barkla using soft X-rays has obtained scattering without change of wave-length, it means that the wave-length of his primary rays is greater than the distance between the adjacent scattering electrons, and hence is scattered by groups of electrons.

The thin radiators and absorbers which Barkla requires to obtain the small change in the character of the secondary rays are from this point of view necessary only because thicker screens would filter out the very soft X-rays of which the wave-length is unmodified. In the published experiments showing this effect (Barkla and Sale, *Phil. Mag.*, Apr. 1923), unfiltered X-rays direct from the X-ray tube have been employed. Under these conditions, rays varying in wave-length continuously from 0·2 or 0·3 to about 1·5 Å are present. To illustrate the effect of this heterogeneity, let us suppose that 80 per cent. of the primary beam consists of the wave-length 0·4 Å, which is changed by the theoretical amount when scattered, but that 20 per cent. is of wave-length 1·2 Å, and is unmodified when scattered. It can then be shown that the effective wave-length of the scattered ray as measured by its absorption coefficient in thin sheets of aluminium is changed by only $\frac{1}{3}$ of the theoretical amount. When, however, matter of appreciable thickness is traversed by the radiation, these great wave-lengths are filtered out, leaving only those wave-lengths for which the modified as well as the unmodified line occurs. Barkla's experiments therefore present no difficulty from the point of view of the quantum theory.

According to Barkla's interpretation of his experiments, the fact that thin screens show effects different from thick ones indicates that the transformation of the X-rays to a softer type does not occur until the scattered rays have traversed layers of matter of appreciable thickness. In the published experiments (Barkla and Sale) thin sheets of paper, of thickness presumably of the order of 0·05 mm., were used to show the effect of scattering by thin films. Barkla himself notes, however, that Wilson's " fish " tracks seem to be associated with the J transformation. But if the transformation occurred after the scattering process, since 0·05 mm. of paper is equivalent to about 5 cm. of air, the fish tracks should appear well outside of the path of the primary rays. The fact that these tracks occur in the path of the primary beam thus indicates that the transformation occurs at the moment of scattering rather than later as Barkla assumes.

Regarding Prof. Barkla's attempt to discredit the evidence obtained from experiments on the scattering and absorption of γ-rays, may I merely point out that the γ-ray experiments decide definitely at least one important question—that the total absorption of very short waves is much less than is permissible on the classical theory ? Any transformation of the scattered ray to a softer type, after it has been scattered, cannot put additional energy into the primary beam ; and any transformation of the primary beam would only represent an additional method of absorption, leaving a still smaller part of the observed absorption to be accounted for by the classical scattering process. As measured, the absorption coefficient agrees satisfactorily with the formula supplied by my form of the quantum theory, and is unquestionably less than that predicted by Thomson's theory.

In view of the small intensity and the longitudinal asymmetry of the scattered γ-rays, and of the observed change of wave-length of scattered X-rays, I cannot see how Thomson's classical theory of scattering is longer tenable, except as an approximation for great wave-lengths. On the other hand, the experiments described by Barkla, as well as the others which have been considered, receive a satisfactory interpretation on the basis of the quantum theory of scattering.

ARTHUR H. COMPTON

Ryerson Physical Laboratory,
   University of Chicago,
      December 16.

A MEASUREMENT OF THE POLARIZATION OF SECONDARY X-RAYS*

By Arthur H. Compton and C. F. Hagenow

ABSTRACT

It was observed by Barkla that the intensity of X-rays scattered from the second radiator perpendicular to the theoretical plane of polarization was about 1/3 as great as the intensity of the radiation in the plane. Since then evidence has accumulated that secondary X-rays even from light elements are not of the same wave-length as the primary rays. If this difference in wave-length is due to fluorescent radiation, it should result in an incomplete polarization of the scattered rays. The object of this experiment was to test this point by measurement of the degree of polarization of the scattered rays.

Heterogeneous X-rays of an average wave-length of about 0.25 A were scattered by paper, aluminium and sulphur. A geometric correction was made for the lack of complete polarization due to the solid angles subtended by the radiators, and the results were extrapolated to zero thickness of the radiators. The experiments indicate that, within a probable error of 1 or 2 per cent, the radiation is completely polarized. This precludes the possibility of there being any considerable amount of fluorescent radiation emitted from the radiator.

In his classic work on the polarization of scattered X-rays, Barkla observed that the intensity of the rays scattered from the second radiator perpendicular to the theoretical plane of polarization was about one third as great as the intensity of the radiation in the plane.[1] Had the polarization been complete, he should have observed zero intensity perpendicular to the plane of polarization. This result has been confirmed by later experimenters.[2] Barkla attributed this observed lack of complete polarization as due chiefly to the large cross section of the beams employed, so that most of the radiation was not scattered at exactly 90°. The evidence has recently become convincing that the secondary X-rays radiated even from light elements such as

---

* Based on a paper read before the American Physical Society, April 22, 1921.
[1] C. G. Barkla, Proc. Roy. Soc. 77, p. 247; 1906.
[2] e.g. Haga, Ann. d. Phys. 23, p. 439; 1907.

carbon are not of the same wave-length as the primary rays,[3] a result inconsistent with the classical theory of scattering. While this difference may be explained by a quantum theory of scattering,[4] it has also been given the alternative interpretation that a large part of the secondary X-radiation is fluorescent rather than scattered.[5] If this fluorescent radiation is similar to the fluorescent K and L radiations, we should not expect it to be polarized, and a considerable lack of polarization of the secondary beam should result. If, on the other hand, the secondary beam is completely polarized at 90° with the primary beam, the hypothesis of fluorescence becomes difficult to defend. The present measurements of the degree of polarization of the secondary X-rays were undertaken primarily to test this point.

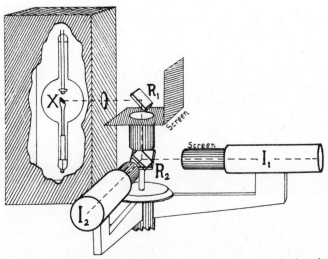

FIG. 1. *For completely polarized X-rays the intensity with the ionization chamber at I_2 should be very small compared with the intensity observed at I_1.*

The arrangement of the apparatus used is shown in Fig. 1. X-rays from the tungsten target X of a Coolidge tube were scattered downward at right angles with the primary beam by the radiator R_1 to the radiator R_2. The radiator R_2 was placed upon the crystal table of a Bragg spectrometer, and scattered the rays into the ionization chamber. According to the usual theory, the intensity when the chamber was

---

[3] A. H. Compton, Bulletin Nat. Res. Council, No. 20, p. 16; 1922; P. A. Ross, Proc. Nat. Acad. Sci.; July, 1923.

[4] A. H. Compton, Phys. Rev. *21*, p. 483; 1923.

[5] e.g. J. A. Crowther, Phil. Mag. *42*, p. 719; 1921; A. H. Compton, Phil. Mag. *41*, p. 749; 1921.

placed at I_2 should be very small compared with the intensity with the chamber at I_1. The X-ray tube was operated at about 130 kilovolts peak and at a maximum of about 10 milliamperes, no absorbing screen being used. The effective wave-length was therefore about 0.25 A. The chamber was filled with methyl iodide, and the ionization current was measured with an electrometer whose sensitiveness was about 2000 divisions per volt. Under these conditions the deflections were conveniently large except when thin radiators were used.

On the basis of the usual view of scattering, there are two chief reasons for a lack of complete polarization when apparatus of this character is employed. The first is the fact that, since in order to obtain sufficient energy the radiators must subtend appreciable solid angles, not all the scattered rays which are used proceed at exactly right angles with the primary beam. The second is that an appreciable portion of the X-rays is scattered two or more times before leaving the radiator, and is hence incompletely polarized. This effect is especially pronounced when the radiator consists of a thick block.

An approximate calculation of the lack of polarization due to the solid angle subtended by the radiators can be made from the dimensions of the apparatus employed. The result of such a calculation of the ratio R_G of the intensity at the position I_2 to that at I_1 is given by the expression

$$R_G = a^2 \left\{ \frac{.16}{k^2} - \frac{.52}{kl} + \frac{1.72}{l^2} - \frac{.52}{lm} + \frac{.16}{m^2} \right\}, \tag{1}$$

where k is the distance from the target to the radiator R_1, l is the distance from R_1 to R_2, and m is the distance from R_2 to the window of the ionization chamber. In deriving this expression, the radiators are supposed to consist of thin, square plates of edge $2a$, mounted at 45° as shown in Fig. 1, and the radiator R_2 is imagined to turn so that its upper face is always towards the ionization chamber. The effect of the solid angle subtended by the focal spot of the X-ray tube and by the window of the ionization chamber was neglected, since auxiliary calculation showed that the error from these sources was not appreciable.

In our experiments the dimensions were approximately, $k = 34$ cm, $l = 11.9$ cm, $m = 15$ cm, and $a = 2.5$ cm. Thus from equation (1) we find $R_G = 5$ per cent. Radiation of this intensity should therefore appear at I_2, even though the radiation scattered at 90° is completely polarized, and though no compound scattering occurs. We shall call this the *geometric correction*.

The error due to compound scattering is difficult to estimate quantitatively from the geometry of the problem. It can however be allowed for by using radiators of different thicknesses, and extrapolating the result to zero thickness. For small thicknesses, the ratio of the energy twice scattered to that scattered once is proportional to the thickness of the radiator. For somewhat greater thicknesses, the relative amount of energy scattered twice is approximately proportional to se^{-bs}, where s is the mass per unit area of the plate, and b is a constant depending upon the absorption coefficient of the rays in the plate. If R_0 is the ratio of the intensity at I_2 to that at I_1 for infinitesimal thickness of the radiators, the ratio for any surface density s should be given approximately by the expression,

$$R = R_0 + ase^{-bs}. \qquad (2)$$

The constants R_0, a and b can be determined from the experimental data. If the polarization is complete for rays scattered at exactly 90 degrees, the experimental value of R_0 thus determined should be identical with the value of R_G calculated from equation (1). Since the form of equation (2) becomes more nearly exact as the surface density s becomes smaller, it serves as a satisfactory formula for extrapolation to zero surface density.

The results of our experiments are shown in table 1, which gives the ratio $R = I_2/I_1$ of the intensities for different thicknesses and surface densities of the various radiators. Most of the figures are the averages of several series of readings. In taking the readings, the intensity was determined by taking the difference between the rate of deflection of the electrometer when both radiators were in place and the rate when radiator R_2 was removed. We thus avoided errors due to stray X-rays.

TABLE 1. *Polarization for Radiators of Different Thicknesses*

	Paper		Aluminium			Sulphur		
Thickness, t in cm	Surface density, s in g/cm²	I_2/I, R in o/o	t	s	R	t	s	R
.25	.25	7	.2	.5	7	.16	.3	7
.5	.5	9	.4	1.1	10	.25	.5	9
1.0	1.0	12	.8	2.1	16	.5	1.0	14
2.0	2.0	13				1.0	2.	13
4.0	4.0	18				2.0	4.	16

The effect due to the compound scattering for the greater thicknesses shows up very prominently in Fig. 2, in which are plotted the experimental values of R for different surface densities s of the radiators. We find the most satisfactory agreement between equation (2) and the experiments if we choose $R_0 = 4.8$ per cent, $a = 8.4$, and $b = 2.57$. The agreement between the experimental value of $R_0 = 4.8$ and the calculated value of $R_G = 5$ per cent is thus very close. We feel justified in concluding from these experiments, therefore, that if the X-rays were scattered at exactly 90 degrees, and if both radiators were of infinitesi-

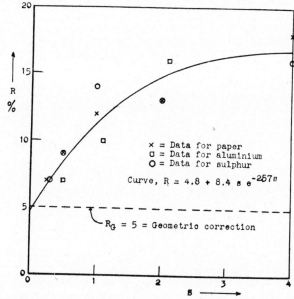

FIG. 2. *The fact that the experimental curve approaches the geometric correction for thin radiators indicates that within experimental error the rays scattered at 90 degrees are completely polarized.*

mal thickness, the intensity of the rays entering the ionization chamber at I_2 would probably be not more than 1 or 2 per cent as great as with the ionization chamber at I_1.

The almost perfect polarization which this result implies leaves no room for any considerable amount of fluorescent radiation in the secondary rays similar to the characteristic fluorescent radiations observed with the heavier elements.[6] Our experiments are on the other hand in complete accord with the view that all the rays from these light elements are truly scattered.

    A. H. C., University of Chicago,
        Chicago, Illinois.
    C. F. H., Washington University, Saint Louis, Missouri.

[6] Note added March 4th: It is inconsistent also with the existence in our experiments of any appreciable "tertiary" x-radiation, of the type suggested by Clark, Duane and Stifler to account for the increased wave-length of the secondary radiation.

THE PHYSICAL REVIEW

THE RECOIL OF ELECTRONS FROM SCATTERED X-RAYS

BY ARTHUR H. COMPTON AND J. C. HUBBARD

ABSTRACT

Quantum theory of the recoil of electrons from scattered x-rays.—This is an extension of the quantum theory of scattering suggested by Compton, which assumes that each directed x-ray quantum is scattered by a single electron. Expressions for the *distribution of recoil velocities, of energies and of ranges* are developed for each of two postulates, assuming (1) the scattered radiation consists of directed quanta, and (2) the scattered radiation proceeds as spherical waves. On the first postulate the maximum recoil energy is shown to be $E_m = h\nu_0 \times 2\alpha/(1+2\alpha)$, where $\alpha = h/mc\lambda_0$; the recoil electrons are shown to be concentrated at angles near the direction of the primary beam; and from the distribution of energy, using a relation given by C. T. R. Wilson, the distribution of ranges is found to be such that two-thirds have tracks shorter than half the maximum range. The maximum range increases rapidly with frequency. The values for the maximum ranges in the case of x-rays (.34 to .48 A) are computed to be about one-third of those observed by Wilson for his fish tracts, but the difference may be due to the lack of homogeneity of the rays used. The relative number of recoil electrons to photo-electrons increases with the frequency and is in agreement with observations by Wilson. The second postulate, however, leads to a value for E_m only one-fourth that given above, a value which is inconsistent with that derived from a consideration of radiation pressure and which leads to values for maximum ranges one-fiftieth of those observed by Wilson. Other experimental observations are cited which also lead to the conclusion that the first postulate is much more likely to be true than the second, hence, that each quantum of scattered radiation is probably emitted in a definite direction.

IN recent papers one of the writers has developed a quantum theory of the scattering of x-rays,[1,2,3] designed primarily to account for the change in wave-length observed when x-rays are scattered. The postulate upon which this theory is based is that x-rays are scattered quantum by quantum, each from a single electron. The change in momentum of the x-ray quantum on being scattered results in a recoil of the scattering

---

[1] Compton, Bull. Nat. Res. Council, No. 20, p. 19 (1922)
[2] Compton, Phys. Rev. **21**, 207 and 483, 1923
[3] Cf. also P. Debye, Phys. Zeits. April 15, 1923

electron with a velocity which may be a considerable fraction of the speed of light. When recoiling from the x-ray quanta which they have scattered, these electrons should appear as a type of secondary β-rays. The "recoil electrons" are, however, sharply distinguished from those secondary β-rays known as "photo-electrons," which are ejected with an energy comparable with $h\nu$, in that their energy is less by a factor of the order of γ/λ, where $\gamma = h/mc = 0.0242$ A.[4] In view of their relatively small energy, it is not surprising that at the time this theory was proposed the production of such recoil electrons by x-rays had not been observed. Evidence was, however, presented[5] to show that the secondary β-rays excited by hard γ-rays in the lighter elements are of this type.

Recently a new type of track has been observed almost simultaneously by C. T. R. Wilson[6] and by W. Bothe,[7] in photographs of the passage of x-rays through moist air. These tracks are very short compared with the usual photo-electron tracks, and occur in rapidly increasing numbers as the wave-length diminishes. A tentative suggestion is made by Bothe that these tracks are due to H particles ejected from the water vapor with an energy of about $h\nu$.* This hypothesis leaves unexplained, however, the fact noticed by Wilson that the short-range tracks always proceed in the initial direction of the x-ray beam. Wilson concludes that both the direction and range of the short-range tracks are in agreement with the suggestion that a single electron scatters an x-ray quantum and in so doing receives the momentum of the quantum. Evidence is given below which strongly supports this conclusion. Wilson's discovery of these recoil tracks, following upon the other successes of the theory, makes the evidence very convincing that the postulate of the scattering of whole quanta by individual electrons is sound.

There are, however, two essentially different methods by which an electron may scatter a quantum. In the postulate as first presented it was supposed that an electron receives the radiation quantum from a definite direction and scatters it in a different but equally definite direction. On this view the velocity and direction of recoil of the scattering electron will depend upon the angle at which the quantum is scattered. It may be imagined, on the other hand, that while the energy and momen-

[4] Cf. Compton, l. c.[1] p. 27
[5] Compton, l. c.[1] p. 71
[6] C. T. R. Wilson, Proc. Roy. Soc. A, **104** (Aug. 1, 1923)
[7] W. Bothe, Zeits. f. Phys. **16**, 319 (July 19, 1923)

*Note added March 6: In a second paper, in which Bothe studies these new rays by an ionization method (Zeits. f. Phys. **20**, 237, 1923), he shows that they are electrons instead of H particles. Their range, as he measures it, is slightly less than the theoretical value, Eq. (22), instead of slightly greater, as measured by Wilson.

tum of the primary quantum are received from a definite direction, the energy thus received is scattered in spherical waves in all directions. In this case every scattering electron will recoil in the direction of the primary ray with a momentum equal to the difference between the momentum of the primary ray and the resultant momentum of the spherical scattered ray. While the first form of the postulate is perhaps a more obvious consequence of the general quantum principle, the second form is in better accord with the interpretation of the quantum suggested by C. G. Darwin,[8] and has been used by C. T. R. Wilson in accounting for the short tracks observed in his photographs.[9] By studying the motions of the recoil electrons it should be possible to choose between these two forms of the quantum hypothesis.

Theory of the Recoil Electrons

Their energy. Using the assumption that each quantum of the primary radiation is scattered in a definite direction by a single electron, it has been shown[10] that the relative velocity of the recoil electron is

$$\beta = 2\alpha \sin \tfrac{1}{2}\varphi \frac{\sqrt{1+(2\alpha+\alpha^2) \sin^2 \tfrac{1}{2}\varphi}}{1+2(\alpha+\alpha^2) \sin^2 \tfrac{1}{2}\varphi}, \quad (1)$$

where $\beta = v/c$ and $\alpha = \gamma/\lambda_0$, γ being $h/mc = 0.0242$ A, and λ_0 being the wave-length of the incident x-rays. φ is the angle between the primary and the scattered x-ray quanta. The expression for the kinetic energy corresponding to this velocity, as derived first by Debye,[3] is somewhat simpler, being

$$E = h\nu_0 \frac{2\alpha \sin^2 \tfrac{1}{2}\varphi}{1+2\alpha \sin^2 \tfrac{1}{2}\varphi}. \quad (2)$$

Debye shows further that the angle θ between the primary ray and the path of the recoil electron is given by the expression

$$\tan \theta = -1/(1+\alpha) \tan \tfrac{1}{2}\varphi. \quad (3)$$

Combining these two expressions, it follows that the energy of the recoil electron ejected at an angle θ with the incident ray is

$$E = \frac{2\alpha\, h\nu_0}{1+2\alpha+(1+\alpha)^2 \tan^2\theta} = \frac{2\alpha\, h\nu_0 \cos^2\theta}{(1+\alpha)^2 - \alpha^2 \cos^2\theta}. \quad (4)$$

The energy of the recoil electron is thus, for small values of α, nearly proportional to $\cos^2\theta$. Its maximum value is at $\theta = 0$, where

$$E_m = h\nu_0 \cdot 2\alpha/(1+2\alpha). \quad (5)$$

[8] C. G. Darwin, Nat. Acad. Sci. Proc. 9, 25 (1923)
[9] Wilson, loc. cit.[6], p. 15
[10] Compton, loc. cit.[2] p. 487

We may calculate the energy of recoil on the second scattering postulate if we notice that the total energy as well as the total momentum of the system (radiation+electron) is the same before and after scattering. The energy equation thus becomes

$$h\nu_0 + 0 = \epsilon_s + mc^2 \,(1/\sqrt{1-\beta^2} - 1), \tag{6}$$

where ϵ_s is the energy of the scattered radiation; and the momentum equation is

$$h\nu_0/c + 0 = \mu_s + m\beta c/\sqrt{1-\beta^2}, \tag{7}$$

where μ_s is the momentum of the scattered radiation. To secure a relation between ϵ_s and μ_s we may now make the assumption that the incident radiation is scattered when the electron has a relative velocity $\bar{\beta} = a/(1+a)$, since this is the velocity which the electron must have in order to give the observed change of wave-length according to the Doppler principle.[11] To an observer moving forward with the scattering electron at the velocity βc, the scattered radiation would appear distributed symmetrically in the backward and forward directions, and its total momentum would therefore be zero. The effective velocity of the radiant energy ϵ_s is hence $\bar{\beta} c$, and its resultant momentum is

$$\mu_s = (\epsilon_s/c^2)\bar{\beta} c = (\epsilon_s/c)\, a/(1+a), \tag{8}$$

where, as before, $a = h\nu_0/mc^2$.

By eliminating ϵ_s and μ_s from Eqs. (6), (7), and (8), we find for the relative velocity of the recoil electron, $\beta = a/(1+a) = \bar{\beta}$. Thus the final velocity of recoil of the scattering electron is just that required on the Doppler principle to give rise to the observed change in wave-length. The kinetic energy of a recoil electron with this velocity is

$$E' = h\nu_0 \cdot \frac{1}{a}\left\{\frac{1+a}{\sqrt{1+2a}} - 1\right\} = h\nu_0 \cdot \frac{\tfrac{1}{2}a}{1+2a}\,(1 - \tfrac{1}{4}a^2 + \ldots). \tag{9}$$

Since a is usually small compared with unity, the energy of recoil according to this form of the quantum postulate is almost exactly ¼ of its maximum value (5) according to the first form of the postulate. This result (9) is in accord with the approximate values of the velocity and energy calculated on similar assumptions by one of the writers[4] and by C. T. R. Wilson.[6]

Attention should be called to a difficulty connected with this view of the scattering process. Eqs. (6), (7), and (8) state that the energy and momentum principles are satisfied and that the wave-length change shall be that which is experimentally observed. The equations do not, however, result in kinetic energy of the recoiling electron identical with that

[11] Compton, loc. cit.[2]

which we should calculate from the work done upon the electron by the radiation pressure. The impulse imparted to the electron by the radiation is obviously equal to the difference in momentum of the primary ray and of the scattered ray, i.e.,

$$\int f dt = h\nu_0/c - \mu_s;$$

or substituting the value of μ_s from Eqs. (6), (7), and (8),

$$\int f dt = (h\nu_0/c)(1-a+\ldots).$$

It is clear, however, that this impulse is imparted while the radiation is being scattered, that is, according to our assumptions, while the electron is moving forward with a velocity $\bar{v} = ac/(1+a)$. The work done on the scattering electron by the radiation pressure is hence

$$W = \bar{v}\int f dt = h\nu_0 [a/(1+a)](1-a+\ldots). \qquad (10)$$

Instead of being equal to the final kinetic energy of the recoiling electron, as given by Eq. (9), this amount of work is about twice as great. It seems

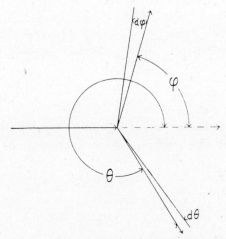

Fig. 1. Electrons which scatter the x-rays in directions between φ and $\varphi + d\varphi$ recoil in directions between θ and $\theta + d\theta$.

impossible to develop a scattering theory on the second form of the quantum postulate (that each scattered quantum proceeds in all directions) without encountering some inconsistency of this character. From the theoretical standpoint we should therefore prefer expression (4) to expression (9) as a statement of the energy of the recoil electrons.

Distribution of the recoil electrons. Let us now determine, on the view that the scattered rays proceed in definite directions, the relative number of electrons which will recoil at different angles. We may suppose, as in Fig. 1, that if the scattered ray proceeds at an angle between φ and $\varphi + d\varphi$ with the incident ray, the recoil electron moves at an angle between

θ and $\theta+d\theta$. Then if $P_\varphi d\varphi$ is the probability that a scattered quantum will lie between φ and $\varphi+d\varphi$, and if $P_\theta d\theta$ is the probability that a recoil electron will be ejected between the angles θ and $\theta+d\theta$, it is clear that

$$P_\theta d\theta = P_\varphi d\varphi. \tag{11}$$

In an earlier paper it was shown that[12]

$$P_\varphi d\varphi = \tfrac{3}{8} \sin \varphi \, d\varphi \frac{(1-\beta^2)\{(1+\beta^2)(1+\cos^2\varphi)-4\beta\cos\varphi\}}{(1-\beta\cos\varphi)^4},$$

where $\beta = a/(1+a)$. In terms of a this becomes,

$$P_\varphi d\varphi = \tfrac{3}{8} \sin \varphi \, d\varphi \frac{(1+2a)\{1+\cos^2\varphi+2a(1+a)(1-\cos\varphi)^2\}}{(1+a-a\cos\varphi)^4}. \tag{12}$$

It follows from Eq. (3) that

$$\tan \tfrac{1}{2}\varphi = -\frac{1}{1+a} \cdot \frac{1}{\tan\theta},$$

whence

$$\left.\begin{aligned}
\sin \varphi &= -\frac{2(1+a)\tan\theta}{(1+a)^2 \tan^2\theta + 1} \\
\cos \varphi &= \frac{(1+a)^2 \tan^2\theta - 1}{(1+a)^2 \tan^2\theta + 1} \\
d\varphi &= \frac{2(1+a)\, d\theta}{\cos^2\theta[(1+a)^2 \tan^2\theta + 1]}
\end{aligned}\right\} \tag{13}$$

Substituting these values in Eq. (12) we obtain for Eq. (11),

$$P_\theta d\theta = -\frac{3(1+a)^2(1+2a)\{(1+a)^4 \tan^4\theta + (1+2a)^2\}}{\{(1+a)^2 \tan^2\theta + (1+2a)\}^4} \cdot \frac{\sin\theta}{\cos^3\theta} d\theta.$$

When we write $a = (1+a)^2$ and $b = (1+2a)$, this becomes

$$P_\theta d\theta = -\frac{3ab(a^2 \tan^4\theta + b^2)}{(a \tan^2\theta + b)^4} \frac{\sin\theta}{\cos^3\theta} d\theta. \tag{14}$$

The probability that a recoil electron will strike unit area placed at a distance R and at an angle θ is

$$P_{R,\theta} = -\frac{P_\theta d\theta}{2\pi R^2 \sin\theta d\theta} = \frac{3}{2\pi R^2} \frac{ab(a^2 \tan^4\theta + b^2)}{(a \tan^2\theta + b)^4 \cos^3\theta}. \tag{15}$$

The total number of recoil electrons is, however, equal to the total number of scattered quanta, which has been shown to be,[13]

$$n = (8\pi/3) I N e^4 / b \, h\nu_0 \, m^2 c^4,$$

where I is the energy per square cm of the incident ray whose frequency is ν_0, and N is the number of electrons effective in scattering the x-rays.

[12] Compton, loc. cit.[2] p. 492
[13] Compton, loc. cit.[2] p. 493

Combining this with Eq. (15), we find for the number of recoil electrons per unit area,

$$nP_{R,\theta} = 4\frac{IaNe^4}{h\nu_0 \, R^2m^2c^4} \cdot \frac{a^2 \tan^4 \theta + b^2}{(a \tan^2 \theta + b)^4 \cos^3 \theta}. \quad (16)$$

Multiplying this by the energy of each recoil electron (Eq. 4), we find for the total energy of the recoil electrons which, if undeviated after scattering the x-ray, should traverse unit area at a distance R and an angle θ with the primary x-ray beam,

$$I_r = \frac{8Ia\alpha \, Ne^4}{R^2m^2c^4} \cdot \frac{a^2 \tan^4 \theta + b^2}{(a \tan^2 \theta + b)^5 \cos^3 \theta}. \quad (17)$$

The concentration of the effect due to the recoil electrons at angles near the direction of the primary beam becomes apparent when we plot from this Eq. (17) the energy per unit solid angle of the recoil electrons ejected in different directions. This is done in curve A of Fig. 2 for such great

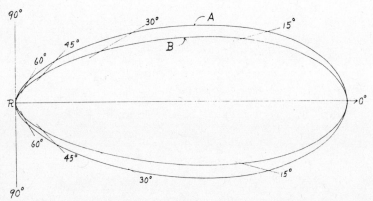

Fig. 2. Spatial intensity distribution of the recoil electrons calculated: A for long waves and B for very short waves ($\lambda = .024$ A) showing a strong concentration near the direction of the incident x-rays.

wave-lengths that a and b are sensibly equal to 1, and in curve B for $\lambda = 0.024$ A, corresponding to hard γ-rays. It will be seen that the form of the distribution curve varies but slightly with the wave-length. It is easy to see from this figure, if it is these recoil electrons which constitute the secondary β-rays excited by γ-rays in light elements, how one might conclude with Rutherford[14] and Wilson[6] that the β particles are ejected nearly in the direction of the incident x-rays or γ-rays.

Energy and range of recoil electrons. It remains to determine the probability that a recoil electron will be ejected with a definite energy. If an

---

[14] E. Rutherford, "Radioactive Substances etc.," p. 276

electron recoiling at an angle θ has an energy E, the probability that the energy of recoil will lie between E and $E+dE$ is

$$P_E dE = P_\theta d\theta. \tag{18}$$

But according to Eq. (4),

$$\tan^2 \theta = k/aE - b/a,$$

where $k = 2ah\nu_0$. Thus

$$\frac{\sin \theta}{\cos^3 \theta} d\theta = \tfrac{1}{2} d(\tan^2 \theta) = -\frac{kdE}{2aE^2}.$$

Substituting these values in Eq. (14), and noting according to Eq. (5) that the maximum energy of a recoil electron is $E_m = k/b$, Eq. (18) becomes

$$P_E dE = \tfrac{3}{2}\left(1 - 2\frac{E}{E_m} + 2\frac{E^2}{E_m^2}\right)\frac{dE}{E_m}. \tag{19}$$

This expression is of the same simple form whatever the frequency of the primary rays. In view of the fact, however, that E_m increases with the frequency, the formula can be applied strictly only in case the incident x-rays are homogeneous. Even in this case it is obvious that a correction will usually have to be made for the energy required to remove the recoil electron from the atom.

C. T. R. Wilson has found[6] that the length of his β-ray tracks is proportional to the square of the energy of the β-ray. Writing s for the length of the track and $1/p^2$ as the constant of proportionality, this fact may be expressed as $s = E^2/p^2$, or $E = ps^{\frac{1}{2}}$. When this value of E is substituted in Eq. (19), the probability that the length of the track of a given recoil electron will lie between s and $s+ds$ is found to be,

$$P_s ds = \tfrac{3}{4}\left(1 - 2\sqrt{(s/s_m)} + 2(s/s_m)\right)ds/\sqrt{s_m s}, \tag{20}$$

whence,

$$P_s s_m = \tfrac{3}{4}\left[\sqrt{(s_m/s)} - 2 + 2\sqrt{(s/s_m)}\right], \tag{21}$$

where s_m is the maximum length of the recoil electron tracks.

If we calculate the relative number of tracks of different lengths s/s_m, we find, according to Eq. (21), the values plotted in Fig. 3. It is found that more than two thirds of the tracks are of less than half of the maximum range, and more than one third are of less than one tenth the maximum range. The value of this maximum range may be calculated from the expression for the maximum energy, Eq. (5). Combining this with $s = E^2/p^2$, and writing $\alpha = h\nu_0/mc^2$, we have

$$s_m = \frac{1}{p^2} \cdot \frac{4h^4 \nu_0^4}{(mc^2 + 2h\nu_0)^2}. \tag{22}$$

It is thus seen that the maximum range increases rapidly with the frequency. These results may be subjected to experimental test.

Another experimental test can be made by comparing the lengths of the photo-electron tracks with those of the recoil electrons. If homogeneous x-rays of frequency ν_0 are used, the highest speed photo-electrons will possess energy $h\nu_0$, whereas the highest speed recoil electrons will possess energy $h\nu_0 \times 2\alpha/(1+2\alpha)$ according to Eq. (4), or approximately $h\nu_0 \times \alpha/2(1+2\alpha)$ according to Eq. (10). The ratio of their energies will thus be either $2\alpha/(1+2\alpha)$ or $\alpha/2(1+2\alpha)$ respectively. The corresponding

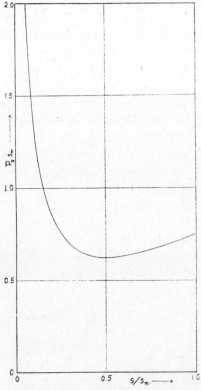

Fig. 3. Most of the recoil electrons have ranges in air less than half the maximum range s_m.

ratios of the lengths of the paths of the photo-electrons and the recoil electrons are

$$R = 4\alpha^2/(1+2\alpha)^2 \qquad (23)$$

and

$$R' = \alpha^2/4\,(1+2\alpha)^2. \qquad (23')$$

Number of recoil electrons. Each recoil electron represents the loss of one quantum of energy from the primary beam, just as does each photo-electron. It follows that the ratio of the number of recoil electrons to the number of photo-electrons should be equal to the ratio of the x-ray energy spent in scattering to that spent in exciting photo-electrons. Thus if the total absorption coefficient of the x-rays in the medium is written as $\mu = \tau + \sigma$, where τ represents the energy spent in exciting photo-electrons and σ that dissipated by scattering, the ratio of the number of recoil electrons to the number of photo-electrons is

$$n_r/n_p = \sigma/\tau. \qquad (24)$$

This ratio may be estimated approximately from absorption measurements. Since it is found that τ is proportional to λ^3, whereas σ is nearly independent of the wave-length, it follows that for great wave-lengths the photo-electrons will predominate, whereas for small wave-lengths the recoil electrons will be greater in number.

Comparison with Wilson's Cloud Expansion Experiments

Number of tracks. Using the experimental data of Hewlett,[15] we estimate for the wave-length 0.5 A at which Wilson begins to observe the tracks which he attributes to recoil electrons, that $\tau/\rho = 0.3$ and $\sigma/\rho = 0.2$ per gram, where ρ is the density of the air. Thus there should be about 1.5 times as many photo-electrons as recoil electrons for this wave-length. If his shortest x-rays were about 0.3 A, they should, by similar calculation, have produced about 3.5 times as many recoil electrons as photo-electrons. These numbers are in satisfactory accord with Wilson's observation[6] that the relative number of short range tracks increases rapidly as the wave-length decreases, being greater in number than the long range tracks for wave-lengths shorter than about 0.45 A. This agreement affords a strong confirmation of his conclusion that these short tracks are due to recoil electrons.

Wilson observes in general a predominance of "sphere" tracks over the short range or "fish" tracks. As the frequency increases the sphere tracks increase rapidly in number,[6] and are accompanied by the development of fish tracks. These observations coincide in detail with what would be expected in view of Eqs. (21) and (22). The great predominance of points or sphere tracks in Wilson's photographs may be attributed to the relatively great probability of tracks of vanishingly small range, and on the other hand, the rapid development with increasing frequency of these points into tracks of measurable length is in agreement with the expression for the maximum range of a single particle as a function of the frequency, given in Eq. (22). The close general agreement between the experiments and the theory leads to the conclusion that the sphere tracks as well as those of definite range must be considered together in any study of x-ray scattering. In order that quantitative comparisons be-

[15] C. W. Hewlett, Phys. Rev. **17**, 284 (1921).

tween experiment and theory may be made, it is desirable that data be secured on the tracks produced by homogeneous x-rays.

Range of tracks. Eqs. (23) show that the recoil electrons with the longest paths should go 16 times as far on the directed quantum hypothesis as on the spherical radiation view. Wilson observes that "When the x-rays are hard enough to eject β-particles of 1.5 cm range, fish tracks of ranges up to about 0.4 mm appear; their range increases as the frequency of the incident radiation is increased, but rarely exceeds 1.5 mm, even when the long tracks have a range exceeding 3 cm."[6] The wave-length required to produce a photo-electron track of 1.5 cm length, according to Wilson's data, is about 0.48 A, whence $\alpha = .0242/.48 = .050$. According to Eq. (23) the longest recoil tracks should thus be $1.5 \times 4\alpha^2/(1+2\alpha)^2 = 0.12$ mm. While this is considerably shorter than the observed tracks of 0.4 mm, it is at least of the correct order of magnitude. Eq. (23), however, would predict a track of only 0.008 mm length, which is very much too short.

Similarly, corresponding to the long tracks of 3 cm range, for which the wave-length is 0.34 A, Eq. (23) predicts recoil tracks of 0.5 mm length and Eq. (23') of 0.03 mm. The difference between the theoretical range of 0.5 mm and the observed range of 1.5 mm is perhaps no greater than might result from the fact that heterogeneous x-rays were used by Wilson in these experiments. For the number of photo-electrons excited in air increases rapidly with increasing wave-length whereas the prominence of the recoil electrons decreases with increasing wave-length. Thus the effective wave-length for the photo-electrons must have been greater than that for the recoil electrons. This consideration certainly accounts for a part of the difference between the theoretical and the experimental values. In order to obtain a more exact test of Eq. (23) it will be necessary to excite the recoil electrons by more nearly homogeneous x-rays.

The present experiments of Wilson suffice to show, however, that Eq. (23'), which leads to a range differing from the experimental value by a factor of about 50, is not correct. This indicates that we must abandon the assumption upon which the equation is based, that the scattered radiation is emitted in spherical waves. Both from the standpoint of the experimental evidence and from the internal consistency of the theory *we therefore seem forced to the conclusion that each quantum of scattered x-rays is emitted in a definite direction.* It would appear but a short step to the conclusion that all radiation occurs as definitely directed quanta rather than as spherical waves.

THE UNIVERSITY OF CHICAGO (A. H. C.)
NEW YORK UNIVERSITY (J. C. H.)
October 25, 1923.

THE WAVE-LENGTH OF MOLYBDENUM Kα RAYS WHEN SCATTERED BY LIGHT ELEMENTS

By Arthur H. Compton and Y. H. Woo

Ryerson Physical Laboratory, University of Chicago

Communicated, May 6, 1924

A paper by Clark, Stifler and Duane in the April number of these Proceedings describes measurements of the wave-length of the X-rays from a molybdenum target after they have been scattered by certain substances of low atomic number. The conclusion drawn from these experiments is that no secondary radiation occurs whose wave-length is increased by the amount $0.024\ (1-\cos\theta)$ A. U. predicted by the quantum theory of scattering. They find, on the other hand, evidence for modified secondary radiation whose minimum wave-length is $\lambda\lambda_k/(\lambda_k-\lambda)$, where λ is the wave-length of the incident rays and λ_k is the critical K absorption wave-length of the radiating element. The experiments described in the present paper were undertaken to examine this question in greater detail.

The apparatus used was identical in general design with that employed

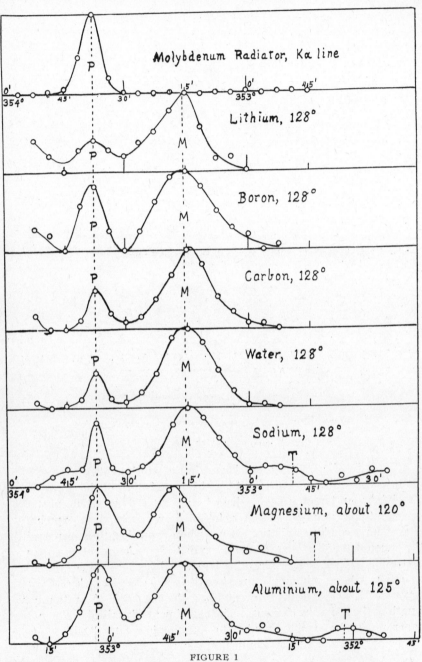

FIGURE 1

Spectra from calcite of the secondary radiation from various elements, traversed by X-rays from a molybdenum target. P marks the position of the primary $K\alpha$ peak, M the theoretical position of the modified peak, and T the peak of the "tertiary radiation" according to Clark, Stifler and Duane's experiments.

by one of the writers for measuring the change of wave-length of X-rays scattered by carbon.[1] In the present work, however, instead of a pair of slits to direct the beam falling on the crystal, a collimator was employed composed of a pile of sheets of lead foil separated by strips of lead foil.[2] This device, due to W. Soller, enabled us to secure greater intensity than in the earlier measurements. The water cooled X-ray tube was operated at about 65 kilovolts peak and 40 milliamperes.

The samples of lithium, carbon and sodium were in the form of cylinders of 8 mm. diameter, and the boron and water were held in a very thin walled, waxed paper cylinder of the same diameter. In order to secure greater intensity, the samples of magnesium and aluminium were in the form of flat plates. These radiators were clamped in turn approximately 2.5 cm. from the focal spot of the X-ray tube, in such a position that they would scatter the rays into the collimating slits at about 125°. A slight surface oxidation of the lithium and sodium occurred during the course of the experiments. The boron used was in the amorphous form, which contains 4 or 5 per cent of oxygen. Our sample also had traces of silica, and the paper container introduced small amounts of carbon and hydrogen. The rays scattered from the water may be considered characteristic of oxygen, with hydrogen and some carbon from the paper container as impurities. The carbon (graphite), magnesium and aluminium presumably had impurities only in relatively small amounts.

The results of our measurements with radiators of molybdenum, lithium, boron, carbon, oxygen (water), sodium, magnesium and aluminium are shown in figure 1. The spectra from the first six radiators were taken without changing the adjustments. For the last two elements, the width of the slits of the diaphragm was increased from 0.1 to 0.2 mm. in order to secure greater intensity. The readjustment resulted in an altered zero point, as is indicated in the figure.

The important point in this figure is that the spectra obtained from the various elements are almost identical in character. In every case an unmodified line P occurs at the same position as the fluorescent Mo $K\alpha$ line, and there is a modified line whose peak is within experimental error at the position M, calculated from the quantum change of wave-length formula given above. There is also perhaps some evidence in the cases of sodium and aluminium for a hump at the position T, where according to the experiments of Clark, Stifler and Duane the peak of the line due to "tertiary radiation" should appear.

In view of the consistency of the results for the different elements, we feel that these experiments show beyond question the reality of the spectrum shift predicted by the quantum theory of scattering.

[1] A. H. Compton, *Physic. Rev.*, **22,** 409 (1923).
[2] *Cf.* W. Soller, *Ibid.*, **23,** 292 (1924).

THE SCATTERING OF X-RAYS.*

BY

ARTHUR H. COMPTON, Ph.D.

Professor of Physics, Ryerson Physical Laboratory, University of Chicago.

The present paper will be confined to a discussion of some of the points which present to us a revolutionary change in our ideas regarding the process of scattering of electromagnetic waves. No attempt will be made to describe the great amount of experimental and theoretical work which has been done on the subject.

The theoretical researches of J. J. Thomson and the experimental work of Barkla and others had shown so striking an agreement between experiment and the theory of the scattering of X-rays based upon the usual electrodynamics that scattering has been classed with dispersion and interference as a phenomenon that is completely explicable according to our classical ideas. Within the past two years, however, scattering phenomena have been observed which are so directly contrary to the predictions of the usual electrodynamics that we apparently have to reverse our attitude almost completely. We thought we could explain the scattering of X-rays on the assumption that radiation proceeds in spherical waves, spreading in all directions in space. We now find that to retain this assumption, if our recent results are correct, we must abandon both the principle of the conservation of momentum and the principle of the conservation of energy—a hard choice, indeed; but that our observations are explained simply if we are willing to imagine the rays as consisting of discrete quanta proceeding in definite directions.

Just as a piece of paper held in the sunlight becomes a source of scattered light, so a piece of paraffin placed in the path of a beam of X-rays becomes itself a source of scattered X-rays. These rays may be examined with a fluoroscope, a photographic plate or an ionization chamber in the same manner as the primary beam. Just as the light coming from zinc sulphide exposed to sunlight consists of a mixture of fluorescent and scattered light,

---

* Based on a paper read before Section B of the American Association for the Advancement of Science, Dec. 28, 1923.

so, in general, the rays from a radiator placed in a beam of X-rays consist of a mixture of fluorescent and scattered rays. There are the fluorescent K and L radiations, X-rays whose wave-length is characteristic of the elements from which they come, and truly scattered rays whose wave-length is independent of the material of the radiator, but is dependent upon the wave-length of the primary rays. From the heavier elements the fluorescent rays usually predominate, but from the lighter elements under ordinary conditions the scattered rays only are present in the secondary beam. It is these truly scattered rays with which we are now concerned.

The classical electron theory has a very straightforward explanation of the scattering of radiation. When an electromagnetic wave traverses matter, the electric vector of the wave gives each electron in the matter an acceleration. In virtue of this acceleration these electrons themselves radiate energy. The frequency of the forced oscillations of the electrons will have to be that of the incident wave, and the frequency of the reradiated waves will, of course, be the same as that of the electrons which produce them. The frequency, and hence the wave-length, of the scattered rays is thus necessarily the same as that of the primary X-rays.

This theory of the scattering of X-rays was first developed in detail by J. J. Thomson.[1] Knowing the charge and mass of the electron, its acceleration when traversed by a wave of known intensity can be calculated, and hence the amount of scattered energy which it will reradiate can be estimated. You will remember that it was by comparing the results of such a calculation with the experimental value of the intensity of the scattered X-rays that Barkla first reached the conclusion that the number of electrons in an atom is equal to about half the atomic weight.[2] The agreement of this estimate with the atomic number, as determined by later methods, indicates that the classical theory as developed by Thomson is at least in approximate accord with the facts.[3]

The earliest experiments on secondary X-rays and γ-rays showed, however, a difference in the penetrating power of the

---

[1] J. J. Thomson, "Conduction of Electricity through Gases," 2nd Ed., p. 321 *et seq.*

[2] C. G. Barkla, *Phil. Mag.*, **7**, 543, 1904; **21**, 648, 1911.

[3] *Cf.* also C. W. Hewlett, *Phys. Rev.*, **19**, 266, 1922.

primary and the secondary rays. In the case of X-rays, Barkla and his collaborators showed that the secondary rays from many of the elements consisted largely of fluorescent radiations characteristic of the radiating element, and that it was the presence of these softer rays which was chiefly responsible for the greater absorption of the secondary rays. When later experiments showed a measurable difference in penetration, even in the case of light elements such as carbon, from which no fluorescent K or L radiation appears, it was only natural to ascribe this difference to a new type of fluorescent radiation, similar to the K and L types,

Fig. 1.

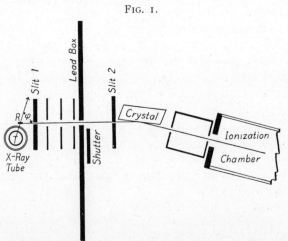

The spectrum of the rays from the radiator R, at an angle φ with primary beam is investigated.

but of shorter wave-length. The careful absorption measurements of Richtmyer and Grant and of Hewlett failed, however, to show any critical absorption limit for these assumed "J" radiations similar to those corresponding to the K and L radiations. Moreover, direct spectroscopic observations by Duane and Shimizu failed to reveal the existence of any spectrum lines under the conditions for which the supposed "J" rays should appear. It thus became evident that the softening of the secondary X-rays from the lighter elements was due to a different kind of process than the softening of the secondary rays from heavy elements where fluorescent X-rays are present.

It was at this stage, about two years ago, that I began the spectroscopic examination of the scattered rays from light elements. The reason that such experiments had not been performed before was doubtless because the low intensity of the scattered rays makes their spectroscopic examination relatively much more difficult than that of the primary X-rays. Instead of showing scattered lines of the same wave-length as the primary rays, these spectra revealed lines in the secondary rays corresponding to those in the primary beam, but with each line displaced slightly toward the longer wave-lengths.

A diagram of the apparatus employed, such as is shown in

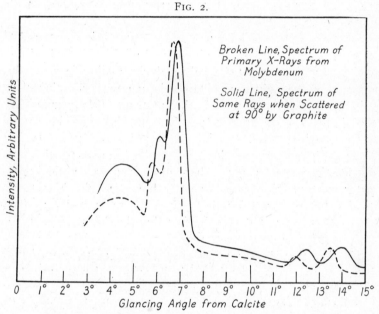

Fig. 2.

Spectrum of molybdenum X-rays scattered by graphite compared with the spectrum of the primary X-rays, showing an increase in wave-length of the scattered rays.

Fig. 1, may help in understanding the significance of the result. There X-rays proceed from the molybdenum target T of the X-ray tube to the carbon radiator R, and are thence scattered at 90° with the primary beam through the slits 1 and 2 to the crystal of the Bragg spectrometer. We thus measure the wave-length of the X-rays that have been scattered at an angle of 90°. This angle may be altered by shifting the radiator and the X-ray tube, and the spectrum of the primary beam may be obtained by merely shifting the X-ray tube, without altering the slits or the crystal.

In Fig. 2 are shown complete spectra of both the primary and the scattered beam. Here the broken line represents the first and second order spectra of the direct rays, and the solid line, drawn of course on a much larger scale, represents the spectrum of the scattered rays obtained under the same conditions. Though the difference between the first order spectra is not great, in the second order the difference in wave-length between the primary and the scattered lines is readily measurable. It is clear from this spectrum that the secondary X-rays do not have a fixed wave-length characteristic of the radiator as do fluorescent rays, but that each line in the secondary spectrum corresponds to a line of the primary spectrum.

Spectra of the molybdenum $K\alpha$ line after being scattered by carbon at different angles are shown in Fig. 3. The only difference between the two sets of values shown in this figure is that the curves on the right were obtained by narrower slits than those on the left. The upper curve is the spectrum of the primary ray, and the curves below are the spectra, using the same slits, of the rays scattered at 45°, 90° and 135°, respectively. It will be seen that though in each case there is one line of exactly the same wave-length as the primary, there also occurs a second line of greater wave-length. These spectra show very clearly not only the fact that the wave-length of the scattered ray differs from that of the primary, but also the fact that the difference increases rapidly at large angles of scattering.

The fact that the secondary rays are of greater wave-length when scattered at large angles with the primary beam suggests at once a Doppler effect as from particles moving in the direction of the primary radiation. According to the classical idea of the scattering process, however, every electron in the matter traversed by the primary X-rays is effective in scattering the rays. Thus in order to account for such a Doppler effect on this view, all of the electrons in the radiating matter would have to be moving in the direction of the primary beam with a velocity comparable with that of light—an assumption obviously contrary to fact. It was clear that if any electrons were moving in this manner, it was only a very small fraction of the whole number in the scattering material, and that it must be this small fraction which was responsible for the scattering. The idea thus presents itself that an electron, if it scatters at all, scatters a complete quantum of the incident radiation; for thus the number of electrons which move forward would just be equal to the number of scattered quanta.

This suggestion that each quantum of X-rays is scattered by a single electron supplies a simple means of accounting on quantum principles for the observed change in wave-length. For if we con-

FIG. 3.

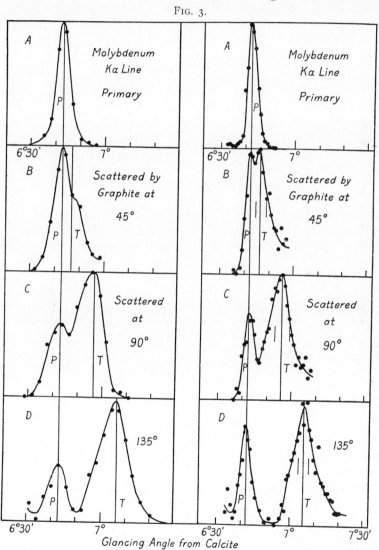

When an X-ray spectrum line is scattered, it is broken up into two lines, one of unmodified wave-length P and a line T whose wave-length is increased by the amount $h/mc \times (1-\cos \varphi)$.

sider the primary rays to proceed in quanta so definitely directed that they can be scattered by individual electrons, along with their energy $h\nu$ they will carry momentum $h\nu/c$. The scattered quantum, however, proceeding in a different direction from the primary, carries with it a different momentum. Thus by the principle of the conservation of momentum, the electron which scatters the ray must recoil with a momentum equal to the vector difference

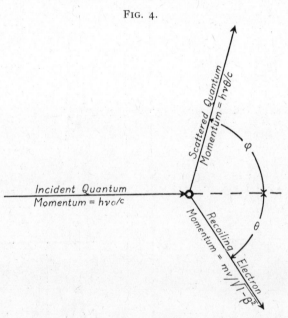

Fig. 4.

When an X-ray quantum is scattered by an electron at an angle φ, the electron recoils an angle θ, removing some of the energy from the quantum and hence reducing its frequency.

between that of the primary and that of the scattered quanta (Fig. 4). But the energy of this recoiling electron is taken from that of the primary quantum, leaving a scattered quantum which has less energy and hence a lower frequency than has the primary quantum. A simple calculation on this basis shows that the scattered ray is of greater wave-length than the primary ray by the amount

$$\lambda^1 - \lambda = \delta\lambda = \frac{h}{mc}(1 - \cos\varphi),$$

where h is Planck's constant, m the mass of the electron, c is the velocity of the light and φ is the angle between the primary and the scattered beams. The electron which scatters this quantum should recoil with an energy which is to a close approximation,

$$E_\nu = h\nu \cdot \frac{h}{mc\lambda} \cos^2 \theta$$

where ν and λ are the frequency and wave-length of the primary wave, and θ is the angle between the primary beam and the direction of recoil of the electron.[4]

These results are subject to direct experimental test. In the spectra shown in Fig. 3, the position of the line T is calculated in each case from the theoretical formula for the change in wave-

FIG. 5

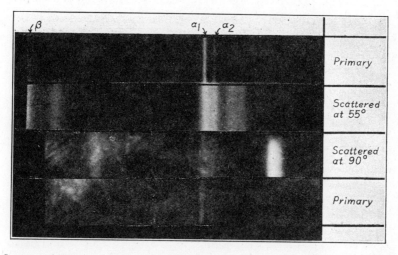

Spectrum of K-rays from molybdenum, scattered from paraffine and reflected from calcite (P. A. Ross).

length. As you can see, in every case these lines fall within experimental error at the maximum of the modified line.

These ionization spectra are completely confirmed by the beautiful photographic spectra obtained by P. A. Ross, using prolonged exposures as great as 250 hours. Fig. 5 shows two of his hitherto unpublished spectra which he has been so kind as to loan me for this occasion. At top and bottom are the spectra of the primary beam, showing the $K\alpha$ doublet resolved and the $K\beta$ line just off the edge of the photograph. The second photograph

---

[4] For a more complete discussion of the theory, cf. A. H. Compton, *Phys. Rev.*, **21**, 483, 1923; P. Debye, *Phys. Zeitschr.*, Apr. 15, 1923.

shows the spectrum of the rays scattered by paraffin at 55° with the primary beam, and the third photograph that of the rays scattered at 90°. Within an experimental error which is very small indeed, the wave-length of the modified line shown in these photographs is the same as that of the primary, and the wave-length of the modified line is greater by the amount predicted by the quantum formula.

Clark and Duane, in certain recent spectroscopic measurements, using the K-rays from tungsten and a carbon radiator, have failed to detect a change in wave-length of this character.[5] Nevertheless, absorption and spectroscopic measurements, performed independently by Ross[6] and myself,[7] have shown such a change of wave-length occurring for the rays scattered by heavy as well as light elements, and for primary rays varying in quality from soft X-rays to hard gamma rays.

The simple theory as I have outlined it accounts only for the existence of the line whose wave-length is modified. It does not seem reasonable to suppose that this part of the radiation is scattered by quanta and the unmodified portion according to the classical theory. We can, however, account for the unmodified portion if we suppose that the incident quantum has an appreciable size, of the order of magnitude of a wave-length, and that if this quantum falls simultaneously on several electrons it will be scattered by the group. Thus a light wave falling on a silver mirror would be many times as long as the distance between two adjacent "free" electrons in the silver, and so would be reflected by a large group of electrons. But the total mass of the group would be so great that the energy lost in its recoil would be insignificant, and the corresponding change in wave-length would be negligible. This conclusion is in agreement with experiments published by Ross[8] and unpublished ones by Sir Ernest Rutherford. But if, as in the case of hard γ-rays, the wave-length is much less than the distance between the electrons in the scattering material, we should thus expect each quantum to be scattered by individual electrons and to give rise only to a ray of modified wave-length; a result also in accord with experiment. For intermediate radia-

---

[5] Clark and Duane, *Proc. Nat. Acad. Sci.*, Dec., 1923.
[6] P. A. Ross, *Phys. Rev.*, Nov., 1923.
[7] A. H. Compton, *Phys. Rev.*, Nov., 1923; *Phil. Mag.*, Nov., 1923.
[8] P. A. Ross, *Science*, **57**, 614, 1923.

tions we should expect, on this view, that some of the quanta would fall on electrons relatively far removed from other electrons in the matter, which would scatter them as modified rays, and that other quanta would fall upon relatively compact groups of electrons which would scatter as units, giving rise to rays of unmodified wave-length. The problem of obtaining a quantitative criterion which will predict when modified scattering will occur, and when unmodified, is a most important one, for it is this criterion which determines whether or not interference is possible.

From the quantitative agreement between the experimental and the observed wave-lengths of the scattered rays, we may look with some confidence for the recoil electrons which result from the quantum theory of scattering. At the time that this theory was proposed there was no direct evidence for the existence of such electrons, though indirect evidence suggested that the secondary β-rays ejected from matter by hard γ-rays are mostly of this type. During the past summer, however, two experimenters, C. T. R. Wilson[9] at Cambridge, and W. Bothe[10] at Charlottenburg, have independently discovered by Wilson's cloud expansion method the existence of a new type of secondary β-rays which they have identified as these recoil electrons. Let me show you some of Wilson's photographs in which these new rays appear. Fig. 6 is a photograph obtained using X-rays that are not very hard. You will see four long tracks, originating in the path of the primary beam, which are produced by electrons which have absorbed the energy of a quantum of the incident rays and are ejected with a kinetic energy of about $h\nu$. These correspond to the photo-electrons. But there are also within the primary beam several very short tracks which appear as spheres. It is tracks of this character which Wilson has identified with the recoil electrons. When harder X-rays are employed, as in Fig. 7, these short tracks increase in length, and develop "tails" on the side of the incident X-rays. For this reason Wilson has dubbed them "fish" tracks. It is a significant characteristic of these fish tracks that their heads are all pointed in the direction of the incident X-ray beam, as is to be expected if they are due to recoil electrons. Let me show one more of Wilson's photographs, Fig. 8, showing both the long and the short tracks, which illus-

---

[9] C. T. R. Wilson, *Proc. Roy. Soc.*, **104**, 1, 1923.
[10] W. Bothe, *Zeitsch. f. Phys.*, **16**, 319, 1923.

FIG. 6.

FIG. 7.

FIG. 8.

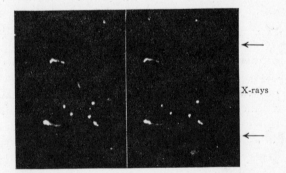

Cloud expansion photographs (Wilson) showing long β-ray tracks produced by photo-electrons and shorter tracks which are identified with the electrons recoiling from scattered X-rays.

trates the fact that the "fish" tracks which start directly forward are longer than those which move at an angle with the primary ray, corresponding to the difference in energy at different angles as predicted by our theoretical formula.

Wilson and Bothe both conclude that the direction and the range of these fish tracks is in agreement with the view that they are due to electrons which have recoiled from the quanta which they have scattered. Professor Hubbard and I have found also that the relative number of the long and the short tracks is in close agreement with what we should expect if each long track represents the fluorescent absorption of a quantum and each short track represents the scattering of a quantum of the incident X-rays. In view of the fact that these recoil electrons were unknown at the time this theory was presented, their existence and the close agreement with the predictions as to their number, direction and velocity supplies strong evidence in favor of the fundamental hypotheses of the quantum theory of scattering.

Regarding the intensity of the scattering of X-rays, I have already mentioned the fact that under certain favorable conditions Barkla and others have shown that the experiments are in good accord with Thomson's classical theory. But the experiments show conclusively, I believe, Professor Barkla to the contrary [11] notwithstanding, that when X-rays of short wave-length are employed the intensity falls distinctly below the minimum value predicted on the usual electrical theory. A number of unsuccessful attempts have been made to modify the electron theory in such a manner as to account for the low intensity and the unsymmetrical distribution of the scattered rays when hard X-rays and γ-rays are employed. These attempts have only served to emphasize that a very radical departure, indeed, is required to account for these phenomena.

It has not been found difficult, however, to account for this dissymmetry and feeble scattering on the basis of the quantum postulate. Both Professor Jauncey [12] and I [13] have obtained, by widely different though somewhat arbitrary methods, formulas for the intensity of X-ray scattering from the quantum standpoint. The resulting formulas, which are nearly identical,

---

[11] C. G. Barkla, *Nature*, Nov. 17, 1923.
[12] C. E. M. Jauncey, *Phys. Rev.*, **22**, 233, 1923.
[13] A. H. Compton, *Phys. Rev.*, **21**, 491, 1923.

approach Thomson's theoretical expression for great wave-lengths; but for shorter wave-lengths they indicate a reduced and unsymmetrical scattering. In this region where the classical formulas fail, the quantum expressions are found to fit the data probably within experimental error. In view of the fact that no new constants are introduced to secure this result, this agreement gives strong support to the quantum scattering postulates.

I have already remarked that experiments to detect the shift in wave-length when light is scattered or reflected have given negative results. This is in line with the fact that for X-rays of great wave-length part of the scattered ray is of the unmodified kind. This has been explained as due to scattering by groups of electrons if the wave-length is much greater than the distance between the adjacent electrons. In one interesting case, however, the scattering electrons are far apart even as compared with the wave-length of light. I refer to the electron atmosphere of the sun, which is present due to the dissociation by heat of the gases in the photosphere. Here we might look for a change of wave-length for the scattered rays, which should be greater near the sun's limb, where the light rays reach us after traversing a thicker layer of free electrons. A difference in wave-length of spectrum lines from the edge and from the centre of the sun, known as the "limb" effect, of origin unknown, has been recognized by solar spectroscopists for many years. Its sign is right and its magnitude is of the proper order to be accounted for as being due to the same process of quantum scattering as occurs with X-rays.

In a recent number of the *Zeitschrift für Physik*, Pauli [14] has considered the interaction of radiation and electrons in a reflecting enclosure, and has found by a thermodynamical argument that if the wave-length of the rays is altered by scattering, the energy absorbed by an electronic oscillator in a field of radiation depends not only upon the radiation reaching the electron, but also upon the radiation which the electron is about to emit. In other words, the action of the electron is conditioned not only by present and past events, but also by events which have not yet happened.

Unquestionably the most important result of this work, however, is the information which it gives regarding the nature of

---

[14] Pauli, *Zeitsch. f. Phys.*, 18, 272, 1923.

electromagnetic radiation. We find that the wave-length and the intensity of the scattered rays are what they should be if a quantum of radiation bounced from an electron, just as one billiard ball bounces from another. Not only this, but we actually observe the recoiling billiard ball, or electron, from which the quantum has bounced, and we find that it moves with just the speed it should if a quantum had bumped into it. The obvious conclusion would be that X-rays, and so also light, consist of discrete units, proceeding in definite directions, each unit possessing the energy $h\nu$ and the corresponding momentum h/λ. So in a recent letter to me Sommerfeld has expressed the opinion that this discovery of the change of wave-length of radiation, due to scattering, sounds the death knell of the wave theory of radiation.

If we wish to avoid this conclusion and to retain the idea that energy proceeds in all directions from a radiating electron, we are presented with an alternative which is perhaps even more radical, namely, that when dealing with interactions between radiation and electrons, the principles of the conservation of energy and of momentum must be abandoned. It is indeed difficult to see how the idea of directed quanta can be reconciled with those experiments in which interference is secured between rays that have moved in different directions, as for example in the interferometer. The conviction of the truth of the spherical wave hypothesis produced by such interference experiments has led Darwin and Bohr in conversation with me to choose rather the abandonment of the conservation principles.

The manner in which this alternative presents itself is very clear-cut. If the energy radiated by an electron striking the target of an X-ray tube is distributed in all directions, only a very small fraction of it will fall upon any particular electron in the scattering material. But this small fraction of the original radiated energy is sufficient to enable the scattering electron to emit a whole quantum of radiant energy. That this is true is shown by the fact that the number of recoiling electrons observed in the Wilson photographs is no larger than the number of scattered quanta, while the change in wave-length of the scattered rays shows that they have all come from these rapidly moving electrons. Thus on the spherical wave hypothesis, when a scattering electron radiates, it emits many times as much energy as it receives from the incident radiation. In the corresponding case of the photo-electric effect, we sometimes try to explain this difficulty on the view that energy is temporarily stored up by the electron. In the present case, this view is even more difficult to defend than in the case of the photo-electric effect, for here the loosely bound electrons, and apparently even the free electrons as well, would have to be able to store up energy as readily as those tightly bound within the atom.

The lack of conservation of momentum on the spherical wave view is even more clearly evident than is the sudden appearance of energy. For just as in the case of the energy received by the scattering electron, so also the impulse received by the electron from the incident radiation is on this view insignificant. We find, however, that the scattering electrons move with a velocity comparable with that of light, an electron suddenly acquiring a momentum in the forward direction which is incomparably greater than the impulse which it receives from the incident ray on the usual wave theory. To retain the conservation of momentum, we might suppose that the remaining part of the atom recoils with a momentum equal and opposite to that of the scattering electron. But the experiments indicate that the momentum may be equally well acquired whether the electron is loosely or tightly bound, and there is even evidence that the free electrons in an ionized gas acquire momentum in the same manner. It is thus clear that the momentum acquired depends only upon the scattering electron and the radiation, and has nothing to do with the remaining part of the atom. If, therefore, as the spherical wave hypothesis requires, radiation does not impart to the electron an impulse as great as it is found to acquire, the momentum is not conserved in the process, that is, action and reaction are no longer equal and opposite.

If this work on the scattering of X-rays is correct, we must therefore choose between the familiar hypothesis that electromagnetic radiation consists of spreading waves, on the one hand, and the principles of the conservation of energy and momentum on the other. We cannot retain both.

It seems to me that the very fact that the energy and momentum principles may be applied to the problem of the scattering of radiation with results in accord with experiment, constitutes a test of their validity for phenomena of this type. For this reason I am inclined toward the choice of these principles even at the great cost of losing the spreading wave theory of radiation. I am by this choice confined to the view that radiation consists of directed quanta. Following perhaps the trail blazed by Duane [15] for interference of plane waves, I still have hope that a path may be found for the quantum theory to pass also through the difficulties presented by other interference experiments.

A GENERAL QUANTUM THEORY OF THE WAVE-LENGTH OF SCATTERED X-RAYS

By Arthur H. Compton

Abstract

Corpuscular quantum theory of the scattering of x-rays.—The theory previously presented[1] gave for the change of wave-length due to scattering, assuming each quantum scattered by a single free electron, $\delta\lambda_F = (h/mc)$ vers φ = .0242 vers φ, where φ is the angle between the primary and scattered ray. This theory is now *extended to scattering by bound electrons*. If the scattering electron is not ejected from the atom, no energy is transferred and no change of wave-length occurs; but if the electron is removed, the change of wave-length must lie between $\delta\lambda_B = \lambda^2(\lambda_s - \lambda)$ and ∞, where λ is the incident wave-length and λ_s the critical ionization wave-length for the scattering electron in its original orbit. If we restrict the theory by assuming that the final momentum possessed by the residual atom is that acquired during the absorption from the incident beam, of the energy hc/λ_s required to remove the electron from the atom, and that the electron, now free, receives the impulse resulting from the deflection of the quantum, the resulting change of wave-length is $\delta\lambda = \delta\lambda_B + \delta\lambda_F$. *Comparison with experimental results* shows that the restricted theory accounts satisfactorily for scattering by the lighter elements and also for the scattering of tungsten rays by Mo if for heavy elements $\delta\lambda_F$ is taken to be zero. *Criticism of the "tertiary radiation" hypothesis*, which leads to the expression $\delta\lambda = \lambda^2(\lambda_s - \lambda)$, shows that it does not account for the large percentage of polarization, for the large relative intensity and for the homogeneity of the scattered x-rays.

IN recent papers the writer[1] and independently P. Debye[2] have developed a quantum theory of the scattering of x-rays by light elements. This theory is based on the idea that an x-ray quantum proceeds in a definite direction, and is scattered by a single electron. The quantum loses energy due to the recoil of the electron, and is thus reduced in frequency. Measurement of the wave-length of scattered x-rays by the writer,[3] Ross[4] and Davis[5] have shown that at least in many cases scattered x-rays occur whose wave-length is changed by the theoretical amount. Moreover, electrons moving in the direction of the primary beam with about the velocity predicted for the recoiling electrons have been found

---

[1] A. H. Compton, Bulletin Nat. Res. Council, No. 20, p. 19 (1922); Phys. Rev. **21**, 207 and 483 (1923)

[2] P. Debye, Phys. Zeits., Apr. 15, 1923

[3] A. H. Compton, Bulletin N. R. C., No. 20, p. 16; Phys. Rev. **22**, 409 (1923)

[4] P. A. Ross, Proc. Nat. Acad. Sci., July, 1923; Phys. Rev. **22**, 524 (1923)

[5] Bergen Davis, Paper before section B of A. A. A. S., at Cincinnati, December, 1923

by C. T. R. Wilson[6] and by Bothe.[7] There is thus strong evidence that in some cases x-rays are scattered in essentially the manner described by the quantum theory.

Very recently, however, Clark and Duane[8] and Clark, Duane and Stifler[9] have shown the existence of a type of secondary radiation whose wave-length is altered more than this quantum theory demands. The lower frequency limit of this modified radiation is in many cases approximately $\nu-\nu_s$, where ν is the frequency of the primary ray and ν_s is a critical ionization frequency of the radiator. Moreover, when a series of radiators of increasing atomic numbers is used, there is a gradual increase in the displacement of the modified line from about the value given by the quantum formula when radiators of low atomic number are used, to a much larger value for radiators of higher atomic number. This result suggests that the writer's quantum formula for the change of wave-length holds only in the case of scattering electrons which are effectively free,[10] and it becomes important to extend the theory to the scattering of x-ray quanta by electrons that are bound within the atom.

In extending the theory it seems desirable to present it first in a very general form; we shall then impose such restrictions as seem to be warranted by the experiments, and shall compare the resulting formulas with the data given by Duane and his collaborators. Finally, Clark and Duane's interpretation of these modified lines as due to "tertiary radiation" excited by the photo-electrons will be briefly discussed.

General Theory of Scattering by Individual Electrons

If we retain the conception used in the original theory, that each x-ray quantum is scattered by an individual electron, two cases are to be considered, that in which the electron is not ejected from its atom, and that in which the scattering electron receives an impulse sufficient to eject it from the atom.

In the first case, evidence from x-ray spectra indicates that there is no resting place for the electron within the atom after it has scattered the quantum unless it returns to its original orbit. The final energy of the atomic system is thus the same after the quantum is scattered as it was before (the kinetic energy imparted by the deflected quantum to a body

---

[6] C. T. R. Wilson, Proc. Roy. Soc. A **104**, 1 (1923)
[7] W. Bothe, Zeits. f. Phys. **16**, 319; **20**, 237 (1923)
[8] G. L. Clark and W. Duane, Proc. Nat. Acad. Sci. **9**, 413, 419 (1923); **10**, 41 (1924)
[9] G. L. Clark, Stifler and W. Duane, paper presented to Am. Phys. Soc. Feb. 23, 1924, privately communicated to the writer (see abstract in Phys. Rev. **23**, 551, 1924)
[10] This was indeed emphasized in the original paper.

as massive as an atom being negligible), implying that the frequency of the scattered ray is unaltered. Scattering by this process would thus give rise to an unmodified line.

In the second case, however, part of the energy of the incident quantum is spent in removing the scattering electron from the atom, part is used in giving the electron and the ionized atom their final motions, and the remainder appears as the scattered ray. If we suppose that the primary beam is propagated along OX, and if the direction of the scattered ray defines the XOY plane, the following equations suffice to define the condition of the quantum, the electron and the ionized atom before and after the scattering occurs.

Energy equation,

$$\frac{hc}{\lambda} = \frac{hc}{\lambda'} + \frac{hc}{\lambda_s} + mc^2\left(\frac{1}{\sqrt{1-\beta^2}} - 1\right) + \frac{1}{2}MV^2 ; \qquad (1)$$

momentum equations,

$$\frac{h}{\lambda} = \frac{hl_1}{\lambda'} + pl_2 + Pl_3 , \qquad (2)$$

$$0 = \frac{h}{\lambda'}m_1 + pm_2 + Pm_3 , \qquad (3)$$

$$0 = 0 + pn_2 + Pn_3 ; \qquad (4)$$

supplementary equations,

$$p = m\beta c/\sqrt{1-\beta^2} , \qquad (5)$$

$$l_2^2 + m_2^2 + n_2^2 = 1 . \qquad (6)$$

In these equations,

λ = wave-length of incident x-ray quantum;
λ' = wave-length of scattered quantum;
λ_s = critical ionizing wave-length for scattering electron;
h = Planck's constant;
c = velocity of light;
m = rest mass of electron;
βc = final velocity of recoiling electron;
p = final momentum of electron;
M and V = mass and final velocity of recoiling atom;
$P = MV$;
$l_1, m_1, 0$ = direction cosines of scattered quantum;
l_2, m_2, n_2 = direction cosines of p;
l_3, m_3, n_3 = direction cosines of P.

We shall use also the following abbreviations: $\nu = c/\lambda$ = frequency of incident radiation; φ = angle between incident and scattered quanta;

θ = angle between incident quantum and p; $\gamma = h/mc = 0.0242$ A;
$a = h/mc\lambda$; $s = \lambda/\lambda_s$; $b = p/mc$; $B = P/mc$; $D = \lambda^2/(\lambda_s - \lambda)$;
$F = \gamma(1 - l_1) = \gamma$ vers ϕ; $A = s(1 + a - \tfrac{1}{2}as) - B(l_3 - B/2a)$.

A straightforward solution of these equations, noting that $\tfrac{1}{2}MV^2$ is always negligible compared with the other terms of Eq. (1), gives for the change in wave-length,

$$\delta\lambda \equiv \lambda' - \lambda = [\lambda/(1-A)][a(1-l_1) + s(1-\tfrac{1}{2}as) + B(l_1l_3 + m_1m_3 - l_3 + B/2a)]. \qquad (7)$$

Instead of solving directly for β it is more convenient to calculate the kinetic energy of the recoiling electron, which is given by these equations as

$$E = h\nu\left\{1 - \frac{1 - a(l_1s + \tfrac{1}{2}s^2) + B(l_3 + l_1l_3s + m_1m_3s - B/2a)}{1 + a(1 - l_1 - s) + B(l_1l_3 + m_1m_3)}\right\}. \qquad (8)$$

If the scattering electrons are free, the critical ionization wave-length is $\lambda_s = \infty$, and the momentum imparted to the atom is $P = 0$, in which case these equations reduce to

$$\delta\lambda = F \equiv a\lambda(1 - l_1) = \gamma \text{ vers } \varphi, \qquad (9)$$

and

$$E_0 = h\nu\{1 - 1/(1 + a - al_1)\} = h\nu\frac{a \text{ vers } \varphi}{1 + a \text{ vers } \varphi}. \qquad (10)$$

These results are identical with those given by Debye and the writer for the scattering by a free electron.

For a definite value of s the change in wave-length according to Eq. (7) is a minimum when all the impulse from the deflected quantum is absorbed by the atom. In this case

$$\delta\lambda = D \equiv \lambda^2/(\lambda - \lambda_s), \qquad (11)$$

which is identical with the minimum wave-length change given by the tertiary radiation theory of Clark and Duane. Eq. (11) may be obtained by substituting the appropriate values of B, l_3 and m_3 in Eq. (7), but can be got more easily directly from Eq. (1), noting that the kinetic energy of the scattering electron is zero if the impulse all goes to the ionized atom.

In the general case there is nothing to prevent all of the energy being taken up by the ejected electron, leaving in Eq. (1) $hc/\lambda' = 0$, or $\lambda' = \infty$. The complete quantum theory of scattering thus gives a possible wave-length range for the scattered ray between $\lambda + \lambda^2/(\lambda_s - \lambda)$ and ∞. We thus have precisely the same wave-length range for the secondary radiation as is predicted by the tertiary radiation theory of Clark and Duane, which also assigns a definite lower limit to the wave-length, but supplies no finite upper limit.

Restrictions Suggested by Experiment

In view of the sharpness of the lines or bands observed in the spectrum of the secondary x-rays, it is clear that the momentum which the atom acquires is defined within rather narrow limits, that is B in Eq. (7) has a rather definite value. An exact prediction of the value of this momentum requires some knowledge of the atom's internal dynamics and of the mechanism of interaction between the quantum and the electron. Thus a sudden impulse applied to the electron would result in a smaller momentum imparted to the atom than would an interaction lasting for a considerable time interval, just as jerking a sheet of paper from under a book disturbs the book less than removing the paper more slowly. Lacking a sufficient knowledge of this mechanism to make a definite prediction, it is nevertheless of interest to study the result of certain plausible assumptions regarding the impulse imparted to the atom.

Let us suppose for example that in removing the electron from the atom all of the work is done by the incident ray. The energy absorbed in this process is hc/λ_s, and since the absorption of this energy leaves the electron at rest outside the atom, the whole impulse, h/λ_s, accompanying the absorption of the energy, must be imparted to the remainder of the atom. Since the electron is now free, any further action of the radiation on the electron will not affect the atom. At the end of the process, the atom therefore retains the momentum $P = h/\lambda_s$ in the direction of the primary beam.

In Eq. (7) we have, therefore, $B = P/mc = h/mc\lambda_s = as$, $l_3 = 1$, and $m_3 = 0$. On substituting these values, Eq. (7) reduces to

$$\begin{aligned}
\delta\lambda &= [\lambda/(1-s)][a(1-s)(1-l_1)+s] \\
&= \lambda^2(\lambda_s-\lambda) + \gamma(1-l_1) = D+F \\
&= \lambda^2(\lambda_s-\lambda) + (h/mc)\text{ vers }\varphi = \lambda^2(\lambda_s-\lambda) - .0242 \text{ vers }\varphi.
\end{aligned} \qquad (12)$$

From Eq. (8) we find that the kinetic energy of the recoil electron is

$$E = h\nu \cdot \frac{a(1-s)^2 \text{ vers }\varphi}{1+a(1-s)\text{ vers }\varphi}. \qquad (13)$$

Solving the original equations, we find that the angle θ between the primary beam and the motion of the electron which recoils from a quantum scattered at an angle φ is given by

$$\tan\theta = -\frac{\cot\tfrac{1}{2}\varphi}{1+a-as}. \qquad (14)$$

It will be seen that according to these equations the motion of the recoil electrons is not much affected by the constraining forces until $s = \lambda/\lambda_s$

becomes comparable with 1, in which case their energy is reduced. This result is supported qualitatively by the recent experiments of Bothe[11] on the ranges of the recoil electrons ejected from different substances.

Experimental Test

The experiments of Clark and Duane[8] and of Clark, Duane and Stifler[9] have shown that the modified line or band excited by electrons of the s group always occurs at wave-lengths greater than $\lambda+\lambda^2/(\lambda_s-\lambda)$. This is in complete accord with our general Eq. (7). It will be of interest, however, to compare also the displacement of the peak of the modified line with the value predicted by Eq. (12).

Perhaps the most precise experimental data referring to the scattering of x-rays by electrons which are loosely bound are those[12] for the scattering of molybdenum Kα rays by carbon. In these experiments, for the rays scattered at 45°, 90° and 135°, the peak of the modified line was displaced (0.0242 vers φ) A within a probable error of about ±0.001 A. This is exactly the displacement F predicted by Eqs. (7) and (12) for free electrons.

The extensive data of Clark, Duane and Stifler are presented in Table I. This table includes only those lines for which the experimental curves are available and for which the positions of the peaks of the modified lines can accordingly be determined. Clark and Duane[8] have published also the short wave-length limits of certain other modified lines, which agree very well with Eq. (11). In the column describing the origin of the modified line, the symbol MoKα-f indicates that the line is due to molybdenum Kα rays scattered by free electrons, the symbol MoKα-AlK indicates that molybdenum Kα rays are scattered by electrons in the K energy level of aluminium, etc. $\delta\lambda$(obs) is in every case the observed wave-length difference between the peaks of the modified line and of the unmodified line excited by the same primary ray.

It will be seen that every observed line except one is accounted for by this form of the quantum scattering theory. The one exception is found in the case of rock-salt, where a line ($\lambda=0.823$ A) occurs between the theoretical positions of the MoKα−NaK and the MoKα−AlK lines. The fact that this line does not alter its position as the angle of scattering is changed from 90° to 135° in the manner characteristic of the modified lines indicates that it is not a true modified line but has some other origin.

When the lighter elements are used as radiators, it will be found from this table that Eq. (12), $\delta\lambda=D+F$, predicts the peak of the modified line

[11] W. Bothe, Zeits. f. Phys. **20**, 237 (1923).
[12] A. H. Compton, Phys. Rev. **22**, 409 (1923).

within experimental error. The only serious departure from this value is observed in the case of the tungsten K rays scattered by molybdenum. In this case the displacement approaches the value $\delta\lambda = D$, given by Eq. (11). As we have seen, this means that when the scattering electron is ejected from an atom in which it is tightly bound, most of the impulse

TABLE I

Wave-length change of modified lines
(Data of Clark, Duane and Stifler)[8,9]

Radiator	Angle	Origin	$\delta\lambda$(obs) peak	$\delta\lambda$(calc) peak($D+F$); limit D		Nature of line
Li	135°	MoKα-f	.035A	.041	0	unresolved from Kα
C	90	MoKα-f	.030	.024	0	partially resolved
Ice	90	MoKα-f	.025	.024	0	partially resolved
Al	90	MoKα-AlK	.094	.093	.069	rather sharp
Al	90	MoKβ-AlK	.067	.078	.054	unresolved from Kα
NaCl	90	MoKα-NaK	.058	.070	.046	unresolved doublet
		MoKβ-ClK	.137	.130	.106	
NaCl	135	MoKα-NaK	.073	.087	.046	unresolved doublet
		MoKβ-ClK	.152	.147	.106	
NaCl	90	unknown				$\lambda = .823$A*
NaCl	135	unknown				$\lambda = .823$A*
C	90	WKα-f	.023	.024	0	faint, unresolved
Cu	90	WKα-f	.028	.024	0	rather broad line
		WKβ-CuK	.053	.051	.027	
Mo	90	WKα-f	.024	.024	0	unresolved
Mo	90	WKαMoK	.106	.130	.106	rather broad
Mo	90	WKβ-MoK	.081	.103	.079	rather broad
La	90	WKα-f	.021	.024	0	faint, unresolved

*Wave-length unaltered as φ changed.

is transferred to the atom before the electron escapes—a result which might have been anticipated. The quantum theory of scattering in its general form is therefore adequate to account completely for the wave-lengths of the modified lines observed from the heavier as well as the lighter elements.

The "Tertiary Radiation" Hypothesis

To account for the wave-length of the modified lines which they observed, Clark and Duane have suggested[13] the hypothesis, mentioned above, that these rays are produced by the collision with the surrounding atoms of the photo-electrons ejected by the primary x-rays. A similar

---

[13] G. L. Clark and W. Duane, Proc. Nat. Acad. Sci. **9**, 422 (1923).

view was at one time defended also by the writer.[14] Since the kinetic energy of the photo-electron is $h(\nu-\nu_s)$, the maximum frequency of the x-rays which they can excite is $\nu'=\nu-\nu_s$. This corresponds to a minimum wave-length greater than the primary by $D\equiv\lambda^2/(\lambda_s-\lambda)$, identical with that given by Eq. (11).

Although we have seen that this value of the minimum wave-length is in satisfactory accord with experiment, there are other considerations which seem to make the hypothesis untenable.

(1) *Polarization of the secondary x-rays.* Under the best conditions, for the rays emitted at 90° with the motion of the impinging electron, experiment[15] shows that not more than about 25 per cent of the x-rays produced when a cathode electron traverses matter are polarized. Since the photo-electrons responsible for the tertiary radiation are ejected through a wide range of angles, the x-rays which they excite should be even less strongly polarized. It is found, however, that the secondary x-rays from light elements at right angles with the primary beam contain not more than perhaps 2 per cent of unpolarized x-rays.[16] Since the spectrum of the secondary rays from these substances shows a large fraction of the energy in the modified ray, it follows that the modified ray at right angles with the incident beam is nearly completely polarized. This fact is not consistent with the hypothesis of tertiary radiation.

(2) *Energy of the secondary x-rays.* In the case of the secondary x-rays from carbon, excited by the Kα line from molybdenum, the spectra show that about 2/3 of the energy lies in the modified ray. Hewlett has shown,[17] however, that in this case the total energy in the secondary beam corresponds to a mass scattering coefficient of 0.20, in accurate agreement with the simple electron theory of scattering. Since the mass absorption coefficient of the molybdenum Kα line in carbon is about .055, this means that 36 per cent of the energy removed from the primary beam reappears as secondary x-rays, or about 24 per cent as modified rays. Even if we suppose that all of the energy removed from the primary beam is initially transformed into photo-electrons, this means that the efficiency of production of x-rays by the impact of these photo-electrons must be 24 per cent. Experiment shows,[18] however, that the efficiency of production of x-rays by electrons traversing carbon

---

[14] A. H. Compton, Phil. Mag. **41**, 762 (1921).

[15] C. G. Barkla, Phil. Trans. **204**, 467 (1905); et al.

[16] A. H. Compton and C. F. Hagenow, Phys. Rev. **18**, 97 (1921); Journ. Opt. Soc. Am. April 1924

[17] C. W. Hewlett, Phys. Rev. **20**, 688 (1922)

[18] Cf. summary by Bergen Davis, Bull. Nat. Research Council No. 7, p. 415 (1920)

with the velocity of these photo-electrons is considerably less than 0.1 per cent. It follows that no appreciable fraction of the modified rays in this case can be due to tertiary radiation. On the other hand, the agreement of Hewlett's experiments with the theoretical scattering coefficient loses its significance unless the modified as well as the unmodified rays are truly scattered.

(3) *Width of the modified lines.* In Ross's spectrum[19] of the Kα line of molybdenum as scattered by paraffine at 55°, the modified α_1 and α_2 can be distinguished, though they differ in wave-length by only .004 A. This implies a sharpness of the modified line which is quite inconsistent with the view that the peak is merely the maximum of a band of general radiation produced by the impact of cathode rays.

These considerations are, however, all in agreement with the idea that the modified line is a type of truly scattered rays. In view of the fact that the wave-length of these lines can be satisfactorily accounted for by the quantum theory of scattering, and especially in light of the experimental evidence for the existence of the recoil electrons, it is very difficult to avoid the conclusion that the modified rays observed in the spectra of secondary x-rays result from the scattering of whole quanta by individual electrons.

RYERSON PHYSICAL LABORATORY,
UNIVERSITY OF CHICAGO
March 29, 1924

---

[19] Reference is made to an unpublished spectrum shown before the A. A. A. S. and Am. Phys. Soc. in December, 1923; see Phys. Rev. **23**, 290 (1924)

THE EFFECT OF A SURROUNDING BOX ON THE SPECTRUM OF SCATTERED X-RAYS

By A. H. Compton and J. A. Bearden

Ryerson Physical Laboratory, University of Chicago

Communicated January 7, 1925

In the September number of these Proceedings Professor Duane and his collaborators have described experiments in which differences occurred in the ionization spectra obtained from sulphur traversed by X-rays from an X-ray tube with a molybdenum target depending upon the presence of a wooden box surrounding the X-ray tube and the sulphur radiator.[1] The fact that the spectra obtained in the presence of the box were similar to those obtained in this laboratory, whereas those obtained in the absence of the box did not show what we call the "modified lines" suggested to these experimenters that the presence of these lines was dependent upon the presence of the box. If this is true, the origin of these lines must be different from that postulated in the quantum theory of X-ray scattering. It was suggested rather that the modified lines were due to a "tertiary radiation" from the carbon in the wood box.

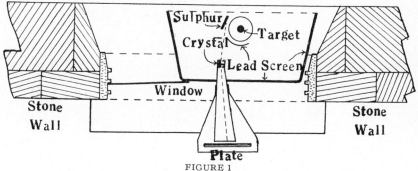

FIGURE 1

The X-ray tube and the sulphur scattering block were placed outside of a window, with a lead screen partially surrounding the X-ray tube to prevent any X-rays from entering the window.

Before our experiment was performed, Y. H. Woo compared the spectra obtained by the ionization method when a lead-lined box surrounded his X-ray tube and radiator with those he previously obtained on using a wood-lined box.[2] The comparison was made for magnesium and aluminium radiators, and new measurements were made on silicon and sulphur radiators. In every case both the modified and unmodified lines appeared with substantially the same intensity as when the wood-lined box was used, and in approximately the positions demanded by the quantum theory.

In order to satisfy ourselves completely, we set up a water cooled molybdenum target X-ray tube and a sulphur radiator outside a third floor window, with no surrounding box, as shown diagrammatically in figure 1. The face of the target pointed upward, so that nearly all of the rays which did not strike the lead screen shielding the window went to the open sky. The spectrum was obtained photographically, using a modified Seeman spectrograph. The tube was operated continuously at about 25 m. a. and 47 k. v. peak.

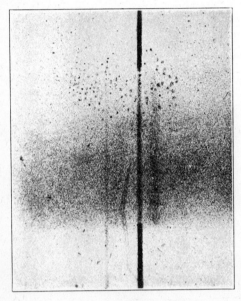

FIGURE 2

Spectrum of X-rays from molybdenum target scattered by sulphur at 118° (middle) compared with spectrum of the fluorescent rays from molybdenum (edges). The presence of the modified as well as the unmodified lines is evident for the scattered rays.

To obtain suitable reference lines on our film, and to be sure that no false lines appeared due to faults in the crystal, the edges of the film were first exposed for 8 hours to the fluorescent rays from a sheet of molybdenum slipped over the face of the sulphur radiator. Thus the $K\alpha$ and β lines of molybdenum were registered, as shown in figure 2. The middle of the film was then exposed for about 170 hours to the secondary rays coming from the sulphur itself. In addition to the unmodified α and β lines, this spectrum shows plainly the modified α line and very faintly the modified β line.

The angle of scattering of the α line was 118 ± 3 degrees, whence according to the quantum formula the wave-length change should be $\delta\lambda = 0.036 \pm 0.001\ A$. Measurements on the $K\alpha$ lines showed $\delta\lambda = 0.037\ A$, which is a satisfactory agreement.

The relative energy in the modified and unmodified lines of this photographic spectrum did not differ much from that shown by Woo's ionization spectra taken with a surrounding box. We are thus unable to find any considerable effect on the character of the spectrum obtained due to the presence of a surrounding box.

[1] Armstrong, A. H., Duane, W., and Stifler, W. W., *Proc. Nat. Acad. Sci.*, **10,** 374 (1924).

Allison, S. K., Clark, G. L., and Duane, W., *Ibid.*, p. 379 (1924).

[2] Woo, Y. H., this number of these PROCEEDINGS.

MEASUREMENTS OF β-RAYS ASSOCIATED WITH SCATTERED X-RAYS

By Arthur H. Compton and Alfred W. Simon

Abstract

Stereoscopic photographs of beta-ray tracks excited by strongly filtered x-rays in moist air have been taken by the Wilson cloud expansion method. In accord with earlier observations by Wilson and Bothe, two distinct types of tracks are found, a longer and a shorter type, which we call P and R tracks, respectively. Using x-rays varying in effective wave-length from about 0.7 to 0.13 A, the ratio of the observed number of R to that of P tracks varies with decreasing wave-length from 0.10 to 72, while the ratio of the x-ray energy dissipated by scattering to that absorbed (photo-electrically) varies from 0.27 to 32. This correspondence indicates that about 1 R track is produced for every quantum of scattered x-radiation, assuming one P track is produced by each quantum of absorbed x-radiation. The *ranges* of the observed R tracks increase roughly as the 4th power of the frequency, the maximum length for 0.13 A being 2.4 cm at atmospheric pressure. About half of the tracks, however, had less than 0.2 of the maximum range. As to *angular distribution*, of 40 R tracks produced by very hard x-rays (111 kv), 13 were ejected at between 0 and 30° with the incident beam, 16 at between 30° and 60°, 11 at between 60° and 90° and none at a greater angle than 90°. The R electrons ejected at small angles were on the average of much greater range than those ejected at larger angles. These results agree closely in every detail with the theoretical predictions made by Compton and Hubbard, and the fact that in comparing observed and calculated values, no arbitrary constant is assumed, makes this evidence particularly strong that the assumptions of the theory are correct, and that whenever a quantum of x-radiation is scattered, an R electron is ejected which possesses a momentum which is the vector difference between that of the incident and that of the scattered x-ray quantum.

IN recently published papers, C. T. R. Wilson[1] and W. Bothe[2] have shown the existence of a new type of β-ray excited by hard x-rays. The range of these new rays is much shorter than that of those which have been identified with photo-electrons. Moreover, they are found to move in the direction of the primary x-ray beam, whereas the photo-electrons move nearly at right angles to this beam.[3] Wilson, and later Bothe,[4] have both ascribed these new β-rays to electrons which recoil from scattered x-ray quanta in accordance with the predictions of the quantum theory

---

[1] C. T. R. Wilson, Proc. Roy. Soc. A **104**, 1 (1923)
[2] W. Bothe, Zeits. f. Phys. **16**, 319 (1923)
[3] See, e.g., F. W. Bubb, Phys. Rev. **23**, 137 (1924)
[4] W. Bothe, Zeits. f. Phys. **20**, 237 (1923)

of x-ray scattering.⁵ In support of this view, they have shown that the direction of these rays is right, and that their range is of the proper order of magnitude. The present paper describes stereoscopic photographs of these new rays which we have recently made by Wilson's cloud expansion method. In taking the pictures, sufficiently hard x-rays were used to make possible a more quantitative study of the properties of these rays.

The cloud expansion apparatus used in our work was patterned closely after Wilson's well-known instrument except that all parts other than the glass cloud chamber itself were made of brass. The timing was done by a single pendulum, which carried a slit past the primary beam and actuated the various levers through electric contacts. The Coolidge x-ray tube, enclosed in a heavy lead box, was excited by a transformer and kenotron rectifiers capable of supplying 280 peak kilovolts. For illumination we used a mercury spark, similar to that of Wilson, through which discharged a 0.1 microfarad condenser charged by a separate transformer and kenotron to about 40 kv. The photographs were made by an "Ontoscope" stereoscopic camera, equipped with Zeiss Tessar $f/4.5$ lenses of 5.5 cm. focal length. Eastman "Speedway" plates (45×107 mm) were found satisfactory.

A typical series of the photographs[6] obtained are reproduced in Plate I, (a) to (f), which show the progressive change in appearance of the tracks as the potential across the x-ray tube is increased from about 21 to about 111 kv.

Especially in view of the fact that the original photographs are stereoscopic, the negatives of course show much more detail than do the reproductions. These suffice to show, however, the two types of tracks, the growth of the short tracks with potential, and the fact that while the long tracks are most numerous for the soft x-rays, the short tracks are most in evidence when hard rays are used. These results are in complete accord with Wilson's observations.

Number of tracks. It has been shown[7] that if the above interpretation of the origin of the two classes of β rays is correct, the ratio of the number of short tracks (type R) to that of long tracks (type P) should be

$$N_R/N_P = \sigma/\tau \qquad (1)$$

where σ is the scattering coefficient, and τ the true absorption coefficient of the x-rays in air; for σ is proportional to the number of scattered

---

[5] A. H. Compton, Bulletin Nat. Res. Council, No. 20, p. 19 (1922); and P. Debye, Phys. Zeits. (Apr. 15, 1923)

[6] These photographs were shown at the Toronto meeting of the British Association in August 1924.

[7] A. H. Compton and J. C. Hubbard, Phys. Rev. **23**, 448 (1924)

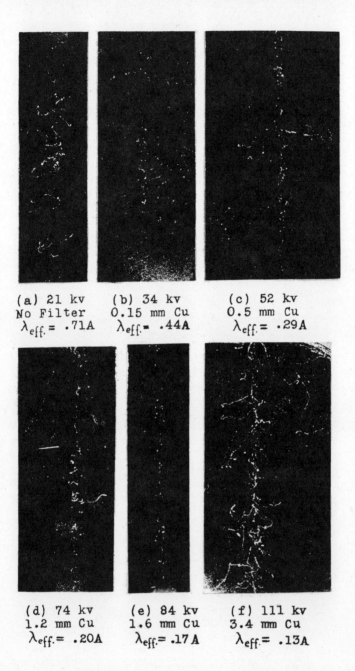

(a) 21 kv
No Filter
$\lambda_{eff.} = .71 A$

(b) 34 kv
0.15 mm Cu
$\lambda_{eff.} = .44 A$

(c) 52 kv
0.5 mm Cu
$\lambda_{eff.} = .29 A$

(d) 74 kv
1.2 mm Cu
$\lambda_{eff.} = .20 A$

(e) 84 kv
1.6 mm Cu
$\lambda_{eff.} = .17 A$

(f) 111 kv
3.4 mm Cu
$\lambda_{eff.} = .13 A$

Plate I. The x-rays pass from top to bottom. In addition to the copper filter, they traverse glass walls 4 mm thick. For the short waves the shorter (R) tracks increase rapidly in length and number. Thus while in (a) nearly all are P tracks, in (f) nearly all are R tracks.

quanta, and τ to the number of quanta spent in exciting photo-electrons, per centimeter path of the x-rays through the air.

In Table I we have recorded the results of the examination of the best 14 of a series of 30 plates taken at different potentials. The potentials given in column 1 of this table are based on measurements with a sphere gap. The potential measurements required corrections due to a slight warping of the frame holding the spheres, and to the lowering of the line voltage when the condenser was charged for the illuminating spark. The latter error was eliminated in the later photographs, at 34, 21, and 74 kv, and the former error was corrected by a subsequent measurement of the sphere gap distances, checked by a measurement of the lengths of the P tracks obtained at the lowest potential. The probable errors of potential measurements are thus unfortunately large, amounting to perhaps 10 percent in every case except that of 74 kv, which is probably accurate to within 5 per cent.

TABLE I
Number of tracks of types R and P.

Potential	Effective wave-length	Total tracks N	R tracks N_R	P tracks N_P	N_R/N_P	σ/τ
21kv	.71A	58	5	49	0.10	0.27
34	.44	24	10	11	0.9	1.2
52	.29	46	33	12	2.7	3.8
74	.20	84	74	8	9	10
84	.17	73	68	4	17	17
111	.13	79	72	1	72	32

The effective wave-lengths as given in column 2 are the centers of gravity of the spectral energy distribution curves after taking into account the effect of the filters employed. Because of the strong filtering, the band of wave-lengths present in each case is narrow, and the effective wave-length is known nearly as closely as the applied potential.

All the tracks originating in the path of the primary beam are recorded in column 3. Of these, the nature of some was uncertain. At the lower voltages it was difficult to distinguish the R tracks from the "sphere" tracks which Wilson has shown are often produced near the origin of a β-ray track by the fluorescent K rays from the oxygen or nitrogen atoms from which the ray is ejected. At the highest voltage the length of some of the R tracks is so great as to make it difficult to distinguish them from the P tracks. The numbers of R and P tracks shown in columns 4 and 5 are those of the tracks whose nature could be recognized with considerable certainty, the uncertain ones not being counted. This procedure probably makes the values of N_R/N_P in column 6 somewhat too small for the lower potentials and somewhat too great for the higher potentials.

The values of σ and τ given in column 7 are calculated from Hewlett's measurements[8] of the absorption of x-rays in oxygen and nitrogen. We have taken from his data the value of τ for 1 A to be 1.93 for air, and to vary as λ^3. The difference between the observed value of $\mu = \tau + \sigma$ and this value of τ gives the value of σ which we used.

The surprisingly close agreement between the observed values of N_P/N_R and the values of σ/τ we believe establishes the fact that the *R tracks are associated with the scattering of x-rays*. In view of the evidence that each truly absorbed quantum liberates a photo-electron or P track,[9] the equality of these ratios indicates that *for each quantum of scattered x-rays about one R track is produced*.

The fact that for the greater wave-lengths the ratio N_R/N_P seems to be smaller than σ/τ may mean that not all of the scattered quanta have R tracks associated with them. This would be in accord with the interpretation which has been given of the spectrum of scattered x-rays. The modified line has been explained by assuming the existence of a recoil electron, and the unmodified line as occurring when the scattering of a quantum results in no recoil electron. On this view the fact that the unmodified line is relatively stronger for the greater wave-lengths goes hand in hand with the observation that N_R/N_P is less than σ/τ for the greater wave-lengths. In view, however, of the meager data as yet available on this point, we do not wish to emphasize this correspondence too strongly.

Ranges of the R tracks. The range of the recoil electrons has been calculated on the basis of two alternative assumptions.[10] First, assuming that the electron recoils from a quantum scattered at a definite angle, its energy is found to be

$$E = h\nu \frac{2a \cos^2\theta}{(1+a)^2 - a^2 \cos^2\theta}, \quad (2)$$

where $a = h\nu/mc^2$, and θ is the angle between the primary x-ray beam and the direction of the electron's motion. This energy is a maximum when $\theta = 0$, and is then,

$$E_m = h\nu \frac{2a}{1+2a}. \quad (3)$$

[8] C. W. Hewlett, Phys. Rev. **17**, 284 (1921)
[9] See, e. g., A. H. Compton, Bull. Nat. Res. Council No. 20, p. 29, 1922
[10] See Compton and Hubbard, loc. cit.[7]

The second assumption is that the R electron moves forward with the momentum of the incident x-ray quantum. In this case the energy acquired is

$$E' = h\nu \cdot \tfrac{1}{2}\frac{a}{1+2a}(1-\tfrac{1}{4}a^2+ \cdots) . \qquad (4)$$

Eq. (3) was found to agree considerably better than Eq. (4) with Wilson's experimental results.

The lengths of the tracks shown on our photographs could be estimated probably within 10 or 20 per cent. These measured values, reduced to a final pressure of 1 atmosphere, are summarized in Table II. In column 2 are recorded the lengths of the longest tracks observed at each potential. S_m is the range calculated from Eq. 3, using C. T. R. Wilson's result[1] that the range of a β-particle in air is $V^2/44$ mm, where V is the potential in kilovolts required to give the particle its initial velocity, and the frequency ν employed is the maximum frequency excited by the voltage applied to the x-ray tube. S' is similarly calculated from Eq. (4).

TABLE II
Maximum lengths of R tracks.

Potential	Observed	Calc. (S_m)	Calc. (S')
21kv	0mm	0.06mm	0.004mm
34	0	0.3	0.02
52	2.5	1.8	0.1
74	6	6	0.4
88	9	12	0.7
111	24	25	1.5

It is evident that the observed lengths of the R tracks are not in accord with the quantity S' calculated from Eq. (4). They are, however, in very satisfactory agreement with the values of S_m given by Eq. (3). This result agrees with the conclusion drawn from Wilson's data,[11] but is now based upon more precise measurements. It follows that *the momentum acquired by an R particle* is not merely that of the incident quantum, but *is the vector difference between the momentum of the incident and that of the scattered quanta.*[12]

This conclusion is supported by a study of the relative number of tracks having different ranges. If the maximum range of the recoil electrons is S_m, Compton and Hubbard find[7] that the probability that the length of a given track will be S is proportional to

$$(2\sqrt{S/S_m}+\sqrt{S_m/S}-2) . \qquad (5)$$

[11] Compton and Hubbard, loc. cit.,[7] p. 449.
[12] That this is true for the β-rays excited by γ-rays has been shown in a similar manner by D. Skobeltzyn, Zeits. f. Phys. **28**, 278 (1924).

This expression assumes that the exciting primary beam has a definite wave-length. To calculate the relative number of tracks for different relative lengths to be expected, we have averaged this expression by a rough graphical method over the range of wave-lengths used in our experiments. These calculated values are given in the last column of Table III, for the relative ranges designated in column 1. A comparison of these

TABLE III
Relative lengths of R tracks.

Range of S/S_M	Per cent of R tracks within this range					Calc.
	52kv	74kv	88kv	111kv	Mean	
0- .2	44	66	60	54	56	53
.2- .4	34	20	26	32	28	22
.4- .6	19	8	4	8	10	14
.6- .8	0	3	5	3	3	8
.8-1.0	3	3	5	3	3	3

calculated values with the observed relative ranges shows a rather satisfactory agreement throughout. It will be noted further that the probabilities of tracks of different relative ranges is found to be about the same for x-rays excited at different potentials. This is in accord with the theoretical expression (5) for the probability, which is independent of the wave-length of the x-rays employed.

Angles of ejection of R tracks. On the view that the initial momentum of an R electron is the vector difference between the momenta of the incident and the scattered quantum, it is clear that these electrons should start at some angle between 0 and 90° with the primary beam. The probability that a given track will start between the angles θ_1 and θ_2 is on this hypothesis,[13]

$$\int_{\theta_1}^{\theta_2} P_\theta d\theta = 3ab \int_{\theta_1}^{\theta_2} \frac{a^2 \tan^4\theta + b^2}{(a \tan^2\theta + b)^4} \frac{\sin\theta}{\cos^3\theta} d\theta, \qquad (6)$$

where $a = (1 + h\nu/mc^2)^2$, and $b = (1 + 2h\nu/mc^2)$.

In our photographs only those taken at 111 kilovolts have tracks long enough to determine the initial direction with sufficient accuracy to make a reliable test of this expression. In all, the directions of 40 tracks were estimated, with the results tabulated in the second column of Table IV. In view of the fact that the photographs were stereoscopic, it was possible to estimate the angles in a vertical plane roughly, though not closer perhaps than within 10 or 15°. The values in the third column are calculated from Eq. (6). It is especially to be noted that, in accord with the

---

[13] See Compton and Hubbard, loc. cit.,[7] Eq. (14).

theory, no R tracks are found which start at an angle greater than 90° with the primary x-ray beam. In view of the small number of tracks observed and the approximate character of the angular estimates, the agreement between the two sets of values is as close as could be expected.

A more searching test of the assumption that the R tracks are electrons which have recoiled from scattered quanta is a study of the relative ranges of the tracks starting at different angles. (See columns 4 and 5 of Table IV.) The calculated ranges in column 5 are based on Eq. (2) for

TABLE IV
Number and range of R tracks at different angles, for 111 kv x-rays.

Angle of emission	Per cent of total number		Average range	
	(obs.)	(calc.)	(obs.)	(calc.)
0°–30°	34	28	9 mm	11 mm
30°–60°	39	50	4	4
60°–90°	27	22	0.9	0.3

the energy at different angles. In this calculation the effective wave-length, as estimated in connection with Table I, is employed. It will be seen that the observed ranges of the tracks ejected at small angles are much greater than that of those ejected at large angles, in substantial agreement with the theory.

It is worth calling particular attention to the fact that in comparing the theoretical and experimental values in these tables, no arbitrary constants have been employed. The complete accord between the predictions of the theory and the observed number, range, and angles of emission of the R tracks is thus of especial significance.

The evidence is thus very strong that there is about one R track or recoil electron associated with each quantum of scattered radiation, and that this electron possesses, both in direction and magnitude, the vector difference of momentum between the incident and the scattered x-ray quantum. Our results therefore afford a strong confirmation of the assumptions used to explain the change in wave-length of x-rays due to scattering, on the basis of the quantum theory.

RYERSON PHYSICAL LABORATORY,
UNIVERSITY OF CHICAGO.
November 15, 1924.

THE DENSITY OF ROCK SALT AND CALCITE

By O. K. DeFoe and Arthur H. Compton

Abstract

Crystals as perfect as could be secured were obtained from a variety of sources, and densities were measured by weighing them when suspended in paraffine oil or water respectively. The weights were calibrated and correction was made for the buoyancy of the air. The measurements are accurate to .2 mg cm^{-3} for rock salt and to .1 mg cm^{-3} for calcite. Variations in density were observed between different parts of the same crystal, as well as between different crystals, which are larger than the experimental error. The estimated uncertainty of the mean density is taken as the probable variation of the individual samples from the mean density. The mean result for 7 good crystals of rock salt is 2.1632±.0004 gm cm^{-3}, and for 6 crystals of calcite, 2.7102±.0004 gm cm^{-3}, at 20°C.

IN order to increase the precision of absolute x-ray wave-lengths, we have undertaken the redetermination of the densities of rock salt and calcite. For this purpose we have secured samples as perfect as we could obtain, from a variety of different sources. Some of these were loaned to us by the National Museum at Washington, to whom we wish to express our grateful acknowledgment.

Our experimental method was in its essentials that described by D. C. Miller.[1] The weights used were carefully calibrated, and the weighings were corrected for the buoyancy of the air. The density of calcite was determined by suspension in freshly boiled distilled water, whose density was used as the standard. For rock salt, paraffine oil was used in order not to dissolve the crystal. The density of the oil was compared with that of water through the medium of a plumb bob, which consisted in one set of measurements of fused quartz, in another of Pyrex glass and in a third of a calcite crystal. Because of the comparatively large thermal expansion coefficient of the oil, the density determinations for rock salt were not as precise as those for calcite.

The results of the measurements on nearly perfect crystals are given in the following table. We have not counted measurements on crystals in which serious imperfections were observed. In reducing the results to 0°C, we have taken Benoit's values for the thermal expansion coefficient of calcite[2] as +0.0000251 deg.$^{-1}$ parallel with the crystal axis and −0.0000056 deg.$^{-1}$ perpendicular to the axis. It follows that $\rho_0 =$

---

[1] D. C. Miller, "Laboratory Physics" (1903) pp. 120-123.
[2] Cf Kaye and Laby, "Physical and Chemical Constants" (1911) p. 53.

$\rho(1+.0000139t)$ where t is the centigrade temperature. Similarly for rock salt,[3] $\rho_0 = \rho(1+.0001212\,t)$. The average temperature at which the measurements were made was about 23°.

TABLE I

Densities of rock salt and calcite
I and II refer to two parts of the same crystal.

Rock salt		Calcite	
Origin	Density (0°C)	Origin	Density (0°C)
Michigan I	2.1687	Missouri	2.7114
Michigan II	2.1683	U. S. A.	2.7110
Arizona I	2.1687	Iceland	2.7111
Arizona II	2.1683	Iceland	2.7098*
Germany I	2.1687	Unknown	2.7109
Germany II	2.1676*	Unknown	2.7112
Unknown	2.1687		
Mean	2.1685	Mean	2.7110

* Weighted 1/2 in calculating mean.

In the case of calcite the determinations of the density of the individual samples were made within a probable error of $\pm.0001$ gm cm^{-3}, and for rock salt within $\pm.0002$ gm cm^{-3}. The experiments thus indicate that different samples of apparently perfect crystals vary in density by amounts greater than the experimental error of our measurements. No chemical analysis was attempted to search for possible impurities which might account for these variations. One of the samples of rock salt (Michigan I) was however tested by Mr. C. C. Van Voorhis for occluded water, by melting in an evacuated Pyrex tube and drawing the liberated gases over phosphorous pentoxide. A similar test with no salt present showed that, if any moisture was occluded in the rock salt, it was an amount small compared with that which escaped from the Pyrex container, even though this had previously been baked out at a high temperature.

In view of these variations in density from one part of an apparently perfect crystal to another, we are unable to determine which value represents the true density of a perfect crystal. As an estimate of the probable error in the value of the density we are therefore on safer ground if we choose the probable variation of the density of any individual crystal from the mean value for all crystals. Thus we determine the following densities:

Calcite: $\rho = 2.7110 \pm .0004$ gm cm^{-3} at 0° C.
$= 2.7102 \pm .0004$ gm cm^{-3} at 20° C.

[3] Cf Landolt, Bornstein, Roth, "Tabellen" (1912), p. 336.

Rock salt: $\rho = 2.1685 \pm .0004$ gm cm^{-3} at 0°C.
$= 2.1632 \pm .0004$ gm cm^{-3} at 20°C.

These values are compared with those given by other authorities, in Table II.

TABLE II

Values for the density of calcite and rock salt

	Authority	Date	Density
Calcite	Clarke[4]	1873	2.723
	Schroeder[5]	1874	2.702
	Retgers[6]	1890	2.712
	Dana[7]	1892	2.65 ± .05
	Bragg[8]	1915	2.71
	Compton[9]	1916	2.7116 ± .0004 (at 18°)
	Ledeaux, Lebard and Dauvillier[10]	1919	2.7125 ± .0015
	This determination	1925	2.7102 ± .0004 (at 20°)
Rock salt	Clarke[9]	1873	2.135
	Retgers[11]	1889	2.167 (at 17°)
	Dana[7]	1892	2.15 ± .05
	Krickmeyer[12]	1896	2.174 (at 20°)
	Haigh[13]	1912	2.170 (at 20°)
	Moseley[14]	1913	2.167
	Baxter and Wallace[15]	1916	2.161
	This determination	1925	2.1632 ± .0004 (at 20°)

A calculation of the grating spaces of rock salt and calcite based on this determination of their densities is described elsewhere in this journal.[16]

WASHINGTON UNIVERSITY (O. K. D.)
UNIVERSITY OF CHICAGO (A. H. C.)
January 31, 1925

---

[4] Clarke, "Constants of Nature" (1873) v. 1, p. 30.
[5] Schroeder, "Neue Handbuch f. Minerologie" (1874) p. 805.
[6] Retgers, Zeits. Phys. Chem. **6**, 193 (1890).
[7] Dana, "System of Minerology" (1892).
[8] W. H. Bragg and W. L. Bragg, "X-rays and Crystal Structure" (1915) p. 115.
[9] A. H. Compton, Phys. Rev. **7**, 646 (1916). In this measurement no correction was made for the buoyancy of the air. Such a correction makes this value agree well with the present determination.
[10] Ledeaux, Lebard and Dauvillier, Compt. Rend. Nov. 24, 1919. This value is based chiefly on that by A. H. Compton,[9] and is hence also subject to a correction for the buoyancy of the air.
[11] Retgers, Zeits. Phys. Chem. **3**, 289 (1889)
[12] Krickmeyer, Zeits. Phys. Chem., **21**, 53 (1896).
[13] Haigh, J. Am. Chem. Soc. **34**, 1137 (1912)
[14] H. G. J. Moseley, Phil. Mag. **26**, 1024 (1913).
[15] Baxter and Wallace, J. Am. Chem. Soc. **38**, 26 (1916).
[16] A. H. Compton, H. N. Beets and O. K. DeFoe, in this issue, p. 625.

THE GRATING SPACE OF CALCITE AND ROCK SALT

By A. H. Compton, H. N. Beets* and O. K. DeFoe

Abstract

Grating spaces of calcite and rock salt are recalculated using the new values for the crystal densities obtained by DeFoe and Compton, and come out for calcite $3.0291 \pm .0010$A and for rock salt, $2.8147 \pm .0009$A at 20°C. The values for each degree from 15° to 25° are given in a table. These values are slightly greater than those usually given. The remaining uncertainty is due chiefly to that in Avogadro's number. The ratio $D(\text{CaCO}_3)/D(\text{NaCl})$ comes out $1.0762 \pm .0002$ (18°C) which agrees satisfactorily with Siegbahn's value, 1.0764. The uncertainty in this value is chiefly associated with the molecular weights.

Absolute wave-length values corrected for refraction.—The correction is made by using an effective grating space slightly smaller than the true value. For the first order from calcite the effective value is 3.0287 and for rock salt 2.8144A, both at 20°C. Using these values the *wave-length of Mo Kα_1* line is found (Siegbahn) to be $.70749 \pm .00023$A.

ACCORDING to Bragg's law the wave-length λ of a beam of x-rays reflected from a crystal at a glancing angle θ is given by

$$n\lambda = 2D\sin\theta, \tag{1}$$

or more precisely by

$$n\lambda = 2D\sin\theta(1 - \delta/\sin^2\theta), \tag{2}$$

where n is the order of the spectrum, D is the grating space or distance between successive layers of atoms in the crystal grating, and $\delta = 1 - \mu$, μ being the index of refraction of the x-rays in the crystal. At present it is possible to measure θ to a degree of accuracy such that by far the greater part of the error in determining the absolute wave-length of a beam of x-rays is due to the uncertainty of D. It is thus of obvious importance to determine this grating space with precision.

From very fundamental considerations of the principles underlying crystal structure, it can be shown that the grating space of a rhombohedral crystal is given by

$$D = (nM/\rho N \varphi(\beta))^{1/3}, \tag{3}$$

where n is the number of molecules per elementary rhombohedron, M is the molecular weight, ρ the density, N the number of molecules per gram molecule, and $\varphi(\beta)$ is the volume of a rhombohedron the distance

---

* Coffin Foundation Fellow.

between whose opposite faces is unity and the angle between whose edges, β, is that between the axes of the crystal. For the two most commonly used crystals, rock salt and calcite, $n=1/2$, and M is known with rather high precision. For rock salt, $\beta=90°$, whence $\varphi(\beta)=1$; but for calcite the uncertainty of the value of β has been considered an appreciable source of error.[1] The density ρ does not appear to have been measured for either crystal with the care its importance would warrant, so that it also contributes appreciably to the probable error. In estimating D, the uncertainty of the value of N introduces the greatest single source of error, though no greater than that which Uhler estimates[1] as due to the combined error of ρ and β in the case of calcite.

By introducing the values of ρ and β which we have measured as described elsewhere in this journal, we are able to reduce somewhat the probable error of the calculation of the grating spaces of calcite and rock salt from Eq. (3). We have:

For calcite,
$n = 1/2$ [2]
$M = 100.075 \pm .03$ [3]
$\rho = 2.7102 \pm .0004$ at $20°$ [4]
$N = (6.0594 \pm .006) \times 10^{23}$ [5]
$\varphi(\beta) = 1.09630 \pm .00007$ at $20°$ [6]

For rock salt,
$n = 1/2$ [2]
$M = 58.46 \pm .02$ [3]
$\rho = 2.1632 \pm .0004$ at $20°$ [4]
$N = (6.0594 \pm .006) \times 10^{23}$ [5]
$\varphi(\beta) = 1$.

Substituting these values in equation (3) we obtain
$$D(\text{CaCO}_3) = (3.0291 \pm .0010) \times 10^{-8} \text{ cm at } 20° \text{ C};$$
$$D(\text{NaCl}) = (2.8147 \pm .0009) \times 10^{-8} \text{ cm at } 20° \text{ C}.$$

The thermal expansion coefficient of these crystals, though not large, is sufficient to make corrections necessary for precise measurements. Perpendicular to the cleavage faces, this coefficient is for calcite[7] 0.0000104 and for rock salt[8] 0.0000404 per degree centigrade. If we use the value 3.02910 as the value of the grating space of calcite, and 2.81470 as that of rock salt at 20°C, their values at other temperatures are given in the following table.

---

[1] H. S. Uhler, Phys. Rev. **12**, 42 (1918).

[2] W. H. Bragg and W. L. Bragg, "X-Rays and Crystal Structure," (1915).

[3] International Atomic Weights, 1921. The estimates of the accuracy are our own, made chiefly on the basis of the data collected by F. W. Clarke, "A Recalculation of Atomic Weights" (1920).

[4] O. K. DeFoe and A. H. Compton in this issue, p. 618.

[5] R. T. Birge, Phys. Rev. **14**, 365 (1919).

[6] H. N. Beets, preceding paper.

[7] M. Siegbahn, "Spektroskopie der Roentgenstrahlen," p. 86 (1924).

[8] Fizeau. Cf Landolt, Bornstein, Roth "Tabellen" p. 336 (1912).

Table I
Grating space of calcite and rock salt at different temperatures

T	$D(CaCO_3)$	$D(NaCl)$
15°C	3.02894	2.81413
16	3.02898	2.81425
17	3.02901	2.81436
18	3.02904	2.81447
19	3.02907	2.81459
20	(3.02910)	(2.81470)
21	3.02913	2.81481
22	3.02916	2.81493
23	3.02919	2.81504
24	3.02923	2.81515
25	3.02926	2.81527

It is interesting to compare these values of the grating spaces with those previously employed.

For calcite

D	Authority	Date
3.04×10^{-8} cm	W. H. Bragg[9]	1914
3.028	W. S. Gorton[10]	1916
3.0279	A. H. Compton[11]	1916
3.0307	Uhler and Cooksey[12]	1917
3.030	Millikan[13]	1917
3.0281	A. H. Compton[14]	1918
$3.0283 \pm .0022$ at 18°	M. Siegbahn[15]	1919
3.02904		
3.028	Duane[16]	1920
3.02855	McKeehan[17]	1922
$3.0291 \pm .0010$ at 20°	This determination	1925

For rock salt

D	Authority	Date
2.80×10^{-8} cm	W. L. Bragg[18]	1913
2.814	Moseley[19]	1913
2.814	E. Wagner[20]	1916
2.810	Davey[21]	1921
2.814	Siegbahn[22]	1924
2.815	Davey[23]	1924
$2.8147 \pm .0009$ at 20°	This determination	1925

[9] W. H. Bragg, Proc. Roy. Soc. A **89**, 468 (1914).
[10] W. S. Gorton, Phys. Rev. **7**, 209 (1916).
[11] A. H. Compton, Phys. Rev. **7**, 655 (1916).
[12] H. S. Uhler and C. D. Cooksey, Phys. Rev. **10**, 645 (1917); by comparison, assuming $D(NaCl) = 2.8140$ A.
[13] R. A. Millikan, Phil. Mag. **34**, 13 (1917).
[14] A. H. Compton, Phys. Rev. **11**, 431 (1918).
[15] M. Siegbahn, Phil. Mag. **37**, 601 (1919); 3.02904 obtained by comparison, assuming $D(NaCl)$ at 18° = 2.81400 A.
[16] W. Duane, Bull. Nat. Res. Council No. 7 (1920).
[17] L. W. McKeehan, Science, N. S. **56**, 757 (1922); average by comparison with rock salt.
[18] W. L. Bragg, "X-Rays and Crystal Structure," (1915) p. 110.
[19] H. G. J. Moseley, Phil. Mag. **26**, 1024 (1913).
[20] E. Wagner, Ann. der Phys. **49**, 625 (1916).
[21] W. P. Davey, Science, **54**, 497 (1921).
[22] [23] See footnotes on the next page.

There has heretofore been a discrepancy between the observed and the calculated values of the ratio $D(CaCO_3)/D(NaCl)$, which has seemed larger than the probable error of the determinations.[24] It will be noticed that this ratio does not involve N, which is responsible for the greatest part of the probable error of D. Our calculated value of this ratio is $1.0762 \pm .0002$ at 18°, in which the greatest part of the uncertainty is due to the molecular weights. This is to be compared with Uhler and Cooksey's[12] value 1.07701 and Siegbahn's[15] value 1.076417 at 18°, both of which were determined by direct spectrometric methods. It will be seen that the agreement with Siegbahn's value, which is the more accurate of the two experimental determinations, is almost within the estimated probable error. There thus remains no significant discrepancy between the calculated and the observed ratios of the grating spaces of rock salt and calcite.

In view of the fact that calcite crystals are much more nearly perfect than are those of rock salt, it is natural to choose calcite for our basis of wave-length measurements. It is a fortunate, though largely fortuitous circumstance that the value which we obtain for the grating space of calcite at 18° is identical with that employed by Siegbahn, namely 3.02904 A. Thus as far as the grating space is concerned our results suggest no correction whatever to his values.

In order to get absolute wave-length measurements it is of course necessary to take into account the refraction of the x-rays in the diffracting crystal. As Ewald and others have shown, if the x-rays are of considerably higher frequency than the maximum critical frequency of the diffracting crystal, this correction can be made by merely altering slightly the grating space.[25] In this case[26]

$$\delta = ne^2\lambda^2/2\pi mc^2, \tag{4}$$

where n is the number of mobile electrons per unit volume of the crystal, e, m and c have their usual meaning, and λ is the wave-length of the incident rays. If the number of effective electrons per atom is equal to the atomic number, this becomes for calcite[27]

$$\delta = 3.67\lambda^2 \times 10^{-6}$$

if λ is expressed in angstroms. If this value is substituted in Eq. (2),

---

[22] M. Siegbahn, "Spektroskopie der Roentgenstrahlen," p. 20 (1924).

[23] W. P. Davey, General Electric Review, **27**, 744 (1924).

[24] Cf A. H. Compton, Phys. Rev. **11**, 431 (1918); M. de Broglie, "Les Rayons X" p. 34 (1922); and especially M. Siegbahn, "Spektroskopie der Roentgenstrahlen," p. 87 (1924).

[25] P. P. Ewald, Phys. Zeit. **21**, 617 (1920).

[26] A. H. Compton, Phil. Mag. **45**, 1123 (1923).

[27] A. H. Compton, l.c.[26] p. 1129.

noting that $\sin \theta = n\lambda/2D$ to a sufficient approximation, we obtain
$$n\lambda = 2D \sin\theta(1 - 0.0000147\ D^2/n^2). \qquad (5)$$
The true wave-length may then be calculated from Eq. (1) if, for the first order of reflection from calcite, we use an effective grating space of $3.0291\ (1-0.000135) = 3.0287$ A.

For rock salt the corresponding correction factor is $(1-0.000096)$. That is, to find the true wave-length we must use for the first order spectrum an effective grating space of 2.8144 A at 20°C.

As a typical case of the determination of absolute wave-length we may consider the $K\alpha_1$ line of molybdenum. Using 3.02904 A as the grating space of calcite at 18°, Siegbahn gives for this wave-length[28] 0.70759 A. Though our present work gives the same value for the true grating space, the correction due to refraction changes[29] the effective grating space for the first order spectrum to 3.02864, reducing the wave-length to 0.70749 A.

We accordingly recommend the fol owing values as the most probable in light of this investigation:

$D(CaCO_3) = (3.0291 \pm .0010) \times 10^{-8}$ cm at 20° C.

$D(NaCl) = (2.8147 \pm .0009) \times 10^{-8}$ cm at 20° C.

Mo $K\alpha_1$ line, $\lambda = 0.70749 \pm .00023$ A.

UNIVERSITY OF CHICAGO (A. H. C. and H. N. B.)
WASHINGTON UNIVERSITY (O. K. D.)
January 31, 1925

---

[28] M. Siegbahn, "Spektroskopie der Roentgenstrahlen," p. 102 (1924).

[29] Professor Siegbahn informs us that he has applied a correction of this character to some of his most recent wave-length determinations.

ON THE MECHANISM OF X-RAY SCATTERING

By Arthur H. Compton

Department of Physics, University of Chicago

Read before the National Academy April 28, 1925

It is now well known that whereas the usual form of the wave theory is inadequate to account for the alteration of the wave-length of X-rays when scattered, this may be explained simply on the hypothesis that the X-rays consist of quanta, each of which interacts with an individual electron. Strong support of this hypothesis is found in the discovery that there appear electrons each with about the momentum it should have acquired if it had deflected an X-ray quantum. That these electrons are associated with the scattered X-rays is evident from the fact that there exists on the average about one such electron for each quantum of scattered X-rays. Nevertheless the great weight of the evidence from many sources for the wave theory of radiation makes the acceptance of this radiation quantum hypothesis difficult.

It is improbable that the idea of spreading waves of radiant energy can be successfully reconciled with these studies of X-ray scattering without abandoning the principles of the conservation of energy and momentum. However, by making this bold step, Bohr, Kramers and Slater have apparently been able to incorporate them within the general scheme of the wave theory.

While discussing this matter in November 1923, Professor Swann called the attention of Professor Bohr and myself to the fact that a crucial test between the two theories could be made if it were possible to detect simultaneously a scattered quantum and the recoiling electron associated with it. For on any spreading wave theory, including that of Bohr, Kramers and Slater, there should be no correlation between the direction of ejection of the recoil electron and that of the effect of the scattered quantum. On the quantum view, however, if the recoiling electron is ejected at an angle θ with the incident X-ray beam, the scattered quantum should appear at an angle ϕ such that

$$\tan \frac{1}{2} \phi = - \frac{1}{1 + \alpha} \cot \theta,$$

where $\alpha \equiv h/mc\lambda$.

Two methods of making this test suggest themselves. One is to use two point counters of the type developed by Geiger, Kovarik and others, one to receive the recoil electron and the other to catch the scattered quantum. The quantum hypothesis would predict simultaneous impulses in both chambers when set at the proper angles. This method has been employed by Mr. R. D. Bennett in our laboratory, and has recently been suggested also by Bothe and Geiger[1] as a means of testing Bohr's theory.

The second method, which Mr. Simon and I have been using, is to take cloud expansion photographs showing simultaneously a recoil electron and a secondary β-ray track produced by the scattered X-ray quantum. To increase the probability of emission of a β-ray by a scattered quantum, diaphragms of thin lead foil are placed inside the chamber. Thus about 1 in 50 of the recoil electrons is found to have a secondary β particle associated with it.

Both methods have yielded provisional results. Mr. Bennett has connected each of a pair of head phones with one of the counting chambers

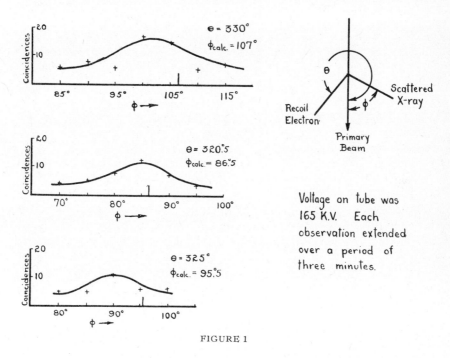

FIGURE 1

through a 3 stage amplifier, and has listened for simultaneous impulses in the two phones. He uses a thin strip of mica to scatter the X-rays, and places the counting chambers inside an evacuated vessel to avoid effects due to the curving of recoil electrons in air. His results for three different settings of the electron counter are shown in Figure 1. The fact that the coincidences are more frequent when the quantum counter is near the theoretical angle is in support of the quantum theory. Mr. Bennett hopes to be able to publish soon results obtained by a recording method, to avoid the uncertainties inherent in auditory observations of this type.

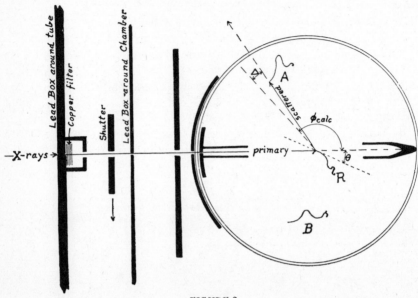

FIGURE 2

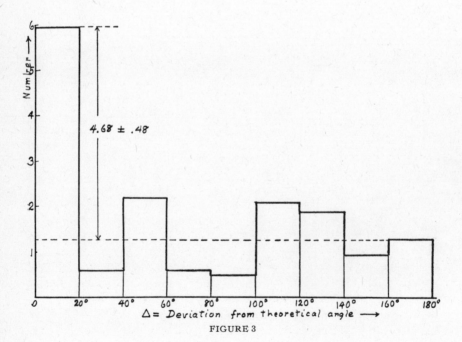

FIGURE 3

Mr. Simon and I have taken about 750 stereoscopic cloud expansion photographs, using apparatus such as that shown diagrammatically in figure 2. Our chief difficulty has been to eliminate stray X-rays, which give rise to meaningless tracks often indistinguishable from those due to scattered quanta. The results of the last 350 plates, in which these rays were reduced to a relatively low intensity, are shown in figure 3. Here I have plotted the deviation Δ of the observed tracks from the theoretical angle. On the spreading wave theory, the values of Δ should be nearly uniformly distributed between 0 and 180 degrees. On the quantum theory they should be concentrated near 0, as is obviously the case. The occurrence of tracks at other angles is explicable as due in part to stray X-rays and in part to the method of plotting the results.

A more detailed account of the work will be published when further experiments, which are now in progress have been completed. The results already obtained, however, permit us to state, with very little uncertainty, that the direction in which a quantum of scattered X-rays can produce an effect is determined at the moment it is scattered, and can be predicted from the direction of motion of the recoil electrons. In other words, *scattered X-rays proceed in directed quanta.*

It is possible to clothe this statement in the language of the wave-theory, if we keep in mind that a wave with a single quantum of energy can produce an effect in only one direction.[2]

[1] W. Bothe and H. Geiger, *Zeitschr. Physik.*, **26** (1924) 44.

[2] Since this paper was read before the Academy, I have received a letter from Dr. Bothe informing me that H. Geiger and he have also observed the coincidences demanded by the quantum theory but contrary to the theory of Bohr, Kramers and Slater. Their work will be described soon in *die Naturwissens haften*.

THE PHYSICAL REVIEW

DIRECTED QUANTA OF SCATTERED X-RAYS

By Arthur H. Compton and Alfred W. Simon[1]

Abstract

Relation between the direction of a recoil electron and that of the scattered x-ray quantum.—It has been shown by cloud expansion experiments previously described, that for each recoil electron produced, an average of one quantum of x-ray energy is scattered by the air in the chamber. If the quantum of scattered x-rays produces a β-ray in the chamber, then a line drawn from the beginning of the recoil track to the beginning of the β-track gives the direction of the ray after scattering. Using a chamber 18 cm in diam. and 4 cm deep traversed by a carefully shielded narrow beam of homogeneous x-rays, with exploded tungsten wires as sources of light, nearly 1300 stereoscopic cloud expansion photographs were taken. Of the last 850 plates, 38 show both recoil tracks and β-tracks. The angles projected on the plane of the photographs were measured and it was found that in 18 cases, the direction of scattering is within 20° of that to be expected if the x-ray is scattered as a quantum so that energy and momentum are conserved during the interaction between the radiation and the recoil electron. This number 18 is four times the number which would have been observed if the energy of the scattered x-rays proceeded in spreading waves, that is if the direction of production of a β-ray was unrelated to the direction of the recoil track. The chance that this agreement with theory is accidental is about 1/250. The other 20 β-rays are ascribed to stray x-rays and to radioactivity. This evidence seems a direct and conclusive proof that at least a large proportion of the scattered x-rays proceed in directed quanta of radiant energy.

AN increasingly large group of phenomena has recently been investigated which finds its simplest interpretation on the hypothesis of radiation quanta, proposed by Einstein to account for heat radiation and the photo-electric effect.[2] It has not been possible, however, to show that any of these phenomena necessarily demand this hypothesis for its explanation. Thus, for example, the photo-electric effect is not inconsistent with the view that the light energy proceeds from its source in expanding waves, if we postulate the existence within atoms of a

---

[1] National Research Fellow.
[2] An interesting summary of this work has been presented by K. K. Darrow in the Bell Technical Journal **4**, 280 (1925).

mechanism for storing energy until a quantum has been received. It is true that no such mechanism is known; but until our knowledge of atomic structure is increased it would be premature to assert that such a storing mechanism cannot exist. The change of wave-length of x-rays when scattered and the existence of recoil electrons associated with scattered x-rays, it is true, appear to be inconsistent with the assumption that x-rays proceed in spreading waves if we retain the principle of the conservation of momentum.[3] Bohr, Kramers and Slater,[4] however, have shown that both these phenomena and the photo-electric effect may be reconciled with the view that radiation proceeds in spherical waves if the conservation of energy and momentum are interpreted as statistical principles.[5]

A study of the scattering of individual x-ray quanta and of the recoil electrons associated with them makes possible, however, what seems to be a crucial test between the two views of the nature of scattered x-rays.[6] On the idea of radiation quanta, each scattered quantum is deflected through some definite angle ϕ from its incident direction, and the electron which deflects the quantum recoils at an angle θ given by the relation[7]

$$\tan \tfrac{1}{2}\phi = -1/[(1+a)\tan \theta], \tag{1}$$

where $a = h/mc\lambda$. Thus a particular scattered quantum can produce an effect only in the direction determined at the moment it is scattered and predictable from the direction in which the recoiling electron proceeds. If, however, the scattered x-rays consist of spherical waves, they may produce effects in any direction whatever, and there should consequently be no correlation between the direction in which recoil electrons proceed and the directions in which the effects of the scattered x-rays are observed

To make the test it is, of course, necessary to observe the individual recoil electrons and to detect the individual scattered quanta.[8] This we have done by means of a Wilson cloud expansion apparatus, in the manner shown diagrammatically in Fig. 1. In a recent statistical study,

[3] Cf A. H. Compton, Jour. Franklin Inst. **198**, 71 (1924).

[4] N. Bohr, H. A. Kramers and J. C. Slater, Phil. Mag. **47**, 785 (1924).

[5] It seems that there still remains a difficulty in accounting on this view for the intensity of scattered x-rays. See Y. H. Woo, Phys. Rev. **25**, 444 (1925).

[6] The possibility of such a test was suggested by W. F. G. Swann in conversation with Bohr and one of us in November 1923.

[7] Cf P. Deybe, Phys. Zeits. (Apr. 15, 1923); A. H. Compton and J. C. Hubbard, Phys. Rev. **23**, 444 (1924).

[8] A preliminary account of this work has been given in Proc. Nat. Acad. Sci. **11**, 303 (June, 1925).

we have found[9] that on the average there is produced about one recoil electron for each quantum of scattered x-rays. Each recoil electron produces a visible track, and occasionally a secondary track is produced by the scattered x-ray. When but one recoil electron appears on the same plate with the track due to the scattered rays, it is possible to tell at once whether the angles satisfy Eq. (1). If the photographs and the measurements can be made with sufficient definiteness, the experiment should thus give an unequivocal answer to the question whether the energy of a scattered x-ray quantum is distributed over a wide solid angle or proceeds in a definite direction.[10]

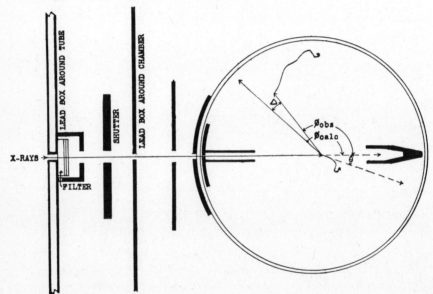

Fig. 1. Diagram of apparatus. On the hypothesis of radiation quanta, if a recoil electron is ejected at an angle θ, the scattered quantum must proceed in a definite direction ϕ_{calc}. In support of this view, many secondary β-ray tracks are found at angles ϕ_{obs} for which Δ is small.

Experimental procedure. In the final experiments we used a high voltage Coolidge x-ray tube excited by an unrectified alternating cur-

[9] A. H. Compton and A. W. Simon, Phys. Rev. **25**, 306 (1925).

[10] W. Bothe and H. Geiger have proposed (Zeits. f. Phys. **26**, 44, 1924) and carried out (Naturwissenschaften **20**, 440, May 15, 1925) a rather similar test. Using two point counters, one to receive the scattered x-rays and the other to receive the recoil electrons, they have found that many of the recoil electrons occur simultaneously with β-rays excited by the scattered x-rays. While such an experiment affords less definite evidence than does the present one regarding the directive nature of the scattered x-rays, it is equally incompatible with Bohr, Kramers and Slater's statistical view of the production of photo and recoil electrons.

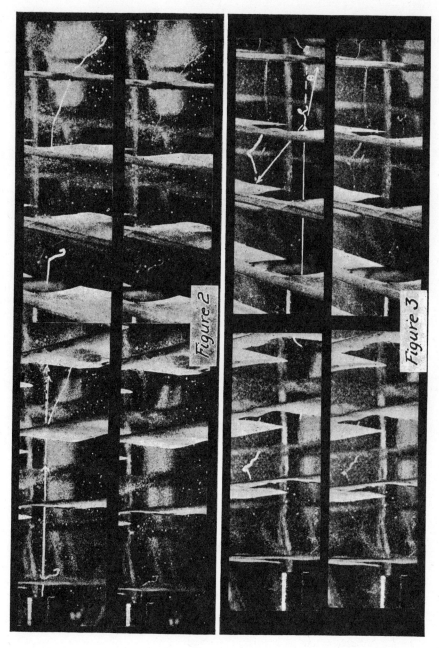

Fig. 2. A square hit. Plate 597. 1 recoil track, 1 secondary. $\theta = -2°$; $\phi_{calc} = +175°$; $\phi_{obs} = +177°$; $\Delta = 2°$.
Fig. 3. Plate 560. 2 recoil tracks, 1 secondary. $\theta = -25°$; $\phi_{calc} = +120°$; $\phi_{obs} = +120°$, $\Delta = 0°$. Recoil track (2) lies in wrong plane to be associated with secondary track.

rent at 140 peak kilovolts. Potentials as high as 250 kv were tried, but the resulting tracks of the recoil electrons were inconveniently long, and it was difficult to shield the expansion chamber adequately from the very penetrating direct rays. The x-rays were rendered approximately homogeneous by filtering through 6 mm of brass and about 2 mm each of copper and aluminium. Effects of stray x-rays were reduced to a minimum by surrounding the x-ray tube by a box of 9 mm lead, surrounding the expansion chamber with a box of 3 mm lead, and interposing suitable additional lead screens as shown diagrammatically in Fig. 1. The illumination was produced by exploding tungsten wires,[11] 0.1 mm in diameter and 14 cm long, by condenser discharges. The condenser, of about 0.1 microfarad capacity, was charged to about 70 kv by means of a separate transformer and "kenotron" thermionic rectifier. It was found necessary to surround this rectifier also with a box of 3 mm lead in order to avoid stray x-rays. The light entered the expansion chamber horizontally, at right angles to the primary beam, and the photographs were taken through the top of the expansion chamber in such a manner that the light was scattered by the water droplets at an angle of about 40°. We used an Ontoscope stereoscopic camera of 5.5 cm focal length, with the lenses stopped down to $f/8$. This gave ample exposure and a focus so deep that a β-ray track in any part of the chamber could be identified. Both the camera and the illuminating spark were enclosed within the lead box surrounding the expansion chamber.

In order to increase the probability that the scattered x-rays would produce β-ray tracks, a comparatively large chamber, about 18 cm in diameter by 4 cm high, was used. Diaphrams of thin lead foil were also suspended inside the chamber (see Figs. 2 to 5), so that the scattered x-rays might make themselves evident by ejecting photo-electrons. We allowed the primary rays to enter through a collimating lead tube, and absorbed them in a hollow lead cone. This eliminated almost completely the effect due to the scattering of the primary rays by the glass walls. When the expansion occurred, however, a slight blast of air proceded from each of these lead tubes, and distorted the tracks for a distance of about 2 cm from the collimator and about 1 cm from the absorbing cone. It was accordingly possible to make accurate measure-

---

[11] The illumination obtained using the tungsten wires was roughly 10 times as brilliant photographically as that from a mercury spark at atmospheric pressure, using the same electrical energy.

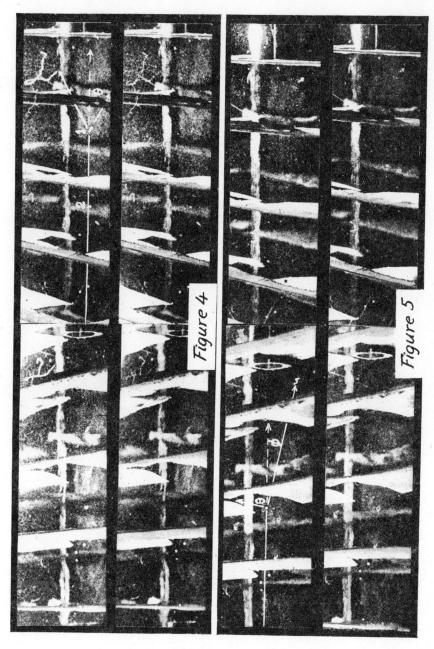

Fig. 4. Plate 1018. 2 recoil tracks, 1 secondary. $\theta = -55°$; $\phi_{calc} = +59°$, $\phi_{obs} = +50°$; $\Delta = 9°$.
Fig. 5. A glancing blow. Plate 725. 1 recoil (sphere) track, 1 secondary. $\theta = \pm 85°$ as estimated from length of track; $\phi_{calc} = \pm 8°$, $\phi_{obs} = -22°$; $\Delta = 14°$ or $30°$.

ments only on those recoil electrons which were ejected from a column of air about 4 cm long near the middle of the expansion chamber.

An examination of the photographs indicates that there was 1 β-ray produced by x-rays scattered from this air column for about every 50 recoil electrons originating in this region, in satisfactory accord with a rough calculation based on the absorption coefficients of the x-rays and β-rays. There were in addition a considerable number of stray β-rays, due in part to stray x-rays and to radioactivity, as was shown by preventing the direct rays from entering the chamber, and probably in part also to the incompletely shielded scattered rays from the walls of the expansion chamber. The presence of these stray rays does not, however, present a serious difficulty in interpreting the photographs, since obviously there can be no correlation between their positions and the directions of ejection of the recoil electrons. Thus the effect of the scattered rays is a definite one superposed upon a random effect due to the stray β-rays.

The photographs. In making the preliminary adjustments and for auxiliary experiments we took about 140 photographs. The pictures which were useful in the final test were divided into three series. In the first series there were 302 plates. The second series of 338 plates and the third series of 511 plates were taken under improved conditions of x-ray shielding, thus reducing the number of stray β-ray tracks. There appear on the average two or three recoil tracks in each picture.

Typical photographs in which the secondary tracks appear are shown stereoscopically ($\times 1.3$) in Figs. 2, 3, 4 and 5. In each case a retouched photograph with arrows marking the direction of the primary, recoil and scattered rays is placed above the untouched photograph. Unfortunately there appear on these plates also other marks due to water drops, pieces of lint, bubbles in the glass, etc. While these mar the beauty of the photographs, they do not impair their value, since such marks can always be identified by comparing successive plates.

Analysis of the photographs. An angular scale was placed on top of the expansion chamber and photographed with the camera raised half the height of the chamber. A transparency print of the resulting photograph was superposed on the negative of the tracks to be measured. By this means we measured the angles approximately as they would be projected on the plane of the top of the cylindrical chamber. We did not find it possible by means of the stereoscopic effect to make reliable estimates of angles in a plane including the line of sight. The measurements were made on the original negatives, using a stereoscope with lenses of the same focal length as those of the camera.

When only one recoil electron and one secondary electron appeared on a photograph, the procedure was to record first the angle θ at which the track of the recoil electron begins. The end representing the origin was rarely doubtful, since these tracks started within a narrow cylinder of air only about 1 mm in diameter. The angle between the incident ray and the line joining the origin of the recoil track and the origin of the secondary track was noted (when the origin of the secondary track

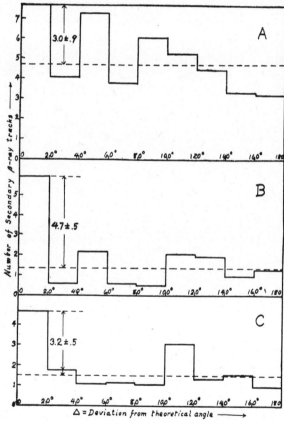

Fig. 6. The weighted number of secondary tracks at different angles Δ from the theoretical position. The exceptionally large number between 0 and 20° indicates that many secondary tracks are due to rays scattered as directed quanta. Curve A: Some stray x-rays present. Curves B and C: Most of stray rays eliminated by lead screens.

could not be identified the measurement was made to a point midway between its two ends). The difference between this angle and the angle ϕ calculated from θ by equation (1) was called Δ, and this value of Δ

was assigned a weight of unity. When a number n of recoil tracks appeared on the same plate with a secondary track, the value of Δ was thus determined for each recoil track separately, and assigned a weight $1/n$. Following this procedure there are values of Δ which are distributed approximately at random between 0 and 180° due to the $n-1$ recoil electrons which are not associated with the secondary track. This is in addition to the random values of Δ resulting from the presence of occasional stray tracks. We discarded the plates on which more than 3 recoil tracks appeared.

The results of this analysis are summarized in Figs. 6 and 7. Figure 6A shows the results for the first 302 plates, in which there appeared about 1 stray track for every 4 photographs. As will be seen, this was almost enough to hide the presence of the comparatively rare tracks due to scattered rays, which nevertheless show themselves by the exceptional concentration of Δ values between 0 and 20°. Figs. 6B and

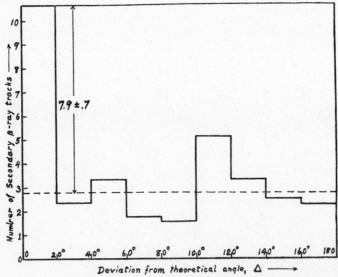

Fig. 7. Sum of curves 6B and 6C.

6C represent respectively the second and third series of plates, in which, by improving the lead shielding, the total number of secondary tracks was reduced to about 1 for every 20 plates. It will be seen that the result of this is to lower the general level of the random Δ values and thus to exhibit much more prominently the concentration of these values at small angles. It is just such a concentration which is to be expected if Eq. (1) holds, that is if individual quanta are scattered in definite directions from individual electrons. In Fig. 7 we have collected the results of the second and third series of plates. This figure shows the distribution of 38 secondary β-rays observed on these plates, 18 of which originate within 20° of the position anticipated from one of the recoil tracks present. The number 18 is about 4 times as many as would be expected if the distribution of the secondary tracks were a matter of chance. An idea of the degree of definiteness of the results can be obtained by calculating the probable variation from the mean of each of the 8 values between 20° and 180° by the usual formula, $\pm.67\sqrt{\overline{\Sigma\delta^2/(n-1)}}$, where δ is the deviation of each value from the mean, and n is the number (8) of values. Thus we find that the number of pairs occurring with values of Δ less than 20° is greater than the mean value for the other angles by 7.9 ± 0.7. The probability that so great an accidental deviation would occur in the positive direction is accordingly $(1/2)\times(.7^2/7.9^2)=1/250$. It is thus highly improbable that so many pairs of tracks should as a matter of chance have been found to fit Eq. (1) satisfactorily.

Two other possibilities remain, (1) that the observed coincidences are the result of an unconscious tendency to estimate the angles falsely, making consistently favorable errors in measurement, and (2) that the agreement with Eq. (1) is real. Regarding the first possibility, we may note that using our methods of measurement it was hardly possible to make an error in determining the angle of ejection of the recoil electron of more than 10° nor in the angle at which the secondary electron appeared of more than 5°. It will be seen that errors of this magnitude could not alter the general form of the curve. The evidence therefore seems unescapable that Eq. (1) describes to a close approximation the angles at which many of the secondary electrons appear.

The question arises, does the presence in Fig. 7 of the considerable number of values of Δ greater than 20° indicate the existence of scattered rays which do not obey the quantum law? It can be shown that about half of these random values of Δ are to be expected merely from the fact that in most of the cases where Eq. (1) describes accurately the relation between the position of the secondary electron and the direction of motion of one recoil electron, there are one or two other recoil electrons which are in no way associated with the secondary β-ray. We are unable to assign definitely the origin of the remaining half of the random Δ values. Undoubtedly some are due to stray x-rays which could not be completely eliminated and some are probably due to β-rays of radioactive origin. Our experiments cannot therefore be taken to afford any evidence for the production of β-rays by scattered x-rays which do not conform to the quantum rule described by Eq. (1).

It is our conclusion, therefore, that at least a large part of the scattered x-rays under investigation produce β-rays at an angle connected by equation (1) with the angle of ejection of an associated recoil electron.

Since the only known effect of x-rays is the production of β-rays, and since the meaning of energy is the ability to produce an effect, our result means that there is scattered x-ray energy associated with each of these recoil electrons sufficient to produce a β-ray and proceeding in a direction determined at the moment of ejection of the recoil electron. In other words, at least a large part of the *scattered x-rays proceed in directed quanta of radiant energy*.

Since other experiments have shown that these scattered x-rays can be diffracted by crystals, and are thus subject to the usual laws of interference, there is no reason to suppose that other forms of radiant energy possess an essentially different structure. It thus becomes highly probable that all electromagnetic radiation is constituted of discrete quanta proceeding in definite directions. It is not impossible to express this result in terms of waves if we suppose that a wave train possesing a single quantum of energy can produce an effect only in a certain predetermined direction.

These results do not appear to be reconcilable with the view of the statistical production of recoil and photo-electrons proposed by Bohr, Kramers and Slater. They are, on the other hand, in direct support of the view that *energy and momentum are conserved during the interaction between radiation and individual electrons*.

RYERSON PHYSICAL LABORATORY,
UNIVERSITY OF CHICAGO.
June 23, 1925.

X-RAY SPECTRA FROM A RULED REFLECTION GRATING

By A. H. Compton and R. L. Doan

Ryerson Physical Laboratory, University of Chicago

Communicated September 5, 1925

We have recently obtained spectra of ordinary X-rays by reflection at very small glancing angles from a grating ruled on speculum metal. Typical spectra thus obtained are shown in the accompanying figures. From some of these spectra it is possible to measure X-ray wave-lengths with considerable precision.

In order to reflect any considerable X-ray energy from a speculum surface it is necessary to work at small glancing angles, within the critical angle for total reflection. (See A. H. Compton, *Phil. Mag.*, **45**, 1121 (1923).) Within this critical angle, which in our experiments, using wave-lengths less than 1.6 angstroms, was less than 25 minutes of arc, the diffraction grating may be used in the same manner as in optical work. The wave-length is given by the usual formula,

$$n\lambda = D(\sin \phi + \sin i)$$

where i is the angle of incidence and ϕ is the angle of diffraction for the nth order. For small glancing angles this may be more conveniently written as

$$n\lambda = D\{\cos \theta - \cos(\theta + \alpha)\} \quad \ldots \ldots \ldots \ldots (1)$$

in which θ is the glancing angle and α is the angle between the zero order and the nth order. For the small angles employed this may be written to a very close approximation as

$$\lambda = \frac{D}{n}\left(\alpha\theta + \frac{1}{2}\alpha^2\right) \quad \ldots \ldots \ldots \ldots (2)$$

In order that several orders of the spectrum should appear inside the critical angle, we had a grating ruled with a comparatively large grating space, $D = 2.000 \times 10^{-3}$ cm. Special pains were taken to obtain a well polished surface, and the ruling was rather light, so as to obtain good reflection from the space between the lines. The reflected beam thus obtained was just as sharply defined as the direct beam.

In our first trials the X-rays direct from the target of a water-cooled Coolidge tube were collimated by fine slits 0.1 mm. broad and about 30 cm. apart. There was some difficulty at first in determining the zero position of the grating but this was solved in the following manner. The primary X-ray beam was first allowed to fall directly upon the film. Then after a brief exposure the speculum grating was brought into the path of the beam by means of a slow motion screw and a longer exposure sufficed to record the reflected image of the zero order, together with the associated first and higher orders. We were able, from the lines thus obtained on the film, to measure both θ and α. Photographs of this type are shown in figure 2 for a copper tube and in figure 3 for a molybdenum tube.

We were not able, with the grating used, to separate sharply the different X-ray spectrum lines. Therefore in order to get a precise measurement of one particular line we reflected the $K\alpha_1$ line of molybdenum from a calcite crystal and studied this beam with the ruled grating. The experimental arrangement is shown diagrammatically in figure 1. Typical diffraction patterns are shown in figures 4 and 5 for two different angles of incidence

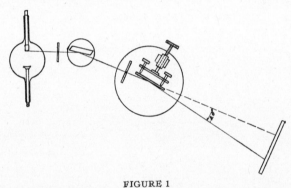

FIGURE 1

of the X-rays on the grating. It was found that the intensity of the spectrum obtained increased with the glancing angle, θ. Thus in figure 4, where $\theta = 0.00095$ radians, only the first order spectrum appears; whereas in figure 5, where $\theta = 0.00308$, there appear the first inside order and three outside orders. The exposure was in each case about 9 hours.

By solving equation 2 for α it will be seen that the inside order cannot appear unless θ is greater than a certain limiting angle. For precise wavelength measurements, however, it is important to have both inside and outside orders because an accurate setting is impossible on the broad zero order line. The image due to the direct beam can of course be made of any desired intensity by controlling the length of its exposure. If β_{-1} is the angle from the image of the direct beam to the first inside order, and β_{+1} that to the first outside order, the glancing angle θ is given by

$$\theta = \frac{\beta_{-1}^2 + \beta_{+1}^2}{2(\beta_{-1} + \beta_{+1})} \tag{3}$$

Using this value of θ, the wave-length can be calculated from equation 2. Following are some examples of measurements and calculations. From the film shown in figure 5 we measured $\beta_{-1} = 0.004815$, $\beta_{+1} = 0.00725$.

Thus from equation 3, $\theta = 0.003140$, and $\alpha_{-1} = -0.001462$ and $\alpha_{+1} = 0.000972$. Substituting these values in equation 2, we get $\lambda = 0.704\text{A}$.

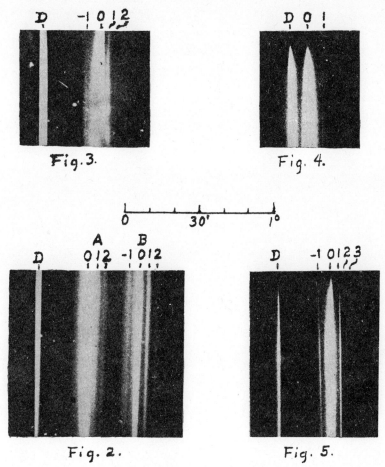

Fig. 3.

Fig. 4.

Fig. 2.

Fig. 5.

Figure 2. Spectrum of X-rays from copper target, excited at 20 kv. D = image of direct beam; for group A, glancing angle $\theta = 9'$, for B $\theta = 20'$. Numbers $-1, 0, 1, 2$ indicate the order of the spectrum. The absence of order -1 in group A is predicted by equation 2.

Figure 3. Spectrum of X-rays from molybdenum target, excited at about 35 kv.

Figures 4 and 5. Spectra obtained using the $K\alpha_1$ line of molybdenum.

The weighted mean value of our measurements on five films showing from 1 to 4 orders of the spectrum of the molybdenum $K\alpha_1$ line is

$$\lambda = 0.707 \pm 0.003\text{A}.$$

From crystal measurements this wave-length is determined as

$$\lambda = 0.7078 \pm 0.0002\text{A}.$$

The agreement is well within the probable error of our experiments. Our measurements of the spectra, obtained using a copper target, give in a similar manner wave-lengths intermediate between the α and β lines of copper, i.e., about 1.4 to 1.5A.

We see no reason why measurements of the present type may not be made fully as precise as the absolute measurements by reflection from a crystal, in which the probable error is due chiefly to the uncertainty of the crystalline grating space.

Light Waves or Light Bullets?

Recent Experiments Show that Light Rays Act Like Projectiles in Knocking Electrons Out of Matter. A Revision of the Time-honored "Wave Theory" of Light Is Suggested

By Arthur H. Compton
Professor of Physics at the University of Chicago

AS LONG ago as the Seventeenth Century, Newton defended the view that light consists of streams of little particles, shot with tremendous speed from a candle or the sun or any other source of light, the sensation of vision being produced by the mechanical impact of these particles or corpuscles on the retina of the eye. At the dawn of the Nineteenth Century, however, experiments were performed which were thought to give positive evidence that light consists of waves. Maxwell interpreted them as electromagnetic waves, and in such terms we have ever since been explaining light rays, X rays and radio rays. We have measured the length of the waves, their frequency and other characteristics, and have felt that we know them intimately. Very recently, however, a group of electrical effects of light have been discovered, for which the idea of light waves suggests no explanation, but whose interpretation is obvious according to a modified form of Sir Isaac Newton's old theory of corpuscular light projectiles or particles.

Einstein and the Photoelectric Effect

When light falls on certain metals, such as sodium, they give off electrons, just as does the filament of a radio tube when heated. This is known as the photoelectric effect. It was to account for this effect that Einstein, about twenty years ago, suggested the reversion to Newton's corpuscular theory of radiation.

This photoelectric effect can be studied especially well with X rays, which are of the same nature as light though of much shorter "wavelength." X rays are produced when a stream of electrons strikes a block of metal, just as sound is produced when a stream of bullets from a rapid-fire gun strikes a target. Suppose an electron is shot with a speed of 100,000 miles a second (these little particles certainly move at a tremendous rate). When it hits a "target" of platinum, an X ray will be produced which may pass, almost undiminished in strength, through a log of wood. But if this X ray hits one of the electrons in the wooden log just right, it will throw it out at nearly 100,000 miles a second, the speed of the first electron. A most surprising thing for a wave to do!

There was once a sailor on a vessel in New York harbor who dived overboard and splashed into the water. The resulting wave, after finding its intricate way out of the harbor, at last reached the other side of the ocean, and a part of it entered the harbor at Liverpool. In this harbor there happened to be a second sailor swimming beside his ship. When

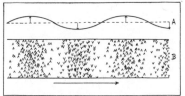

THE TWO CONCEPTIONS OF LIGHT
In A, the usual conception, the arrows indicate the direction of the electric field. In B, each checkmark represents an individual quantum moving from left to right but capable of producing an electric force at right angles to its motion. The new conception of light fits empirical observations like the usual conception, and therefore does not upset our actual laboratory practice

the wave reached him, he was surprised to find himself knocked by the wave up to the deck. Impossible? No more so than that an ether wave should perform this task which an X ray is found to do!

Concentrated Bundles of Energy

In order to account for this photoelectric effect, then, Einstein suggested that light and X rays do not consist of waves at all, but of concentrated bundles of energy which he calls "quanta." These quanta shoot in all directions with the speed of light, but each remains an undivided unit. Thus, when the electron strikes the target of the X-ray tube, its energy is transformed into one of these X-ray quanta, which goes out with tremendous speed until it finds an electron to which it can impart its energy. The X-ray quantum spends itself on this electron, which then shoots off with the speed of the first. On this view, a beam of light or X rays would be less like a ripple in a pond and more like a shower of shot from a shot-gun, each shot representing a light quantum.

We thus have a new picture of light as consisting of streams of little particles. Therefore, a light ray is no longer distinguished by its frequency or wavelength, but by the energy of its quantum. Einstein's relativity theory showed him that due to its energy each quantum would also have a certain definite inertia, or mass. Because of this mass, which is considered the measure of matter, these particles are in the usual sense material, so that we may speak of light as a form of matter. But there is now no need to imagine an ether, such as was necessary to propagate waves, for the inertia of these particles will carry them with undiminished speed from the remotest ends of the universe without any such conducting medium.

The Scattering of X Rays

Since the idea of light quanta was invented primarily to explain the photoelectric effect, the fact that it does so very well is no great evidence in its favor. The wave theory explains so satisfactorily such things as the reflection, refraction and interference of light that the rival quantum theory could not be given much credence unless it was found to account for some new thing for which it had not been especially designed. This is just what the quantum theory has recently accomplished in connection with the scattering of X rays.

Just as the moon, placed in the path of a beam of sunlight, becomes a source of scattered or diffusely reflected light, so one's finger, placed in the path of an X-ray beam, becomes in turn a source of scattered X rays. If a whistle is blown in front of a brick wall, the echo comes back with the same pitch as the original sound. Similarly we should expect, if X rays are waves, that the X-ray "echo" or scattered X rays should have the same wavelength as the rays which produced them. Spectra, however, prove that a part of the scattered X rays is of distinctly greater wavelength than the parent primary rays. Thus, for each line in the primary spectrum there are two lines in the scattered spectrum, one of the same and the other of greater wavelength than the primary line. This observed change of wave-

PROFESSOR ALBERT EINSTEIN
revived the old Newtonian idea of light corpuscles in form of quanta. It was for this work rather than for work on relativity that Professor Einstein was recently awarded the Nobel prize

PROFESSOR A. A. MICHELSON
The leading exponent of light waves and their uses, and one of America's foremost physicists. It was Professor Michelson's original "Michelson-Morley" experiments on whose results Einstein based his theory of relativity

TRAILS OF RECOILING ELECTRONS
X rays coming from the right have knocked the electrons nearly in the same direction

KNOCKED AT 100,000 MILES PER SECOND
If waves are not like projectiles, how can they knock electrons around like this?

length is directly contrary to the predictions of the wave theory.

The quantum theory, however, gives a direct explanation of the change of wavelength accompanying the scattering of X rays. We now think of the X ray once more as a little particle which is deflected or scattered when it collides with an electron of the matter through which it passes. If the electron is loose, it will recoil from the quantum which it deflects, so that part of the quantum's energy is spent in setting the electron in motion. Thus, the energy of the scattered quantum will be less than it was before it collided with the electron. But as we have seen, such a decrease in energy of the quantum corresponds to what in the wave theory we call an increase in wavelength, which is just what the experiments show. When a calculation is made on this basis of the amount of the wavelength change, we find exact agreement with the experiments.

The Recoiling Electrons

The part of the scattered rays which have the same spectrum as the primary rays is due, according to this view, to quanta which have bounced from electrons and which are held so tightly by their parent atoms that they are unable to recoil.

In view of the success of the quantum theory in explaining the change of wavelength of the scattered X rays, we look with interest to see whether the electrons that are supposed to recoil from the scattered quanta really exist. When the theory was first proposed, no one had ever observed such electrons. C. T. R. Wilson had, however, developed a method for photographing the trails left by individual electrons passing through air, by condensing water droplets on the ions produced when the electrons shoot through the air, and photographing the droplets under bright illumination. Wilson at once used this method to search for the predicted recoiling electrons, and found indisputable evidence of their existence.

In our laboratory at Chicago we have taken a series of similar pictures, in which the trails of even the shorter type are long enough to measure with accuracy. We find that the number of these tracks, as well as their direction and range, agrees very well with the predicted characteristics of the electrons recoiling from scattered quanta. Since their very existence was unknown before they were predicted by the quantum theory, these recoil electrons must be taken as a strong support of the theory of radiation quanta.

Waves or Projectiles?

We have seen that the spectrum of the scattered X rays is what it should be if the rays consist of little projectiles which have bounced from the electrons in the scattering material, just as one billiard ball bounces from another. Not only this, but we have photographed the recoiling billiard ball, or electron, from which the quantum has bounced. The obvious interpretation would be that X rays, and so also light rays or radio rays (for they are all the same kind of thing) consist of streams of little particles.

The question immediately arises, what then about radio waves? Are there no such things? The answer is undoubtedly yes. That there are electric waves of some sort can be shown directly in the laboratory, where standing waves of brush discharge can be produced along conducting wires. Therefore, there is no reason to doubt but that similar waves go out into space, substantially as the electromagnetic theory states.

An attempt at the reconciliation of these two viewpoints, which seems to be growing in favor, is to suppose that electric waves are not ether waves, but successive sheets of radiation quanta, somewhat as one sometimes sees sheets of drops in rainstorms. Where according to the usual theory the electric field is strong, there will be a concentration of quanta, and these quanta will be moving forward just as the wave does. Also there will be some quanta directed upward and some down, just as part of the electric field in the wave is directed upward and part down. From a 1,000-watt broadcasting station of 300 meters wavelength, it can be shown that, at a distance of 1,000 kilometers, there would be more than 400 such quanta in each cubic centimeter of the wave. Thus, the wave would be to all appearances continuous, though actually made up of discrete particles, just as water seems continuous though made up of molecules.

What Will Be the Solution?

The manner in which we are to reconcile these new ideas with the long established facts of interference and refraction of light is very obscure. A suggestion has been made by Duane which may ultimately lead to a solution, but as yet the way is very dark. It seems that the most rapid advances in the study of the nature of radiation have recently been made by those who have adopted the idea of light corpuscles.

I believe, however, that most physicists look forward to a final solution of the problem of the nature of light in some combination of the wave and quantum theories.

ONE 10,000,000TH MOSQUITO POWER FROM 60 HORSEPOWER
A 40-kilowatt, 200,000-volt transformer producing a single X-ray quantum, amplified three stages and heard as a click in the headset. A mosquito climbing up one inch of screen does the work of 10,000,000 of these X-ray quanta

PHOTOGRAPHING ELECTRON TRACKS LIKE THE ABOVE
The chamber in which the electrons are photographed is enclosed in the square lead box the foreground. The condenser, made of ordinary glass milk bottles which may be seen top of the apparatus, supplies the illuminating sparks for taking the photographs

10. Electron distribution in sodium chloride. ARTHUR H. COMPTON, University of Chicago.—Extending Duane's method of Fourier analysis, a series expression is obtained for the mean electron density at different distances from the centers of the atoms in a crystal. Every term in the series may be determined by measurements of the intensities of the reflection of x-rays from the crystal, whereas in Duane's method a constant term remains undetermined. Using the data of Bragg, James and Bosanquet for the reflection from rock-salt, the electron density is found to fall definitely to zero for the sodium ion 1.1A, and for chlorine at 2.0A. from the center. Sodium has a group of eight electrons (probably K and $L2_2$) near the center with two electrons (probably $L2_1$) at .9A from the center. Near the center of the chlorine ion are 10 electrons (K and L) outside of which are eight others, resolvable into $4(3_3?)$ at .74A, $2(3_2?)$ at 1.14A and $2(3_1?)$ at 1.60A. These distances are uncorrected for thermal agitation. The electron distribution curves are similar to but show more detail than those obtained by Bragg and Havighurst from the same data.

X-RAYS AS A BRANCH OF OPTICS

By Arthur H. Compton

One of the most fascinating aspects of recent physics research has been the gradual extension of the familiar laws of optics to the very high frequencies of x-rays. until at present there is hardly a phenomenon in the realm of light whose parallel is not found in the realm of x-rays. Reflection, refraction, diffuse scattering, polarization, diffraction, emission and absorption spectra, photoelectric effect, all of the essential characteristics of light have been found also to be characteristic of x-rays. At the same time it has been found that some of these phenomena undergo a gradual change as we proceed to the extreme frequencies of x-rays, and as a result of these interesting changes in the laws of optics we have gained new information regarding the nature of light.

It has not always been recognized that x-rays is a branch of optics. As a result of the early studies of Röntgen and his followers it was concluded that x-rays could not be reflected or refracted, that they were not polarized on traversing crystals, and that they showed no signs of diffraction on passing through narrow slits. In fact, about the only property which they were found to possess in common with light was that of propagation in straight lines. Many will recall also the heated debate between Barkla and Bragg, as late as 1910, one defending the idea that x-rays are waves like light, the other that they consist of streams of little bullets called "neutrons." It is a debate on which the last word has not yet been said!

THE REFRACTION AND REFLECTION OF X-RAYS

We should consider the phenomena of refraction and reflection as one problem, since it is a well known law of optics that reflection can occur only from a boundary surface between two media of different indices of refraction. If one is found, the other must be present.

In his original examination of the properties of x-rays, Röntgen[1] tried unsuccessfully to obtain refraction by means of prisms of a variety of materials such as ebonite, aluminium and water. Perhaps the experiment of this type most favorable for detecting refraction was one by Barkla.[2] In this work x-rays of a wave length which excited strongly the characteristic K radiation from bromine were passed through a crystal of potassium bromide. The precision of his experiment was such that he was able to conclude that the refractive index for a wave length of 0.5A probably differed from unity by less than 5 parts in a million.

Although these direct tests for refraction of x-rays were unsuccessful, Stenström observed[3] that for x-rays whose wave lengths are greater than about 3A, reflected from crystals of sugar and gypsum, Bragg's law, $n\lambda = 2D \sin \theta$, does not give accurately the angles of reflection. He interpreted the difference as due to an appreciable refraction of the x-rays as they enter the crystal. Measurements by Duane and Siegbahn and their collaboraters[4] showed that discrepancies of the same type occur, though they are very small indeed, when ordinary x-rays are reflected from calcite.

The direction of the deviations in Stenström's experiments indicated that the index of refraction of the crystals employed was less than unity. If this is the case also for other substances, total reflection should occur when x-rays in air strike a polished surface at a sufficiently sharp glancing angle, just as light in a glass prism is totally reflected from a surface between the glass and air if the light strikes the surface at a sufficiently sharp angle. From a measurement of this critical angle for total reflection, it should be possible to determine the index of refraction of the x-rays.

When the experiment was tried,[5] the results were strictly in accord with these predictions. The apparatus was set up as shown in Fig. 1, reflecting a very narrow sheet of x-rays from a polished mirror onto

---

[1] W. Röntgen, Sitzungber. der Würzburger Phys. Med. Ges. Jahrg. 1895. These papers are reprinted in German in Ann. d. Phys., *64*, p. 1; 1898, and in English translation by A. Stanton in Science, *3*, p. 227; 1896.

[2] C. G. Barkla, Phil. Mag., *31*, p. 257; 1916.

[3] W. Stenström, Dissertation, Lund, 1919.

[4] Duane and Patterson, Phys. Rev., *16*, p. 532; 1920. M. Siegbahn, C. R., *173*, p. 1350; 1921; *174*, p. 745; 1922.

[5] A. H. Compton, Phil. Mag., *45*, p. 1121; 1923.

the crystal of a Bragg spectrometer. It was found that the beam could be reflected from surfaces of polished glass and silver through several minutes of arc. By studying the spectrum of the reflected beam, the critical glancing angle was found to be approximately proportional to

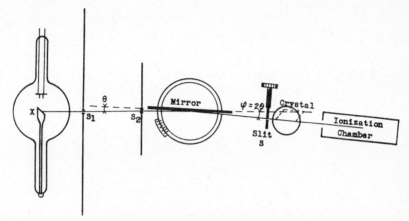

FIG. 1. *Apparatus for studying the total reflection of x-rays.*

the wave length. For ordinary x-rays whose wave length is half an Angström, the critical glancing angle from crown glass was found to be about 4.5 minutes of arc, which means a refractive index differing from unity by a little less than 1 part in a million.

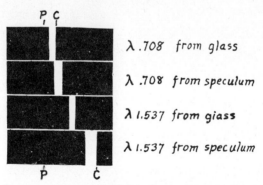

FIG. 2. *Total reflection of x-rays from polished glass and speculum metal (Doan). P = direct beam, C = critical angle of the totally reflected beam.*

Fig. 2 shows some photographs of the totally reflected beam and the critical angle for total reflection taken recently by Dr. Doan[6] working at Chicago. From the sharpness of the critical angles shown in this

[6] R. L. Doan, Phil. Mag., 20, p. 100; 1927.

figure, it is evident that a precise determination of the refractive index can thus be made.

You will recall that when one measures the index of refraction of a beam of light in a glass prism it is customary to set the prism at the angle for minimum deviation. This is done primarily because it simplifies the calculation of the refractive index from the measured angles. It is an interesting comment on the psychology of habit that most of the earlier investigators of the refraction of x-rays by prisms also used their prisms set at the angle for minimum deviation. Of course, since the effect to be measured was very small indeed, the adjustments should have been made to secure not the minimum deviation but the maximum deviation possible. After almost thirty years of attempts to refract x-rays by prisms, experiments under the conditions to secure maximum refraction were first performed by Larsson, Siegbahn and Waller,[7] using the arrangement shown diagrammatically in Fig. 3. The x-rays

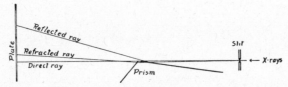

FIG. 3. *Refraction of x-rays by a glass prism. Arrangement of Larsson, Siegbahn and Waller.*

struck the face of the prism at a fine glancing angle, just greater than the critical angle for the rays which are refracted. Thus the direct rays, the refracted rays, and the totally reflected rays of greater wave length were all recorded on the same plate.

Fig. 4 shows one of the resulting photographs. Here we see a complete dispersion spectrum of the refracted x-rays, precisely similar to the spectrum obtained when light is refracted by a prism of glass. The presence of the direct ray and the totally reflected ray on the same plate make possible all the angle measurements necessary for a precise determination of the refractive index for each spectrum line.

For a generation we have been trying to obtain a quantitative test of Drude and Lorentz's dispersion theory in the ordinary optical region. But our ignorance regarding the number and the natural frequency of the electron oscillators in the refractive medium has foiled all such attempts. For the extreme frequencies of x-rays, however, the problem becomes greatly simplified. In the case of substances such as

[7] Larsson, Siegbahn and Waller, Naturwiss, 1924.

glass, the x-ray frequencies are much higher than the natural frequencies of the oscillators in the medium, and the only knowledge which the theory requires is that of the number of electrons per unit volume in the dispersive medium. If we assume the number of electrons per atom to be equal to the atomic number, we are thus able to calculate at once the refractive index of the medium for x-rays. In the case of glass this calculation gives agreement with experiment within the experimental error, which is in some cases less than one per cent. So we may say

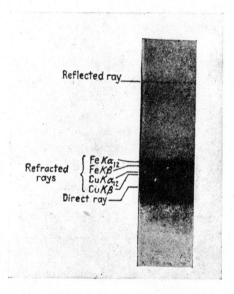

FIG. 4. *Prism spectrum of x-rays obtained by Larsson, Siegbahn and Waller.*

that the laws of optical dispersion given by the electron theory are first established on a quantitative basis by these experiments on the refraction of x-rays.

Another way of looking at the problem is to assume the validity o the dispersion equation developed from the electron theory, and to use these measurements of the refraction of x-rays to calculate the number of electrons in each atom of the refracting material. This affords us what is probably our most direct as well as our most precise means of determining this number. The precision of the experiments is now such that we can say that the number of electrons per atom effective in refracting x-rays is within less than one half of one per cent equal to the atomic number of the atom.

Thus optical refraction and reflection are extended to the region of x-rays, and this extension has brought with it more exact knowledge not only of the laws of optics but also of the structure of the atom.

THE DIFFRACTION OF X-RAYS

Early in the history of x-rays it was recognized that most of the properties of these rays might be explained if, as suggested by Wiechert,[8] they consist of electromagnetic waves much shorter than those of light. Haga and Wind performed a careful series of experiments[9] to detect any possible diffraction by a wedge shaped slit a few thousandths of an inch broad at its widest part. The magnitude of the broadening was about that which would result[10] from rays of 1.3A wave length. The experiments were repeated by yet more refined methods by Walter and Pohl,[11] who came to the conclusion that if any diffraction effects were present, they were considerably smaller than Haga and Wind had estimated. But on the basis of photometric measurements of Walter and Pohl's plates by Koch,[12] using his new photoelectric microphotometer, Sommerfeld found[13] that their photographs indicated an effective wave length for hard x-rays of .4A, and for soft x-rays a wave length measurably greater.

It may have been because of their difficulty that these experiments did not carry as great conviction as their accuracy would seem to have warranted. Nevertheless it was this work perhaps more than any other which encouraged Laue to undertake his remarkable experiments on the diffraction of x-rays by crystals.

Within the last few years Walter has repeated these slit diffraction experiments, making use of the $K\alpha$ line of copper, and has obtained perfectly convincing diffraction effects[14]. Because of the difficulty in determining the width of the slit where the diffraction occurs, it was possible to make from his photographs only a rough estimate of the wave length of the x-rays. But within this rather large probable error the wave length agreed with that determined by crystal spectrometry.

While these slit diffraction experiments were being developed, and long before they were brought to a successful conclusion, Laue and his

---

[8] E. Wiechert, Sitz. d. Phys-okon Ges. zu Königsberg, 1894.
[9] Haga and Wind, Wied. Ann., *68*, p. 884; 1899.
[10] A. Sommerfeld, Phys. ZS., *2*, p. 59; 1900.
[11] Walter and Pohl, Ann. der Phys., *29*, p. 331; 1909.
[12] P. P. Koch, Ann. der Phys., *38*, p. 507; 1912.
[13] A. Sommerfeld, Ann. der Phys., *38*, p. 473; 1912.
[14] B. Walter, Ann. der Phys., *74*, p. 661; 1924; *75*, Sept. 1924.

collaborators discovered the remarkable fact that crystals act as suitable gratings for diffracting x-rays. You are all acquainted with the history of this discovery. The identity in nature of x-rays and light could no longer be doubted. It gave a tool which enabled the Braggs to determine with a definiteness previously almost unthinkable the manner in which crystals are constructed of their elementary components. By its help Moseley and Siegbahn have studied the spectra of x-rays, we have learned to count one by one the electrons in the different atoms, and we have found out something regarding the arrangement of these electrons. The measurement of x-ray wave lengths thus made possible gave Duane the means of making his precise determination of Planck's radiation constant. By showing the change of wave length when x-rays are scattered, it has helped us to find the quanta of momentum of radiation which had previously been only vaguely suspected. Thus in the two great fields of modern physical inquiry, the structure of matter and the nature of radiation, the discovery of the diffraction of x-rays by crystals has opened the gateway to many new and fruitful paths of investigation. As the Duc de Broglie has remarked, "if the value of a discovery is to be measured by the fruitfulness of its consequences, the work of Laue and his collaborators should be considered as perhaps the most important in modern physics."

These are some of the consequences of extending the optical phenomenon of diffraction into the realm of x-rays.

There is, however, another aspect of the extension of optical diffraction into the x-ray region, which has also led to interesting results. It is the use of ruled diffraction gratings for studies of spectra. By a series of brilliant investigations, Schumann, Lyman and Millikan, using vacuum spectrographs, have pushed the optical spectra by successive stages far into the ultraviolet. Using a concave reflection grating at nearly normal incidence, Millikan and his collaborators[15] found a line, probably belonging to the L series of aluminium, of a wave length as short as 136.6A, only a twenty-fifth that of yellow light. Why his spectra stopped here, whether because of failure of his gratings to reflect shorter wave lengths, or because of lack of sensitiveness of the plates, or because his hot sparks gave no rays of shorter wave length, was hard to say.

Röntgen had tried to get x-ray spectra by reflection from a ruled grating, but the task seemed hopeless. How could one get spectra from

---

[15] Millikan, Bowen, Sawyer, Shallenberger, Proc. Nat. Acad., 7, p. 289; 1921; Phys. Rev., 23, p. 1; 1924.

a reflection grating if the grating would not reflect? But when it was found that x-rays could be totally reflected at fine glancing angles, hope for the success of such an experiment was revived. Carrara,[16] working at Pisa, tried one of Rowland's optical gratings, but without success. Fortunately we at Chicago did not know of this failure, and with one of Michelson's gratings ruled specially for the purpose, Doan found that he could get diffraction spectra of the K series radiations from both copper and molybdenum.[17] Fig. 5 shows one of our diffraction spectra, giving several orders of the $K\alpha_1$ line of molybdenum, obtained by reflection at a small glancing angle. This work was quickly

FIG. 5. *Spectrum of the $K\alpha_1$ line of molybdenum, $\lambda = .708A$, from a grating ruled on speculum metal (Compton and Doan). D marks the direct beam, and O the directly reflected beam.*

followed by Thibaud,[18] who photographed a beautiful spectrum of the K series lines of copper from a grating of only a few hundred lines ruled on glass. That x-ray spectra could be obtained from the same type of ruled reflection gratings as those used with light was now established.

The race to complete the spectrum between the extreme ultraviolet of Millikan and the soft x-ray spectra of Siegbahn began again with renewed enthusiasm. It had seemed that the work of Millikan and his coworkers had carried the ultraviolet spectra to as short wave lengths

[16] N. Carrara, N. Cimento, *1*, p. 107; 1924.
[17] A. H. Compton and R. L. Doan, Proc. Nat. Acad., *11*, p. 598; 1925.
[18] J. Thibaud, C. R., Jan. 4, 1926.

as it was possible to go. On the x-ray side, the long wave length limit was placed, theoretically at least, by the spacing of the reflecting layers in the crystal used as a natural grating. De Broglie, W. H. Bragg, Siegbahn and their collaborators were finding suitable crystals of greater and greater spacing, until Thoraeus and Siegbahn,[19] using crystals of palmitic acid, measured the $L\alpha$ line of chromium, with a wave length 21.69A. But there still remained a gap of almost three octaves between these x-rays and the shortest ultraviolet in which, though radiation had been detected by photoelectric methods, no spectral measurements has been made.

Thibaud, working in de Broglie's laboratory at Paris, made a determined effort to extend the limit of the ultraviolet spectrum, using his glass grating at glancing incidence.[20] His spectra, however, stopped at 144A, a little greater than the shortest wave length observed in Millikan's experiments.

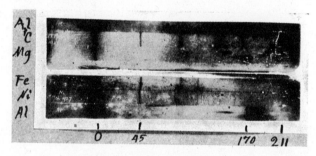

FIG. 6. *Osgood's grating spectra of soft x-rays from Al, C, Mg, Fe and Ni, showing lines from $\lambda = 45A$ to $\lambda = 211A$. These are the first spectra bridging the gap between the soft x-rays and the ultraviolet.*

But meanwhile Dauvillier, also working with de Broglie, was making rapid strides working from the soft x-ray side of the gap. First,[21] using a grating of palmitic acid, he found the $K\alpha$ line of carbon of wave length 45A. Then[22] using for a grating a crystal of the lead salt of mellissic acid, with the remarkable grating space of 87.5A, he measured a spectrum line of thorium as long as 121A, leaving only a small fraction of an octave between his longest x-ray spectrum lines and Millikan's shortest ultraviolet lines. The credit for filling in the greater part of the remaining gap must thus be given to Dauvillier.

[19] Siegbahn and Thoraeus, Arkiv f M. o F., *19*, p. 1; 1925.
[20] J. Thibaud, J. de Phys. et Rad., *8*, p. 15; 1927.
[21] A. Dauvillier, C. R., *182*, p. 1083; 1926.
[22] A. Dauvillier, J. de Phys. et Rad., *6*, p. 1; Jan. 1927.

The final bridge between the x-ray and the ultraviolet spectra has however been laid by Osgood,[23] a young Scotchman working with me at Chicago. He also used soft x-rays as did Dauvillier, but instead of a crystal grating, he did his experiments with a concave glass grating in a Rowland mounting, but with the rays at glancing incidence. Fig. 6 shows a series of Osgood's spectra. The shortest wave length here shown is the $K\alpha$ line of carbon, 45A, and we see a series of lines up to 211A. An interesting feature of these spectra is an emission band in the aluminium spectrum at about 170A, which is probably in some way associated with the L series spectrum of aluminium. These spectra

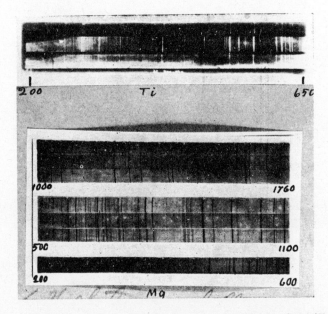

FIG. 7. *Spectra of the extreme ultraviolet, from Mg and Ti, 200A to 1760A (Hoag).*

overlap, on the short wave length side, Dauvillier's crystal measurements, and on the side of the great wave lengths, Millikan's ultraviolet spectra.

In the September number of the Physical Review, Hunt[24] describes similar experiments, using however a plane ruled grating at glancing incidence, in which he has measured lines from 2A down to the carbon line at 45A, thus meeting the shortest of Osgood's measurements. On the other hand, Fig. 7 shows some beautiful spectra of the extreme ultraviolet obtained recently by Dr. Hoag, working with Professor Gale

[23] T. H. Osgood, Nature, *119*, p. 817; June 4, 1927; Phys. Rev., November, 1927.
[24] F. L. Hunt, Phys. Rev., Sept. 1927.

at Chicago, using a concave grating at grazing incidence. These spectra extend from 200A to 1760A, overlapping Osgood's x-ray spectra on the short wave length side, and reaching the ordinary ultraviolet region on the side of the great wave lengths. Thus from the extreme infrared to the region of ordinary x-rays we now have a continuous series of spectra from ruled gratings.

Whatever we may find regarding the nature of x-rays, it would take a bold man indeed to suggest, in light of these experiments, that they differ in nature from ordinary light.

It is too early to predict what may be the consequences of these grating measurements of x-rays. It seems clear, however, that they must lead to a new and more precise knowledge of the absolute wave length of x-rays, and thus to direct determinations of the grating spaces of crystals. This will in turn afford a new means of determining Avogadro's number and the electronic charge, which should be of precision comparable with that of Millikan's oil drops.

THE SCATTERING OF X-RAYS AND LIGHT

The phenomena that we have been considering are ones in which the laws which have been found to hold in the optical region apply equally well in the x-ray region. This is not the case, however, for all optical phenomena.

The theory of the diffuse scattering of light by turbid media has been examined by Drude, Lord Rayleigh, Raman and others, and an essentially similar theory of the diffuse scattering of x-rays has been developed by Thomson, Debye and others. Two important consequences of these theories are, (1) that the scattered radiation shall be of the same wave length as the primary rays, and (2) that the rays scattered at 90 degrees with the primary rays shall be plane polarized. The experimental tests of these two predictions have led to interesting results.

A series of experiments performed during the last few years[25] have shown that secondary x-rays are of greater wave length than the primary rays which produce them. This work is too well known to require description. On the other hand, careful experiments to find a similar increase in wave length in light diffusely scattered by a turbid medium have failed to show any such effect.[26] An examination of the spectrum

---

[25] For an account of this work, see e.g. the writer's "X-Rays and Electrons," Chap. IX, Van Nostrand, 1926.

[26] E. g., P. A. Ross, Proc. Nat. Acad., *9*, p. 246; 1923.

of the secondary x-rays shows that the primary beam has been split into two parts, as shown in Fig. 8, one of the same wave length and the other of increased wave length. When different primary wave lengths are used, we find always the same difference in wave length between these two components; but the relative intensity of the two components changes. For the longer wave lengths the unmodified ray has the greater energy, while for the shorter wave lengths the modified ray is predominant. In fact when hard γ-rays are employed, it is not possible to find any radiation of the original wave length.

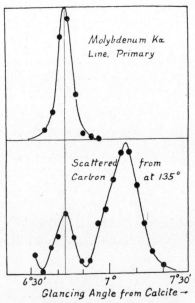

FIG. 8. *A typical spectrum of scattered x-rays, showing the splitting of the primary ray into a modified and an unmodified ray.*

Thus in the wave length of secondary radiation we have a gradually increasing departure from the classical electron theory of scattering as we go from the optical region to the region of x-rays and γ-rays.

The question arises, are these secondary x-rays of increased wavelength to be classed as scattered x-rays or as fluorescent rays? An important fact bearing on this point is the intensity of the secondary rays. From the theories of Thomson, Debye and others it is possible to calculate the absolute intensity of the scattered rays. It is found that this calculated intensity agrees very nearly with the total intensity of the modified and unmodified rays, but that in many cases the observed intensity of the unmodified ray taken alone is very small compared with the calculated intensity. If the electron theory of the intensity of scattering is even approximately correct, we must thus include the modified with the unmodified rays as scattered rays.

Information regarding the origin of these secondary rays is also given by a study of their state of polarization. We have called attention to the fact that the electron theory demands that the x-rays scattered at 90 degrees should be completely plane polarized. If the rays of increased wave-length are fluorescent, however, we should not expect them to be strongly polarized. You will remember the experiments performed by Barkla[27] some twenty years ago in which he observed strong polarization in x-rays scattered at right angles. It was this experiment which gave us our first strong evidence of the similar character of x-rays and light. But in this work the polarization was far from complete. In fact the intensity of the secondary rays at 90 degrees dropped only to one third of its maximum value, whereas for complete polarization it should have fallen to zero. It might have seemed that the remaining third was due to really unpolarized rays of a fluorescent type.

The fact that no such unpolarized rays exist was established by repeating Barkla's experiment[28] with scattering blocks of different

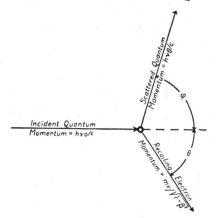

Fig. 9. *An x-ray photon is deflected through an angle ϕ by an electron, which in turn recoils at an angle θ, taking up a part of the energy of the photon.*

sizes. When very small blocks were used, we found that the polarization was nearly complete. The lack of complete polarization in Barkla's experiments was due chiefly to the multiple scattering of the x-rays in the large blocks that he used to scatter the x-rays. It would seem that

[27] C. G. Barkla, Proc. Roy. Soc. A., 77, p. 247; 1906.
[28] A. H. Compton and C. F. Hagenow, J.O.S.A. and R.S.I., *8*, p. 487; 1924.

the only explanation of the complete polarization of the secondary rays is that they consist wholly of scattered rays.

According to the classical theory, an electromagnetic wave is scattered when it sets the electrons which it traverses into forced oscillations, and these oscillating elecrons reradiate the energy which they receive. In order to account for the change in wave-length of the scattered rays, however, we have had to adopt a wholly different picture of the scattering process—that shown in Fig. 9. Here we do not think of the x-rays as waves, but as light corpuscles, quanta, or, as we may call them, photons. Moreover, there is nothing here of the forced oscillation pictured on the classical view, but a sort of elastic collision, in which the energy and momentum are conserved.

This new picture of the scattering process leads at once to three consequences that can be tested by experiment. There is a change of wave-length

$$\delta\lambda = \frac{h}{mc}(1-\cos\phi), \qquad (1)$$

which accounts for the modified line in the spectra of scattered x-rays. Experiment has shown that this formula is correct within the precision of our knowledge of h, m and c. The electron which recoils from the scattered x-ray should have the kinetic energy,

$$E_{\text{kin}} = h\nu \cdot \frac{h\nu}{mc^2}\cos^2\theta, \qquad (2)$$

approximately. When this theory was first proposed, no electrons of this type were known; but they were discovered by Wilson[29] and Bothe[30] within a few months after their prediction. Now we know that the number, energy and spatial distribution of these recoil electrons are in accord with the predictions of the photon theory. Finally, whenever a photon is deflected at an angle ϕ, the electron should recoil at an angle θ given by the relation,

$$\cot\tfrac{1}{2}\phi = \tan\theta, \qquad (3)$$

approximately.

This relation we have tested[31] using the apparatus shown diagrammatically in Fig. 10. A narrow beam of x-rays enters a Wilson expansion

---

[29] C. T. R. Wilson, Proc. Roy. Soc., *109*, p. 1; 1923.
[30] W. Bothe, ZS. f. Phys., *16*, p. 319; 1923; *20*, p. 237; 1923.
[31] A. H. Compton and A. W. Simon, Phys. Rev., *26*, p. 289; 1925.

chamber. Here it produces a recoil electron. If the photon theory is correct, associated with this recoil electron, a photon is scattered in the direction ϕ. If it should happen to eject a β-ray, the origin of this

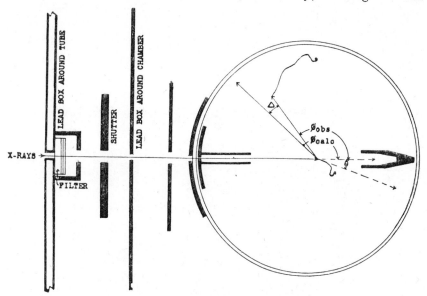

FIG. 10. *An electron recoiling at an angle θ should be associated with a photon deflected through an angle ϕ.*

β-ray tells the direction in which the photon was scattered. Fig. 11 shows a typical photograph of the process. A measurement of the angle θ at which the recoil electron on this plate is ejected and the angle ϕ of the origin of the secondary β-particle, shows close agreement with the photon formula. This experiment is of especial significance, since it shows that for each recoil electron there is a scattered photon, and that the energy and momentum of the system photon plus electron are conserved in the scattering process.

The evidence for the existence of directed quanta of radiation afforded by this experiment is very direct. The experiment shows that associated with each recoil electron there is scattered x-ray energy enough to produce a secondary beta ray, and that this energy proceeds in a direction determined at the moment of ejection of the recoil electron. Unless the experiment is subject to improbably large experimental errors, therefore, *the scattered x-rays proceed in the form of photons.*

Thus we see that as a study of the scattering of radiation is extended into the very high frequencies of x-rays, the manner of scattering changes. For the lower frequencies the phenomena could be accounted for in terms of waves. For these higher frequencies we can find no interpretation of the scattering except in terms of the deflection of corpuscles or photons of radiation. Yet it is certain that the two types of radiation, light and x-rays, are essentially the same kind of thing. We are thus confronted with the dilemma of having before us convincing evidence that radiation consists of waves, and at the same time that it consists of corpuscles.

FIG. 11. *Photograph showing recoil electron and associated secondary β-ray. The upper photograph is retouched.*

It would seem that this dilemma is being solved by the new wave mechanics. De Broglie[32] has assumed that associated with every particle of matter in motion there is a wave whose wave length is given by the relation,

$$mv = h/\lambda,$$

where mv is the momentum of the particle. A very similar assumption was made at about the same time by Duane,[33] to account for the diffraction of x-ray photons. As applied to the motion of electrons, Schrödinger has shown the great power of this conception in studying atomic structure.[34] It now seems, through the efforts of Heisenberg, Bohr and others, that this conception of the relation between corpuscles

[32] L. de Broglie, Thèse, Paris, 1924.
[33] W. Duane, Proc. Nat. Acad., 1924.
[34] E. Shrödinger, Ann. der Phys., *79*, pp. 361, 489, 734; *80*, 437; *81*, 109; 1926; Phys. Rev., *28*, p. 1051; 1926.

and waves is capable of giving us a unified view of the diffraction and interference of light, and at the same time of its diffuse scattering and of the photoelectric effect. It would however take too long to describe these new developments in detail.

We have thus seen how the essentially optical properties of radiation have been recognized and studied in the realm of x-rays. A study of the refraction and specular reflection of x-rays has given an important confirmation of the electron theory of dispersion, and has enabled us to count with high precision the number of electrons in the atom. The diffraction of x-rays by crystals has given wonderfully exact information regarding the structure of crystals, and has greatly extended our knowledge of spectra. When x-rays were diffracted by ruled gratings, it made possible the study of the complete spectrum from the longest to the shortest waves. In the diffuse scattering of radiation, we have found a gradual change from the scattering of waves to the scattering of corpuscles.

Thus by a study of x-rays as a branch of optics we have found in x-rays all of the well known wave characteristics of light, but we have found also that we must consider these rays as moving in directed quanta. It is these changes in the laws of optics when extended to the realm of x-rays which have been in large measure responsible for the recent revision of our ideas regarding the nature of the atom and of radiation.

UNIVERSITY OF CHICAGO,
CHICAGO, ILLINOIS,
OCTOBER 10, 1927.

Coherence of the Reflected X-Rays from Crystals.

On Jauncey's theory (*Phys. Rev.*, **25**, 314; 1925) of the unmodified line in the Compton effect, an unmodified X-ray is scattered when the energy of the impulse imparted to an electron is insufficient to eject the electron from the parent atom. In this case the impulse is presumably imparted to the atom itself. The change of wave-length of the unmodified ray should thus be of the order of [(mass of the electron)/(mass of the atom)] × (change of wave-length of the modified line). It is generally assumed that no coherence occurs for modified scattering on account of the change of wave-length. In the case of unmodified scattering, however, it is assumed that coherence does occur, as, for example, in regular crystal reflection (see papers by Williams, *Phil. Mag.*, **2**, 657; 1926; and Jauncey, *Phys. Rev.*, **29**, 757; 1927). But how can coherence occur in unmodified scattering if there is a change of wave-length, however small? Perhaps there is no change of wave-length at all in unmodified scattering.

Following the idea underlying Jauncey's interpretation of the unmodified line, the atom should not by itself receive the impulse of the scattered quantum unless the energy acquired from this impulse is sufficient to give at least one quantum of vibrational heat energy to the atom. According to Einstein's theory of specific heats, this energy is $h\nu$, where ν is a frequency of the order of that of the *reststrahlen* from the substance in question. The wave-lengths of two bands of these *reststrahlen* from rock-salt are 47μ and 54μ (Wood, "Physical Optics," p. 412), or, let us say, of the order of 50μ. A quantum of this wave-length has an energy of 0·024 electron. If we consider the K_a line of molybdenum reflected from the (100) planes of rock-salt in the nth order, the energy of recoil given to a sodium atom by the impulse imparted by the reflected quantum is $0·0252 \sin^2 \theta$ volt-electrons (see Compton, "X-Rays and Electrons," p. 267). Replacing $\sin \theta$ by $n\lambda/2d$, the energy of recoil is $4·02 \times 10^{-4} \times n^2$ volt-electrons. Hence the ratio of the energy of recoil for each order to the energy of a quantum of *reststrahlen* is as follows:

n	1	2	3	4	5	6	7
ratio	0·017	0·067	0·15	0·27	0·42	0·60	0·82

The highest possible order of reflection according to Bragg's law is the seventh. The fact that the ratio is always less than unity indicates that the energy of recoil is always less than that of a quantum of thermal agitation, and this implies that the thermal agitation will not be excited. Presumably, therefore, the impulse is imparted to the crystal as a whole. There is thus no reason to anticipate an absence of coherence in the reflected rays.

This point of view leads naturally to the quantum interpretation of crystal diffraction as suggested by Duane.

G. E. M. Jauncey.
Washington University, St. Louis.
A. H. Compton.
University of Chicago, Sept. 1.

ON THE INTERACTION BETWEEN RADIATION AND ELECTRONS

By Arthur H. Compton

Abstract

In the production of recoil electrons we have an example of the action of radiation on free electrons, whereas the photoelectric effect with x-rays is an example of the action of radiation on a pair of positive and negative charges. In both effects experiment indicates that the whole momentum absorbed from the radiation is imparted to the electron that is set in motion by the radiation, showing that the duration of the action of the radiation is short compared with the natural period of the electron in the atom. It is assumed that the action is sensibly instantaneous. In contrast with the prediction of Lorentz's force equation, which would predict an impulse imparted to an isolated electron almost in the direction of the electric vector, the experiments show that the preferred direction of motion of the recoil electrons is perpendicular to the electric vector. An impulse on a free electron in the direction of the electric vector would not be consistent with the conservation of momentum. The photo-electrons on the other hand have the electric vector of the incident wave as their preferred direction of motion (neglecting radiation pressure), though the experiments show that the impulse imparted to the electron by the radiation may make a considerable angle with the electric vector. In this case the conservation of linear momentum permits motion in any direction, since equal and opposite impulses are applied to the positive and negative parts of the atom by the electric vector; but the conservation of the angular momentum of the system requires that the impulse shall be imparted in a direction determined by the instantaneous position of the electron in the atom. The experiments of Auger and Bubb are consistent with this requirement, but indicate that Lorentz's force equation is only statistically valid in defining the direction of the action of the electric vector on the photo-electron.

WE MAY distinguish between the actions of radiation upon electrons in which the electrons are ejected from the matter traversed and those in which the electrons affected remain in the matter. In the first group are the photoelectric effect and the production of recoil electrons, or as we may call it, the recoil effect. The motions of these electrons after leaving the matter may be studied, and the information which such a study affords regarding the mode of action of the radiation is the chief subject of this paper. Included in the second group of actions is the production of excited atoms by the absorption of radiation, an action which is doubtless similar in character to the photoelectric effect, and such phenomena as the exciting of high frequency currents in conductors by electric waves, and the polarization of dielectric media when traversed by electric waves. It is not possible in phenomena of the latter type to observe the motions of the individual electrons, but our large scale measurements are consistent with the view that each electron in the medium is subject to the force per unit charge

$$F = E + [vH]/c, \qquad (1)$$

given by Lorentz's force equation.

The number of recoil and photo-electrons. The experiments indicate that when x-rays traverse different elements, the number of recoil electrons ejected is proportional to the number of electrons traversed by the rays,[1] except that for soft x-rays a correction must be applied for the electrons which are bound to tightly in the atom to be ejected by the recoil process.[2] This proportionality suggests strongly that the action is one in which the radiation and the electrons only are concerned, the positive part of the atom playing no essential part. That is, the recoil effect seems to be the action of radiation on electrons which are effectively *free*.

In support of this suggestion we may point out: 1. The recoil electrons have been identified with those which scatter x-rays,[3] and according to the classical electron theory the scattering process is one in which we can consider the electrons alone, without taking into account the positive part of the atom. 2. It is found that the energy and momentum of the system photon plus electron are conserved, within a rather small experimental error,[4] without taking into account any action on the positive part of the atom.

The photoelectric action of x-rays is, however, apparently an action between radiation and a pair of associated positive and negative charges. Experiments such as those of de Broglie with the magnetic spectrograph[5] show that the large majority of the photo-electrons ejected by x-rays come from the K energy level of the atom, supporting the view that it is these photo-electrons which have received the energy "truly absorbed" from the x-ray beam. Owen's observation[6] that the true absorption of x-rays per atom is proportional to the fourth power of the atomic number, and Moore's observation[7] that the number of photo-electrons is likewise proportional to its fourth power, when interpreted in terms of Moseley's law, means that the probability that a photo-electron ejected from the K shell of an atom traversed by x-rays is approximately proportional to the square of the energy required to remove a K electron from the atom. That is to say, the photoelectric effect becomes a very improbable event for loosely bound electrons, and for free electrons should not occur at all.

This conclusion is confirmed by the fact that if a free electron takes all the energy of the photon which it absorbs it must acquire more momentum than that possessed by the photon, so that the energy and momentum of the system photon plus electron cannot both be conserved in the photoelectric process.[8] The motion of the atomic core must also be considered to make conservation possible.

[1] A. H. Compton and A. W. Simon, Phys. Rev. **25**, 306 (1925).

[2] J. M. Nuttall and E. J. Williams, Manchester Memoirs **70**, 1 (1926).

[3] C. T. R. Wilson, Proc. Roy. Soc. **A104**, 1 (1923); A. H. Compton and J. C. Hubbard, Phys. Rev. **23**, 439 (1924).

[4] A. H. Compton and A. W. Simon, Phys. Rev. **26**, 289 (1925).

[5] M. de Broglie, Jour. de physique **2**, 265 (1921).

[6] E. A. Owen, Proc. Roy. Soc. **A94**, 522 (1918).

[7] H. Moore, Proc. Roy. Soc. **A91**, 337 (1915).

[8] Cf. the writer's "X-Rays and Electrons," p. 265, note 1.

Short duration of the recoil and photoelectric actions. Statistical studies of the motion of recoil and photo-electrons by the cloud expansion method have indicated that in both cases the forward momentum of the electron is on the average approximately equal to the momentum of the incident quantum.* The predominant motion of the recoil electrons is forward, though the experiments support the theory in showing a transverse component resulting from the deflection of the motion of the scattered photon.[1] For the photo-electrons the predominant motion is transverse, though, on the average, with a forward component which is stronger for the shorter wavelengths as is indicated in the three curves of figure 1, representing data due to Auger.[9]

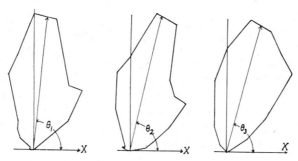

Fig. 1. Longitudinal distribution of photo-electrons for x-rays of three different wavelengths, according to Auger. A photo-electron ejected at the angle θ has a forward momentum equal to that of the incident photon.

If the experimental result is correct that the recoil and photo-electrons retain all the forward impulse imparted by the photon, it means that the impulse has been imparted in a time short compared with the natural period of the electron in its parent atom. For if the duration were longer than this, most of the impulse would be transferred from the electron to the more massive parent atom. This means that the duration of the impulse due to the photon is also short compared with the period of the associated wave. Thus the action of the radiation on the electron cannot be an oscillatory one with the frequency of the wave associated with the photon. The experiments are on the other hand consistent with the view that both the recoil and the photo-electrons are ejected by sensibly instantaneous impulses from

[9] P. Auger, J. de phys. et rad. **6**, 205 (1925).
[10] F. Kirchner, Ann. d. Physik. **81**, 1113 (1926).

* Since this was written, experiments by Loughridge (Phys. Rev. **30**, 1927) have been published which show a forward component to the photo-electrons' motion which seems to be greater than that predicted by equation (2). Williams, in experiments as yet unpublished, finds that the forward component is almost twice as great as that predicted by this theory. These results indicate that the mechanism of interaction between the photon and the atom must be more complex than that here postulated. The fact that the forward momentum is found to be of the same order of magnitude as that of the incident photon, however, suggests that the momentum of the photon is acquired by the photo-electron, while an additional forward impulse is imparted by the atom. Thus these more recent experiments also support the view that the photo-electron acquires both the energy and the momentum of the photon.

the photons. It is in fact difficult to imagine any other type of action which would not impart to the positive core of the atom some of the forward momentum of the photon.

Direction of the impulse imparted to recoil electrons. We have noticed that the predominant motion of the recoil electrons is forward, that is, perpendicular to the electric vector of the incident wave. Recent experiments by Kirchner[10] have shown that even the transverse component of the motion of the recoil electrons is on the average greater in the direction of the magnetic than in that of the electric vector of the incident wave. These results are precisely what we should expect from an application of energy and momentum conservation to the system photon plus electron, if the distribution of the scattered photons is to be in approximate accord with Thomson's classical theory of the distribution of the scattered x-rays. Such a motion of the electrons is however in striking contrast with that predicted by Lorentz's force equation (1). Since the speed of these electrons is at all times small compared with that of light, this equation predicts a motion of the electron almost parallel (or anti-parallel) with the electric vector of the x-ray wave, that is, in a direction perpendicular to the preferred motion of the recoil electrons as shown by the experiments.

Our attention is thus forcibly called to the fact that if the field of an electromagnetic wave acts on a free electron in the manner indicated by Lorentz's equation, the momentum of the system radiation plus electron is not in general conserved in the process. For if the Poynting vector expresses the momentum of the radiation, we find that this momentum is wholly in the direction of propagation of the electromagnetic wave, whereas the impulse imparted to an electron by the electric vector is according to equation (1) perpendicular to the direction of propagation. This means that the electron acquires a transverse momentum which is not balanced by any transverse momentum lost by the radiation. Thus if the momentum is to be conserved, this equation cannot represent the action of radiation on a free electron.

An experimental test of this point is not easy. In experiments with steady or slowly changing electric and magnetic fields, the applied fields are due to the presence of electrically charged or magnetized bodies, which receive the reaction from the force applied to any charge in the field. It is only with radiation fields that the test can be made, since only in this case can the electromagnetic field be considered separate from the charges which give rise to the field. Apparently the only example of the action of radiation on isolated electrons that has been studied experimentally is the production of recoil electrons when x-rays are scattered. In this case, as we have seen, the impulse imparted to the electron by the radiation is in the direction required by the conservation of momentum, which is almost perpendicular to that suggested by the classical force equation.

Direction of the impulse imparted to photo-electrons. When a radiation field acts upon a pair of positive and negative charges, there is no difficulty with the conservation of linear momentum, for the impulses imparted by the electric vector of the radiation to the positive and negative charges will presumably be equal and opposite. From the standpoint of momentum conservation, therefore, it would not be surprising if the greater part of the momentum of the photo-electron were transverse, as suggested by equation (1). We must consider also, however, the conservation of angular momentum.

If the photon acts instantaneously upon a photo-electron, as the experiments suggest, in order that it may impart to the atom all of its energy and at the same time its linear and angular momentum, there is only one definite direction in which the impulse may act. If the angular momentum of the photon is zero, it must not impart any angular momentum to the atom. That is, neglecting the effect of radiation pressure, the impulse imparted to the electron must be along the line joining the electron and the atomic core. If the photon possesses angular momentum, as may be the case with circularly polarized light, the impulse must be in the direction necessary to give this angular momentum to the dissociated atom and electron.

There is thus a single line in the atom on which an electron can lie where the photon can impart to it a photoelectric impulse along the electric vector of the associated wave, and still conserve the angular momentum of the

Fig. 2. Lateral distribution of photo-electrons for incompletely polarized x-rays, according to Bubb.

system. If we were to suppose that the impulse must be exactly in the direction of the electric vector, this would mean that if the electron were in any other position in the atom, the photon could not act upon it photoelectrically. A more plausible assumption would seem to be that the photon may act on the electron in any position in the atom, with an impulse in the direction demanded by the conservation of angular momentum, but that the probability that such action shall occur is greater the nearer its direction approaches the electric vector. This assumption would be consistent with the conservation of angular momentum for each individual event, and would be statistically in accord with the force equation.

The experimental evidence is in complete accord with the latter assumption. It is found that the impulse imparted to the photo-electrons is not always in the same direction, but may occur in a wide variety of directions. This is illustrated by Auger's experiments shown in figure 1. Except for the angle θ, which as we have seen is due to the radiation pressure or momentum of the photon, the most probable direction of emission is that of the electric

[11] F. W. Bubb, Phys. Rev. **23**, 137 (1924).

vector; but many photo-electrons are also observed at other angles. The experiments of Bubb,[11] summarized in Fig. 2, show in a similar manner how the directions of ejection are distributed when polarized x-rays are used. We see here clearly that though the probability is greatest for ejection in the plane of the electric vector, some electrons are ejected at all possible angles. In these experiments the polarization of the x-rays was not complete. A correction for the unpolarized x-rays reduces the probability of emission perpendicular to the electric vector approximately to zero. Even with this correction, however, photo-electrons are observed at all other angles. Both this experiment of Bubb and that of Auger have been confirmed by a number of different investigators.[12]

Attempts have been made[13] to account for the variation in the direction of emission of the photo-electrons as due to the initial motions of the electrons in their orbits; but these have failed to account for the fact that the probability of emission in the different directions is practically the same for all atoms from which the electrons come, whereas the initial motions of the electrons may differ widely for electrons in the different atoms. There does not seem to be any way of accounting for this wide distribution of the directions of emission other than to suppose that the impulse applied to the electron by the radiation is variable in direction.[14] That is, the impulse is not necessarily in the direction of the electric vector; this (neglecting the effect of radiation pressure) is only the most probable direction for the impulse to act.

In the treatment of photoelectric emission from the standpoint of wave mechanics, Wentzel[15] has concluded that the probability of photoelectric ejection at an angle α with the electric vector (neglecting radiation pressure) is proportional to $\cos^2 \alpha$. This is precisely the result to which Auger and Perrin[16] had been led empirically in order to account for Auger's experiments such as those shown in Fig. 1. The distribution of the directions of emission of the photo-electrons is thus of exactly the type which we should expect from considerations of conservation of angular momentum.

In the case of the photo-electric effect with visible or ultra-violet light, the magnitude of the work function required to remove the photo-electrons from the metal suggests that we may be dealing with "conductivity" electrons associated with the whole mass of metal. If this is the case, there should be no difficulty with the conservation of the angular momentum of the system, and we might expect the photo-electrons to be ejected almost exactly in the direction of the electric vector. An experimental test of this point in the case of the selective photoelectric effect would be of great interest.

[12] W. Bothe, Zeits. f. Physik. **26**, 59 (1924); D. H. Loughridge, Phys. Rev. **26**, 697 (1925); F. Kirchner, Zeits. f. Physik. **27**, 385 (1926).

[13] F. W. Bubb, Phil. Mag. **49**, 824 (1925); W. Bothe, Zeits. f. Physik. **26**, 74 (1924).

[14] For a more detailed discussion of this point, see the writer's "X-Rays and Electrons," p. 25 (1926).

[15] G. Wentzel, Zeits. f. Physik. **40**, 574 (1926).

[16] P. Auger and F. Perrin, C. R. **180**, 1742 (1925).

Both in the case of the action of x-rays on isolated electrons (recoil effect) and that of their action on a pair of positive and negative charges (photoelectric effect) we find evidence of radiation pressure, which means a force of the type indicated by the second term of Lorentz's force equation. We find, however, that for isolated electrons there is no evidence that there exists any force associated with the electric field of the x-ray wave. At least the favored direction of recoil is at right angles with the electric field. For an electron associated with a positively charged atom, the first term of equation (1) may be taken to represent the most probable direction in which the electric field will act on the electron. The direction of the photoelectric action in each individual case is however apparently determined by the requirements of the conservation of the angular momentum of the system.

UNIVERSITY OF CHICAGO,
   October 5, 1927.

Journal of The Franklin Institute

Devoted to Science and the Mechanic Arts

| Vol. 205 | FEBRUARY, 1928 | No. 2 |

SOME EXPERIMENTAL DIFFICULTIES WITH THE ELECTROMAGNETIC THEORY OF RADIATION.

BY

ARTHUR H. COMPTON, Ph.D.

Professor of Physics, University of Chicago.

DURING the last few years it has become increasingly evident that the classical electromagnetic theory of radiation is incapable of accounting for certain large classes of phenomena, especially those concerned with the interaction between radiation and matter. It is not that we question the wave character of light—the striking successes of this conception in explaining polarization and interference of light can leave no doubt that radiation has the characteristics of waves; but it is equally true that certain other properties of radiation are not easily interpreted in terms of waves. The power of the electromagnetic theory as applied to a great variety of problems of radiation is too well known to require emphasis. It is, however, only by acquainting ourselves with the failures of this powerful theory that we can hope to develop a more complete theory of radiation which will describe the facts as we know them.

The more serious difficulties which present themselves in connection with the theory that radiation consists of electromagnetic waves, propagated through space in accord with the demands of Maxwell's equations, may be classified conveniently under four heads:

(1) How are the waves produced? The classical electro-

(Note.—The Franklin Institute is not responsible for the statements and opinions advanced by contributors to the JOURNAL.)

COPYRIGHT, 1928, by THE FRANKLIN INSTITUTE.

dynamics requires as a source of an electromagnetic wave an oscillator of the same frequency as that of the waves it radiates. Our studies of spectra, however, make it appear impossible that an atom should contain oscillators of the same frequencies as the emitted rays.

(2) The photo-electric effect. This phenomenon is wholly anomalous when viewed from the standpoint of waves.

(3) The scattering of X-rays, and the recoil electrons, phenomena in which we find gradually increasing departures from the predictions of the classical wave theory as the frequency increases.

(4) Experiments on individual interactions between quanta of radiation and electrons. If the results of the experiments of this type are reliable, they seem to show definitely that individual quanta of radiation, of energy $h\nu$, proceed in definite directions.

The Photon Hypothesis.—In order to exhibit more clearly the difficulties with the classical theory of radiation, it will be helpful to keep in mind the suggestion that light consists of corpuscles. We need not think of these two views as necessarily alternative. It may well be that the two conceptions are complementary. Perhaps the corpuscle is related to the wave in somewhat the same manner that the molecule is related to matter in bulk; or there may be a guiding wave which directs the corpuscles which carry the energy. In any case, the phenomena which we have just mentioned suggest the hypothesis that radiation is divisible into units possessing energy $h\nu$, and which proceed in definite directions with momentum $h\nu/c$. This is obviously similar to Newton's old conception of light corpuscles. It was revived in its present form by Einstein, it was defended under the name of the "Neutron Theory" by Sir William Bragg, and has been given new life by the recent discoveries associated with the scattering of X-rays.

In referring to this unit of radiation I shall use the name "photon," suggested recently by G. N. Lewis.[1] This word avoids any implication regarding the nature of the unit, as contained for example in the name "needle ray." As compared with the terms "radiation quantum" and "light quant," this name has the advantages of brevity and of avoiding any implied

---

[1] G. N. Lewis, *Nature*, Dec. 18, 1926.

dependence upon the much more general quantum mechanics or quantum theory of atomic structure.

Virtual Radiation.—Another conception of the nature of radiation which it will be desirable to compare with the experiments is Bohr, Kramers and Slater's important theory of virtual radiation.[2] According to this theory, an atom in an excited state is continually emitting virtual radiation, to which no energy characteristics are to be ascribed. The normal atoms have associated with them virtual oscillators, of the frequencies corresponding to jumps of the atom to all of the stationary states of higher energy. The virtual radiation may be thought of as being absorbed by these virtual oscillators, and any atom which has a virtual oscillator absorbing this virtual radiation has a certain probability of jumping suddenly to the higher state of energy corresponding to the frequency of the particular virtual oscillator. On the average, if the radiation is completely absorbed, the number of such jumps to levels of higher energy is equal to the number of emitting atoms which pass from higher to lower states. But there is no direct connection between the falling of one atom from a higher to a lower state and a corresponding rise of a second atom from a lower to a higher state. Thus on this view the energy of the emitting atoms and of the absorbing atoms is only statistically conserved.

THE EMISSION OF RADIATION.

When we trace a sound to its origin, we find it coming from an oscillator vibrating with the frequency of the sound itself. The same is true of electric waves, such as radio waves, where the source of the radiation is a stream of electrons oscillating back and forth in a wire. But when we trace a light ray or an X-ray back to its origin, we fail to find any oscillator which has the same frequency as the ray itself. The more complete our knowledge becomes of the origin of spectrum lines, the more clearly we see that if we are to assign any frequencies to the electrons within the atoms, these frequencies are not the frequencies of the emitted rays, but are the frequencies associated with the stationary states of the atom. This result cannot be reconciled with the electromagnetic theory of radiation, nor has

---

[2] N. Bohr, H. A. Kramers and J. C. Slater. *Phil. Mag.*, **47**, 785 (1924); *Zeits. f. Phys.*, **24**, 69 (1924).

any mechanism been suggested whereby radiation of one frequency can be excited by an oscillator of another frequency. The wave theory of radiation is thus powerless to suggest how the waves originate.

The origin of the radiation is considerably simpler when we consider it from the photon viewpoint. We find that an atom changes from a stationary state of one energy to a state of less energy, and associated with this change radiation is emitted. What is simpler than to suppose that the energy lost by the atom is radiated away as a single photon? It is on this view unnecessary to say anything regarding the frequency of the radiation. We are concerned only with the energy of the photon, its direction of emission, and its state of polarization.

The problem of the emission of radiation takes an especially interesting form when we consider the production of the continuous X-ray spectrum.[3] Experiment shows that both the intensity and the average frequency of the X-rays emitted at angles less than 90 degrees with the cathode-ray stream are greater than at angles greater than 90 degrees. This is just what we should expect due to the Doppler effect if the X-rays are emitted by a radiator moving in the direction of the cathode rays. In order to account for the observed dissymmetry between the rays in the forward and backward directions, the particles emitting the radiation must be moving with a speed of the order of 25 per cent. that of light. This means that the emitting particles must be free electrons, since it would require an impossibly large energy to set an atom into motion with such a speed.

But it will be recalled that the continuous X-ray spectrum has a sharp upper limit. Such a sharp limit is, however, possible on the wave theory only in case the rays come in trains of waves of considerable length, so that the interference between the waves in different parts of the train can be complete at small glancing angles of reflection from the crystal. This implies that the oscillator which emits the rays must vibrate back and forth with constant frequency a large number of times while the ray is being emitted. Such an oscillation might be imagined for an electron within an atom; but it is impossible for an electron moving

---

[3] The difficulty here discussed was first emphasized by D. L. Webster, *Phys. Rev.*, **13**, 303 (1919).

through an irregular assemblage of atoms with a speed comparable with that of light.

Thus the Doppler effect in the primary X-rays demands that the rays shall be emitted by rapidly moving electrons, while the sharp limit to the continuous spectrum requires that the rays be emitted by an electron bound within an atom.

The only possible escape from this dilemma on the wave theory is to suppose that the electron is itself capable of internal oscillation of such a character as to emit radiation. This would, however, introduce an undesirable complexity into our conception of the electron, and would ascribe the continuous X-rays to an origin entirely different from that of other known sources of radiation.

Here again the photon theory affords a simple solution. It is a consequence of Ehrenfests's adiabatic principle that photons emitted by a moving radiator will show the same Doppler effect, with regard to both frequency and intensity, as does a beam of waves.[4] But if we suppose that photons are radiated by the moving cathode electrons, the energy of each photon will be the energy lost by the electron, and the limit of the X-ray spectrum is necessarily reached when the energy of the photon is equal to the initial energy of the electron, $i.e.$, $h\nu = eV$. In this case, if we consider the initial state as an electron approaching an atom with large kinetic energy and the final state as the electron leaving the atom with a smaller kinetic energy, we see that the emission of the continuous X-ray spectrum is the same kind of event as the emission of any other type of radiation.

Absorption of Radiation.—According to the photon theory, absorption occurs when a photon meets an atom and imparts its energy to the atom. The atom is thereby raised to a stationary state of higher energy—precisely the reverse of the emission process.

On the wave theory, absorption is necessarily a continuous process, if we admit the conservation of energy, since on no part of the wave front is there enough energy available to change the atom suddenly from a state of low energy to a state of higher energy. What evidence we have is, however, strongly against the atom having for any considerable length of time an energy intermediate between two stationary states; and if such inter-

---

[4] *Cf., e.g.*, A. H. Compton, *Phys. Rev.*, **21**, 483 (1923).

mediate states cannot exist, the gradual absorption of radiation is not possible. Thus the absorption of energy from waves is irreconcilable with the conception of stationary states.

We have seen that on the theory of virtual radiation the energy of the emitting atoms and of the absorbing atoms is only statistically conserved. There is according to this view therefore no difficulty with supposing that the absorbing atom suddenly jumps to a higher level of energy, even though it has not received from the radiation as much energy as is necessary to make the jump. It is thus possible through virtual oscillators and virtual radiation to reconcile the wave theory of radiation with the sudden absorption of energy, and hence to retain the idea of stationary states.

THE PHOTO-ELECTRIC EFFECT.

It is well known that the photon hypothesis was introduced by Einstein to account for the photo-electric effect.[5] The assumption that light consists of discrete units which can be absorbed by atoms only as units, each giving rise to a photo-electron, accounted at once for the fact that the number of photo-electrons is proportional to the intensity of the light; and the assumption that the energy of the light unit is equal to $h\nu$, where h is Planck's constant, made it possible to predict the kinetic energy with which the photo-electrons should be ejected, as expressed by Einstein's well-known photo-electric equation,

$$mc^2 \left(\frac{1}{\sqrt{1-\beta^2}} - 1 \right) = h\nu - w_p. \qquad (1)$$

Seven years elapsed before experiments by Richardson and Compton[6] and by Hughes[7] showed that the energy of the emitted electrons was indeed proportional to the frequency less a constant, and that the factor of proportionality was close to the value of h calculated from Planck's radiation formula. Millikan's more recent precision photo-electric experiments with the alkali metals[8] confirmed the identity of the constant h in the photo-electric equation with that in Planck's radiation formula. De Broglie's beautiful experiments[9] with the magnetic spectrograph

---

[5] A. Einstein, *Ann. d. Phys.*, **17**, 145 (1905).
[6] O. W. Richardson and K. T. Compton, *Phil. Mag.*, **24**, 575 (1912).
[7] A. L. Hughes, *Phil. Trans.*, **A. 213**, 205 (1912).
[8] R. A. Millikan, *Phys. Rev.*, **7**, 355 (1916).
[9] M. de Broglie, *Jour. de Phys.*, **2**, 265 (1921).

showed that in the region of X-ray frequencies the same equation holds, if only we interpret the work function w_p as the work required to remove the electron from the Pth energy level of the atom. Thibaud has made use of this result [10] in comparing the velocities of the photo-electrons ejected by γ-rays from different elements, and has thus shown that the photo-electric equation (1) holds with precision even for β-rays of the highest speed. Thus from light of frequency so low that it is barely able to eject photo-electrons from metals to γ-rays that eject photo-

FIG. 1.

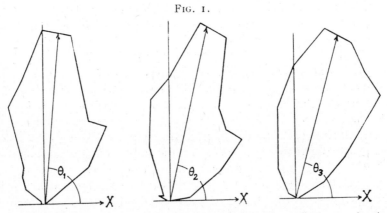

Longitudinal distribution of photo-electrons for X-rays of three different effective wave-lengths, according to Auger.

electrons with a speed almost as great as that of light, the photon theory expresses accurately the speed of the photo-electrons.

The direction in which the photo-electrons are emitted is no less instructive than is the velocity. Experiments using the cloud expansion method, performed by C. T. R. Wilson [11] and others,[12] have shown that the most probable direction in which the photo-electron is ejected from an atom is nearly the direction of the electric vector of the incident wave, but with an appreciable forward component to its motion. There is, however, a very considerable variation in the direction of emission. For example,

---

[10] J. Thibaud, *C. R.*, **179**, pp. 165, 1053, and 1322 (1924).

[11] C. T. R. Wilson, *Proc. Roy. Soc.*, **A, 104**, 1 (1923).

[12] A. H. Compton, *Bull. Natl. Res. Coun. No. 20*, p. 25 (1922); F. W. Bubb, *Phys. Rev.*, **23**, 137 (1924); P. Auger, *C. R.*, **178**, 1535 (1924); D. H. Loughridge, *Phys. Rev.*, **26**, 697 (1925); F. Kirchner, *Zeits. f. Phys.*, **27**, 385 (1926).

if we plot the number of photo-electrons ejected at different angles with the primary beam we find, according to Auger, the distribution shown in Fig. 1. Each of these curves, taken at a different potential, represents the distribution of about 200 photo-electron tracks. It will be seen that as the potential on the X-ray

FIG. 2.

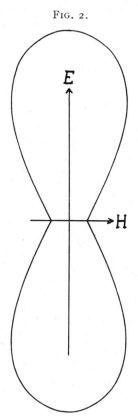

Lateral distribution of photo-electrons for incompletely polarized X-rays, according to Bubb.

tube increases, the average forward component of the photo-electron's motion also increases.

When polarized X-rays are used, there is a strong preponderance of the photo-electrons in or near the plane including the electric vector of the incident rays. Thus Fig. 2 shows the distribution found by Bubb of the direction of the photo-electrons ejected from moist air when traversed by X-rays that have been polarized by scattering at right angles from a block of paraffin.

Because of multiple scattering in the paraffin, the scattered rays are not completely polarized, and this is probably sufficient to account for the fact that some photo-electrons appear to start at right angles with the electric vector. This effect with X-rays is doubtless similar in character to the selective photo-electric effect discovered many years ago by Pohl and Pringsheim, in which the number of electrons ejected by light from the liquid surface of sodium-potassium alloy is greater when the electric vector is in a plane perpendicular to the surface than when parallel to the surface.

Recent experiments have shown that the direction in which the photo-electrons are ejected by X-rays is at least very nearly independent of the material from which the electrons come.[13]

Can Electromagnetic Waves Produce Photo-electrons?—Before discussing the production of photo-electrons from the standpoint of radiation quanta, let us see what success meets the attempt to account for them on the basis of electromagnetic waves. The fact that they are emitted approximately in the direction of the electric vector would suggest that the photo-electrons are ejected by the direct action of the electric field of the incident rays. If this were the case, however, we should expect the speed of the ejected electrons to be greater for greater intensity of radiation, whereas experiment shows that for the same wave-length intense sunlight ejects an electron no faster than does the feeble light from a star. Furthermore, the energy available from the electromagnetic wave is wholly inadequate. Thus in a recent experiment performed by Joffe and Dobronsawov,[14] X-rays were produced by the impact on a target of 10^4 to 10^5 electrons per second. Since on the electromagnetic theory an X-ray pulse is of the order of 10^3 waves in length or 10^{-16} seconds in duration, the X-ray pulses must have followed each other at widely separated intervals. It was found, however, that photo-electrons were occasionally ejected from a bismuth particle which subtended a solid angle not greater than 10^{-5}. It is clearly impossible that all the energy of an X-ray pulse which has spread out in a spherical wave should spend itself on this bismuth particle. Thus on the wave theory the ejection of the photo-

---

[13] E. A. Owen, *Proc. Phys. Soc.*, **30**, 133 (1918); Auger, Kirchner, Loughridge, *loc. cit.*

[14] A. Joffe and N. Dobronsawov, *Zeits. f. Phys.*, **34**, 889 (1925).

electron, which has almost as much energy as the original cathode electron, could not have been accomplished by a single pulse. It cannot therefore be the direct action of the electric vector of the wave, taken in the usual sense, which has ejected the electron.

We may assume, on the other hand, that the energy is gradually absorbed in the bismuth particle of Joffe's experiment until an amount $h\nu$ has accumulated, which is then spent in ejecting the photo-electron. We have already called attention to the fact that this gradual absorption hypothesis implies the existence of stationary states in the atom having infinitesimal gradations of energy, whereas the evidence is very strong that atoms cannot endure except in certain definitely defined stationary states. But new difficulties also arise. Why do the photo-electrons tend to start in the direction of the electric field of the incident wave? If we suppose that it is the gradual absorption of energy from a wave which liberates the electron, why does there exist a tendency for the electron to start with a large component of its motion in a forward direction? The forward impulse due to the radiation pressure as the energy is gradually absorbed will be transferred to the atom and not left with absorbing electron. The accumulation hypothesis is thus difficult to defend.

Photons and Photo-electrons.—On the photon theory it is possible to account in a simple manner for most of the properties of the photo-electrons. We have seen how Einstein was able to predict accurately the velocity of the photo-electrons, assuming only that energy is conserved when a photon acts on an electron. In order to account for the direction of emission we must ascribe to the photon some of the properties of an electromagnetic pulse. Bubb introduced the suggestion [15] that we ascribe to the photon a vector property similar to the electric vector of an electromagnetic wave, so that when the photon traverses an atom the electrons and the nucleus receive impulses in opposite directions perpendicular to the direction of propagation. Associated with this electric vector, we should also expect to find a magnetic vector. Thus if an electron is set in motion by the electric vector of the photon at right angles to the direction of propagation, the magnetic vector of the photon will act on the moving electron in the direction of propagation. This is strictly analogous to the radiation pressure exerted by an electromagnetic wave on an elec-

---

[15] F. W. Bubb, *Phys. Rev.*, **23**, 137 (1924).

tron which it traverses, and means that the forward momentum of the absorbed photon is transferred to the photo-electron.

In the simplest case, where we neglect the initial momentum of the electron in its orbital motion in the atom, the angle between the direction of the incident ray and the direction of ejection is found from these assumptions to be,

$$\theta = \tan^{-1} \sqrt{2/\alpha}, \qquad (2)$$

where $\alpha = \gamma/\lambda$, and $\gamma = h/mc = 0.0242$ A. The quantity α is small compared with unity, except for very hard X-rays and γ-rays. Thus for light, equation (2) predicts the expulsion of photo-electrons at nearly 90 degrees. This is in accord with the rather uncertain data which have been obtained with visible and ultra-violet light.[16]

The only really significant test of this result is in its application to X-ray photo-electrons. In Fig. 1 are drawn the lines θ_1, θ_2 and θ_3 for the three curves, at the angles calculated by Auger from equation (2). It will be seen that they fall very satisfactorily in the direction of maximum emission of the photo-electrons. Similar results have been obtained by other investigators.[17] This may be taken as proof that a photon imparts not only its energy, but also its momentum to the photo-electrons.[18]

If the angular momentum of the atomic system from which the photo-electron is ejected is to be conserved when acted upon by the radiation, the electron cannot be ejected exactly in the direction of θ, but must receive an impulse in a direction determined by the position of the electron in the atom at the instant

---

[16] *Cf.* A. Partsch and W. Hallwachs, *Ann. d. Phys.*, **41**, 247 (1913).

[17] W. Bothe, *Zeits. f. Phys.*, **26**, 59 (1925); F. Kirchner, *Zeits. f. Phys.*, **27**, 385 (1926).

[18] Since this was written, experiments by Loughridge (*Phys. Rev.*, **30**, 1927) have been published which show a forward component to the photo-electron's motion which seems to be greater than that predicted by equation (2). Williams, in experiments as yet unpublished, finds that the forward component is almost twice as great as that predicted by this theory. These results indicate that the mechanism of interaction between the photon and the atom must be more complex than here postulated. The fact that the forward momentum of the photo-electron is found to be of the same order of magnitude as that of the incident photon, however, suggests that the momentum of the photon is acquired by the photo-electron, while an additional forward impulse is imparted by the atom. Thus these more recent experiments also support the view that the photo-electron acquires both the energy and the momentum of the photon.

it is traversed by the photon.[19] Thus we should probably consider the electric vector of the X-ray wave as defining merely the most probable direction in which the impulse should be imparted to the electron. This is doubtless the chief reason why the photo-electrons are emitted over a wide range of angles instead of in a definite direction, as would be suggested by the calculation just outlined.

It will readily be seen that if the time during which the photon exerts a force on the electron is comparable with the natural period of the electron in the atom, the impulse imparted to the electron will be transferred in part to the positive nucleus about which the electron is moving. The fact that the photo-electrons receive the momentum of the incident photon means that no appreciable part of the photon's momentum is spent on the remainder of the atom. This can only be the case if the time of action of the photon on the electron is short compared with the time of revolution of the electron in its orbit. Such a short duration of interaction is a natural consequence of the photon conception of radiation, but is quite contrary to the consequences of the electromagnetic theory.

The Photo-electric Effect and Virtual Radiation.—It is to be noted that none of these properties of the photo-electron is inconsistent with the virtual radiation theory of Bohr, Kramers and Slater. The difficulties which applied to the classical wave theory do not apply here, since the energy and momentum are conserved only statistically. There is nothing in this theory, however, which would enable us to predict anything regarding the motion of the photo-electrons. The degree of success that has attended the application of the photon hypothesis to the motion of these electrons has come directly from the application of the conservation principles to the individual action of a photon on an electron. The power of these principles as applied to this case is surprising if the assumption is correct that they are only statistically valid.

PHENOMENA ASSOCIATED WITH THE SCATTERING OF X-RAYS.

As is now well known, there is a group of phenomena associated with the scattering of X-rays for which the classical wave theory of radiation fails to account. These phenomena may be

---

[19] *Cf.* A. H. Compton, *Phys. Rev.*, 1928.

considered under the heads of: (1) The change of wave-length of X-rays due to scattering, (2) the intensity of scattered X-rays, and (3) the recoil electrons.

The earliest experiments on secondary X-rays and γ-rays showed a difference in the penetrating power of the primary and the secondary rays. In the case of X-rays, Barkla and his collaborators showed that the secondary rays from the heavy elements consisted largely of fluorescent radiations characteristic of the radiator, and that it was the presence of these softer rays which was chiefly responsible for the greater absorption of the secondary rays. When later experiments [20] showed a measurable difference in penetration even for light elements such as carbon, from which no fluorescent K or L radiation appears, it was natural to ascribe [21] this difference to a new type of fluorescent radiation, similar to the K and L types, but of shorter wave-length. Careful absorption measurements [22] failed, however, to reveal any critical absorption limit for these assumed "J" radiations similar to those corresponding to the K and L radiations. Moreover, direct spectroscopic observations [23] failed to reveal the existence of any spectrum lines under conditions for which the supposed J-rays should appear. It thus became evident that the softening of the secondary X-rays from the lighter elements was due to a different kind of process than the softening of the secondary rays from heavy elements where fluorescent X-rays are present.

A series of skilfully devised absorption experiments performed by J. A. Gray [24] showed, on the other hand, that both in the case of γ-rays and in that of X-rays an increase in wave-length accompanies the scattering of the rays of light elements.

It was at this stage that the first spectroscopic investigations of the secondary X-rays from light elements were made.[25]

---

[20] C. A. Sadler and P. Mesham, *Phil. Mag.*, **24**, 138 (1912); J. Laub, *Ann. d. Phys.*, **46**, 785 (1915).

[21] Barkla and White, *Phil. Mag.*, **34**, 270 (1917); J. Laub, *Ann. d. Phys.*, **46**, 785 (1915), *et al.*

[22] *E.g.*, Richtmyer and Grant, *Phys. Rev.*, **15**, 547 (1920).

[23] E. G. Duane and Shimizu, *Phys. Rev.*, **13**, 288 (1919); **14**, 389 (1919).

[24] J. A. Gray, *Phil. Mag.*, **26**, 611 (1913); Jour. Frank. Inst., Nov., 1920, p. 643.

[25] A. H. Compton, *Bull. Natl. Res. Coun. No. 20* (1922); *Phys. Rev.*, **22**, 409 (1923).

According to the usual electron theory of scattering it is obvious that the scattered rays will be of the same frequency as the forced oscillations of the electrons which emit them, and hence will be identical in frequency with the primary waves which set the electrons in motion. Instead of showing scattered rays of the same wave-length as the primary rays, however, these spectra revealed lines in the secondary rays corresponding to those in the primary beam, but with each line displaced slightly toward the longer wave-lengths.

This result might have been predicted from Gray's absorption measurements; but the spectrum measurements had the advantage of affording a quantitative measurement of the change in wavelength, which gave a basis for its theoretical interpretation.

The spectroscopic experiments which have shown this change in wave-length are too well known [26] to require discussion. The interpretation of the wave-length change in terms of photons being deflected by individual electrons and imparting a part of their energy to the scattering electrons is also very familiar. For purposes of discussion, however, let us recall that when we consider the interaction of a single photon with a single electron the principles of the conservation of energy and momentum lead us [27] to the result that the change in wave-length of the deflected photon is

$$\delta\lambda = \frac{h}{mc}(1 - \cos\varphi), \qquad (3)$$

where φ is the angle through which the photon is deflected. The electron at the same time recoils from the photon at an angle of θ given by,

$$\cot\theta = -(1+x)\tan\frac{1}{2}\varphi; \qquad (4)$$

and the kinetic energy of the recoiling electron is,

$$E_{\text{kin}} = h\nu \frac{2\alpha \cos^2\theta}{(1+\alpha)^2 - \alpha^2 \cos^2\theta}. \qquad (5)$$

The experiments show in the spectrum of the scattered rays two lines corresponding to each line of the primary ray. One

---

[26] *Cf., e.g.,* A. H. Compton, *Phys. Rev.,* **22**, 409 (1923); P. A. Ross, *Prov. Nat. Acad.,* **10**, 304 (1924).

[27] A. H. Compton, *Phys. Rev.,* **22**, 483 (1923); P. Debye, *Phys. Zeits.,* **24**, 161 (1923).

of these lines is of precisely the same wave-length as the primary ray, and the second line, though somewhat broadened, has its centre of gravity displaced by the amount predicted by equation (3). According to experiments by Kallman and Mark [28] and by Sharp,[29] this agreement between the theoretical and the observed shift is precise to within a small fraction of 1 per cent.

The Recoil Electrons.—From the quantitative agreement between the theoretical and the observed wave-lengths of the scattered rays, the recoil electrons predicted by the photon theory of scattering were looked for with some confidence. When this theory was proposed, there was no direct evidence for the existence of such electrons, though indirect evidence suggested that the secondary beta-rays ejected from matter by hard γ-rays are mostly of this type. Within a few months of their prediction, however, C. T. R. Wilson [30] and W. Bothe [31] independently announced their discovery. The recoil electrons show as short tracks, pointed in the direction of the primary X-ray beam, mixed among the much longer tracks due to the photo-electrons ejected by the X-rays.

Perhaps the most convincing reason for associating these short tracks with the scattered X-rays comes from a study of their number. Each photo-electron in a cloud photograph represents a quantum of truly absorbed X-ray energy. If the short tracks are due to recoil electrons, each one should represent the scattering of a photon. Thus the ratio N_r/N_p of the number of short tracks to the number of long tracks should be the same as the ratio σ/t of the scattered to the truly absorbed energy when the X-rays pass through air. The latter ratio is known from absorption measurements, and the former ratio can be determined by counting the tracks on the photographs. The satisfactory agreement between the two ratios [32] for X-rays of different wave-lengths means that on the average there is about one quantum of energy scattered for each short track that is produced.

This result is in itself contrary to the predictions of the classical wave theory, since on this basis all the energy spent on

---

[28] H. Kallman and H. Mark, *Naturwiss.*, **13**, 297 (1925).

[29] H. M. Sharp, *Phys. Rev.*, **26**, 691 (1925).

[30] C. T. R. Wilson, *Proc. Roy. Soc.*, **104**, 1 (1923).

[31] W. Bothe, *Zeits. f. Phys.*, **16**, 319 (1923).

[32] A. H. Compton and A. W. Simon, *Phys. Rev.*, **25**, 306 (1925); J. M. Nuttall and E. J. Williams, *Manchester Memoirs*, **70**, 1 (1926).

a free electron (except the insignificant effect of radiation presrure) should reappear as scattered X-rays. In these experiments, on the contrary, 5 or 10 per cent. as much energy appears in the motion of the recoil electrons as appears in the scattered X-rays.

That these short tracks associated with the scattered X-rays correspond to the recoil electrons predicted by the photon theory of scattering becomes clear from a study of their energies. The energy of the electron which produces a track can be calculated from the range of the track. The ranges of tracks which start in different directions have been studied,[33] using primary X-rays of different wave-lengths, with the result that equation (5) has been satisfactorily verified.

In view of the fact that electrons of this type were unknown at the time the photon theory of scattering was presented, their existence, and the close agreement with the predictions as to their number, direction and velocity, supply strong evidence in favor of the fundamental hypotheses of the theory.

Interpretation of These Experiments.—It is impossible to account for scattered rays of altered frequency, and for the existence of the recoil electrons, if we assume that X-rays consist of electromagnetic waves in the usual sense. Yet some progress has been made on the basis of semi-classical theories. It is an interesting fact that the wave-length of the scattered ray according to equation (3) varies with the angle just as one would expect from a Doppler effect if the rays are scattered from an electron moving in the direction of the primary beam. Moreover, the velocity that must be assigned to the electron in order to give the proper magnitude to the change of wave-length is that which the electron would acquire by radiation pressure if it should absorb a quantum of the incident rays. Several writers [34] have therefore assumed that an electron takes from the incident beam a whole quantum of the incident radiation, and then emits this energy as a spherical wave while moving forward with high velocity.

This conception that the radiation occurs in spherical waves, and that the scattering electron can nevertheless acquire suddenly

---

[33] Compton and Simon, *loc. cit.*

[34] C. R. Bauer. *C. R.*, **177**, 1211 (1923); C. T. R. Wilson, *Proc. Roy. Soc.*, **104**, 1 (1923); K. Forsterling, *Phys. Zeits.*, **25**, 313 (1924); O. Halpern, *Zeits. f. Phys.*, **30**, 153 (1924).

the impulses from a whole quantum of incident radiation is inconsistent with the principle of energy conservation. But there is the more serious experimental difficulty that this theory predicts recoil electrons all moving in the same direction and with the same velocity. The experiments show, on the other hand, a variety of directions and velocities, with the velocity and direction correlated as demanded by the photon hypothesis. Moreover, the maximum range of the recoil electrons, though in agreement with the predictions of the photon theory, is found to be about four times as great as that predicted by this semi-classical theory.

There is nothing in these experiments, as far as we have described them, which is inconsistent with the idea of virtual oscillators continually scattering virtual radiation. In order to account for the change of wave-length on this view, Bohr, Kramers and Slater assumed that the virtual oscillators scatter as if moving in the direction of the primary beam, accounting for the change of wave-length as a Doppler effect. They then supposed that occasionally an electron, under the stimulation of the primary virtual rays, will suddenly move forward with a momentum large compared with the impulse received from the radiation pressure. Though we have seen that not all of the recoil electrons move directly forward, but in a variety of different directions, the theory could easily be extended to include the type of motion that is actually observed.

The only objection that one can raise against this virtual radiation theory in connection with the scattering phenomena as viewed on a large scale, is that it is difficult to see how such a theory could by itself predict the change of wave-length and the motion of the recoil electrons. These phenomena are directly predictable if the conservation of energy and momentum are assumed to apply to the individual actions of radiation on electrons; but this is precisely where the virtual radiation theory denies the validity of the conservation principles.

We may conclude that the photon theory predicts quantitatively and in detail the change of wave-length of the scattered X-rays and the characteristics of the recoil electrons. The virtual radiation theory is probably not inconsistent with these experiments, but is incapable of predicting the results. The classical theory, however, is altogether helpless to deal with these phenomena.

The Origin of the Unmodified Line.—The unmodified line is probably due to X-rays which are scattered by electrons so firmly held within the atom that they are not ejected by the impulse from the deflected photons. This view is adequate to account for the major characteristics of the unmodified rays, though as yet no quantitatively satisfactory theory of their origin has been published.[35] It is probable that a detailed account of these rays will involve definite assumptions regarding the nature and the duration of the interaction between a photon and an electron; but it is doubtful whether such investigations will add new evidence as to the existence of the photons themselves.

A similar situation holds regarding the intensity of the scattered X-rays. Historically it was the fact that the classical electromagnetic theory is unable to account for the low intensity of the scattered X-rays which called attention to the importance of the problem of scattering. But the solutions which have been offered by Breit,[36] Dirac[37] and others[38] of this intensity problem as distinguished from that of the change of wave-length, seem to introduce no new concepts regarding the nature of radiation or of the scattering process. Let us therefore turn our attention to the experiments that have been performed on the individual process of interaction between photons and electrons.

INTERACTIONS BETWEEN RADIATION AND SINGLE ELECTRONS.

The most significant of the experiments which show departures from the predictions of the classical wave theory are those that study the action of radiation on individual atoms or on individual electrons. Two methods have been found suitable for performing these experiments, Geiger's point counters, and Wilson's cloud expansion photographs.

(1) *Test for Coincidences with Fluorescent X-rays.*—Bothe has performed an experiment[39] in which fluorescent K radiation

---

[35] *Cf.*, however, G. E. M. Jauncey, *Phys. Rev.*, **25**, 314 and 723 (1925); G. Wentzel, *Zeits. f. Phys.*, **43**, 14,779 (1927); I. Waller, *Nature*, July 30, 1927. It is possible that the theories of the latter authors may be satisfactory, but they have not yet been stated in a form suitable for quantitative test.

[36] G. Breit, *Phys. Rev.*, **27**, 242 (1926).

[37] P. A. M. Dirac, *Proc. Roy. Soc.*, **A.** (1926).

[38] W. Gordon, *Zeits. f. Phys.*, **40**, 117 (1926); E. Schrodinger, *Ann. d. Phys.*, **82**, 257 (1927); O. Klein, *Zeits. f. Phys.*, **41**, 407 (1927); G. Wentzel, *Zeits. f. Phys.*, **43** (1927).

[39] W. Bothe, *Zeits. f. Phys.*, **37**, 547 (1926).

from a thin copper foil is excited by a beam of incident X-rays. The emitted rays are so feeble that only about five quanta of energy are radiated per second. Two point counters are mounted, one on either side of the copper foil, in each of which an average of one photo-electron is produced and recorded for about twenty quanta radiated by the foil. If we assume that the fluorescent radiation is emitted in quanta of energy, but proceed in spherical waves in all directions, there should thus be about 1 chance in 20 that the recording of a photo-electron in one chamber should be simultaneous with the recording of a photo-electron in the other.

The experiments showed no coincidences other than those which were explicable by such sources as high-speed beta particles which traverse both counting chambers.

This result is in accord with the photon hypothesis. For if a photon of fluorescent radiation produces a beta-ray in one counting chamber it cannot traverse the second chamber. Coincidences should therefore not occur.

According to the virtual radiation hypothesis, however, coincidences should have been observed. For on this view the fluorescent K radiation is emitted by virtual oscillators associated with atoms in which there is a vacancy in the K shell. That is, the copper foil can emit fluorescent K radiation only during the short interval of time following the expulsion of a photo-electron from the K shell, until the shell is again occupied by another electron. This time interval is so short (of the order of 10^{-15} sec.) as to be sensibly instantaneous on the scale of Bothe's experiments. Since on this view the virtual fluorescent radiation is emitted in spherical waves, the counting chambers on both sides of the foil should be simultaneously affected, and coincident pulses in the two chambers should frequently occur. The results of the experiment are thus contrary to the predictions of the virtual radiation hypothesis.

(2) *Bothe and Geiger's Coincidence Experiments.*—We have seen that according to Bohr, Kramers and Slater's theory, virtual radiation is being continually scattered by matter traversed by X-rays, but only occasionally is a recoil electron emitted. This is in sharp contrast with the photon theory, according to which a recoil electron appears every time a photon is scattered. A crucial test between the two points of view is afforded by an experiment devised and brilliantly performed by Bothe and

Geiger.[40] X-rays were passed through hydrogen gas, and the resulting recoil electrons and scattered rays were detected by means of two different point counters placed on opposite sides of the column of gas. The chamber for counting the recoil electrons was left open, but a sheet of thin platinum prevented the recoil electrons from entering the chamber for counting the scattered rays. Of course not every photon entering the second counter could be noticed, for its detection depends upon the production of a β-ray. It was found that there were about ten recoil electrons for every scattered photon that recorded itself.

The impulses from the counting chambers were recorded on a moving photographic film. In observations over a total period of over five hours, sixty-six such coincidences were observed. Bothe and Geiger calculate that according to the statistics of the virtual radiation theory the chance was only 1 in 400,000 that so many coincidences should have occurred. This result therefore is in accord with the predictions of the photon theory, but is directly contrary to the statistical view of the scattering process.

(3) *Directional Emission of Scattered X-rays.*—Additional information regarding the nature of scattered X-rays has been obtained by studying the relation between the direction of ejection of the recoil electron and the direction in which the associated photon proceeds. According to the photon theory, we have a definite relation (equation 4) between the angle at which the photon is scattered and the angle at which the recoil electron is ejected. But according to any form of spreading wave theory, including that of Bohr, Kramers and Slater, the scattered rays may produce effects in any direction whatever, and there should be no correlation between the directions in which the recoil electrons proceed and the directions in which the secondary β-rays are ejected by the scattered X-rays.

A test to see whether such a relation exists has been made,[41] using Wilson's cloud apparatus, in the manner shown diagrammatically in Fig. 3. Each recoil electron produces a visible track, and occasionally a secondary track is produced by the scattered X-ray. When but one recoil electron appears on the same plate with the track due to the scattered rays, it is possible to tell at once whether the angles satisfy equation (4). If two or three

---

[40] W. Bothe and H. Geiger, *Zeits. f. Phys.*, **26**, 44 (1924); **32**, 639 (1925).
[41] A. H. Compton and A. W. Simon, *Phys. Rev.*, **26**, 289 (1925).

recoil tracks appear, the measurements on each track can be approximately weighted.

Out of 850 plates taken in the final series of readings, thirty-eight show both recoil tracks and secondary β-ray tracks. On eighteen of these plates the observed angle φ is within 20 degrees of the angle calculated from the measured value of θ, while the other twenty tracks are distributed at random angles. This ratio 18 : 20 is about that to be expected for the ratio of the rays

FIG. 3.

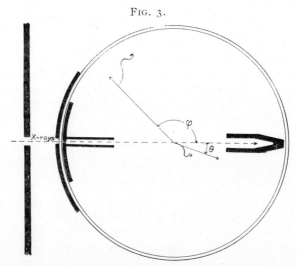

If the X-rays excite a recoil electron at an angle θ, the photon theory predicts a secondary β-particle at an angle φ.

scattered by the part of the air from which the recoil tracks could be measured to the stray rays from various sources. There is only about 1 chance in 250 that so many secondary β-rays should have appeared at the theoretical angle.

If this experiment is reliable, it means that there is scattered X-ray energy associated with each recoil electron sufficient to produce a β-ray, and proceeding in a direction determined at the moment of ejection of the recoil electron. In other words, the scattered X-rays proceed in directed quanta of radiant energy.

This result, like that of Bothe and Geiger, is irreconcilable with Bohr, Kramers and Slater's hypothesis of the statistical production of recoil and photo-electrons. On the other hand, both of these experiments are in complete accord with the predictions of the photon theory.

RELIABILITY OF EXPERIMENTAL EVIDENCE.

While all of the experiments that we have considered are difficult to reconcile with the classical theory that radiation consists of electromagnetic waves, only those dealing with the individual scattering process afford crucial tests between the photon theory and the statistical theory of virtual radiation. It becomes of especial importance, therefore, to consider the errors to which these experiments are subject.

When two point counters are set side by side, it is very easy to obtain coincidences from extraneous sources. Thus, for example, the apparatus must be electrically shielded so perfectly that a spark on the high-tension outfit that operates the X-ray tube may not produce coincident impulses in the two counters. Then there are high-speed alpha- and beta-rays, due to radium emanation in the air and other radio-active impurities, which may pass through both chambers and produce spurious coincidences. The method which Bothe and Geiger used to detect the coincidences, of recording on a photographic film the time of each pulse, makes it possible to estimate reliably the probability that the coincidences are due to chance. Moreover, it is possible by auxiliary tests to determine whether spurious coincidences are occurring—for example, by operating the outfit as usual, except that the X-rays are absorbed by a sheet of lead. It is especially worthy of note that in the fluorescence experiment the photon theory predicted absence of coincidences, while in the scattering experiment it predicted their presence. It is thus difficult to see how both of these counter experiments can have been seriously affected by systematic errors.

In the cloud expansion experiment the effect of stray radiation is to hide the effect sought for, rather than to introduce a spurious effect. It is possible that due to radio-active contamination and to stray scattered X-rays beta particles may appear in different parts of the chamber, but it will be only a matter of chance if these beta particles appear in the position predicted from the direction of the ejection of the recoil electrons. It was in fact only by taking great care to reduce such stray radiations to a minimum that the directional relations were clearly observed in the photographs. It would seem that the only form of consistent error that could vitiate the result of this experiment would be the psychological one of misjudging the angles at which the beta particles appear. It hardly seems possible, however, that errors in the measurement of these angles could be large enough to account for the strong apparent tendency for the angles to fit with the theoretical formula.

It is perhaps worth mentioning further that the initial publications of the two experiments on the individual scattering process were made simultaneously, which means that both sets of experimenters had independently reached a conclusion opposed to the statistical theory of the production of the beta rays.

SUMMARY.

The classical theory that radiation consists of electromagnetic waves propagated in all directions through space affords no adequate picture of the manner in which radiation is emitted or absorbed. It is inconsistent with the experiments on the photo-electric effect, and is entirely helpless to account for the change of wave-length of scattered radiation or the production of recoil electrons.

The theory of virtual oscillators and virtual radiation which are associated statistically with sudden jumps of atomic energy and the emission of photo-electrons and recoil electrons, does not seem to be inconsistent with any of these phenomena as viewed on a macroscopic scale. This theory, however, seems powerless to predict the characteristics of the photo-electrons and the recoil electrons. It is also contrary to Bothe's and Bothe and Geiger's coincidence experiments and to the ray track experiments relating the directions of ejection of a recoil electron and of emission of the associated scattered X-ray.

According to the photon theory, the production and absorption of radiation is very simply connected with the modern idea of stationary states. It supplies a straightforward explanation of the major characteristics of the photo-electric effect, and it accounts in the simplest possible manner for the change of wave-length accompanying scattering and the existence of recoil electrons. Moreover, it predicts accurately the results of the experiments with individual radiation quanta, where the statistical theory fails.

Unless the three experiments on the individual events are subject to improbably large experimental errors, the conclusion is, I believe, unescapable that radiation consists of directed quanta of energy, *i.e.*, of photons, and that energy and momentum are conserved when these photons interact with electrons or atoms.

Let me say again that this result does not mean that there is no truth in the concept of waves of radiation. The conclusion is rather that energy is not transmitted by such waves. The power of the wave concept in problems of interference, refraction, etc., is too well known to require emphasis. Whether the waves serve to guide the photons, or whether there is some other relation between photons and waves is another and a difficult question.

THE SPECTRUM AND STATE OF POLARIZATION OF FLUORESCENT X-RAYS

By A. H. Compton

Ryerson Physical Laboratory, University of Chicago

Communicated June 11, 1928

A brilliant series of experiments by Barkla and Sadler,[1] begun in 1907, showed the now well-known phenomenon of fluorescent x-rays, excited in the heavier elements when traversed by primary x-rays of shorter wavelength. They identified two series of such fluorescent rays, which they called K and L radiations. With the advent of crystal spectrometry, the Braggs[2] and Moseley and Darwin[3] showed that the absorption coefficients of the lines in the x-ray spectra were the same as the absorption coefficients which Barkla and Sadler had found for their fluorescent radiations, thus identifying the fluorescent K and L series radiations with the characteristic line radiation which comes directly from the target of an x-ray tube. This identification was made more definite when spectra of the fluorescent rays were obtained, which showed the same lines as those present in the direct rays. Such spectra have been published, for example, by Duane and Shimizu,[4] Clark and Duane,[5] Woo[6] and D. L. Webster.[7]

In Sadler's earliest studies of the fluorescent rays,[8] he finds by absorption measurements that under favorable conditions not more than 1 per cent of the secondary rays from a fluorescing radiator are scattered. The spectra of Clark and Duane,[5] however, indicate that with fluorescent radiators of barium, lanthanum, molybdenum and silver, excited by x-rays produced at about 90 kv., less than 4 per cent of the secondary radiation consists of the homogeneous fluorescent rays. Woo's spectra, on the other hand, show only a strong line spectrum with no indication of any continuous spectrum mixed with the fluorescent rays.

In figure 2 is shown a spectrum of the fluorescent x-rays from silver, excited by x-rays from a tube with a tungsten target, operated at 53 kv. The apparatus was arranged as shown in figure 1, which is drawn approximately to scale. In order to detect, if possible, any continuous spectrum, the slits collimating the beam incident on the crystal were made rather wide, as is indicated by the width of the α and β lines. Perhaps the most striking feature in the spectrum is the complete absence of any continuous spectrum. In fact an estimate of the ratio of the energy in the continuous

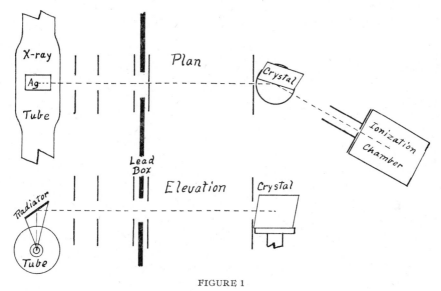

FIGURE 1

spectrum to that in the line spectrum from these data gives 0.007 ± 0.01. New absorption measurements indicate also that the line radiation constitutes more than 99 per cent of the total radiation. These results are in complete agreement with the early findings of Sadler, and suggest that the greater part of the radiation studied by Clark and Duane was not really the secondary rays from the radiator under investigation.

It is worthwhile calling attention again, as Barkla has done long ago, to the value of these fluorescent rays as a source of homogeneous x-rays. If the presence of the β and γ lines is objectionable, they can be removed by a suitable filter, leaving practically nothing except the $K\alpha$ radiation. The homogeneity thus secured is more complete than that usually got by crystal reflection. When homogeneity is sought by filtering the direct beam from a molybdenum target through a zirconium filter, under the best conditions only about 25 per cent of the transmitted radiation is of the $K\alpha$ type.[9] By filtering the fluorescent beam, on the other hand, the homogeneity may be made as great as 98 per cent.

Under the conditions of the present experiment, the intensity of the fluorescent rays as measured at the ionization chamber was approximately $1/200$ of that of the primary beam. The rays were easily visible with the fluorscope.

Ratio of the α to the β Lines.—The ratio of the area of the α peak to that of the β peak from these experiments is 4.58. When this is corrected for:

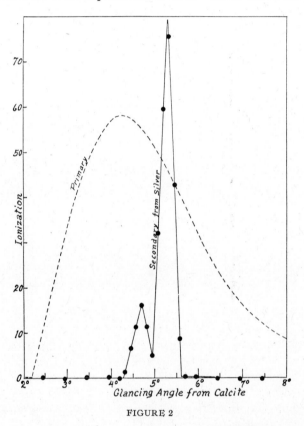

FIGURE 2

(1) the difference in ionization per unit intensity for the two wave-lengths (the method of calculating this correction will be discussed in a later paper), a factor of 0.825; (2) the difference in the absorption of the α and β lines in the air and in the fluorescent radiator, factor 1.10; and (3) the difference in the reflecting power of the crystal for the two types of radiation, a factor of 1.08, the ratio of the energies of the two lines as emitted from the silver atoms in the radiator becomes 4.5. This agrees within the errors of experiment and correction with Unnewehr's ratio[10]

of 4.65 for the α and β lines from a silver target. Thus the relative intensity as well as the positions of the spectrum lines is approximately the same in the fluorescent as in the primary x-rays.

Unpolarized Character of Fluorescent X-Rays.—In their earliest experiments on the characteristic secondary radiation, Barkla and Sadler tested for a possible polarization of the rays by using a radiator which emitted strong fluorescent radiation as an "analyzer" for the direct rays from an x-ray tube.[11] A carbon scattering block used as analyzer reveals in this case some 15 per cent polarization of the primary beam. They were unable, however, to detect any polarization when a fluorescing radiator was used as analyzer. If the probable error of their measurements was 1 or 2 per cent, this would mean that the polarization of the fluorescent rays is probably less than about 10 per cent.

More recently, Mark and Szilard,[12] in a study of the x-rays polarized by reflection from a crystal, have found no indication of analysis of these rays by a fluorescing radiator except insofar as scattered rays were mixed with the fluorescent rays. Their method was a photographic one.

The question of the polarization of the fluorescent x-rays has acquired new interest as a result of Bishop's conclusion that the line radiation from the target of an x-ray tube, which is partly of fluorescent origin, is partially polarized.[13] Moreover, Raman has recently found a type of secondary light rays which are similar to fluorescent rays in that their wave-length differs from that of the primary rays, but which are strongly polarized.[14] It thus becomes important to make a more careful test than has hitherto been made for possible polarization of fluorescent x-rays.

The method employed was that used by Hagenow and the writer in examining the polarization of scattered x-rays.[15] A horizontal beam of x-rays struck a radiator R_1 placed above the axis of a Bragg x-ray spectrometer. The secondary rays proceeding downward fell on a second radiator R_2, also placed over the spectrometer axis, but on a level with the window of the ionization chamber. When both radiators R_1 and R_2 were plates of graphite, the ionization current with the chamber in the perpendicular position was only 0.09 of that in the parallel position, indicating almost complete polarization. When R_1 was silver and R_2 graphite,

$$\frac{i_\perp}{i_\parallel} = 0.968. \qquad (a)$$

When R_1 was silver and R_2 a plate of compressed ZrO, the observed ratio was

$$\frac{i_\perp}{i_\parallel} = 0.964. \qquad (b)$$

An auxiliary experiment showed that the zirconium oxide analyzer gave almost the same intensity in the $\parallel$ and $\perp$ positions even when struck by rays polarized by scattering from carbon. The ratio 0.964 observed in case b is thus to be ascribed to the lack of geometric symmetry in the || and ⊥ positions. The corrected value of the intensities in the two positions is thus

$$\frac{\perp}{||} = \frac{a}{b} = \frac{0.968}{0.964} = 1.004$$

with a probable error, estimated from the variations of the readings, of ±0.005.

This means that the characteristic fluorescent x-rays from silver are completely unpolarized, within a probable error of 0.5 per cent. It follows that any polarization in the line spectrum of the primary x-rays must be due to the portion of the rays resulting from the direct action of the cathode rays.

[1] C. G. Barkla and C. A. Sadler, *Phil. Mag.*, **16**, 550 (1908); C. A. Sadler, *Ibid.*, **17**, 739 (1909); **18**, 107 (1909); **19**, 337 (1910). C. G. Barkla, *Phil. Trans.*, **217**, 315 (1917).
[2] W. H. Bragg and W. L. Bragg, *Proc. Roy. Soc.*, **A88**, 428 (1913).
[3] H. G. J. Moseley and C. G. Darwin, *Phil. Mag.*, **26**, 210 (1913).
[4] W. Duane and T. Shimizu, *Phys. Rev.*, **14**, 391 (1919).
[5] G. L. Clark and W. Duane, *Proc. Nat. Acad. Sci.*, **9**, 415 (1923); **10**, 41 and 92 (1924).
[6] Y. H. Woo, *Phys. Rev.*, **27**, 119 (1926), et al.
[7] D. L. Webster, *Proc. Nat. Acad. Sci.*, **14**, 330 (1928).
[8] C. A. Sadler, *Phil. Mag.*, **17**, 739 (1909).
[9] A. W. Hull, *Phys. Rev.*, **10**, 666 (1917).
[10] E. C. Unnewehr, *Ibid.*, **22**, 529 (1923).
[11] C. G. Barkla and C. A. Sadler, *Phil. Mag.*, **16**, 550 (1908).
[12] H. Mark and L. Szilard, *Zeits. Phys.*, **35**, 743 (1926).
[13] J. B. Bishop, *Phys. Rev.*, **28**, 625 (1926).
[14] C. V. Raman, *Indian Journ. Phys.*, **2**, part 2 (1928).
[15] A. H. Compton and C. F. Hagenow, *J. O. S. A. & R. S. I.*, **8**, 487 (1924).

THE CORPUSCULAR PROPERTIES OF LIGHT

By Arthur H. Compton

University of Chicago

THE development of the modern conception of light quanta, or photons, began with Planck's ideas concerning heat radiation. Newton indeed had defended the hypothesis of light corpuscles, but the facts which he cited to support this view were later reconciled by Fresnel with the wave theory of light. It was not until new problems were studied, such as the intensity heat radiation and the electrical effects of light, that any real need arose for corpuscles as alternative or supplementary to the wave theory of light.

Planck's Theory of Heat Radiation

Planck was confronted[1] with the fact that the only theory of emission of radiation from hot bodies to which the classical mechanics and electrodynamics would lead, predicted rays much more intense than are actually observed, and of the wrong color. It is a matter of common experience that as a body gradually becomes hotter it first glows a dull red, then orange, and bright gold and finally white. According to the formula developed from the usual kinetic theory, however, the light emitted should always be of the same blue color, differing only in intensity as the temperature changes. Such a conclusion followed necessarily from the fact that all oscillators in thermal equilibrium with each other should have on the average the same kinetic energy, whatever their natural frequency of oscillation. But the oscillators of higher frequency will be subject to greater acceleration if their kinetic energy is the same, and hence, according to electrodynamical principles, should radiate more energy than those of lower frequency. Thus at all temperatures the theory predicted that the high frequency radiation should be more intense than the low frequency radiation.

Planck saw a possible way of escape from this difficulty if he were to suppose that at low temperatures only the oscillators of low frequencies could emit radiation, whereas at high temperatures those of higher frequencies could also radiate. In order to accomplish this result he introduced the simple assumption that the oscillators in the hot body can emit radiation only in units, or quanta, whose energy is proportional to the frequency of the radiation, i.e.,

$$E = h\nu, \qquad (1)$$

where h is the constant of proportionality between the frequency and the the energy E of the unit. With this limitation it is possible for only those

---

[1] Planck, Verh. d. Deut. Phys. Ges. **2**, 237 (1900). A complete account of Planck's studies of this problem is given in his "Wärmestrahlung" (1915), published by Blackiston's in English translation by Masius.

oscillators which have energy greater than $h\nu$ to emit a unit of radiation. Thus at low temperatures, where the average energy of the oscillators is low, only low frequency rays can be emitted. At higher temperatures the higher frequency oscillators will have enough energy to emit their larger units of radiant energy, and so as the temperature rises the center of gravity of the radiation will shift to higher and higher frequencies. Thus with one bold assumption regarding the unitary nature of the emitted light, Planck was able to arrive at a reasonable explanation of the hitherto insoluble problem of the color of the light emitted by hot bodies.

It would take us too far afield to describe how Planck developed this idea of energy quanta to account quantitatively for the intensity as well as the spectral energy distribution of heat rays. In his hands and those of others the theory has assumed a variety of forms, but it has always retained the essential feature that the rays from the hot body must be emitted in units whose energy is proportional to the frequency. The introduction of this idea has marked the opening of an important epoch in the development of theoretical physics.

Einstein and the Photoelectric Effect

The units of radiant energy introduced by Planck were not corpuscular. He supposed that the radiation from an oscillator, though having a definite amount of energy, would spread itself through all space after the manner of a spherical electromagnetic wave. It remained for Einstein[2] to introduce the conception of a corpuscular unit of radiation, or photon, in his effort to account for the photoelectric effect.

When Einstein approached this problem it was recognized that the speed with which photoelectrons are ejected from a surface increases with increasing frequency of the light, and it was generally supposed that the number of photoelectrons emitted was proportional to the intensity of the light striking the photoelectric surface. He saw that this proportionality would follow from the assumption that the light which excited the photoelectrons occurs as a stream of particles, each of which would spend its energy in ejecting an electron fron an atom of the photoelectric material. If each of these particles had energy $h\nu$, as might be inferred from Planck's theory of heat radiation, this picture of the process would account also for the increase of speed with higher frequencies. If a certain amount of work w_0 is required to remove the electron from the atom, Einstein supposed that all the rest of the photon's energy is spent in giving kinetic energy to the electron, thus deriving his famous photoelectric equation,

$$E_{kin} = h\nu - w_0. \qquad (2)$$

It was years before this theory received an adequate test. Experiments by Ladenburg[3] favored the view that the velocity rather than the kinetic energy

---

[2] A. Einstein, Ann. d. Physik **17**, 145 (1905).
[3] E. Ladenburg, Phys. Zeits. **8**, 590 (1907).

was proportional to the frequency of the incident light, and different results were obtained with different metals. Richardson and Compton[4] and independently Hughes[5] showed that the differences found for different metals were due to their different contact potentials, and to the fact that the value of w_0 is different from metal to metal. They were indeed able to show that Einstein's equation was of the right form, and that the constant of proportionality h is approximately the same as Planck's constant. A few years later Millikan,[6] using greater care in securing strictly monochromatic light, was able by means of Einstein's equation to secure from photoelectric measurements one of the best experimental determinations that we have of Planck's constant.

The photoelectric effect is especially prominent with x-rays, for these rays eject photoelectrons from all kinds of substances. The velocities of these x-ray photoelectrons have been measured by means of their curvature in a magnetic field, using the so-called magnetic spectrograph. M. de Broglie[7] showed that even for these very high frequencies Einstein's equation holds, if by w_0 we now mean the work required to remove the electron from the o level of the atom. In fact Robinson[8] has applied this equation to his measurements of the speed of x-ray photoelectrons from various substances as a powerful method of studying the energy levels of the different atoms. In a similar way, Ellis,[9] Meitner,[10] Thibaud[11] and others have used equation (2) as a means of determining γ-ray frequencies from the speed of the secondary β-rays. Recent measurements of these frequencies by crystal methods[12] show that even for these exceedingly great energies Einstein's law holds. Over a range of kinetic energies corresponding to a drop through potential differences from 1 volt to 2 million volts Einstein's photoelectric equation has thus been verified to within an experimental error of 1 percent. It is thus one of the most adequately tested laws in the realm of physics.

Before these photoelectric experiments had been carried to a successful conclusion, Duane and Hunt[13] observed a closely related phenomenon which is frequently called the inverse photoelectric effect. They found that when an x-ray tube is operated at a constant potential, there is a definite lower limit to the wave-length of the x-rays from the tube, and that this limiting wave-length is inversely proportional to the voltage. This result may be written in the form,

$$Ve = hc/\lambda_{min} = h\nu_{max}, \qquad (3)$$

---

[4] O. W. Richardson and K. T. Compton, Phil. Mag. **24**, 575 (1912).
[5] A. L. Hughes, Phil. Trans. Roy. Soc. **A212**, 205 (1912).
[6] R. A. Millikan, Phys. Rev. **7**, 18 and 355 (1916).
[7] M. de Broglie, J. de Phys. et Radium **2**, 265 (1921).
[8] H. R. Robinson, Phil. Mag. **50**, 241 (1925).
[9] C. D. Ellis, Proc. Roy. Soc. **A100**, 1 (1922); Proc. Camb. Phil. Soc. **22**, 374 (1924), et al.
[10] L. Meitner, Zeits. f. Physik **11**, 35 (1922), et al.
[11] J. Thibaud, Comptes rendus **178**, (1924), et al.
[12] L. T. Steadman, Phys. Rev. **33**, 1069 (1929).
[13] W. Duane and F. L. Hunt, Phys. Rev. **6**, 166 (1915).

where V is the applied potential, and the other letters have their usual significance. Duane and Hunt's quantitative measurements, confirmed and extended by a number of other investigators[14] have shown that the factor of proportionality in this equation is the same quantity h as that which appears in Planck's theory, equation (1). In fact the measurement of this limiting x-ray wave-length is perhaps our best direct method of determining Planck's constant.

The significance of this work will perhaps be more obvious if we imagine the following experiment: Let two x-ray tubes, A and B, be placed side by side. Tube A is operated at a constant potential of say 100,000 volts. A cathode electron with a kinetic energy Ve strikes the target of tube A and gives raise to an x-ray of frequency $\nu = Ve/h$. This ray strikes the target of tube B and there ejects a photoelectron whose kinetic energy according to equation (2) is $Ve - w_0$. This means that all of the energy of the cathode electron in tube A has been transmitted to the photoelectron ejected from the target of tube B. How is it possible for such a complete transfer of energy to be effected?

A precisely similar difficulty arises in connection with Bohr's picture of radiation and absorption by atoms, which was developed[15] while these studies of the photoelectric effect were going on. According to this picture, radiation is emitted by an atom only when it changes from one state to another having less energy, in which case the frequency is given by the expression,

$$h\nu = \delta E, \qquad (4)$$

where δE is the loss in energy by the atom. When an atom absorbs energy, it changes from one state to another of higher energy and the frequency of the absorbed radiation is again given by equation (4), where δE now means the increase in the energy of the atom. Thus we see again that if one system suddenly radiated an amount of energy δE, another atomic system, which may be as far away as the earth is from a distant star, may suddenly have its energy increased by the same amount when the radiation reaches it.

The impossibility that an electromagnetic wave whose energy spreads in all directions should effect such a sudden and complete transfer of energy is obvious. It is equally clear that Einstein's photon conception affords a simple and adequate method of making the transfer. There have not been lacking, however, attempts to explain these phenomena without resorting to assumptions departing so completely from the electromagnetic waves of Maxwell.

One such attempt is the accumulation hypothesis, according to which the light energy is gradually accumulated by the atom, and the photoelectron is finally ejected when the accumulated energy exceeds a certain critical value. This process requires the existence of stored energy of all possible amounts within the atom, since the kinetic energy of the ejected photo-

---

[14] E. g., Duane, Palmer and Chi-Sun-Yeh, J. Opt. Soc. Am. **5,** 376 (1921); E. Wagner, J. d. Rad. Elek. **16,** 212 (1919).

[15] N. Bohr, Phil. Mag. **26,** 1,476 and 857 (1913).

electron may have any value, depending upon the frequency of the radiation which traverses its parent atom. Furthermore this energy must remain stored for indefinitely long periods of time, for otherwise emission of photoelectrons would not occur at once upon exposure to the light—time would be required for the atom to accumulate sufficient energy. We are thus led to imagine an atom which may possess any energy whatever, and whose energy may gradually increase as radiation is absorbed. Such a picture is wholly inconsistent with Bohr's idea of an atom with definite stationary states and which changes only suddenly from one such state to another. It is true that recent developments in quantum mechanics have led us to revise considerably Bohr's conception of electron orbits; but this hypothesis of stationary states seems more firmly established than ever, and continues to be the fundamental principle of spectral analysis. We thus find it difficult to consider seriously an accumulation hypothesis which would mean atoms having all possible amounts of energy.

There is another apparently fatal difficulty with this explanation of the photoelectric effect, in that it fails to account for the direction in which the photoelectrons are emitted. Experiments by Wilson[16] Bubb[17] and others[18] have shown that the most probable direction in which a photoelectron is ejected from an atom by x-rays is nearly that of the electric vector of the incident wave, but with an appreciable average forward component to its motion. This forward component is about equal to the momentum $h\nu/c$ of the incident photon, as would be expected if the electron suddenly absorbs a photon of energy $h\nu$ and escapes before any appreciable impulse has been transferred to the atom.[19] On the other hand, if the energy is gradually accumulated by the electron, the forward impulse received from the radiation would be transferred to the whole atom, and no reason appears for the strong forward component to the photoelectron's motion. Thus the accumulation hypothesis does not seem to be tenable.

If the atom cannot gradually accumulate energy, since a spherical electromagnetic wave cannot give up its whole energy to a single atom, the occurrence of photoelectrons with the energy $h\nu$ means that we must either give up our old view that light comes in spherical waves or abandon the doctrine of the conservation of energy. Bohr, Kramers and Slater[20] at one time preferred to assume that energy is not conserved when an individual photoelectron is produced. They supposed that on the average the energy appearing in the photoelectrons is equal to that absorbed from the radiation, but under

---

[16] C. T. R. Wilson, Proc. Roy. Soc. **A104,** 1 (1923).

[17] F. W. Bubb, Phys. Rev. **23,** 137 (1924).

[18] *E.g.*, Auger, Comptes rendus **178,** 1535 (1924); D. H. Loughridge, Phys. Rev. **26,** 697 (1925); F. Kirchner, Zeits. f. Physik **27,** 285 (1926); E. J. Williams, Nature **121,** 134 (1928).

[19] The average forward component is found in certain recent experiments to approach a value $9/5 \times h/\lambda$, where h/λ is the momentum of the photon. This value has been derived on the basis of wave mechanics (cf. A. Sommerfeld "Atombau Ergänzungsband" 1929 p. 222, and E. J. Williams, Nature **123,** 565, (1929).

[20] N. Bohr, H. A. Kramers and J. C. Slater, Phil. Mag. **47,** 785 (1924).

the stimulus of the incident waves any particular electron might suddenly escape at high speed without any corresponding loss in energy by the remainder of the system. That is, the conservation of energy, and similarly the conservation of momentum, would become statistical laws. The authors of this theory assume that, though the rays are propagated as spherical waves the motion of the photoelectrons would be the same as if they were ejected by photons. It has thus been difficult to devise a photoelectric experiment which would distinguish between this "virtual radiation" hypothesis and that of photons. The degree of success that has attended the application of the photon hypothesis to the motion of photoelectrons has however come directly from the application of the conservation principles to the individual action of a photon on an electron. The power of these principles as applied to this case is surprising if the assumption is correct that they are only statistically valid.

Quantum Phenomena Associated with the Scattering of X-Rays

We have seen that Einstein's hypothesis of corpuscular units of radiant energy gives a satisfactory account of the photoelectric effect. As Jeans has significantly remarked, however, Einstein invented the photon hypothesis just to account for this one effect, and it is not surprising that it should account for it well. In order to carry any great weight the hypothesis should also be found applicable to some phenomena of widely different character. Just such phenomena have recently been found associated with the scattering of x-rays—the change in wave-length of the scattered rays, and the recoil electrons associated with them.

The earliest experiments on secondary x-rays and γ-rays showed a difference in the penetrating power of the primary and the secondary rays. Barkla[21] and his collaborators showed that the secondary rays from the heavy elements consisted largely of fluorescence radiations characteristic of the radiator, and that it was the presence of these softer rays which was chiefly responsible for the great absorption of the secondary rays. When later experiments showed a measureable difference in penetration even for light elements such as carbon, from which no fluorescence radiation appears, it was natural to ascribe this difference to a new type of fluorescence radiation, similar to the K and L types, but of shorter wave-length.[22] Careful absorption measurements[23] failed however, to reveal any critical absorption limit for these assumed "J" radiations similar to those corresponding to the fluorescence K and L radiations. Moreover, direct spectroscopic observations[24] failed to reveal the existence of any spectrum lines under conditions with which the supposed J rays should appear. It thus became evident that the softening of the secondary x-rays from the lighter elements was due to a different

---

[21] C. G. Barkla and C. A. Sadler, Phil. Mag. **16**, 550 (1908).

[22] C. G. Barkla and Miss White, Phil. Mag. **34**, 270 (1917); J. Laub, Ann. d. Physik **46**, 785 (1915); J. A. Crowther, Phil. Mag. **42**, 719 (1921).

[23] Richtmyer and Grant, Phys. Rev. **15**, 547 (1920).

[24] Duane and Shimizu, Phys. Rev. **13**, 288 (1919); **14**, 389 (1919).

kind of process from the softening of the secondary rays from heavy elements where fluorescence x-rays are present.

According to the usual electron theory of x-ray scattering, the primary waves set the electrons into forced oscillation, and they, because of their accelerations, reradiate x-rays in all directions. It is thus obvious that the scattered rays will be of the same frequency as the primary rays which set the electrons in motion. A series of skillfully devised absorption experiments, performed by J. A. Gray,[25] showed however that both in the case of x-rays and γ-rays an increase in wave-length accompanies the scattering of the rays by light elements. When spectroscopic studies were made[26] they likewise revvealed lines in the spectrum of the scattered rays corresponding to those in the primary beam, but with each line displaced slightly toward the longer wave-lengths. These spectra had the advantage over the absorption measurements of affording a quantitative determination of the change in wavelength, which gave a basis for its theoretical interpretation.

The photon conception gives a simple interpretation of this phenomenon. If we suppose that each x-ray photon is deflected by a single electron the electron will recoil from the impact. That is, part of the photon's energy is spent in setting the electron in motion, so the photon has less energy after deflection than before. The problem is very similar to that of the elastic collision of a light ball with a heavy one. If we assume that the energy and momentum are conserved in the process we can calculate the loss in energy and hence the increase in wave-length of a photon which is scattered at an angle ϕ with the primary ray. We thus find[27] for the increase in wavelength,

$$\delta\lambda = \frac{h}{mc}(1-\cos\phi),\qquad(5)$$

where h is again Planck's constant, m is the mass of the electron and c is the velocity of light. The electron at the same time recoils from the photon at an angle θ given by,

$$\cot\theta = -(1+\alpha)\tan\tfrac{1}{2}\phi,\qquad(6)$$

where $\alpha = h/mc\lambda$, and the kinetic energy of the recoiling electron is

$$E_{kin} = h\nu\frac{2\alpha\cos^2\theta}{(1+\alpha)^2-\alpha^2\cos^2\theta}.\qquad(7)$$

Until very recently the experiments showed just two lines in the spectrum of the scattered rays corresponding to each line of the primary ray. Of these lines one, the "unmodified line," is of very nearly the same wavelength as the primary ray, whereas the second, or "modified line," though apparently somewhat broadened, has its center of gravity shifted by ap-

---

[25] J. A. Gray, Phil. Mag. **26**, 611 (1913); J. Frank. Inst. Nov. 1920, p. 643.
[26] A. H. Compton, Bull. Nat. Res. Coun. No. 20 (1922); Phys. Rev. **22**, 409 (1923).
[27] A. H. Compton, Phys. Rev. **22**, 483 (1923); P. Debye, Phys. Zeits. **24**, 161 (1923).

proximately the amount predicted by equation (5). According to experiments by Kallman and Mark[28] and by Sharp[29] the agreement between the theoretical and the observed shift is precise to within a small fraction of 1 percent.

Within the last year Davis and Mitchell,[30] using their high resolving power double crystal spectrometer, have found that the "unmodified line" has a complex structure, with one line the same wave-length as the primary rays and with a group of other lines each of whose frequencies differs from that of the primary beam by approximately the limiting frequency of some energy level in the normal atom. Thus there is a line whose frequency is given approximately by the relation,

$$h\nu'' = h\nu - h\nu_k \tag{8}$$

where ν is the primary frequency and $h\nu_K$ is the energy of the K energy level. Such lines may be accounted for by supposing that the incident photon spends enough energy upon the atom to release a K electron (or to transfer it to an outside orbit) and then escapes from the atom with its remaining energy. The process is thus analogous to the photoelectric effect, where, however, the photon instead of the electron escapes with the energy remaining after the electron has been removed from its original orbit. These lines seem to be the x-ray analogues of the Raman lines, which had been discovered[31] a few months earlier in the visible spectrum.

Very recent experiments by Davis and Purks[32] have shown a similar fine structure in the modified line. Such a structure is consistent with the photon conception of the scattering process, and had indeed been predicted on this basis by the writer[33] using certain assumtions regarding the action of the photon on bound electrons. The experiments of Sharp[29] and DuMond,[34] however, seem to indicate a broad and almost structureless modified line, which would seem to be in disagreement with the results of Davis and Purks. The spectra obtained by the latter investigators also show a change in wavelength which is almost 10 percent less than that predicted by equation (5), a result difficult to reconcile with the observations of other recent experiments. It is important to settle these differences in the detailed experimental results, because they are of significance regarding the manner in which a photon acts upon an electron bound within an atom. There is in these experiments, however, no evidence of a disagreement with the basic corpuscular theory from which equation (5) was derived.

[28] H. Kallman and H. Mark, Naturwiss. **13**, 297 (1925).
[29] H. M. Sharp, Phys. Rev. **26**, 691 (1925).
[30] B. Davis and D. P. Mitchell, Phys. Rev. **32**, 331 (1928).
[31] C. V. Raman, Indian J. Phys. **2**, 387 (1928).
[32] B. Davis and H. Purks, Phys. Rev. **34**, 1 (1929).
[33] A. H. Compton, Phys. Rev. **24**, 168 (1924).
[34] J. W. M. DuMond, Phys. Rev. **33**, 643 (1929).

Recoil Electrons

From the close agreement between the theoretical and the observed wave-lengths of the scattered rays, the recoil electrons predicted by the photon theory of scattering were looked for with some confidence. When this theory was proposed, there was no direct evidence for the existence of such electrons, though indirect evidence suggested[35] that the secondary β-rays ejected from matter by hard γ-rays are mostly of this type. Within a few months of their prediction, however, C. T. R. Wilson[36] and W. Bothe[37] independently announced their discovery. The recoil electrons show as short tracks in the cloud expansion photographs, pointed in the direction of the primary beam, mixed among the much longer tracks due to the photo-electrons ejected by the x-rays.

Perhaps the most convincing reason for associating these short tracks with the scattered x-rays comes from a study of their number. Each photo-electron in a cloud photograph represents a quantum of truly absorbed x-ray energy. If the short tracks are due to recoil electrons, each one should represent the scattering of a photon. Thus the ratio N_r/N_p of the number of short tracks to the number of long tracks should be the same as the ratio σ/τ of the scattered to the truly absorbed energy when the x-rays pass through air. The latter ratio is known from absorption measurements, and the former ratio can be determined by counting the tracks on the photographs. The satisfactory agreement between the two ratios for x-rays of different wave-lengths means that on the average there is about one quantum of energy scattered for each short track that is produced.

This result is in itself contrary to the predictions of the classical wave theory, since on this basis all the energy spent on a free electron (except the insignificant effect of radiation pressure) should reappear as scattered x-rays. In these experiments on the contrary, 5 or 10 percent as much energy appears in the motion of the recoil electrons as in the scattered x-rays.

That these short tracks correspond to the recoil electrons predicted by the photon theory of scattering becomes clear from a study of their energies. The energy of an electron which produces a track in an expansion chamber can be calculated from the range of the track. The ranges of the tracks which start in different directions have been studied,[38] using primary x-rays of different wave-lengths, with the result that equation (7) has been satisfactorily verified. A more accurate check on these recoil electron energies has recently been made by Bless,[39] using a magnetic spectrometer, and with results wholly consistent with the theory.

In view of the fact that electrons of the recoil type were unknown when the photon theory of scattering was presented, their existence, and the close

---

[35] A. H. Compton, Bull. Nat. Res. Coun. No. 20, p. 27 (1922).

[36] C. T. R. Wilson, Proc. Roy. Soc. A104, 1 (1923).

[37] W. Bothe, Zeits. f. Physik 16, 319 (1923); 20, 237 (1923).

[38] A. H. Compton and A. W. Simon, Phys. Rev. 25, 306 (1925); J. M. Nuttall and E. J. Williams, Manchester Memoirs, 70, 1 (1926).

[39] A. A. Bless, Phys. Rev. 30, 871 (1927).

agreement with the predictions as to their number, direction and velocity, supply strong evidence in favor of the photon hypothesis.

Interpretation of these Experiments

It is impossible to account for scattered rays of altered frequency, and for the existence of the recoil electrons, if we assume that x-rays consist of electromagnetic waves in the ordinary sense. Yet some progress has been made on the basis of semi-classical theories. It is an interesting fact that the wave-length of the scattered ray according to equation (5) varies with the angle just as one would expect from a Doppler effect if the rays are scattered from an electron moving in the direction of the primary beam. Moreover, the velocity that must be assigned to the electron in order to give the proper magnitude to the change of wave-length is that which the electron would acquire by radiation pressure if it should absorb a quantum of the incident rays. Several writers[40] have therefore assumed that an electron takes from the incident beam a whole quantum of the incident radiation, and then emits this energy as a spherical wave while moving forward with high velocity. There is, however, the difficulty that this theory predicts recoil electrons all moving in the same direction and with the same velocity. The experiments show, on the other hand, a variety of directions and velocities, with the velocity and direction correlated as demanded by the photon hypothesis. Moreover, the maximum range of the recoil electrons, though in agreement with the predictions of the photon theory, is found[38] to be some four times as great as that predicted by this semi-classical theory.

There is nothing in these experiments, however, which is inconsistent with the idea of virtual oscillators continually scattering virtual radiation. In order to account for the change of wave-length on this view, Bohr, Kramers and Slater assumed[20] that the virtual oscillators scatter as if moving in the direction of the primary beam, accounting for the change in wave-length as a Doppler effect. They then suppose that occasionally an electron, excited by the primary virtual rays, will suddenly move forward as if it had received the momentum of a photon. Though we have seen that the electrons move in a variety of different directions, the theory could easily be extended to include the type of motion that is actually observed. It is difficult, however, to see how such a theory could by itself predict the change in wave-length and the motion of the recoil electrons.

We may conclude that the photon theory predicts quantitatively and in detail the change of wave-length of the scattered x-rays and the characteristics of the recoil electrons. The virtual radiation theory is probably not inconsistent with these experiments, but is incapable of predicting the results. The classical theory is altogether helpless to deal with these phenomena.

---

[40] C. R. Bauer, Comptes rendus **177**, 1211 (1923); C. T. R. Wilson, Proc. Roy. Soc. **A104**, 1 (1923); K. Forsterling, Phys. Zeits. **25**, 313 (1924); O. Halpern, Zeits. f. Physik **30**, 153 (1924).

Experiments with Individual Radiation Quanta

We have seen that while these experiments on the photoelectric effect and on the scattering of x-rays give results which cannot be reconciled with the classical picture of electromagnetic waves, they do not suffice to distinguish between the photon theory and the theory of virtual radiation. The latter theory succeeded in avoiding the difficulties of the classical theory by considering the conservation of energy and momentum as only statistically valid. If experiments can be performed on the interaction of individual photons and electrons, it should be possible to make a direct test of the conservation principles, and to distinguish between the virtual radiation hypothesis and that of photons. Three important experiments of this type have been reported.

(1) *Test for Coincidences with Fluorescence X-Rays.* Bothe has performed an experiment[41] in which fluorescence rays from a thin copper foil are excited by a beam of incident x-rays. Two point counters of the type developed by Geiger are mounted, one on either side of the foil, in each of which an average of 1 photoelectron is recorded for about 20 quanta radiated by the foil. If we assume that the fluorescence radiation is emitted in quanta of energy, but proceed in spherical waves in all directions, there should thus be about 1 chance in 20 that the recording of a photoelectron in one chamber should be simultaneous with the recording of a photoelectron in the other. The experiments showed no coincidences other than those which were explicable by such sources as high speed beta particles which traverse both counting chambers.

This result is in accord with the photon hypothesis. For if a photon of fluorescence radiation produces a β-ray in one counting chamber it cannot traverse the second chamber. Coincidences should therefore not occur.

According to the virtual radiation hypothesis, however, coincidences should have been observed. For on this view the fluorescence K radiation is emitted by virtual oscillators associated with atoms in which there is a vacancy in the K shell. That is, the copper foil can emit fluorescence K radiation only during the short interval of time following the expulsion of a photoelectron from the K shell, until the shell is again occupied by another electron. This time interval is so short (less than 10^{-15} seconds) as to be sensibly instantaneous on the scale of Bothe's experiments. Since on this view the virtual radiation is emitted in spherical waves, the counting chambers on both sides of the foil should be simultaneously affected, and coincident pulses in the two chambers should frequently occur. The results of the experiment are thus contrary to the predictions of the virtual radiation hypothesis.

(2) *Coincidences of Scattered X-Rays and Recoil Electrons.* We have seen according to Bohr, Kramers and Slater's theory, virtual radiation is being continually scattered by matter traversed by x-rays, but only occasionally is a recoil electron emitted. This is in sharp contrast with the photon theory, according to which a recoil electron appears every time a photon is scattered.

---

[41] W. Bothe, Zeits. f. Physik **37,** 547 (1926).

A crucial test between these two points of view is afforded by an experiment devised and brilliantly performed by Bothe and Geiger.[42] X-rays were passed through hydrogen gas, and the resulting recoil electrons and scattered rays were detected by means of two point counters on opposite sides of the column of gas. The chamber for counting recoil electrons was left open, but a thin sheet of platinum prevented the recoil electrons from entering the chamber for counting the scattered rays. The impulses from the counting chambers were recorded on a moving photographic film.

In observations over a period of five hours 66 coincidences between the impulses due to recoil electrons and the scattered rays were observed. Bothe and Geiger estimated that according to the virtual radiation theory the chance was only 1 in 400,000 that so many coincidences should have occurred. This result is therefore in accord with the predictions of the photon theory, but is directly contrary to the statistical view of the scattering process.

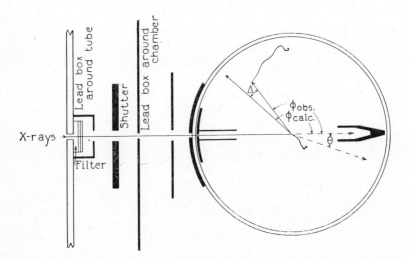

Fig. 1.

(3) *Directional Emission of Scattered X-Rays.* According to the photon theory, we have a definite relation (equation 6) between the angle at which the photon is scattered and the angle at which the recoil electron is ejected. But according to any form of spreading wave theory, including that of Bohr, Kramers and Slater, the scattered rays may produce effects in any direction whatever, and there should be no correlation between the direction in which a recoil electron proceeds and the direction in which the scattered x-ray produces an effect. A test to see whether such a relation exists has been made,[43] using a cloud expansion apparatus, in the manner shown diagrammatically in Fig. 1. Each recoil electron produces a visible track, and oc-

---

[42] W. Bothe and H. Geiger, Zeits. f. Physik **26**, 44 (1924); **32**, 639 (1925).
[43] A. H. Compton and A. W. Simon, Phys. Rev. **26**, 289 (1925).

casionally a secondary track is produced by the scattered x-ray before it escapes from the chamber. When but one recoil electron appears on the same plate with the track due to the scattered rays, it is possible to tell at once whether the angles satisfy equation (6). If two or three recoil tracks appear, the measurements on each track can be appropriately weighted.

Of 850 plates taken in the final series of readings, 38 show both recoil tracks and secondary β-ray tracks. On 18 of these plates the observed angle ϕ is within 20 degrees of the angle calculated from the measured value of θ, while the other 20 tracks are distributed at random angles. This ratio 18:20 is about that to be expected for the ratio of the rays scattered by the part of the air from which the recoil tracks could be measured to the stray rays from various sources. There is only about 1 chance in 250 that so many secondary β-rays should have appeared at the theoretical angle.

This result means that associated with each recoil electron there is scattered x-ray energy sufficient to produce a β-ray, and proceeding in a direction determined at the moment of ejection of the recoil electron. In other words, the scattered x-rays proceed in directed units of radiant energy.

This result, like those of the previous two experiments, is irreconcilable with the virtual radiation hypothesis of the production of recoil and photoelectrons. On the other hand all of these experiments with individual radiation quanta are in complete accord with the predictions of the photon theory.

The Paradox of Waves and Particles

Experiments on the photoelectric effect and on scattered x-rays, taken together with these experiments on the individual interactions of radiation and electrons, show therefore that radiation is emitted in units, is propagated in definite directions, and is absorbed again in units of undiminished energy. Light thus has all the essential characteristics of particles. It is well known however that light has the characteristics of waves. The phenomena of reflection, refraction, polarization and interference, which occur with light, can leave no reasonable doubt about its wave properties. How can these two apparently conflicting conceptions be reconciled?

Electron Waves. Before attempting to answer this question, let us notice that this dilemma applies not only to radiation, but also to other fundamental fields of physics. When the evidence was growing strong that radiation, which we have always thought of as waves, had the properties of particles, L. de Broglie asked, may it not then be possible that electrons, which we have known as particles, may have the properties of waves? He was able to give a mathematical proof[44] that the dynamics of any particle may be expressed in terms of the propagation of a group of waves. That is, the particle may be replaced by a train of waves—the two, so far as their motion is concerned, may be made mathematically equivalent. The motion of a particle in a straight line is represented by a plane wave. The wave-length is determined by the momentum of the particle. Thus just as the momentum of

[44] L. de Broglie, Phil. Mag. **47**, 446 (1924); Thesis, Paris 1924.

a photon is $h\nu/c = h/\lambda$ so the wave-length of a moving electron is given by $mv = h/\lambda$, or

$$\lambda = h/mv. \qquad (9)$$

In C. T. R. Wilson's cloud expansion photographs we have ocular evidence that electrons are very real particles indeed. Nevertheless de Broglie's suggestion that they should act as waves has been subjected to experimental test by Davisson and Germer[45] and later by G. P. Thomson,[46] Rupp,[47] Kikuchi[48] and others.

For our present purpose we may describe Thomson's experiment, which is typical of them all. His experiment is analogous to those in which Debye and Scherrer[49] and Hull[50] secured diffraction patterns of x-rays by passing them through powdered crystals placed some distance in front of a photographic plate. Thomson replaced the x-ray beam with a stream of cathode rays (falling through about 30,000 volts potential difference), and the mass of powdered crystals with a sheet of gold leaf. The resulting photographs showed the same kind of diffraction pattern as that obtained when x-rays pass through gold leaf. Indeed from the size of the diffraction rings the wave-length of the cathode rays could be calculated, and was found to be just that predicted by de Broglie's formula (9). If the diffraction of x-rays by crystals proves that they are waves, this diffraction of cathode rays establishes equally the wave characteristics of electrons.

We are thus faced with the fact that the fundamental things in nature, matter and radiation, present to us a dual aspect. In certain ways they act like particles, in others like waves. The experiments tell us that we must seize both horns of the dilemma.

A Suggested Solution. During the last few years there has gradually developed a solution of this puzzle, which though at first rather difficult to grasp seems to be free from logical contradictions and essentially capable of describing the phenomena which our experiments reveal. A mention of some of the names connected with this development will suggest some of the complexities through which the theory has gradually gone. There are L. de Broglie, Duane, Slater, Schrödinger, Heisenberg, Bohr and Dirac, among others, who have contributed to the growth of this explanation.[51] The point of departure of this theory is de Broglie's proof, mentioned above, that the motion of a particle may be expressed in terms of the propagation of a group of waves. In the case of the photon, this wave may be taken as the ordinary

---

[45] C. J. Davisson and L. H. Germer, Phys. Rev. **30**, 705 (1927).
[46] G. P. Thomson, Proc. Roy Soc. **A117**, 600 (1928); **A119**, 651 (1928).
[47] E. Rupp, Ann. d. Physik **85**, 981 (1928).
[48] S. Kikuchi, Jap. J. of Phys. **5**, 83 (1928).
[49] Debye and Scherrer, Phys. Zeits. **17**, 277 (1916).
[50] A. W. Hull, Phys. Rev. **10**, 661 (1917).
[51] A review of the development of this theory is given in the report of the fifth Solvay Congress, "Electrons et Photons," Brussels, 1928, written chiefly by W. L. Bragg, A. H. Compton, L. de Brogile, E. Schrödinger, W. Heisenberg and N. Bohr.

electromagnetic wave. The wave corresponding to the moving electron is generally called by the name of its inventor, a de Broglie wave.

Consider, for example, the deflection of a photon by an electron on this basis, that is, the scattering of an x-ray.[52] The incident photon is represented by a train of plane electromagnetic waves. The recoiling electron is likewise represented by a train of plane de Broglie waves propagated in the direction of recoil. These electron waves form a kind of grating by which the incident electromagnetic waves are diffracted. The diffracted waves represent in turn the deflected photon. They are increased in wave-length by the diffraction because the grating is receding, resulting in a Doppler effect.

In this solution of the problem we note that before we could determine the direction in which the x-ray was to be deflected, it was necessary to know the direction of recoil of the electron. In this respect the solution is indeterminate; but its indeterminateness corresponds to an indeterminateness in the experiment itself. There is no way of performing the experiment so as to make the electron recoil in a definite direction as a result of an encounter with a photon. It is a beauty of the theory that it is determinate only where the experiment itself is determinate, and leaves arbitrary those parameters which the experiment is incapable of defining.

It is not usually possible to describe the motion of either a beam of light or a beam of electrons without introducing both the concepts of particles and waves. There are certain localized regions in which at a certain moment energy exists, and this may be taken as a definition of what we mean by a particle. But in predicting where these localized positions are to be at a later instant, a consideration of the propagation of the corresponding waves is usually our most satisfactory mode of attack.

Attention should be called to the fact that the electromagnetic waves and the de Broglie waves are according to this theory waves of probability. Consider as an example the diffraction pattern of a beam of light or of electrons, reflected from a ruled grating, and falling on a photographic plate. In the intense portion of the diffraction pattern there is a high probability that a grain of the photographic plate will be affected. In corpuscular language, there is a high probability that a photon or electron, as the case may be, will strike this portion of the plate. Where the diffraction pattern is of zero intensity, the probability of a particle striking is zero, and the plate is unaffected. Thus there is high probability that a photon will be present where the "intensity" of an electromagnetic wave is great, and a lesser probability where this "intensity" is smaller.

It is a corollary that the energy of the radiation lies in the photons, and not in the waves. For we mean by energy the ability to do work, and we find that when radiation does anything it acts in particles.

In this connection it may be noted that this wave-mechanics theory does not enable us to locate a photon or an electron definitely except at the

---

[52] E. Schrödinger, Ann. d. Physik **82**, 257 (1927).

instant at which it interacts with another particle. When it activates a grain on a photographic plate, or ionizes an atom which may be observed in a cloud expansion chamber, we can say that the particle was at that point at the instant of the event. But in between such events the particle can not be definitely located. Some positions are more probable than others, in proportion as the corresponding wave is more intense in these positions. But there is no definite position that can be assigned to the particle in between its actions on other particles. Thus it becomes meaningless to attempt to assign any definite path to a particle. It is like assigning a definite path to a ray of light: the more sharply we try to define it by narrow slits the more widely the ray is spread by diffraction.

Perhaps enough has been said to show that by grasping both horns we have found it possible to overcome the dilemma. Though no simple picture has been invented affording a mechanical model of a light ray, by combining the notions of waves and particles a logically consistent theory has been devised which seems essentially capable of accounting for the properties of light as we know them.

Starting with Planck's epoch-making suggestion that radiation is emitted in discrete units proportional to the frequency, we have thus seen how Einstein was led to suggest corpuscular quanta of radiation or photons in order to account for the photoelectric effect, and how recent experiments with x-rays, especially those with individual x-ray quanta, have seemed to establish this corpuscular hypothesis. Yet we have long known that light has the characteristics of waves. For centuries it has been supposed that the two conceptions are contradictory. Goaded on, however, by the obstinate experiments, we seem to have found a way out. We continue to think of light as propagated as electromagnetic waves; yet the energy of the light is concentrated in particles associated with the waves, and whenever the light does something it does it as particles.

34

An Attempt to detect a Undirectional Effect of X-Rays

By

Arthur H. Compton, K. N. Mathur and H.R. Sarna.

It has been shown by Bubb[2] that a polarized beam of x-rays ejects photoelectrons whose most probable direction of emission lies in the plane of the electric vector of the polarized waves. If we picture the x-rays as consisting of a stream of photons, each of which is capable of ejecting a photoelectron, this may be explained by assigning to each photon certain vector characteristics and a definite orientation. Such was Bubb's interpretation of his experimental results.

In the case of polarized radiation, however, the electric vector though in a definite direction alternates between a positive and negative sense in this direction. Expressing this in terms of vector photons, a plane polarised electromagnetic wave would be represented by successive groups of photons oriented alternately parallel and antiparallel to a fixed line in the plane of the electric vector. The question arises, is it possible to obtain radiation with an excess of photons oriented in one sense in the plane of polarization? If such a unidirectional property could be detected, it would be a qualitatively new phenomenon in optics.

It would seem that the rays coming directly from the target of an x-ray tube might have such unidirectional characteristics. According to Stokes' theory, these x-rays are produced by the sudden stopping of the cathode electrons when

---

[1] An abstract of this paper has appeared in the Physical Review, 31, 159 (January, 1928).

[2] F. W. Bubb, Phys. Rev., 23 137 (1924).

they reach the target, in which case the electric vector and the corresponding photons should be strictly unidirectional. On Kramers' view,[3] the impinging electron radiates while coursing in a hyperbolic orbit about the nucleus of an atom in the target. In this case also there should be a larger component of the electric vector of the resulting ray anti-parallel to the cathode ray stream than parallel to it.

When examined more closely, however, it is by no means certain that theory demands such a unidirectional property of the x-rays. In the case of the part of the rays which gives rise to the line spectrum, it is evident that in order that sharp lines may appear it is necessary that there shall be long trains of waves, that is, alternate groups of photons oriented in opposite directions. But the continuous spectrum also has its sharp upper limit of frequency, which means a complete interference at small glancing angles from the crystal such as can occur only with long trains of waves. Furthermore the studies of electromagnetic pulses by J. J. Thomson and Sommerfeld have shown that when unidirectional pulses traverse a thin layer of matter their asymmetry is rapidly reduced. From the photon standpoint, however, it is not obvious that a sharp frequency limit will mean anything more than photons with a fixed maximum of energy, and the attempt to detect such a unidirectional property of the x-rays does not appear hopeless.

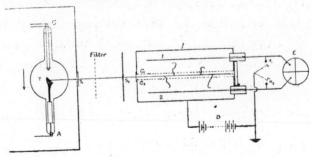

Fig. 1.

[3] H. A. Kramers, Phil, Mag., 46, 836 (1923).

The experiment was performed as shown diagrammatically in Figure 1. X-rays from a "universal type" Coolidge tube, placed in a horizontal position, passed through two narrow vertical slits S_1, S_2, and into an ionization chamber I. The x-rays formed a thin sheet between two gauzes G_1 and G_2. If the x-rays resulting from the stopping of a cathode ray form a pulse, as Strokes suggests, the direction of the electric vector of this pulse at the ionization chamber should be such as to exert a force on an electron in the direction $G_1 \rightarrow G_2$ We might thus anticipate a greater number of photo-electrons in the side 2 of the chamber than in the side 1. As will be seen from the figure, the electrodes 1 and 2 are connected to opposite pairs of quadrants of a Dolezalek electrometer. Thus if more photo-electrons enter one side of the chamber than the other, producing greater ionization, there will be a corresponding deflection of the electrometer.

It was found that the readings were rather sensitive to changes in the location of the x-ray beam between the two gauzes. This required an accurate alignment of the slits and of the chamber itself. Some difficulty was also experienced due to the uneven brilliance of different portions of the focal spot. This resulted in an unsymmetrical sheet of x-rays between the gauzes, the dissymmetry varying with the position of the target of the tube. Errors due to this source were reduced by making the slits narrow. In order to eliminate systematic errors, alternate readings were taken with the tube reversed in direction.

Experiments were tried at tube potentials from about 40 to about 100 kilovolts, with and without filters placed in the path of the x-rays. In some preliminary experiments it appeared that an effect of about 5 per cent. in the predicted direction was present, and this provisional result was described at the meeting of the All-India Science Congress held in Lahore in January, 1927. Later, more refined experiments failed to show any consistent change in the ionization in the two halves of the chamber when the x-ray tube was reversed. It appeared that the positive result of the earlier experiment was due to the dissymmetry of the x-ray beam described in the paragraph above. When the slits were made narrower no such effect was observed.

The results of our final experiments may be summarized by the statement that on reversing the x-ray tube, no dissymmetry in the ionization current appeared which was larger than the probable error of our measurements, about 1 part in 500.

It would be premature to conclude from this negative result that the photon does not possess the vector properties of a line in a definite direction. The possibility is clearly present that the conditions of our experiment were not suitable to ensure any predominance of the photons oriented in one sense. If, however, it is essentially impossible to produce radiation in which the electric vector or the protons directed in one sense can be distinguished from those in the opposite sense, the sense of the electric vector as well as of the photon ceases to have any significance. In this regard the electric field of radiation would be different in quality from a static electric field.

These experiments were performed at the Punjab University in the winter of 1926-27 and the authors are grateful to Dr. S. S. Bhatnagar for the facilities provided for this work at the University Chemical Laboratories, Lahore.

A NEW WAVE-LENGTH STANDARD FOR X-RAYS.

BY

ARTHUR H. COMPTON, Ph.D.,

University of Chicago.

THE first reliable value of the wave-length of an X-ray beam was obtained when Bragg[1] determined the atomic arrangement in a rock salt crystal, and used the calculated distance between its layers of atoms as a standard for measuring the wave-length of the palladium K line. The wave-length is given by the Bragg formula,

$$n\lambda = 2D \sin \theta, \qquad (1)$$

where n is the order of the spectrum, D is the distance between successive layers of atoms, and θ is the glancing angle of diffraction. When one takes into account the refraction of the X-rays as they enter the crystal, this relation becomes,

$$n\lambda = 2D \sin \theta \left(1 - \frac{1-\mu}{\sin^2 \theta}\right), \qquad (2)$$

where μ is the index of refraction. In this equation n is a known integer, θ can be measured with high precision, and $1 - \mu$ can be measured accurately enough so that its uncertainty introduces no appreciable error in the result. The wave-length λ can thus be measured as accurately as D is known.

The grating constant of a crystal whose structure is known can be calculated in terms of Avogadro's number. Thus for a simple cubic crystal,

$$D = (W/N\rho)^{1/3}, \qquad (3)$$

where W is the molecular weight, N is the number of atoms per gram atom, and ρ is the crystal density. For other than simple cubic crystals, the right hand side of equation (3) must

---

[1] W. L. Bragg, *Proc. Roy. Soc. A.*, 89, 248 (1913).

be multiplied by a simple numerical factor. In this equation, the molecular weight is thought to be known with high precision. The density can be precisely measured, though there may be some question whether the measured density of a large crystal is identical with the density of the perfect crystal lattice which is postulated in deriving this equation. The chief uncertainty in the grating space would seem to be that

Fig. 1.

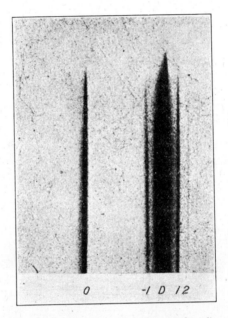

Spectrum of Mo Kα_1 line from grating ruled on speculum metal, 50 lines per mm. (Doan and Compton, 1925) D = direct beam, o = zero order (direct reflection).

due to the uncertainty in Avogadro's number. This is estimated by Millikan to be known to about 1 part in 1,000, which means an uncertainty in D and hence also in λ of about 1 part in 3,000.

About four years ago Dr. Doan and I found that it is possible to secure X-ray spectra from ruled reflection gratings if the X-rays are allowed to strike the reflecting surface at a sufficiently sharp glancing angle. One of our spectra, of the Kα line of molybdenum, is shown in Fig. 1. From such a spectrum it is possible to measure the wave-length in terms of the spacing of the lines ruled on the grating, which can in turn be measured directly. Though the angle of diffraction is much smaller than that occurring when the crystal grating is used, it seemed from the start that the definiteness of the grating space of the ruled grating should make it preferable for making absolute determinations of X-ray wave-lengths.

In these first ruled grating measurements, our probable error was about 0.4 per cent., and to this accuracy we found our results in accord with the crystal values. Since that time similar spectra have been obtained using more refined methods by such investigators as Thibaud,[2] Hunt,[3] Bäcklin,[4] Wadlund[5] and others. There has been a definite tendency for the wavelengths as measured by these men to be greater than those measured on the same spectral lines by the crystal method, though in most cases the difference has hardly been larger than the probable experimental error. During the last year, however, Dr. J. A. Bearden, working at Chicago, has made measurements by the ruled grating method[6] that seem to be of much higher precision than those of the previous investigators. In Fig. 2 are shown spectra obtained by Wadlund (above) and Bearden (below), both using a grating of the same characteristics as that from which Fig. 1 was made. A comparison of these three spectra shows the gradual improvement of technique. Bearden's measurements result in a wave-length 0.24 per cent. greater than that given by Siegbahn from crystal measurements. This difference is about 25 times the probable error of his measurements, and thus seems to indicate a hidden error in our wave-length estimates based on crystal diffraction.

If one calculates from the X-ray wave-lengths as thus determined the crystalline grating space, and thence Avogadro's number and the charge on the electron, one finds values for these constants which differ by a surprisingly large amount from the accepted values. It will therefore be worth while to discuss some of the X-ray wave-length measurements using ruled gratings, and to look for possible errors which may be present.

---

[2] J. Thibaud, *Comptes Rendus*, 182, 55 (1926), *Rev. d'Opt.*, 5, 97 (1926).
[3] F. L. Hunt, *Phys. Rev.*, 30, 227 (1927).
[4] E. Backlin, Thesis, Upsala (1928).
[5] A. P. R. Wadlund, *Phys. Rev.*, 32, 841 (1928).
[6] J. A. Bearden, *Proc. Nat. Acad. Sci.*, June (1929).

In making such standard wave-length measurements, one must choose between the use of soft X-rays of comparatively great wave-length, for which the angles to be measured are relatively large, but for which a vacuum spectrometer is necessary, and the use of ordinary X-rays, for which the angles to be measured are inconveniently small, but for which the

FIG. 2.

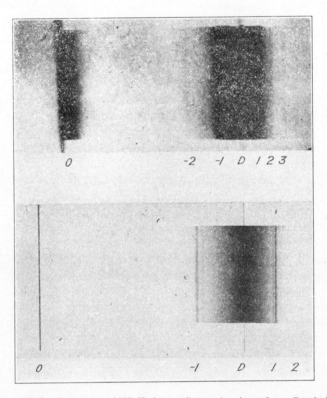

Upper figure, Wadlund's spectrum of NCuKα from 50 line grating; lower figure, Bearden's spectrum of CuKα and Kβ lines from the same grating.

necessary adjustments can be made and continually checked in the open air. Bäcklin's experiments were done using a vacuum spectrometer and relatively long wave-lengths. Bearden and Wadlund chose the latter alternative, having in mind also that for the ordinary X-rays the index of refraction is more accurately known, and the comparison with the crystal wave-lengths is of correspondingly greater significance.

The angles to be measured, using the latter method, are less than 40 minutes of arc. Such angles may be measured by the help of a finely divided circle, by the use of a telescope and scale, or by measuring the separation of the X-ray lines on a photographic plate placed at a known distance. Wadlund, following a suggestion made by Doan, relied upon the telescope and scale. In any case, however, the separation of the lines on a photographic plate must be used in some part of the experiment, and Bearden preferred to measure all of his angles by this means.

It will not be necessary here to describe Bearden's experiment in detail, since this has been done elsewhere.[7] One should perhaps mention, however, that by taking photographs on two plates at a known distance apart, he determines the wave-length from the formula,

$$n\lambda = \frac{D}{2b^2} \frac{(y-x)^2(y \pm z)z}{y^2}. \qquad (4)$$

Here n is the order of the spectrum, D is the grating space, b is the distance between the two plates, and x, y and z are distances between lines on the plates themselves. Thus every quantity that needs to be known is susceptible to precise direct measurement.

It will be interesting to compare the results of different investigators who have used the ruled grating method:

TABLE I.

Investigator.	Spectra Line.	λ	Grating λ − Crystal λ.	Probable Error.	Date.
Compton-Doan...	MoKα_1	.7078A	− .1%	± .4%	1925
Thibaud.........	CuKα	1.540	+ .1	± 1.	1926
Hunt............	PtMα	6.1	+ 1.6		1927
	AlKα	8.5	+ 2.0		
Bäcklin.........	MoLα	5.402	+ .1	± .2	1928
	AlKα	8.333	+ .2	± .1	
	MgKα	9.883	+ .2	± .2	
	FeLα	17.61	+ .1	± .2	
Wadlund........	MoKα_1	.708	+ .1	± .1	1928
	CuKα_1	1.537	0	± .08	
	FeKα_1	1.938	+ .3	± .2	
Howe...........	ZnLα	12.24	+ .2	± .3	1929
	CuLα	13.31	+ .6	± .3	
	FeLα	17.61	+ .5	± .3	
Bearden........	CuKα	1.5422	+ .23	± .01	1929
	CuKβ	1.3926	+ .24	± .01	

[7] J. A. Bearden, *Proc. Nat. Acad.*, June 1929.

It will be noted that there is a consistent tendency for the ruled grating wave-lengths to have higher values than the wave-lengths measured from crystals. Though in most cases the difference is hardly greater than the probable error of the measurements, Bearden's experiments, which appear to be much the most precise, show an outstanding discrepancy from the crystal values.

The only case in which there is an apparent conflict between Bearden's experiments and the results of other grating measurements is Wadlund's value for the $K\alpha_1$ line of copper. To see whether this difference is significant, let us note the individual values from which Wadlund's mean result is taken. He found:

TABLE II.

Plate No.	Order.	Wave-length.	Grating λ − Crystal λ
105	1	$1.5394 \pm .0016$	$+.14\%$
	2	$1.5361 \pm .0011$	$-.08$
	3	$1.5378 \pm .0008$	$+.03$
106	1	$1.5312 \pm .0028$	$-.40$
	2	$1.5330 \pm .0008$	$-.28$
	3	$1.5427 \pm .0011$	$+.34$
	4	$1.5423 \pm .0009$	$+.32$
	6	$1.5359 \pm .0005$	$-.09$

It will be seen that orders 3 and 4 from his plate 106, which would seem to be two of his best measurements, give wave-lengths larger than Bearden's mean result. (Bearden used the unresolved $\alpha_1\alpha_2$ doublet, whereas Wadlund used only the α_1 component). In order to be certain that the difference was not a real one, Bearden repeated the measurement on the copper $K\alpha$ line using Wadlund's own grating, and with the advantage of the knowledge of the possible sources of error in Wadlund's experiments. The result was entirely consistent with Bearden's measurements using his other gratings.

It is perhaps worth noting that the sharpness of the best lines on Bearden's plates is such that an error in setting great enough to make his results agree with the crystal measurement would mean placing the cross hair of the comparator microscope beside the line instead of over it.

CONSEQUENCES OF THIS ABSOLUTE WAVE-LENGTH MEASUREMENT.

Using the value of the wave-length of the copper Kβ line as found by Bearden, the grating space D of a calcite crystal can be obtained from equation (2). Siegbahn and Dolejsek[8] find for θ in this case $12° \, 15' \, 28.2''$ for $n = 1$, with the calcite crystal at $18°$ C.; whence $\sin \theta = 0.229334$. The writer has given[9] for the value of $(1 - \mu)/\sin^2 \theta$ for calcite, 1.46×10^{-4}. These data give for the grating space of the calcite (100) planes,

$$D_{\text{calcite}} = 3.0366 \pm .0004 \, A. \; (18° \text{ C.}),$$

and

$$D_{\text{calcite}} = 3.0367 \pm .0004 \, A. \; (20° \text{ C.}). \tag{5}$$

From a relation similar to equation (3) we find for Avogadro's number,

$$N = \frac{nM}{\rho\varphi} \frac{1}{D^3}, \tag{6}$$

where for calcite,

$$n = \tfrac{1}{2},$$
$$\rho = 2.7102 \pm .0004 \text{ g. cm.}^{-3}, [10]$$
$$M = 100.078 \pm .006, [11]$$
$$\varphi = 1.09630 \pm .00007, [12]$$
$$D = 3.0367 \pm .0004 \times 10^{-8} \text{ cm. (Eq. 5).}$$

Thus,

$$N = 6.0142 \pm .0026 \times 10^{23} \text{mole}^{-1}. \tag{7}$$

Knowing Avogadro's number, the electronic charge e is given in terms of Faraday's constant of electrolysis F by the relation,

$$e = F/N. \tag{8}$$

Using the value[2] $F = 9648.9 \pm .7$ e.m.u., and $c = 2.99796 \times 10^{10}$ cm./sec., we thus obtain

$$e = 4.810 \pm .002 \times 10^{-10} \text{e.s.u.} \tag{9}$$

---

[8] M. Siegbahn and V. Dolejesek, *Zs. f. Phys.*, 10, 160 (1922).
[9] A. H. Compton, "X-rays and Electrons" (1926), p. 212.
[10] O. K. DeFoe and A. H. Compton, *Phys. Rev.*, 25, 618 (1925).
[11] R. T. Birge, *Phys. Rev.*, Sup. 1 (1929).
[12] H. N. Beets, *Phys. Rev.*, 25, 621 (1925).

Perhaps the most direct method of obtaining a value of Planck's constant h on the basis of this work is through the measurement of the limit of the X-ray spectrum by Duane and his collaborators. Writing the Duane-Hunt law in the form

$$h = Ve\lambda/c,$$

where V is the potential applied to the X-ray tube, and λ is the minimum wave-length, and expressing λ by equation (2), remembering that for this experiment $n = 1$, we get

$$h = \frac{2}{c} VeD \sin \theta (1 - b), \tag{10}$$

where $b = (1 - \mu)/\sin^2 \theta = 1.46 \times 10^{-4}$. The experiments of Duane, Palmer and Chi-Sun-Yeh [13] give directly the value of $V \sin \theta$ as 2039.9 ± 9 volts. Using the values of e and D obtained above, we thus obtain

$$h = 6.629 \pm .004 \times 10^{-22} \text{ erg sec.} \tag{11}$$

If we write Bohr's expression for the Rydberg constant in the form,

$$R = 2\pi^2 \frac{e^5 m}{h^3 e}, \tag{12}$$

we may solve for the value of e/m. Using [14] $R = 3.28988 \times 10^{15}$ we thus find

$$e/m = 1.769 \pm .003 \times 10^7 \text{ e.m.u./g.} \tag{13}$$

A comparison of the values thus obtained with the accepted values found by other methods, as recently summarized by Birge,[14] is given in the following table.

TABLE III.

Comparison of Values of Fundamental Constants.

	From Bearden's Wave-lengths.	Accepted Values.	Per cent. Difference.
D Calcite	3.0367 A.	3.0283 A.	0.3
N	6.014×10^{23}	6.064×10^{23}	0.8
e	4.810×10^{-10}	4.770×10^{-10}	0.8
h	6.629×10^{-27}	6.547×10^{-27}	1.2
e/m	1.769×10^7	1.769 (deflection)	0
		1.761 (spectrum)	0.5

[13] W. Duane, H. H. Palmer, and Chi-Sun-Yeh, *J. Opt. Soc. Am.*, 5, 376 (1921).
[14] R. T. Birge, *Phys. Rev.* Sup. 1, (1929).

The probable error of every value in the table as estimated by the usual methods is less than 0.1 per cent. The large differences thus point strongly toward a hidden error either in the experiments or in the formulæ relating the experimental data to these fundamental constants. The most significant differences are those in the values of e and of e/m, since these quantities are subject to independent experimental measurement. The accepted values of N, D and h are all based upon the experimental value of e, so the differences in their values give but little additional information. We have been diligently searching for such a hidden error for the past six months without success, and it now seems necessary to appeal for help by calling the discrepancies to your attention.

It is interesting however to note that if one uses these new values of e and h, better agreement is found with Eddington's recent speculations [15] regarding Sommerfeld's α. He has brought forward reasons for believing that $1/\alpha$ should be the integer 136, where $\alpha = 2\pi e^2/hc$. Using the accepted values of e, h and c, we find $1/\alpha = 137.2$, which seems to differ by more than the experimental error. The values given above, however, lead to $1/\alpha = 136.45 \pm .15$ which is not far from Eddington's value. While one cannot say that the theoretical argument leading to this value is cogent, it is suggestive that the one integer which can have significance is in close agreement with these new values of the constants.

Possible Sources of Error.—We have been unable to locate any errors in the technique of the experiments which could introduce an error in the wave-length as large as 0.3 per cent. The use of the wide variety of gratings was suggested by the thought that there might be some error in the grating formula (4). The fact that different gratings gave the same results, and that measurements of different orders of spectra led to the same wave-lengths, confirmed the validity of the formula. One may perhaps however call attention to a very general proof of this equation which indicates that no surface peculiarity of the grating can in any way affect the angle of diffraction. In Fig. 3 is shown diagrammatically the simple case of diffraction of two rays from adjacent lines of the reflection grating.

---

[15] A. S. Eddington, *Proc. Roy. Soc.*, A, 122, 358 (1929).

The condition for an interference maximum at the angle r is that $\overline{CB} - \overline{AD} = n\lambda$, where n is an integer. In Fig. 4 we have supposed that due to refraction or some other cause the ray diffracted from the slit A traverses and irregular path

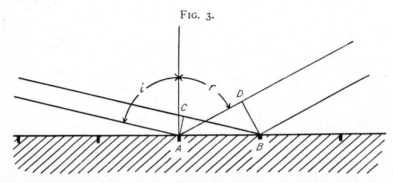

FIG. 3.

$A'AA''$ while in the neighborhood of the grating, but that when beyond some arbitrary plane $A'B'A''B''$ it is propagated in a straight line. If the ray diffracted from the adjacent slit B traverses a path of the same shape, and through media of the same refractive index, the phase lag along $B'BB''$ will be the

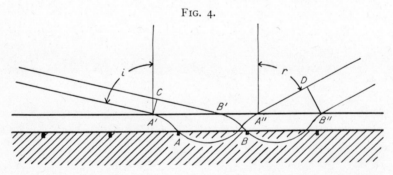

FIG. 4.

same as that along $A'AA''$. Thus here again the interference maximum occurs when $\overline{CB'} - \overline{A''D} = n\lambda$. Since $A'B' = AB = A''B''$, this means that the angles i and r are the same as those of the simple case in Fig. 3, from which equation (4) follows.

I can accordingly find no basis on which to doubt the reliability of the wave-length values resulting from Bearden's experiments.

Nov., 1929.] NEW WAVE-LENGTH STANDARD FOR X-RAYS. 615

The passage from the wave-length to the crystal grating space through equation (2) seems so direct as to admit little chance of error. A question might occur regarding the effect of refraction. The correction for refraction introduced in this equation has been verified by thorough theoretical studies by Darwin[16] and Ewald,[17] Moreover, the fact that the equa-

FIG. 5.

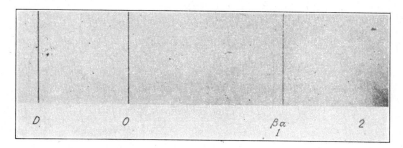

Bearden's spectrum of CuK series from glass grating of 600 lines per mm.

tion as thus corrected leads to the same wave-length for the different orders of reflection is a strong experimental confirmation. This fact also makes untenable the suggestion that the spacing of the crystal lattice may change near the surface of the crystal.

Avogadro's number it would seem involves no additional uncertainty except that of the crystal density. The existence of crystal faults is well known, and Joffe[18] and Zwicky[19] among others have emphasized the existence of resulting gaps in the crystal lattice. It should follow that the minute crystal elements would have a higher density than the large scale crystals. Unfortunately such a difference is in the wrong direction to account for the difference between the crystal and the ruled grating wave-lengths. A large impurity of high molecular weight might account for the difference; but it is improbable that such an impurity exists. It should be noted also that the difference is nearly the same whether a

---

[16] C. G. Darwin, *Phil. Mag.*, 27, 318 (1914).
[17] P. Ewald, *Phys. Zeits.*, 21, 617 (1920).
[18] A. Joffe, "The Physics of Crystals" (1928).
[19] F. Zwicky, *Proc. Nat. Acad.*, 15, 253 (1929).

comparatively perfect crystal such as calcite is used or whether one employs the relatively imperfect crystal, rock salt.

From Avogadro's number the relation through Faraday's constant of electrolysis to the electronic charge seems straightforward.

Such directly conflicting experimental evidence regarding these fundamental constants is not a new problem. Thus for example there is the outstanding difference in the values of e/m as determined by deflection experiments as compared with that resulting from spectroscopic observations. It would seem that the best mode of attack would be the redetermination of these constants by whatever independent methods may be capable of giving sufficient precision.

From the Philosophical Magazine, vol. viii. *December* 1929.

The Efficiency of Production of Fluorescent X-Rays.
By Arthur H. Compton, *Professor of Physics at the University of Chicago.*

THERE exists a simple theory of the mechanism by which fluorescent X-rays are produced, which gives a definite prediction of the intensity of these rays. According to this theory, when an electron in the K energy level of an atom absorbs a quantum of X-rays, it is ejected as a photoelectron. An electron from the L or outer levels will then fall into the K level, resulting in the emission of a quantum of K series X-rays. For each quantum of energy absorbed by the K electrons there should thus be emitted one quantum of fluorescent X-rays of the K type. The ratio of the absorbed to the emitted rays can thus be directly calculated.

This simple theory has, however, predicted two important results which are inconsistent with experiment. The intensity of the fluorescent X-rays is found not to be as great as this theory indicates; and the β-rays excited by the X-rays are found to be more plentiful than it predicts. Both of these facts are indicated by the data accumulated by Sadler* and have been emphasized by Barkla † and by Barkla and Dallas ‡.

The Compound Photoelectric Effect and Fluorescent Yield.

A source of difficulty with the theory was made evident by Auger's discovery in 1925 § of the "compound photoelectric effect," *i. e.*, the simultaneous ejection of several β-particles from the same atom. By measuring their ranges ‖ he found that in addition to the photoelectron whose velocity is given by Einstein's equation

$$E_{kin} = h\nu - h\nu_K$$

for absorption by a K electron, there might be emitted from the same atom an electron of energy

$$E = h\nu_K - 2h\nu_L$$

and others of a similar type. The total energy of all the β-particles from a single atom was never more than $h\nu$. This observation showed that there are two different processes by which the atom can return to its normal condition. The first process is that in which the energy liberated when the atom returns to its normal state is radiated as a fluorescent ray. In the second process the energy absorbed by the atom is spent in ejecting from the atom one or more electrons. This may be considered either as a kind of internal photoelectric absorption, or, as Auger thinks more reasonable, a radiationless energy transfer of the type which Klein and Rosseland describe as an "event of the second kind." Such an event—the transfer of the energy of an excited atom to an electron—has been recognized for some time in optics, and Rosseland had opined ¶ that it should also appear in the X-ray domain. It is clear that the occurrence of events of this second type will result in less intense fluorescent radiation,

---

* C. A. Sadler, Phil. Mag. xvii. p. 739 (1909); xviii. p. 107 (1909); xix. p. 337 (1910).
† C. G. Barkla, Phil. Trans. ccxvii. p. 315 (1917).
‡ C. G. Barkla and Miss Dallas, Phil. Mag. xlvii. p. 1 (1924).
§ P. Auger, *C. R.* clxxx. p. 65 (1925); *J. de Phys. et Radium*, vi. p. 205 (1925).
‖ P. Auger, *Ann. de Physique*, vi. p. 183 (1926).
¶ Rosseland, *Zeits. f. Phys.* xiv. p. 173 (1923). The suggestion that such an event might account for the anomalous ionization and fluorescence by X-rays was suggested by Kossel, *Zeits. f. Phys.* xix. p. 333 (1923), and by Barkla and Dallas, Phil. Mag. xlvii. p. 1 (1924).

and more intense β radiation, which is precisely what is needed to satisfy the demands of Sadler's and Barkla's work.

Auger * has determined the probability that an event involving fluorescence will occur by the direct method of counting the number n of photoelectrons ejected from the K shell of an element (these can be identified for the heavier elements, because the range is shorter than that of a photo-electron ejected from an outer level), and comparing this with the number n_2 of atoms from which go associated tracks of the second kind. The difference $n-n_2$ is the number n_1 of fluorescent quanta emitted. He thus obtains what is known as the " fluorescence yield,"

$$w = n_1/n. \quad \ldots \quad \ldots \quad (1)$$

His results are summarized in the following table:

TABLE I.

Element.	Voltage.	w.
19 A	70 kv	·07
36 Kr	70	·5
	22	·51
54 Xe	43	·71

The experiments on krypton, using two different wave-lengths of primary X-rays, indicate that the fluorescence yield is a property of the atom and is independent of the wave-length of the exciting radiation. It is evident from these results that especially for the lighter elements the compound photoelectric effect is of relatively great importance.

Fluoresence Yield from Measurements of Fluorescence.

It is also possible to determine the fluorescence yield from measurements of the intensity of the fluorescent rays. If the compound photoelectric effect is the sole cause of the difficulty with the simple theory outlined above, the yield thus calculated should be identical with that deter-

* P. Auger, *Ann. de Physique*, vi. p. 183 (1926).

mined by Auger. Kossel* and Bothe† have calculated approximate values of this factor using the fluorescence data given by Sadler and Barkla respectively. Kossel's results are :—

TABLE II.

Element	24 Cr	26 Fe	27 Co	29 Cu	30 Zn
w	·23	·32	·39	·42	·51

These values are of the same order of magnitude as those of Auger, but vary with atomic number at a surprising rate. Kossel himself evidently does not have much confidence in their reliability, since he remains uncertain whether the experiments really show that w is less than 1. His work is of especial interest, however, in that he was led from these figures to give the first suggestion of atomic transformations of the second kind in order to account for a fluorescence yield less than unity.

The values of w calculated by Bothe, 0·5 for Br and 0·23 for Cr, fit more acceptably than do Kossel's with Auger's data. Jauncey and De Foe‡ have also made a rough measurement of the ratio of the fluorescent energy from copper to the absorbed rays, leading to a value of the fluorescence yield less than unity.

Since Auger's discovery of the compound photoelectric effect, two measurements of the fluorescence yield have been made, one by Balderston§ and one by Harms ||. Balderston compares the total number of absorbed quanta with the number of fluorescent K quanta, calling the ratio u. Since only about $\frac{7}{8}$ of all the absorbed quanta are absorbed by the K shell, his ratio u must be divided by $\frac{7}{8}$ in order to get the fluorescence yield, w, which is comparable with Auger's data. As thus corrected, his values are :—

* W. Kossel, *Zeits. f. Phys.* xix. p. 333 (1923).
† W. Bothe, *Physik. Zeits.* xxvi. p. 410 (1925).
‡ G. E. M. Jauncey and O. K. De Foe, Proc. Nat. Acad. xi. p. 520 (1925).
§ L. Balderston, Phys. Rev. xxvii. p. 695 (1926).
|| M. I. Harms, *Ann. d. Phys.* lxxxii. p. 87 (1926).

TABLE III.

Element	26 Fe	28 Ni	29 Cu	30 Zn	42 Mo	47 Ag
w	·38	·45	·50	·57	·95	·86

The experimental error in these determinations is probably large. Balderston calls attention to an uncertainty of 13 per cent. in the solid angle subtended by the window of the ionization chamber. Errors which are probably of the same order of magnitude are introduced also by uncertainty in the absorption in the ethyl bromide vapour, and by the method of calculation, which involves an extrapolation, assuming a linear variation over a range of wave-lengths in some cases more than 3 times the range over which observations were taken. More serious errors of a consistent nature are apparently introduced by the assumption that the ionization of different wave-lengths in air is proportional to the absorption in air, whereas for the shorter wave-lengths a large part of the absorption is due to scattering, and does not result in ionization. Also, since in his experiments the fluorescent rays struck the side walls of the ionization chamber, whereas the primary rays did not, it was not possible to compare the relative intensities, even of the same wave-length, by comparing the ionizations. These values can therefore be assigned but little weight.

Harms* has made a more careful analysis of the problem. In the following table his results are given in columns 2 and 3.

TABLE IV.

Element.		w'.	w.	w (corrected).
26	Fe	·530	·303	·282
29	Cu	·589	·394	·378
30	Zn	·598	·418	·403
34	Se	·647	·530	·517
38	Sr	·598	·590	·615
42	Mo	·754	·754	(·730)

* M. I. Harms, *loc. cit.*

In calculating w' Harms has made use of Kossel's relation between the wave-length and the ionization current per erg of β-ray energy. As he himself notes, Kulenkampff* and Kircher and Schmitz† have independently reached the conclusion that from ·5A to 1·5A there is a strict proportionality between absorbed energy and ionization. Yet Kossel considers the older work of Lenard and Holthusen more reliable, and Harms has used Kossel's correction factor to calculate the energy from the ionization. The conclusion of Kulenkampff and of Kircher and Schmitz has recently been confirmed by Crowther and Bond‡, and is supported by the results about to be described. Harms's data should thus not have been corrected by Kossel's factor, and he should have obtained the values of w given in column 4.

There are some minor corrections that should be applied to Harms's results :—(1) The mass scattering coefficient of air for λ ·71 should be, from Hewlett's data on carbon §, not less than 0·20, whereas Harms has used the value 0·17 ; (2) a part of the secondary X-rays is scattered, whereas Harms assumes that it is wholly fluorescent. From my measurements this fraction varies from 4 per cent. in the case of λ ·71 A exciting secondary rays in iron to 0·4 per cent. in selenium ; (3) the value for strontium was obtained using a surface of strontium sulphate instead of the element, and a correction of 5 per cent. must be applied to make it comparable with the values for the other elements ; (4) the value for molybdenum is an extrapolated one, assuming that the variation of w with the atomic number is linear. This assumption probably is not justified. These minor corrections give us the values in the last column.

Harms's measurements are subject to some uncertainty, because the intensity of the primary and fluorescent beams were measured by two ionization chambers of very different design. The measurements thus involved a comparison of the capacities of the two systems, the sensitivity of the two electrometers, and the effective path of the X-rays in the two ionization chambers. Moreover, he used a beam from a molybdenum target, filtered through

---

* H. Kulenkampff, *Ann. der Phys.* lxxix. p. 97 (1926).
† H. Kircher and W. Schmitz, *Zeits. f. Phys.* xxxvi. p. 484 (1926).
‡ J. C. Crowther and W. N. Bond, *Phil. Mag.* vi. p. 401 (1928).
§ C. W. Hewlett, *Phys. Rev.* xvii. p. 284 (1921).

a zirconium screen, and assumed that the effective wavelength of the transmitted rays was that of the K lines. A spectral analysis of the X-rays under these conditions * shows that ordinarily only about 25 per cent. of the filtered beam consists of the Kα lines, and that there is a band of the continuous spectrum in the neighbourhood of 0·5A which has more energy than has the Kα radiation. This heterogeneity makes it impossible to estimate the absorption accurately from tables of absorption coefficients, as Harms found it necessary to do. It would, nevertheless, seem that Harms's values of the fluorescence yield, after applying the corrections noted above, are more reliable than those of the other investigators.

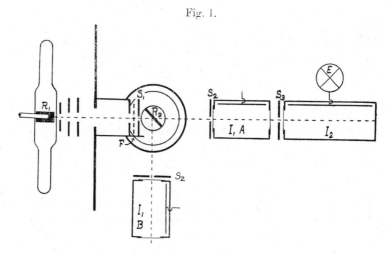

Fig. 1.

New Measurements of the Fluorescent Yield.

In view of the unsatisfactory status of these fluorescence measurements, I have undertaken some new measurements of the fluorescence yield. These measurements were meant to be of only a preliminary character; but as they have led to results that seem more reliable than those now in the literature, and since it is uncertain when the experiments can be continued, it seems worth while to publish them at this time.

The apparatus used in my experiments is shown in fig. 1, which is drawn approximately to scale.

* A. W. Hull, Phys. Rev. x. p. 666 (1917).

Nearly homogeneous X-rays were obtained by taking the fluorescent rays from plates of various elements placed at R_1, just over the X-ray tube*. These homogeneous rays were limited by a diaphragm S_1, and fell on the radiator R_2 whose fluorescence yield was to be determined. R_2 was a flat thick plate placed at 45 degrees on the crystal table of a Bragg spectrometer, and of area large enough to intercept all the rays coming through the diaphragm S_1. The fluorescent rays from R_2 entered the ionization chamber I_1 at its position B, through a lead diaphragm S_2. The ionization current thus obtained was compared with that observed when the chamber was turned to the position A, with the radiator R_2 removed, and with a second diaphragm at S_1 of opening small enough so that the ionization current was of the same order of magnitude as that found in position B. The water-cooled tungsten target X-ray tube was usually operated at about 40 milliamperes and 53 peak kilovolts.

The ionization chamber was 8 cm. diameter and 11 cm. long, inside. The dimensions of the diaphragm used at S_2 were such that no X-rays reached the side walls in either position A or B. The front and back faces of the chamber were covered with thin celluloid, and the chamber was filled with methyl bromide vapour. The fraction of the X-rays absorbed by the gas in the chamber was measured for each wave-length employed by the use of a second ionization chamber I_2. The absorption by the celluloid windows was corrected for by measuring the direct beam through an equal thickness of celluloid placed at F. For the longer-wave-lengths it was necessary to apply a correction for the absorption in the 11 cm. column of air displaced by the ionization chamber.

Calculation of the Fluorescence Yield.

The intensity of the fluorescent X-rays may be calculated as follows: If μ' is the total absorption coefficient of the primary X-rays in the radiator, we have

$$\mu' = \sigma' + \tau', \quad \ldots \ldots \ldots (2)$$

---

* The high degree of homogeneity thus obtained was recognized long ago by Barkla, and has recently been investigated by the writer (Proc. Nat. Acad. xiv. p. 549, 1928). In the latter paper I failed to mention the work of Allison and Duane (Proc. Nat. Acad. xi. p. 486, 1925), who also call attention to this remarkable homogeneity.

where τ' is the photoelectric absorption, and σ' is the absorption due to scattering. If, as is the case in these experiments, the primary wave-length λ' is shorter than the critical K absorption wave-length of the radiator λ_K, the greater part of the photoelectric absorption is that due to the electrons in the K shell, τ_K'. Let us suppose that a beam of intensity I' and of cross-section A' traverses a thickness ds of the radiator. The number of quanta absorbed by the K electrons, and hence the number of photoelectrons ejected from the K energy levels of the atoms will then be

$$dn = I'A'\tau_K' ds \cdot \frac{1}{h\nu'}, \quad \ldots \quad (3)$$

where ν' is the effective frequency of the primary rays. By equation (1), the number of fluorescent quanta is then $dn_1 = w\,dn$, where w is the fluorescence yield, and the total power in the fluorescent beam will be $h\nu'' dn_1$, where ν'' is the effective frequency of the fluorescent ray. Barkla and Sadler have shown that the intensity of the fluorescent ray is the same in all directions. Thus the power in the fluorescent ray entering the ionization chamber from the thin layer ds (uncorrected for absorption) is

$$dP'' = h\nu'' dn_1 \frac{A''}{4\pi r^2} = \frac{w}{4\pi} P' \frac{A''}{r^2} \tau_K' \frac{\nu''}{\nu'} ds, \quad . \quad . \quad (4)$$

where A'' is the area of the diaphragm S_2, r is the distance from R_2 to S_2, and $P' = A'I'$ is the power in the primary beam striking the radiator.

If the fluorescent ray leaves the surface at the same angle with the normal as that at which the primary beam enters, the paths of the primary beam and of the secondary beam in the radiator are equal, having the value, let us say, s. Writing μ' and μ'' for the absorption coefficients of the two beams *, we thus have for the power in the fluorescent beam from a thick radiator,

$$\begin{aligned}P'' &= \frac{w}{4\pi} P' \frac{A''}{r^2} \tau_K' \frac{\nu''}{\nu'} \int_0^\infty e^{-(\mu'+\mu'')s} ds \\ &= \frac{w}{4\pi} P' \frac{A''}{r^2} \frac{\tau_K'}{\mu'+\mu''} \frac{\nu''}{\nu'}\end{aligned} \quad \right\} \quad \ldots \quad (5)$$

Strictly speaking, one should use an absorption coefficient intermediate between μ' and τ', since part of the primary rays scattered in the radiator are reabsorbed before leaving the radiator. However, μ' and τ' are so nearly equal that this difference can be neglected.

Writing instead of ν''/ν' its equivalent λ'/λ'', we obtain for the value of the fluorescence yield,

$$w = 4\pi \frac{r^2}{A''} \frac{\mu' + \mu''}{\tau_K'} \frac{\lambda''}{\lambda'} \frac{P''}{P'}. \quad \ldots \quad (6)$$

The quantities r and A'' can be measured directly, λ' and λ'' are the weighted mean wave-lengths of the fluorescent radiations from the radiators R_1 and R_2 respectively*, and μ' and μ'' are the absorption coefficients in the radiator of the wave-lengths λ' and λ''. The values of μ' and μ'' have been interpolated from the tables of absorption data compiled by the writer †. According to Richtmyer and Warburton ‡, the coefficient k in Owen's formula for the photoelectric absorption per atom,

$$\tau_a = KZ^4\lambda^3,$$

is 0·0224 on the short wave-length side of the K absorption limit and 0·0033 on the long wave-length side. The fraction $(224-33)/224 = 0\cdot 85$ of τ is thus due to the K electrons. For absorption in the radiators here considered, the scattering term σ is less than 1 per cent. of μ and may be neglected in the calculation. We thus have,

$$\tau_K' = 0\cdot 85\mu'. \quad \ldots \quad (7)$$

In order to obtain the ratio P''/P' from the experimental data, let us assume in accord with the results of Kulenkampff §, Kircher and Schmitz ‖, and Crowther ¶ that the ionization is proportional to the energy spent in producing β-rays. Of the total energy absorbed in the methyl bromide vapour, the fraction τ/μ is spent in exciting photo-electrons (the part spent in exciting recoil electrons may be neglected). As we have seen, if the frequency is greater than the K limit of bromine, 85 per cent. of this is absorbed in the K shell, and 15 per cent. in the outer shells. If n_a is the total number of absorbed quanta, a number

$$n_f = 0\cdot 85 w_B \frac{\tau}{\mu} n_a$$

---

* On the basis of the results of Unnewehr (Phys. Rev. xxii. p. 529, 1923) and the writer (Proc. Nat. Acad. xiv. p. 549, 1928) the ratio of the a lines to the β lines has been taken as 5 : 1 for the elements used.

† A. H. Compton, 'X-Rays and Electrons' (Van Nostrand, 1926), p. 184.

‡ F. K. Richtmyer and F. W. Warburton, Phys. Rev. xxii. p. 539 (1923).

§ H. Kulenkampff, Ann. der Phys. lxxix. p. 97 (1926).

‖ H. Kircher and W. Schmitz, Zeits. f. Phys. xxxvi. p. 484 (1926).

¶ J. C. Crowther and W. N. Bond, Phil. Mag. vi. p. 401 (1928).

reappears as fluorescent K rays of frequency ν_{Br}''. The energy of the K rays escaping to the walls is thus

$$0.85 w_{Br} \frac{\tau}{\mu} n_a h \nu_{Br}'' e^{-\mu'' \bar{x}}, \quad \ldots \quad (8)$$

where μ'' is the absorption of the bromine K rays in the methyl bromide vapour, and $\bar{x}$ is the effective distance traversed by the rays before reaching the walls. An approximate calculation of $e^{-\mu''\bar{x}}$ gives 0·937. Noting that the absorbed energy is $n_a h \nu$, and that $\nu''/\nu = \lambda/\lambda''$, this means that the fraction of the absorbed energy lost as K radiation is

$$0.85 \times 0.937 w_{Br} \frac{\tau_{Br}}{\mu_{Br}} \frac{\lambda}{\lambda_{Br}''} = 0.796 \left(w \frac{\tau}{\mu} \frac{\lambda}{\lambda''} \right)_{Br}.$$

Any fluorescent L or M radiation will be absorbed before reaching the walls of the ionization chamber, and hence will appear as β-ray energy. There are, however, also scattered rays from the methyl bromide vapour which escape from the ionization chamber without producing ionization. The number of such scattered quanta is $n_s = n_a \sigma / \mu$, of which a fraction $e^{-\mu' \bar{x}}$ will escape from the chamber. Thus the total fraction of the absorbed energy which escapes from the ionization chamber is

$$0.796 \left(w \frac{\tau}{\mu} \frac{\lambda}{\lambda''} \right)_{Br} + e^{-\mu' \bar{x}} \frac{\sigma}{\mu}.$$

If we call R the ratio of the energy spent in producing ionization to the absorbed energy, we thus have, for rays shorter than the K limit of bromine,

$$R = 1 - 0.796 \left(w \frac{\tau}{\mu} \frac{\lambda}{\lambda''} \right)_{Br} - e^{-\mu' \bar{x}} \frac{\sigma}{\mu}. \quad \ldots \quad (9)$$

When λ is greater than $\lambda_{K\,Br}$ the second term in this expression becomes negligibly small.

Let i' and i'' be the ionization currents due to the primary and secondary rays respectively, f' and f'' the fractions of the two beams absorbed by the methyl bromide, and S' and S'' be the areas of the slits S_1 used in the two cases. Then,

$$\frac{i''}{i'} = \frac{P'' f'' S''}{P f' S'} \frac{R''}{R'} \frac{e^{-(\mu_a'' r + x'')}}{e^{-(\mu_a' r + x')}}.$$

Here μ_a' and μ_a'' are the absorption coefficients in air o the two beams, r is as before the distance from R_2 to the ionization chamber, and x' and x'' account for the absorption in the celluloid window. We thus have,

$$\frac{P''}{P'} = \frac{i''}{i'} \frac{f S'R'}{f''S''R''} \frac{e^{y''}}{e^{y'}}, \quad \ldots \ldots \quad (10)$$

where
$$y = \mu_a r + x.$$

Making use of equations (7) and (10), expression (6) now becomes

$$w = 14 \cdot 78 \frac{r^2}{A''} \frac{\mu' + \mu''}{\mu'} \frac{\lambda''}{\lambda'} \frac{i''}{i'} \frac{S'}{S''} \frac{f'}{f''} \frac{R'}{R''} \frac{e^{y''}}{e^{y'}}. \quad . \quad (11)$$

The Measurement of Relative Intensities for different Wave-lengths.

It will be seen that an expression similar to equation (10) affords a means of comparing the power in two X-ray spectrum lines of different wave-lengths in terms of the ionization resulting from them. The appropriate form of the expression is

$$\frac{P_1}{P_2} = \frac{i_1 f_2 R_2}{i_2 f_1 R_1} \frac{e^{-y_2}}{e^{-y_1}}, \quad \ldots \ldots \quad (12)$$

where the significance of the various terms is as in equation (10). This expression, of course, does not correct for the difference in reflecting power of the grating for the two wave-lengths.

Evaluation of the Fluorescence Yield.

In the present experiment, referring to equation (11), $r = 15 \cdot 78$ cm., $A'' = 7 \cdot 70$ cm.^{2}, and the other quantities are given in the following table:

TABLE V.

Experiment.	λ'.	λ''.	$\frac{\mu'+\mu''}{\mu'}$.	$\frac{i''}{i'}$.	$\frac{S'}{S''}$.	$\frac{f'}{f''}$.	$e^{(y''-y')}$.	$\frac{R'}{R''}$.	w.
1. Sn→Mo	·482	·696	1·425	·370	·00374	·440	1·016	1·136	·69
2. Ag→Mo	·549	·696	1·290	·352	·00374	·588	1·013	1·093	·67
3. Sn→Se	·482	1·085	2·24	·727	·000472	·742	1·091	·788	·53
4. Mo→Se	·696	1·085	1·427	·866	·000472	1·69	1·073	·694	·55
5. Sr →Se	·859	1·085	1·334	·970	·000472	2·33	1·048	·622	·56
6. Sn→Ni	·482	1·629	4·69	·261	·000472	·343	1·316	·788	·33
7. Mo→Ni	·696	1·629	2·30	·441	·000472	·781	1·299	·694	·38
8. Sr →Ni	·859	1·629	1·75	·579	·000472	1·08	1·267	·622	·37
9. Se →Ni	1·085	1·629	1·40	1·39	·000472	·463	1·210	1·000	·37
10. Zn→Ni	1·410	1·629	1·292	1·42	·000472	·805	1·102	1·000	·42

The diaphragm apertures S' and S" can be measured directly. The ratios i''/i' are the ratios of ionization currents as read from the electrometer, except that corrections have been applied to take account of the scattered primary rays mixed with the fluorescent rays from R_1 and also for the scattered rays mixed with the fluorescent rays from R_2. These corrections were based on absorption measurements of the mixed rays. Only in the cases of tin rays falling on nickel and of the primary rays falling on zinc were these corrections large—that is, in cases 6 and 10. The absorbed fractions f' and f'' were measured directly by means of the auxiliary ionization chamber I_2. The absorption coefficients $\mu a'$ and $\mu a''$ used in calculating y' and y'' were interpolated from a table given by Siegbahn *.

In order to evaluate R'/R'', it was at first assumed that, since the atomic numbers of selenium and bromine are nearly the same, the fluorescence yield w will be the same for both. Expressing R in terms of w, as in equation (9), equation (11) can thus be solved for w in experiments 3, 4, and 5, giving the mean value $w = 0.549$. In making this calculation, since under the conditions of these experiments σ was less than 1 per cent. of μ, the last term in equation (9) was neglected. Thus for wavelengths greater than the K limit of bromine R may be taken to have the value 1. Using this value 0.549 for the fluorescent yield in bromine, provisional values of R for the other cases could be calculated, and provisional values of w determined for molybdenum and nickel. This showed the rate of change of w with atomic number, and indicated that if 0.542 is the mean value of w for bromine and selenium, its value for bromine should be .565. Using this value for w_{Br}, and taking $\tau = \mu$, the values of R'/R'' shown in column 9 are calculated from equation (9). From equation (11) we then obtain the values of w given in the last column.

Discussion of Results.

There is no significant variation in the values of the fluorescence yield w with the wave-length of the exciting radiation. The only apparent departure from this statement is for the nickel radiator when excited by rays from

* M. Siegbahn, 'Spectroscopy of X-Rays,' p. 246.

tin and from zinc. As we have noted, however, in these two cases the corrections due to the presence of scattered X-rays were so large as to make the results considerably less reliable than in the other cases. The experiments thus indicate (in support of the conclusions of all previous investigators) that the fluorescent yield is a constant characteristic of the radiator, but independent of the exciting radiation.

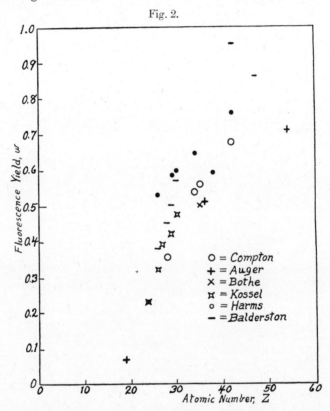

Fig. 2.

If we had assumed that the number of ions produced per unit energy by the β-rays in the ionization chamber were a function of the wave-length of the incident X-rays, as was assumed by Kossel and Harms, this independence of w of the exciting wave-length would not have appeared. For the efficiency of ion production by the fluorescent rays from any one radiator would have remained constant, while that due to the exciting rays would on this assumption have varied with their wave-length, resulting in a corresponding variation in w. In so far as the values of w here obtained are constant for a given radiator R_2 therefore, this work supports the conclusions reached by Kulenkampff, Kircher and Schmitz, and Crowther that the ionization by β-rays per unit energy is independent of their energy.

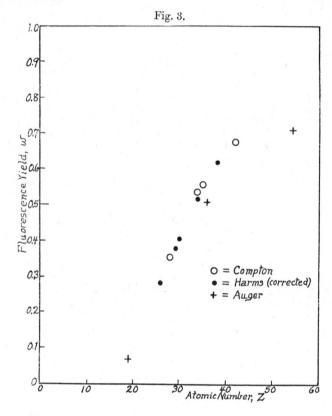

Fig. 3.

In fig. 2 are plotted the values of the fluorescence yield given by the different investigators. Of these, as we have seen, Auger's values have been obtained by the direct method of comparing the number of single photoelectrons with the number of compound photoelectrons. The results of Bothe, Kossel, and Balderston can be considered only as rough approximations, and Harms's values should be corrected for certain unjustified assumptions in the calculations. In fig. 3 Harms's values of w as thus corrected are compared with those of Auger and the writer. Their agreement with the writer's results is seen to be good.

If the assumption of Auger and Kossel is correct, that the compound photoelectric effect is the whole explanation of the fact that the fluorescence yield is less than unity, Auger's values should fall on the same curve with those of Harms and the writer. Though this agreement is not exact, it is probably within the limits of experimental error. The present work thus confirms Auger and Kossel's theory that the compound photoelectric effect is responsible for the fact that the fluorescence yield is less than unity.

It will be noted that these experiments show a rapid increase of the fluorescence yield with increasing atomic number. For elements as light as aluminium, fluorescence should thus be alm non-existent. On the other hand, for a heavy element such as iodine, a larger portion of the absorbed energy is spent in producing fluorescence and a smaller portion in producing ionization. This gives a lighter element, such as argon, a certain advantage for use in an ionization chamber, though in this case it is, of course, more difficult to secure complete absorption.

Of practical significance is the fact that, having a knowledge of the fluorescence yield, it is now possible to make a reliable calculation of the relative intensity of X-ray beams of different wave-lengths, in terms of the ionization which they produce. The necessary formula is given in equation (10).

SUMMARY.

A review of the published data describing the intensity of fluorescent X-rays shows that the various observers agree that the number of quanta of fluorescent X-rays emitted by a radiator is considerably less than the number of quanta which it absorbs from the primary beam. The values of the efficiency of fluorescence that appear in the literature are for the most part, however, found not to be quantitatively reliable.

New experiments give values of the "fluorescence yield," or ratio of the number of fluorescent K quanta to the number of photoelectrons ejected from the K shell, of 0·68 for a molybdenum radiator, 0·56 for bromine,

0·54 for selenium, and 0·37 for nickel. The values seem to be independent of the wave-length of the exciting rays.

These measurements agree within experimental error with Auger's values of the fluorescent yield, based on his count of the frequency of occurrence of the compound photoelectric effect, and thus indicate that it is this effect which makes the fluorescent yield less than unity.

An expression is derived for calculating the relative intensity of two X-ray beams of different wave-length, in terms of the ionization currents obtained and other factors.

Ryerson Laboratory,
University of Chicago,
December 13, 1928.

What Things Are Made Of—I

The New Physical Concept of the Universe. It is a Composite of Three Things—Protons, Electrons, and "Photons"

By ARTHUR H. COMPTON, Ph.D.
*Professor of Physics, University of Chicago.
Member American Philosophical Society; National Academy of Sciences;
National Research Council. Nobel Physics Prize Winner, 1927*

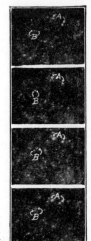

MOLECULAR JAZZ

Figure 1: Four successive "movie" pictures of oil particles suspended in water, magnified 2700 diameters

WHO is there who has not asked himself, "What is this world around me?" Rocks, trees, people—what are the parts of which they are made, and how are these parts put together? Light, radio waves, X rays—what are these radiations?

Upon such questions men have pondered since the beginning of thought. Until recently, opinions on these matters were speculative; but during the last heroic generation many of these questions have been answered with a definiteness which was unthinkable in an earlier era. It will not be possible in this brief survey to present adequately the great mass of evidence which has been collected regarding the nature of things. We shall rather discuss only a few typical experiments that tell us something about how things are made

When we take apart this infinitely complex mechanism which we call dirt, or perhaps a diamond, or it may be a flower, we find it made up of a myriad of tiny molecules. Each of these molecules is itself complex but is more perfectly formed than the wheels of a watch, and has continued to run for billions of years without winding and without wear. We find that the molecules which make up matter in all its endless variety of forms are themselves built up of a few hundred kinds of atoms. Some of these atoms differ from each other by their chemical properties, others only in one respect, by their weights.

But we cannot stop here. Our few hundred atoms are themselves made of yet more tiny particles. It is as sung by the poet Pope:

"The larger fleas have smaller fleas
  Upon their backs to bite 'em.
These smaller fleas have other fleas,
  And so ad infinitum."

These tiny particles show themselves through their electric charges. There are found to be two kinds of them, carrying positive and negative electric charges, respectively. But as nearly as we can tell, all those of any one kind are exactly alike. The positively charged particles, called *protons*, have most of the weight of the atom, while the negatively charged particles, or *electrons*, are the lively little bodies responsible for chemical combinations, electrical conductivity, and the like. By grouping themselves in various ways these little particles form the various atoms.

ROCKS and trees and people do not, however, make up the whole of the universe. How about the sunlight that makes the trees grow and gives us warmth? Newton wrote of light as consisting of little particles or corpuscles. Huyghens and Fresnel explained its properties in terms of waves, and Maxwell, by introducing the idea that these waves are electromagnetic, foreshadowed the recent developments in X rays and radio. But recently Einstein has brought back a modified conception of Newton's light corpuscles, now called *photons*, and evidence for their existence has become convincing.

At the present moment, light is the darkest of the physicist's problems. It seems that we can prove that it consists of waves; but we can also show that it consists of little particles. The two conclusions cannot really be inconsistent; but how light can consist of waves and at the same time of tiny particles is as yet an unsolved riddle.

Just a century ago a botanist named Brown was examining some tiny objects suspended in water under a high power microscope. "They're

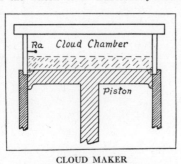

CLOUD MAKER

Figure 2: When piston is pulled down, the air in the chamber expands and cools

alive," he exclaimed. But to make sure, he turned his glass on some wax particles suspended in the same way, and found that they also moved. Everything that is small enough moves, he found, unless it is held fast to something solid. These are the "Brownian movements." It is as if a swarm of little "Brownies" were pushing the globules hither and thither. In Figure 1 is shown a series of moving pictures of these motions, taken with the ultra-microscope. Notice the motion of the particles A and B. The smaller particles move faster than the larger ones. Under the microscope the Brownian motion has the appearance of a jazz dance on a crowded floor.

WHAT is the cause of this motion? For many years physicists had explained the pressure of a gas on the walls of a vessel which holds it by supposing that the gas is made of little particles, the "molecules," which dart about in the gas at high speed. They could calculate how fast these molecules must go to account for the observed pressure, and found that if the energy of their motion was proportional to the temperature, they could account accurately for the increase of pressure with temperature. The calculation showed too that the energy of motion of the little molecules should be the same as that of the big molecules, which would mean that the little ones must move faster.

Now a careful study of these Brownian movements, such as those shown in Figure 1, reveals the fact that the energy of motion of these particles under the microscope is just what the kinetic theory says a molecule should have. And the speed of their motion increases at higher temperatures, just as the molecular theory predicted. At one instant more molecules strike a globule on one side, and the next instant more strike on another. In

Reprinted with permission. Copyright © 1929 by Scientific American, Inc. All rights reserved.

fact, in a very real sense, one may correctly consider the motion of these little particles as true molecular motions. We must remember, however, that these globule "molecules" consist of perhaps a million atoms each.

Such things as this show us that our idea of "dead matter" is indeed far from the truth. We have a glimpse of the continual activity in things we call dead.

Not long ago a group of us were camping beside a frozen lake in the Himalaya Mountains, 13,000 feet above the sea. The lake was surrounded on three sides by mountains. At about the middle of the day, when the wind was blowing over the mountains, one could see clouds gathered about their peaks, clouds which evaporated almost as soon as they were formed. On one side of the mountains the clouds could be seen rising with the air, and becoming more dense as the air expanded at the higher levels. On the other side, where the air was coming down, the clouds, being compressed at the lower levels, gradually evaporated into thin air. The air was so cooled by expansion when it was rising that the moisture condensed into clouds, which would then evaporate as soon as the air sank to a lower level and became warmed by compression.

Dr. C. T. R. Wilson tells me that it was while watching a phenomenon of this kind in his native hills of Scotland, that he conceived the idea of making artificial clouds by allowing moist air to expand. This study quickly revealed the fact that it is very difficult to make a cloud form unless there is something in the air, such as dust particles, which may act as nuclei for condensation. Even at a high mountain peak, the fact that a cloud forms indicates that there are some dust particles, although perhaps excessively minute, floating in the clear blue air.

IT is possible, however, to remove the dust from air. The air may be filtered through cotton, or if the water droplets in the cloud are allowed to fall as rain they carry the dust particles down with them. When the air has thus been cleaned, Wilson found that the cloud would condense on ions if any were present. Ions are the electrically charged pieces of broken atoms or molecules. The "alpha rays" from radium were known to produce ions in the air through which they pass. So Wilson tried the experiment of making clouds condense on the ions produced by alpha rays.

The apparatus which he used in this experiment is shown diagrammatically in Figure 2. There is a chamber with glass top and sides, which is closed at the lower end by a movable piston. This chamber is filled with air, saturated with moisture. The piston is suddenly moved downward to give the proper degree of expansion, but no cloud forms, because the air has been carefully freed from dust. Then alpha rays are allowed to shoot from a speck of radium placed in the chamber, and a cloud forms on the ions produced.

Two photographs which Wilson took of the clouds thus formed are shown in Figure 3. Instead of a diffuse cloud surrounding the radium, we see a group of sharply defined lines radiating from the source. We see that the alpha rays do not spread uniformly in all directions but are concentrated along definite, straight lines. The interpretation is clear. These rays are little particles which shoot through the air at such high speed that they break in pieces the molecules through which they pass, leaving ions to mark their trails. The ions are in turn made visible when droplets of moisture condense upon them. Each of these linear clouds thus marks the path along which traveled one of the particles ejected from the speck of radium. For want of a better name we shall call the thing which is responsible for such a track an "alpha" particle.

BUT what are these alpha particles? The question has been answered in a striking manner by Rutherford. He surrounded a strong radium preparation by a glass vessel with walls so thin that the alpha particles could pass through. After a long period of time he found that a gas was accumulating where the alpha particles were collecting. This gas was compressed into a small tube and its spectrum examined. Its spectrum was found to be exactly like that of helium gas. In other words, these alpha particles were found, when collected in large numbers, to constitute helium. We may call them, then, *atoms* of helium.

Using this method of collecting alpha particles or helium atoms it is possible to estimate in a very direct manner the number of atoms in a given volume of gas. We can count directly the number of alpha particles ejected from a weak preparation of radium. This might be done by counting the trails of the alpha particles in photographs of the type shown in Figure 3.

A BETTER method, however, is to count them electrically, using a sensitive electroscope which records each alpha particle as it enters a small chamber. Figure 4 shows a record which Rutherford obtained in this manner of the alpha particles escaping from a weak source of radium. Of course the number of alpha particles is proportional to the amount of radium, so by comparing the strength of the radium preparations the number of alpha particles in the gas collected in the spectrum tube can be estimated.

The number thus determined is a perfectly tremendous one. It means little to say that in a thimble full of helium there are atoms to the number of 3 with 19 ciphers after it. Let us suppose that we could paint each of the molecules in a thimble full of water

MORE ARTIFICIAL CLOUDS

Figure 3: Two photographs of the cloud formed in air ionized by alpha rays shot out from radium (Wilson)

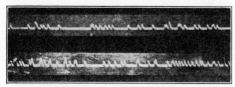

COUNTING ALPHA PARTICLES

Figure 4: A record of the alpha particles entering a counting chamber connected to an electroscope (Rutherford)

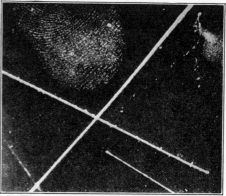

From Eddington, "Stars and Atoms". Courtesy of Yale University Press

TWO CLUES

Figure 5: Would you deny the existence of the maker of the finger print because the print was all you could see?

THREE TRAILS

Figure 6: Trails of an alpha particle (helium atom) and two beta particles (that is, electrons) shot out from radium

green. Let us now spill this water on the ground and wait for thousands of years until it has flowed to the sea, has re-evaporated and rained over the earth so that our water molecules are distributed all over the world. Then wherever we should go each drop of water that we should examine would probably have in it one of the green molecules from the original thimble.

But perhaps you will say, "Show us the atom and we will be satisfied." Let me ask you, then, what is the spot in the corner of Figure 5? If you answer, "It is a man's finger," I shall reply, "Then the heavy white lines across the paper are helium atoms."

BUT if you are more cautious and tell me that the spot is the print of a man's finger I shall be equally careful and state that the white line is merely the print left by the helium atom.

The story is told that toward the close of the 19th Century one of Lord Kelvin's students approached him with the question, "What do you think of this new theory that the atom has structure?" "What!" said Kelvin, "the atom has a structure? The very word 'atom' means the thing that can't be divided. How then can it have a structure?" The insubordinate youth is said to have replied, "That shows the disadvantage of knowing Greek."

In Figure 6 is shown the trail of an alpha particle, and above it two fainter trails, one crooked and the other straight. From their appearance one would guess that these fainter trails are due to objects much smaller than the alpha particle. But the alpha particle itself, as we have seen, is an atom of helium, which is next to the smallest atom that we know. What then can these other particles be? Let us call them "beta particles" to avoid any suggestion as to their nature. In Figure 7 are shown such beta particles knocked out of air by the action of X rays. It is found that they can be ejected from anything. They are, that is, a common component of all different forms of matter of whatever nature.

The photograph shown in Figure 8 was taken with a strong magnetic field applied to the expansion chamber when a cloud was formed. It will be seen that the trails of the beta particles are curved in circles. Such a bending of paths is just what we should expect if the moving particles were electrically charged. For a magnetic field produces a force on a moving electric charge just as it does on a wire carrying an electric current. From the direction of the curvature one can show that the charge carried by the particle is negative. If we can measure the magnitude of this charge it will be possible to tell from the curvature of the trails what is the mass of the beta particles.

No one has been able to devise an electroscope sensitive enough to measure the very minute charge carried by one of these beta particles. But Millikan has measured their charge in a most interesting manner. He

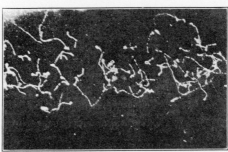

RESULT OF COLLISIONS

Figure 7: Beta particles (electrons) produced by X rays passing through air and knocking them out of atoms

caught a tiny droplet of mercury between the electrified plates of a condenser and threw a beam of ultra-violet light on the droplet. The effect was to eject a beta ray from the droplet. The droplet was left with a positive charge equal to the negative charge carried away by the beta particle, and the electric field of the condenser could be so adjusted that the force on the droplet's charge would just balance its weight. Knowing the weight of the droplet, it was thus possible to determine its charge. It was found that each beta ray which left the droplet carried with it the same negative charge, a charge equal in magnitude to that carried by a hydrogen ion in electrolysis.

Having thus found that the charge carried by a beta particle is the same as that carried by a hydrogen ion, magnetic deflection experiments such as those shown in Figure 8 indicate that the mass of the beta particle is only 1/1845 that of a hydrogen atom. We were thus right in our guess that these beta particles are things much smaller than the smallest known atom. Since we find that such particles can be removed from every kind of matter, it follows that they must be one of the components of which atoms themselves are built. We shall now give to these beta particles the name of *electrons*.

IT is not possible to suppose that atoms are constituted solely of electrons, for electrons have negative charges, and atoms are electrically neutral. The atoms must, therefore, have some positive charge as an essential part of their structure. An examination of the trails of alpha particles such as those shown in Figures 9 and 10 give an important clue to the distribution of the positive charge within the atom. Let me call attention to the pair of tracks in Figure 9, one of which goes nearly straight and the other of which has two rather sharp breaks. It is really a most remarkable thing that this one track is so straight, since a simple calculation shows that in the portion of the track shown in this figure the alpha particle has passed through some 20,000 atoms. But we have seen that the alpha particle itself is nothing but a helium atom with a double positive charge. This photograph therefore means that while the helium atom is passing through the oxygen or nitrogen atoms of air, we have repeated 10,000 times the most unusual phenomenon of two bodies occupying the same space at the same time.

Occasionally, however, the alpha particle strikes something so hard and immovable that it must change its course. From the fact that there is only about one such collision for every 10,000 atoms traversed, it is clear that the object struck is much

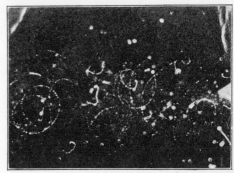

CURVED PATHS

Figure 8: A magnetic field attracts a moving electric charge, whether it moves in a wire or in free space

smaller than the atom itself. Further, the manner in which the impinging helium atom glances off shows that the object is heavier than an atom of helium. It is not, however, a collision with a stone wall. You see in Figure 9 that the particle struck by the alpha ray itself rebounded and left a little track.

Figure 10 shows a similar event more clearly. From the relative length of these two tracks it seems probable that the object struck has a mass several times that of the helium atom, or about that of the oxygen atom. Collisions of this character indicate that there must be something within the atom hard and impermeable, very much smaller in size than the atom itself, yet possessing practically the whole mass and weight of the atom. This something, whatever it may be, has been named the atomic *nucleus*.

IT has been shown by Rutherford that the atomic nucleus deflects an alpha particle as if the force between them were one of repulsion between two electric charges. On this view the paths of alpha particles passing a nucleus should be as shown diagrammatically in Figure 11. All of the particles are bent slightly, but only those coming close to the nucleus are bent through a large angle. It is obvious that if the charge on the nucleus is large its effect will extend to a greater distance. That is, the nucleus will act as a larger obstacle and the number of collisions will be

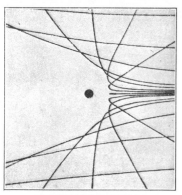

THEORETICAL PATHS

Figure 11: Reflections of charged particles shot toward a charge of the same sign

greater. So by counting the number of collisions occurring when a group of alpha particles passes through a known number of atoms we can determine the charge on the nucleus. Measurements of this kind have shown that the nucleus of the hydrogen atom has a positive charge equal to that of one electron, helium that of two, lithium three, and so on down the list of chemical elements to uran-

A PARADOX?

Figure 9: The "paradox" is stated in the text. The answer is: the atoms consist mainly of emptiness

ium with a charge equal to 92 electrons on its nucleus.

This discovery suggests that the nucleus of the atom may be built up of units carrying a positive charge equal to the negative charge of the electron. Such a unit we find in the nucleus of the hydrogen atom. It is perhaps surprising that the positive unit of electric charge should be associated with a mass almost 2,000

TWO CARROMS

Figure 10: Double photograph of a collision of alpha particle with oxygen atom

times as great as that associated with the negative unit.

Rutherford has, however, performed a series of experiments which gives good reason to believe that our guess is correct. These experiments consist in shooting alpha rays from radium through various substances. It is found that particles having the same charge and mass as the hydrogen nucleus can be knocked out of some of the lighter elements. An event of this kind is shown in Figure 12, a remarkable photograph taken by Mr. Blackett. We see here the impact of an alpha particle with the nucleus of a nitrogen atom. There is a thin trail left by the hydrogen nucleus escaping from the nitrogen nucleus.

Similar results are obtained when alpha particles traverse boron, aluminum, phosphorus and certain other light elements. It would seem that the only reason that hydrogen cannot

MODERN ALCHEMY

Figure 12: A double photograph of an alpha particle knocking a hydrogen nucleus (proton) out of a nitrogen atom (Blackett photo)

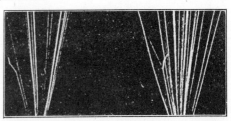

be liberated from other elements by such methods is that our hammer, the alpha particle, does not strike a sufficiently powerful blow.

The evidence thus seems very strong that the nuclei of the various atoms are indeed built up of an aggregate of hydrogen nuclei, which we shall now call *protons*, cemented together with electrons. In the case of oxygen, for example, the nucleus presumably consists of 16 protons, corresponding to the fact that the oxygen atom weighs 16 times as much as hydrogen. To hold these protons together, there will be eight electrons, leaving a resultant positive charge on the nucleus of eight units.

IT is this nuclear charge of eight electronic units which is detected by the alpha particle deflections, such as those in Figure 10. Around this nucleus, arranged as a sort of atmosphere, will be distributed eight more electrons, making a neutral atom. These outer eight electrons are responsible for the ordinary physical and chemical properties of oxygen.

What Things Are Made Of—II

The Paradox of Light; Einstein and the Photoelectric Effect; Peculiar X-Ray Echoes; Photons and Electrons; The Paradox of Particles and Waves

By ARTHUR H. COMPTON, Ph.D.

Professor of Physics, University of Chicago.
Member American Philosophical Society; National Academy of Sciences;
National Research Council. Nobel Physics Prize Winner, 1927

(Concluded from February)

THE tangible things with which we are acquainted, rocks and mountains, trees and people, are built of atoms and electrons and protons. In stating the subject of this article, however, the word "things" was chosen because it included an important something which we are apt to overlook. I refer to

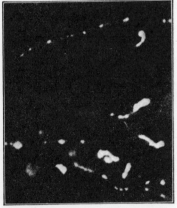

ELECTRON TRAILS

Figure 13: Beta particles recoiling from the impact of X-ray particles or "photons"

light and similar radiations such as wireless waves and X rays which may be grouped with light under the term "radiations." What is light?

As long ago as the 17th Century, Newton defended the view that light consists of little particles shot with tremendous speed from a candle or the sun or any other source of light. At the dawn of the 19th Century, however, experiments were performed which gave very direct evidence that light consists of waves. Maxwell interpreted them as electromagnetic waves and in such terms we have ever since been explaining light rays, X rays and radio rays. We have measured the length of the waves, their frequency, and other characteristics and have felt that we know them intimately. Very recently, however, a group of electrical effects of light has been discovered for which the idea of light waves gives no explanation, but whose interpretation is obvious according to a modified form of Newton's old theory of light particles.

The evidence that these radiations consist of waves is so familiar that I shall not take time to outline it. Let me, rather, present some of the reasons why we feel that light consists of corpuscles.

When a beam of light falls upon the surface of certain metals, such as metallic sodium, electricity in the form of electrons is found to be emitted from the surface. This effect is especially prominent with X rays, for these rays eject electrons from all sorts of substances, as was shown for example in Figure 7 (see February issue). X rays are produced when a stream of electrons hits a block of metal inside an X-ray tube. It is as if one were shooting at a steel plate with a rapid fire gun. The stream of bullets represents the electrons and the racket produced when the bullets strike the steel plate corresponds to the X rays emitted at the metal target inside the tube.

LET us suppose that an electron strikes the target of an X-ray tube at a speed of say 100,000 miles a second. (These little particles certainly move tremendously fast!) The X ray produced by this electron may be knocked out of the metal, and the speed of this electron will be almost as great as that of the original electron which gave the rise to the X ray.

The surprising nature of this phenomenon may be illustrated by an experience which I had when camping in my early boyhood. My older brother, with several of the older boys, built a diving pier around the point a half mile away from camp, where the water was deep, while we younger boys built a diving pier of our own in the shallower water near the camp. One hot, July day my brother dived from his diving board into the deep water. By the time the ripples from the splash had gone around the point to where I was swimming a half mile away, they were of course much too small to notice. You can imagine my surprise therefore when these insignificant ripples, striking me as I was swimming under our diving pier, suddenly lifted me bodily from the water and set me on the diving board!

Does this sound impossible? It is no more impossible than for an ether ripple sent out when an electron dives into the target of an X-ray tube, to jerk an electron out of a second piece of metal with a speed equal to that of the first electron.

CONSIDERATIONS of this type showed to Einstein the futility of trying to account for the photoelectric effect on the basis of waves. He suggested, however, that this effect might be explained if light or X rays moved in particles. These particles we now call "photons." The picture of the X-ray experiment on this view would be that when the cathode electron strikes the target of an X-ray tube, its energy of motion is trans-

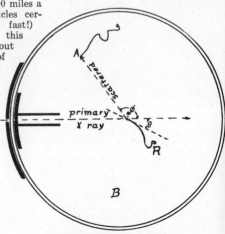

TRAILING AN X RAY

Figure 14: Photographing two electrons, R and A, struck by an X ray. See the text

rmed into a photon, that is, a particle of X rays which goes with the speed of light to the second piece of metal. Here the photon gives up its energy to one of the electrons of which the metal is composed, and throws it out with an energy of motion equal to that of the first electron.

Thus Einstein was able to account in a very satisfactory way for the phenomenon of the emission of photoelectrons. But his theory had been devised for just this purpose. It was not surprising that it should work well for this one fact. It would naturally carry much greater weight if it could be shown that his theory accounted for other facts for which it had not been originally intended. This is what it has recently done in connection with certain properties of scattered rays.

If you hold your hand in the light of a lamp, your hand scatters light from the lamp into your eyes. This is the way by which your hand is made visible. In the same way, if the lamp were an X-ray tube, your hand would scatter X rays to your eyes. If you had a blue light in the lamp, your hand would appear blue. If the light were yellow, your hand would appear yellow and so on. But some five years ago, we noticed that when one's hand or anything else scatters X rays, the "color" or wavelength of the rays is changed. The corresponding effect with light would be for one's hand to appear green when illuminated with a blue light, to appear yellow when illuminated with green light, red when lighted by a yellow lamp and so on. This change in wavelength is now known to physicists as the "Compton Effect," after the author of the present article.—Editor.]

If X rays are waves, scattered X rays are like an echo. When one whistles in front of a barn, the echo comes back with the same pitch as the original tone. This must be so, because each wave of the sound is reflected from the barn, and as many waves return as strike, so the frequency or pitch of the echoed wave is the same as that of the original wave. Similarly an X-ray echo should be thrown back by the electrons in the scattering material, and should have the same pitch or frequency as the incident rays. Thus the wave theory does not account for the lowered pitch which the scattered X rays are found to have.

The corpuscular idea revived by Einstein suggests, however, a simple explanation of this effect. On this view, we may suppose that each X-ray photon is deflected by a single electron, just as, for example, a golf ball might bounce from a football.

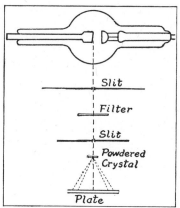

HULL'S APPARATUS

Figure 16: Used for performing X-ray diffraction experiments with crystals. Thomson uses an analogous apparatus.

The football will recoil from the golf ball, and part of the energy is spent in setting the football in motion. Thus the golf ball bounces off having less energy than when it struck. In the same way, the electron from which the X-ray photon bounces will recoil, taking part of the photon's energy, and the deflected photon will have less energy than before it struck the electron. The reduction in energy of the X-ray photon corresponds on Einstein's view to a decrease in frequency of the scattered X rays, just as the experiments show. In fact, the theory is so definite that it is possible to calculate just how great a change in pitch should occur, and the calculation is found to correspond accurately with the experiments.

Within a few months after this new theory of the scattering of X rays had been proposed, Dr. C. T. R. Wilson was able to photograph the trails left

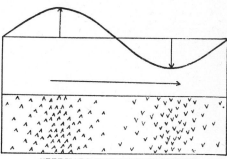

"PERHAPS THE BEST PICTURE"

Figure 15: Sheets of photons may make up electromagnetic waves. The wave is represented above; the sheets of photons corresponding to it are below

when electrons in air recoiled from X rays which they scattered. In Figure 13 there are two high-speed beta particle trails at top and bottom. These are photo-electrons. Between them are shorter trails left by electrons recoiling from X rays which they have scattered. Notice how those which are knocked straight ahead go farther than those which received a glancing blow. Thus we have observed not only the loss in energy of the deflected golf balls, but also the footballs, or electrons, from which they have bounced.

Finally it was found possible to follow not only the electron which recoiled from the impact of the X ray, but also the path of the deflected X-ray particle as it bounced from the electron. Through the air in a cloud expansion chamber (Figure 2, February issue) was passed a beam of X rays so faint that we would find only one or two electrons recoiling from the scattered X rays, as in Figure 14.

NOW if the recoiling electron moves downward, the X-ray particle must have glanced upward, just as when the golf ball bounces to one side the football recoils to the other. If, however, the scattered X ray goes as a wave, spreading in all directions, there is no more reason to expect it to affect a second electron on one side than on the other. The photographs show that the second electron struck by the scattered X ray is on the side corresponding to A of Figure 14, opposite to the direction of recoil of the first electron. An X ray is thus scattered in a definite direction, as it should be if it is a particle.

But if X rays consist of particles, so also must light and heat rays, for they are all the same kind of thing. For centuries it has been thought that the corpuscular and wave conceptions of the nature of light are contradictory; but when we are confronted with apparently convincing evidence that light consists of particles, the two conceptions must in some way be reconcilable. The theoretical physicists are hard at work on a reconciliation of the two theories. One suggestion is

DIFFRACTION OF X RAYS

Figure 17: Pattern obtained by passing a pencil of X rays through powdered aluminum crystals (Hull). See Figure 18

DIFFRACTION OF ELECTRONS
Figure 18: Diffraction pattern obtained by passing a pencil of electrons through powdered crystals of gold. See Figure 17

that the energy of radiation is carried by the particles and that the waves serve merely to guide the particles. According to the second view, the particles of radiation exist in any true sense only when the radiation is acting on atoms or electrons, and that between such events the radiation moves as waves. These ideas, however, are difficult to state in any satisfactory form.

Perhaps the best picture that one can give of the relation between waves and particles is the analogy of sheets of rain which one sometimes sees in a thunderstorm. We may liken the waves to the sheets of rain that one sees sweeping down the street or across the fields. The radiation particles or photons would correspond to the rain drops of which the sheet is composed. Such an idea is pictured in the diagram, Figure 15.

THIS conception is probably fairly accurate when we are thinking of radio rays. For in the case of radio rays, even a feeble signal, such as one broadcast from Los Angeles and heard in New York, would have waves consisting of thousands of photons per cubic inch. But in the case of X rays, a single photon carries enough energy to detect—and one particle is difficult to arrange in sheets.

The fact remains that the evidence before us seems to demand that light and other forms of radiations consist both of waves and of particles.

If then, light, which has long been known as waves is now found to consist of particles, may it not be that such things as atoms and electrons which have long been known as particles may have the characteristics of waves? Thus reasoned the French physicist de Broglie. He went so far as to calculate what the wavelength of an electron should be when moving at a certain speed. The calculation indicated that the wavelength of an electron moving at moderate speed is about the same as the wavelength of an X ray.

Now it is not many years since the wave characteristics of X rays were demonstrated by finding that they may be diffracted by crystals. De Broglie's suggestion was accordingly tested during the last year by two Americans, Davisson and Germer, by diffracting an electron stream from a crystal in the same way. They found in these experiments the same kind of interference effects that Laue and the Braggs had observed with X rays.

About a year ago I had the pleasure of calling on Sir J. J. Thomson, who did so much to establish the corpuscular nature of the electron. His son, G. P. Thomson, was home on a visit, and was telling his enthusiastic father and myself of his new experiments on the diffraction of electrons. This experiment was the analogue of Hull's powder method of diffracting X rays by crystals.

HULL'S apparatus may be illustrated by Figure 16. A beam of rays passes through a pair of slits and traverses a mass of powdered crystals which throw a diffraction pattern on the photographic plate. Using a beam of X rays of a definite wavelength traversing a mass of aluminum crystals, Hull obtained the photograph shown in Figure 17.

In Thomson's experiment, the X-ray tube was replaced by the cathode of a vacuum discharge tube. The electrons in the cathode ray stream were shot through a thin sheet of gold leaf (which replaces the powdered crystals in the X-ray experiment) and then fell on the photographic plate. Figure 18 shows the result. By the close similarity between Figures 17 and 18, Professor G. P. Thomson was able to convince his father, as well as the rest of us, that we now have precisely the same kind of evidence for believing in the wave characteristics of electrons that we have for believing in the wave characteristics of X rays.

Our paradox of waves and particles is thus not confined to the nature of light, but applies to electrons as well. Atoms and molecules are now also being treated as complex bundles of waves. Light which we have long thought of as waves has the properties of particles; and electrons which Figures 7 and 8 show so clearly as particles have the properties of waves. There seems to be a dualistic aspect to these fundamental entities. The distinction between the conceptions of waves and particles may not be as sharp as we have thought.

HOW then does the matter stand? The tangible objects with which we are familiar we find constituted of molecules. These, in turn, are composed of atoms and these of positively charged and massive protons and the negatively charged and mobile electrons. The light which makes plants grow and which gives us warmth has the double characteristics of waves and particles, and is found to consist ultimately of photons.

Having carried the analysis of the universe as far as we are able, there thus remain the proton, the electron, and the photon—these three. And, one is tempted to add, the greatest of these is the photon, for it is the life of the atom.

We sometimes think of standardization as being the distinctive keynote of modern industry. But even a Ford car has hundreds of parts that differ from each other. What, then, shall we say of the Workman who by using only three different parts, protons, electrons, and photons, has made a universe with its infinite variety of beauty and life?

THE END

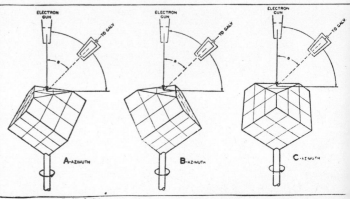

Courtesy Journal of The Franklin Institute

APPARATUS FOR INVESTIGATING ELECTRON DIFFRACTION

Electrons from a tungsten filament are accelerated by an electron gun; they strike a crystal of nickel and are received by an adjustable collector. The three positions permit measuring intensity of scattering of electrons not only at various angles but three azimuths, derived from the molecular structure of the crystal. An experiment performed by Davisson and Germer

COMPSA—COMPTON EFFECT

COMPTON EFFECT. The Compton Effect is the change in quality of a beam of X-rays when it is scattered. Imagine

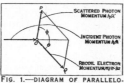

FIG. 1.—DIAGRAM OF PARALLELOGRAM OF MOMENTA IN COMPTON EFFECT

that a piece of paper when held between the eyes and a green light appears green, but that when the paper is moved to a position at right angles with the light its colour changes to yellow, and when turned to the opposite side from the light its colour becomes red. Such a change in colour would correspond to the increase in wave-length which X-rays undergo when they are scattered, a small change when scattered at a small angle, but a larger difference for the rays scattered at a large angle. This phenomenon owes its chief interest to the fact that it indicates a corpuscular structure for X-rays.

History.—The earliest experiments on secondary X-rays showed a difference in the penetrating power of the primary and the secondary rays. Barkla and his collaborators found (1908) that the secondary X-rays from heavy elements consist mostly of fluorescent radiations characteristic of the radiating element, and that it is the presence of these fluorescent rays which is chiefly responsible for the greater absorbability of the secondary rays. Later experiments, showed, however, a measurable difference in penetration even for the rays coming from light elements, such as carbon, from which no such fluorescent rays are emitted. It was established by J. A. Gray (1920) that in such cases the change in quality was an accompaniment of the process of scattering or diffuse reflection of the primary X-rays. A spectroscopic study of the scattered X-rays by A. H. Compton (1923) revealed the fact

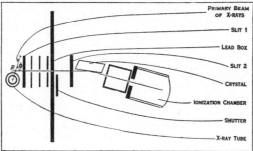

FIG. 2.—DIAGRAM ANALYZING THE SPECTRUM OF THE RAYS

that different primary wave-lengths are increased in wave-length by the same amount when the rays are scattered, and he showed at the same time that this change could be explained if the X-rays are corpuscular in nature.

The Experiment and Its Explanation.—According to the theory that X-rays consist of electromagnetic waves, scattered X-rays are similar to an echo. When an X-ray wave passes through a piece of paper composed of electrons, each electron is set in vibration by the wave and, because of its forced vibrations, emits a new wave which goes in all directions as a scattered X-ray. The number of vibrations of these new waves per second is the same as the number of vibrations of the electron, which is in turn the same as the frequency of the original X-rays. Experiment, however, shows that the frequency of the scattered rays is less

Reprinted with permission. Copyright © by *Encyclopaedia Britannica*, 14th edition, 1929. All rights reserved.

than that of the primary rays. This prediction of the wave theory of X-rays is thus incorrect.

The corpuscular theory of the scattering process supposes that each X-ray particle, or "photon" may collide with an electron of the scattering material and bounce off. In fig. 1 is shown a diagram of such a collision. The photon strikes the electron at O, and bounces off toward P, while the electron recoils from the impact in the direction OQ. The collision is supposed to be elastic; but a part of the energy of the photon is spent on the recoiling electron. It follows that the deflected or scattered photon must have less energy than it had before the collision. Such a decrease in the energy of the photon would be described in the language of the wave theory as a decrease in frequency or an increase in wave-length of the scattered X-ray. (See QUANTUM THEORY.)

As we shall show later, the photon theory can be put in a quantitative form, in which it predicts an increase in wave-length of the X-rays due to the scattering process of $2.42 \times 10^{-10} \times (1 - \cos \phi)$ centimetre, where ϕ is the angle between the primary and the scattered rays.

A diagram of the apparatus used for testing this prediction is shown in fig. 2. X-rays pass through a radiator R, which may be for example a block of carbon or paraffin. Some of the rays are scattered through slits 1 and 2 into the X-ray spectrometer. In this instrument a crystal of calcite takes the place of the prism or the grating of an optical spectrometer and spreads the rays into a spectrum, which is examined by the ionization chamber. (See SPECTROSCOPY: ROENTGEN RAY.) By placing the X-ray tube before the slits in place of the radiator, the spectrum of the primary X-rays can be compared with that of the scattered rays.

Fig. 3 compares the spectrum of the primary X-rays with the spectrum of these rays after they have been scattered by a block of graphite. The upper curve shows a prominent line in the X-ray spectrum of molybdenum. The lower curves show the spectrum of these rays after being scattered from graphite at three different angles. In each case, in addition to a line of the original wave-length, there appears a more prominent line of increased wave-length. Measurements on spectra of this type have shown that the difference in wave-length between the two sets of lines is given accurately by the formula $2.42 \times 10^{-10} \times (1 - \cos \phi)$ cm. as predicted by the photon theory.

The line whose wave-length has not been changed is called the "unmodified" line. It may be accounted for as due to photons deflected by electrons that are too tightly held in the atom to recoil from the impact of the photon.

The Recoil Electrons.—We have seen that, according to the photon theory, when an X-ray particle collides with an electron, the electron recoils from the impact unless held too tightly by its atom. Electrons recoiling in this manner were discovered independently by C. T. R. Wilson and W. Bothe (1923) a few months after their prediction. Fig. 4 is a photograph of the trails of four such recoil electrons, taken by Ikeuti, using Wilson's method. It will be seen that the tracks of these electrons start nearly in the direction of the X-ray beam, as they should if they are recoiling from deflected X-ray photons. In fact a detailed study of such photographs shows that the number of these trails is about equal to the number of photons of scattered X-rays, and that their directions and ranges are in good accord with the predictions of the photon theory.

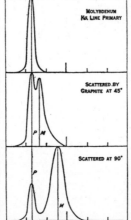

FROM "PHYSICAL REVIEW," 1923
FIG. 3.—SCATTERED X-RAY SPECTRUM
Line P is shown and one of increased wave-length, its position marked by M

The corpuscular character of the scattered X-rays is shown most clearly by tracing the path of a photon after it has collided with an electron. This has been done (Compton and Simon, 1925) in the manner shown diagrammatically in fig. 5. A feeble beam of X-rays is admitted into a cloud expansion chamber of the type devised by Wilson to show the trails left by fast moving electrons. A photon is scattered by an electron at O, and the trail of the electron as it recoils is visible. If it starts along the line OQ, the X-ray particle must have proceeded in the direction OP, determined by the usual mechanical laws of elastic collisions. The deflected photon can make itself visible by exciting a second high-speed particle before it escapes through the wall of the chamber. The track at A represents such an occurrence. When such a second track appears it is possible to trace the path followed by the X-ray particle after its collision with the first electron. If the scattered X-rays did not consist of particles, but were propagated as waves spreading in all directions, when a second electron appears, there is no more reason why it should occur at A than at some other position such as B. The fact that in the experiments the scattered ray excited secondary electrons near the line OP, determined by the angle of recoil Θ, means that the X-rays go in definite directions.

Unless there is some improbably large error in the experiments, we may therefore infer that scattered X-rays go as discrete particles in definite directions. At the same time, experiments on the diffraction, interference and polarization of light and X-rays, and on electrical oscillations associated with electric waves, can leave no doubt but that electromagnetic radiation has the properties of waves. No satisfactory explanation has as yet been offered of how radiation may have at the same time the properties of waves and those of particles. Such a reconciliation does not, however, seem impossible.

The Photon.—The experiments associated with the Compton Effect thus seem to establish the existence of a particle of radiation. This particle, the photon, may be classified with the electron and the proton as one of the three fundamental units of matter. It does not possess an electric charge as do the electron and the proton, but it does have an electric "field," that is, it exerts a force on an electron in its neighbourhood. It also has mass, the essential characteristic of matter, its mass being $2.19 \times 10^{-38} \lambda$ grams, where λ is the wave-length of the radiation expressed in centimetres. For a hard gamma ray, of wave-length 2.4×10^{-10} cm., its mass is equal to that of an electron at rest; but for or-

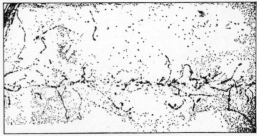

FROM "PROCEEDINGS OF THE ROYAL SOCIETY," 1923, BY COURTESY OF THE COUNCIL, AND OF PROFESSOR C. T. R. WILSON
FIG. 4.—TRAILS OF ELECTRONS RECOILING FROM SCATTERED X-RAYS

dinary light its mass is only about 0.000005 that of an electron. The photon seems to disappear when absorbed by an atom, and to be created again when the atom emits radiation. However, the suggestion has been made by G. N. Lewis (1926) that the photon is really retained by the atom and does not lose its identity. The motion of the photon is always with the speed of light, which in free space is about 3×10^{10} cm. per second.

Calculation of the Change of Wave-length of Scattered X-rays.—The photon theory can be put in quantitative form by making use of Einstein's postulate (1905) that the energy of the photon is proportional to the frequency of the corresponding wave. Einstein assumes that the energy of a light particle is $E=h\nu$, where ν is the number of vibrations per second of the corresponding wave and h is a universal constant which has the value 6.55×10^{-27} erg seconds. For a photon moving with the velocity of light, the theory of relativity demands that its momentum shall be E/c, where c is the velocity of light, i.e., the momentum of a

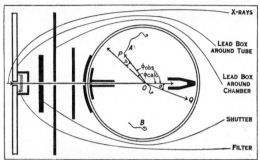

FIG. 5.—DIAGRAM SHOWING THAT A SECOND ELECTRON MAY BE EJECTED BY THE SCATTERED RAY AT A, BUT NOT AT B

photon is $h\nu/c$, or its equivalent h/λ, where λ is the wave-length of the corresponding wave.

The mathematical statement that the total energy after the collision between the photon and the electron is the same as before is,

$$h\nu = h\nu' + \tfrac{1}{2}mv^2, \qquad (1)$$

where ν' is the "frequency" of the photon after collision, and $\tfrac{1}{2}mv^2$ is the kinetic energy with which the electron recoils (neglecting higher powers of v^2/c^2, which become important only when the electron's speed v is comparable with that of light).

The statement that the total momentum of the photon and electron along the X axis remains equal to the $h\nu/c$ after the collision is, to the same degree of approximation,

$$\frac{h\nu}{c} = \frac{h\nu'}{c}\cos\phi + mv\cos\theta \qquad (2)$$

Similarly, along the Y axis the momentum is

$$0 = \frac{h\nu'}{c}\sin\phi - mv\sin\theta \qquad (3)$$

In these three equations we have three unknown quantities, $=\nu'$, v and θ (in the experiments ϕ is usually known), for which the equations may be solved. It is more convenient, however, to express the results of the solution thus:

$$\delta\lambda = \lambda' - \lambda = \frac{h}{mc}(1-\cos\phi) = 2.42 \times 10^{-10}(1-\cos\phi) \qquad (4)$$

$$E_{kin} = \tfrac{1}{2}mv^2 = h\nu \times 2a\cos^2\theta (\text{approx.}) \qquad (5)$$

$$\cot\theta = -(1+a)\tan\tfrac{1}{2}\phi, \qquad (6)$$

where $a = h/mc\lambda$. These equations represent the solutions of equations (1), (2) and (3) except for higher powers of v^2/c^2. Equations (4) and (6) are exact solutions if the relativity expressions for the kinetic energy and momentum of the electron are used.

Equation (4) expresses the difference in wave-length between the two sets of lines shown in fig. 3. It has been found to be as accurate as our knowledge of the constants, h, m, and c. Equation (5) describes the motion of the recoil electrons and has been found to agree with the experiments. The last equation (6) has been verified by experiments such as that pictured in fig. 5.

BIBLIOGRAPHY.—A. H. Compton, *X-Rays and Electrons* (1926); E. N. da C. Andrade, *The Structure of the Atom* (1927); see also A. H. Compton, in *Physical Review* (1923); C. T. R. Wilson, in *Proc. Roy. Soc. A.* (1923); A. H. Compton and A. W. Simon, in *Physical Review* (1925), Bothe and Geiger, in *Zeits. für Physik* (1925).

(A. H. C.)

THE DETERMINATION OF ELECTRON DISTRIBUTIONS FROM MEASUREMENTS OF SCATTERED X-RAYS

BY ARTHUR H. COMPTON

RYERSON PHYSICAL LABORATORY, UNIVERSITY OF CHICAGO

(Received March 5, 1930)

ABSTRACT

A calculation based on classical electromagnetic theory is made of the intensity of the x-rays scattered by an atom in which the electrons are arranged with random orientation and with arbitrary radial distribution. Conversely *an expression is derived for the radial distribution of the electrons in an atom*, assuming that they have random orientation. This expression has the form of a Fourier integral, which can be evaluated from observed intensities of the scattered x-rays for different wave-lengths and angles.

A comparison of this calculation with Wentzel's quantum theory of x-ray scattering suggests the introduction of a certain correction factor to express more nearly the intensity of the modified rays. It is also noted that the interpretation of $\psi\bar\psi$ as a *probability of the occurrence of an electron* leads to the correct value for the intensity of total scattered x-rays.

As an example of the application of the new method of calculation, Barrett's experimental data for the scattering of x-rays by helium are analyzed to give *the distribution of the electrons in the helium atom*. The resulting distribution is in close agreement with the value calculated by Pauling on the basis of wave mechanics, but differs by more than the probable experimental error from the electron orbits given by Bohr's theory.

1. INTRODUCTION

IT IS well known that the intensity of the x-rays scattered at small angles may be considerably greater than is anticipated on the assumption that each electron in the scattering material acts independently of the other electrons. When the scattering of x-rays by solids and liquids is considered, at least a part of this "excess scattering" may be ascribed to the interference between the rays scattered by neighboring atoms. In the case of gases, however, such interference is negligible, since the phases of the rays scattered by neighboring molecules are random. It has nevertheless long been recognized[1] that groupings of the electrons in the atoms themselves should result in some excess scattering in the forward direction. Calculations of the intensity of the scattered x-rays for typical electron distributions have in fact been made by Debye,[2] Schott,[3] the writer,[4] Glocker[5] and others. The converse problem of determining the electron distribution corresponding to an observed angular

---

[1] D. L. Webster, Phil. Mag. **25**, 234 (1913); C. G. Darwin, Phil. Mag. **27**, 325 (1914).
[2] P. Debye, Ann. d. Physik **46**, 809 (1915).
[3] G. A. Schott, Proc. Roy. Soc. **96**, 695 (1920).
[4] A. H. Compton, Washington University Studies, **8**, 98 (1921).
[5] R. Glocker, Zeits. f. Physik **5**, 54 (1921).

distribution of scattered x-rays has not however been attempted. We shall in the present paper obtain a solution of this problem which applies to certain important cases, and illustrate its application by determining the electron distribution in atoms of helium.

Because of the very important information which can thus be obtained regarding atomic structures, the problem would doubtless have been long ago pressed to a solution had it not been for an obstinate theoretical difficulty. Calculations of the effect of interference on the intensity of x-ray scattering are based upon the classical electron theory and electrodynamics. In the course of these x-ray diffraction studies, however, it became evident[6] that these classical theories are inadequate to supply a complete solution of the problem of the intensity of scattered rays. The problem was accordingly "laid on the table" until a new quantum dynamics should be developed which would be able to supply a more reliable solution. Recently Wentzel[7] has shown how the wave mechanics may be applied to this problem, and from his discussion it appears that the classical electron theory itself should give results which are not greatly in error.

In the meantime, closely allied problems have been successfully attacked on the basis of classical electron theory. In our studies of the diffraction of x-rays by crystals, which is of course only a special case of the general problem of x-ray scattering, application of the usual wave theory has enabled us to arrive at satisfactory arrangements of the atoms in the crystals, and has recently been used to determine also electron distributions in the atoms.[8] We have every reason to believe that the information supplied by this work regarding atomic arrangements is reliable, and even the electron distributions found by its use are too satisfactory to admit any major error in the method of analysis. Similarly the classical wave diffraction theory has been successfully applied to the x-ray study of molecular shapes and sizes of liquids,[9] and very recently also to the study of interatomic distances in gaseous molecules.[10] We are thus encouraged to undertake again a more detailed analysis of the scattering of x-rays by gases, on the basis of classical theory. The results of this analysis will then be compared with Wentzel's conclusions, to see what modifications are necessary in light of quantum mechanics.

2. Intensity of the X-rays Scattered by a Group of Electrons having a Random Angular Distribution

Let us suppose that an atom has Z electrons whose distances from the nucleus are at any instant $r_1, r_2 \cdots r_s$, and whose angular distribution is random. We imagine that this atom is traversed by an x-ray wave propagated

[6] A. H. Compton, Bull. Nat. Res. Council No. 20, p. 10 (1922).

[7] G. Wentzel, Zeits. f. Physik **43**, 1 and 779 (1927).

[8] For summaries of the latter work, cf. e.g., A. H. Compton, "X-rays and Electrons," Chapter V, or W. L. Bragg, "Electrons et Photons," report of the Fifth Solvay Congress, Paris (1928).

[9] For a summary of this work, cf. e.g., G. W. Stewart, Phys. Rev. Supp., Jan., 1930.

[10] P. Debye, L. Bewilogua and F. Ehrhardt, Phys. Zeits. **30**, 84 (1929); Ber. Sächsischen Ak. d. Wiss. zu Leipzig **81**, 29 (1929).

along the X axis, and that the forced oscillations of the electrons give rise to a scattered wave at an arbitrary distant point P at an angle ϕ. If A_e is the amplitude of the electric vector and δ the phase at P of the wave scattered by an electron coincident with the nucleus, the electric vector due to the nth electron in the group is (Fig. 1),

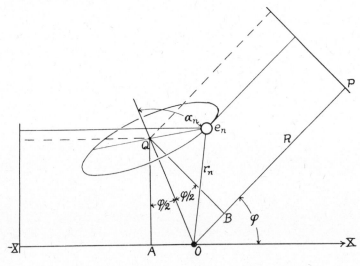

Fig. 1.

$$E_n = A_e \cos \{\delta - (2\pi/\lambda)2r_n \cos \alpha_n \sin (\phi/2)\}, \qquad (1)$$

where $2r_n \cos \alpha_n \sin (\phi/2)$ is the total difference in path[11] between the ray scattered from e_n and that scattered from O, and α_n is the angle between Oe_n and the line OQ which bisects the angle $-XOP$. Equation (1) may be written

$$E_n = A_e \cos (\delta - x_n), \qquad (2)$$

where

$$x_n = (4\pi r_n/\lambda) \cos \alpha_n \sin (\phi/2). \qquad (3)$$

The total electric vector due to all the electrons in the atomic group is then,

$$E = \sum_1^Z E_n = A_e \sum_1^Z \cos (\delta - x_n). \qquad (4)$$

Let us choose the origin of time such that the phase of the wave scattered from O is $\delta = pt$, where $p = 2\pi\nu$ is the phase frequency of the incident wave. The electric vector at the instant t is then, from Eq. (4),

$$E = A_e \sum_1^Z \cos (pt - x_n)$$

$$= A_e \sum_1^Z (\cos pt \cos x_n + \sin pt \sin x_n). \qquad (5)$$

[11] Cf. e.g., A. H. Compton, "X-rays and Electrons," p. 385.

The intensity of the scattered ray at this instant is however proportional to E^2, say bE^2, or,

$$I_i = bA_e^2 \left\{ \sum_1^Z (\cos pt \cos x_n + \sin pt \sin x_n) \right\}^2. \quad (6)$$

When this expression is averaged over a complete cycle, from $t=0$ to $t=2\pi/p$, all the terms in the summation disappear except those of the form

$$\cos x_m \cos x_n + \sin x_m \sin x_n,$$

and we find for the intensity averaged over a cycle,

$$I_\alpha = \frac{1}{2} bA_e^2 \sum_1^Z{}_m \sum_1^Z{}_n (\cos x_m \cos x_n + \sin x_m \sin x_n). \quad (7)$$

For a single electron, this becomes

$$I_e = \tfrac{1}{2} bA_e^2. \quad (8)$$

As Thomson has shown,[12] for unpolarized x-rays

$$I_e = \frac{Ie^4}{2m^2R^2c^4}(1 + \cos^2 \phi), \quad (9)$$

where I is the intensity of the primary beam traversing the electron, e, m and c have their usual significance, and R is the distance from O to P. Equation (7) may thus be written,

$$I_\alpha = I_e \sum_1^Z \sum_1^Z (\cos x_m \cos x_n + \sin x_m \sin x_n). \quad (10)$$

Since we have assumed that the electrons have random angular distribution, we must now average this intensity over all angles α_n. The probability that any α will lie between α and $\alpha+d\alpha$ is for random orientation $\tfrac{1}{2} \sin \alpha \, d\alpha$. Writing then

$$x_n = z_n \cos \alpha_n, \quad (11)$$

where

$$z_n \equiv (4\pi r_n/\lambda) \sin (\phi/2), \quad (12)$$

the probable contribution to the intensity due to the orientations α_n is

$$dI_\alpha = I_e \left\{ \sum_1^Z \sum_1^Z{}_{m \neq n} \tfrac{1}{4} [\cos(z_m \cos \alpha_m) \cos(z_n \cos \alpha_n) \right.$$
$$+ \sin(z_m \cos \alpha_m) \sin(z_n \cos \alpha_n)] \times \sin \alpha_m \sin \alpha_n d\alpha_m d\alpha_n$$
$$\left. + \sum_1^Z \tfrac{1}{2} [\cos^2(z_n \cos \alpha_n) + \sin^2(z_n \cos \alpha_n)] \sin \alpha_n d\alpha_n \right\}.$$

[12] J. J. Thomson, Conduction of Electricity through Gases, 2nd Ed., p. 325; or cf. "X-rays and Electrons," p. 60.

Integrating over all values of α_m and α_n this takes the simple form,

$$I_r = I_e \left\{ Z + \sum_1^Z \sum_1^Z {}_{m \ne n} \frac{\sin z_m}{z_m} \frac{\sin z_n}{z_n} \right\}. \tag{13}$$

Equation (13) represents the scattering by electrons arranged at fixed distances $r_1, r_2 \cdots$ from the nucleus, but with random orientations.

As an example of the application of this formula, consider the case of an atom with two electrons, both at a distance $r = a$ from the center, but with random orientations. We may write equation (13) in the form

$$S \equiv \frac{I_r}{ZI_e} = 1 + \frac{1}{Z} \sum \sum {}_{m \ne n} \frac{\sin z_m \sin z_n}{z_m z_n}, \tag{14}$$

which in the present case becomes,

$$S = 1 + \left(\frac{\sin z_a}{z_a} \right)^2. \tag{15}$$

A graph of this expression is shown in Fig. 2 by the solid line. This may be compared with scattering by two electrons separated by a fixed distance $2a$, which is given by the expression[13]

$$S = 1 + \frac{\sin 2z_a}{2z_a}, \tag{16}$$

and is represented in the figure by the broken line.

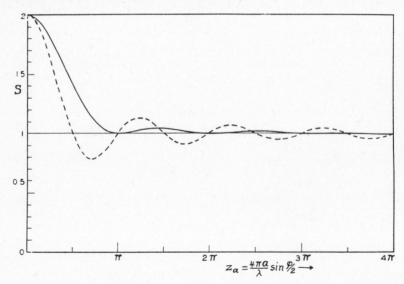

Fig. 2. Relative scattering per electron for an atom of two electrons. Solid line, both electrons at radius a and random orientation. Broken line, electrons at opposite ends of diameter $2a$.

[13] P. Debye, reference 2, or "X-rays and Electrons," p. 72.

If the probability that any one electron shall lie between r and $r+dr$ is $u(r)dr$, and if this probability is the same for every electron, we have,

$$dI_s = I_e \left\{ Z + \sum \sum_{m \neq n} \frac{\sin kr_m \sin kr_n}{k^2 r_m r_n} u(r_m) u(r_n) dr_m dr_n \right\},$$

where
$$k \equiv z_m/r_m = (4\pi/\lambda) \sin(\phi/2). \tag{17}$$

Since $u(r_m)$ assumed the same for all electrons, the integral of this expression may be written,

$$I_s = I_e Z + I_e \sum \sum_{m \neq n} \left\{ \int_0^a u(r) \frac{\sin kr}{kr} dr \right\}^2,$$

where a is the maximum radius of the atom.
Since

$$\sum_1^Z \sum_1^Z {}_{m \neq n} 1 = Z^2 - Z,$$

$$I_s = I_e \left\{ Z + (Z^2 - Z) \left[\int_0^a u(r) \frac{\sin kr}{kr} dr \right]^2 \right\}. \tag{18}$$

For the relative scattering per electron we thus have

$$S = \frac{I_s}{ZI_e} = 1 + (Z-1) \left\{ \int_0^a u(r) \frac{\sin kr}{kr} dr \right\}^2. \tag{19}$$

Expressions (13) or (18) may be applied to calculate the intensity of the rays scattered by an electron group, according as the electrons are at fixed distances from the center of the atom, or as they have a continuous radial distribution.

According to equation (18), I_s should never fall below ZI_e, since the term representing the interference is always positive. In this respect our calculation differs from that of Debye,[2] who considers electrons at fixed distances from each other, of which equation (16) is the simplest example.

3. Comparison with Results of Quantum Mechanics.

Wentzel's equation (3a) for the intensity of the modified scattered rays[14] may be written in the form

$$I_{unm} = I_e \left\{ \int \sum_n 4\pi r_n^2 \rho_n \frac{\sin kr}{kr} dr \right\}, \tag{20}$$

where $\sum \rho_n = \sum u_n{}^2$, the electrical charge distribution in electronic units, the subscript indicating the nth quantum number. Noting that $\sum 4\pi r^2 \rho_n$ is numerically equal to our $Zu(r)$, this may be written

$$I_{unm} = I_e \cdot Z^2 \left\{ \int u(r) \frac{\sin kr}{kr} dr \right\}^2. \tag{21}$$

---

[14] G. Wentzel, reference 7, p. 781.

His equation (4a) for the intensity of the modified scattered rays (uncorrected for the change of wave-length) may similarly be written

$$I_{\text{mod}} = I_e \left\{ Z - \sum \left[\int 4\pi r^2 \rho_n \frac{\sin kr}{kr} dr \right]^2 \right\}. \quad (22)$$

The total intensity of the scattered rays thus becomes,

$$I_s = I_{\text{unm}} + I_{\text{mod}} \quad (23)$$
$$= I_e \left\{ Z - \sum \left[\int 4\pi r^2 \rho_n \frac{\sin kr}{kr} dr \right]^2 + Z^2 \left[\int u(r) \frac{\sin kr}{kr} dr \right]^2 \right\}$$

This expression becomes identical with equation (18) if

$$4\pi r^2 \rho_n = u(r). \quad (24)$$

We have noted above that

$$\sum_{1}^{Z} 4\pi r^2 \rho_n = Z u(r), \quad (25)$$

whence relation (24) holds if

$$\sum_{1}^{Z} 4\pi r^2 \rho_n = Z \cdot 4\pi r^2 \rho_n, \quad (26)$$

i.e. if the charge distribution for every electron is the same. This is precisely the assumption on which equation (18) is derived. Wentzel, in his numerical calculation of equation (22) takes

$$\rho_n = u_n^2 \quad (27)$$

as the charge density for the nth electron, instead of

$$\rho_n = (1/Z) \sum_{1}^{Z} u_n^2, \quad (28)$$

which is the equivalent of (26). This introduces a slight difference between the results of his calculation and that of ours. It would seem however that relation (28) is in better accord with present interpretation of quantum mechanics than is (27), and if its validity is admitted, our classical equation (18) becomes identical with Wentzel's quantum equation (23).

This comparison shows that if we interpret $\sum u_n^2 dxdydz$ as the electric charge in the volume element, the scattering which we calculate is the *unmodified* scattering (eq. 21). If, however, we intrepret it as the probability that a discrete electron will be present in the volume element, as we have done in deriving equation (18), we calculate the *total* scattering. Since the total scattering is experimentally observed, it would seem that the latter interpretation has the better physical justification.

In his derivation of equation (22), Wentzel has assumed the limiting case of very long wave-lengths, for which the scattering by a free electron is identical with that calculated on the classical theory. For shorter wave-lengths

Breit[15] and Dirac[16] have shown that the intensity of the modified rays from free electrons is reduced in the ratio

$$\frac{I_{mod}}{I_{class}} = \left(\frac{\lambda}{\lambda'}\right)^3 = (1 + \gamma \text{ vers } \phi)^3, \tag{29}$$

where λ and λ' are the wave-lengths of the primary and the modified ray respectively, and $\gamma = h/mc\lambda$. We may accordingly expect to get a closer approximation to the intensity of the modified scattering if we multiply equation (22) by equation (29), or using the equivalent part of equation (18),

$$I_{mod} = ZI_e(1 - F^2/Z^2)(1 + \gamma \text{ vers } \phi)^{-3}. \tag{30}$$

Similarly[17]

$$I_{unm} = I_e F^2, \tag{31}$$

where

$$F \equiv Z \int_0^a u(r) \frac{\sin kr}{kr} dr, \tag{32}$$

which is identical with the so-called "atomic structure factor."

A convenient method of comparing the experiments with the theoretical calculations is thus to multiply the observed intensity of the modified rays by the factor $(1 + \gamma \text{ vers } \phi)^3$, and add to the observed intensity of the unmodified rays. The resulting value

$$I_s' = I_{mod}(1 + \gamma \cos \phi)^3 + I_{unm} \tag{33}$$

may then be compared directly with the value of I_s derived by the classical equation (18).

4. Analysis of Scattering Data to Determine Radial Electron Distribution

If the distribution of the electrons is spherically symmetrical, as we have assumed, we may represent the probability that an electron will lie between r and $r + dr$ by a Fourier sine series of the form,

$$u(r) = A_1 r \sin \pi r/a + A_2 r \sin 2\pi r/a + \cdots + A_n r \sin n\pi r/a + \cdots \tag{34}$$

Substituting this value of $u(r)$ in equation (19) we get,

$$S = 1 + (Z - 1)\left[\sum_1^\infty \int_0^a \frac{A_n}{k} \sin\left(n\pi \frac{r}{a}\right) \sin(kr) dr\right]^2. \tag{35}$$

If the scattering I_s is evaluated for $k = n\pi/a$, i.e., by equation (17) for

$$\sin(\phi/2)/\lambda = n/4a, \tag{36}$$

---

[15] G. Breit, Phys. Rev. **27**, 242 (1926).

[16] P. A. M. Dirac, Proc. Roy. Soc. **111**, 405 (1926). This relation (29) presumably does not hold for wave-lengths so short that the velocity of the recoil electron approaches c. In this case the formula of Klein and Nishina presents a closer approximation.

[17] It is interesting that the ratio I_{mod}/I_{unm} is expressed by equations (30 and 31) in terms of interference. It was early suggested by the writer (Phil. Mag. **46**, 910 (1923)) that this ratio might be thus expressed, as an alternative to the more obvious description developed later by Jauncey, in terms of the ratio of the energy of recoil of the scattering electron to its binding energy in the atom. Wentzel[7] shows that equivalent expressions of the ratio I_u/I_m may be made in terms of either interference or energy of recoil.

where a is an assumed maximum radius, all integrals in the sum of equation (35) vanish except the nth, giving

$$S_n = 1 + (Z-1)\left[\frac{1}{4}\frac{a^2}{k^2}A_n^2\right]. \tag{37}$$

Thus

$$A_n = \pm\frac{2k}{a}\left\{\frac{S_n-1}{Z-1}\right\}^{1/2}. \tag{38}$$

Corresponding to each value of S_n we thus determine the nth term of the Fourier series (34), and thus eventually the value of $u(r)$.

Our series (34) has in it an arbitrary radius a, and in evaluating the series the data for only certain arbitrarily chosen values of k are employed. If this arbitrary radius is made large, the values of k which are used come closer together, and our series approaches the Fourier integral,

$$u(r) = r\int_0^\infty B\sin(\pi r x)dx, \tag{39}$$

where

$$x \equiv n/a = (4/\lambda)\sin(\phi/2), \tag{40}$$

according to equation (36), and

$$B \equiv A_x a = 2\pi x\left\{\frac{S_x-1}{Z-1}\right\}^{1/2}. \tag{41}$$

If instead of the probable position of a single electron, we wish to find the probable number of electrons between r and $r+dr$, we have only to multiply $u(r)$ by the number of electrons per atom, giving by equation (34)

$$U(r) = Zu(r) = Zr\sum_1^\infty A_n \sin n\pi r/a, \tag{42}$$

or by (39),

$$U(r) = Zr\int_0^\infty B\sin(\pi r x)dx. \tag{43}$$

It is interesting to compare equation (42) with the similar series expressing the radial distribution of electrons in the atoms of a crystal,[1]

$$U(r) = 8\pi\frac{r}{D^2}\sum_1^\infty nF_n \sin 2\pi n\frac{r}{D}. \tag{44}$$

We note that a of equation (42) corresponds to $D/2$ of (44), since both quantities represent the assumed outer limit of the atom. The series are accordingly identical if $2\pi(r/a^2)nF_n = ZrA_n$. Using the value of A_n given by (38), and noting that $D = 2a = (n\lambda/2)\sin\frac{1}{2}\phi$, this means that

$$F_n = Z\left\{\frac{S_x-1}{Z-1}\right\}^{1/2}. \tag{45}$$

This expression enables us to compare the "F" curves obtained from crystal reflection with the data given by scattering experiments.

[18] A. H. Compton, "X-rays and Electrons," p. 164. An integral identical in form with (43), but representing the electron distribution in atoms of a crystal, has been given by G. E. M. Jauncey and W. D. Claus, Phys. Rev. **32**, 20 (1928).

5. Tests of the Method of Analysis

Before applying equation (43) to the interpretation of experimental data, it will be of interest to study its application to certain cases where the solution is known.

a. Consider the intensity distribution described by equation (15). From equations (15) and (41) we have

$$B = \pm (2/a) \sin \pi x a. \tag{46}$$

Substituting this value in (43), since $Z = 2$,

$$U = 4(r/a) \int_0^\infty \sin(\pi a x) \sin(\pi r x) dx. \tag{47}$$

This integral is zero,[19] except when $r = a$, in which case its value becomes infinite, indicating a concentration of the electrons at the distance $r = a$ from the nucleus, in accord with the original assumption on which (15) was based.

b. An atom of four electrons, each of whose probability of lying between r and $r + dr$ is $u(r) = 2r/a^2$ between $r = 0$ and $r = a$, and is zero beyond $r = a$.

By equation (19) we find,

$$S = 1 + 3\left\{ \frac{2(1 - \cos \pi x a)}{\pi^2 x^2 a^2} \right\}^2. \tag{48}$$

From (41) then,

$$B = \pm \frac{4}{\pi x a^2}(1 - \cos \pi x a),$$

and equation (39) becomes,

$$u(r) = 4\frac{r}{a}\left\{ \int_0^\infty \frac{\sin \pi r x}{\pi a x} dx - \int_0^\infty \frac{\cos \pi a x \sin \pi r x}{\pi a x} dx \right\}. \tag{49}$$

The value of the integrals is[20] $1/2a$ for $[r < a]$, and 0 for $[r > a]$, whence

$$u(r) = 2r/a^2 \qquad [r < a], \tag{50}$$
$$= 0 \qquad [r > a],$$

as initially assumed.

These tests check the accuracy of the mathematical analysis. They of course say nothing, however, regarding the validity of our physical assumptions of spherical symmetry and of independence of the positions of the various electrons in the atomic groups.

6. Electron Distribution in Helium

The formulas that have been developed above are directly applicable only to the scattering of x-rays by gases, in which case the interference effect due to neighboring molecules is negligible. In the case of the noble gases we are also free from interference between adjacent atoms, since the gases are mona-

---

[19] At any point when $r \neq a$ the integral is strictly speaking indeterminate; but its average value over a finite range of x is zero.

[20] B. O. Peirce, "A Short Table of Integrals" (1910) nos. 484 and 485.

tomic, and according to current theories the probable electron distributions should be spherically symmetrical as we have assumed in our calculations. Fortunately recent experiments by Barrett[21] supply sufficient information regarding the scattering by helium to yield valuable information.

In Barrett's Fig. 7 he compares the scattering by helium with that by hydrogen, which he finds identical with that calculated from the Breit-Dirac quantum formula for the range investigated. With an effective wave-length of 0.49A, he finds that helium and hydrogen scatter equally, within experimental error, at angles greater than 60°, but that at 40°, 30° and 20° the scattering by helium is greater by the ratios 1.025, 1.08 and 1.26 respectively. These values are indicated by the circles in Fig. 3, where $S \equiv I_s/ZI_e$ is plotted against x. At sufficiently small angles the phase difference between the rays from the two electrons in helium must be negligible, in which case our theory demands that the value of S must approach 2. For small values of x the phase

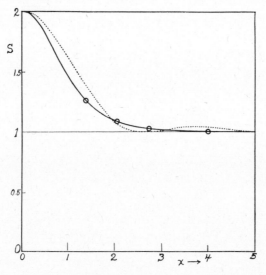

Fig. 3. Solid line, relative scattering by helium, based on Barrett's data (circles). Broken line, calculated scattering by Bohr type helium atom.

differences are small quantities of the first order; but the amplitudes, being proportional to the cosines of the phase differences, are affected only in the second order of small quantities.[22] Thus the S curve must leave the $x=0$ axis parallel to the x axis, and must initially be of a parabolic form.[23] We can thus interpolate the S curve between $x=0$ and $x=1.4$ with some degree of assurance as indicated by the solid line.

This S curve can be transformed into a B curve by the help of equation (41), giving the result shown in Fig. 4. Here again the values given by the experimental data are shown by the circles.

[21] C. S. Barrett, Phys. Rev. **32**, 22 (1928).

[22] This may be seen by finding the maximum value of E from equation (5) for small values x_n. This maximum is unaffected to the first power of x_n, but is reduced by terms containing x_n^2.

[23] These conclusions are valid only if the atom is not of infinite extent.

For values of x greater than 3 the experiments suggest that B gradually approaches zero.[24] The values of the integral U are not much affected by the exact manner of this approach as long as it is slow and continuous. For con-

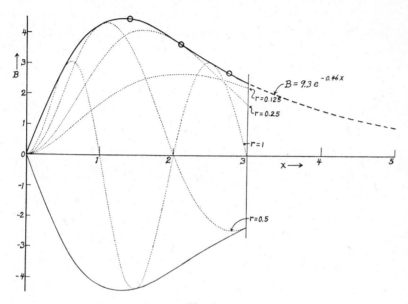

Fig. 4.

venience, therefore, we may assume that beyond some arbitrary value of x, say x_1, B may be expressed by

$$B = be^{-ax} \quad [x > x_1]. \tag{51}$$

In order that at $x = x_1$ the values of B and dB/dx shall be continuous, we must have

$$a = -\left(\frac{1}{B}\frac{dB}{dx}\right)_{x_1} \tag{52}$$

and

$$b = B_1 e^{ax_1} \tag{53}$$

In order to evaluate $U(r)$ for a definite value of r we must determine the integral,

$$\Phi = \int_0^\infty B \sin(\pi r x) dx. \tag{54}$$

This may be separated into two parts,

$$\Phi_1 \equiv \int_0^{x_1} B \sin(\pi r x) dx, \tag{55}$$

and

$$\Phi_2 \equiv \int_{x_1}^\infty B \sin(\pi r x) dx. \tag{56}$$

[24] B must approach zero for large values of x unless the electron density at the center of the atom is infinite.

The first part Φ_1 may be evaluated graphically, by plotting $B \sin(\pi r x)$ for various values of r, as indicated by the dotted lines of Fig. 4, and integrating from O to x_1 with a planimeter. Φ_2 may be determined by substituting in equation (56) the value of B given in equation (51) and integrating, which gives

$$\Phi_2 = B_1 \frac{a \sin \pi r x_1 + \pi r \cos \pi r x_1}{a^2 + \pi^2 r^2}. \qquad (57)$$

From Fig. 4 we find for helium, if $x_1 = 3$, $B_1 = 2.36$ and $a = 0.46$, whence the value of Φ_2 may be determined for any desired value of r.

As typical examples, we have the following values (Table I):

TABLE I.

r	Φ_1	Φ_2	$\Phi = \Phi_1 + \Phi_2$	$U = Zr\Phi$
1.125A	5.75	1.64	7.39	1.85
0.25	8.16	−0.42	7.74	3.87
0.5	3.14	−0.35	2.79	2.79
1.0	0.86	−0.70	0.16	0.32

The resulting values of U plotted against r are shown in Fig. 5 by the solid line.

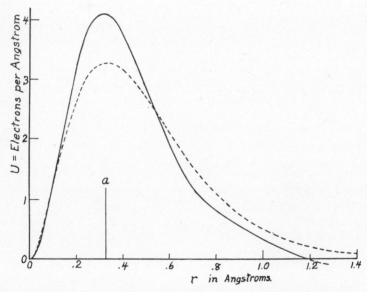

Fig. 5. Radial electron distribution in helium. Solid line, based on Barrett's x-ray scattering data. Broken line, Pauling's calculation from wave mechanics. a = radius of Bohr orbits.

It is of great interest to compare the "observed" distribution with that calculated theoretically. According to the Bohr-Sommerfeld theory, the electrons in helium should both traverse approximately circular orbits with unit angular momentum, the radius of the orbits being given by

$$a = h^2/[4\pi^2 e^2 m(Z - s)]. \tag{58}$$

where h, e, m and Z have their usual significance, and s is the "screening constant" of each electron on the other, having a value[25] of approximately $s = 0.39$. Using the usual values of the constants, we thus find $a = 0.33$A. It will be seen that this value falls very close to the radius of maximum electron density as shown in Fig. 5.

If we assume that the two electrons are on a spherical surface of fixed radius $r = 0.33$A, the intensity of the scattered rays should be given by equation (15). The values of s thus calculated are shown by the dotted line Fig. 3. The differences between this dotted curve and the experimental points are considerably greater than the probable experimental error. Yet it is not impossible that a combination of heterogeneous x-rays such as Barrett used and the presence of incoherent rays (Compton scattering) at the large angles might flatten out the dotted curve to resemble the experimental one.

The distribution found from this analysis of Barrett's scattering data is however in striking agreement with that calculated on Schrödinger's wave mechanics. Thus Pauling[25] has shown that the radial electron distribution in helium can be expressed to a close approximation by

$$U(r) = Zr^2 X^2, \tag{59}$$

where for helium in the normal state he finds,

$$X = u_{1,0} = -2\left(\frac{Z-s}{a_0}\right)^{3/2} e^{-\xi/2} \tag{60}$$

$a_0 = h^2/4\pi^2 e^2 m = 0.53A$,

$Z = 2$

s = screening const = 0.39

$\xi = 2(Z - s)r/a_0$.

Substituting these values in equation (59) we get the U curve shown by the dotted line of Fig. 5. The striking similarity between this distribution predicted by the quantum theory and that coming from our interpretation of the scattering experiments is the more convincing when it is noted that there are no arbitrary constants available to make the two curves correspond. This agreement is a strong argument in favor of a continuous electron distribution, as predicted by the wave-mechanics, as opposed to the Bohr quantum theory of definite orbits.

---

[25] Cf. e.g. L. Pauling, Proc. Roy. Soc. **A114**, 181 (1927).

50. Scattering of x-rays and the distribution of electrons in helium. ARTHUR H. COMPTON. *University of Chicago.*—An analysis of the theory of scattering of x-rays by monatomic gases makes it possible to express the probable distribution of the electrons in the atom as a Fourier integral. To evaluate this integral it is necessary to know the intensity of scattering of x-rays of known wave-length at different angles. With data recently obtained by C. S. Barrett the distribution of electrons in atoms of helium gas is thus determined. This distribution agrees satisfactorily with that calculated by Pauling from wave-mechanics but differs by more than the experimental error from that predicted from Bohr's theory.

23. Electron distribution in argon, and the existence of zero point energy. ARTHUR H. COMPTON, *The University of Chicago*.—Data on the scattering of x-rays by argon, reported by Wollan are analyzed by the method recently described by the writer (Phys. Rev. **35**, 925 (1930)) The resulting electron distribution curve (U curve) shows a maximum electron density at about 0.10A from the center of the atom. The data of James and Firth (Proc. Roy. Soc. **A114**, 181 (1927)) on rock salt at 0°K give a maximum electron density of chlorine atoms at about 0.19A from the center, for 900°K at about 0.58A. The difference between the latter two values is due to the thermal motion of the atoms in the crystal. Similarly, the large difference between the radius of maximum electron density in the atoms of argon gas and of chlorine in rock salt at 0°K must mean motion of the chlorine ions in the crystal lattice. The amplitude of this motion is of the order of magnitude predicted on the quantum theory of zero point energy. This confirms the conclusion of James Waller and Hartree, who however based their conclusion on a theoretical rather than an experimental electron distribution for argon and chlorine.

JOURNAL
of the
OPTICAL SOCIETY
OF AMERICA

VOLUME 21
No. 2

FEBRUARY
1931

THE OPTICS OF X-RAYS*

BY ARTHUR H. COMPTON

[UNIVERSITY OF CHICAGO, CHICAGO, ILLINOIS.
RECEIVED OCTOBER 31, 1930]

In opening this symposium on the optics of the extremes of the spectrum, it may be well to review briefly the immense range of the electromagnetic radiations. In Fig. 1 these various rays are graphically

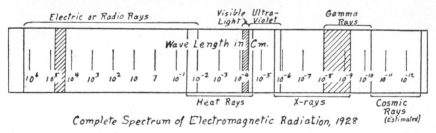

FIG. 1. *Complete spectrum of electromagnetic radiation on a logarithmic scale. Visible light is only a small but very important part of this spectrum.*

compared. At the extreme left I have arbitrarily started the spectrum at a wave length of eighteen kilometers, which is the wave length of certain transatlantic wireless signals. There is no reason why longer waves could not be produced if desired. The electric waves continue in

---

* Invited paper, Charlottesville Meeting of the Optical Society.

an unbroken spectrum down to 0.1 mm, rays recently studied at Cleveland by the late Dr. Nichols and Mr. Tear. Overlapping these electric waves are the heat rays, which have been observed from about .03 cm to .00003 cm, including the whole of the visible region. The heat rays in turn are overlapped by the ultraviolet rays, produced by electric discharges; and these reach well into the region described as x-rays. The bridge between these ultraviolet and the x-rays was closed only a few years ago by Osgood's soft x-ray studies made at Chicago. Beyond these are in turn the gamma rays and the cosmic rays. Thus over a range of wave lengths of from 2×10^{-13}cm to 2×10^{6}cm there is found to be a continuous spectrum of radiations, of which visible light occupies only a very narrow band.

The great breadth of this wave length range will perhaps be better appreciated if we expand the scale until the wave of a cosmic ray has a length equal to the thickness of a post card. The longest wireless wave would on this scale extend from here to the nearest fixed star.

When the physicist speaks of light, he refers to all the radiations included in this vast range. We believe that they are all the same kind of thing, and that anything which may be said about the nature of the rays in one part of this region is equally true of the rest.

Review of Evidence of X-rays As a Kind of Light

In the early studies of the properties of x-rays there appeared only one truly optical characteristic, namely that of straight line propagation. In the hands of Roentgen and his contemporaries, x-rays were found neither to be reflected nor refracted. They showed no polarization, nor could they be diffracted as light waves are. As a result, many theories were put forward as to their nature. Roentgen himself favored the idea of longitudinal ether waves. Some favored corpuscles, other electric waves of the Herzian type. Wiechert proposed, almost immediately upon their discovery, the idea that x-rays consist of rays similar to light but of much shorter wave length; yet it was many years before this suggestion found confirmation in experiment.

The first strong evidence that x-rays are really similar to light in their fundamental nature appeared when Barkla found that they could be polarized by scattering in the same way that sun-light is polarized when it is scattered by the sky. This was in 1905, ten years after the discovery of the rays.

It was eight years more before the discovery of Laue showed that x-rays could be diffracted by crystals in the same way that light is diffracted by a ruled grating, thus establishing the wave characteristics of these rays. The striking consequences of Laue's discovery, leading as it did to undreamed of knowledge about the interior of crystals and of atoms themselves, are a matter of common knowledge.

The ordinary optical properties of refraction and reflection from polished surfaces had, however, not yet shown themselves. Though a few years after Laue's discovery evidence arose of the refraction of x-rays within a crystal, it was not until 1922 that our experiments showed the reflection of x-rays from polished glass surfaces when the rays strike the surface incident at sufficiently fine glancing angles. Only five years ago Siegbahn and his collaborators performed the first successful experiments showing the refraction of x-rays on passing through a prism of glass.

The parallelism between x-rays and light is now complete. All the distinctly optical characteristics—reflection, refraction, diffraction, and polarization, and even interference, are now recognized as properties of x-rays as well as light. Yet, because of their great difference in wave length, there remains a notable difference between these types of radiation. It will be recalled that the resolving power of an optical instrument depends upon the wave length of the light that is used. We make use of this fact in examining minute organisms under a microscope with blue light instead of white. The fact that x-rays have a wave length 10,000th that of ordinary light, makes it possible with these rays to study the detailed structure of minute bits of matter in a much more effective way than can be done with the comparatively coarse waves of light. It is partly for this reason that it becomes of especial interest to review some of the important findings of our x-ray studies and to acquaint ourselves with some of the optical problems which are now being faced in the study of x-rays.

Some Fruits of the Optics of X-rays

"If, as Henri Poincare has said, the value of a discovery is to be measured by the fruitfulness of its consequences, the work of Laue and his collaborators should be considered as perhaps the most important of modern physics." Thus the Duke de Broglie describes the significance of the discovery that x-rays may be diffracted by crystals. Not only did this discovery show the identical character of x-rays and light, which had previously been questioned, but it also afforded a tool which has made it possible to examine the interior structure of matter with a definiteness previously almost unthinkable. By the help of this discovery the Braggs analysed the atomic structure of crystals. Moseley studied the spectra of x-rays, finding from the beautiful regularity of the changes in the spectra as we pass from one element to the next, that each element has in its structure one more electron than that of the element next lighter. The measurement of x-ray wave lengths in the hands of Duane and Hunt have supplied us with what is probably our most precise method of determining Planck's constant h, which plays so important a part in the photoelectric effect, and is fundamental in the

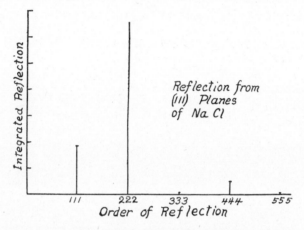

FIG. 2. *Intensities of reflection from the (111) planes of rock salt.*

quantum theory. In fact it is hard to find an aspect of atomic or molecular structure which has not been affected by the consequences of Laue's discovery of the diffraction of x-rays.

It is not, however, my desire to dwell at length on this work today, because to most of you the principle features of these investigations are already known. It may be more profitable to discuss some of the less familiar aspects of the application of the optics of x-rays.

The work of the Braggs in applying Laue's discovery to the study of the distribution of atoms in crystals is well known. The recent elaboration of their methods that has made possible a direct investigation of the arrangement of electrons in atoms is not however so familiar. The problem is essentially similar to that of the study of an optical grating. When one has a knowledge of the wave length of the light employed, a measurement of the position of different orders of the diffracted lines affords exact knowledge of the spacing between the lines of the grating. This is the optical analogue of the method of finding the distance between the layers of atoms in a crystal. It is found, however, that if the lines of a grating are broad, the higher orders of diffraction rapidly diminish in intensity. Thus, from a study of the intensities of the orders, it is possible to estimate the breadth or diffuseness of the lines ruled on the grating. In a precisely similar manner, measurements of the relative intensity of different orders of x-rays reflected from crystals afford a means of measuring the diffuseness of the atomic layers of which the

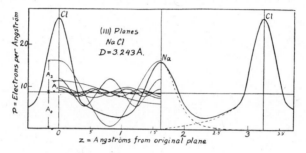

FIG. 3. *Density of electron distribution in layers parallel to the (111) planes of rock salt.*

crystal is built. This diffuseness is that due to the distribution of the electrons within the atom.

In Fig. 2 is shown, above, the relative intensity of the various orders of reflection from the (111) face of rock salt as measured by W. L. Bragg

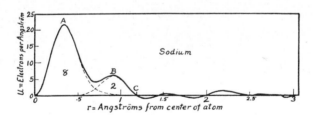

FIG. 4. *Radial distribution of electrons in the sodium ion of rock salt.*

and his collaborators. By a Fourier analysis of these intensities one finds the distribution of the electrons in the atomic layers of this crystal shown in Fig. 3. Alternate lines of sodium and chlorine atoms appear, corresponding to the alternately high and low intensities of the even and odd orders of diffraction. A more detailed examination, making use of the data from other faces of the crystal, gives the distribution of the electrons in the atoms of sodium shown in Fig. 4. It will be seen here that the radius of these atoms is of the order of 1A, which is comparable with the distance between the layers of the successive atoms. This is one of our most direct methods of studying the distribution of electrons in atoms.

Very recently similar information has been obtained from a study of the diffraction of x-rays scattered by the molecules of a gas. You are all familiar with the time-honored prediction of rain following a ring around the moon. Such a ring is frequently due to the diffraction of the moonlight by droplets of water suspended in the air. From the size of the

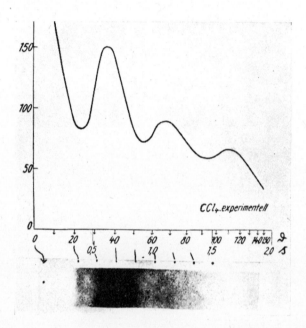

FIG. 5. *Diffraction pattern of x-rays traversing carbon tetrachloride vapor.*

ring, knowing the wave length of the light, it is possible to estimate the diameter of the water drops. In a similar way x-rays may be scattered by molecules of a gas, diffraction rings are formed, and from their size it is possible to measure the diameter of the ring. Fig. 5 shows one of these diffraction rings obtained by Debye and his colleagues, on passing x-rays through carbon tetrachloride vapor. From the size of these diffraction rings they were able to show that the distance between the atoms of chlorine in the carbon-tetrachloride molecules is 3.45A. Thus our x-ray diffraction affords a powerful means of studying the structure of molecules.

In monatomic gases the study of x-ray diffraction has also given important information. In the case of helium for example, the form of the diffraction curves observed by Barrett and shown in Fig. 5 has revealed the form of the atom itself. When put through the mathematical mill,

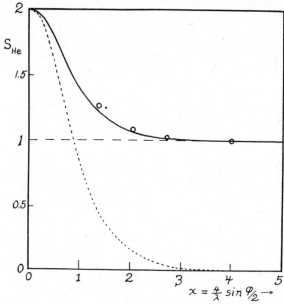

FIG. 6. *Intensity of x-rays scattered by helium for different values of x. Data from Barrett's experiments; solid line from Heisenbert atom; dotted line from Schrödinger atom.*

from these data is ground out the electron distribution in helium shown by the solid curve of Fig. 6. This result may be compared with the curve for sodium, shown in Fig. 4, obtained by a different method.

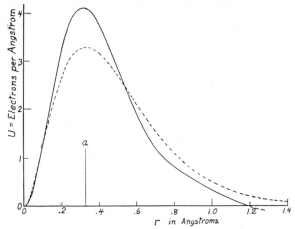

FIG. 7. *Radial distribution of electrons in atom of helium. Solid curve calculated from Barrett's experiments; broken line, Heisenberg-Schrödinger theory.*

Studies such as these come as near as is now possible to showing the inside of matter as it would be seen by an x-ray microscope. Though the nature of things seems to make impossible the invention of this mythical instrument, yet by the careful analysis of systematic experi-

FIG. 8. *"Appearance"* of a helium atom, Heisenberg-Schrödinger theory (after Langer).

ments it has been possible to find practically the same information that such an instrument would reveal. When we thus "look" at the atom of helium we find it, as illustrated in Figs. 7 and 8, composed of a diffuse electron cloud, concentrated near the center of the atom.

Current Problems of X-ray Optics

Closely associated with the analysis of the structure of crystals is the counting of the number of molecules in a cubic centimeter of the crystal. Such a measurement affords us, likewise, a knowledge of Avogadro's number and of the charge on the electron.

The method of making such a measurement is straight-forward. Bragg's law,

$$n\lambda = 2D \sin \theta, \qquad (1)$$

expresses the angle θ at which an x-ray beam of wave length λ is diffracted from a crystal, in terms of the distance D between successive layers of atoms and the order of diffraction n. The quantity n is a small integer, usually 1, while θ is an angle which can be measured with high precision. It is thus possible to determine the distance D between the layers of atoms almost as accurately as the wave length λ of the x-rays is known. The most precise as well as the most direct method of determining this wave length is by the help of a diffraction grating ruled on glass. The fact that it has recently become possible to make precise

FIG. 9. *Bearden's spectrum of CuK series from glass grating of 600 lines per mm.*

wave length measurements by this method is the more interesting since the older physicists were unable to observe any diffraction effects whatever when x-rays strike an optical grating. We now find that if the x-rays are allowed to fall on the grating at a fine glancing angle, a strong reflected beam is observed which has in it the various orders of diffraction.

Careful measurements of x-ray wave lengths by this method have been made by a number of investigatiors, including Wadlund, Bäcklin, Bearden, and Cork. Typical of their results are those of Bearden, one of whose spectra is shown in Fig. 9. He finds for the $K\alpha$ line of copper a wave length of $\lambda = 1.5422$ A. This spectrum line is diffracted in the first order from a crystal of calcite at an angle of 12° 15′ 28.2″. In order to obtain a precise estimate from these data of the distance between the layers of atoms in calcite, it is necessary to apply a small correction to Equation 1 to take into account the refraction of the x-rays as they enter the surface of the crystal. When this is done, it is found from Equation (1) that the grating space of calcite is,

$$D = 3.0367 \pm .0004 \text{A } (20°C), \tag{2}$$

where the probable error is due chiefly to the uncertainty of the measurements of the wave length.

W. H. Bragg has shown that the distance between layers of atoms in the crystal of calcite is given by the relation,

$$D = \left\{ \frac{1}{2} \frac{M}{N\rho\varphi} \right\}^{1/3}, \qquad (3)$$

where M is the molecular weight of calcium carbonate, N is Avogadro's number, i.e. the number of molecules per gram molecule, ρ is the density of the crystal, and φ is a geometric factor depending on the angle between the faces of calcite. Using the values,

$$M = 100.078,$$
$$\rho = 2.7102 \text{ g cm}^{-3},$$
$$\varphi = 1.09630,$$

which result from measurements of the usual type, and the value of D given above, we may solve Equation 3 for Avogadro's number. This gives

$$N = (6.0142 \pm .003) \times 10^{23} \text{ per mole.} \qquad (4)$$

Knowing Avogadro's number, the electronic charge e is given in terms of Faraday's constant of electrolysis F by the relation,

$$e = F/N. \qquad (5)$$

Using the value $F = 9648.9 \pm .7$ emu., and $c = 2.99796 \times 10^{10}$ cm./sec., we thus obtain

$$e = 4.810 \pm .002 \times 10^{-10} \text{ esu.} \qquad (6)$$

It will be recognized that this result differs appreciably from the value of the charge on the electron

$$4.770 \pm .002 \times 10^{-10} \text{ esu.}$$

based on Millikan's oil drop measurements. In searching for an explanation of this difference two difficulties have come to light. First the wave length measurements made by different investigators using different ruled gratings have given results that differ by more than the apparent probable error of the individual measurements. At the present writing it would seem that this difference is due chiefly to irregularities in the rulings of the gratings, which become especially prominent in experiments of this kind, because only a comparatively small number of lines of the grating are exposed to the incident x-ray beam. It would seem that this difficulty should be removed by making a more precise grating, and by designing the experiment in such a way that a larger portion of the grating will be exposed to the x-rays.

A second difficulty, which is more fundamental, is due to the imperfections of the crystal itself. It is known that crystals such as rock salt are much less perfect than crystals of calcite; but even calcite cannot be considered as an ideally perfect geometric arrangement of atoms in a lattice. Zwicky has recently brought forward theoretical considerations which strongly suggest that a crystal such as calcite may have extra concentrations of atoms at rather regularly spaced intervals in its lattice, sufficient to make the average density of the crystal a fraction of a percent greater than the density of the small portions of the crystal which can be considered approximately perfect. Imperfections such as those supposed by Zwicky render invalid a precise estimate of Avogadro's number according to the method as we have just outlined. Before such an application of our x-ray wave length measurements can be made with precision, it will be necessary for us to re-examine the nature of the crystal lattice and to assure ourselves that we are using a crystal that is really perfect.

This difficulty with the crystal lattice does not, however, affect the ruled grating measurements of the wave length of the x-rays. In fact it appears that by this straight-forward application of optical methods to x-ray diffraction we obtain our most reliable standard for x-ray wave lengths.

Index of Refraction and the Specific Charge of the Electron

Recently Birge and others have called attention to the fact that one of the fundamental constants of physics, the ratio of the charge to the mass of the electron e/m, appears to have slightly different values according to the method by which this constant is measured. Spectroscopic data such as measurements of the Rydberg constant and studies of the Zeeman Effect agree in giving a value of

$$e/m_s = 5.279 \times 10^{17} \text{ esu/g};$$

measurements of the diffraction of cathode rays by electric and magnetic fields, however, have agreed in giving a value of about

$$e/m_d = 5.303 \times 10^{17} \text{ esu/g}.$$

This difference is considerably greater than the apparent experimental error of the two types of measurement. It might be supposed that this difference represents a true difference in the properties of the electron when it is attached to an atom. Such a proposition is however very difficult to justify on theoretical grounds, and it becomes highly desirable to find additional methods of determining this ratio which will be independent of those just considered. It appears that such a method is available, if it is possible to make a precise determination of the index of refraction of x-rays.

Acording to the Drude-Lorentz theory of optical dispersion, the index of refraction of a medium containing electric oscillators is given by the expression

$$\mu = 1 + \frac{e^2}{2\pi m} \sum_1^N \frac{n_s}{(\nu_s^2 - \nu^2)} \quad (7)$$

where n_s is the number of electrons which have a natural frequency of ν_s, N is the number of such groups of electrons and ν is the frequency of light.

If the frequency is very high, as in the case of x-rays, we may reach the condition where ν_s^2 is negligible compared with ν^2, giving, in the limiting case for very high frequency and elements of low atomic weight in which ν_s is not large,

$$\mu = 1 - \frac{ne^2}{2\pi n \nu^2} \quad (8)$$

where n is now the number of electrons per cm³ of the refracting material. Solving this expression for the ratio e/m, and noting that

$$\nu = c/\lambda$$

we obtain,

$$\frac{e}{m} = \frac{2\pi c^2}{ne\lambda^2(1-\mu)}. \quad (9)$$

It will be noted, however, that the quantity

$$ne = \frac{Ne}{W\rho} = \frac{F}{W\rho}.$$

where N is now Avogadro's number, W the molecular weight, and ρ the density, and $F = Ne$ is Faraday's constant of electrolysis. We thus have in equation 9 a value of e/m expressed in terms of the index of refraction μ and of other quantities which are subject to precise measurements by the usual methods. We should thus be able to determine the specific charge of the electron almost as precisely as we can measure $1-\mu$.

In view of the fact that all the early attempts failed to measure any refraction of x-rays whatever, it is not surprising that existing measurements have failed to give the precision necessary to give valuable results by this method. During the past year, however, Dr. H. E. Stauss, working at Chicago, has developed a precision method of determining x-ray refractive indices which has given results of great interest. Fig. 10 shows the arrangement of his apparatus. X-rays, after passing through

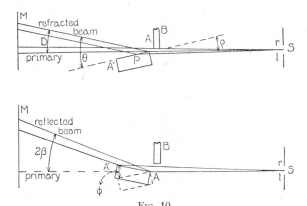

FIG. 10

a pair of slits strike one face of a right-angled prism of quartz at almost normal incidence, and are transmitted to the second face at a very fine glancing angle. Thence the rays are refracted out of the prism at an angle θ. This disposition of the apparatus is that which secures the maximum deviation of the beam, thus differing from the usual arrangement used with visible light, where the prism is placed so as to give the minimum deviation. Otherwise the problem of determining the index of refraction from the angles of incidence and refraction is precisely the same as that in the usual optical problem. In Fig. 11 is shown a dispersion spectrum of the molydenum K series which he has thus obtained.

Stauss has made measurements on the $K\alpha$ and $K\beta$ lines of molybdenum, for which he finds

$$(1 - \mu)_{K\alpha} = (1.804 \pm .001) \times 10^{-6}$$

and

$$(1 - \mu)_{K\beta} = (1.436 \pm .001) \times 10^{-6}.$$

Inserting these values into equation 9, and using Bearden's wave-length standard to calculate λ, Stauss finds from the α lines,

$$e/m = 5.290 \pm .003 \times 10^{17}$$

and from the β line
$$5.293 \pm .003 \times 10^{17}.$$

The probable error here estimated is that due merely to the uncertainty in $1-\mu$.

It will be seen that these values of the specific charge of the electron lie between those obtained from the spectroscopic and from the deflec-

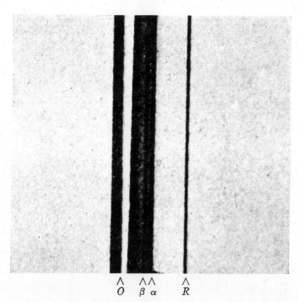

FIG. 11. *Dispersion spectrum of K series of molybdenum. O = primary beam; β and α = refracted spectrum lines; R = totally reflected beam.*

tion measurements, and it would seem that the new values are of at least comparable precision. It will be noted, however, that the value of e/m thus observed depends upon λ^2, and we have seen that λ is itself uncertain because of the probable error in the standard wave length found by different investigators. These differences in the wave length are of the order of .1%, which would mean differences in e/m of the order of .2%. A variation of this magnitude, however, would make Stauss' value of e/m fit with either of the usual values of the specific charge. It will therefore be necessary to refine our measurements of x-ray wave lengths before it is possible by this method to choose between the two alternative values of the specific electronic charge. Stauss' experiments have however indicated that we have in such an application of the methods of x-ray optics a valuable method of determining another of nature's fundamental constants.

Conclusion

Of the various methods of obtaining information regarding the world around us, probably the knowledge observed through the light which enters our eyes is the most important. It is surprising that so much information can be obtained through the medium of a single octave of electro-magnetic waves. It was perhaps to be expected, when x-rays were discovered, that because of their much shorter wave length they should be able to give us finer and more detailed knowledge of our world. We see how they are fulfilling this promise. We must consider x-rays as a kind of light, so that all x-ray problems are essentially problems of optics. Yet because of the differences in wave length they form a supplement to visible light which is of enormous value.

Review of Scientific Instruments

Vol. 2   JULY, 1931   No. 7

A PRECISION X-RAY SPECTROMETER AND THE WAVE LENGTH OF Mo $K\alpha_1$

By ARTHUR H. COMPTON

[UNIVERSITY OF CHICAGO, CHICAGO, ILLINOIS. RECEIVED APRIL 10, 1931]

ABSTRACT

An x-ray spectrometer was designed with the first crystal mounted on an arm supported by the frame of the spectrometer, and with the second crystal mounted on the central table of the instrument, whose position is read from a precision circle. The ionization chamber (of 28 cc capacity, filled with krypton) is on an arm whose position is read by a second precision circle. The instrument was built by the Societe Genevoise de Physique.

The first crystal was adjusted to throw the $K\alpha_1$ line of molybdenum over the main axis of the spectrometer, and measurements were made with the second crystal in the $(1, -1)$, $(1, +1)$, $(1, -4)$ and $(1, +4)$ positions (Allison's notation). The reflection maxima from calcite (corrected to 18° C) occur at $\theta_1 = 6°42'35.''5$ and $\theta_4 = 27°51'33.''0$ with a probable error of $0.''25$ due chiefly to errors in reading the circle. Using an apparent grating space for the first order of 3.02904A at 18° C, we get $\lambda = 707.830 \pm .002$ mA. Larsson obtained $707.831 \pm .003$ mA using Siegbahn's photographic method. Comparison of θ_1 and θ_4 gives for the index of refraction in calcite, $1 - \mu = (2.10 \pm .15) \times 10^{-6}$, as compared with Hatley's value of $(2.04 \pm .09) \times 10^{-6}$.

In order to increase the precision of x-ray diffraction measurements, with both crystals and ruled gratings, it was desirable to have available an ionization spectrometer which could be used with two or more crystals in series, with the final crystal over the axis of the precision circle. The Societe Genevoise collaborated in the design, and has constructed three of the instruments. Several spectrometers of the same general type have since been built by other instrument makers. Since this form of spectrometer can be used for certain types of measurement to which others are not readily adaptable, it is perhaps worth while to describe some of its special features.

PRINCIPLE OF OPERATION

A diagrammatic sketch of the spectrometer is shown in Fig. 1. The crystal A serves to select the ray which is to be studied in detail by crystal B. The circle on which crystal A is mounted is carried by an arm

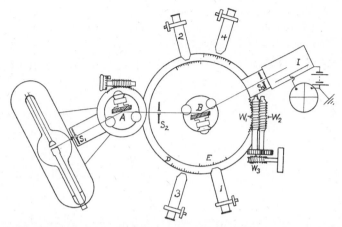

Fig. 1. *Diagrammatic sketch of double crystal spectrometer, showing second crystal over spectrometer axis, and x-ray tube mounted to turn around axis of first crystal.*

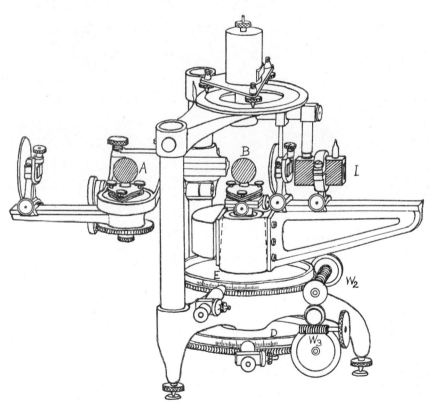

Fig. 2. *Perspective drawing of spectrometer with cover removed.*

July, 1931]   X-Ray Spectrometer   367

supported by the frame of the spectrometer, and is turned by a worm gear, the head of which is divided to minutes of arc. The arm supporting the slit S is driven by a similar worm gear, and turns about the same axis as does the crystal A. The angle of the crystal table B can be read to seconds of arc from circle D by two microscopes, as can also the position

FIG. 3. *Spectrometer in use, with x-ray tube (in lead box) and electrometer.*

of the ionization chamber[1], using circle E. The worms which are used to drive these two circles may be geared together, with a speed ratio of 2:1, and may be turned by a third worm W_3. Each revolution of worms W_1 or W_2 means an angular rotation of 1 degree, whereas a revolution of worm W_3 turns the crystal through 72 seconds and the ionization chamber through 144 seconds of arc.

The arm carrying the ionization chamber may be counterpoised for any chamber weighing less than 8 kg. Since most of the weight is carried by a ball bearing, even a heavy chamber can be turned smoothly and easily. The chamber supplied with the instrument is 12 cm in diameter by 32 cm long, with a window of aluminum, and fused quartz insulation. It may thus be used with methyl bromide or methyl iodide. This chamber is designed for absolute intensity measurements, where it is important that the β rays produced within the chamber should not reach the walls. For most purposes a chamber of only 28 cm capacity and filled with krypton has however been found more satisfactory. This chamber is shown in the figures, and is described in detail below.

The electrometer is of the Compton quadrant type, as built by the Cambridge Scientific Instrument Company, and is mounted about 18 cm above the crystal.

With this arrangement it is necessary to move the x-ray tube in order to reflect different wave lengths from crystal A. This has been accomplished by mounting the lead box, which holds the tube, on a lathe tool carrier. This is done in such a way that the tube swings about the axis of crystal table A. The tube is in a small lead tank, filled with oil, so that the weight to be moved is not excessive. A perspective drawing, partly diagrammatic, with the cover of the circles removed, is shown in Fig. 2, and Fig. 3 is a photograph of the instrument in use.

Some Construction Details

The cone which forms the axis of the instrument is supported directly by the frame of the spectrometer. The outside of this cone is the bearing about which revolves the arm carrying the ionization chamber. Inside the cone is the shaft carrying the crystal table B at one end and the circle D at the other. Thus the rotation of the crystal is entirely unaffected by that of the ionization chamber. As will be seen from the photograph, both circles are protected by a covering which is a part of the frame of the instrument. The circles have graduations at every 10 minutes of arc, and 1 scale division of the micrometer corresponds to 1 second.

Both crystal tables are fitted with levelling screws, and the crystal holders have micrometer adjustments in a horizontal plane. The arm which carries the first crystal table has room also for a third table between A and B if this is needed. The slits are symmetrical, and may be rotated about a horizontal axis. Their jaws are square rather than V shaped, and are faced with gold, 2 mm thick.

The Ionization Chamber

Special attention has been paid to the design of an ionization chamber which will give maximum sensitiveness and minimum electrical leakage. Krypton gas at 72 cm pressure is used in a copper walled chamber whose internal dimensions are $1.6 \times 3.2 \times 5.5$ cm. This column of gas is sufficient to absorb more than one third of any x-ray beam of wave length greater than 0.4A, and about 85 percent of the K radiation of molybdenum. The window, 0.6×2.0 cm, is covered with thin cellu-

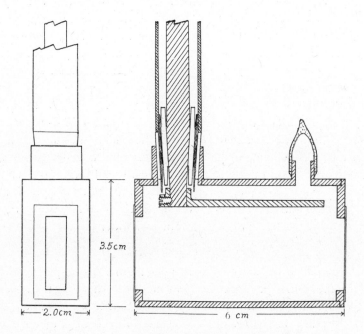

Fig. 4. *Details of ionization chamber.*

loid, cemented to the copper with "Duco household cement." The arrangement of the electrode, guard ring, and insulators is shown in Fig. 4, which represents a longitudinal section of the chamber. Amber insulation is used. All soldering was done with pure tin, in order to avoid the effect of radioactive lead. The chamber is normally operated at about 50 volts, and for convenience is insulated with tape, which is then covered with a grounded sheath of 1/32 inch lead.

The rod coming through the insulator is rigid enough to support a cross rod, which extends to a fine coiled wire suspended from the insulated quadrants of the electrometer. Thus there are only two insulators supporting the insulated system, reducing the chance of electrical leakage.

The effect of ionization within the tubes shielding the connection between the ionization chamber and the electrometer is minimized by using small shielding tubes of the same material as the lead wire. This gives a minimum contact potential difference, so that any ions formed may recombine before they find the wire.[1]

The effect of the small shielding tubes and rather large leads is to increase slightly the capacity. However the leads are relatively short, so that the total capacity of the system, including the electrometer, is about 30 cm. Using a sensitiveness of 3×10^{-4} volts per scale division (half period 4.5 seconds), this means 60,000 ions per mm deflection. If each β-ray produces an average of 600 ion pairs, this means 100 particles per division. In a 100 mm deflection, due to 10^4 particles, there is thus a mean error of about 10^{-2} or 1 percent, due solely to the limited number of β particles included in the measurement. Variations of this kind were found to be the largest source of error in many of the intensity measurements. They of course cannot be reduced by improving the measuring instruments, nor by greater steadiness of the source of x-rays. They introduce rather a natural limit to the precision of ionization measurements.

There is also occasionally (every fifteen minutes or so) an alpha particle emitted from the walls of the ionization chamber, which produces a deflection of about one scale division.

A test with the chamber exhausted showed no perceptible drift of the spot of light. When filled with krypton, the deflection due to penetrating radiation and other natural sources was about 1 mm in 4 minutes. This means about 10 ions per cm³ per second. On the other hand, when set on the Mo $K\alpha$ line reflected from the first crystal, with 1 milliampere flowing through the tube, the deflection was 360 mm in 1 second. Thus the ionization chamber-electrometer system is suitable for reading currents of from 4×10^{-17} to 4×10^{-12} amperes. At the sensitiveness employed the zero point of the electrometer does not shift more than a few centimeters in a week.

MEASUREMENT OF ANGLE OF REFLECTION
THE Mo $K\alpha_1$ LINE

A test of the instrument was made by measuring the angle of reflection of the $K\alpha_1$ line of molybdenum from a cleaved surface of calcite.

[1] Cf. A. H. Compton, Phys. Rev. 7, 649; 1916.

For this purpose the positions of the x-ray tube and of crystal A were so adjusted that the $K\alpha_1$ line of molybdenum passed over the main axis of the spectrometer. This was done with the aid of a magnifying fluoroscope, consisting of a lens of 2.5 cm focal length with a fluorescent screen placed at its focus. The angles of crystal B were then measured when reflecting the $K\alpha_1$ line in the -1 and $+1$, and the -4 and $+4$ orders. Half the difference between these angles gave θ_1 and θ_4.

Adjustments

The crystals A and B were freshly cleaved from an optically perfect sample of Iceland spar. The portions of the surfaces that were used

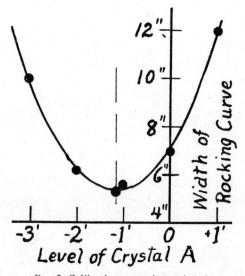

Fig. 5. *Calibration curve of crystal circle.*

showed microscopic "steps," but gave sufficiently good optical definition that their position could be observed with a telescope and Gauss eyepiece. Crystal B was set with its face accurately parallel to the axis of the spectrometer, in the following manner. First a plane-parallel plate of glass was mounted on the crystal table, and was so levelled that the images of the cross hairs of the Gauss eyepiece were reflected back upon themselves when the glass plate was turned to reflect from either face. Thus the axis of the telescope was adjusted accurately perpendicular to the spectrometer axis. The crystal was then placed on position, and levelled until it in turn reflected the cross hairs back upon themselves. The error in this adjustment, due chiefly to the optical imperfection of the crystal face, was probably not more than 5 seconds of arc.

Crystal A was placed approximately parallel to the main spectrometer axis by turning the telescope and Gauss eyepiece alternately on crystals A and B. Due to the lack of parallelism between the spectrometer axis and the axis about which the telescope turned, the error of this adjustment was of the order of 1 minute of arc. The final adjustment was made by levelling crystal A until the width of the $(1, -1)$ rocking curve is a minimum.[1] A typical curve, showing the width of the rocking curve at half maximum for different settings of crystal B, is shown in Fig. 5. In this manner, the proper adjustment of the level can be made to within a few seconds of arc.

Two slits only, those next to the x-ray tube and to the ionization chamber respectively, need accurate levelling. This was done as well as possible with the help of a telescope placed perpendicular to the spec-

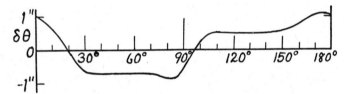

Fig. 6. *Effect of level of crystal A on width of rocking curve (widths taken at half maximum).*

trometer axis. The final adjustment was however made by finding the position of the slit next to the chamber at which the angle between the $+4$ and -4 reflections was a minimum.

As is well known, the height of the slits affects both the width and the position of the minimum of the lines. This effect was made small by using slits only 2 mm high and 40 cm apart, and the remaining effect was corrected for as described below.

The circle was calibrated at 22.5 degree and at 4.5 degree intervals, using two distant illuminated cross hairs at a fixed distance apart, which were reflected from a mirror into a large telescope. The mean calibration curve is shown in Fig. 6. There are in addition random variations in the position of neighboring lines which are of the order of ± 0.4 second of arc. The largest errors in this calibration came in reading the position of the circle graduations. It will be seen that the circle can be relied upon to within about 1 second without applying any corrections.

In order further to reduce the chance of consistent errors, three sets of readings were taken using different portions of the circle.

[1] Cf. S. K. Allison and J. H. Williams, Phys. Rev. *35*, 1476; 1930.

The Angle Measurements

In Figs. 7 and 8 are plotted typical data, forming one set of measurements of θ_1 and θ_4. By taking the observations in the order indicated by the letters, errors due to gradual shifts of the apparatus are eliminated. The principle errors are probably due to uncertainty in reading the circle.

After correcting for the circle calibration, the angle measurements are as follows.

$$\theta_1 = \tfrac{1}{2}(\theta_{1,-1} - \theta_{1,1}) = 6° \ 42' \ 34.''0 \text{ at } 22.30°C$$
$$= 6° \ 42' \ 35.''2 \quad 21.05°C$$
$$= 6° \ 42' \ 36.''0 \quad 21.25°C$$
$$\theta_4 = \tfrac{1}{2}(\theta_{1,-4} - \theta_{1,4}) = 27° \ 51' \ 30.''4 \quad 22.60°C$$
$$= 27° \ 51' \ 31.''2 \quad 21.35°C$$
$$= 27° \ 51' \ 32.''0 \quad 20.73°C$$

It can be shown that if $\delta D/D$ is the relative expansion of the grating space of the crystal due to a rise in temperature δT, the corresponding change in the angle of reflection is

$$\delta\theta = -\tan\theta \frac{\delta D}{D}. \tag{1}$$

Using Siegbahn's value[1] of

$$\frac{\delta D}{D} = 1.04 \delta T \times 10^{-5},$$

these angles have been reduced to 18°C, giving

$\theta_1 = 6° \ 42' \ 35.''1$ $\theta_4 = 27° \ 51' \ 35.''6$
$ 6° \ 42' \ 36.''0$ $ 27° \ 51' \ 34.''9$
$ 6° \ 42' \ 36.''8$ $ 27° \ 51' \ 35.''1$

Mean $6° \ 42' \ 36.''0$ Mean $27° \ 51' \ 35.''2$

with a probable error of the mean in each case of about $\pm 0.''25$.

There is also a correction to be applied due to the height of the slits. If a ray which strikes the crystal is inclined at a small angle α with the perpendicular to the spectrometer axis, it will be reflected not at the theoretical angle θ but at $\theta + \delta\theta$, where

$$\delta\theta = \tfrac{1}{2}\alpha^2 \tan\theta.$$

[1] M. Siegbahn, "The Spectroscopy of X-rays," (1925) p. 85.

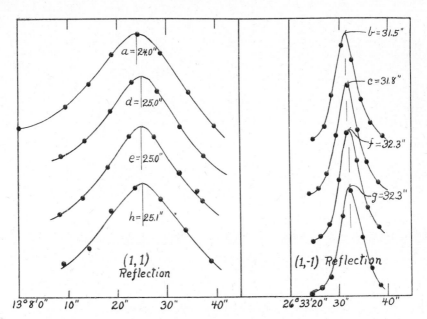

Fig. 7. *Successive readings, taken in order a, b, c · · ·, of the positions of the (1, 1) and (1, −1) reflections.*

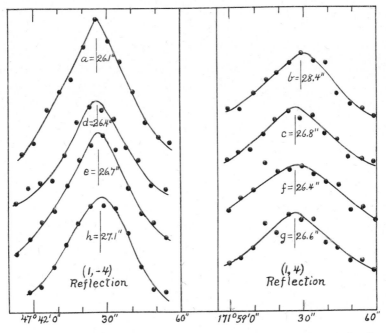

Fig. 8. *Readings of the 4th order reflections. Note the great width of the (1, 4) reflection as compared with (1, −1).*

Now if a is the height of one slit, b that of the other, and if L is their distance apart, the maximum value of α (if the adjustment is accurate) is

$$\alpha_{max} = \frac{1}{2}\frac{(a+b)}{L}, \text{ and}$$

$$\delta\theta_{max} = \frac{1}{8}\frac{(a+b)^2}{L^2}\tan\theta. \tag{2}$$

By integrating the effect over both slits it can be shown that the effective value of $\delta\theta$ is

$$\delta\theta_{eff} = \sqrt{\frac{1}{3}}\frac{\sqrt{a^2+b^2}}{a+b}\delta\theta_{max}. \tag{3}$$

In the present case, where $a = b$, this becomes

$$\delta\theta_{eff} = \sqrt{\frac{1}{6}}\delta\theta_{max} = \sqrt{\frac{1}{6}\frac{1}{2}\frac{a^2}{L^2}}\tan\theta. \tag{4}$$

Using $a = 0.2$ and $L = 40$ cm, this gives $\delta\theta_1 = 0.''5$, and $\delta\theta_4 = 2.''2$. Thus our corrected angles become,

$$\theta_1 = 6° 42' 35.''5 \pm .''25$$
$$\theta_4 = 27° 51' 33.''0 \pm .''25$$

at 18°C.

Index of Refraction and Wave Length

Because of the refraction of x-rays in the crystal, it is well known that Bragg's law

$$n\lambda = 2D\sin\theta \tag{5}$$

must be replaced by

$$n\lambda = 2D\sin\theta\left\{1 - \frac{\delta}{\sin^2\theta}\right\}, \tag{6}$$

if a precise value of λ is to be calculated. Here $\delta \equiv 1-\mu$, where μ is the index of refraction. On the other hand, by comparison of the angles of reflection of the various orders, it is possible to calculate the refractive index from the expression,[1]

$$\delta = \frac{\lambda_1 - \lambda_n}{\lambda_1}\frac{n^2}{n^2-1}\sin^2\theta_1, \tag{7}$$

where

$$\lambda_n = \frac{2D}{n}\sin\theta_n. \tag{8}$$

---

[1] Cf. *e.g.*, A. H. Compton, "X-rays and Electrons," p. 213.

Placing n = 4 in these expressions, we thus find from the above values of θ_1 and θ_4,

$$\delta = (2.10 \pm .15) \times 10^{-6}. \tag{9}$$

This is in satisfactory agreement with Hatley's value[1] of $\delta = (2.04 \pm .09) \times 10^{-6}$ for the Mo $K\alpha$ line in calcite, which he obtained by the crystal wedge method. It does not correspond so well, however, with the value 1.85×10^{-6} to be expected from the Drude Lorentz dispersion theory.

Siegbahn's wave lengths are calculated from Bragg's law (eq. 5) using for first order reflection $D' = 3.02904$A at 18°C, where as compared with equation (6),

$$D' = D\left(1 - \frac{\sigma}{\sin^2 \theta}\right).$$

Using the value of $\delta = 1.85 \times 10^{-6}$, when measurement is made in the first order this becomes equivalent to calculation from equation (6) using a value of $D = 3.02945$A.

The calculation of λ should be made from θ_4, since the value from θ_1 is by comparison of negligible weight. Using $D = 3.02945$A in eq. (6), we thus obtain

$$\lambda = 707.830 \pm .002 \text{ milliAngstroms}.$$

This is to be compared with other recent values, recalculated on the same basis,[4] as follows,

Observer		Date
Leide[2]	707.80 m A	1925
Allison and Armstrong[3]	707.865 ± .07	1925
Larsson[4]	707.831 ± .003	1927
Present Result	707.830 ± .002	1931

The remarkable agreement between the present results and those of Larsson is of course in part fortuitous. But considering the fact that his measurements were made using a photographic method, and a wholly different technique of angle determination, the agreement serves to emphasize the reliability of modern x-ray measurements.

The writer wishes to thank Mr. Y. Tu for his valuable help in taking the readings.

[1] C. C. Hatley, Phys. Rev. *24*, 486; 1924.
[2] A. Leide, Comptes Rendus, *180*, 1203; 1925.
[3] S. K. Allison and A. H. Armstrong, Phys. Rev. *26*, 701; 1925.
[4] A. Larsson, Phil. Mag. *3*, p. 1136; 1927.

THE UNCERTAINTY PRINCIPLE AND FREE WILL

In his very excellent presentation of the uncertainty principle, published in a recent number of SCIENCE,[1] Professor Darwin concludes with a comment regarding the significance of this principle in connection with the problem of "free will," which should not be allowed to pass without comment. He may be correct in his view that "the question is a philosophic one outside the thought of physics." Yet the reason that he offers to show that the uncertainty principle does not help to free us from the bonds of determinism is inadequate.

Darwin's argument is that "physical theory confidently predicts that the millions of millions of electrons concerned in matter-in-bulk will behave . . . regularly, and that to find a case of noticeable departure from the average we should have to wait for a period of time quite fantastically longer than the estimated age of the universe." He apparently overlooks the fact that there is a type of large-scale event which is erratic because of the very irregularities with which the uncertainty principle is concerned. I refer to those events which depend at some stage upon the outcome of a small-scale event.

As a purely physical example, one might pass a ray of light through a pair of slits which will so diffract it that there is an equal chance for a photon to enter either of two photoelectric cells. By means of suitable amplifiers it may be arranged that if the first

[1] C. G. Darwin, SCIENCE, 73, 653, June 19, 1931.

2

photon enters cell A, a stick of dynamite will be exploded (or any other large-scale event performed); if the first photon enters cell B a switch will be opened which will prevent the dynamite from being exploded. What then will be the effect of passing the ray of light through the slits? The chances are even whether or not the explosion will occur. That is, the result is unpredictable from the physical conditions.

Professor Ralph Lillie has pointed out[2] that the nervous system of a living organism likewise acts as an amplifier, such that the actions of the organism depend upon events on so small a scale that they are appreciably subject to Heisenberg uncertainty. This implies that the actions of a living organism can not be predicted definitely on the basis of its physical conditions.

Of course this does not necessarily mean that the living organism is free to determine its own actions. The uncertainty involved may merely correspond to the organism's lack of skill. Yet it does mean that living organisms are not subject to physical determinism of the kind indicated by Darwin.

ARTHUR H. COMPTON

UNIVERSITY OF CHICAGO

[2] Ralph Lillie, SCIENCE, 66, 139, 1927. Lillie draws much the same conclusion as that found here.

ASSAULT ON ATOMS[1]

By Arthur H. Compton

[With 2 plates]

Twenty-five hundred years ago, Thales, the first true scientist of ancient Greece, undertook to solve the problem, "Of what and how is the world made?" Almost a hundred generations have passed and the problem is not yet solved.

Democritus and his followers thought they had found the solution. Everything is made of atoms. "According to convention there is a sweet and a bitter, a hot and a cold, and according to convention there is color. In truth there are atoms and a void." Thus, in terms of motions of minute particles the ancient Atomists accounted for their world. Mountains and seas, trees and people, even life and thoughts, were thus explained.

But Socrates and Plato would have none of their atoms. Did they not in Democritus' hands rob men of their personality? Atoms are thus worse than useless, for they destroy the basis of morality. Here in Athens, around the question of atoms, was staged the first great battle between science and religion. Epicurus and Lucretius took up the cudgels on behalf of the atomists, but Plato carried the day, and atoms were forgotten until the revival of scientific thought during the Renaissance. Though our present day atomic theories are based on much firmer foundations than those of Democritus, they owe their origin to his ideas, transmitted down through the centuries.

A few years ago we were camped beside a mountain lake in the foothills of the Himalayas, studying cosmic rays. The warm air from the plains of India was carried up over a range of mountains, and came down again into the beautiful Vale of Kashmir. Clouds were continually forming as the air, cooled by expansion as it came up the mountain side, became supersaturated with moisture. But after passing the peak of the range, the air was warmed by com-

---

[1] Read before the American Philosophical Society, Apr. 23, 1931. Reprinted by permission from Proceedings of the American Philosophical Society, vol. 70, No. 3, 1931.

pression as it sank to lower levels, and the clouds evaporated into thin air.

It was while watching such clouds in his native hills of Scotland that C. T. R. Wilson conceived his beautiful laboratory experiments on clouds. Of course he couldn't bring the mountains into his laboratory, but he could expand his moist air in a cylinder with a piston at one end. He made his cylinder of glass in order to see what was going on. I have one patterned after his design here in my hand. Here are the glass top and sides, with the whole vessel partially filled with inky water. There is a lamp beside the glass cylinder so we can see better what is going on. I can compress the air in the glass chamber by squeezing the bulb. We let the air remain under this pressure for a moment, until it becomes saturated with moisture, and then allow it to expand. As it expands the air cools and a cloud forms in the chamber just as it did on the mountain top.

Did it ever occur to you that, when a cloud forms, each little drop of moisture in the cloud must condense on something? Usually it condenses on a speck of dust floating in the air, and after a rainstorm these dust particles are carried to the ground and the air is beautifully clear. But when the dust has been removed, what can the drops condense on? There are always in the air some broken bits of atoms and molecules, which we call ions. These ions are produced by rays from radioactive substances in the ground and other sources. So, Mr. Wilson tried the experiment of placing a speck of radium in his expansion chamber, to see what kind of clouds would be formed. Let's see what happens when we repeat his experiment. Those of you who are near enough will see the little white lines radiating out from the tip of the glass rod which carries the radium. These little white lines are tiny clouds of water drops, condensed on the ions left along the paths of particles shot out by the radium. It is clear that particles of some kind are coming from the radium. What are they?

A series of photographs will illustrate what is happening in this chamber. A picture taken from above (pl. 1, fig. 1) shows the glass walls of the chamber, and the rod on which the speck of radium is placed. The more or less diffuse lines are the clouds of water drops that mark the paths of the particles ejected from the radium.

What are these particles? Let us call them alpha particles, in order not to imply anything about what they are, and look into their properties. Plate 1, Figure 2, shows a sharper photograph, each line a thin straight cloud, marking the path of an alpha particle. Rutherford (recently made Lord Rutherford in recognition of his work with atoms) caught a large number of these particles to find out what they were when there are enough of them to handle. Niton is a radio-active gas, a hundred thousand times as active as radium. He compressed some of this gas into a fine glass tube with walls so thin that the alpha particles would pass right through. After a few days he noticed gas collecting in the space surrounding this tube, and this gas he forced into a fine tube above. On passing an electric discharge through the tube and looking through a spectroscope at the light emitted, he saw the brilliant spectrum characteristic of the gas helium.

Many of you know the romance of helium. Observed many years ago by Lockyer in the spectrum of the sun, it remained unknown on the earth for a generation until Rayleigh and Ramsay, making a precise measurement of the density of the nitrogen in the air, found it different from the nitrogen prepared in the laboratory. Search for the cause of the discrepancy revealed a whole series of new gases—argon with which our incandescent lamps are filled, neon with which we advertise our wares in blazing red, helium with which we now fill our dirigibles, and two others, krypton and xenon, which are now of great value in certain laboratory experiments. Thus was helium found, and here we see it being formed—the birth of helium atoms. For these alpha particles are none other than atoms of helium gas.

We can count these atoms one by one as they come from a preparation of radium. It might be done using an expansion chamber of this type, and counting the tracks as they appear. A better method is to allow the atoms to enter an electrical counting chamber. Each particle then can make its record on a moving film, as we see in Plate 1, Figure 3. Every little peak here marks the birth of a helium atom from its parent radium.

Imagine that we have thus counted all the atoms of helium that come through the walls of Rutherford's glass tube, and make the gas that he observed in his spectroscope. How many atoms would we have? In a little glass bulb the size of a large pea, filled with helium at atmosphere pressure, the number of atoms is about 1 with 19 ciphers after it. Perhaps that doesn't mean much to you. Let me put it this way. Two thousand years ago Julius Caesar gave a dying gasp, " Et tu Brute?" In the intervening millenniums the molecules of air that he breathed out with that cry have been blown around the world in ocean storms, washed with rains, warmed by the sunshine, and dispersed to the ends of the earth. Of course only a very small fraction of these molecules are now in this room; but at your next breath each of you will probably inhale half a dozen or so of the molecules of Caesar's last breath.

Molecules and atoms are very tiny things; but there are so many of them that they make up the world in which we live.

The story is told of Lord Kelvin, a famous Scotch physicist of the last century, that after he had given a lecture on atoms and molecules, one of his students came to him with the question, " Professor, what is your idea of the structure of the atom? " " What," said Kelvin, " the *structure* of the atom? Why, don't you know, the very word ' atom ' means the thing that can't be cut. How, then, can it have a structure? "

" That," remarks the facetious young man, " shows the disadvantage of knowing Greek."

Does the atom have parts?

THE ELECTRON

Do you see the faint little trail at the bottom of Plate 1, Figure 4? It appears to be due to something much smaller than the particle which made the broad bright trail above it. If we called the one an alpha particle, let us call the other a beta particle, and try to find out what kind of thing it is.

Plate 1, Figure 5 shows a large number of these beta particles, that have been knocked out of air molecules by the action of X rays. You can see where the X rays passed through the middle of the chamber. Now every substance has its own peculiar kind of atoms. Iron atoms differ from oxygen atoms, and these from atoms of carbon and so on. But these beta particles are all alike, as far as we can tell, and they can be knocked out of anything. Had we put into the chamber fried eggs or a platinum wrist watch, the same kind of beta particles would have been observed. Thus beta particles are things which go to make up all kinds of matter. They are more fundamental even than atoms.

But what are these beta particles? In the first place they carry an electric charge. Notice in Plate 2, Figure 1 how their trails are curled up if a magnet is held near the expansion chamber. This is because the moving electric charge acts like a wire carrying an electric current, and the particles form the armatures of tiny electric motors.

Professor Millikan, a member of our society, spent years at the University of Chicago in measuring the charge carried by one of these little particles. He built himself an electroscope in which a tiny drop of oil took the place of the usual gold leaf, and he would catch these beta particles on his oil drop. Every particle carried the same charge, he found. It was also the same charge that a hydrogen ion carries when water is dissociated into oxygen and hydrogen by the passage of an electric current.

Smithsonian Report, 1931.—Compton PLATE 1

1

2

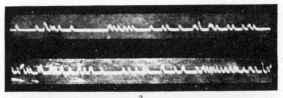
3

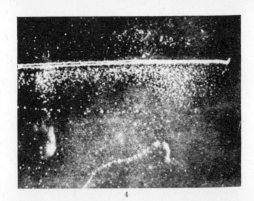
4

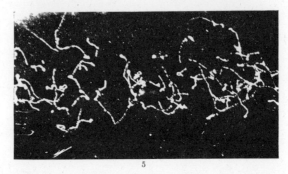
5

1, Tracks of alpha particles (helium atoms) made visible by condensing clouds along their paths (Wilson); 2, helium atoms ejected from radium (Wilson); 3, counting atoms. Each peak marks the entrance of one helium atom into the counting chamber (Geiger and Rutherford); 4, trails of alpha and beta particles (Wilson); 5, beta particles ejected from air by X rays (Wilson)

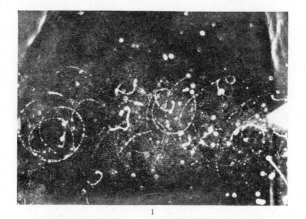

1, Beta particles (electrons) curved by magnetic field (Skobeltzyn); 2, alpha particle knocking hydrogen nucleus out of nitrogen atom (Blackett); 3, "appearance" of a helium atom, as found by X rays (Langer)

Because it carries this unit of electric charge, which seems to be an indivisible unit, these beta particles were called *electrons*, and by that name they have become familiar.

These electrons have been weighed, too, and their weight is found to be very small indeed. The atom of hydrogen is the smallest atom we know, and as we have seen, it is a very tiny thing. But an electron weighs only 1/1845 as much as does a hydrogen atom. Thus we were correct in guessing that the beta particle which made the faint trail was much smaller than the alpha particle that made the broad bright streak on an earlier photograph.

The electron is indeed one of the components of which the atom is built. We can in fact count the number of electrons that each atom has. Hydrogen has 1 electron, helium 2, lithium 3, and so on. Oxygen has 8 electrons in each atom, iron 26, and uranium, the heaviest atom of all, has 92 electrons.

THE NUCLEUS AND THE PROTON

But this is only a part of the story. The electrons are all particles of *negative* electricity. The atom itself is electrically neutral, and must therefore have in it some positive electricity to neutralize the negative electrons. If time were available, I should describe for you the beautiful experiments carried out by Rutherford and Aston in Cambridge, Dempster at the University of Chicago, and others, which have shown that this positive electricity is concentrated in a very small nucleus, which though much smaller in size that the atom has yet nearly all the atom's weight.

The careful experiments of Dempster and Aston have shown that the weights of the nuclei of the various atoms, such as oxygen, nitrogen, sodium, and the rest, are whole multiples of a unit which is nearly equal to the weight of the hydrogen nucleus. This suggested that the various atomic nuclei are built up of hydrogen nuclei. This idea was supported by the fact that the electric charge carried by the various atomic nuclei is always an integral multiple of the charge carried by hydrogen nucleus.

Many attempts have been made to make one element out of another. This is in fact the old problem of alchemy, to make gold out of lead. The first success was got by Rutherford. He didn't get gold out of lead; but he did get hydrogen out of nitrogen and out of aluminum and other elements.

The experiment can best be shown using again our cloud expansion apparatus, as has been done for example by our fellow member, Professor Harkins. Plate 2, Figure 2 shows a group of alpha particles shooting through nitrogen gas. Most of them go straight to the end of the path, and this is remarkable, for each alpha passes right through tens of thousands of nitrogen atoms before its flight is stopped. But here we see a really surprising occurrence. The alpha particle dives into a nitrogen atom, and out of it emerges a smaller particle, which goes out leaving a thin straight trail. The nitrogen nucleus with the alpha particle now attached moves heavily along in a different direction. The alpha particle has served as a hammer to knock a hydrogen nucleus out of a nitrogen atom.

Similar experiments have been done with many other elements, and most of the lighter ones have thus been disintegrated, expelling always a hydrogen nucleus. Thus we may take this nucleus, like the electron, as a component of which the various atoms are built. We give to the hydrogen nucleus now the name of *proton*, i. e., the original or fundamental thing. Out of protons and electrons we believe all the 92 different kinds of atoms are built.

HOW THE ATOM IS BUILT

Old Ptolemy, the ancient Greek astronomer, knew that there was a sun and a moon, the earth, and the planets, but he didn't know what the solar system is. When Copernicus and Galileo showed, however, that there is a sun, around which revolve planets in definite orbits, then men felt that they had become acquainted with their world. So, though we have found the parts of which the atom is made, we really don't know the atom until we know how these parts are put together.

Perhaps the best way to find out how something is made is to look at it. If it is something like a watch, which we can hold in our hands, this is comparatively easy. If it is the cell structure of a muscle that we wish to examine, we put it under a microscope. But some things are too small to see, even in a microscope. By using ultra-violet light of wave length shorter than ordinary light, we can photograph such things as typhoid bacilli with increased sharpness. But atoms are too small even for this.

Now, X rays have a wave length only a ten-thousandth that of light, and if we could use them in a microscope it should be possible for us to observe even the tiny atoms. Unfortunately, we can not make lenses that will refract X rays, and even if we could, our eyes are not sensitive to X rays. So it would seem that we shall never be able to see an atom directly.

It is nevertheless possible for us in the laboratory to get by more round-about methods precisely the same information about an atom that we should if we could look at it with an X-ray microscope. I have spent a large part of the last 16 years trying to find what the atom looks like, and it has become something of a game with me.

Last summer while spending a brief vacation in northern Michigan, I noticed a fuzzy ring, not very large, around the moon. Half an hour later the ring was perceptibly smaller, and within an hour we had to come in out of the rain.

This ring was due to the diffraction of the moonlight by tiny water droplets that were beginning to form a cloud. The size of the ring depends upon the size of the water drops—if the drops are small, the ring is big, and vice versa. So when the ring grew smaller it meant that the drops were growing larger. Soon they would fall as rain.

Our method of studying atoms is very similar to this method of finding out the size of the droplets in a cloud. Instead of the moon we use an X-ray tube, and in place of the cloud of water droplets we use the atoms in air or helium. For the wave length of the X rays bears about the same ratio to the size of a helium atom that a light wave bears to a droplet of water in a fog. The helium atoms spread the X rays out into a halo. This halo, now of X rays scattered by the helium atoms, corresponds precisely to the ring around the moon diffracted by the cloud droplets. Likewise here, from the diameter of this halo, we can estimate the size of the helium atom. We can also tell pretty much what it looks like, just as if the atom were under the microscope.

Plate 2, Figure 3, shows how the helium atom would look if we were to see it with an X-ray microscope. The picture is drawn carefully from the data we have got from the diffraction halos. Of course, it is highly magnified, about a thousand million times. Such a magnification would make a pea appear as big as the earth.

In the middle of this fuzzy ball somewhere is the nucleus of the helium atom, which has in it the protons. This fuzzy atmosphere is due to the electrons. We noted above that the helium atom has only two electrons in it. You may wonder how with only two electrons the atom can seem so diffuse. Did you ever see the boys on the Fourth of July waving the sparklers to make circles or figures eight? Of course the sparklers weren't in the form of circles; they appeared that way because they moved so fast. So here, the electrons give this continuous, diffuse appearance to the atom because they are now here and now there, and we have caught a "time exposure" of their average positions. This is, of course, what we would see if we could look at the atom.

There have been 57 varieties of atomic theories proposed. Lord Kelvin thought the atom was something like a smoke ring; J. J. Thomson said it was a sphere of jelly. Rutherford called it a miniature solar system, while Bohr and Sommerfeld calculated precisely the orbits of the planetary electrons revolving about the central nucleus. Lewis and Langmuir objected, and said the atom is a cube. "Not so, it's a tetrahedron," claimed Lande. "Quite a mistake; it's a diffuse atmosphere of electricity around a central core," says Schrödinger. "Only it isn't diffuse electricity," complains Heisenberg, "It's electrons moving now here, now there, which make up this atmosphere."

Each of these theories has found support in that it has explained certain physical or chemical or spectroscopic properties of atoms. For the most part, each theory has been better than the one before, because it has explained the things which the earlier one described and some new thing as well. It may seem over-optimistic to suppose that there is anything final about the most recent theory. Yet the fact remains that there is one and only one such picture, namely, that of Heisenberg, that describes what we find when with our "X-ray eyes" we look into the atom.

Does this mean that the problem of the structure of the atom is solved? Not yet! We feel that we know in general outline what this electron atmosphere of the atom is like; but there's the nucleus of the atom. What is it like?

"What's the idea of bringing that up?" you ask me. "Surely that little nucleus isn't big enough to amount to anything!"

It is the nucleus of the radium atom from which the alpha particles came. Did it occur to you that those alpha particles carry a tremendous amount of energy? It is about a million times as much as is released when a molecule of TNT explodes. It is only because they are liberated one at a time that the alpha particles make so little impression.

Did you ever pause to wonder where all the energy of the sun comes from which it is pouring out as heat? If it were made of pure coal burning in oxygen, the sun could shine with its present brilliance for only a few thousand years, less than the era of history, before it would be reduced to a cinder. Even if it were composed of uranium or radium, and got its heat from their disintegration, it would last only for a few billion years, which is about the age of our own earth; yet our geological records indicate no change in the sun's brightness over this vast period. The best astronomical evidence indicates that the sun must be at least a thousand billion years old. What is the enormous supply of energy which has kept it hot for so long a time? Professor McMillan has pointed out that apparently the only way to explain the sun's long life is to suppose that the sun is consuming itself. If under the extreme pressure and temperature of the sun's interior the electrons and protons in an atom should come together and neutralize each other, all of their energy would be liberated and add to the sun's heat. Such a process would release energy almost beyond belief. From five drops of water, if we could thus squeeze out all the energy, we should be able to run all the power stations in Philadelphia for 24 hours.

Is it possible for man to tap these great stores of energy? We do not know. We know the energy is there, and the evidence is strong that it is being liberated in the sun and stars. But under what conditions? Perhaps we can not realize the proper conditions here on the earth. In any case it is our job—the physicists' job, that is—to find out whether this energy can be used, and, if so, how.

If we are to find the conditions for the release of these vast stores of energy, we must acquaint ourselves with the atomic nucleus, for it is there that the energy lies. Studies of the band spectra of molecules have shown us something about the rotation of the nucleus. The masses of the nuclei and their electric charges have been measured by the help of magnetic spectrographs and scattered X rays. Attempts have been made to disintegrate atomic nuclei by bombardment with high speed electrons shot by high voltages. But by far the most fruitful tool for studying the nucleus has been radioactivity.

Experiments with scattered alpha rays have shown the minute size and relatively large mass of the nucleus. They have enabled us to measure its charge and even to estimate the field of electric force in its neighborhood. Further information on the latter point is given by the speed with which the alpha particles are ejected from the radioactive nucleus. Combining the evidence from these alpha ray experiments, it becomes evident that surrounding the nucleus there is a "potential wall" which prevents alpha particles that are outside from entering the nucleus and those on the inside from escaping. We are thus afforded a basis for developing a quantum theory of radioactive disintegration according to which the probability of an alpha particle jumping this wall is greater if it has large energy, and a qualitative explanation of one of the fundamental laws of radioactivity is obtained. Studies of the sharpness of gamma ray lines suggest a nucleus in which planetary alpha particles correspond to the electrons of the outer atom; though how these particles are held together remains unknown. Similarly the condition of the electrons in the nucleus remains unsolved. There is no gamma radiation that can be traced to these electrons, and when they appear as beta particles their energies are distributed over broad bands. Though much new light is shed by these studies in radioactivity, the nucleus of the atom, with its hoard of energy, thus continues to present us with a fascinating mystery.

Thus our assault on atoms has broken down the outer fortifications. We feel that we know the fundamental rules according to which the outer part of the atom is built. The appearance and properties of the electron atmosphere are rather familiar. Yet that inner citadel, the atomic nucleus, remains unconquered, and we have reason to believe that within this citadel is secreted a great treasure. Its capture may form the main objective of the physicists' next great drive.

The Appearance of Atoms as Determined by X-Ray Scattering

E. O. WOLLAN AND A. H. COMPTON, *University of Chicago*
(Received July 23, 1934)

Photographs are shown which are the images of atoms of helium, neon and argon as obtained by x-ray diffraction, magnified about 2×10^8 times. The images are obtained by photographing a rotating template whose shape is calculated by a mathematical transformation of our measured values of the x-rays scattered by the respective gases. This mathematical-mechanical procedure corresponds to the lens which forms the image when a microscope is used. The images formed by our procedure should be true representations of the electron distributions in the atom, except for the limited resolving power and certain minor aberrations. The photographs show the helium atom as a diffusely continuous region filled with electricity. In neon, the inner group of K electrons is clearly distinguishable from the L electron group. The resolving power is insufficient to distinguish the K and L groups of electrons in argon, but does separate these from the M electrons. The appearance of these atoms is in good accord with modern quantum theory of atomic structure.

THE study of the diffraction of x-rays by crystals and gases has made it possible to obtain a great deal of information regarding the arrangement and spacing of the atoms in gas molecules and in complicated crystals. It has also given us what is probably the most direct method of obtaining the distribution of the electrons within an atom. In the present paper it will be shown how one can obtain, with the aid of x-rays, what, in effect, corresponds to a photograph of an atom. It is, of course, evident that one cannot photograph an atom by the straightforward method of projecting its image on a photographic plate by means of lenses, since there are no lenses which will focus x-rays. The details of the appearance of atoms are obtained from x-ray diffraction data and the photograph is produced by a combined mathematical and mechanical procedure.

Many examples could be cited in which the details of the appearance of an object cannot be obtained by direct photography, but a study of its diffraction pattern can be made to give the desired results. Our knowledge of star diameters for example, is based on such information. A simple illustration of the way a photograph of an object can be derived from its diffraction pattern is given by Michelson.[1] By means of calculation based on the diffraction pattern produced by a single slit, one can construct a new diffraction screen, and if parallel light is passed through this, an image of the original slit is obtained. W. L. Bragg[2] has used a method

[1] A. A. Michelson, *Studies in Optics* (University Chicago Press, 1928), p. 60.
[2] W. L. Bragg, Zeits. f. Krist. **70**, 489 (1929); *Crystalline State* (Macmillan, 1934), p. 227.

The photographs obtained by Bragg do not show great detail in the structure of the atoms as those presented. This is doubtless due in part to the fact that photographic method of obtaining the image which uses must result in considerable loss of detail; but the also the essential limitation of the crystal method

229

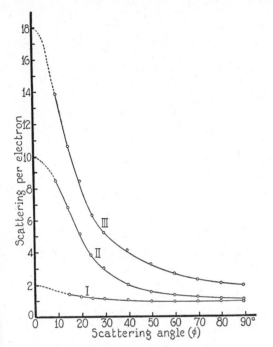

FIG. 1. Intensity of x-rays scattered by gases. I, helium; II, neon; III, argon.

almost identical in principle for securing a photographic image of the projection on a plane of the atoms in a crystal.

We have used a somewhat similar procedure in obtaining pictures of atoms. The atom is the diffraction screen for which we measure the intensity of the diffraction pattern. The photograph is a result of a mathematical and mechanical procedure which, in effect, does the work which the lens does in a camera.

In the experimental set-up a parallel beam of x-rays falls on an atom which gives its characteristic diffraction pattern or in other words an angular distribution of intensity of scattering. We cannot, however, isolate a single atom, nor could we measure the very feeble scattering from it. Actually, we observe the x-ray diffraction from a gas consisting of n atoms distributed at random, which diffraction pattern we know is the same as that due to one atom except that it is n times as intense. The photograph which we obtain then represents an average picture of these n atoms.

Fig. 1 shows the diffraction data obtained[3] by scattering x-rays of wave-length $\lambda = 0.71$A from helium, neon and argon. The quantity S represents the contribution to the total scattered intensity made by each electron measured in units of the Thomson scattering by a single free electron. From these data one can calculate the distribution of the diffracting medium (the electrons) within the atom as a function of its radius. If we represent the number of electrons which lie at distances between a and $a+da$ from the center of the atom by $U(a)$ we have previously shown[4] that it is related to the experimental quantity S by the Fourier integral,

$$U(a) = \frac{2Za}{\pi} \int_0^\infty \left\{ \frac{S-R}{Z-R} \right\}^{\frac{1}{2}} k \sin (ak) dk, \quad (1)$$

where Z is the atomic number, $S = I_\varphi/ZI_e$, I_e being the Thomson scattering by a single free electrons, $k = (4\pi/\lambda) \sin (\varphi/2)$, and $R = (1 + (h/mc\lambda) \text{vers } \varphi)^{-3}$. Using the data for S given in Fig. 1, $U(a)$ can be calculated from Eq. (1) by means of a graphical integration. The result of

The atoms which are photographed are in thermal motion about their neutral positions. This, of course, makes a blurred image. Our method, being based on diffraction by gaseous molecules, assumes a random arrangement, and hence unaffected by the thermal motion. Thus our practical resolving power almost equals the theoretical value.

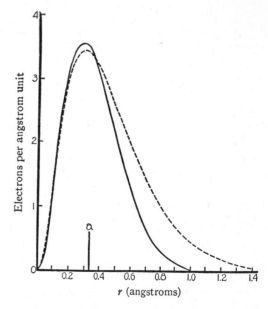

FIG. 2. Radial electron distribution in helium. Solid line from experiment, broken line from wave mechanics according to Hartree.

[3] E. O. Wollan, Phys. Rev. **37**, 862 (1931).
[4] A. H. Compton, Phys. Rev. **35**, 925 (1930).

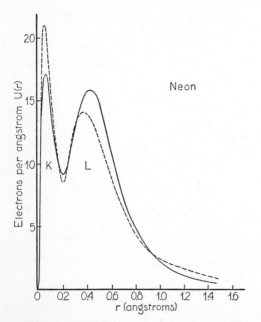

FIG. 3. Radial electron distribution in neon. Solid line from experiment, broken line from wave mechanics according to Hartree.

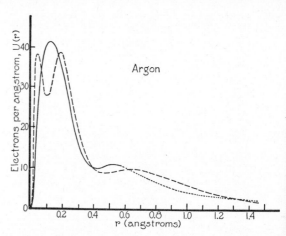

FIG. 4. Radial electron distribution in argon. Solid line from experiment, broken line from wave mechanics according to Hartree.

such a calculation for helium is shown by the solid line in Fig. 2. The dotted line is a similar curve calculated on the basis of wave mechanics and one sees that the electron distribution calculated in this way is in good accord with the experimental curve. On the basis of the older Bohr theory all the electrons in the atom would lie at a distance $a=0.33$A from the center of the atom, which does not accord with the experimental results.

Fig. 3 shows the electron distribution curve for a neon atom calculated in a similar way from Eq. (1) and the data of Fig. 1. An especially interesting feature of this curve is the fact that the K and L electrons are resolved. A similar analysis of argon as shown in Fig. 4 fails to separate the K and L groups, the approximate position of which can be seen from the dotted curve obtained from wave mechanics. This is, of course, due to the fact that these details of the atom become smaller and more difficult to resolve as the atom becomes heavier.[5]

It is interesting in this connection to make a calculation of the resolving power of our apparatus. To do this, let us assume that two electrons can be distinguished if the waves scattered by them traverse path differences of more than half a wave-length. At the most favorable orientation of the two electrons, this will occur when

$$\lambda/2 = 2\delta \sin(\varphi/2)$$

or

$$\delta = \lambda/4 \sin(\varphi/2),$$

where δ is the minimum distinguishable distance. Taking the value of λ and the maximum value of φ for which the data of Fig. 1 were obtained, we get $\delta = 0.25$A. This value of δ shows why the K and L electrons were resolved in the neon atom and not in the argon atom.

Although the curves of Figs. 2, 3 and 4 represent, in one way, our knowledge of these atoms, it is not easy on the basis of them to construct in one's imagination a picture of the atoms. An image of an atom as it would appear through an x-ray microscope would correspond to the projection of the electron atmosphere on a plane perpendicular to the direction of observation. To do this, let us represent the atom in cylindric

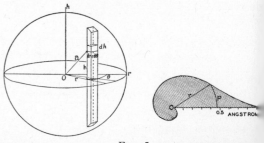

FIG. 5.

[5] For a more detailed account of these results, cf. E. O. Wollan, Rev. Mod. Phys. **4**, 243 (1932).

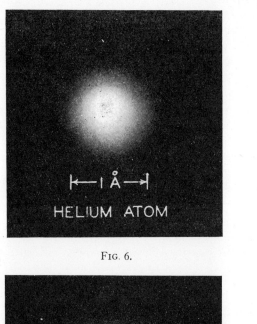

FIG. 6.

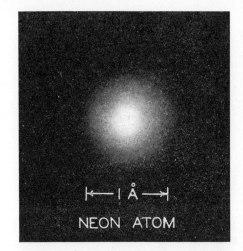

FIG. 7.

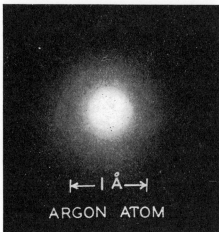

FIG. 8.

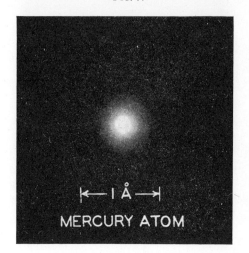

FIG. 9.

FIGS. 6, 7 AND 8 show "appearance" of atoms of helium, neon and argon, respectively, as observed with x-rays of wave-length 0.71A and a spectrometer working to an angle of 90°.

FIG. 9 shows "appearance" of mercury atom, based on Thomas-Fermi electron distribution, as supported by x-ray scattering experiments. This picture represents closely, except for differences in scale, the "appearance" of any atom heavier than argon.

coordinates as in Fig. 5, and let us project the electron density on the plane $h=0$. The number of electrons projected on the element of area $d\theta \cdot dr$ will then be

$$P(r) \cdot r \cdot d\theta \cdot dr = 2\int_0^\infty \rho \cdot r \cdot d\theta \cdot dr \cdot dh, \quad (2)$$

where $\rho = U(a)/4\pi a^2$ is the volume density of distribution of the electrons and $P(r)$ represents the number of electrons per square Angstrom in the plane of projection. Putting $h=(a^2-r^2)^{\frac{1}{2}}$, (2) becomes

$$P(r) = \frac{1}{2\pi}\int_r^\infty \frac{U(a)}{a(a^2-r^2)^{\frac{1}{2}}} da. \quad (3)$$

Values of $P(r)$ for a given atom can be determined from Eq. (3) by using the values of $U(a)$ given in curves of the form shown in Figs. 2, 3 and 4. On the other hand, $P(r)$ can be represented directly in terms of the experimentally observed quantity S by combining Eqs. (1) and (3). This gives

$$P(r) = \frac{Z}{\pi^2}\int_r^\infty \int_0^\infty \left\{\frac{S-R}{Z-R}\right\}^{\frac{1}{2}} \frac{k \sin(ak)}{(a^2-r^2)^{\frac{1}{2}}} dk da, \quad (4)$$

where one sees that the right-hand side of this equation depends only on the experimentally observed intensity S and other known or measurable quantities, and not on any theory of atomic structure.

From the Fourier integrals represented by Eq. (3) or (4), one can calculate $P(r)$ as a function of r by graphical integration. Making such a calculation on the basis of the experimental values of S in Fig. 1, we proceed to represent the function photographically. A simple method of doing this is the following: We plot $P(r)$ in polar coordinates against r, as shown by the shaded figure in Fig. 5 for the case of helium. Since $P(r)$ represents the number of electrons per square Angstrom, the arc shown in the figure will correspond to the total number of electrons at a distance r from the center which is also proportional to the scattering power of the atom at this value of r. Hence, a template of this form cut from a piece of white paper will scatter light in the same proportion that the corresponding atom scattered x-rays. If this template is placed on a black background and rotated about its center, a photograph of it will, in effect, be a photograph of the corresponding atom.

Figs. 6, 7 and 8 show the photographs which have been obtained according to this procedure for the atoms of helium, neon and argon. These photographs represent the atoms as observed with x-rays of wave-length 0.71 Angstrom, in which, instead of a lens, the image has been formed, as just described, by means of a mathematical and mechanical method. The magnification, which is equal to the ratio of the unit on which r is plotted in Fig. 5, to the unit in which the atom is measured, is in the present instance, equal to about 2×10^8 times. The resolving power of the apparatus as we have noted above was sufficient to separate the K and L electrons in neon, but in argon, due to its greater compactness, these are not resolved. The K electrons in neon are shown in the photograph by the very bright center, and the L electrons are represented by the diffuse outer ring. In argon the bright center corresponds to the combined K and L electrons, and the outer diffuse part to the M electrons.

Fig. 9 shows a similar picture for an atom of mercury.[6] The fact that the atom appears so concentrated is to be expected for so heavy an atom. This fact is however overemphasized in the picture, due to the inadequacy of the photographic process to represent large variations of density. For mercury the rays scattered from the center of the atom are so very intense that it becomes impossible to represent photographically the very much less scattering from the outer part of the atom. This shortcoming of the photographic process applies also to the case of the lighter atoms but to a very much less extent. It is a difficulty with which any photographer must contend.

There are other inaccuracies which may be introduced into our photographs due to certain approximations in our mathematical-mechanical method. These inaccuracies can hence be considered as analogous to the aberrations obtained with a lens. The photographs shown in Figs. 6–9 are free from these aberrations if one can assume (1) that the atoms are spherical and (2) that every electron has the same chance of occurring at any place in the atom. Auxiliary evidence indicates that assumption (1) is valid for all atoms of the noble gas group, and that (2) is exact for helium, very approximately valid for argon, and not seriously in error for neon. Thus the appearance of the helium, neon, and argon atoms, as shown by these photographs, should correspond very closely to their true form.

The x-ray diffraction experiments under discussion thus supply us with just the same information that would be obtained if we could look at the atom with an x-ray microscope. The size of the atom is clearly observed, and, as we have seen, the resolving power of our apparatus is also sufficient to distinguish various groups of electrons within the atom.

Photographs representing the *theoretical* electron density (Schrödinger's function $\psi\psi^*$) for various states of hydrogen-like atoms have been published by White.[7] These bear a close resemblance to the photographs of the *experimentally observed* atoms pictured here. It is significant that our pictures are in good accord with predictions of the new quantum mechanics.

---

[6] The x-ray scattering data on which the mercury atom photograph is based, were obtained by P. Scherrer and Stäger, Helv. Phys. Acta **1**, 518 (1928).

[7] H. E. White, Phys. Rev. **37**, 1423 (1931).

Incoherent Scattering and the Concept of Discrete Electrons*

ARTHUR H. COMPTON, *University of Chicago and Oxford University*
(Received December 21, 1934)

Experiment has shown that scattered x-rays contain both a coherent and an incoherent portion. Classical electron theory provides a formula for scattering closely similar to that based on wave mechanics, but includes both coherent and incoherent components only if the atom consists of discrete electrons. From a comparison of the classical and wave-mechanics theories of scattering it is concluded that the most accurate classical analog of Schrödinger's $\psi\psi^*$ is the probability of occurrence of discrete electrons, and that a particular electron is associated with each particular function $\psi_n\psi_n^*$.

EXPERIMENT shows that two types of scattered x-rays occur, one of which is of the same wavelength as the primary rays and the other of greater wavelength. Definite phase relations occur between the unmodified rays scattered by neighboring portions of matter, whence these rays are described as *coherent*. The modified rays, differing from each other in wavelength, can have no fixed phase relations, and are described as *incoherent*. Theoretical descriptions of the origin of the two types of rays have been given on the basis of (1) photons, (2) wave mechanics and (3) classical electromagnetic theory. It is a point of no little interest that whereas according to wave mechanics both types of scattered rays are interpreted in terms of wave functions distributed continuously throughout space, according to classical electromagnetic theory discrete electrons are required if incoherent rays are to occur. A comparison of the various theories of scattering is thus helpful in understanding the realm within which the concept of the electron is applicable.

It was first recognized from a consideration of the interaction of atoms and photons[1] that in addition to coherent scattering, collisions should occur involving changes of the photon's energy, which would mean a change of frequency and hence incoherence. Whereas Compton and Debye concerned themselves primarily with electrons which were knocked free from the atom by the photon's impact, Smekal noted also that any possible energy change of the atom might give to the scattered photon a frequency described by

$$h\nu_s = h\nu + W_i - W_f, \quad (1)$$

where W_i and W_f are the initial and final energies. In the hands of Jauncey and others[2] the concept of photons colliding with electrons moving within the atom has shown itself capable of describing in quantitative form the distribution of energy between the coherent and the incoherent rays.

According to the wave-mechanics theory,[3] under the influence of the field of the incident electromagnetic wave the characteristic functions for higher energy states of an atom assume finite values, and the radiation which it emits has the frequencies described by Eq. (1). If the final state of the atom is identical with the initial state i, the frequency is unchanged, and coherent radiation is emitted. In calculating this part of

---

* Based on an address presented at the Symposium on Ray Scattering at the St. Louis meeting of the Am. Phys. Soc., Dec. 1, 1934.

[1] A. H. Compton, Bull. Nat. Res. Council No. 20, 19 (1922); Phys. Rev. 21, 483 (1923); 24, 168 (1924). P. Debye, Physik. Zeits. 24, 161 (1923). A. Smekal, Naturwiss. 11, 873 (1923).

[2] G. E. M. Jauncey, Phil. Mag. 49, 427 (1925); Phys. Rev. 25, 314 (1925). J. W. M. Dumond, Phys. Rev. 33, 643 (1929). S. Chandrasekhar, Proc. Roy. Soc. A125, 231 (1929).

[3] This wave-mechanics theory includes the main features of the theory of dispersion by virtual oscillators introduced by H. A. Kramers, Nature 113, 673 (1924); 114, 310 (1924), and extended by H. A. Kramers and W. Heisenberg, Zeits. f. Physik 31, 681 (1925), whose theory included the incoherent radiation of the frequency given by Eq. (1). E. Schrödinger's first wave-mechanical theory of scattering, Ann. d. Physik 81, 109 (1926), was approximately equivalent to that of Kramers, giving only the coherent term. O. Klein, Zeits. f. Physik 41, 407 (1927), in sections 5 and 6 of his paper, gives a detailed and lucid account of the origin of the coherent and incoherent radiation according to wave mechanics. This theory was extended by Wentzel, Zeits. f. Physik 43, 1, 779 (1927), who first derived Eq. (6) from wave-mechanical principles. Eq. (3) is due to I. Waller, Phil. Mag. 4, 1228 (1927); Zeits. f. Physik 41, 213 (1928); and especially good for clarity of interpretation, I. Waller and D. R. Hartree, Proc. Roy. Soc. A124, 119 (1929), who corrected Wentzel's theory by taking into account the limitations of Pauli's exclusion principle.

the scattering, only the ψ functions of the normal state o of the atom are therefore concerned. That is, the coherent scattering is identical with that from an atom having a continuous distribution of electric charge of density

$$\rho = -e\psi_0\psi_0^*. \qquad (2)$$

If the final state differs from the initial state, the scattered ray is by Eq. (1) incoherent with the primary. The possible final states which may occur are all of those permitted by Pauli's exclusion principle, which is equivalent to the statement that an electron may be transferred to any level allowed by the selection rules, and not already occupied in the normal atom. Since the most probable transitions would in any case be those corresponding to complete ionization, this limitation is not stringent. It has the effect, however, of making the inner electrons of a Bohr atom less effective in incoherent scattering than are the outer ones.[4]

The formula thus developed by Waller[3] from wave mechanics may be written as

$$R_w \equiv \frac{I_a}{I_e} = \underbrace{(\sum_1^z f_n)^2}_{\text{content}} + \underbrace{\sum_1^z (1-f_n^2) - \sum_{mn}{}'' (f_{mn}^2)}_{\text{incoherent}}. \qquad (3)$$

[$\sum_{mn}{}''$ = sum over all pairs of electrons with the same spin, and $m \neq n$.]

Here I_a is the scattering by an atom, I_e that by an electron according to classical electron theory, f_n and f_{mn} are defined by the expressions,

$$f_n = \iiint \psi_n \psi_n^* e^{i\mathbf{k}\cdot\mathbf{r}} d\tau, \qquad (4)$$

$$f_{mn} = \iiint \psi_m \psi_n^* e^{i\mathbf{k}\cdot\mathbf{r}} d\tau, \qquad (5)$$

where ψ_n is Schrödinger's wave function corresponding to the nth electronic state in the normal atom, $\mathbf{k} = 2\pi(\mathbf{s}'-\mathbf{s})/\lambda$, where $\mathbf{s}'$ and $\mathbf{s}$ are unit vectors in the directions of the scattered and the primary rays respectively, $\mathbf{r}$ is the vector distance of the volume element $d\tau$ from the center of the atom and Z is the atomic number. Thus f_{mn} is the amplitude of the scattered wave due to the m to n transition, in terms of that due to a point charge electron as unity. The first term thus represents the scattering for the unchanged atom, which is coherent. The second term, as Wentzel shows, represents the scattering for all transitions for which $m \neq n$, and the third term takes account of those transitions which are disallowed by the Pauli exclusion principle. These terms thus describe the incoherent scattering.

In this theory, both radiation and atom are treated as distributed continuously through space. Its relation to the photon-electron theory described above corresponds to de Broglie's theorem of the equivalence of waves and particles. They may be considered as alternative views of the same phenomenon.[5]

It is noteworthy that classical electrical theory leads to a formula (Eq. (6)) almost identical with that derived from wave mechanics (Eq. (3)), but only if the atom is assumed to consist of discrete electrons. Woo's extension of Raman's classical theory[6] gives the expression,

$$R_c = (\sum_1^z f_n)^2 + \sum_1^z (1-f_n^2), \qquad (6)$$

where f_n has the same significance as in Eq. (4), except that $\psi_n \psi_n^*$ is replaced by p_n, where $p_n d\tau$ is the probability that the nth electron will lie in the volume element $d\tau$. If we should assume on the other hand that $\sum p_n d\tau$ is the portion of the continuously distributed electric charge in the volume element $d\tau$, and that this volume element has the same ratio of charge to mass as does an electron, we should obtain merely[7]

---

[4] In most cases the effect of the negative term representing the Pauli exclusion principle is practically negligible. G. G. Harvey, P. S. Williams and G. E. M. Jauncey, Phys. Rev. **46**, 365 (1934), have recently shown, however, that in the diffuse scattering from crystals this term may become of experimental importance.

[5] The photon theories of Jauncey and others are not in their present forms precisely equivalent to Eq. (3), though there is no apparent reason why they could not be made. It seems possible from wave mechanics, however, to arrive more easily at a rigorously derived formula.

[6] The classical theory of coherent scattering by the electrons in an atom was first treated extensively by Debye, Ann. d. Physik **46**, 809 (1915). That incoherent scattering must occur was pointed out from classical principles by A. H. Compton, X-Rays and Electrons, p. 1 (1926), and the theory was developed independently C. V. Raman, Ind. J. Phys. **3**, 357 (1928) and A. Compton, Phys. Rev. **35**, 925 (1930), leading to Eq. (6). Eq. (6) was derived by Y. H. Woo, Phys. Rev. **41**, (1932), following an extension of Raman's theory G. E. M. Jauncey, Phys. Rev. **37**, 1193 (1931).

[7] This result is implicit in Wentzel's wave-mechan theory (reference 5), and is derived explicitly by A.

$$R_c' = (\sum_1^z f_n)^2. \qquad (7)$$

Since definite phase relations exist between the rays scattered by the various volume elements of charge, this expression represents coherent radiation. Raman has shown[6] that the second term of Eq. (6) arises from the fact that in an atom composed of discrete electrons the positions of these electrons are continually changing, giving variable phase relations. There is thus a portion of the radiation from each electron (represented by the nth term in the summation) which is incoherent with that from the rest of the atom.

To the last term in Eq. (3) there seems to be no exact classical analog. It represents constraints upon each electron's motion due to the presence of the other electrons. If an attempt were made to build an atom out of electrons according to classical principles, such constraints would necessarily arise, but they could not be expected to introduce a term in the scattering formula identical with that resulting from Pauli's exclusion principle. According to classical electron theory, electrons should partially shield each other from the action of the electric field of the primary wave due to (1) the electrostatic field of the displaced neighboring electrons, and (2) the radiation field resulting from this acceleration. The former effect is an aspect of refraction, the latter is an increase of the electrical inertia due to proximity of the electrons. Both effects are dimensionally different from the Pauli exclusion effect, and of a smaller order of magnitude.

Further differences arise if the wave-mechanical theory is developed in sufficient detail to take into account the recoil of the scattering electrons and the corresponding increase in wavelength of the incoherent rays. Here the classical analog of radiation pressure gives a term of entirely the wrong order of magnitude. Nevertheless even here a treatment of the change in wavelength as a Doppler effect has, in the hands of Jauncey and DuMond[2] shown that momenta can be properly ascribed to the electrons within the atom, thus emphasizing their discrete existence.

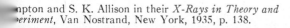
...mpton and S. K. Allison in their *X-Rays in Theory and Experiment*, Van Nostrand, New York, 1935, p. 138.

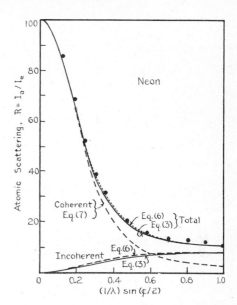

FIG. 1. Scattering of x-rays by neon. Experiments, Wollan. Theories based on Hartree electron distribution.

To a first approximation (Eq. (7)), a classical counterpart of the wave-mechanics atom is thus given by supposing the atom to consist of a continuously distributed charge of density given by Eq. (2). As will be seen from Fig. 1 this represents the most important part of the scattering at small angles. A second and much closer approximation (Eq. (6)) is obtained by replacing the continuous charge distribution with discrete electrons, the probability of occurrence of the nth electron being given by,

$$p_n d\tau = \psi_n \psi_n^* d\tau. \qquad (8)$$

Attempts to arrive at a still closer approximation by taking into account the constraints on the electron's motions cannot be expected to give the third term of Eq. (3).

It is noteworthy also that in order to get the closest classical analog to the wave-mechanical atom, we must assume that each electron has its own characteristic region within which it moves, i.e., $p_n \neq p_m$ (except where n and m differ only in the direction of the electron's spin). If each electron in the atom is assumed to have the same probability of occurrence at any point as every other electron, all the p_n's and hence all the f_n's will be alike (Eq. (4)), and Eq. (6) reduces to the form derived by Raman and

Compton,[6]

$$R_c = Z^2 f^2 + Z(1-f^2), \qquad (9)$$

where f is the common value of the structure factor for all of the electrons. This is no longer identical with the first two terms of Eq. (3), and is found[8] to be in somewhat less satisfactory agreement with the experiments.

Though we are accustomed to think of the Schrödinger ψ_n functions as distributed continuously throughout space, these functions are strictly speaking in $3n$-dimensional space. The apparent overlapping of the functions is merely a convenient 3-dimensional approximation. Thus the discreteness of the electrons in the classical theory corresponds on the wave-mechanics theory to equally discrete ψ_n functions, which are completely separated by being in different dimensions.

The closest classical analog to the wave-mechanical atom is thus one composed of discrete electrons, each of which has its own characteristic probability $\psi_n \psi_n^* d\tau$ of occurring within a given volume element.[9] It is accordingly proper not only to speak of electrons occurring within the atom, but also to distinguish each individual electron by the name of the corresponding quantum state. That is, K electrons are distinguishable from L electrons, etc. Thus for treatment of scattering problems, individual electrons grouped within the atom form the most nearly adequate classical picture corresponding to the continuous de Broglie waves of the wave-mechanics atom.

---

[8] G. Herzog, Zeits. f. Physik **69**, 207 (1931); E. O. Wollan, Rev. Mod. Phys. **4**, 241 (1932).

[9] Cf. A. H. Compton, Phys. Rev. **35**, 931 (1930), and Tech. Rev. Mass. Inst. **33**, 19 (1930). The conclusion there drawn was that the observed presence of incoherent scattering indicated that the discrete electron interpretation of $\psi\psi^*$ was necessary to bring agreement between theory and experiment. This argument was criticized by G. Herzog (reference 8), because of a supposed lack of agreement between the classical and quantum formulas, and because the concept of discrete electrons does not enter into the quantum theory of scattering. The statement that the best *classical* interpretation of $\psi\psi^*$ is the probability of occurrence of discrete electrons seems to be the legitimate conclusion from the considerations here advanced.

Scattering of X-Rays by a Spinning Electron

ARTHUR H. COMPTON, *University of Chicago, Chicago, Illinois*
(Received October 5, 1936)

A theory based upon classical electrodynamics is presented for the scattering of x-rays by an electron which is spinning and has a magnetic moment. The radiation scattered by such a magnetic doublet is found to be almost completely unpolarized. Being proportional to ν^2, it is negligible for ordinary x-rays, but should comprise the major part of the scattering according to classical theory for wave-lengths as short as hard gamma-rays. The rays thus scattered by one electron should be incoherent with those from every other. In all of these features the radiation thus magnetically scattered is closely similar in properties to that described by the added term of Klein and Nishina's quantum theory of scattering. Only in the distribution of scattered rays with angle does there appear a fundamental difference between the results of the two theories. Insofar as the two theories agree, we may consider the classical interpretation of Klein and Nishina's added term to be scattering due to the electron's spin. In particular, an interpretation according to classical electron theory of Rodger's experimental discovery of an important unpolarized component in scattered x-rays of very high frequency is that the electron is a magnetic doublet, which for these frequencies is comparable in importance with its electric charge.

THE concept of a spinning electron[1] which because of its spin has a magnetic moment has been useful in accounting for several physical phenomena, especially the fine structure of spectral lines.[2] When such an electron is traversed by an electromagnetic wave it should scatter radiation in a distinctive manner. The characteristics of the radiation thus scattered, as calculated according to classical electrodynamics, are similar to those of the scattered radiation described by Klein and Nishina's quantum-mechanics theory of the phenomena.[3] This theory, it will be recalled, added a term, which to the second power of α is

$$I_{KN} = I_e \alpha^2 \frac{(1-\cos \varphi)^2}{1+\cos^2 \varphi}, \quad (1)$$

to that given by the earlier formula of Breit, Gordon and Dirac.[4] Here I_e is the scattering per electron as calculated by Thomson's classical electron theory

$$I_e = I_0 \frac{e^4}{m^2 r^2 c^4}(1+\cos^2 \varphi), \quad (2)$$

φ being the angle between the primary and the scattered rays, r the distance from the electron to the observer, e, m, c and h have the usual significance, and

$$\alpha \equiv h\nu/mc^2 = h/mc\lambda, \quad (3)$$

where λ is the wave-length of the incident rays. This term arose from the use of Dirac's relativistic form of quantum mechanics, whose classical analog contains a spin term and an imaginary electric moment term. Presumably a part at least of the added scattering predicted by Klein and Nishina should thus be attributable to electron spin. The fact that classical electrodynamics, when applied to a spinning electron gives rise to a very similar expression (Eq. (9)), supports this presumption.

Following the successes of the Klein and Nishina formula in describing the intensity of scattering and the absorption of x-rays of short wave-length, Rodgers has recently[5] shown also that for these short wave-lengths an unpolarized component of the scattered x-rays appears. According to the Klein and Nishina theory,[6] it is only the component described by Eq. (1) which is incompletely polarized when scattered at 90°. On classical theory, as is well known, the scattered rays due to the electron's charge are completely polarized, while, as we see below, those due to its magnetic moment have an important unpolarized component. Rodger's experimental discovery of an unpolarized component thus gives specific confirmation of the new term in the Klein-Nishina formula, and affords new and direct evidence for the spin and the magnetic moment of the electron.

Precession of a Spinning Electron

Let **p** be the angular momentum of the electron's spin, and **γ** be its magnetic moment. The torque exerted on the electron by the electromagnetic wave which traverses it is then

$$\mathbf{T} = \boldsymbol{\gamma} \times \mathbf{H},$$

where **H** is the magnetic vector of the wave. The result is a precession of the spin axis about **H** such that

$$d\mathbf{p}/dt = \mathbf{T},$$

and a change of magnetic moment at the rate

$$\dot{\boldsymbol{\gamma}} = (\gamma/p)[\boldsymbol{\gamma} \times \mathbf{H}].$$

If now the magnetic vector of the primary is expressed by

$$\mathbf{H}_0 = \mathbf{A}_0 \cos 2\pi\nu t,$$

we have, assuming that the rate of rotation of the spinning electron is high compared with ν,

$$\ddot{\boldsymbol{\gamma}} = (\gamma/p)[\boldsymbol{\gamma} \times (d\mathbf{H}/dt)]$$
$$= -2\pi\nu(\gamma/p)[\boldsymbol{\gamma} \times \mathbf{A}_0] \sin 2\pi\nu t.$$

But the magnitude of the magnetic vector of the wave radiated by this changing magnetic doublet is

$$H_s = (\ddot{\gamma}/rc^2) \sin \theta,$$

---

[1] Cf. A. H. Compton, Phil. Mag. **41**, 279 (1921).
[2] E.g., L. Pauling and S. Goudsmidt, *The Structure of Line Spectra* (1930), p. 54.
[3] O. Klein and Y. Nishina, Zeits. f. Physik **52**, 853 (1929).
[4] For a summary of this work, cf. e.g. A. H. Compton and S. K. Allison, *X-Rays in Theory and Experiment* (1935), p. 234.
[5] E. Rodgers, preceding paper, this issue.
[6] A detailed discussion of the polarization characteristics of the rays scattered according to their theory has been given by Y. Nishina, Zeits. f. Physik **52**, 869 (1929).

where θ is the angle between $\ddot{\gamma}$ and $\mathbf{r}$, i.e., between $\mathbf{T}$ and $\mathbf{r}$. The instantaneous intensity of the scattered wave is then

$$I' = \frac{c}{4\pi} H_s^2 = \frac{c}{4\pi} \frac{4\pi^2 \nu^2}{r^2 c^4} \frac{\gamma^2}{p^2} [\gamma \times \mathbf{A}_0]^2 \sin^2\theta \sin^2 2\pi\nu t.$$

The average value of this intensity taken over a complete cycle, assuming that the period of the electron's precession is long compared with the period of the wave, is

$$I'' = I'/2 \sin^2 2\pi\nu t.$$

The corresponding average intensity of the primary wave traversing the electron is

$$I_0 = (c/8\pi) A_0^2.$$

We may thus write

$$I'' = I_0 \frac{4\pi^2 \nu^2 \gamma^4}{r^2 c^4 \, p^2} \sin^2\xi \sin^2\theta, \qquad (4)$$

where ξ is the angle between γ and $\mathbf{A}_0$.

We next proceed to average I'' over all possible orientations of the spin axis. Let us refer to Fig. 1, in which the X axis is the direction of propagation of the primary beam, and the XY plane is chosen to include OP, the direction of the scattered ray under consideration. The primary magnetic vector $\mathbf{H}$ is then normal to OX and at an arbitrary angle ζ with the Y axis, and the magnetic doublet γ is at an angle ξ with $\mathbf{H}$. The torque $\mathbf{T}$ is perpendicular to γ and $\mathbf{H}$. The angle θ between $\ddot{\gamma}$ and $\mathbf{r}$ is given by

$$\cos\theta = -\cos\varphi \sin\psi - \sin\varphi \cos\psi \sin\zeta.$$

Thus, $\sin^2\theta = 1 - \cos^2\varphi \sin^2\psi - 2 \sin\varphi \cos\varphi$

$$\times \sin\psi \sin\zeta - \sin^2\varphi \cos^2\psi \sin^2\zeta. \qquad (5)$$

The average intensity for all orientations of the spin axis is,

$$I_s = \frac{1}{4\pi} \int_0^\pi \int_0^{2\pi} I'' \sin\xi \, d\psi \, d\xi.$$

Substituting the value of I'' and of $\sin^2\theta$ given in Eqs. (4) and (5), we obtain on integration

$$I_s = I_0 \frac{4}{3}\pi^2 \frac{\nu^2 \mu^4}{r^2 c^4 p^2} [1 + \sin^2\varphi \cos^2\zeta]. \qquad (6)$$

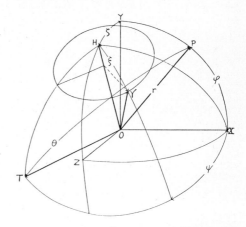

Fig. 1.

State of Polarization

For the ray scattered at $\varphi = 90°$, if the primary ray is polarized with the magnetic vector along the Y axis, $\zeta = 0$, and the factor in brackets becomes equal to 2. In this case reference to Fig. 1 shows that the direction of $\mathbf{T}$ and hence of $\ddot{\gamma}$ is perpendicular to the Y axis, which is the direction of scattering, and that the probability distribution of this direction is axially symmetrical about Y. Thus the scattered ray is in this case completely unpolarized.

If the primary ray has its magnetic vector along the Z axis, $\zeta = 90°$, and the bracketed factor becomes 1. In this case $\ddot{\gamma}$ lies in the XY plane. This gives rise to a component of the ray scattered along the Y axis which is polarized in the same plane as is the ray scattered by the electric charge of the electron.

Since for an unpolarized primary beam the average magnitudes of the Y and Z components of the magnetic vector are equal, this means that when an unpolarized ray is scattered at 90° the scattered ray may be thought of as consisting of an unpolarized component of intensity 2 upon which is superposed a polarized component of intensity 1. If this ray is scattered again at 90° by a magnetic doublet, as in the usual technique for studying polarized x-rays, it can be shown that the ratio of $I_{\parallel}$ to $I_{\perp}$ is 5/4. Thus the magnitude of the polarization of rays scattered only by the magnetic moment of the electron, as defined in the usual manner, should be

$$P \equiv (I_{\parallel} - I_{\perp})/(I_{\parallel} + I_{\perp}) = 1/9. \qquad (7)$$

Nishina has shown[6] that the corresponding polarization of the added term in the Klein-Nishina formula is 0.

Intensity

In the theory of spectra, it is found necessary to assume that the effective value of the electron's angular momentum and magnetic moment are, respectively,[2]

$$p = \sqrt{(\tfrac{3}{4})} h/2\pi$$

and $\quad \gamma = \sqrt{(\tfrac{3}{4})}(h/2\pi)(e/mc).$ (8)

Substituting these values in Eq. (6), and taking the average value of $\sin^2 \zeta$ as $\tfrac{1}{2}$ for unpolarized radiation, we obtain,

$$I_\gamma = \frac{1}{8} \frac{e^4}{m^2 r^2 c^4} \frac{h^2 \nu^2}{m^2 c^4}(3-\cos^2 \varphi).$$

Using relations (2) and (3), this may be written,[7]

$$I_\gamma = I_c \alpha^2 \frac{3-\cos^2 \varphi}{4(1+\cos^2 \varphi)}.$$ (9)

Noting that $\alpha = 0.024/\lambda$, if λ is expressed in angstroms, this means that for wave-lengths shorter than about 0.024A, in the middle gamma-ray region, the magnetic moment of an electron is responsible for more scattering than is its electric charge.

---

[7] This result, except for an error of a factor of 2, has been given in footnote 166, p. 237, of A. H. Compton and S. K. Allison's *X-Rays in Theory and Experiment* (1935), where a comparison with the results of Klein and Nishina's theory is also given. The derivation of the expression has, however, not been presented.

It will be seen that this magnetic scattering is of the same order of magnitude as the added term (1) of Klein and Nishina's theory. It is noteworthy however that while both theories give the same type of variation of intensity with frequency, the variation with angle of scattering is fundamentally different. The factor $(1-\cos \varphi)$ present in Klein and Nishina's formula suggests on classical theory an incoherent scattering which vanishes at $\varphi = 0$ because of exactness of phase relationships. Such a process has no place in the present theory of scattering by a magnetic doublet.

Phase Relationships

It will be seen from Fig. 1 that for every orientation of **T** there is an equally probable orientation in the opposite sense. It follows that if the squares of the sum of the amplitudes of the waves scattered by a group of spinning electrons is averaged over all possible orientations, all the product terms must vanish, leaving only the sum of the squared terms. Thus each doublet scatters a wave which is incoherent with that from every other doublet. Because of this incoherence it is legitimate to assume, as we have done, that the scattering by an electron's magnetic doublet is independent of that by its electric charge.

This feature is also characteristic of the scattering represented by Klein and Nishina's added term. In their case incoherence follows from the fact that the initial and final states of the scattering atoms are of different energy, implying an increased wave-length, so that no exact phase relationships can exist.

An Alternative Interpretation of Jauncey's "Heavy Electron" Spectra

In the February first number of *The Physical Review*,[1] Jauncey has shown two photographs of magnetic spectra of electrons, in both of which the main deflected image is taken as evidence of a heavy electron of rest-mass about 3 times normal. Following the suggestion of C. T. Zahn[2] that this image may be caused by scattered electrons, I have calculated that in both of Jauncey's photographs the image in question corresponds with that to be expected from electrons once scattered by the lower plate of his velocity selector.

Jauncey has shown that his velocity selector, with parallel plates 5 cm long and 0.105 cm apart, and with the magnetic and electric fields used in his experiments, will not transmit β-particles coming directly from Ra E, except in the chosen low range of velocities. When these β-rays are however bent by the magnetic field to strike the lower plate, this part of the plate becomes the source of secondaries of only slightly reduced energy. The rays moving forward may then pass through the selector if they do not again strike the plate. Thus for β-rays once scattered, the resolving power is no greater than for a velocity selector of half the length if primary particles only were considered. From the dimensions of Jauncey's apparatus it can thus be shown that while for primary particles the minimum transmissable radius of curvature is 30 cm, the corresponding minimum for particles once scattered is 7.5 cm.

Numerical calculation shows that with the magnetic and electric fields used in Jauncey's experiments any electron with curvature low enough to reach the film will have a radius of curvature of more than 7.5 cm between the plates. That is, the apparatus is not effective in selecting velocities of electrons that have been once scattered. That it does, however, act as an effective velocity selector for the direct β-rays is shown by the sharpness of Jauncey's line C, due to electrons with velocities close to the chosen value of $\beta = 0.33$.

Calculations of the position of the images to be expected from such once-scattered rays are also in satisfactory agreement with the observed positions of the diffuse images in Jauncey's photographs.

It would thus appear that the heavy electron interpretation of Jauncey's photographs is not required. His new lines may alternatively be ascribed to a kind of "second-order" magnetic spectrum caused by once-scattered particles. Higher order scattering might also, as suggested by Zahn, be supposed to produce detectable effects. The present calculation indicates, however, that only single and not multiple scattering processes need be considered to account for Jauncey's results.

ARTHUR H. COMPTON

University of Chicago,
Chicago, Illinois,
February 16, 1938.

[1] G. E. M. Jauncey, Phys. Rev. **53**, 265 (1938).
[2] Made in discussion of Jauncey's results at the meeting of the American Physical Society, Indianapolis, Dec. 30, 1937, and soon to appear in the *Physical Review*.

Journal of The Franklin Institute

Devoted to Science and the Mechanic Arts

| Vol. 230 | AUGUST, 1940 | No. 2 |

WHAT WE HAVE LEARNED FROM SCATTERED X-RAYS.*

BY

ARTHUR H. COMPTON, Ph.D., Sc.D., LL.D.,

Ryerson Physical Laboratory, University of Chicago.

Just as sunlight is scattered by sky and clouds, so x-rays are scattered by any object that they strike. Most of our information about the world has come through light scattered by objects into our eyes. It is thus not surprising that important additions to our knowledge should come from the scattering of the light of super frequency which we call x-rays. The new information thus gained has concerned the nature of x-rays and light, and the nature of the objects which scatter the x-rays.

Only a few months after Rœntgen's announcement of the discovery of x-rays, Michael Pupin, working with an x-ray tube of unusual power, and using Edison's fluoroscope as a sensitive detector, discovered scattered x-rays. He noted that when any object, such as a book or a man's hand, was placed in the x-ray beam, the object itself became a source of radiation which would light up a fluoroscope placed in the neighborhood. It was like brightening a room by placing a sheet of white paper in the path of a sunbeam that comes through a window.

An interpretation of this phenomenon of scattering was

---

* Read by Dr. Karl Compton, President, Massachusetts Institute of Technology, at Medal Day Meeting, Wednesday, May 15, 1940.

(Note—The Franklin Institute is not responsible for the statements and opinions advanced by contributors in the JOURNAL.)

given some years later by J. J. Thomson in terms of electromagnetic waves acting on electrons. He had already calculated the energy that should be radiated when an electric charge is moving with changing velocity. When x-rays are scattered, he supposed that the electric field of the impinging wave changes the velocity of the electrons in the scattering material. Thus Thomson was able to calculate how intense the scattered rays should be in terms of the number of electrons present.

The calculation was a bold one with far-reaching consequences. Up to this time no satisfactory quantitative test of the theory of radiation from an accelerated charge had been possible. Though the charge and mass of the electron were roughly known, no one knew how many electrons were to be found in each atom. It was, however, correctly assumed that most of the electrons had natural frequencies comparable with those of light, and hence that with regard to x-ray waves of a thousand fold greater frequency they would act almost as if free from constraining forces. A favorable test of Thomson's calculations would thus at once establish the theory that x-rays were electromagnetic waves, that an accelerated charge emits radiation, and, assuming the validity of the theory, serve to estimate the number of electrons in the scattering material.

In 1905 Charles Barkla, stimulated by Thomson's calculations, found a slight polarization of the x-rays coming directly from an x-ray tube, which strongly suggested the electromagnetic character of the rays. Within a year he had performed the then very delicate experiment of measuring x-rays scattered twice at right angles. The first scattering should theoretically polarize the rays completely, and the second scattering of these already polarized rays should serve to analyze their polarization. His experiments showed 50 per cent. polarization as compared with a theoretical 100 per cent., which was adequate to show the strong probability of the correctness of the electromagnetic wave theory of x-rays.

Seventeen years later Dr. Hagenow and I repeated Barkla's work with the more refined equipment then available, and found that the lack of complete polarization which he had observed was due chiefly to multiple scattering in his large polarizing and analyzing blocks, and that when smaller blocks are used the polarization does indeed approach 100 per cent. as the theory demands.

It is interesting to note that this first verification of Thomson's theory indicates a motion of the electrons in the scattering material along the direction of the electric vector of the x-ray wave. This shows that the electrons carry electric charges. A few years ago Dr. Rodgers, working in our laboratory, extended the polarization experiments to show that the electron has likewise a magnetic moment. Assuming that an electron is spinning, it should act as a gyroscopic magnet. The magnetic field of an incident x-ray wave should set the resulting magnet in motion in such a way as to produce a scattered ray which has no polarization. This magnetically scattered ray should however be appreciable only for x-rays produced at potentials greater than 200 kilovolts. Using x-rays up to 800 kilovolts potential, where the theoretical value of the scattering by the magnetic doublet of the electron should be comparable with that by its electric charge, Rodgers observed just the expected amount of unpolarized x-rays. Thus scattered x-rays have shown also that the electron is magnetic.

Following his polarization experiments, Barkla measured the intensity of the scattered x-rays. This was found to follow a much more complicated rule than was indicated by Thomson's simple theory. Part of the difference was traced to the fact that in the heavier atoms the electrons are grouped so closely that the scattered x-rays are nearly in phase with each other. This leads to an intensity of scattering greater than Thomson's normal value. On the other hand, for very short x-rays waves, the scattering was found to be less than Thomson predicted, a departure whose origin was not revealed until later.

By carefully selecting the experimental conditions, however, and assuming the essential validity of Thomson's calculations, Barkla was able to make a reliable estimate of the number of electrons in the scattering material. This he found to be a number in each atom equal approximately to half of its atomic weight.

It was one of those striking coincidences of physics that

this first reliable determination of the number of electrons in an atom should have appeared in the same 1911 issue of *The Philosophical Magazine* as that in which Rutherford published his first measurement of the charge on the nucleus of the atom based on the scattering of alpha particles. This charge was likewise estimated as equal to half the atomic weight. At once it became evident that the atom consists of a positively charged nucleus, surrounded by a number of electrons sufficient to neutralize the charge. With Moseley's remarkable x-ray spectra, two years later, we found how the number of these electrons changes merely by one as we go from element to element, the basis of our concept of atomic number, a result which might indeed have been inferred from the earlier experiments of Barkla and Rutherford.

A complete account of what scattered x-rays have taught us would include at this point a consideration of the diffraction of x-rays by crystals and of x-ray spectra, for these are merely especial aspects of scattering. With the work of Laue, the Braggs, Mosely, Siegbahn and many others, however, this work has become so well known that in spite of its outstanding importance we may properly leave it with this passing mention.

My own active interest in scattered x-rays came first when I realized that here was a method of determining the arrangement of electrons in atoms. You are familiar with the ring around the moon caused by the diffraction of moonlight by droplets of fog. In this phenomenon the size of the ring depends upon the ratio of the length of the light wave to the size of the drop. When the size of the corona becomes smaller we know that the droplets are growing, and that very probably they will fall as rain.

In a similar manner x-rays are diffracted by the atoms in a gas, and it should be possible by studying the diffraction halos to estimate the size of the atoms. Since it is the electrons in an atom which scatter x-rays, this would mean finding the distances of the electrons from each other.

It was in 1914 that I began by this direct method to investigate the electronic structure of atoms. The Braggs had been using x-ray diffraction as an effective method of learning the arrangement of atoms in crystals. Why should not a more refined study lead to a knowledge of the electron distributions in the atoms themselves. The most promising approach seemed at first to study the intensity of the various orders of x-ray diffraction from crystals. C. G. Darwin had made some calculations of intensity of this diffraction in terms of the electron distribution, and I was able to show how it should be possible to calculate backward from observed intensity measurements to arrive at the electron arrangements. The necessary experiments were delicate, and involved many corrections, but were at last completed, only to find that the resulting electron distribution contained an important unknown factor. This factor was the magnitude of the heat motions of the atoms in the crystal that was diffracting the x-rays. Until more was known about the heat motions of atoms, we could not reliably use experiments with crystals to determine the electronic structure of their component atoms.

The obvious alternative to studying atoms arranged in the regular order of a crystal lattice was to use atoms distributed entirely at random. Whereas for the crystal it is necessary to take account of the diffraction pattern of the regularly arranged atoms themselves, for an amorphous substance the diffraction by each moecule should be independent of that of every other, and no such correction for their mutual effects need be considered.

It was from such considerations that in 1917 Debye and Scherrer, working in Switzerland, began to study the diffraction of x-rays by such supposedly amorphous materials as sheets of metal. The result was diffraction patterns showing that these materials were really masses of tiny crystals. Instead of developing a method of determining electronic distributions, therefore, they discovered the powerful and convenient powdered crystal method of studying atomic arrangements in metals.

My own objective was turned toward the scattering of x-rays by monatomic gases, where if under any conditions, one might rely on the random arrangement of the scattering atoms. The first major problem was the experimental one of obtaining enough scattered x-rays to measure. By certain improvements in technique, Barkla had succeeded in measuring reliably the scattering of the total x-ray beam from a large block of solid material. We needed to measure the scattering of a single wave-length of x-rays from a volume of gas small enough to determine precisely the angle at which the rays were scattered. As compared with Barkla's experiments a factor of roughly a million fold had to be gained.

The problem was precisely similar to that of striving to photograph spiral nebulæ at greater and greater distances. The telescope is made larger, the plates more sensitive, the time of exposure increased, and the aim of the telescope toward the star is improved. So we introduced water-cooled x-ray tubes, ionization chambers with more strongly absorbing gas, an electrometer (which my brother who reads this paper helped to develop) of greatly increased sensitivity, and finally reached the required factor of a million by using P. A. Ross's difference absorption method of obtaining the effects of a homogeneous x-ray beam a hundred times more intense than that reflected from a crystal.

Before this technical difficulty had been overcome, however, we were faced by an unforeseen theoretical problem. When a spectroscope is turned toward the sky, the same Fraunhofer spectral lines are observed as are to be seen when the spectroscope is pointed directly at the sun. This must be so. For every light wave that strikes an air molecule is scattered, and the number of scattered waves will thus be the same as the number of incident ones. This means that the frequency and wave-length of the scattered light must be the same as that of the light that comes from the sun. It had gradually become evident, however, that the diffused x-rays and gamma rays coming from a scattering block were more readily absorbed than the incident rays. Various attempts to explain this phenomenon were put forward, most of them based on the idea that an important part of the diffused rays were of fluorescent origin. When, however, we turned the x-ray spectrometer on the diffused rays we found that almost all of the x-rays were increased in wave-length. Fluorescent rays should not be polarized, but these rays of increased wave-length were. If we were to consider those rays to be scattered whose intensity was given by Thomson's classical theory, certainly we had found a change in wave-length of scattered x-rays. Here was a fundamental difficulty that needed to be resolved before any diffraction theory could reliably be applied which would assume that the diffracted ray and the primary ray were of the same wave-length.

We soon found that the observed change of wave-length could be quantitatively explained by assuming that the x-ray beam consists of a stream of photons, particles carrying a quantum of energy each, as had been suggested previously by Einstein in his theory of the photoelectric effect. We supposed that each photon is deflected by a single electron, which recoils from the impact, taking away some of the photon's energy. This theory involved thus both the energy and the momentum of the photon, whereas to explain the photoelectric effect only its energy was important.

The theory thus used to account for the change in wave-length implied that scattered x-rays should be accompanied by electrons recoiling with an energy only a few per cent. that of photoelectrons. Such electrons had not yet been identified but were now immediately observed by C. T. R. Wilson, using his cloud chamber, and our studies showed that their number and energy corresponded accurately with the values predicted from the photon theory.

In a desparate effort to save the wave theory of radiation, it was then pointed out that these phenomena were consistent with the view that x-rays are waves if we would consider that the conservation of energy and momentum is only statistically conserved in the scattering process, but not in the motion of the individual electrons. This proposal to abandon the exact validity of the conservation principle stimulated tests by Bothe and Geiger in Germany and by ourselves of the individual process of electron-photon interaction. It was found that whenever a recoil electron appears a photon is scattered in just the direction and with just the energy required for the exact conservation of their energy and the momentum. Thus not only did the conservation principles receive an important new confirmation, but we had actually traced the path of an x-ray from place to place, showing that it travels as a compact bundle of energy. In other words, x-rays were particles.

As these studies were in progress, Louis de Broglie brought forward his now famous theory that the motions of waves and particles are really indistinguishable, if we suppose that associated with each particle is a wave whose length is inversely proportional to its momentum. His results were immediately applicable to our experiments on the wave-length change of scattered x-rays. A year or two later they were shown by Davisson and Germer and by G. P. Thomson, to predict accurately the wave properties of electrons. Soon it was found that moving hydrogen and helium atoms and hydrogen molucules can be diffracted by crystal gratings, and that their wave-lengths are also given by de Broglie's theory.

So here was the duality of particles and waves, introduced first in the case of x-rays when our scattering experiments revealed the particle properties of rays long known to be waves, and then extended by de Broglie's theory to moving particles of every type.

It would take us too far afield to describe how this duality became the basis of the uncertainty principle, philosophically one of the most significant developments of modern physics. Nor can we stop to describe the development of the new quantum mechanics, which, as Heisenberg showed, could be derived from this principle as a starting point. For our present purpose the significant fact is that within a half dozen years from the first measurement of the change in wave-length of scattered x-rays a new quantum theory of x-ray scattering had been developed in which electrons had only probable, not actual, positions within the atom. Now it was evident that when x-rays are scattered with changed wave-length they act independently of each other. It is only the scattered rays whose wave-length remains unchanged that are affected by the arrangement of the scattering electrons. The theory was now in shape for reliable application to the data on x-ray scattering.

New work on the scattering of x-rays by helium, neon and argon was now undertaken by Charles Barrett and E. O. Wollan in our laboratory and by G. Herzog in Zurich. In helium the electrons were found to be distributed diffusely in a region whose average size is about that specified by Bohr's old orbital theory. In neon, two groups of electrons are identified, and in argon three groups, though the resolving power of the x-ray waves was not sufficient to distinguish clearly between the two inner groups. The results are indeed as definite as if we were looking directly at the atoms with an x-ray microscope. Though perhaps not so exciting, it was however very satisfying that the electron distributions thus observed are in complete agreement with modern quantum theories of atomic structure, at least if we interpret the wave-function in terms of the probability of electron occurrence rather than as an actual volume density of electricity.

Comparing the new electron distributions from gases with the older ones from crystals it was now possible to estimate the thermal motions of the crystal atoms. We were thus able directly to show that even at the zero point of temperature a motion of the atoms occurs equal to half a quantum of energy in each mode of vibration. This served to distinguish between two forms of quantum theory of heat which had been a vexing question since the early theories of Planck.

So finally, after some twenty years of study by experimenters in this country and Europe, the problem which I had set myself for my doctor's thesis has received an answer. I am happy to have had a part in its solution. Fortunately, my professors did not demand the completion of the problem before they released me from the university halls. It gives me great pleasure that the Franklin Medal committee has seen in my share of this work a worthy contribution to science. I thank you.

AUGUST, 1940

PHYSICAL DIFFERENCES BETWEEN TYPES OF PENETRATING RADIATION*

By ARTHUR H. COMPTON

The University of Chicago
CHICAGO, ILLINOIS

THE unexpected discovery of the value of roentgen rays and radium for treatment of cancer emphasizes the importance of examining the characteristics of the more recently discovered types of radiation. Experience indicates that radiations that are to be useful for tumor treatment must be ionizing rays. We may thus neglect those rays, such as infra-red and radio rays, which have a wave-length longer than light and consider only those of shorter wave-length. Ultraviolet light is too rapidly absorbed to be considered as an agent for other than surface treatment. Of the more penetrating ionizing rays, roentgen rays, i.e., light of very short wave-length, are the most familiar. We must accordingly consider the relative effects of roentgen rays of different wave-lengths or voltages. In addition, however, there are various types of corpuscular rays which are likewise penetrating when they have sufficient energy. Among these corpuscular rays are the familiar electrons, which we know as cathode rays and beta rays, and the less familiar alpha rays, protons, neutrons, deuterons, positrons and mesotrons, not to mention the neutrinos and neutrettos whose existence is perhaps still questionable. Our present problem is to interpret the physical properties of these various rays in such a manner as to indicate their probable therapeutic value.

We can quickly dispose of most of the less familiar radiations. There is one group which consists of fast moving atomic nuclei. These are the protons, deuterons and alpha particles, the nuclei of hydrogen, heavy hydrogen, and helium respectively. In order that these rays may have a penetrating power comparable with roentgen rays, their energy needs to be immensely greater than present technique can impart, of the order of hundreds of thousands of kilovolts. It is probable that at such energies their properties might make them of practical interest. But this lies in the distant future.

Positrons, or positive electrons, are found to act in a manner very similar to the familiar negative electrons, except that they may attach themselves to negative electrons and spend themselves in the production of a penetrating gamma ray. Positrons thus have ionizing and penetrating powers intermediate between those of electrons and gamma or roentgen rays.

Most penetrating of all known forms of radiation at very high energy is the mesotron, a newly discovered component of cosmic rays. The mesotron is a particle with either a positive or a negative charge, of the same magnitude as that of an electron, and with a mass when at rest of about one-tenth that of a hydrogen atom. The mesotron's exceptional penetration appears only when its energy is over a hundred million electron volts. At present we know of no artificial method of producing these rays, and their intensity in cosmic rays is far too feeble for therapeutic use. Furthermore, their ionizing properties are found to be so similar to those of high speed electrons that they would seem to be of no unusual therapeutic interest.

Of the available types of penetrating radiations, this leaves us with roentgen rays (or gamma rays), electrons (or beta rays), and neutrons. The two chief physical characteristics of these rays with which we are concerned are their penetration and their method of ionization. The meaning of dif-

---

* Read at the Fortieth Annual Meeting, American Roentgen Ray Society, Chicago, Ill., Sept. 19-22, 1939.

ferences in penetration, as measured, for example, by their absorption coefficients, is too familiar to require elaboration. It may be worth while, however, to explain what is meant by differences in the method of ionization.

I presume you have all read of, if you have not seen, the cloud chambers with which photographs are made of the tracks of ionizing particles. Figure 1 shows such a glass-walled cloud chamber filled with a fog condensed on ions produced by a mixture of gamma rays and neutrons. Two distinct types of tracks will be noted, heavy white tracks in which the individual droplets are too close to be distinguished, and thinner, more diffuse tracks, where the individual droplets can be seen. These tracks represent respectively the protons which are formed by the neutrons, and the electrons which the roentgen rays excite. Neither neutrons nor roentgen rays produce ionization directly. It is thus evident that neutrons produce ionization in much more concentrated regions than do roentgen rays. In view of this marked physical difference in the mode of ionization, therefore, it would not be surprising to find a distinctive biological action of the two types of rays.

There is a similar, though less striking contrast between the concentration of ions produced by slowly and rapidly moving electrons. The number of ions per centimeter path of the electron is in fact inversely proportional to the square of the speed with which it moves. This means that low voltage electrons produce more concentrated ionization than do high voltage electrons. When the electron's energy is as great, however, as 200 electron kilovolts, it is already moving with a speed so close to that of light that additional energy does not appreciably change its speed, and the ions along the path do not spread out any more. It follows that with regard to ion density, no important changes in the properties of electrons occur for energies greater than about 200,000 electron volts.

Cathode electrons at voltage available in the laboratory are too easily absorbed to be classed as penetrating rays. The tubes would need to be operated at potentials of the order of 100 times those now available for them to compare favorably in penetration with the roentgen rays now in use. Nor should we expect such immensely energetic electrons to show ionization characteristics differing appreciably from those of ordinary high voltage roentgen rays. There

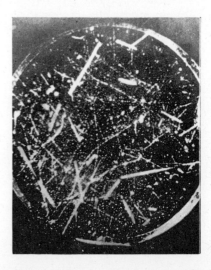

FIG. 1. Cloud photographs of tracks of protons (dense white lines) and electrons (thin lines) excited respectively by neutrons and gamma rays (Lawrence).

would thus appear no reason for developing supervoltage electron rays for deep therapy.

Neutrons and roentgen rays show much the same penetration in human tissue. Both are sufficiently penetrating to be useful in deep treatment. At present the only type of neutron generator capable of producing neutrons in quantity adequate for human therapy is the so-called "cyclotron," developed by Professor E. O. Lawrence. This consists essentially of a powerful magnet and an electrical oscillator, which serve to produce a high energy beam of deuterons. When this beam strikes a target of beryllium, gamma rays and neutrons are produced, much as cathode rays striking a tungsten target produce roentgen rays. The gamma rays may be removed by a filter of lead, which the neutrons readily penetrate, leaving a nearly pure neutron beam. Fig-

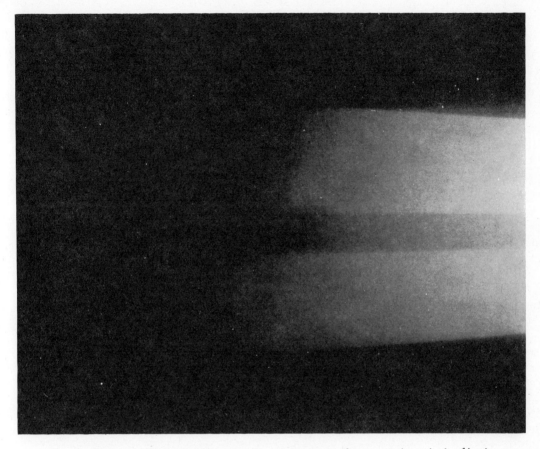

Fig. 2. Photographic impression made by neutron rays after traversing 1 inch of lead. Central shadow cast by 3 cm. of paraffin (Lawrence).

ure 2 is a photograph made by such a neutron beam after passing through the diaphragm of Lawrence's first therapeutic cyclotron. With his latest instrument, shown in Figure 3, weighing some 220 tons and operating at about 100 kilowatts, a beam of neutrons is produced whose biological effect is of the same order of intensity as that of filtered roentgen rays from the usual 200 kilovolt, therapeutic roentgen tube.

While the penetration and the intensity of such neutron rays are closely similar to the high voltage roentgen rays in common use, the neutrons show two significant physical differences. The first of these is that of the much greater concentration of ions along the paths of the protons which the neutrons excite in the irradiated material. The second is that neutrons act much more strongly upon hydrogen-containing substances than upon any other, while for hard roentgen rays hydrogen, having the fewest electrons, is least affected. This marked absorption of neutrons by hydrogen is illustrated by the shadow in the middle of the neutron beam shown in Figure 2. This shadow is cast by a block of paraffin 3 cm. thick, which would hardly affect a roentgen-ray beam which, like these neutron rays, had been filtered through an inch of lead. Since various organic materials differ in their hydrogen content, this may lead to important biological distinctions in the effect of the rays.

It is as yet too early to say whether neutron rays are really preferable to roentgen rays for any kind of treatment. Because of the physical differences just mentioned, however, it would be surprising if some corresponding differences in their biological ef-

Fig. 3. University of California's 220 ton cyclotron, built for therapeutic purposes, with operating staff (Lawrence).

fects did not appear. Different investigators are not yet in agreement, however, with regard to the magnitude of these biological differences.

There remains the important question as to the physical properties of roentgen rays of higher voltage than now in use. Commercial roentgen-ray equipment of 400 kv. is readily available. A few studies extending over several years have been performed with roentgen rays up to 1,000 kv. Within the last year, Van Atta at Massachusetts Institute of Technology and Bouwers at Eindhoven, Holland, have reported physical measurements with roentgen rays up to about 3 million volts. In cosmic rays we have identified and studied the production, absorption and ionization of roentgen rays excited by electrons of energies between 10 million and 1,000 million electron volts. These studies have made it possible to check and refine the theory of roentgen-ray action. Following are some of the more significant conclusions.

Up to at least several million electron volts, roentgen-ray production continues to increase approximately as the square of the voltage. Over 500 kv. the roentgen rays become strongly concentrated in the general direction of the roentgen-ray beam (Fig. 4). Thus, at 2,000 kv., Van Atta and Northrup find about six times as strong radiation in line with the cathode rays as at right angles. This indicates that the supervoltage tube should be designed in a different manner from the tubes of more moderate voltage.

Van Atta find that in lead the penetrating power of the roentgen rays continues to increase up to 2.5 million volts, whereas in organic material, Bouwers finds a maximum penetration at a little more than 1 million volts. That is, in human tissue it seems that 3 million volt roentgen rays are less penetrating than 2 million volt rays, and hardly more penetrating than 1 million volt rays. The physical reason for this upward rise in the absorption for very high voltages is well

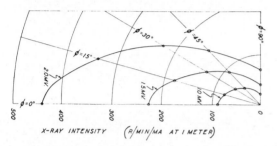

FIG. 4. Variation of roentgen-ray intensity with azimuthal angle at 1.0, 1.5, and 2.0 million volts (Van Atta and Northrup).

understood, but must not be discussed now. Packard of Columbia has recently reported that when the effect of scattered rays is taken into account, he finds the penetration of 700 kv. roentgen rays almost as satisfactory as those at 900 kv. This is clearly indicated by his curves shown in Figure 5. Stone of California goes yet farther, showing that if a patient is irradiated from both sides, 400 kv. roentgen rays give just as high a ratio of depth to surface dosage as do his 1,100 kv. roentgen rays. These studies make it clear that with regard merely to penetrating power, little if anything is to be gained in human therapy by using voltages over half a million, and that even here the advantage over 200,000 volts is less than has frequently been supposed.

It has already been noted that at the higher voltages the ionization along the electron paths is less concentrated, but that the spacing between the ions does not increase much for electron energies over 200 kv. We must remember, though, that the recoil electrons have a much lower average energy than the photons which produce them, and these in turn have less energy than the cathode electrons in the roentgen tube. Thus, to produce recoil electrons with an average energy of 200 kv. requires cathode rays of about 2,000 kv. This means

that there is a physically measurable increase in the diffuseness of ion production as the roentgen-ray voltage used increases from 200 to 2,000 kv.

In view of this difference, an observation of Stone becomes of special significance. He showed that when roentgen rays of 200 and 1,200 kv. are accurately controlled to the same dosage, their physiological effects on opposite halves of the same patient could not be distinguished. That is, changes in the diffuseness of the ions, which I estimate to be in the ratio of about 6:1, do not produce detectably different physiological effects. It, of course, remains possible that the 100 fold difference between the concentration of ions produced by neutrons and hard roentgen rays might be very appreciable. This would appear at present to be the most hopeful suggestion from the physical considerations that I have been presenting.

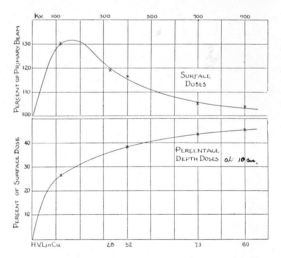

FIG. 5. Curves showing the surface and depth dosage with roentgen rays at potentials of from 100 to 900 kv. (Packard).

While it does not come strictly within the subject of penetrating radiations, this discussion would not be complete without mentioning the possibilities that have recently been opened by the production of artificial radioactive elements. Most interesting from the therapeutic standpoint are perhaps radio sodium, potassium, iodine, and phosphorus. These can be injected into the blood stream, or taken as food, and are in certain cases absorbed by the affected tissue. Where this can be done, it makes possible local irradiation under ideal conditions. Qualitatively, however, the effects of the irradiation from such substances will presumably be identical with those of roentgen rays or radium as ordinarily applied.

Of the available forms of penetrating radiation, therefore, the only ones whose physical properties give promise of valuable therapeutic properties are roentgen rays and neutrons. Studies made with roentgen rays up to very high voltages indicate that nothing is to be gained with regard to penetration or nature of the ionization by going to potentials over about 500 kv. Neutron rays are of about the same penetrating power as such roentgen rays, but differ in that the ions they produce occur in highly concentrated groups, and in their preferred action on hydrogen bearing matter. Because of these physical differences, differences in the biological effects of the two types of rays should appear; but it is not yet known under what condition, if any, neutrons are superior in their therapeutic action to roentgen rays. This, however, is evidently a hopeful field of experimentation.

Modern Physics and the Discovery of X-Rays

ARTHUR H. COMPTON, D.Sc.

University of Chicago

AMONG THE MOST striking practical developments resulting from the discoveries of modern physics are radio and the atomic bomb. Both are fruits of the discovery of x-rays. In our effort to understand the nature of the world around us, prominent recent advances include knowledge of the arrangement of atoms in crystals and molecules, recognition of the several elemental particles of which atoms are built, the electron and the nucleus, the proton and the neutron, the positron, the mesotrons, and so on, and something of the way in which these particles combine to form atoms. We have learned in the theory of relativity the laws of motion of stars and atoms moving at very high speeds and more precise laws of gravitation. In the quantum theory we have greatly improved our understanding of the nature of light and x-rays and have learned how to describe the motions of atoms and the parts of atoms. All of these new findings stem from Röntgen's discovery of x-rays fifty years ago, and in their development x-rays themselves have been used as a vitally important tool.

We might point out, also, how chemistry, geology, biology, and philosophy have been enriched by Röntgen's discovery. We could show how the electronic tools have stimulated the growth of industry, how the electron tube has made possible not only the radio but also the long-distance telephone and greatly improved telegraphic communication. We could describe the use of x-radiation and radium in the diagnosis and treatment of diseases. All of these have come from the discovery and use of x-rays. They are, however, part of a larger story that we cannot here take time to tell.

Two years before the discovery of x-rays, in his statement of the purpose for which the new Ryerson Physical Laboratory of the University of Chicago was built, Professor A. A. Michelson noted that the fundamental principles of physics had been well established. The future of physics research, he explained, lay in making more precise measurements of the known physical constants. It was for such precision measurements that the new building was designed. This attitude toward physics was common to the leading thinkers of the period, who from the time of Galileo, through Newton, Faraday, Maxwell, and Helmholtz had developed an elegantly organized description of how events in the physical world happen. Ours was a determined world, precisely predictable according to laws that were clearly understood.

X-RAYS

As typical of the scientific work of the period, Wilhelm Conrad Röntgen was then engaged in a careful study of the densities of various crystals. It seems that the immediate occasion for turning his interest to new fields was a publication by Lenard of an experiment with cathode rays striking a thin window from which rays (which came to be called "Lenard rays") were observed to emerge into the surrounding air. Lenard assumed that these were the cathode rays themselves, which penetrated the thin window of the discharge tube and could go a few centimeters further through the air. Röntgen was not so sure. He surmised that perhaps the rays outside the tube were of a different nature, produced possibly by the cathode rays, but of a considerably more penetrating character. He accordingly set up equipment similar to Lenard's but with walls too thick for the cathode rays to penetrate, surrounded his discharge tube with black paper to keep the light from getting out, and had a crystal such as was commonly used to observe ultraviolet light to see what would happen.

How Röntgen saw the fluorescing crystal when the electrical discharge was passed through the evacuated tube is now a matter of familiar history. Otto Glasser sets the probable date of this event as Nov. 8, 1895. From there on, developments were rapid. Before announcing his discovery, Röntgen himself made so thorough an investigation that for the next several years the investigators who rushed into the field added little more than refinement of detail to the statements about the properties of the new rays made in his own initial publications. The effect on a photographic plate, the electrical conductivity of the surrounding air, the slight effect on the retina of the eye, the remarkable penetration but partial absorption of the rays traversing various materials, the sharpness of shadows, unsuccessful attempts to refract, reflect and diffract the rays, even a try at reflecting the rays from a crystal of calcite, were described in Röntgen's publications of 1896. His was a triumph of individually conducted research.

ELECTRONS

It was, however, the uses to which the new rays were put that made this discovery of such extraordinary importance. Within a few months, with the help of x-rays, the existence of ions was demonstrated. For years the idea of electrically charged atoms and molecules had been used in the effort to explain the electrical conductivity of flames and of salt solutions. At a time, however, when one could not be sure even of the existence of atoms and molecules, the theory of ions gained little acceptance. It was when J. J. Thomson and E. Rutherford showed that air made conducting by x-rays could carry just so much current but no more, and that when the exposed air was passed through a strong electric field it was no longer conducting, that people were ready to accept the ionic hypothesis. These were properties predicted on the assumption of ions and for which there appeared no other explanation.

From this discovery of ions came in turn a long line of scientific and practical consequences. Combining the concept of ions with Faraday's laws of electrolysis gave to Arrhenius the firm basis for the theory of electrolytes, which gave impetus to new developments in physical chemistry. Johnstone Stoney noted that this theory required that the charges on each ion should be small multiples of a definite unit, for which he suggested the name "electron." J. J. Thomson surmised that cathode rays consist of tiny "corpuscles" carrying negative charges of this magnitude, and by a brilliant series of experiments measured approximately the charge and the mass of these "electrons." Here was a particle 2,000 times smaller in mass than the lightest atom. What had been named "the thing that can't be cut" is therefore itself composed of smaller parts. Thus came our knowledge of the electron and the beginning of our effort to learn the structure of the atom.

What the discovery of the electron has meant is in itself a story worth many volumes. Its exploitation in the electron tube, through the radio, sound-movies, radar, etc., has changed our social life, our economy, our political development, and has played a crucial part in the outcome of the recent war. On the scientific side, one of the many uses of the electron has been as an object for study while moving at speeds approaching that of light. It was the new phenomena thus presented that led to the theory of relativity, with its far-reaching implications concerning the relations of time and space and of matter and energy.

RADIUM

The early x-ray tubes of the type used by Röntgen glowed with a green fluorescence while emitting x-rays. Though Röntgen himself knew the two phenomena were of entirely different origins, Becquerel started from this observation to search for possible penetrating radiation that might be emitted by natural salts that show fluorescence. Among such fluorescing materials are various compounds of uranium. When these materials were placed on a black paper-covered photographic plate, they left their images. It was immediately noted that the non-fluorescent compounds of uranium were just as effective as were those that showed fluorescence. However, the discovery of radioactivity had been made. Here were natural materials which of themselves emit rays having effects like x-rays.

There followed an intensive study of the materials that show this remarkable characteristic of radioactivity. In addition to the then familiar elements uranium and thorium, polonium and radium and a score of other new radioactive elements were discovered. Rutherford, Soddy, Mme. Curie, and many others shared in showing how one atom emits a positively charged helium atom or a negatively charged electron and becomes an atom of different chemical properties, a natural transmutation from one chemical element to another. The energies involved in these radioactive changes were a million times greater per atom than those in ordinary chemical processes such as combustion. But no way could be found whereby the rate of transformation from one element to another could be changed. This rate was one of nature's established facts.

By the use of rays emitted by radioactive materials many important discoveries were made. Among them was the fact that each atom has within it a tiny "nucleus," only a ten-thousandth the diameter of the atom itself, which possesses nearly all of the atom's mass. By bombarding various substances with the alpha particles (charged atoms of helium) thrown off by radium, and noting how these particles were deflected by the materials they traversed, Rutherford was able to show that the nucleus of each atom has a positive charge which in electron units is equal to about half of its atomic weight. Around this nucleus circulates an equal number of negative electrons, forming a kind of atmosphere. Thus was blazed a trail which has led to the complete solution of the electronic structure of the atom.

Then came the remarkable discovery of atomic "fission." As a result of shooting an alpha particle from radium against an element of low atomic weight, such as lithium or beryllium, a new kind of particle called a "neutron" is produced. This particle is like the nucleus of a hydrogen atom except that it has no electric charge. If a neutron in turn falls on an atom of the special kind of uranium that has atomic weight 235, the nucleus of the atom is split into two roughly equal pieces and in the process emits further neutrons. If these neutrons are in turn caught by other atoms of U-235 the process is repeated, emitting still further neutrons, and so on indefinitely. Each such "fission" liberates a hundred times as much energy as is given out in the already highly energetic process of radioactive disintegration. This is the atomic chain reaction which goes on explosively in the atomic bomb or in a controllable manner in the chain-reacting piles used to make the plutonium used in such bombs.

Not only did the x-ray tube fortuitously guide us to the hidden store of atomic energy. At many stages, also, studies of x-rays and radioactivity have been intertwined, so that the growth of each subject has been connected intimately with the other. Now in bringing the greatest of all wars to a dramatic close, and in making available a source of energy vastly greater than the fuel at man's disposal, atomic fission has justified all the hopes of the Fermis, Rutherfords, Curies, and Röntgens, whose labors have brought us this Promethean gift.

THE NATURE OF THINGS

For understanding the nature of the world we live in, however, the use of x-rays themselves has during the last fifty years been perhaps the most effective of our methods of research. In 1912 von Laue and his collaborators found that x-rays could be diffracted by crystals just as light is diffracted on passing through a finely woven cloth. This showed at once that x-rays act like waves, just as light does, and also that crystals are indeed composed of regularly spaced layers of particles.

Investigations by the Braggs and others demonstrated that these particles are, in fact, the chemical atoms, and from the manner in which they diffract x-rays it was possible to learn the exact arrangement of these atoms in all the ordinary crystals. On the other hand, led by Moseley, the study of the spectra of elements used as the targets of x-ray tubes showed remarkable results. A regularity appeared in these spectra that made it possible to arrange the elements in a simple series of atomic numbers, starting with hydrogen as one and continuing to uranium as 92. Combined with the theory put forward at the same time by Bohr, it became evident that this atomic number is the number of electrons belonging in the atmosphere of each atom. Thus another important step was taken in learning how atoms are made.

With regard to its effect on our attitude toward the world, the most unexpected and far-reaching discovery of modern physics is perhaps that all the elementary pieces of which things are made have the characteristics of both waves and particles. This paradoxical duality came to light first in experiments on the nature of x-rays, which can be diffracted as waves from a ruled grating or can collide with an electron as one billiard ball bounces from another. This supplied the final evidence needed to establish a general "quantum" theory of mechanics that would apply to particles of atomic size as well as to ordinary things. The theory based on these experiments with x-rays concludes that all particles should show certain characteristics of waves, and that, as a result, the future of events on an atomic scale is predictable only as a statistical probability. Electrons and neutrons and atoms, in fact all kinds of things on which it has been possible to make the tests, show the predicted properties of both waves and particles, and the actions of these things are found to be indeterminate to just the predicted degree.

According to the mechanics that Röntgen knew, as Laplace once said, an intelligence so comprehensive that it would know completely the existing physical state of the world should see both the past and the future as if they were present. The result of the new quantum mechanics is, on the contrary, to show that there is a necessary range of uncertainty in any prediction of the future. For atomic events, and all larger happenings, such as the explosion of an atomic bomb, in which the result depends upon some individual atomic event, the uncertainties thus introduced into future actions are so great that only statistical statements as to what will probably happen have any significance. In the case of astronomical phenomena we deal with occurrences in which so many actions are concerned that the statistical predictions amount to practical certainty. As to the uncertainty of human actions, which depend upon nerve currents involving small numbers of molecules, the degree of uncertainty is unknown. We can merely say that if physics is the sole determiner of human events and our intentions are of no effect, then our future actions are not determined by the present situation, but probably within rather wide limits are a matter of chance. Since the day that Lucretius, in his *De Rerum Natura*, asked how free will was to be reconciled with a world whose atoms are governed by pushes and pulls, men have felt that the determined world of science presents a formidable barrier to belief in the effectiveness of purpose. For those who have grasped the meaning of the quantum mechanics, this barrier has ceased to exist.

HISTORICAL SIGNIFICANCE OF RÖNTGEN'S DISCOVERY

For the person who is concerned with the growth of the distinctively human attributes of man, it is such consequences as these that are the true measure of the greatness of Röntgen's work. If, indeed, science is the great intellectual quest of the modern world, his discovery is one of the greatest achievements of the age. There are those, however, who want to measure importance in dollars, or in the shaping of national destiny, or in terms of human life

Even in such practical terms the discovery of x-rays should be reckoned as an outstanding event in man's history.

As to human life, one could show that the direct effect of the use of x-rays and radium in diagnosis and therapy has saved a number of lives that is comparable with the number of soldiers killed in a world war. In terms of dollars, the money spent by the United States in building and using radio, radar, and atomic bombs, to mention only a few of the industrial consequences of Röntgen's discovery, has during the recent war been several per cent of the total national income. As to the shaping of political events, it is such factors as the radio with its nation-wide broadcasts of news and music and advertisements, that make our nation a cohesive unit.

Now as we face the future of civilization, what are the great factors that shape our thinking? As one commentator has expressed it, the single fact of the atomic bomb in the hands of America and Britain dwarfs the rest of the war situation into relative insignificance. As a result of the search into the nature of things which x-rays initiated, man has found a new basis on which to organize his world.

It may be correctly said that if Röntgen had not discovered x-rays, someone else would probably have found them within a year. It is true, likewise, that this discovery was itself based on a foundation of painstaking and brilliant researches made by many others in previous years. The discovery nevertheless is properly recognized as marking the beginning of a great new era. First it was the era of modern physics. This then developed into a period of vigorous industrial and social growth such as could not have arisen without the stimulus of the new scientific discoveries.

Perhaps, after all, the greatest human meaning of Röntgen's work lies in the increasing interdependence of people's lives in the world that he has helped to bring into being. This interdependence means a greater need for co-operation. This in turn means that the post-Röntgen world is one in which love of one's neighbor, as expressed in the willingness of each to work for the other, becomes a matter of rapidly increasing value. It is, indeed, such events as these that shape the destiny of man.

Washington University,
St. Louis 5, Mo.

The Scattering of X-Ray Photons

Arthur H. Compton
Washington University, Saint Louis 5, Missouri

WITH the recent celebration by the American Physical Society of the fiftieth anniversary of the discovery of x-rays, my thoughts go back to the autumn of 1920—some 25 years ago—when G. E. M. Jauncey, C. F. Hagenow and I began a series of experiments at Washington University on the scattering of x-rays. To us, x-rays were light rays of very short wavelength. We polarized them, we diffracted them, we reflected them from polished surfaces. But our chief concern was that some of the secondary rays which seemed to be scattered were more easily absorbed than the primaries from which they came. Besides, the rays of shorter wavelength were not scattered as strongly as the theories said they should be.

Those were exciting times. But if one were interested in confirming his pet theory, the work was disappointing; for, time after time, experiments gave results that were contrary to expectation. Finally, in desperation, we tried a crucial experiment. A scattered ray should at least have the same frequency and wavelength as the primary ray of which it was the echo. The test would be to compare, with a crystal x-ray spectrometer, the wavelengths of the primary and the scattered x-rays.

In those days it was no small task to obtain scattered x-rays of high enough intensity to measure their wavelength. We had to build special x-ray tubes without help of glass-blowers, for there were none in Saint Louis who knew the art, and x-ray manufacturers were not interested in our unorthodox designs. But the homemade tubes as finally built worked well enough. The x-rays from them seemed to show a definite increase in wavelength on scattering.

Fearful of personal bias, I had one of the graduate students take and record the readings of the electrometer while I shifted the crystal angles. Not knowing what we were looking for, he felt that the changing readings as we moved past one line after another were very erratic. "Too bad the apparatus wasn't working so well today," was his final comment. The data he had just taken were the ones that have since been published most frequently as showing the typical changes in wavelength of x-ray for different angles of scattering.

We knew now that x-rays were indeed increased in wavelength in the process of scattering. Was it possible that this was some kind of Doppler effect, even though the graphite block that scattered the rays was sitting stationary besides the x-ray tube? Perhaps each x-ray quantum was scattered by a single, moving electron. How much impulsive momentum would an electron receive if it scattered one quantum of x-rays? The answer came quickly:

$$mv = h/\lambda.$$

And, if the electron was moving forward with that momentum while the rays were being scattered, how much increase in wavelength would result for rays scattered at 90°? Again the answer was straightforward:

$$\lambda - \lambda_0 = h/mc = 2.4 \times 10^{-10} \text{ cm}.$$

Amazing! This was precisely the value of the wavelength change that was shown by our spectrometer measurements. It looked like hot lead.

Then followed months of refinement of the theory and of the experiments. Apparently the Doppler effect idea was capable of describing also the relatively low intensity of scattering x-rays of the shorter wavelengths. It was surprising how the concept of particles, or photons, simplified the theory of radiation emitted by objects moving with speeds comparable with that of light. Use of the principles of conservation of energy and of momentum as a basis for calculating the interaction of particles was a natural one, and led to the prediction of the so-called "recoil" electrons.

Actually these ideas were for the most part not new. They acquired importance because the theory became so well confirmed by experiment that it could be considered firm; and the experiments, having now a theoretical basis, took on greater significance. For many years Eve and Florance and Barkla and Gray had known that scattered x-rays and gamma-rays did not behave as theory said they should. Twenty years earlier Einstein had introduced the idea of the needle-ray quantum of radiation as a basis for explaining the photoelectric effect. Indeed, while our work was under way in Saint Louis, Debye, working then at Zurich, developed independently the same theory of the scattering of x-ray photons by electrons, and was looking for someone who would put his results to the test.

Most readers today will not remember the hot debates and intensive experiments that were carried on during the next few years. First, the experiments were not believed. Then their interpretation was questioned. Better give up the principles of conservation of energy and of momentum than the beautiful simplicity of the electromagnetic wave theory of light!

The definitive experiments that showed effects of individual photons of x-rays associated with individual recoiling electrons were the final step that showed that x-rays do indeed act as particles. First, Bothe and Geiger showed by delicate tests with counters that when a recoil electron is ejected from one side of a thin foil, a scattered photon may simultaneously appear on the opposite side. A. W. Simon and I followed this with a cloud-chamber test, repeated since by several others with similar results, which showed that when a recoil electron appears, a photon is scattered in precisely the direction required by the simple theory of collision of particles.

X-rays, which by this time had shown all the wave properties exhibited by light, were thus clearly capable of acting as particles.

This result came at a most fortunate time for the development of physics. Louis de Broglie had just published his paper showing that wave groups moving in a medium of variable refractive index are equivalent to particles moving in a field of force. From this he obtained his famous expression for the wavelength of a moving particle,

$$\lambda = h/mv.$$

The fact that this expression applied perfectly to the momentum and wavelength of the x-ray photons gave life to his theory. In the hands of Schrödinger, it became the starting point of quantum wave mechanics.

Heisenberg, who had been approaching quantum mechanics through matrix theory, saw in the experimental duality of waves and particles a new method of approach, through the principle of uncertainty. This principle merely combines the limitation in determining position because of the finite resolving power associated with wavelength and the uncertainty in determining momentum because of the recoil from the emission of the particle which signals the position of the object. By a simple argument Heisenberg thus found that a limit to experimental precision was given by the expression,

$$\delta p \delta q \approx h.$$

This theory, like de Broglie's, assumed that all things have the dual properties of waves and particles. The x-ray tests showed this to be true for electromagnetic rays. Now came in quick succession the experiments of Davisson and Germer and of G. P. Thomson, showing that electrons are diffracted by crystals, and similar experiments by many others showing diffraction effects with such well-known particles as protons, hydrogen and helium atoms, hydrogen molecules and, very recently, neutrons.

A rather striking illustration of what this revolution meant in the thinking of physicists occurred when I called on J. J. Thomson in September 1927 at his home in Cambridge. He showed me with enthusiasm the photographs his son George had taken of the diffraction of cathode rays by films of aluminum and gold. Those photographs showed circular diffraction patterns closely comparable with those obtained when x-rays pass through similar films. Jauncey[1] has referred to the vigorous argument between J. J. Thomson and Lenard in the early days over the nature of cathode rays. Lenard thought of them as a type of wave; Thomson considered them to be particles. Because they could be deflected with electric and magnetic fields and because the charge on each ion could be measured, the physics world agreed that Thomson

---

[1] G. E. M. Jauncey, *Am. J. Physics* **13**, 362 (1945).

had proved his point; he had shown that cathode rays were particles of electricity. But here was now the old physics hero rejoicing at the achievement of his son in establishing that these same cathode rays were waves. I wonder what would have happened if George Thomson had taken his diffraction photographs at the time of the controversey between Lenard and "J. J."

Today we take it for granted that all kinds of particles have wave characteristics and that all waves have corpuscular characteristics. Which properties are predominant is only a question of the conditions under which tests are being made.

It is worth noting that the scattering of x-rays helps us in our understanding of the precise significance of particles and of waves. Briefly it is this. Whenever a ray produces a physical effect it acts as a particle; the effect is discrete and occurs in a definite position. The wave, on the other hand, serves as a convenient means of predicting where the action of the particle will probably occur. The wave is thus something of a conceptual device, while the particle has a more precise significance in terms of physical action.

When many particles are concerned in an action, for example, in the blackening of a photograph plate, the effect of the individual particles may be hidden by the action of the mass. In such cases it may be preferable to describe the action in terms of waves than of particles. An illuminating example is that of the electron structure of an atom. Instead of the older concept of electrons moving in orbits, we now think of an electron atmosphere about the nucleus whose density at any point may be expressed by the value of the appropriate Schrödinger wave function. The question sometimes arises whether in this case the electron considered as a particle continues to have any meaning. Experiments fail to determine its position or even to locate its orbit.

Experiments on the scattering of x-rays from atoms, however, show that even here the particle concept has fundamental reality. It is found that the x-rays scattered by atoms or by groups of atoms in molecules or crystals are much more intense than is to be expected if the electricity is distributed continuously within the atom.

In addition to the x-rays which enter into interference relations between one atom and the next and which can correspondingly be described as "coherent" x-rays, there is an equally important part which we call, "incoherent" rays. To explain this "incoherent" radiation, we need to assume that the electricity in the atoms is not continuously distributed but consists of discrete particles whose positions are distributed at random. The wavelength of these scattered rays differs from that of coherent rays because of a kind of Doppler effect attributable to the velocity of the electron as a particle within the atom. This shows itself in the width of the modified line in the spectrum of scattered x-rays. The work of Jauncey, Dumond and others has shown that the electron velocities within the atom as thus estimated agree well with those calculated from atomic theory.

Such studies of the scattering of x-ray photon confirm the theory of the meaning of waves that Heisenberg and Bohr have emphasized. On this view, a region where a wave is of large amplitude is one for which there is a high probability of occurrence of a particle. The amplitude of wave therefore is a measure of the probability of the presence of the particle. If, however, we are concerned with the physical actions that may occur, we must concentrate our attention on the particles themselves, for these are the so entities that produce physical effects.

The study of the scattering of x-ray photon has likewise increased our knowledge of the characteristics of the elemental particles themselves. Consider, for example, the properties the electron. For a long time the only characteristics of this particle revealed by experiment were its charge and its mass. Early calculations of its size turned out to have little meaning. The development of the "exclusion" principle as applied to spectra gave reason to believe that the electron likewise has a spin and should consequently have the characteristics of a tiny magnet. One of the most convincing lines experimental evidence in this connection is scattering of x-rays of very short wavelength. It is possible to show, following the consideration of Heisenberg's principle of uncertainty, that a particle having a mass comparable that of an electron has a natural diameter which value is roughly h/mc, which in the case of an electron is equal to the wavelength of an x-ray excited by about 500 kv. If one thinks of this natural size as being the distance through which the electron charge is distributed, it is apparent that, considered from the point of view of electromagnetic theory, intereference will occur between the waves scattered from different parts of this electron. On this view, when the waves which traverse the electron have a wavelength of the same order of magnitude as the natural size, little scattering is to be expected.

If, however, the electron is spinning and has a magnetic field, the magnetic component of the traversing wave will produce torques on the electron which will change the direction of its rotation. This should give rise also to scattered x-rays. Calculation shows that for waves of the length of gamma-rays, such magnetic scattering should be more important than the well-known electrical scattering from the electron. This is perhaps the best interpretation that can be given on classical principles of the term in the formula for scattered x-rays which was added by Klein and Nishina in their quantum mechanical treatment of the problem.

Experiments of two kinds have verified this added term in the scattering formula. The first one dealing strictly with the energy of the scattered rays. Rather refined experiments of this kind with rays of very high frequency have, as is well known, confirmed the need for the added term. The second line of evidence is considerably more specific and definite. It comes from the fact that the electric component of the rays scattered at 90° is completely polarized, while the portion that results from the magnetic dipole of the electron should be unpolarized. Shortly before the war this prediction was tested by Dr. Mc Rogers, who showed that when x-rays of very short wavelengths are scattered, they exhibit little polarization, precisely in accord with Nishina's theory.

Such results make it clear that the electron is to be considered as not only an electric charge but also a tiny magnet.

No attempt will be made here to discuss the transformation of high energy photons into positive and negative electrons. This remarkable phenomenon of pair production, revealed first by the cosmic ray experiments of C. D. Anderson and interpreted especially by J. R. Oppenheimer and his colleagues, suggests that we may be on the threshold of a new period in physics where it will be possible to understand much more thoroughly the relations among the various elemental particles. Here lies perhaps the great future of physics.

Perhaps even more important in its general human implication is the principle of uncertainty. This stems directly, as we have seen, from the experimental duality of waves and particles. In my philosophy course as a college student, I was impressed by the impossible human situation of living in a world in which we had to assume that our choices and purposes are effective while at the same time scientists seemed to have established the fact that all physical actions had been determined from the beginning of time. The problem was an old one. Lucretius in his *De Rerum Natura* asks the puzzled question, "How then are we to wrest freedom of the will from a nature in which the motions of each particle are produced by the impact of other particles upon them in an order that is fixed by natural law?" The classical expression of the verdict of the older physics is that of Laplace in the introduction to his *Celestial Mechanics:* he points out that a being with knowledge and understanding of the present state of the world, knowing the position and motion of each of its particles, would know both the past and the future as if they were the present.

Some years before the development of quantum mechanics, Schrödinger pointed out that such a statement, though it is an accurate conclusion from Newtonian mechanics, nevertheless assumes a knowledge of the reliability of our mechanical theories that goes far beyond possible experimental tests. To Bohr and Heisenberg we owe the more precise awareness of the degree of uncertainty to which our knowledge is subject.

The uncertainties to which we refer are those that necessarily become of importance in dealing with actions on a very minute scale. A single x-ray photon would be an example. We do not know at all in which direction a particular x-ray photon will be scattered when it falls upon a block of graphite. No refinement of our experiments is in principle capable of making this direction predictable.

The fact that the distribution of scattered x-rays can be predicted statistically with precision is sometimes taken to mean that all large-scale events are thus predictable. But what is overlooked is that many large scale events are based at some stage upon processes on a very minute scale. Human actions are themselves examples of such events. The uncertainty in the minute event is reflected in the large scale event that follows.

A typical example is the explosion of an atomic bomb. It may be set off by the capture of a neutron that has in turn been released by some radioactive process. If, at the right moment, the initial neutron is produced, an effective explosion will occur. The type of explosion which will occur thus becomes a matter that is predictable, not precisely, but only as to its statistical probability. Here is an event of distinctly large magnitude whose occurrence is uncomfortably subject to uncertainty of the Heisenberg type.

It is considerations of this kind that have made it necessary for us to modify sharply our ideas of cause and effect. It is not appropriate here to discuss their implications with regard to the vital human problem of freedom in a world of law. It is perhaps pertinent to say, however, that no longer should a physicist try to argue, as Laplace might well have done, that effort to achieve a result can have no meaning. We are well aware now that our physical laws are not adequate for predicting a definite future, and that there is ample room within our system of physical laws to admit the effectiveness of purpose.

It was far from my thoughts when we started our experiments with x-rays 25 years ago that such work would have a bearing on such age-old problems of philosophy. We are still far from having solved these problems, but it is perhaps justifiable to feel that the way has been opened to approach a solution.

The contribution to physics of the studies of scattering of x-rays have been more concrete. They have introduced the duality of waves and particles, and have played a part in opening to physicists a new study of the relations between the elementary praticles of which the world is made.

Man's Awareness and the Limits of Physical Science

A. H. Compton, *Washington University*

Attention is called to the conflict between the views of Bohr and Schrödinger regarding the nature of the physical world. Bohr considers this to include only that which can in principle be verified by observations using material instruments. Schrödinger introduces the hypothesis of a physical continuum, typified by the concepts of wave mechanics, whose changes follow a strictly causal determinism, but of which observations with material instruments can give only partial information.

Making use of Bohr's evidence that one's awareness of his intentions makes possible the prediction of his actions more definitely than is in principle possible from physical observations, a category of ideological reality is introduced. This category includes feelings, ideas, intentions, etc., of which one is immediately aware. It is distinguished from the physically real world as defined by Bohr. In terms of this distinction, Schrödinger's determined continuum has ideological but not physical reality. The question remains whether this continuum may have objective reality outside the physical world.

The usefulness of one's awareness of his intentions in forecasting physical events suggests the following rather obvious hypothesis: that there exists an objective world regarding which physical observations reveal one aspect and man's awareness reveals another. Such a world is not evidently related to Schrödinger's continuum. The present hypothesis can, however, be used in reconciling the "outer" physical indeterminateness of man's actions with a high degree of "inner" determinateness. Here seems to be an answer to Schrödinger's criticism of the concept of an undetermined world as violating the reasonable inference of self-determined actions associated with the sense of moral responsibility.

THE WORLD OF SCIENCE IN THE LATE EIGHTEENTH CENTURY AND TODAY

ARTHUR H. COMPTON

Professor of Natural Philosophy, Washington University

(*Read April 19, 1956*)

It is my privilege to speak about the world of science in the late eighteenth century and today. It is evident that those who arranged our program did not want a catalogue of scientific achievements. Rather they have asked us to consider the setting in which science was studied when Benjamin Franklin was doing his important work.

I shall discuss this matter in two ways, first by telling something of the relation of Benjamin Franklin's studies of electricity to other similar investigations before and after his time. Then I want to draw attention to the development of mechanics, which during the eighteenth century was the main line of professional physics research. Galileo, Newton, and Laplace replaced Aristotle's view of the world as a God-impelled system with that of a world of material objects moving and acting according to simple mechanical laws. Our century has now replaced their deterministic world-machine with a new system where again physics permits us to suppose that events have human meaning.

It is this century's development of the understanding of the electricity that Franklin was studying which has been largely responsible for shaking the foundations of the mechanics that Franklin's contemporaries believed they were establishing so firmly. Thus the studies of eager amateurs such as Franklin, carried on with a burning interest in finding new truth, have added greatly to our knowledge of nature.

We may think of Franklin as the prototype of the typical American scientist and engineer—the enthusiastic, resourceful, and ingenious searcher for new facts and relationships, who relies largely upon others for their systematic elucidation, but is alert to put the new knowledge to useful work. Closely parallel would be the relation of Michelson and Morley's famous experiment to Einstein's relativity theory, or the discovery and application of thermoelectricity by Thomas Edison and Lee DeForest to its electron interpretation by J. J. Thomson and O. W. Richardson.

BENJAMIN FRANKLIN AND THE UNDERSTANDING OF ELECTRICAL PHENOMENA

Let me then first indicate the relation of Franklin's study of electricity to that of others working in the same field. Franklin began his experiments in 1746 at the age of forty, and continued his most active investigation of electrical phenomena for about seven years. During this period he found his experiments of the most engrossing interest. "For my own part," he writes in his first letter to Peter Collinson, Fellow of the Royal Society at London, "I never was before engaged in any study that so totally engrossed my attention as this has lately done. . . ." He ceased his studies, as so many others have been forced to do, because of the demands of responsible citizenship, when he went to England as representing the American colonies.

Franklin stepped into a field of study that was of very live interest among amateurs in science but was for the most part thought of as a novel form of parlor entertainment. The professional physicists of the day were concentrating their attention on the elaboration of Newton's principles of mechanics and on the development of precision astronomical instruments or were exploring quantitatively the phenomena of heat. One reads of the Burgomeister of Danzig lighting with an electric spark a candle that he had just blown out, and of igniting alcohol vapor by sparks drawn from an electrified jet of water. People line up to receive the jolt of an electric discharge. It would seem that the appearance of improved machines for producing frictional electricity was a profitable business chiefly because of orders from those for whom these experiments were a fascinating hobby. One is reminded of the development of short wave radio a generation ago by enthusiastic amateurs who were working in an area discarded by the professionals as of negligible importance.

It is however true that some of this amateur work was of strictly high scientific merit. Thus Stephen Gray in 1729 reported in the *Philosophical Transactions* of the Royal Society a series systematic experiments on all kinds of materials to see which ones were insulators and which conductors. The effects of frictional electricity he found could be transmitted up to 765 feet over wires supported by silk thread. Also Charles François Du Fay, of Paris, reported during the years 1733–1737 a series of observations in which he established the existence of two kinds of electricity. The account of his results as reported in the *Philosophical Transactions* for 1734 is worth repeating:

Chance has thrown in my way another Principle, more universal and remarkable ... and which casts a new Light on the Subject of Electricity. This Principle is, that there are two distinct Electricities, very different from one another; one of which I call *vitreous Electricity* and the other *resinous Electricity*. The first is that of Glass, Rock-Crystal, Precious Stones, Hair of Animals, Wool, and many other Bodies. The Second is that of Amber, Copal, Gum-Lack, Silk, Thread, Paper, and a vast Number of other Substances. The characteristick of these two Electricities is, that a Body of the vitreous Electricity, for example, repels all such as are of the same Electricity; and on the contrary, attracts all those of the resinous Electricity. (*Phil. Trans.* **38**: 258.)

One of the developments that attracted Franklin's early interest was the Leiden jar. It was a clergyman of Pomerania, E. G. von Kleist, who seems first to have noticed by accident the properties of what we now call an electrical condenser. The phenomenon was, however, first reported to the French Academy of Sciences in 1746 as a letter from Musschenbroek of Leiden, communicated by the distinguished naturalist, Ferchault de Réaumur. As translated by Professor A. Wolf of London,[1] a portion of this letter will give a sense of the level at which such experiments were being done:

I wish to report to you a new but terrible experiment, which I advise you on no account to attempt yourself. ... I was carrying out some researches on the force of electricity; for that purpose I had suspended by two cords of blue silk, an iron gun-barrel, AB, which was receiving electricity by conduction from a glass globe which was being rapidly rotated on its axis, and rubbed meantime by the application of the hands. From the other end, B, there hung freely a brass wire, the end of which was immersed in a round glass vessel D, partly filled with water, which I was holding in my right hand F, while with the other hand E, I tried to draw sparks from the electrified gun barrel. Suddenly my right hand, F, was struck with such violence that my whole body was shaken as by a thunderbolt. The vessel, although made of thin glass, does not break as a rule, and the hand is not displaced by this disturbance, but the arm and the whole body are affected in a terrible manner which I cannot express; in a word, I thought it was all up with me.

One Dutch physicist, writing shortly afterward, states that this effect was first noted by a man named Cunaeus, a wealthy amateur, living in Leiden. It was William Watson, a Fellow of the Royal Society, who in 1748 described a modification of the Leiden jar, with metal lining inside and out, essentially the form that is still used.

Franklin obtained from England one of the best frictional electrical machines that was then made, and began his experiments at Philadelphia. He described his results in letters to Peter Collinson, a London merchant through whom he had obtained the electrical equipment. One of his earlier letters (Letter III) gives his interpretation of the action of the Leiden jar in terms of his "one fluid" theory of electricity. As he explains,[2]

At the same time that the wire and top of the bottle, &c. is electrised *positively* or *plus,* the bottom of the bottle is electrised *negatively* or *minus,* in exact proportion: *i.e.* whatever quantity of electrical fire is thrown in at the top, an equal quantity goes out of the bottom.

By way of commentary he writes some years later to Collinson[3] expressing some of his "Opinions and Conjectures concerning the Properties and Effects of the electrical Matter, arising from Experiments and Observations, made at Philadelphia, 1749." He writes,

Thus common matter is a kind of spunge to the electrical fluid. ... But in common matter there is (generally) as much of the electrical as it will contain within its substance. If more is added, it lies without upon the surface, and forms what we call an electrical atmosphere; and then the body is said to be electrified.

Franklin's idea of a single electrical fluid stimulated other hypotheses, such as the two-fluid theory suggested by Robert Symmer in 1759. The result was further active experiments. It was not possible, however, to suggest any true dif-

---

[1] A. Wolf, *A history of science, technology, and philosophy in the eighteenth century,* 222–223, New York, Macmillan, 1939.

[2] *Cf.* I. Bernard Cohen, *Benjamin Franklin's experiments,* 180, Cambridge, Harvard Univ. Press, 1941.

[3] *Op. cit.,* 213 and 214.

ference that might be expected to appear according to the two alternative hypotheses.

Franklin's fame as a scientist arose chiefly from his identification of lightning as an electrical phenomenon. This was not a new suggestion. Thus, among others, Isaac Newton, thirty-three years earlier, had described the audible sparks that can be drawn from a body electrified by friction, and compared them with lightning and thunder. What Franklin did was first to show one after another that the effects of electric sparks are identical in kind with those of lightning. He then gave convincing direct evidence by his kite experiment, as described in his Letter XI addressed to Peter Collinson in 1752:[4]

As soon as any of the thunder clouds come over the kite, the pointed wire will draw the electric fire from them, and the kite, with all the twine, will be electrified, and the loose filaments of the twine will stand out every way, and be attracted by an approaching finger. And when the rain has wet the kite and twine, so that it can conduct the electric fire freely, you will see it stream out plentifully from the key on the approach of your knuckle. At this key, the phial may be charged; and from electric fire thus obtained spirits may be kindled, and all the other electric experiments be performed, which are usually done by the help of a rubbed glass globe or tube, and thereby the sameness of the electric matter with that of lightning completely demonstrated.

Franklin's practical mind turned to the use of sharp pointed conductors as lightning rods for protection of buildings. His writings show that he was well acquainted with the conditions needed to make these rods effective. Quite independently and at almost the same time the use of pointed conductors as lightning rods was being tried in Moravia by Father Procopius Diviš.

In passing it is worthy of note that the use of persons in these electrical experiments was not primarily for dramatic or entertainment effect. The fact was that the sensation of a shock was the most sensitive means then known for detecting the electrification of a charged condenser. The intensity of the shock was a rough measure of the degree of electrification. Similarly a low capacity spark was most readily detected by the ignition of alcohol vapor.

While Franklin was engaged upon these and other electrical experiments, a group of Russian, English, and Swedish experimenters, Wilcke, Canton, and Æpinus, discovered electrical induction. Qualitatively the chief phenomena of electrostatics were now known. The next stage was that of placing the knowledge of electricity on a quantitative basis. In this phase of the study the work of Joseph Priestley, Henry Cavendish, and Charles Augustin Coulomb was outstanding.

The use of balls of cork or pith suspended on thread had been long used as a means of observing electrification. Franklin, at this time living in England, told Priestley, who was twenty-seven years his junior, of an unpublished experiment that he had tried.[5] This was an attempt to electrify a pair of cork balls suspended *inside* a metal vessel. He had been unable to detect any effect on the balls when the vessel was charged. To Priestley, familiar with the consequences of the inverse square law of force as developed in the theory of gravitational phenomena, this was a very suggestive result. He accordingly repeated Franklin's experiment, using a pair of pith balls hanging entirely within a tin quart cup. The very slight effects that were caused when the cup was electrified or discharged he explained as a result of using a vessel that was not completely closed. Thus Priestley asked the question, "May we not infer from this experiment, that the attraction of electricity is subject to the same laws with that of gravitation, and is therefore according to the squares of the distances?"

Such were the first suggestions of the famous experiment carried out in a more refined manner by Cavendish, which remains today as our chief reliance for accepting the inverse square law of force between electric charges. Cavendish enclosed an insulated metal ball, 12.1 inches in diameter, inside of a pair of hinged conducting hemispheres. Having charged the hemispheres, he connected them momentarily to the sphere inside, and then removed the enclosing conductor. Using a pith-ball electroscope, he was unable to detect any charge on the inner metal sphere. If this result was exactly true he reasoned that the repulsion between like electric charges must follow precisely an inverse square law of force From the precision of his experiments he concluded that the exponent must lie between 1.98 and 2.02. James Clerk Maxwell, who long afterward published this work of Cavendish, repeated the experiment with sufficient precision to reduce the uncertainty to one part in 40,000.

Not knowing of this work of Cavendish, Coulomb in 1784 arrived at the same result by th

---

[4] *Op. cit.*, 266.

[5] *Cf.* A. Wolf, *op. cit.*, 242.

more direct means of measuring the force between two pith balls similarly charged. Using his newly developed torsion balance he compared directly the force between the balls when separated at different distances. Coulomb also performed an experiment similar to that of Cavendish, but with less refinement, from which he also correctly concluded that in an electrified conductor the charge does not penetrate the interior. It was these experiments of Coulomb that gave the basis for the subsequent mathematical development of electrostatic theory.

To complete the story of electricity as known during the eighteenth century we must refer to the well-known work of Luigi Galvani at Bologna and of Alessandro Volta at Como. It is noteworthy that Galvani was engaged in a systematic study of the effect of electrical impulses in causing contraction of animal muscle. It was accidental that he had a frog leg in contact with both iron and brass when about 1780 he made his famous discovery, but it was because he was alert to the convulsive effects of electric discharges that the contractions of his frog legs appeared significant. Nor was it thus surprising that he should have recognized the contractions as an electrical phenomenon. The fact that Galvani ascribed the effect mistakenly to electricity in the animal organism did not prevent the very fruitful development of his observation.

It was Volta who correctly identified the Galvanic current as a flow of electricity arising from the contact of the two dissimilar metals. The electrical nature of the current he demonstrated by use of a condenser of widely variable capacity and an electrometer of greatly improved sensitivity that he had invented. Then in 1800, at the close of the century, by means of his Voltaic pile the potential differences became large enough that all kinds of electrical phenomena could be readily produced. Animal electricity had no part in such experiments. Electricity could now be made to flow in a continuous current. The way was open for the study of electrodynamics which was carried on so effectively during the nineteenth century, and was to affect so greatly our whole picture of the physical world.

I have spent this time on the development of our understanding of electricity during the eighteenth century because it was in this field that Benjamin Franklin made his most important contributions to science. He also added to our understanding of radiant heat, of meteorology, of hydrodynamics, and of many other aspects of science. But time will not permit me to discuss this work if I am to tell in a balanced way what the world of science in the late eighteenth century was like. In concentrating on the subject of electricity we have seen how one man contributed to an important new field of knowledge and stimulated its rapid development. As noted before, however, this was not the field where most of the great scientists of Franklin's day were devoting their efforts.

DEVELOPMENT OF NEWTONIAN MECHANICS

Isaac Newton died when Benjamin Franklin was twenty-one years old. It was Newton's great generalizations regarding mechanics and gravitation, formulated in his *Philosophiae Naturalis Principia Mathematica* of 1687, that continued during the eighteenth century as the focus of attention for the professors of natural philosophy. D'Alembert's Principle, Maupertuis' formulation of the Principle of Least Action, Lagrange's use of generalized coordinates in his elegant statement of the laws of mechanics, and Laplace's application of these principles to the stability of the solar system in his *Mécanique Céleste*—it was studies such as these that occupied the chief scholarly scientific thought of the period, especially on the continent of Europe.

At this distance it is difficult to appreciate the importance attached to the development of mechanics in reshaping man's outlook on the world. The change that was going on was from a Ptolemaic description of the universe to a Copernican system, which was in turn simply interpreted according to the laws of mechanics. It was Laplace's work, published from 1796 to 1805, that put the finishing touches on this development with a completeness that carried full conviction as to its truth. Why did this seem so important?

The answer lies in the world view accepted by Europe in the Middle Ages. In its essence this view was not so much Christian as it was Aristotelian, but it had become basic to the thinking of scholars in both the Christian and the Islamic lands. As far as my reading goes, it appears that until the time of Galileo, this Aristotelian interpretation of order in the world was universally accepted by Western scholars. The new mechanics completely undermined his view. According to the older system, order in the universe was maintained by the continual free action of a God on whose wisdom and generally beneficent attitude one could rely. Now order in the universe was seen to follow as a necessary result of simple and immutable properties of matter. The world was thus indifferent to the needs that man might feel. It was the firm establishment of this revised point of view that was the great accomplishment of the natural philosophers of the eighteenth century. Because Aristotle's physics had been incorporated so implicitly in Christian and Islamic thought, this revolution in mechanics meant that the whole idea of God had to be thought through again from the beginning. Those of little faith dropped religion as being inconsistent with science. Those who knew the basic realities of religion from their own experience sought a framework for religion that was consistent with science.

The central difficulty was simple but fundamental. According to Aristotle the normal state of undisturbed matter is that of rest. When motion was observed it implied a cause. For animals and man the cause of their own motion was their volition. Similarly the stars were kept moving in their courses by the God of the universe. The reliability of God's will was demonstrated by the great regularity of these motions.

Even in Copernicus' writings and those of Kepler there is nothing to disturb this point of view. Only now it is the earth instead of the sun that moves with regularity. God is still the Prime Mover.

The revolution against this view was started by Galileo when he postulated that a body would continue in uniform motion unless acted upon by some force. No longer was a Mover needed to keep the body in motion. The next big step was taken when Newton showed that if gravitation between bodies followed an inverse square law of force, the motion of the planets was just what was to be expected from the laws of mechanics as observed in the laboratory. There was nothing here that required special attention from the Master Hand. Such orbital motion followed from the simple properties of matter.

There remained however certain perturbations in the motions of the planets, irregular departures from the elliptic orbits that Newton's calculations predicted. These might be disturbances caused by the attraction of other planets. What was to prevent the whole solar system from thus flying apart? Did not the heavens still need God's guiding hand? It was this lingering question that was answered so completely by the work of Laplace, when he showed that the orbits were stable under gravitational forces. Here was the point of his oft-quoted reply to Napoleon's remark, "M. Laplace, they tell me you have written this large book on the system of the universe, and have never even mentioned its Creator." As the story is told by Professor Rouse Ball of Cambridge,[6] Laplace answered bluntly, "I had no need for that hypothesis."

It was the fact that he could now truly make this reply that gave the epochal significance to Laplace's work. Now it could be said with confidence that the regular motion of the planets follows in detail the predictions based solely on the assumption that matter has certain simple mechanical properties. Perhaps the original creation of matter with such properties was a divine act; but there is no indication whatever of continued divine intervention in the physics of mechanical or celestial affairs. The Aristotelian argument for an all-powerful Creator continuously at work was gone. It was easy to infer, as stated for example by G. S. Brett in his book *Sir Isaac Newton,* that "In this way a severe rationalism was put forward in opposition to all the romantic forms of religion that went by the name of 'enthusiasm.' The seat of religious belief was thus moved from the heart to the head; mysticism was excommunicated by mathematics . . . the way was opened for a liberal Christianity which might ultimately supersede traditional beliefs," and for the "religion within the limits of reason" sought by Kant.

It should be noted in passing that the common interpretation here given by Brett is precisely the reverse of the facts. It was the professors of systematic theology who believed they were establishing religion on a rational foundation who were caught off base by the mechanical revolution. It was those whose religion was based rather on intuition and experiences of the heart, which they knew first hand as reliable facts, who found themselves undisturbed by the mechanists' arguments. These were those whose guide was not Aristotle's philosophy but the Hebraic writings that had little to do with cosmology. Immanuel Kant, though a member of a theological faculty, counted himself in this latter group when he noted that the true evidence for God is to be found in the majesty of the heavens above and of the moral law within. Such evidence was not to be sought, that is, in intellectual arguments, though as a true philosopher Kant recognized that the idea which one holds of God must be consistent with the limits of reason.

To make his point completely clear Laplace ex-

---

[6] Sir William Dampier, *A history of science,* 193, Cambridge, Cambridge Univ. Press, 1944.

plained that if there were a mind of such infinite capacity that it could know at any time the positions and motions of all the particles in the universe and the forces between them "both the past and the future would be present to his eyes." In other words, the world is a completely determined mechanical system. Note that this statement not only rules out the effectiveness of an assumed *Divine* Agent, but rules out also the effectiveness of the *human* will in determining the course of physical events. That is, man's feeling of responsibility for what happens is an illusion. This is an accurate statement of the implications of Newton's mechanics.

As of the mid-twentieth century no physicist would now subscribe to such a statement, if for no other reason than that the precise knowledge of the position and motion of a particle is meaningless or impossible according to the quantum concepts that replace Newtonian mechanics when dealing with small-scale phenomena. Thus determinism is no longer considered a valid interpretation of the physical world.

An accurate statement would be that while on the one hand observation of physical events reveals nothing that requires for their interpretation the operation of intelligence; on the other hand what we can observe, and the physical laws as they are now known, are not inconsistent with the effectiveness of purpose in shaping the course of the events in nature. This balanced statement applies equally to one's own actions with reference to his responsibility for what he does, and to events occurring in the external world as related to other intelligences, either of men or of God. That is to say, we recognize now that we cannot call on physics and astronomy to give evidence for the effective action of free minds, either human or divine. But at the same time we recognize also that we cannot, on the basis of any kind of physical observation, deny that either human or divine minds may be effective in determining the course of certain types of events, in particular the actions of living organisms. Whether mind may participate in determining the course of events simply cannot be answered by physical observations.

Perhaps we of the post-Laplacian period have not taken seriously enough the moral implications of a strict belief in Newton's Laws. Immanuel Kant, by juggling an unknowable world of things-in-themselves about which observed phenomena give us no information, wrested to his own satisfaction a ghostly kind of freedom. But it could not mean real responsibility for what happens. Actual physical events, according to Newton's laws, have been determined from the beginning of time. In practice men have resorted to the common sense assurance that they are responsible for their acts, and have shaped their lives accordingly.

At other times and places people have not been so complacent. Dampier in his *A History of Science* points out that the decline of Greek science began when Socrates urged that it is one's will, not the laws of physics, that determines action.[7] Waiting in prison to drink the hemlock, with the door left open by a friendly jailer, he asks his students why they suppose he is sitting there in a cramped position on his bed when the door has been left open for him to escape. Is it, as Anaxagoras would have him believe, because of the tension of tendons over his joints? Not so, says Socrates. It is because I have been condemned by the people of Athens, and as a man of honor I will not run stealthily away.

Fifteen hundred years later the religious authorities of Islam, under the leadership of al Ghazali, became concerned over the influence of science. Tolerance gave way to persecution. Science must no longer be studied because it leads "to loss of belief in the origin of the world and in the creator."

A few years ago I had the opportunity to ask Professor Habib of the department of history at Aligarh University whether it was in fact this religious antagonism to science in the early twelfth century that was responsible for its decline, or should we look for other factors? He answered that he had just been lecturing on this point to his classes. He told me the counts that the religious leaders of that period had against science. In brief, science reduced the course of events in the world to natural law. If they are thus determined, what becomes of the effectiveness of God's will? Is not God all powerful? How can this be if He is bound by the laws that science proclaims? Here was indeed the chief reason that science was abandoned in Islam.

Shortly afterward, at Lahore, I was talking with a young Muslim Oxford graduate who was working ardently to establish Pakistan as a Muslim state. I asked him, "How, in a nation that is committed to the concept of God, are you going to develop the scientific technology that you see so necessary?" He understood at once what I meant.

[7] *Op. cit.*, 27–28.

This was to him and his associates a live problem. But he had his answer ready: "You Christians have shown that science can be consistent with belief in God. If you can solve this problem so also can we."

It is worth noting that the great leaders of eighteenth-century thought who were reshaping the outlook on the universe did not themselves see in what they were doing any reason for abandoning the idea of God. Their ideas were merely moving from the traditional pattern to a new view that was to them more satisfactory. It was the lesser minds, or those reacting against a dominating religious authority, who shied away from religion when the new science showed that old interpretations were inadequate. Copernicus, Galileo, Newton, Kant, and even Laplace himself counted themselves firmly as theists. In placing Laplace in this group I am using as authority our late associate George Sarton, who called my attention to letters that Laplace wrote to his son. In these letters Laplace urges the young man to maintain his belief in God, for this, he says, is the best basis for a stable attitude toward life.

So likewise the scientists of our century who have sought to orient our thinking regarding man's place in his world. Representative among these are such persons as the mathematician-philosopher, A. N. Whitehead; the astronomer, Arthur Eddington; the physicists, Albert Einstein, Erwin Schrödinger, and Werner Heisenberg; and the physiologist, Charles Sherrington. These men, like their eighteenth-century counterparts, have no need for introducing the concept of purpose, either human or divine, in accounting for the events that they see happening in the physical world. They are however unanimous in recognizing the importance of ideas, ideals, and purpose in understanding the meaning of what happens, and most of them introduce the concept of God into their world picture.

The idea of God used by today's men of science is far different from that of the followers of Aristotle, but in a form fitted to science as now known the idea remains alive among them. We of the present day are however far from agreement as to the form that the idea of God should take.

EIGHTEENTH-CENTURY PHILOSOPHY AND THE GROWTH OF SCIENCE

Time will not permit us to consider adequately certain other important aspects of eighteenth-century scientific thought. Published summaries of the science of this period are available which treat these matters in detail. A more adequate picture would include the marked development of precision instruments, especially for use in astronomy and navigation. It would show how advances were made in the measurement of temperature and of quantity of heat, and how meteorology became a subject of systematic study. The great advances of chemistry in which Lavoisier played so important a part, the extension of our geographic knowledge to include most of the earth, the classification of plants and animals by Linnaeus, Buffon, and many others, the introduction of inoculation against small pox, a concern with psychological phenomena, and a beginning of the science of demography, with Wallace and Malthus calling sharp attention to the threat of over-population—these would all be included.

But perhaps more significant, because it reflects the spirit of the age, was the strong development of philosophy. It is perhaps correct to describe the eighteenth century as the modern philosopher's great period. The basic reason for this growth was the impact of scientific thought on the traditional ideas that had been incorporated into religion. Thus philosophy affected people where they lived.

A mere mention of names will indicate what I mean. Thomas Aquinas, in the mid-thirteenth century, had performed the remarkable feat of synthesizing the heritage of classical scientific and philosophical thought with Christian doctrine. Descartes, Spinoza, von Leibnitz, and Locke, during the seventeenth century, had begun to wrestle with the new problems posed by science. Chief among these problems was that of understanding how we can know anything.

It was at this point that early in the eighteenth century Bishop George Berkeley entered, calling sharp attention to the fact that one can know only what is in his own mind, and that in one's ideas of which he is directly aware, is the essence of reality. David Hume was skeptical even of the reality of ideas. By general consent, however, was Immanuel Kant who, toward the close of the century, went as far as one could then go toward understanding how knowledge is reached.

Kant was himself a physicist of considerable stature. He anticipated Laplace in formulating nebular hypothesis. He was the first to point out that tidal friction must retard the earth's rotation. He explained the trade winds as an effect of the earth's rotation on the atmosphere. He was thoroughly conversant with Newtonian mechanics and accepted its complete validity. Coming before the establishment of the principle of biological evolution and the physicist's proof of the limitation of precision to our knowledge of physical events, his views of the place of ultimate ends in the interpretation of what happens in the world show remarkable balance.

In the view of J. B. S. Haldane and many others, as described by Dampier,[8] "of all the older philosophies, Kant's metaphysics best represents the position to which physical and biological science [now] point. . . . Relativity and the quantum theory; biophysics, biochemistry and the idea of purposeful adaptation; all these latest developments of science . . . have brought scientific philosophy back to Kant." Bertrand Russell and certain others have expressed themselves strongly in opposition to this high rating of Kant. But until the twentieth-century philosophers, notably A. N. Whitehead, brought into the total world picture such concepts as that of organism, it is perhaps fair to say that Kant supplied us with the most suitable systematic working philosophy that men of science had available. In fact, one may question whether without Kant's analysis of our concepts of space and time such a development as Einstein's theory of relatively could have occurred. It is equally doubtful whether our ideas of physical causality would have been revised without Kant's criticism of the basic meaning of this concept.

In sum, we see the eighteenth century as a period of completion of the classical analytical mechanics, establishing on a firm basis our understanding of the motion of matter under force. This achievement carried with it a revolutionary change in attitude toward the place of God and men in the world of nature, arriving at a position which has again been fundamentally altered during the twentieth century. During the eighteenth century nearly all of the main fields of science showed marked advance. Only biological science, following Harvey and waiting for Darwin, was relatively stagnant. The great new field of science that was opening was electricity, in whose investigation Benjamin Franklin and other amateurs took a leading part. It was these studies that led directly to modern electrical technology and the electron theory of matter, and played an important part in making possible our new view of the universe.

The dominant spirit of the eighteenth century was shown in the release of scientific thought from authoritarian theology and in the strong growth of a type of philosophic thought that was fitted to the new world of science. It was these developments that cleared the field and laid the foundation on which later generations of science, including our own, could build substantial structures.

---

[8] *Op. cit.,* 211.

The Scattering of X Rays as Particles*

A. H. COMPTON
Washington University, St. Louis, Missouri
(Received March 22, 1961)

> The experimental evidence and the theoretical considerations that led to the discovery and interpretation of the modification of the wavelength of x rays as a result of scattering by electrons are reviewed, as is the controversy between Duane and the author that took place in 1923–24. The confirmatory evidence obtained by Bothe, Geiger, Simon, and Compton is summarized.

I HAVE been asked to say something about how the study of the scattering of x rays has led to the concept of x rays acting as particles.

In the interest of conserving time I shall summarize the first part of the story by noting that, beginning in 1917, I spent five years in an unsuccessful attempt to reconcile certain experiments on the intensity and distribution of scattered x rays with the electron theory of the phenomenon that had been developed by Sir J. J. Thomson. Then a series of experiments that I performed at Washington University, beginning in 1922, confirmed an observation by J. A. Gray[1] of Queen's University of Kingston, Ontario, that the secondary rays produced when x rays pass through matter are in fact of the nature of scattered rays, showing the same polarization and approximately the intensity predicted by Thomson's electron theory and, further, that in the process of scattering, these rays are in some way altered to increase their absorbability. From my absorption measurements I was able to estimate that over a wide range of wavelengths of the primary rays the increase in the absorbability of scattered rays was what it should be if their wavelength was increased by about 0.03 A over the wavelength of the primary ray. This result I checked with an x-ray spectrometer, measuring an increase in the wavelength of approximately 0.02 A.

At this point I found myself engaged, as a member of a committee of which William Duane of Harvard was the chairman, in preparing a report for the National Research Council on secondary radiations produced by x rays. When it came to publication of the report, Duane objected to including my revolutionary conclusion that the wavelength of the rays was increased in the scattering process just described because he felt that the evidence was inconclusive. At the insistence of A. W. Hull, however, this portion of my report was included in the publication.[2]

At this point I paused in my experiments in order to concentrate on their theoretical interpertation. I found at once that the change of wavelength that I observed for scattering at 90° was what should be expected if the scattering electrons were moving in the direction of the primary beam at about half the speed of light which would mean that each electron had a momentum equal to that of a quantum of energy of the frequency of the primary x rays. It was obvious, however, that not all of the electrons in the scattering material, which was fixed in my apparatus, could be moving forward at such a velocity; yet according to the theory all of the electrons should participate in the scattering

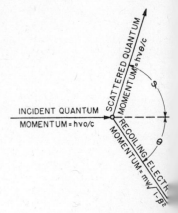

FIG. 1. Recoil of an electron upon scattering of an x-ray photon.

---

* Paper delivered as part of a program on "Topics in the history of modern physics" on February 3, 1961, at a joint session of the American Physical Society and the American Association of Physics Teachers during their annual meetings in New York City.
[1] J. A. Gray, J. Franklin Inst. 189, 643 (1920).

[2] A. H. Compton, Bull. Natl. Research Council, No. 19 (1922).

$$\delta\lambda = \lambda' - \lambda = (h/mc)(1-\cos\phi), \quad (1)$$

$$E_{kin} = 2\alpha \cos^2\theta/(1+\alpha)^2 = \alpha^2 \cos\theta, \quad (2)$$

$$\cot\theta = (1-\alpha)\tan(\phi/2), \quad (3)$$

where

$$\alpha = h\nu/mc^2.$$

The change in wavelength I measured repeatedly at Washington University. Figure 2 shows the results of one series of these experiments.

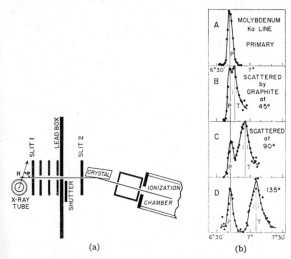

FIG. 2. (a) Schematic arrangement of the apparatus for determination of the spectrum of the scattered x rays. (b) Experimental results.

The results, confirming accurately the theoretical predictions, immediately became a subject of the most lively scientific controversy that I have ever known. I reported the results shown in Fig. 2 before the American Physical Society in April, 1923. At the meeting of the American Physical Society during the Christmas holidays of that year there was arranged a rather formal debate between Duane and myself on the validity of the results. Having frequently repeated the experiments I entered the debate with confidence, but was nevertheless pleased to find that I had support from P. A. Ross of Stanford and M. de Broglie of Paris, who had obtained photographic spectra showing results similar to my own. Duane at Harvard with his graduate students had been able to find not the same spectrum of the scattered rays, but one which they attributed to tertiary x rays excited by photoelectrons in the scattering material. I might have criticized his interpretation of his results on rather obvious grounds, but thought it would be wiser to let Duane himself find the answer. Duane followed up this debate by visiting my laboratory (at that time in Chicago) and invited me to his laboratory at Harvard, a courtesy that I should like to think is characteristic of the true spirit of science. The result was that neither of us could find the reason for the difference in the results at the two laboratories, but it turned out that the equipment that I was using was more sensitive and better adapted than was Duane's to a study of the phenomenon in question.

During the following summer at Toronto there occurred a meeting of the British Association for the Advancement of Science, with Sir William Bragg presiding over the physics section. In the previous decade Sir William, as also Ernest Rutherford, had been greatly impressed by the

process. This led me to examine what would happen if each quantum of x-ray energy were concentrated in a single particle and would act as a unit on a single electron. Thus I was led to the now familiar hypothesis, illustrated in Fig. 1, of an x-ray particle colliding with an electron and bounding elastically from it with reduced energy, the lost energy appearing as the recoil energy of the electron. This idea, of an x-ray quantum losing energy by collision with an electron, must have been already in the mind of Peter Debye, then working at Zürich, for immediately upon the appearance of my report in the Bulletin of the National Research Council, he published a paper[3] in the Physikalische Zeitschrift in which he presented an explanation of the change in wavelength of the scattered rays identical in principle with my own hypothesis, and appearing in print only a few days after my first full publication.[4]

Assuming that the energy and the momentum of the incident quantum and of the electron are conserved in this collision process, one is led to a group of three expressions representing a change of wavelength, the energy of the recoil electron, and the relation between the angle of recoil and the angle of scattering of the photon. Each of these formulas, expressed in Eqs. (1)–(3), is subject to precise experimental test.

P. Debye, Physik. Z. **24**, 161 (1923).
A. H. Compton, Phys. Rev. **21**, 484 (1923).

forward momentum of the secondary electrons ejected from matter by both x rays and gamma rays and had been led thereby to defend a corpuscular theory of the scattering of the rays. This interpretation, however, he had abandoned following the experiments by von Laue and by himself on the reflection of x rays by crystals, which had given him confidence in the wave interpretation of x rays. At this Toronto meeting a full afternoon was set aside for a continuation of the debate. The result was inconclusive. It was summarized by Sir C. V. Raman by this statement to me privately after the meeting. "Compton," he said, "you are a good debater, but the truth is not in you." Nevertheless, it seems to have been this discussion that stimulated Raman to the discovery of the effect which now bears his name. Duane followed up this meeting by a new interpretation of the change in wavelength which he attributed to what he called a "box" effect, explaining that surrounding the scattering apparatus with a lead box had in some way altered the character of the radiation. This interpretation I answered by repeating the experiment out of doors with essentially the same results, and at the same time Duane and his collaborators in a repetition of their own experiments began to find the spectrum line of the changed wavelength in accord with my collision theory. At the next meeting of the American Physical Society they reported a very good measurement of this change in wavelength.

In the meantime other experimenters had not

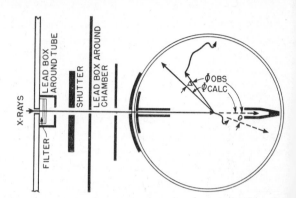

FIG. 4. Apparatus for measurement of scattering angle.

been idle. Within a few months after my first paper C. T. R. Wilson at Cambridge and W. Bothe in Germany had found the recoil electrons predicted by the corpuscular theory. Figure 3 shows one of C. T. R. Wilson's photographs of the cloud tracks left by these electrons in air traversed by x rays. The appearance of the trails led Wilson to call them "fish" tracks, with their tail toward the x-ray tube and their head pointed in the direction of the beam. A. W. Simon and I repeating these experiments, showed that the number of the tracks and their ranges were just what should be expected according to the theory that each recoiling electron was the result of the impact of one photon of x rays that it scattered. I had the opportunity to show some of these fish tracks in a cloud chamber to S. K. Allison, who was at that time working in Duane's laboratory. It is possible that it was these tracks, rather than the evidence of the x-ray spectra, that convinced Duane of the validity of the corpuscular theory. In any case, since that time no one seems to have questioned the correctness of our experimental results.

Immediately following their observation of the recoil electrons, Bothe and Geiger reported an observation of coincidences of recoil electrons and associated scattered photons as observed by a pair of counters. Simon and I were engaged in checking the angles at which the recoil electron and the associated scattered photon would occur. The apparatus that we used is shown diagrammatically in Fig. 4. According to the theory, associated with an electron recoiling at an angle any effect of the associated scattered photon

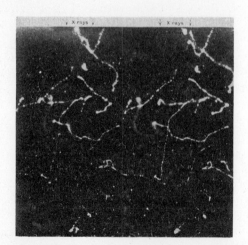

FIG. 3. Cloud chamber tracks produced by recoil electrons (after C. T. R. Wilson).

[5] A. H. Compton and A. W. Simon, Phys. Rev. 25, (1925).

should occur in the direction of ϕ, as given by Eq. (3). With a specially designed cloud chamber, out of 850 photographs 38 showed a β particle resulting from the photon associated with the recoil electron.[6] Figure 5 shows a typical photograph.

This result is of especial interest because it shows that it is possible to follow the path of an x-ray particle or photon by examining the secondary electrons that it ejects along its way. It is clear that the x rays thus scattered proceed in direct quanta of radiant energy; in other words, that they act as photon particles. This test of the relation between the angles θ and ϕ

FIG. 5. Typical cloud chamber photograph of recoiling electron track.

[6] A. H. Compton and A. W. Simon, Phys. Rev. **26**, 289 (1925).

is a crucial test of the conservation of energy and momentum as related to the process of the scattering of photons by electrons. The results of Simon and myself have accordingly been reexamined and refined by a number of experimenters, as summarized recently by Robert Shankland in his *Atomic and Nuclear Physics*.[7] The net result is a full confirmation of the angular relation given by Eq. (3).

Time does not permit me to review the evidence that was accumulating in the meantime that gave full support likewise to the electromagnetic wave character of the x rays: complete polarization of x rays scattered at 90°, the diffraction of x rays from ruled gratings, as well as from crystals, interference phenomena, and refraction phenomena, precisely analogous to results obtained with light. It became evident that though x rays moved and did things as particles, they nevertheless have also the characteristic optical qualities that identify them as waves. Thus we were introduced to the concept of light as having the nature of waves and particles as having a kind of reality, a difficult concept to which L. de Broglie was, however, at the same time giving a theoretical meaning.

It may be fair to say that these experiments were first to give, at least to physicists in the United States, a conviction of the fundamental validity of the quantum theory.

[7] Robert Shankland, *Atomic and Nuclear Physics* (The Macmillan Company, New York, 1960), 2nd ed., p. 204.

Appendix 1

A. H. Compton and O. W. Richardson

In September 1913 Arthur Compton entered the graduate school at Princeton University to begin his advanced work in physics. He was extremely fortunate to have as his principal advisor and research director during his first term Professor O. W. Richardson. At that time Richardson was beginning an experimental program in X-rays, and he had assembled a modern set of equipment for his own experiments. Compton started to work immediately on X-ray problems with Richardson; and he found that the recent experience with his brother Karl, at Wooster, was a great help. Compton also took Richardson's course in electromagnetic theory, which immediately became and continued to be one of the major factors in Compton's development as a research physicist. It is abundantly clear that Richardson was a great inspiration in starting Compton off properly on his research in graduate school. The cordial relations between Richardson and Compton were life-long, as the Richardson papers from the University of Texas library reveal.

Since Richardson returned to England at the end of December 1913 his direct association with Compton was for only one term. However, this was an exceedingly vital period in Compton's education. His Ph.D. thesis, which he completed in 1916, owed much to the initial introduction and inspiration that Richardson had given him. Richardson was so impressed with the abilities of young Compton that he turned over all of his X-ray equipment to him when he left. Richardson recalled this incident, and the nature of their early association, when he introduced Compton as Guthrie Lecturer of the Physical Society of London in 1935.

> In my own case it is a particular personal pleasure to welcome Arthur Compton on account of our long acquaintance [which] extends over 20 years. He was in fact one of my old research students at Princeton. In that capacity he holds 3 records. (1) The shortest time. He came to the Palmer Laboratory in September 1913 and I left for London at the end of December that year. However, it happened that for about a year before that time I had been collecting and assembling a lot of apparatus with a view to an extensive series of researches on X-rays. I could not bring it away with me (it was the property of the laboratory) so when I left I handed it over to C[ompton] with my blessing. I do not think that I told him much about what to do with it. But I got him interested in X-rays and I may have impressed him with the importance of the subject and that it was a field of research likely to be productive of valuable results. Well I would like to regard that achievement as one of my greatest contributions to the progress of physics. You can sometimes do better by handing your apparatus to another than by fumbling about with it yourself.

I said that as a research student he held 3 records. The second is that he is the most famous of them and the third that he is the only one who has been awarded a Nobel Prize. I think there can be no doubt that the discovery of the law of the interaction between radiation and free electrons resulting in the Compton Effect for which he received the N[obel] Prize is one of the greatest discoveries in an age of great discoveries.[1]

During the year 1919–20, which Compton spent in the Cavendish Laboratory at Cambridge, he and Richardson were again in close touch. Richardson was then professor of physics at King's College in London. Toward the end of his year at Cambridge, Compton made arrangements with Richardson to continue work during the summer of 1920 in Richardson's laboratory in London, where he had X-ray equipment. The two problems that Compton requested permission from Richardson to study in his laboratory are set forth in a letter dated March 24, 1920:

> There are two experimental problems which I should like to complete before I return to the United States, for which there is not adequate equipment in the Cavendish laboratory, but which could be carried out satisfactorily in your laboratory. Both of these refer to special tests on the absorption of X-rays under certain special conditions.
>
> The first of these problems is to determine the relative absorption of primary X-rays and of X-rays reflected from a crystal, both of the same wave-length, in material of low atomic weight (carbon or paraffin). A theory of absorption which I have worked out indicates that if X-rays are produced by an electron which is suddenly set into vibration and is damped by its own radiation, there should occur a type of absorption about half as great as the scattering, in addition to the familiar absorption which is proportional to the cube of the wavelength. . . .[2]
>
> The second problem is in connection with the absorption of X-rays in iron when magnetized and when unmagnetized.[3]

It is especially interesting to note that Compton's Minnesota work, directed toward locating the cause of ferromagnetism, was still very much in the forefront of his research programs.

The Richardson papers again show the close relationship between Compton and Richardson in late 1921. In a paper dated December 3, 1921, there are calcula-

1. Quoted by permission of Lady O. W. Richardson, Professor H. O. W. Richardson, and the Humanities Research Center, the University of Texas at Austin.
2. In a letter to the editor dated September 28, 1970, Dr. P. P. Ewald pointed out that, had Compton followed up this experiment in the light of Ewald's dynamical theory of X-ray diffraction, he probably would have discovered what is now known as the Borrmann effect, which, as it turned out, was not discovered until 1948.
3. Letter of A. H. Compton to O. W. Richardson dated March 24, 1920. Quoted by permission of Mrs. Arthur H. Compton and the Humanities Research Center, the University of Texas at Austin.

tions by electromagnetic theory in which Richardson attempts to explain results that Compton had published in the *Washington University Studies*.[4] These studies were part of a sustained effort that Compton made to explain his X-ray and gamma-ray scattering results by classical electrodynamics. In general, the method was to assume an electron of finite size, so that interference effects would cause a falling off of intensity with increased scattering angle. Both Richardson and Compton labored on this problem, but of course neither was successful in explaining the Compton effect by purely classical means.

There are also notes in the Richardson papers discussing Compton's report on his scattering experiments at the 1927 Solvay Conference. An excerpt from this discussion follows.

> Whether the laws of conservation of energy and momentum do or do not apply to a single elementary act [X-rays scattered by electrons] it is quite certain that they hold for the average result of a large number of such acts. Accordingly the process under discussion must be governed by the equations of momentum (a) and energy (b). If for simplicity we neglect the relativity refinements these are
>
> $$\frac{h\nu}{c} = \overline{mv} + \overline{MV} \qquad \text{(a)}$$
>
> and
>
> $$h\nu = \frac{1}{2}\overline{mv^2} + \frac{1}{2}\overline{MV^2} \qquad \text{(b)}$$
>
> where m, M are the masses, v and V the velocities of the electron and the positive residue respectively and the bars denote averages.[5]

Here we note with interest that both the energy $h\nu$ and momentum $h\nu/c$ of the photon are used. However, equations (a) and (b) show that the use of special relativity in these problems was by no means yet standard. It is interesting that this discussion shows less than acceptable agreement between theory and experiment, and it may very well be that the neglect of the theory of relativity was responsible.

A most interesting item in the Richardson papers regarding Compton is his recommendation of Compton for the Hughes Medal of the Royal Society of London. This recommendation was made sometime after Compton had received the Nobel Prize, when the full significance of the Compton effect for quantum mechanics was generally recognized. It is of special interest as it again emphasizes the fact that Compton and Richardson had been led to the momentum of the

---

4. Classical electrodynamics, p. 105 (see Bibliography entry 36). Reproduced on p. 224 of this book.
5. Calculation in O. W. Richardson's notebook made at the 1927 Solvay Conference in Brussels. Quoted by permission of Lady O. W. Richardson, Professor H. O. W. Richardson, and the Humanities Research Center, the University of Texas at Austin.

photon by the route of electromagnetic theory rather than from the 1917 photon hypothesis of Einstein. Even long after the Compton effect was completely accepted on the quantum basis, Richardson clung to his electromagnetic theory. This gives additional strong support to the view that Compton also came to his quantum theory of scattering by analogy with classical electromagnetic theory. The pertinent remarks in Richardson's introduction are these:

> Arthur Holly Compton Professor of Physics in the University of Chicago is an outstanding all round physicist with the emphasis on the experimental side but carrying also a very effective mathematical equipment. From 1913 to about 1932 most of his work was concerned with X-rays. I had the privilege of introducing him to this field. During most of 1913 I had spent a good deal of time and energy in obtaining and assembling a lot of up to date X-ray apparatus. He was a research student of mine at the time, and when I returned to England at the beginning of 1914 I could not take it with me so I handed it over to him with my blessing and some advice. This turned out a very good thing to have done as it is very unlikely that I could have done so much good with it as he did.
>
> His most important work with X-rays was on the Compton effect, which he discovered, and which is now universally called after him. When X-rays fall on matter it emits electrons and secondary X-rays. The bulk of these secondary rays are of two kinds, Barkla's characteristic rays, caused by the reorganization of the electrons in the atoms after one of them has been ejected by the primary X-rays and the unmodified rays. The last most likely originate in various ways but the bulk of them are probably the analogue of ordinary light scattering. An important advance in this subject was made in 1913 by J. A. Gray who showed that in the case of elements of low atomic weights when the characteristic rays were excluded, the secondary rays were measurably less penetrating than the primary; this corresponds to a diminution of the frequency of the waves.
>
> It is clear from his writings that somewhere about 1920 Compton's mind had become much occupied with the conflict, as old as Newton, whether light was made up of waves or particles or both. Of course the experiments of Young and Fresnel early in the 19th century proved that light certainly had undulatory properties. But by 1920 stubborn facts were emerging, notably in connection with photoelectric action, which were impossible to reconcile with the assumption that light was nothing but electromagnetic waves of the classical type. These difficulties disappeared if light of frequency ν were assumed to be dynamically equivalent to a collection of particles of energy $h\nu$ (h = Planck's constant).
>
> It occurred to Compton that from this standpoint the interaction between radiation and a free electron is very simple and in fact is the simplest interaction that radiation can undergo. Associated with the energy $h\nu$, *according to the*

electromagnetic theory,[6] there is momentum $h\nu/c$ (c = velocity of light). The interaction is thus reduced to a very ancient problem, that of the encounter of two infinitesimal billiard balls with known energies and momenta. As the radiation moves with the velocity of light, in most cases the electron can be treated as if it were at rest. It is then obvious that in the collision the electron will acquire energy from the radiation and the conservation of momentum requires that if the electron moves off in a certain direction the radiation will travel in a certain other direction. But reduction of energy of a quantum of radiation means increase in wave length, and this increase will be a predetermined function of the direction of the "scattered" radiation and of the direction of motion of the "recoil" electron.

He published these conclusions in 1922. In 1923 he established the change in wave length, first qualitatively by Barkla's absorption coefficient methods and then quantitatively with the X-ray spectrometer. In the succeeding years he investigated the energies of the recoil electrons as a function of their direction of motion and showed that the correlation, predicted by the theory, between the direction and energy of the recoil electrons on the one hand and the direction and change of wave length of the radiation on the other did in fact occur. This correlation is of fundamental importance in the general theory of the interaction of radiation with matter.

Compton's results met with a surprising amount of opposition when first published, especially when one recalls that they furnished the only satisfactory explanation of the already known anomalous scattering of Gray which had ever been put forward.[7]

R. S. S.

---

6. The italics have been added by the editor to emphasize that Richardson and Compton formed their concepts even for quantum problems largely by analogy with the classical theories they knew so well.
7. Unpublished papers of O. W. Richardson. Quoted by permission of Lady O. W. Richardson, Professor H. O. W. Richardson, and the Humanities Research Center, the University of Texas at Austin.

Appendix 2

An Exchange of Letters between A. H. Compton and Gordon Ferrie Hull

In 1953, thirty years after the discovery of the Compton effect, Professor Gordon F. Hull wrote Compton asking what had led him to the idea of the momentum of the X-ray quantum. Excerpts from Hull's letter and Compton's reply are quoted below. It is interesting to note that Compton recalled that Einstein's needle rays were "in the air," but that he had only seen an abstract of Einstein's 1917 paper. He clearly states that the route that led him to $h\nu/c$ for the momentum was that of classical electromagnetism, as I have also concluded after studying the notebooks, papers, and correspondence of Professor Compton written at the time of his discovery. The following quotations from the two letters give the essential items: [1]

1. Hull to Compton, April 23, 1953:

Recently I have been trying to find the logic which led you to the value of the momentum of a photon as equal to E/c. In your long and excellent article in Bull. Nat. Res. Council of 1922 you obviously were feeling your way for you have put forth numerous proposals regarding the nature of a quantum. You are, for example, particularly concerned as to what happens at the separating surface of an interferometer. Do some quanta go through unchanged while others are reflected? (Without knowing about your discussion I take up the same question in my book in Elementary Modern Physics). But nowhere in that great article do you equate the momentum of a photon to E/c. Then in the convincing article in Phys. Rev. 21, 1923 you suddenly state that relation. There is no intermediate article that I know of.

Did you use the relation $E = mc^2$ and take mc as the momentum? You make no reference to $E = mc^2$. Of course you could have asserted that your experimental results prove the relation $E = mc^2$. Indeed you could have derived that relation.

2. Compton to Hull, May 17, 1953:

I am pleased to hear from you, and to have my attention brought back to questions of physics. During the past two weeks I have re-read my own early papers and have looked up one or two of Einstein's. It becomes evident to me, as you have indicated, that the general significance of the mc^2 relation has been a gradual growth. I believe it is fair to say that its validity has been verified experimentally with regard to radiation, kinetic energy and nuclear

---

1. The quotation from the letter of Gordon Ferrie Hull is presented here with the permission of Gordon Ferrie Hull, Jr.; that from Compton, with the permission of Mrs. Arthur Holly Compton.

potential energy. I know of no test with regard to gravitational energy. If there is any generally valid theoretical demonstration it has not come to my attention. I have not, however, had an opportunity to look into this matter sufficiently thoroughly to satisfy myself as to its present status.

It is evident that the fact that radiant energy "acts as if it" has mass, E/c^2, was a familiar concept before Einstein's studies. This dates from Maxwell's electromagnetic theory of radiation pressure, and the implied radiation mass, as discussed for example by Poincaré (as you note) and by Hasenöhrl (1904). Your own convincing experiments of 1901 and those of Lebedev placed this theory on a reliable foundation. Einstein (*Phys. Zeits., 18*, 122, 1917) in his discussion of the impulse imparted to a molecule when it absorbs or emits energy, also makes use of the effect of radiation pressure. In fact, on p. 127 of his paper, Einstein uses the expression $h\nu/c$ for the impulse, noting that it applies to the recoil from emitted radiation only in case this radiation is directed, and presents a powerful argument that such *directed* radiation in quanta must occur.

It is noteworthy that in this article (which is by the way a masterpiece) Einstein does not refer to the radiation itself as possessing momentum or mass. He is concerned rather with the impulses to which the molecules are subject as they emit and absorb radiation, though he calls special attention to the intimate relation between energy and momentum where radiation is being studied.

To the best of my recollection, when I wrote my papers on the scattering of x-rays to which you refer, the idea of "needle-rays," having energy $h\nu$ and momentum $h\nu/c$, was one which at the time was widespread among physicists and was usually associated with Einstein's name. Probably I had heard of it first from Richardson, with whom I studied in 1913 and had also discussed these problems in 1920. To him the concept was of special importance in connection with the photo-electric effect. In 1922 I was acquainted with only an abstract of Einstein's 1917 paper, and arrived independently at the value of the impulse imparted by a quantum of radiation as $h\nu/c$. My calculation also was based on Maxwell's theory.

In the six months between my writing the two papers of 1922 and 1923, to which you refer, I became impressed by the usefulness of the concept of a particle of radiation that possessed momentum as well as energy. It was in terms of such particles accordingly that I wrote the 1923 paper. I recall having the impression that there was nothing new in this concept, but I was unable to find any specific statement or use of it in the literature. This accounts for my unwillingness either to indicate that the hypothesis was original with me or to make any reference to its origin. My impression now is that the idea was one which was "in the air," but which up to that time had not appeared sufficiently useful for anyone to publish. This impression is supported by the fact that P. Debye published his theory of my effect at practically the same time using the same concept. The idea had apparently been in his mind for more than a year.

At this time (1922) the idea of the photon as a particle of mass $h\nu/c^2$ was a familiar one to me, as derived from its momentum as $h\nu/c$. I do not find any publication of this value for the mass, however, until several years later. At no time in my thinking did I derive this mass from the general relation $m = E/c^2$. Rather, I thought of the demonstration that the momentum is $h\nu/c$ as an independent confirmation of the relation $m = E/c^2$ for the particular case of radiation.

It should be noted that Professor Compton's acquaintance with Einstein's 1917 paper was based on his reading it in 1953 to answer Hull's letter, and that he had not read it earlier. The reader must form his own judgment as to the influence of Einstein's photon hypothesis on Compton in the years 1917–23, but it is my opinion that classical electrodynamics rather than the photon hypothesis was chiefly responsible for providing the concepts that led Compton to the formulation of his theory of X-ray scattering.

R. S. S.

Appendix 3

THE COMPTON EXPERIMENT by Albert Einstein

I want to report in the following discussion on an important experiment with regard to light, or electromagnetic radiation, an experiment carried out about a year ago by the American physicist [Arthur] Compton. In order to recognize fully the significance of this experiment, we must realize the highly remarkable situation of the theory of radiation at the moment.

In the first half of the nineteenth century, work in optics was still primarily concerned with reflection and refraction of light (concave mirrors, lens systems). Up to that time the Newtonian corpuscular or emission theory of light was generally held. According to this theory light consists of corpuscles that move in a straight line and without variation in form so long as they remain in a homogeneous medium, but that in general they undergo a sudden change of direction at surface boundaries. On this assumption a rather complete theory of all phenomena observed up to that time was developed, especially those phenomena connected with the telescope and microscope.

However, approximately a hundred years ago, as the phenomena of interference and diffraction (as well as of the polarization of light) became more exactly known, it was necessary to replace the Newtonian principle of the nature of light by a theory completely different from it, the undulatory theory that had been proposed by Huygens about a century and a half earlier. According to this theory light consists of elastic waves that are propagated in space (or ether) in a like manner in all directions, just as in two dimensions the surface waves in water proceed from one point, the point at which the surface is set into oscillation. Only this theory offered an explanation for the fact that a light ray propagates in all directions after passing through a very narrow opening. Only this theory is able to give an explanation for the fact that in a light-illuminated space dark places are present in the case of interference and diffraction phenomena, or for the fact that several bundles of light rays cancel each other out locally. This undulatory theory was able to present the most complicated phenomena of diffraction and interference with an astronomical precision so great that the belief in its correctness very quickly became unshakeable.

The investigations of Faraday and Maxwell produced a modification of the undulatory theory and at the same time a still firmer foundation for it. Through these investigations the undulatory field of light was divested of its mechanical nature. Maxwell's theory of electricity and magnetism also encompasses the undulatory theory of light without changing its formal structure. His theory estab-

From a paper by Albert Einstein originally published as "Komptonsche Experiment" in *Berliner Tageblatt,* April 20, 1924. The translation is by Lucille B. Pinto and is included here with the permission of the estate of Albert Einstein.

lishes quantitative relationships between the optical and electrical behavior of empty space as well as of measurable bodies and reduces the number of mutually independent hypotheses on which undulatory optics is based. At the turn of the century, physics seemed to have gained by means of this theory a firm base on which one hoped to be able to found all its branches, including mechanics.

But it turned out otherwise. As a result of Planck's work on the law of radiation emitted from hot bodies, it was shown that Maxwell's theory was not able to explain this discovery. It was also impossible to explain the general observation that the effects of radiation do not depend on the intensity, but only on the frequency. This is extremely paradoxical and does not seem possible of reconciliation with the basic concepts of the undulatory theory.

Imagine that somewhere on the open sea gigantic waves are produced that spread out on all sides from a center of agitation. Naturally the height of waves so produced will be lesser the further they have traveled from the point of origin. Imagine now ships of the same size spaced out over this region of the ocean before the postulated waves are produced. What will happen when the waves start coming? The ships near the point of origin will capsize or be destroyed, but those far enough away from this point will suffer no damage; they will only be brought to a harmless pitching and tossing. Consider now that molecules hit by radiation will behave analogously to the ships hit by the waves of the sea. Whether the molecules are chemically changed or not should not depend on the wave length of the effective radiation alone but also on its intensity; it is exactly this that is not confirmed by experience.

The solution to this failure of the general theory was the hypothesis of light quanta. In spite of general respect for the undulatory theory, this working hypothesis gained ground because radiation in an energetic relationship behaves as if it consists of energy projectiles whose quantity of energy depends only on the frequency (color) of the radiation and is proportional to it. Newton's corpuscular theory of light comes to life again although it failed completely in the area of the basically undulatory properties of light.

There are therefore now two theories of light, both indispensable, and—as one must admit today in spite of twenty years of tremendous effort on the part of theoretical physicists—without any logical connections. Quantum theory has made Bohr's theory of the atom possible and has explained so many facts that it must contain a large measure of truth. In view of these facts it becomes of utmost importance to consider how far one ought to go in ascribing the properties of projectiles to the light corpuscles or quanta.

A projectile transmits not only energy to the hit body but also an impulse in the direction of its motion. Is this also true in the case of light quanta? On the basis of theoretical considerations this question was answered affirmatively some time ago, and Compton's experiment has proved the correctness of this assumption. In order to understand the experiment, one must consider more exactly the mechanism of the process known as scattering, which, for example, produces the blue color of the sky.

If an electromagnetic wave strikes a free elementary particle (electron) or one bound to an atom, the electron is set into an oscillating motion by the alternating electric fields of the wave. It then radiates in turn (like an antenna of wireless telegraphy), in all directions, waves of the same frequency whose energy is derived from the original wave. This causes the light to be scattered in all directions by the illuminated medium containing such particles, and indeed the scattering is stronger the shorter the wave length of the primary light. Thus is scattering interpreted by the undulatory theory.

The process is explained differently by the quantum theory. According to this theory a light quantum strikes an electron and, because of this, changes its direction and at the same time imparts velocity to the electron. The kinetic energy that is imparted to the electron by this collision must therefore be taken from the striking quantum so that the scattered quantum has less energy (in undulatory theory a lower frequency) than does the impinging radiation. More exact consideration shows that the frequency loss of the scattered radiation can be calculated exactly. The percentage of the change in frequency for visible light is very slight; but for harder X radiation, which is indeed nothing else than very shortwave light, it is appreciable.

Compton found that X radiation scattered by means of suitable media actually shows the frequency change predicted by quantum theory (not however by the undulatory theory). This can be explained in the following way: according to the Rutherford-Bohr theory every atom possesses a number of electrons that are so loosely bound to the atom that in the case of the quantum collision of an X ray they behave as if they were freely movable. For light that is scattered by such electrons, the above consideration is applicable. The positive result of the Compton experiment proves that radiation behaves as if it consisted of discrete energy projectiles—and not only with regard to the effects of impact.

Bibliography of Compton's Scientific Works

PAPERS

1909

1. A criticism of Mr. C. W. Williams's article, "Concerning aeroplanes." *Fly* 1 (February): 13.
2. Comparison of Wright and Voisin aeroplanes. *Sci. Am.* 100 (February 13): 135 (L).
3. Striving for perfect machine. *Aeronautics* 5 (August): 58ff.

1911

4. Aeroplane stability. *Sci. Am. S.* 72 (August 12): 100–102.

1913

5. A laboratory method of demonstrating the earth's rotation. *Science* 37 (May 23): 803–6.

1914

6. New light on the structure of matter. *Sci. Am. S.* 78 (July 4): 4–6.

1915

7. A determination of latitude, azimuth, and the length of the day independent of astronomical observations. *Phys. Rev.* 5 (February): 109–17; also in *Pop. Astron.* 23 (April): 199–207.
8. Watching the earth revolve. *Sci. Am. S.* 79 (March 27): 196–97.
9. An agglomeration theory of the variation of the specific heat of solids with temperature. *Phys. Rev.* 5 (April): 338–39 (A).
10. What is matter made of? *Sci. Am.* 112 (May 15): 451–52.
11. The distribution of the electrons in atoms. *Nature* 95 (May 27): 343–44 (L).
12. The variation of the specific heat of solids with temperature. *Phys. Rev.* 6 (November): 377–89.

1916

13. A physical study of the thermal conductivity of solids. *Phys. Rev.* 7 (March): 341–48.
14. On the location of the thermal energy of solids. *Phys. Rev.* 7 (March): 349–54.
15. The X-ray spectrum of tungsten. *Phys. Rev.* 7 (April): 498–99 (A).

16. A recording X-ray spectrometer, and the high frequency spectrum of tungsten. *Phys. Rev.* 7 (June): 646–59.

1917

17. The intensity of X-ray reflection, and the distribution of electrons in atoms. *Phys. Rev.* 9 (January): 29–57. (Ph. D. diss., Princeton University.)
18. The reflection coefficient of monochromatic X-rays from rock salt and calcite. *Phys. Rev.* 10 (July): 95–96 (A).
19. The nature of the ultimate magnetic particle (with Oswald Rognley). *Science* 46 (October 26): 415–16.

1918

20. The size and shape of the electron. *J. Wash. Acad. Sci.* 8 (January 4): 1–11.
21. The nature of the ultimate magnetic particle (with Oswald Rognley). *Phys. Rev.* 11 (February): 132–34 (A).
22. The size and shape of the electron. *Phys. Rev.* 11 (April): 330 (A).
23. The non-molecular structure of solids. *J. Franklin Inst.* 185 (June): 745–74.
24. Note on the grating space of calcite and the X-ray spectrum of gallium. *Phys. Rev.* 11 (June): 430–32.

1919

25. An addition to the theory of the quadrant electrometer (with K. T. Compton). *Phys. Rev.* 13 (April): 288 (A).
26. The law of absorption of high frequency radiation. *Phys. Rev.* 13 (April): 296 (A).
27. The size and shape of the electron: I. The scattering of high frequency radiation. *Phys. Rev.* 14 (July): 20–43.
28. A sensitive modification of the quadrant electrometer: Its theory and use (with K. T. Compton). *Phys. Rev.* 14 (August): 85–98.
29. The size and shape of the electron: II. The absorption of high frequency radiation. *Phys. Rev.* 14 (September): 247–59.
30. Radio-activity and gravitation (E. Rutherford with A. H. Compton). *Nature* 104 (December 25): 412 (L).

1920

31. A photoelectric photometer. *Trans. Illum. Eng. Soc.* 15 (February 10): 28–33.
32. Cathode fall in neon (with C. C. Van Voorhis). *Phys. Rev.* 15 (June): 492–97.
33. Radioactivity and the gravitational field. *Phil. Mag.* 39 (June): 659–62.

34. Is the atom the ultimate magnetic particle? (with Oswald Rognley). *Phys. Rev.* 16 (November): 464–76.

1921

35. The absorption of gamma rays by magnetized iron. *Phys. Rev.* 17 (January): 38–41.
36. Classical electrodynamics and the dissipation of X-ray energy. *Wash. Univ. Studies* 8 (January): 93–129.
37. Possible magnetic polarity of free electrons. *Phil. Mag.* 41 (February): 279–81.
38. The elementary particle of positive electricity. *Nature* 106 (February 24): 828 (L).
39. The degradation of gamma-ray energy. *Phil. Mag.* 41 (May): 749–69.
40. The wave-length of hard gamma rays. *Phil. Mag.* 41 (May): 770–77.
41. The magnetic electron. *J. Franklin Inst.* 192 (August): 145–55.
42. Secondary high frequency radiation. *Phys. Rev.* 18 (August): 96–97 (A).
43. The polarization of secondary X-rays (with C. F. Hagenow). *Phys. Rev.* 18 (August): 97–98 (A).
44. The width of X-ray spectrum lines. *Phys. Rev.* 18 (October): 322 (A).
45. A possible origin of the defect of the combination principle in X-rays. *Phys. Rev.* 18 (October): 336–38 (A).
46. The softening of secondary X-rays. *Nature* 108 (November 17): 366–67 (L).

1922

47. The width of X-ray spectrum lines. *Phys. Rev.* 19 (January): 68–72.
48. The spectrum of secondary X-rays. *Phys. Rev.* 19 (March): 267–68 (A).
49. The intensity of X-ray reflection from powdered crystals (with Newell L. Freeman). *Nature* 110 (July 8): 38 (L).
50. Total reflection of X-rays from glass and silver. *Phys. Rev.* 20 (July): 84 (A).
51. Secondary radiations produced by X rays. *Bull. Nat. Res. Council* 4, pt. 2 (October): 56 pp.
52. Radiation a form of matter. *Science* 56 (December 22): 716–17 (L).

1923

53. A quantum theory of the scattering of X-rays by light elements. *Phys. Rev.* 21 (February): 207 (A).
54. The luminous efficiency of gases excited by electric discharge (with C. C. Van Voorhis). *Phys. Rev.* 21 (February): 210 (A).
55. A quantum theory of the scattering of X-rays by light elements. *Phys. Rev.* 21 (May): 483–502.

56. Wave-length measurements of scattered X rays. *Phys. Rev.* 21 (June): 715 (A).
57. The total reflexion of X-rays. *Phil. Mag.* 45 (June): 1121–31.
58. Recoil of electrons from scattered X-rays. *Nature* 112 (September 22): 435 (L).
59. Absorption measurements of the change of wave-length accompanying the scattering of X-rays. *Phil. Mag.* 46 (November): 897–911.
60. The spectrum of scattered X-rays. *Phys. Rev.* 22 (November): 409–13.
61. The quantum integral and diffraction by a crystal. *Proc. Nat. Acad. Sci.* 9 (November 15): 359–62.

1924

62. A quantum theory of uniform rectilinear motion. *Phys. Rev.* 23 (January): 118 (A).
63. Scattering of X-ray quanta and the J phenomena. *Nature* 113 (February 2): 160–61 (L).
64. A measurement of the polarization of secondary X-rays (with C. F. Hagenow). *J. Opt. Soc. Am. and Rev. Sci. Instr.* 8 (April): 487–91.
65. The recoil of electrons from scattered X-rays (with J. C. Hubbard). *Phys. Rev.* 23 (April): 439–49.
66. The wave-length of molybdenum $K\alpha$ rays scattered by light elements (with Y. H. Woo). *Phys. Rev.* 23 (June): 763 (A).
67. A general quantum theory of the wave-length of scattered X-rays. *Phys. Rev.* 23 (June): 763 (A).
68. The wave-length of molybdenum $K\alpha$ rays when scattered by light elements (with Y. H. Woo). *Proc. Nat. Acad. Sci.* 10 (June): 271–73.
69. The scattering of X-rays. *J. Franklin Inst.* 198 (July): 57–72.
70. A general quantum theory of the wave-length of scattered X-rays. *Phys. Rev.* 24 (August): 168–76.
71. The scattering of X-rays. *Radiology* 3 (December): 479–85.

1925

72. Measurements of the beta-rays excited by hard X-rays (with Alfred W. Simon). *Phys. Rev.* 25 (January): 107 (A).
73. Tests of the effect of an enclosing box on the spectrum of scattered X-rays (with J. A. Bearden and Y. H. Woo). *Phys. Rev.* 25 (February): 236 (A).
74. The effect of a surrounding box on the spectrum of scattered X-rays (with J. A. Bearden). *Proc. Nat. Acad. Sci.* 11 (February): 117–19.
75. Measurements of β-rays associated with scattered X-rays (with Alfred W. Simon). *Phys. Rev.* 25 (March): 306–13.
76. The density of rock salt and calcite (O. K. DeFoe with A. H. Compton). *Phys. Rev.* 25 (May): 618–20.

77. The grating space of calcite and rock salt (with H. N. Beets and O. K. DeFoe). *Phys. Rev.* 25 (May): 625–29.
78. On the mechanism of X-ray scattering. *Proc. Nat. Acad. Sci.* 11 (June 15): 303–6.
79. Directed quanta of scattered X-rays (with Alfred W. Simon). *Phys. Rev.* 26 (September): 289–99.
80. X-ray spectra from a ruled reflection grating (with R. L. Doan). *Proc. Nat. Acad. Sci.* 11 (October 15): 598–601.
81. Light waves or light bullets? *Sci. Am.* 133 (October): 246–47.

1926

82. Diffraction of X-rays by a ruled metallic grating (R. L. Doan with A. H. Compton). *Phys. Rev.* 27 (January): 104–5 (A).
83. Electron distribution in sodium chloride. *Phys. Rev.* 27 (April): 510–11 (A).

1927

84. Röntgenstrahlen als Teilgebiet der Optik. *Zeits. für tech. Phys.* 7 (December): 530–37.
    X-rays as a branch of optics. *J. Opt. Soc. Am. and Rev. Sci. Instr.* 16 (February 1928): 71–87; also in *Les Prix Nobel en 1927* (Stockholm: P. A. Norstedt et Fils, 1928). (Nobel lecture, December 12, 1927.)
85. Coherence of the reflected X-rays from crystals (G. E. M. Jauncey with A. H. Compton). *Nature* 120 (October 15): 549 (L).

1928

86. On the interaction between radiation and electrons. *Phys. Rev.* 31 (January): 59–65; also in *Atti del Congresso Internazionale dei Fisici* (September 1927), 1: 161–70. Bologna, 1928.
87. An attempt to find a unidirectional effect of X-ray photons (with K. N. Mathur and H. R. Sarna). *Phys. Rev.* 31 (January): 159 (A).
88. Some experimental difficulties with the electromagnetic theory of radiation. *J. Franklin Inst.* 205 (February): 155–78.
89. The spectrum and state of polarization of fluorescent X-rays. *Proc. Nat. Acad. Sci.* 14 (July 15): 549–53.
90. Discordances entre l'expérience et la théorie electromagnetique du rayonnement. In *Electrons et photons*, report of the 5th Conference on Physics, Solvay Conference, Brussels, 1927, pp. 55–104. Paris: Gautier-Villars, 1928.
    The corpuscular properties of light. *Naturwis.* 17 (June 28, 1929): 507–15; *Phys. Rev. S.* and *Rev. Mod. Phys.* 1 (July, 1929): 74–89.

1929

91. What is light? *Sigma Xi Quar.* 17 (March): 14–34; reprinted in *Sci. Mo.* 28 (April): 289–303; *Proc. Ohio State Educ. Conf.* 35 (September 15, 1930): 401–20 (with alterations); *Smithsonian Rep. for 1929*, pub. no. 3038, pp. 215–28 (Washington, 1930); *J. Chem. Educ.* 7 (December 1930): 2769–87.
92. An attempt to detect a unidirectional effect of X-rays (with K. N. Mathur and H. R. Sarna). *Indian J. Phys.* 3, pt. 4 (May): 463–66.
93. A new wave-length standard for X-rays. *J. Franklin Inst.* 208 (November): 605–16.
94. The efficiency of production of fluorescent X-rays. *Phil. Mag.* 8 (December): 961–77.
95. What things are made of: I. *Sci. Am.* 140 (February): 110–13. II. *Sci. Am.* 140 (March): 234–36.
96. Compton effect. *Encyclopaedia Britannica*, 14th ed.

1930

97. The efficiency of X-ray fluorescence. *Phys. Rev.* 35 (January): 127–28 (A).
98. The determination of electron distributions from measurements of scattered X-rays. *Phys. Rev.* 35 (April 15): 925–38.
99. Scattering of X-rays and the distribution of electrons in helium. *Phys. Rev.* 35 (June 1): 1427–28 (A).
100. Are planets rare? *Science* 72 (August 29): 219 (L).
101. Looking inside the atom: X-ray scattering and the structure of atoms. *Tech. Rev.* 33 (October): 19ff.

1931

102. Electron distribution in argon, and the existence of zero point energy. *Phys. Rev.* 37 (January 1): 104 (A).
103. The optics of X-rays. *J. Opt. Soc. Am.* 21 (February): 75–89.
104. Precision wave-length measurement with the double crystal X-ray spectrometer. *Phys. Rev.* 37 (June 15): 1694 (A).
105. A precision X-ray spectrometer and the wave length of Mo $K\alpha_1$. *Rev. Sci. Instr.* 2 (July): 365–76.
106. The uncertainty principle and free will. *Science* 74 (August 14): 172.
107. Assault on atoms. *Proc. Am. Phil. Soc.* 70, no. 3: 215–29: reprinted in *Smithsonian Rep. for 1931*, pub. no. 3150, pp. 287–96 (Washington, 1932).
108. Ionization as a function of pressure and temperature (with R. D. Bennett and J. C. Stearns). *Phys. Rev.* 38 (October 15): 1565–66 (L).
109. The constancy of cosmic rays (R. D. Bennett with J. C. Stearns and A. H. Compton). *Phys. Rev.* 38 (October 15): 1566 (L).

1932

110. Comparison of cosmic rays in the Alps and the Rockies. *Phys. Rev.* 39 (January 1): 190 (A).
111. Ionization by penetrating radiation as a function of pressure and temperature (with R. D. Bennett and J. C. Stearns): *Phys. Rev.* 39 (March 15): 873–82.
112. Variation of the cosmic rays with latitude. *Phys. Rev.* 41 (July 1): 111–13 (L).
113. Diurnal variation of cosmic rays (R. D. Bennett with J. C. Stearns and A. H. Compton). *Phys. Rev.* 41 (July 15): 119–26.
114. Use of argon in the ionization method of measuring cosmic rays (with John J. Hopfield). *Phys. Rev.* 41 (August 15): 539 (L).
115. Progress of cosmic-ray survey. *Phys. Rev.* 41 (September 1): 681–82 (L).
116. Studies of cosmic rays. In *Carnegie Inst. Wash. yr. bk.*, 1931–32 (December 9), pp. 331–33.
117. Sea level intensity of cosmic rays in certain localities from 46° south to 68° north latitude. *Phys. Rev.* 42 (December 15): 904 (A).

1933

118. A geographic study of cosmic rays. *Sci. Mo.* 36 (January): 75–87.
119. Some evidence regarding the nature of cosmic rays. *Phys. Rev.* 43 (March 1): 382 (A).
120. A geographic study of cosmic rays. *Phys. Rev.* 43 (March 15): 387–403.
121. A positively charged component of cosmic rays (Luis Alvarez with A. H. Compton). *Phys. Rev.* 43 (May 15): 835–36 (L).
122. The significance of recent measurements of cosmic rays. *Science* 77 (May 19): 480–82.
123. Nature of cosmic rays. *Nature* 131 (May 20): 713–15.
124. The secret message of the cosmic ray. *Sci. Am.* 149 (July): 5–7.
125. An improved cosmic-ray meter (with J. J. Hopfield). *Rev. Sci. Instr.* 4 (September): 491–95.
126. Studies of cosmic rays. In *Carnegie Inst. Wash. yr. bk.*, 1932–33 (December 15), pp. 334–39.
127. Nature of cosmic rays. In *The science of radiology*, O. Glasser, ed., pp. 398–411. Springfield, Ill.: Thomas. (Reprinted in part from 118 and 120 above).
128. Progress of world-survey of cosmic rays. *Trans. Am. Geophys. Union*, 14th annual meeting, 1933, pp. 154–58.

1934

129. Scientific work in the "Century of Progress" stratosphere balloon. *Proc. Nat. Acad. Sci.* 20 (January): 79–81.

130. Further geographic studies of cosmic rays (with J. M. Benade and P. G. Ledig). *Phys. Rev.* 45 (February 15): 294–95 (A).
131. Cosmic-ray ionization at high altitudes (with R. J. Stephenson). *Phys. Rev.* 45 (April 1): 441–50.
132. Cosmic-ray ionization in a heavy walled chamber at high altitudes (with R. J. Stephenson). *Phys. Rev.* 45 (April 15): 564 (A).
133. Interpretation of data from world cosmic-ray survey. *Science* 79 (April 27): 378 (A).
134. "Appearance" of atoms as observed with X-rays (with E. O. Wollan). *Science* 79 (April 27): 379–80 (A).
135. A precision recording cosmic-ray meter (with E. O. Wollan, R. D. Bennett, and A. W. Simon). *Phys. Rev.* 45 (May 15): 758 (A).
136. The appearance of atoms as determined by X-ray scattering (E. O. Wollan with A. H. Compton). *J. Opt. Soc. Am.* 24 (September): 229–33.
137. Composition of cosmic rays (with H. A. Bethe). *Nature* 134 (November 10): 734–35 (L).
138. Studies of cosmic rays. In *Carnegie Inst. Wash. yr. bk.*, 1933–34 (December 14), pp. 316–21.
139. Magnitude of cosmic-ray bursts. *Nature* 134 (December 29): 1006 (L).
140. A precision recording cosmic-ray meter (with E. O. Wollan and R. D. Bennett). *Rev. Sci. Instr.* 5 (December): 415–22.

1935

141. Incoherent scattering and the concept of discrete electrons. *Phys. Rev.* 47 (January 15): 203 (A).
142. Incoherent scattering and the concept of discrete electrons. *Phys. Rev.* 47 (March 1): 367–70.
143. The composition of cosmic rays. *Proc. Am. Phil. Soc.* 75 (April 20): 251–74.
144. Cosmic rays. *Nature* 135, no. 3418 (May 4): 695–98; *Sci. Am.* 153 (September): 133.
145. An apparent effect of galactic rotation on the intensity of cosmic rays (with Ivan A. Getting). *Phys. Rev.* 47 (June 1): 817–21.
146. An attempt to analyse cosmic rays. *Proc. London Phys. Soc.* 47 (July 1): 747–73.
147. Studies of cosmic rays. In *Carnegie Inst. Wash. yr. bk.*, 1934–35 (December 13), pp. 336–40.
148. A study of cosmic-ray bursts at different altitudes (with Ralph D. Bennett). In *Papers and discussions of the International Conference on Physics, London, 1934*. Vol. 1, *Nuclear Physics*, p. 225. London: London Physics Society.

1936

149. Recent developments in cosmic rays. *Rev. Sci. Instr.* 7 (February): 71–81.
150. The physics of higher voltage X rays. In *Graduate Course, Tumor Clinic of Michael Reese Hospital, Chicago, Supplement, September 21–27*, pp. 34–41. Chicago: Michael Reese Hospital.
151. Scattering of X-rays by a spinning electron. *Phys. Rev.* 50 (November 15): 878–81.
152. Studies of cosmic rays. In *Carnegie Inst. Wash. yr. bk.*, 1935–36 (December 11), pp. 343–46.
153. Cosmic rays as electrical particles. *Phys. Rev.* 50 (December 15): 1119–30.

1937

154. An energy distribution analysis of primary cosmic rays. *Phys. Rev.* 51 (January 1): 59 (A).
155. Effect of galactic rotation on cosmic rays. *Science* 85 (January 1): 25 (A).
156. Variations of cosmic rays with latitude on the Pacific Ocean (with R. N. Turner). *Phys. Rev.* 51 (June 1): 1005 (A).
157. On the origin of cosmic rays (with P. Y. Chou). *Phys. Rev.* 51 (June 15): 1104 (L).
158. Cosmic rays on the Pacific Ocean (with R. N. Turner). *Phys. Rev.* 52 (October 15): 799–814.
159. Studies of cosmic rays. In *Carnegie Inst. Wash. yr. bk.*, 1936–37 (December 10), pp. 356–58.

1938

160. An alternative interpretation of Jauncey's "heavy electron" spectra. *Phys. Rev.* 53 (March 1): 431 (L).

1939

161. Significance of sidereal time variations of cosmic rays (with P. S. Gill). *Phys. Rev.* 55 (January 15): 233 (A).
162. Cosmic-ray intensity and the thermal expansion of the atmosphere (with M. Schein and P. S. Gill). *Science* 89 (May 5): 398 (A).
163. Time variations of cosmic rays. *J. Franklin Inst.* 227 (May): 607–20.
164. Cosmic rays on the Pacific Ocean (with P. S. Gill). *Rev. Mod. Phys.* 11 (July–October): 136.
165. Recurrence phenomena in cosmic-ray intensity (A. T. Monk with A. H. Compton). *Rev. Mod. Phys.* 11 (July–October): 173–79.
166. Chicago cosmic-ray symposium. *Sci. Mo.* 49 (September): 280–84.

1940

167. What we have learned from scattered X-rays. *J. Franklin Inst.* 230 (August): 149–57.
168. Physical differences between types of penetrating radiation. *Am. J. Roentgenol. and Radium Therapy* 44 (August): 270–75; excerpted in *Radiography* 7 (May 1941): 60–73.
169. Effect of an eclipse on cosmic rays. *Phys. Rev.* 58 (November 1): 841 (L).
170. Report on cosmic-ray research at the University of Chicago. In *Carnegie Inst. Wash. yr. bk.*, 1939–40 (December 13): 116–21.
171. Studies of cosmic rays at high altitudes. In *Am. Phil. Soc. yr bk.*, 1940, pp. 147–49. (Report as recipient of grant from the Penrose Fund.)

1941

172. Recurrence pulses in cosmic-ray intensity (with A. T. Monk). *Phys. Rev.* 59 (January 1): 112 (A).
173. Protons as primary cosmic rays (with Marcel Schein). *Science* 93 (May 9): 436 (A).
174. Recent studies of cosmic rays at high altitudes. *Science* 93 (May 16): 462 (A).
175. Report on cosmic-ray research at the University of Chicago. In *Carnegie Inst. Wash. yr. bk.*, 1940–41 (December 12): 121–26.

1942

176. Report on cosmic-ray research at the University of Chicago. In *Carnegie Inst. Wash. yr. bk.*, 1941–42 (December 18): 90–94.

1943

177. On the fluctuations of cosmic rays. In *Symposium on Cosmic Rays*, pp. 59–66. Rio de Janeiro: Acad. Brasileira Ciênc.

1945

178. Modern physics and the discovery of X-rays. *Radiology* 45 (November): 534–38.

1946

179 The scattering of X-ray photons. *Am. J. Phys.* 14 (March–April): 80–84.

1952

180. Man's awareness and the limits of physical science. *Science* 116 (November 14): 519 (A).

1956

181. The world of science in the late eighteenth century and today. *Proc. Am. Phil. Soc.* 100 (August): 296–303.

1961

182. The scattering of X rays as particles. *Am. J. Phys.* 29 (December): 817–20.

BOOKS

X-rays and electrons. New York: Van Nostrand, 1926.

X-rays in theory and experiment (with S. K. Allison). New York: Van Nostrand, 1935.

Index

Absorption, 187, 358; coefficients, 353, 421; edges, 355; of gamma rays, 220, 247, 265, 367; law of X-ray absorption, 138, 177, 185, 251; quantum theory of, 364; of scattered X-rays, 251, 396, 414
Allison, S. K., xi, xxiv, xxvii
Alvarez, L. W., xi, xxvi, xxix, 724, 769
Atomic structure by X-rays, 59
Auger, P., 547, 550, 558, 585, 614
Avogadro's number, 136, 607

Barkla, C. G., xvii, xxii, 80, 95, 225, 230, 439, 441, 586, 713
Bearden, J. A., xi, xxv, 485, 603, 665
Benade, J. M., xi
Bennett, R. D., xi, xxi, xxix, 505
Bohr, Kramers, Slater, theory of, xx, xxi, xxviii, 554, 568, 572, 584
Bohr, N., xx, 85, 87
Bothe, W., xxi, 589, 591
Bothe-Geiger experiment, xxi, xxiii, 469, 488, 540, 566, 570, 733
Bragg, W. H., xxii, 16, 45, 52, 59, 89, 105, 212, 666
Bragg, W. L. (Sir Lawrence), xi, xv, 45, 59, 105
Bragg equation, xxiii, 77, 601; corrected for refraction, 403
Bragg spectrometer, 209, 416, 430, 462; double-crystal types, 673
Breit, G., 647
Brillouin, L., xi
Broglie, L. de, xxi, 543, 594
Broglie, M. de, 347, 546, 557, 583
de Broglie waves, xxi

Cambridge University, xvi
Chicago, University of, xi, xxi, xxv
Collidge X-ray tube, 50, 88, 187, 599
Compton, K. T., xiii, xviii, xix, 557, 583, 712
Compton effect, xxi, xxvii, 382, 476, 504, 637, 747
Compton electrometer, 163
Compton wave-length, 429
Cosmic rays, xiii, xxvii, xxix
Crystals: density of, 496; grating space of, 135, 499

Darwin, C. G., xiv, 59, 60, 89, 213, 228, 320, 611
Davis, Bergen, 588

Debye, P., xiv, xx, 21, 26, 39, 594, 641, 716, 747
Defoe, O. K., xi, 496
Dempster, A. J., xxiv
Diffraction of X-rays by crystals, 434, 532; by ruled grating, 519
Dirac, P. A. M., xi, xxi, xxiii, xxviii, 647
Doan, R. L., xi, xxv, xxix, 519, 534
Duane, W., xv, xxii, xxiv, 468, 482, 583
DuMond, J. W. M., xxiv

Earth's rotation, 1, 5, 14
Eckart, C., xi
Einstein, A., xiv, xix, xxiii, xxvii, 29, 39, 523, 557, 582, 759
Electromagnetic theory, 224, 552
Electron, 689, 703; charge of, 607, 666; complex, effect on scattering, 139, 247; distribution in atoms, xv, xxvii, 16, 71, 258, 525, 640, 654, 661, 699; free, and scattered quanta, 348, 488, 589; magnetic properties, 93, 219, 261, 294; recoil, 488; size and shape of, 94, 139, 151, 177; spin and ferromagnetism, xvi, 93
Eve, A. S., xvii
Ewald, P., xi, 611

Fermi distribution, xxiv
Ferromagnetism, xxvi, 90, 207
Florance, D. C. H., xvii
Florescent X-rays, 274, 277, 334, 358, 576, 613
Foldy, L. L., xi
Fourier analysis, electron distributions, 661

Gamma-rays: scattering of, 265; wavelength of, 284
Geiger, H., xxi
General Electric Co., xv, xvi
Gingrich, N. S., xi, xxvi
Goudsmit, S., xi, xvi
Gratings, diffraction of X-rays from, xxv
Gray, Prof. J. A., xvii, xviii, xxiii, 336, 564, 587, 746

Hagenow, C. F., xxvi, 306, 441, 713
Havighurst, R. J., xv
Heisenberg, W., xi, xxiii, 594
Hughes, A. L., xi
Hull, A. W., 635
Hull, Gordon F., 756

Jauncey, G. E. M., 616, 711
Jeans, J. H., xiv, xxiv
Johnston, Marjorie, xi
J radiation, 439

Kirkpatrick, H. A., xxiv
Kirkpatrick, Paul, xxiv
Klein, M. J., xi

Lewis, G. N., 553
Lorentz, H. A., 545

Magnetization and X-ray reflection, 207, 294; electron and ferromagnetism, 216
Mathur, K. N., 597
Michelson, A. A., 374, 523, 727
Millikan, R. A., xxi, xxviii, 25, 136, 533, 583
Minnesota, University of, xv, xvii
Moseley, H. G. J., xiv, 45, 80

National Research Council, Bulletin of, xix, xx, 321
Nobel Prize, award to A. H. Compton, xxii, xxvii, 527

Osgood, T. H., xi, xxv, 535

Pauli, W., 472
Photo-electric effect, 557, 582, 614; by X-rays, 341
Photo-electrons, 191, 341, 349, 546, 557, 561; spatial distribution, 547, 558
Photons, 553, 561, 630, 732
Planck's quantum theory, 24, 85, 581
Poincaré, H., 659
Polarization of X-rays, 306, 441, 576
Princeton University, xiii, xiv, xvi, xix

Radiation and matter, 378
Raman, C. V., xxii, 588
Recoil electrons, Compton electrons, 413, 446, 589
Reflection of X-rays: and electron distribution, 59, 71; from grating, 519; from imperfect crystals, 66, 88, 320; and magnetization, 90, 207; from perfect crystals, 63, 544; total reflection, 372, 402, 529
Refraction of X-rays, 369, 530, 667; deviation from Bragg's law, 370, 402; and grating space of crystals, 135; significance of refractive index less than unity, 407; by total reflection, 371, 407

Richardson, O. W., xi, xiii, xix, xx, xxiv, 347, 557, 583, 751
Rodgers, Eric, xi, xxvi, 714
Roentgen, C. W., 533, 727
Rognley, O., xv, xxvi, 93, 103, 207
Ross, P. A., xxiv, 467
Rutherford, E., xiv, xvi, 100, 187, 190, 213, 687

Sargent, B. W., xi
Sarna, H. R., 597
Scattering of X-rays and gamma rays, 146, 323, 460, 538, 746; from bound electrons, 426, 458; change of wave length in, 265, 311, 318, 335, 385, 393, 416, 431, 465, 587; conservation of energy and momentum, xxi, 385, 466, 478, 586; intensity and scattering angle, 146, 290, 390, 392, 398; J-phenomena, 439; polarization of, 441; quantum theory, 237, 382, 414, 446, 476; recoil, or Compton electrons, 390, 395, 413, 488; spectrum of, 318, 429, 431, 463, 465; Thomson theory, 224, 266
Schrödinger, E., xxi, xxiii, 542, 595
Shankland, R. S., xiii, xxi, 749
Shook-Roentgen apparatus, 50
Simon, A. W., xi, xxi, 507, 508, 571, 592, 748
Specific heats, 18, 39, 110
Spectrum lines of X-rays, 313, 318
Stearns, J. C., xxvi, xxix
Stuewer, R. H., xi

Taylor, G. I. (Sir Goffrey), xxiv
Thermal conductivity, 28, 31
Thomson, Sir G. P., xi
Thomson, Sir J. J., xvii, 57, 224, 713, 733
Thomson theory of X-ray scattering, xviii, 59, 94, 147, 383, 643

Uhlenbeck, G., xi, xvi

Van Vleck, J. H., xi

Washington University, St. Louis, iv, xviii, xix, 224
Wave-lengths of X-rays: change on scattering, 286; by crystal diffraction, 503, 601, 605; from gratings, 519, 601, 605
Wave-particle duality, 593
Webster, D. L., xi, xxiv, 17, 65, 74, 137, 233
Wentzel, G., xi, xxviii, 550, 641
Westinghouse Electric Mfg. Co., xvi

Wilson, C. T. R., xxii, xxiii, 349, 455, 470, 488, 540, 566, 589, 718
Wollan, E. O., xi, xxvi, xxix, 698, 705, 719
Woo, Y. H., xxi, 457

Wooster, College of, xiii

X-ray: spectrometer, 45, 672; wave-length standard, 601

125